# Science 2.0
# I Upgraded My Science

## Book One of the Philosophy of Science Series

## By: Mark My Words

Copyright: From October 2017 to September 2019

Original Version: 31OCT2017

Current Version: 09SEP2019

I love science, so I want my science to be at its best. Introducing Science 2.0! This book has been a couple of years in the making.

Science 2.0 allows all of the evidence into evidence, and then it pursues a preponderance of that evidence.

In the final and enhanced edition of this book, I pulled out all the stops and did precisely what they don't want us to do – I falsified the second law of thermodynamics and found its replacement.

Exploring Consciousness, the Mind-Brain Problem, Quantum Consciousness, Non-Local Consciousness, Universal Consciousness, Out-of-Body Experiences (OBEs), Near-Death Experiences (NDEs), Shared-Death Experiences (SDEs), and other Non-Local Non-Physical Transdimensional Experiences. Getting rid of Materialism and Naturalism makes for better and more interesting science.

As an upgraded science, Science 2.0 absolutely needs an Account of Origins and a Model of Reality that matches perfectly with the Observational Evidence and the Lived Experiences of the human race. Consequently, a great deal of time and effort has been placed into finding and/or developing such a thing. You will have to judge for yourself if the process and this project has been a success, or not.

I'm trying to fix everything that's wrong with science. That's impossible, I know, but it's fun to try.

Science 2.0 represents not only a serious upgrade to the Philosophy of Science but also a great introduction to Origin Science in general. Science 2.0 is based upon everything that has ever been experienced and observed. Science 2.0 gives preference to observation and experience over philosophical speculation and wishful thinking.

This book represents a portion of my Doctorate in Origin Science, as well as a small portion of my Doctorate in the Philosophy of Science. Of course, there is no university on this planet that will give a person a PhD in the truth. Currently, they only give Doctorate Degrees in Materialism, Naturalism, Darwinism, Nihilism, Atheism or Creation Ex Nihilo, the Theory of Evolution, the Second Law of Thermodynamics, Classical Realism, materialistic and naturalistic interpretations of Quantum Mechanics, and Creation by Chance. A doctorate degree in creation by chance, materialism, naturalism, or atheism is worthless because these things have been falsified by Science, Observation, and the Experiences of the human race.

Therefore, Science 2.0 will have to stand on its own as a true and realistic replacement for these falsified philosophies, religions, and ideas until the science community catches on to what's currently wrong with science as a whole. In this book, I did a much better job of meeting my burden of proof than the Naturalists and Darwinists have ever done. I will have to be satisfied with that for now.

Mark My Words


The Official Website:   https://science-2-0.com/

The Associated Blog:   https://science-2-0.com/blog/

A Sister Website:   http://www.markme.website/

The Associated Forum:   https://markme.us/forums/forum/science-2-0/


First Edition

# Table of Contents

Dedication ................................................................................................................. 13
Introduction ............................................................................................................. 14
Science 2.0: Redefining Science ............................................................................. 15
PART 0 – PREFACE MATERIALS FOR SCIENCE 2.0 ...................................................... 19
    Introducing Science 2.0 ....................................................................................... 20
        Statistics Falsifies Creation by Chance or Creation by Magic ........................... 21
        Science Is Observation and Experience ............................................................ 25
    The Probability of Protein Synthesis by Chance Alone ........................................ 31
    A Binomial Distribution Table Falsifies Creation by Chance ............................... 34
        Chance Cannot Do Causation! ........................................................................... 38
    The Perpetual Motion Cycle ................................................................................. 42
    Evolution Is Entropy ............................................................................................. 46
    Freedom of Science .............................................................................................. 50
    The Null Hypothesis ............................................................................................. 51
    Origin Science ....................................................................................................... 54
    A New Model for Reality ...................................................................................... 55
        The Transformation of Energy .......................................................................... 55
    Introducing the Law of Psyche ............................................................................ 64
    Falsifying the Second Law of Thermodynamics .................................................. 67
    Heat Transfer Is One Definition for Entropy ....................................................... 77
        Work In = Work Out ........................................................................................... 82
        $S = \Delta Q$ ......................................................................................................... 84
        Examining the Equation for Heat ...................................................................... 87
        $Q = mc\Delta T$ .................................................................................................. 87
    Entropy Is Mass .................................................................................................... 89
        Entropy Is Mass or Resistance to Acceleration ................................................. 91
        Entropy Is Mass's Heat Storage Capacity ......................................................... 91
        Deriving the Mass-Based Definition for Entropy ............................................. 93
    I Finally Found a Couple of True Definitions for Entropy ................................... 97
        Introducing the Ultimate Law of Thermodynamics ....................................... 108
        Introducing the Quantum Law of Thermodynamics ...................................... 117
        The True Version of the Second Law of Thermodynamics ............................ 126
        Reference Material Regarding Entropy and Syntropy ................................... 131
    Science Is in Desperate Need of an Upgrade .................................................... 133
    Dark Matter and Dark Energy – Proof of the Non-Physical .............................. 140

- Dark Energy ... 142
- Dark Matter ... 143
- The Brain Is MAPPED ... 145
  - Source ... 148
  - References ... 149
- We Find the Truth by Living It and Experiencing It ... 150
- Science Is Observation and Experience ... 152
  - Reference ... 153
- Denying the Existence of Observed Facts ... 154
- Proof of God's Existence ... 155
- Entropy Is Death and Syntropy Is Eternal Life ... 157
- The Purpose of Science 2.0 Is to Identify the Falsehoods in Science ... 161
- Science Needs a Philosophy of Science that Is True ... 162

PART I — DEFINING SCIENCE 2.0 ... 165
- The Law of Witnesses ... 167
- The Dedicated Opposition ... 168
- The Epiphany ... 169
- We Need an Evidence-Based Science, Not Scientific Inferences ... 172
- Defining Science ... 175
- Foundational Tenets of Science 2.0 ... 179
- Invisible Forces and Fields ... 180

PART II – FIXING SCIENCE WITH QUANTUM MECHANICS ... 182
  - Reference ... 182
- Syntropy vs. Entropy ... 183
  - Source ... 188
  - References ... 188
- Interpretations of Quantum Mechanics ... 189
  - Source ... 193
  - References ... 193
- Transdimensional Physics ... 194
  - Transdimensional Physics Has NO Physical Limitations ... 198
  - Quantum Weirdness ... 200
  - Transdimensional Physics in Action ... 204
  - Source ... 209
  - References ... 209
- Quantum Field Theory ... 210

PART III – SCIENCE 2.0, ORIGINS, AND SYNTROPY ... 213
- Observational Studies Cannot Establish Causation ... 220

- Let's Run a Few Science Experiments .................................................................. 226
- Creation by Chance.................................................................................................. 227
- Proving Non-Existence ........................................................................................... 229
- Natural Selection Doesn't Work as Advertised......................................................... 230
  - Our Created Solar System ................................................................................... 230
  - Our Created Stars and Galaxies .......................................................................... 230
  - Our Created Universe – Apologetics Symposium .............................................. 230
  - Our Created Universe – Seattle Creation Conference ....................................... 231
  - The Probability of a Protein Forming by Chance ................................................ 231
  - The Probability of Making a Protein by Chance.................................................. 231
  - Evidence for Creation by Outside Intervention .................................................. 231
- ORIGIN: Probability of a Single Protein Forming by Chance ................................. 231
- Science Is Evidence.................................................................................................. 233
- Evolution Produced Your Brain................................................................................ 243
  - Information Exchange at the Quantum Level ..................................................... 247
  - Everything Falsifies the Theory of Evolution ..................................................... 247
  - Information Storage Capacity Falsifies the Theory of Evolution ...................... 249
  - Telepathic Communication Is Physically Impossible......................................... 249
  - Brain Mapping Requires Intelligence.................................................................. 251
  - Someone Psyche Clearly Mapped Your Brain .................................................... 253
  - Quantum Waves Are WiFi at the Psyche Level ................................................... 254
  - The Theory of Evolution is Self-Defeating ......................................................... 255
  - It Took Millions of Years for Evolution to Make Your Brain ............................... 258
  - Source ................................................................................................................... 260
  - References ............................................................................................................ 260
- Syntropy versus Entropy.......................................................................................... 262
- PART IV — THE WEAKNESSES OF THE SCIENTIFIC METHOD .............................. 264
- SCIENTISM ................................................................................................................ 265
- The Scientific Method Is Based Upon Affirming the Consequent ........................... 266
- Scientism.................................................................................................................. 272
  - Defining Scientism............................................................................................... 272
  - The Weaknesses of Naturalism ........................................................................... 273
  - Science and Human Behavior.............................................................................. 274
  - Science Is Based Upon Philosophical Assumptions and Philosophical Interpretations. 275
  - Defining Science................................................................................................... 276
  - The Purpose of Science ....................................................................................... 277
  - Behavioral Sciences ............................................................................................. 278
  - Knowledge............................................................................................................ 280

- Positivism and Realism .................. 281
- Observation .................. 281
- Definition of Behaviorism .................. 282
- Some Limitations and Weaknesses of Scientific Methods .................. 283
- The Affirming the Consequent Logic Fallacy .................. 283
- Science, Logic, and Reason .................. 285
- Psyche and Lived Experiences Are Axiomatic .................. 287
- Unidentified Variables and Unknown Gods .................. 287
- The Problem of Operationalization .................. 288
- Objectivity .................. 289
- Alternative Epistemologies .................. 289
- Qualitative Methods .................. 290
- Lived Experiences .................. 290
- Conclusion and Solution .................. 291
- VERIFICATION VERSUS FALSIFICATION .................. 293
  - Verificationism .................. 293
  - Falsificationism .................. 293
  - Applying Falsification to Materialism .................. 295
  - Reference Articles .................. 297
- Philosophy of Science .................. 299
  - The Essential Philosophy of Science .................. 299
- My Philosophical Proof of God's Existence .................. 302
  - Kalam Cosmological Argument .................. 302
  - My Adjustments to the Kalam Cosmological Argument .................. 302
  - My Philosophical Proof of God's Existence .................. 303
  - Commentary on My Philosophical Proof of God .................. 304
  - Go Out of Your Way to Get the Right Interpretation .................. 308
  - A Proof or Argument Has to Be Logically Sound .................. 309
  - Observational Reality and Common-Sense Logic .................. 310
  - Critiquing Craig's Cosmological Argument .................. 311
  - Conclusions .................. 313
- PART V — MY TOP TEN MOST FAVORITE SCIENCE THEMES AND MY FAVORITE BOOKS SUPPORTING THOSE THEMES .................. 316
  - 1. Scientific Explanation of Psyche, Spirit Matter, Tachyons, and Quantum Non-Locality or Quantum Non-Physicality .................. 317
    - Quantum Mechanics Has Never Been Falsified .................. 318
    - Quantum Non-Locality .................. 320
    - Quantum Law of Complementarity .................. 321

The Non-Local Consciousness Interpretation of Quantum Mechanics ..................... 323
Tachyons............................................................................................................................. 324
Materialism and Naturalism Cause the Mind-Brain Problem ................................. 326
The Quantum Zeno Effect ............................................................................................... 327
No Entropy in the Non-Local Realms ........................................................................... 329
Telepathy ............................................................................................................................ 329
Teleportation and Quantum Tunneling ...................................................................... 330
Interpretations of Quantum Mechanics ..................................................................... 331
David Bohm's Interpretation of Quantum Mechanics ............................................ 332
Three Main Interpretations of Quantum Mechanics............................................... 332
Observational Support for Quantum Consciousness or Non-Local Consciousness...... 336
Another Winner – Henry P. Stapp's Orthodox Interpretation of Quantum Mechanics. 339
The Monistic Idealism Interpretation of Quantum Mechanics ............................ 348
Idealism ............................................................................................................................... 351
The Orthodox Interpretation of Quantum Mechanics ............................................ 353
Integrating the BioPsychoSocial Model of Reality with Quantum Mechanics ............ 364
2. Observational Evidence and Experiential Evidence that Falsifies Materialism and Naturalism ........................................................................................................................ 366
Quantum Field Theory Falsifies the Second Law of Thermodynamics ..................... 370
3. Solving the Mind-Brain Problem ............................................................................... 372
A Winner in the Mind-Brain Solution Category ....................................................... 373
Winners in the Mind-Brain Solution Category ......................................................... 374
The Physical Brain Is a Transceiver ............................................................................. 376
The Raspberry Award and the Rotten Tomato ....................................................... 380
A Clear Winner in this Mind-Brain Category............................................................. 381
Another Winner in this Mind-Brain Category – Brain Wars................................... 384
Another Winner – How the Self Controls Its Brain ................................................. 384
Honorable Mention in this Mind-Brain Category .................................................... 386
Signs of Psyche – BioPsychoSocial .............................................................................. 391
Signs of Psyche – BioPsychoSocial – Chapter 2........................................................ 399
Schizophrenia ................................................................................................................... 405
Alzheimer's Disease ........................................................................................................ 406
Materialism and Naturalism Cause the Mind-Brain Problem ............................... 407
3. Mind-Over-Matter Resolves the Mind-Brain Problem ......................................... 407
The Placebo Effect ........................................................................................................... 409
Neuroplasticity ................................................................................................................. 410
Epigenetics ........................................................................................................................ 410
Prayer.................................................................................................................................. 411

    Hypnosis .................................................................................................. 414

    Psychedelics ........................................................................................... 416

    Mediums and Channeling ....................................................................... 417

    Mysticism and Clairvoyance ................................................................... 419

    Quantum Zeno Effect .............................................................................. 421

    Conclusions Regarding Mind-Over-Matter ............................................. 425

4. Scientific Proof that Evolution (Genetic Drift), Natural Selection, Random Mutations, and Entropy Cannot Design and Create and Build Anything at All ................ 426

    Evolution Is Entropy................................................................................ 427

    Genetic Entropy ...................................................................................... 431

    Kin Selection........................................................................................... 442

5. The Beginning of Anything Always Points Us to the Beginner ................. 445

    Obviously Made ...................................................................................... 445

6. Scientific Proof of the Invisible and the Immaterial ................................. 446

    Falsifying the Philosophy of Mechanism ............................................... 447

7. Scientific Proof that Macroevolution Could Never Have Done the Job of Design and Creation ........................................................................................................... 448

    Mathematical Modeling Proves that Abiogenesis of Proteins and Genomes Is Impossible .................................................................................................................. 450

8. Fine-Tuning, Extra-Dimensionality, Transdimensionality, Non-Locality, and String Theory Point Us Indirectly to God .................................................................. 451

    Fine-Tuning ............................................................................................. 453

    Unifying Quantum Physics, Consciousness, M-Theory, Heaven, Neuroscience, and Transcendence .......................................................................................... 455

9. Using the Scientific Method to Falsify Materialism, Naturalism, Darwinism, and Atheism .................................................................................................................. 458

    Personality Theory – The Philosophy of Psychology ............................. 461

10. The Unexplainable and Unexplained Aspects of the Big Bang Theory Point Us Directly to God, Because He Is the Only One Who Could Have Done All the Science that Needed to Be Done at that Point in Time ............................................................... 464

    Conclusion .............................................................................................. 476

PART VI — SCIENTIFIC PROOF OF GOD'S EXISTENCE ................................. 481

    Scientific Proof of God's Existence .............................................................. 482

    Using the Scientific Methods to Prove God's Existence .............................. 484

        Let's use the Scientific Method ............................................................. 486

        What's the opposite of Materialism? ..................................................... 488

        What's the opposite of Darwinism? ...................................................... 489

        What's the opposite of Naturalism? ...................................................... 489

        What's the opposite of Atheism?........................................................... 490

        Quantum Non-Locality........................................................................... 491

Scientific Proof of God's Existence ................................................................................. 493
The Materialists Don't Want You Using Scientific Theories to Prove the Truth and to Find the Truth ................................................................................................................................ 496
PART VII — THE BOOKS I HAVE WRITTEN IN AN ATTEMPT TO SUPPORT THESE THEMES 499
The Grand Deception and the GRAND TRUTH ....................................................... 507
You Have to Believe in the Theory of Evolution ....................................................... 511
Signs of Psyche – BioPsychoSocial ......................................................................... 513
Science 2.0 Formally and Functionally Opposes the Denialistic Philosophies and the Exclusionary Philosophies ........................................................................................ 518
   Applied Science ..................................................................................................... 519
   My Autobiography ................................................................................................. 521
   Book 13. "Tripping the Light Fantastic: How Prescription Drugs Almost Killed Me" ... 522
Science 2.0 Is Based Upon the Non-Local Interpretations of Consciousness and Quantum Mechanics ............................................................................................... 523
   Book 5. "Quantum Mechanics from a Non-Physical Spiritual Perspective" ............... 524
Science 2.0 Is Based Upon Eye-Witness Evidence of All Kinds ................................. 526
   Book 4. "The Second Comforter: Supping with Our Resurrected Lord Jesus Christ" 527
   Book 8. "I Am Not a Creationist: So What Am I?" ................................................... 528
   Book 10. "Putting Psyche Back into Psychology: Restoring Science to Consciousness" ............................................................................................................................. 529
Science 2.0 Is Based Upon All of the Evidence that Falsifies Materialism, Naturalism, Darwinism, Nihilism, and Atheism ........................................................................... 530
Science 2.0 Is Based Upon All the Observational Evidence that Solves the Mind-Brain Problem ................................................................................................................... 531
   Book 7. "NATURE vs. NURTURE vs. NIRVANA: An Introduction to Reality" ............ 531
   Book 9. "The Ultimate Model of Reality: Psyche Is the Ultimate Cause" ................ 532
   Book 10. "Putting Psyche Back into Psychology: Restoring Science to Consciousness" ............................................................................................................................. 532
   Book 11. "BioPsychoSocial: Including Psyche or Light into our Theoretical Models" 532
Science 2.0 Is Based Upon All the Scientific Evidence that Falsifies Darwinism and the Theory of Evolution ................................................................................................. 534
   Book 1. "Summary Of: The Theory of Evolution Proves that God Exists" ................ 535
   Book 2. "The Theory of Evolution Proved to Me that God Exists: Why I Am No Longer an Atheist and Why I No Longer Believe in the Theory of Evolution" ..................... 535
   Book 3. "The Scientific Method Proves That the Theory of Evolution Is False" ........ 536
   Book 6. "Using the Scientific Method to Eliminate the Usual Suspects and to Prove the Truth" ................................................................................................................. 536
Science 2.0 Allows Evidence for the Non-Local, the Immaterial, the Non-Physical, and the Invisible into Evidence ............................................................................................. 538
   Book 12. "God Is in the Light: God is light, and in Him is no darkness at all" ......... 538

Science 2.0 Is Based Upon All the Evidence that Falsifies Macroevolution, Abiogenesis, Spontaneous Generation, and Creation by Chance .................................................. 539

    Book 1. "Summary Of: The Theory of Evolution Proves that God Exists" ............... 539

    Book 2. "The Theory of Evolution Proved to Me that God Exists: Why I Am No Longer an Atheist and Why I No Longer Believe in the Theory of Evolution" ...................... 539

    Book 3. "The Scientific Method Proves That the Theory of Evolution Is False" ......... 540

    Book 6. "Using the Scientific Method to Eliminate the Usual Suspects and to Prove the Truth" ................................................................................................................. 540

Science 2.0 Allows All of the Evidence for Fine-Tuning, Transdimensionality, Non-Locality, and Light into Evidence ................................................................................................ 542

    Book 12. "God Is in the Light: God is light, and in Him is no darkness at all" ......... 543

Science 2.0 Uses the Scientific Methods to Falsify Materialism, Darwinism, and Naturalism .................................................................................................................................... 544

Science 2.0 Chooses to Believe All of the Evidence that Indicates that Our Physical Universe Had a Beginning and Could Someday Have an End ...................................... 545

    Book 16. "Scientific Proof of God's Existence: Finding God Where the Atheists Refuse to Look for Him" ............................................................................................... 545

    Book 17. "Quantum Neuroscience: The Answer to Life, the Universe, and Everything" ............................................................................................................................... 546

Science 2.0 States That There Is No Such Thing as Creation Ex Nihilo, Because It Has Never Been Experienced nor Observed ........................................................................ 547

    Book 8. "I Am Not a Creationist: So What Am I?" ............................................... 547

Science 2.0 Allows Any Evidence for God's Existence into Evidence ........................... 548

    Book 14. "Scientific Proof of God's Existence: A Primer" ..................................... 550

Science 2.0 Massively Upgrades Science and the Philosophy of Science ..................... 551

    Book 17. "Quantum Neuroscience: The Answer to Life, the Universe, and Everything" ............................................................................................................................... 551

    Book 18. "Syntropy in Defense of Quantum Mechanics: The Answer to Life, the Universe, and Everything" ................................................................................... 552

    Book 5. "Quantum Mechanics from a Non-Physical Spiritual Perspective" ............... 554

    Book 15. "Science 2.0: I Upgraded My Science" ................................................. 555

PART VIII — SCIENCE 2.0 NEEDS AN EYE-WITNESS ACCOUNT OF ORIGINS OR CREATION FROM SOMEONE WHO WAS THERE ............................................................................... 556

Introducing Ultimate Cause and a Psyche Ontology ................................................... 558

A Philosophy of Science for Personality Theory or Psyche Theory ............................... 561

The Creation Drama ................................................................................................... 564

    THE FIRST DAY ................................................................................................. 564

    THE SECOND DAY ............................................................................................. 566

    THE THIRD DAY ................................................................................................ 567

    THE FOURTH DAY ............................................................................................. 569

    THE FIFTH DAY ................................................................................................. 570

- THE SIXTH DAY ........................................................................................................ 572
- THE CREATION OF ADAM ..................................................................................... 573
- THE CREATION OF EVE .......................................................................................... 573
- ADAM AND EVE ARE INTRODUCED INTO THE GARDEN ................................ 574
- THE GARDEN OF EDEN .......................................................................................... 575
- THE GODS WITHDRAW .......................................................................................... 575
- LUCIFER APPROACHES ADAM ............................................................................. 576
- EVE PARTAKES OF THE FRUIT ............................................................................. 577
- ADAM PARTAKES OF THE FRUIT ......................................................................... 577
- THEIR NAKEDNESS ................................................................................................ 578
- THE GODS RETURN ................................................................................................ 579
- LUCIFER IS EXPELLED FROM THE GARDEN .................................................... 580
- CONDITIONS OF THE FALL ................................................................................... 581
- ADAM AND EVE ARE EXPELLED FROM THE GARDEN .................................. 581
- THE TELESTIAL WORLD ........................................................................................ 582
- The Book of Abraham ................................................................................................. 583
- The Book of Moses ..................................................................................................... 589
- Margaret Barker .......................................................................................................... 596
- John Pratt .................................................................................................................... 598
- A Revelation about Origins from the Biblical God Jesus Christ ............................... 599
- Other Scriptures about Origins ................................................................................... 602
- My Personal Interpretation of this Evidence .............................................................. 608

Part VII — Conscious While Outside the Physical Body ................................................ 618
- The Astral Traveler ..................................................................................................... 619
- Experiencing the Light ............................................................................................... 622
- Evidence of Psyche and Signs of Life ........................................................................ 625
- NDE Encounters with Jesus ....................................................................................... 629

Part VIII — Conscious After Dying .................................................................................. 631
- Shared-Death Experiences .......................................................................................... 633
- Reincarnation and Previous Lives .............................................................................. 634

PART IX — QUANTUM CONSCIOUSNESS AND UNIVERSAL CONSCIOUSNESS ................. 636
- Books Supporting Quantum Consciousness ............................................................... 637
- Books Supporting Universal Consciousness or Cosmic Consciousness ................... 642
- Prejudice: The Good, the Bad, and the Ugly ............................................................. 644
  - My Observations and Conclusions .......................................................................... 646

PART X — THE SCIENTIFIC BREAKTHROUGHS FROM SCIENCE 2.0 ............................. 648
- A Physical Atom Is Physically Impossible ................................................................ 649
- Falsifying Creation by Chance ................................................................................... 652

Spirit Matter or Dark Matter Exists below the Ground State ................................. 653
The Conservation Laws Falsify the Second Law of Thermodynamics ........................... 660
Entropy Is the Heat Storage Capacity of Mass .......................................... 661
    Entropy Is Mass or Resistance to Acceleration ...................................... 663
    Deriving the Mass-Based Definition for Entropy ..................................... 663
    The Philosophy of Science Falsifies the Second Law of Thermodynamics ................. 667
The Correlation Coefficient of Zero Falsifies the Second Law of Thermodynamics ........ 695
    Quantum Field Theory Represents the Truth in Science ............................... 699
    Entropy Defined as Disorder Is Self-Defeating ...................................... 702
    The Second Law of Thermodynamics Is Protected and Unfalsifiable .................... 707
    There's NO Correlation between Natural Selection and Our Genes ..................... 719
    Disorder Doesn't Really Exist in Our Part of the Multiverse ........................ 720
The Thermodynamics Version of Entropy Is Also Based upon Fiction ..................... 725
Entropy Defined as Death or Natural Selection ........................................ 732
Quantum Field Theory Falsifies the Second Law of Thermodynamics ...................... 733
    The Truth Is in Plain Sight ........................................................ 737
Using Statistics to Falsify the Second Law of Thermodynamics ......................... 738
    Entropy Defined as Maximum Disorder ................................................ 738
    Entropy Defined as an Infinitely Increasing Superfluid ............................. 742
    Creation by Entropy ................................................................ 743
    The Null Hypothesis ................................................................ 750
    Disorder or Chance Has No Power .................................................... 753
    The Null Hypothesis Falsifies Naturalism, Darwinism, and Physicalism ............... 764
    Identifying Reality and Truth ...................................................... 768
    Inferential Statistics Falsifies the Second Law of Thermodynamics .................. 775
Everything Falsifies the Second Law of Thermodynamics ................................ 789
    The Force in Between the Two Objects ............................................... 791
    Joint Heirs with Christ ............................................................ 796
Falsifying My Materialism, My Naturalism, and My Atheism ............................. 800
Tilting Windmills .................................................................... 806
PART XI — THE MIND OF GOD ............................................................ 811
    Books that the Biblical God Had a Hand in Producing ................................ 812

## Dedication

Dedicated to every open-minded scientist everywhere.

# Introduction

Most people don't know that the Scientific Method is fundamentally flawed; and, that fact or reality is where all of the confusion and contention comes from where science is concerned.

The Scientific Method is based upon a logic fallacy that is called *affirming the consequent*. The whole of modern-day Science, Materialism, and Scientific Naturalism is based upon this logic fallacy; and, most scientists don't even know it, because they have never studied the Philosophy of Science.

The weaknesses of science and the scientific methods explain why science needs to be upgraded to Science 2.0.

Science is Observation.

If you observe carefully, you will notice that the Materialists, Naturalists, Darwinists, Nihilists, and Atheists define science as Materialism, Naturalism, Darwinism, Behaviorism, Determinism, Scientism, Physicalism, Atomistic Reductionism, Nihilism, and Atheism. These philosophies have been falsified by Observational Evidence, Experiential Evidence, Eye-Witness Evidence, and Empirical Evidence. In other words, Materialism, Naturalism, and their derivatives have been falsified by Observations. Notice carefully, and you will observe that the promoters of these false and falsified philosophies rely upon *affirming the consequent* and the suppression of evidence in order to make their case and meet their burden of proof.

Defining science as Materialism, begging the question, suppressing evidence, and similar practices result in a wide variety of different logic fallacies known variously as *circular reasoning, affirming the consequent, begging the question, jumping to conclusions, scientific inferences, personal interpretations, confirmation biases*, and *ad hoc just-so stories*. These are the bread and butter of Scientific Materialism and the foundation upon which modern-day science is built. It's a house of cards – a pyramid scheme.

Science 2.0 resolves the problem by defining Science as Observation, Experience, Experimentation, Knowledge, Fact, Truth, and Evidence. Science 2.0 is a paradigm shift in our way of thinking and in the way that we do Science. Science 2.0 allows all of the evidence into evidence.

Furthermore, under Science 2.0, there is NO officially sanctioned interpretation of the scientific evidence. You are encouraged to make your own interpretation of the evidence. Science 2.0 allows all of the evidence into evidence, and then leaves it up to YOU to interpret the evidence, decide for yourself what to make of the evidence, and choose for yourself how to use that evidence in your own life. You are going to do so anyway, so why not encourage you to do so right from the start?

Science 2.0 is the way that Science should have always been done but wasn't.

# Science 2.0: Redefining Science

https://science-2-0.com/

Modern-day science is in desperate NEED of an upgrade. The "interpretation phase" of the Scientific Method NEEDS to be completely overhauled, fixed, tightened, and upgraded. We need to enhance our Philosophy of Science. All of our stupid, idiotic, and illogical "personal interpretations" of the scientific data ARE the fatal flaw of the Scientific Method. We no longer try to find the true interpretation or the true cause of the scientific data. We simply go with Creation by Chance as our ONLY causal explanation for everything that exists in this universe. It's sloppy science and Bad Science.

Ludwig Boltzmann was the one who pushed for and gave us the "disorder definition" for entropy. Thanks to Boltzmann and some other people, the second law of thermodynamics stopped being about heat or thermodynamics and became all about random disorder or chaos instead. Then Boltzmann and his associates went one step further and decided that "random fluctuations in entropy" or "random fluctuations in disorder" can create Boltzmann Brains, the Big Bang, Planets, Stars, Galaxies, Genomes, Proteins, Eyes, and Life Forms out of thin air from nothing as if by magic.

https://en.wikipedia.org/wiki/Boltzmann_brain

**This is one of the dumbest ideas ever created by the mind of man. It's right up there with the theory of evolution. It is Creation by Disorder, Creation by Chaos, Creation by Entropy, or Creation by Chance. Yet, they all take it seriously, so seriously that they have CHANCE designing and creating everything that exists in our universe. These people want chaos, random disorder, entropy, or chance to be our Designer, our Creator, and our God.**

Boltzmann and his followers decided that everything in our universe was designed and created by "random fluctuations in entropy". In other words, it was created by disorder, created by entropy, or created by chance. Even the world-famous Stephen Hawking jumped onto the bandwagon and decided that the Big Bang, Planets, Stars, and Galaxies were designed and created by "quantum fluctuations". It's the same stupid idea. It's magic! It's Creation Ex Nihilo, Creation by Entropy, Creation by Random Disorder, Creation by Chaos, Creation by Death (natural selection), and Creation by Chance.

**Creation by Chance is the dumbest idea ever created by the mind of man, and we all fell for it, because we desperately wanted it to be true.**

Creation by Chance in every form has to be eliminated from science, if we really want to find and know the truth. Creation by Chance has to be tossed out of science as the ONLY causal explanation for the origin of everything that exists in this universe. Science needs to be upgraded. Currently, Creation by Chance or Chance Causality is the ONLY interpretation and the ONLY explanation that is given to scientific evidence within our science community at-large. We need to do better. We need to reach out and discover the True Cause of everything that we experience and observe.

That's one of the main reasons why I decided to upgrade my science to Science 2.0. Science 2.0 allows ALL of the evidence into evidence, and then it tries to pursue a preponderance of that evidence. I redefined science from "Creation by Chance" to "Observation and Experience" instead. That represents a massive upgrade to science as a whole. For me personally, it changed everything for the better. I'm actively looking for the True Cause of everything that exists, rather than assigning everything to chance.

I'm a scientist more than I am a philosopher, but I'm willing to do what it takes to upgrade the Philosophy of Science while I'm upgrading Science as well. I want to replace all of the stupid ideas and falsified philosophies with the things that have actually been experienced and observed.

We haven't experienced nor observed Creation by Chance, or Creation Ex Nihilo, or Spontaneous Generation, or Abiogenesis, or Macro-Evolution, or Chemical Evolution; but, we have experienced and observed the effects of Psyche, Intelligence, Quantum Consciousness, Conservation of Energy, Action at a Distance, Quantum Information Conservation, and ALL of the massless, heatless, entropyless, non-physical Photons and Quantum Fields that exist. We have observed, experienced, verified, and proven the effects of the massless, entropyless, non-physical Quantum Fields. Gravity, microwaves, radio waves, magnetism, x-rays, light rays, thoughts, after-death memories, and other types of quantum waves ARE massless, entropyless, non-physical quantum entities or quantum objects that we have actually experienced and observed. Each one of them is sustained by a different type of Quantum Field. Thanks to Quantum Field Theory and Quantum Mechanics, the non-physical has been proven to exist. It's been experienced and observed!

Rather than going with "Creation by Chance" as the majority of our science community has done, I choose to go with "Observation and Experience" instead. I choose to upgrade my science to Science 2.0 by abandoning Creation by Chance and by going with observations and lived experiences in its place. I choose to redefine science as observation and experience and knowledge rather than defining science as Creation by Chance and Creation Ex Nihilo as our science community has done. I choose to upgrade my science by choosing to find better and more demonstrable interpretations for the scientific evidence than the Creation by Chance interpretation that is typically given to scientific data in our modern era within our science community.

Most scientists in this world define "science" as Materialism, Naturalism, Physicalism, Physical Reductionism, Darwinism, Nihilism, Behaviorism, Determinism, Classical Physics, Entropy, Atheism, Creation Ex Nihilo, and Creation by Chance. Such an act is called *begging the question* or *jumping to conclusions*, which is a logic fallacy. In the complete absence of evidence, they define science as something that it can't possibly be. It's a *category error* logic fallacy. They define "science" as Creation by Chance and leave it at that. They make no effort to find the True Cause or the True Origin of the things that exist.

Materialism, Physicalism, Darwinism, Behaviorism, and Determinism were carefully and purposefully designed to convince people that God and the Supernatural do not exist. They were designed to convince people that Action at a Distance or Quantum Mechanics does not exist. Materialism, Naturalism, and their derivatives were designed to exclude and eliminate any and all evidence demonstrating or proving that Psyche or Syntropy or Photons exists. Materialism, Darwinism, and Naturalism make the claim that the immaterial, massless, entropyless, non-physical Photons and Quantum fields DO NOT EXIST. These falsified philosophies claim that ONLY physical matter exists. The Materialists, Naturalists, Darwinists, Nihilists, and Atheists ONLY believe in physical matter, entropy, death, and creation by chance. Therefore, all of their "science" is interpreted as Creation by Physical Matter (idolatry), Creation by Disorder or Entropy (the second law of thermodynamics), Creation by Death (natural selection), or Creation by Chance. These falsified philosophies represent everything that is currently wrong with modern-day science. The whole thing is nothing but Creation by Chance.

We'll never be able to use Materialism and Naturalism to prove that God exists or to prove that Action at a Distance exists because these atheistic philosophical assumptions start with the *pre-chosen conclusion* that God, Psyche, Action at a Distance, the massless Photons, the entropyless Quantum Fields, and Quantum Mechanics DO NOT EXIST. The

ONLY way to prove that God exists is through direct observations and lived experiences. If God doesn't reveal Himself to us, then He remains forever unknown. Likewise, the BEST way to prove that Psyche, Action at a Distance, Spirit Matter or Dark Matter, Quantum Tunneling, the Quantum Zeno Effect, and the massless, heatless, entropyless, non-physical Photons and Quantum Fields exist IS to live them and experience them and observe them for ourselves.

I used to be a Materialist, Naturalist, Nihilist, and Atheist; so, I know of what I speak. I have been there and done that. Self-deception works, and it works every time. When it comes time to interpret the scientific evidence, garbage in produces garbage out. Creation by Entropy, Creation by Disorder, or Creation by Chance is the dumbest idea ever produced by the mind of man; and, most of our science community has chosen to define "science" as Creation by Entropy, Creation by Disorder, Creation by Death (natural selection), or Creation by Chance.

One day, I decided that I didn't like where that road was taking me, so I (My Human Psyche) chose to turn around and go the other way. I didn't like My Materialism, My Naturalism, My Nihilism, and My Atheism. I didn't like all the deceptions and the lies. So, I did a 180 and headed off in the other direction; and, my life and my science have been getting better and better ever since.

How we choose to define things makes all the difference in the world. As long as we continue to define "science" as Materialism and Naturalism, then Science's most basic, most fundamental, and most essential questions will remain forever unanswered and unknown to us. It's unavoidable. As long as we continue to define science as Creation Ex Nihilo or Creation by Chance, we will NEVER find the True Cause of anything that exists.

**Creation by Chance is the dumbest way to interpret scientific evidence, because Creation by Chance IS the Null Hypothesis in every science experiment that we conduct. By definition, the Null Hypothesis was produced by chance alone. The Null Hypothesis or Creation by Chance is automatically falsified by True Causal Entities and True Causal Forces. The whole purpose of a science experiment is to find the True Cause of what we are observing and experiencing, which means that the whole purpose of a science experiment is to falsify and reject and eliminate the Null Hypothesis, or Creation by Chance, or Chance Causality.**

We NEED to upgrade our science to include everything that has ever been experienced and observed, and then we NEED to find its True Cause.

Materialism, Naturalism, Behaviorism, and Determinism have NO scientific explanation for something as simple and basic as choice! They can't explain choice, so they say that it does not exist. That's what they do with everything that they can't explain – they say that it doesn't exist, sweep it under the rug, and pretend that it doesn't exist. That's what they tried to do with Quantum Mechanics and Quantum Field Theory. They can't explain it, so they just ignore it. As long as we choose to define "science" as Materialism and Naturalism, then Science's most basic concepts will remain unexplained and unexplainable.

However, if we choose to define "science" as Fine-Tuning, then instantly we have thousands (if not millions) of different Scientific Proofs of God's Existence suddenly spring forth before our eyes as if from nowhere. I KNOW because I have done that as well. Science IS Fine-Tuning; and, Fine-Tuning is an infinitely better definition for Science than Materialism, Naturalism, Darwinism, and Atheism. Whenever we do Science, we are in fact doing Fine-Tuning of some sort, are we not? We are testing things and trying to determine their True Cause.  We are trying to figure out how things really work.  We are trying to eliminate chance from the equation.

How we choose to define things makes all the difference in the universe.  How we choose to interpret the scientific data makes all the difference in a science experiment.

When we choose to define "science" as Fine-Tuning, Science's explanatory power literally goes through the roof!  The same thing happens when we choose to define "science" as Quantum Mechanics, Action at a Distance, Intelligence, Psyche, Quantum Consciousness, Quantum Field Theory, Observation, Conservation of Energy, Conservation of Quantum Information, Lived Experiences, Eye-Witness Evidence, and Syntropy.  Instantly, we can explain EVERYTHING that comes our way.  Suddenly, nothing is beyond our reach.

Mark My Words

# PART 0 – PREFACE MATERIALS FOR SCIENCE 2.0

In these chapters and essays, I try to demonstrate why science is in serious need of an upgrade to Science 2.0. It's not just a gimmick. It's an absolute need. Science has been off-course ever since it was designed and created. Science harbors falsehoods rather than pursuing the truth. Science as it currently stands is in desperate need of an upgrade.

Science 2.0 allows all of the evidence into evidence, and then it pursues a preponderance of that evidence. It's a new and better way of doing science. It's the way that science should have always been done but wasn't.

These introductory essays are an attempt to demonstrate that materialistic and naturalistic Newtonian Science and Classical Physics are in desperate need of a massive upgrade, because they are very limited in their explanatory power.

Over a period of three years, I slowly taught myself to understand physics from a massless, non-physical, quantum perspective. I eventually came to understand that ONLY Nature (or Nature's Psyche) and the Gods can actually DO physics at the massless, non-physical, quantum level. So, understanding physics from a non-physical perspective is not the same thing as DOING physics at the non-physical level; but, it's a step in the right direction where Science is concerned.

The massless and non-physical aspects of Physics and Science are wide open and undiscovered country, because the Materialists, Naturalists, Darwinists, and Atheists who run our public schools won't let us study nor do research into the massless and the non-physical, because they have erroneously chosen to believe that it does not exist. If you want the truth, you have to get it someplace else besides our public schools.

I eventually discovered that it is Nature's Psyche who collapses the wave function, and NOT the Human Psyche or the Human Observer. Nature's Psyche was collapsing the wave function eons BEFORE the first particle of physical matter was designed and made.

These are NEW scientific discoveries that nobody has ever found nor thought of before, because nobody has been willing to look at physics from a massless, non-physical, quantum, transdimensional perspective before now.

Even something as simple as the true definition for entropy has remained undiscovered until now because nobody has ever chosen to look at entropy from a massless and non-physical perspective until now.

In this book, I introduce a new and better way to do science that produces astounding results, especially in the realm of Quantum Theory or the scientific study of the massless and the non-physical. You have to be willing to look at it before you will actually discover it.

Science 2.0 allows all of the evidence into evidence, and then it pursues a preponderance of that evidence. The main benefit of Science 2.0 is that it permits us to falsify Materialism, Naturalism, Creation by Natural Selection (Creation by Death), Darwinism (Creation by Chance), Nihilism, Atheism (Creation Ex Nihilo), Classical Realism, and the Second Law of Thermodynamics (Creation by Chaos and Disorder) using the science and the evidence that has actually been experienced and observed.

The false is falsified by the truth, and the truth is repeatedly and constantly observed.

Mark My Words

# Introducing Science 2.0

Science 2.0 is open-minded science.

Science 2.0 allows ALL of the evidence into evidence, and then it pursues a preponderance of that evidence. It's a NEW and BETTER way to do science. It's the way science should have been done but wasn't.

I'm trying to fix everything that's wrong with science. That's impossible, I know, but it's fun to try.

A philosophy of science is an interpretation of science. The purpose of Science 2.0 is to demonstrate clearly and conclusively that there are BETTER ways to interpret science than the way we are currently interpreting science. Materialism, Naturalism, Darwinism (Creation by Chance), Nihilism, and Atheism (Creation Ex Nihilo) ARE NOT SCIENCE. They are philosophies of science. They are interpretations of science. And, they are the worst way that science can be interpreted. They produce Bad Science because they are false and falsified religions or philosophies of science.

The purpose of Science 2.0 is to demonstrate clearly and conclusively that there are better ways to interpret the scientific evidence than the way that we are currently doing the job. Science 2.0 is not just a gimmick. Science 2.0 is meant to be a true upgrade to both Science and the Philosophy of Science.

Materialism, Naturalism, Darwinism, Nihilism, Determinism, Physical Reductionism, Behaviorism, and Atheism ARE NOT science. They are philosophies of science. They are different ways that science is interpreted. A philosophy of science is a way that science gets interpreted. This is where science falls apart and fails – through all of the different, contradictory, mutually exclusive, and faulty personal interpretations that we give to scientific evidence.

I'm looking for a NEW, BETTER, STANDARDIZED, and MORE REALISTIC philosophy of science or interpretation of science.

Science 2.0 is my attempt to provide the world with a superior Philosophy of Science. I'm trying to find better, more realistic, and more demonstrable ways to interpret science than the faulty, falsified, and outdated ways of interpreting science that we are currently getting from Materialism, Naturalism, Darwinism, Nihilism, Atheism, and their derivatives.

Materialism and Naturalism deny the existence of the non-physical, which means that **they deny the existence** of quantum waves (photons or light), quantum consciousness (intelligence or psyche or life force), quantum fields (gravity, radio waves, microwaves, x-rays, magnetism), and every other invisible non-physical force, field, and entity that exists (dark matter or spirit matter as well as dark energy). Materialism, Naturalism, Darwinism, Nihilism, and Atheism **deny the existence** of every NEW SCIENCE that has been discovered, proven, and verified during the past century and a half. Materialism, Naturalism, and their derivatives are faulty, falsified, and outdated philosophies of science. They have become worthless, and it's time for us to do better.

Rather than interpreting science in terms of Materialism and Naturalism, I'm trying to interpret science in terms of Quantum Mechanics, Quantum Field Theory, the Perpetual Motion Cycle, the Conservation of Energy, the Conservation of Quantum Information, Quantum Consciousness, Spacetime, Action at a Distance, Dark Energy, Dark Matter or Spirit Matter, Quantum Tunneling, the Quantum Zeno Effect, and the other non-physical things in this universe that have actually been experienced and observed.

**The non-existence of these non-physical things has not been experienced nor observed!**

I'm trying to develop a better Philosophy of Science or a better Interpretation of Science than the falsified garbage we are getting from Materialism, Naturalism, Darwinism, Nihilism, and Atheism which have NEVER been experienced NOR observed. I'm trying to base Science upon what has actually been experienced, and observed, and verified. I'm trying to provide Science with a better interpretation or a better Philosophy of Science.

The false is falsified by the truth, and the truth is repeatedly and constantly experienced and observed. Massless, heatless, chargeless, entropyless, non-physical quantum phenomena such as quantum waves or photons are constantly being experienced and observed; and, their verified and proven existence FALSIFIES Materialism, Naturalism, Darwinism, Atheism, and their derivatives.

Of course, Science 2.0 is going to be resisted and rejected by the science community at-large, because it falsifies Materialism, Naturalism, Darwinism, Nihilism, and Atheism – and it eliminates them from science. It replaces them with a better Philosophy of Science that actually matches with what has been experienced and observed. As a new and better Philosophy of Science, Science 2.0 will be rejected and ignored by the people who need it most, because it is the truth and is based upon the truths which falsify Materialism, Naturalism, Darwinism, and their derivatives. These people don't want the truth. They prefer the deceptions and the lies instead.

**Statistics Falsifies Creation by Chance or Creation by Magic**

https://ultimate-model-of-reality.com/statistics-falsifies-creation-by-chance-or-creation-by-magic/

Causation implies correlation. If there is NO correlation between variables – as in the case of perfect disorder, perfect randomness, or pure chance – then there is NO causal relationship between the variables.

Materialism, Naturalism, Darwinism, Nihilism, Atheism, Creation Ex Nihilo, the Theory of Evolution, and the Second Law of Thermodynamics are ALL Creation by Chance or the Null Hypothesis. There can be NO causal relationship between pure chance (or random disorder) and anything else. Chance and causation are mutually exclusive. Whenever chance is your independent variable, there can be NO causal relationship between your independent variable and ANY dependent variable, because there can be NO correlation between pure chance and ANY dependent variable. Chance precludes causation and correlation. No correlation, then NO causation. Creation by Chance, Random Disorder, Creation Ex Nihilo, and Creation by Magic have NO correlation with anything, which means that they can't cause anything. This is the foundational RULE of Science and the Scientific Methods.

Statistics and Science are supposed to eliminate chance, blind luck, disorder, randomness, or entropy as a causal agent or a causal source. Yet, there is a group of scientists in our society who deliberately and knowingly equates correlation with causation, in complete violation of the rules of statistics and science. They literally turn chance, or entropy, or random luck into the cause of everything that exists. We know who they are by what they teach and preach.

Pagano, R.R. (2010). *Understanding Statistics in the Behavioral Sciences* (9th ed.). Belmont, CA: Wadsworth.

> **Statistics uses probability, logic, and mathematics as ways of determining whether or not observations made in the real world or laboratory are due to random happenstance or perhaps due to an orderly effect one variable has on another. Separating happenstance, or chance, from cause and effect is the task of science, and statistics is a tool to accomplish that end.**
>
> **Occasionally, data will be so clear that the use of statistical analysis isn't necessary. Occasionally, data will be so garbled that no statistics can meaningfully be applied to it to answer any reasonable question. But I will demonstrate that most often statistics is useful in determining whether it is legitimate to conclude that an orderly effect has occurred. If so, statistical analysis can also provide an estimate of the size of the effect.**
> (*Understanding Statistics*, p. xxvii.)

According to the Philosophy of Science, the whole purpose for statistics and the scientific method is to ELIMINATE chance as a causal source or a causal agent. Chance is an invalid cause or a false cause according to statistics and the scientific method. Chance falsifies a scientific theory and ruins a science experiment according to the rules of science. Chance or entropy is unreliable. Chance is not replicable in a laboratory nor in a science experiment. Chance as a causal explanation is the very definition of Bad Science. Chance, or entropy, or disorder can't design and create anything, according to statistics and the scientific method. Chance, or entropy, or randomness, or luck is an invalid cause according to the rules of science. There's NO chance that chance can design and create anything reliably.

Chance and causation are mutually exclusive. If we can demonstrate from the results of a science experiment that the Null Hypothesis is true – that the results of the experiment were produced by chance alone – then WE KNOW that the independent variable is NOT the cause of the dependent variable. That's how a science experiment works. We use the presence of chance to eliminate or falsify causation, because chance can never do True Causation. Chance and causation are mutually exclusive. If we detect causation or choice, then WE KNOW that chance had nothing to do with it. If we are detecting nothing but pure chance in our scientific data, then WE KNOW that there is NO causal relationship or NO correlation between the independent variable and the dependent variable. That's how we detect causation – by eliminating chance as the only explanation for the scientific evidence that we are observing. If we eliminate, falsify, or reject Chance or the Null Hypothesis, then we have established that some type of causation has taken place. Chance and causation are mutually exclusive. By definition, in principle or practice, chance cannot do True Causation.

But, what do we see from the Materialists, Naturalists, Darwinists, and Atheists?

We observe these people using chance or entropy or happenstance as the cause of DNA, RNA, genes, proteins, synapses, eyes, brains, life-forms, physical matter, planets, stars, galaxies, and universes. According to these people, everything that exists was caused by chance, in complete violation of the rules of science. These people use chance, disorder, happenstance, or entropy as the causal explanation for everything, in complete violation of the intent behind statistics and the scientific method. These people grant themselves a *special exemption* from the rules of science, and then they use chance, or entropy, or happenstance, or death (natural selection) to explain everything that they encounter in science. They cheat, in order to prove their point.

In Darwinism, Naturalism, Materialism, and Atheism, these people literally turn chance, happenstance, or random fluctuations in entropy into our Designer and our Creator. These people deny the existence of orderly effects and the existence of causal agents; and instead, they attribute everything to happenstance or chance. These people make chance or happenstance the CAUSE of everything that exists.

Science, the scientific method, and statistics were designed to prevent these people from doing precisely what they are doing – making chance or happenstance the cause of everything they see around them. When it comes to creation by entropy or design by chance, these people even go so far as to make the size of the effect infinite. These people literally define entropy as ALL of the heat in the universe, or ALL of the disorder and chance in the universe, or as ALL of the energy in the universe. They make entropy the ultimate causal agent, and they literally make the size of entropy's effect infinite. They cheat. The Materialists, Naturalists, Darwinists, Nihilists, and Atheists violate the rules of science and defeat the purpose of statistics; and then, they make their deceptions and lies axiomatically true instead. Within all of their science, these people literally turn correlation into causation every time. They turn chance into causation and make chance the cause of everything that exists.

Can you believe it? These people literally turn chance or entropy into our Designer, our Creator, and our God. That's what they do, is it not? Isn't that what we have experienced and observed from these people? The observational science, that we call Behaviorism, observes these people using chance or entropy or natural selection to explain everything that they do in science, in complete violation of the rules of science and the purpose of statistics. These people literally turn statistics into the art of lying with math, each time they manipulate the data so that it makes it seem that chance or happenstance is designing and creating everything we see around us, including physical matter and genomes.

The Materialists, Naturalists, Darwinists, and Atheists don't do any experiments to determine if entropy causes things to happen, because they can't. Instead, they *jump straight to the conclusion* that entropy or chance caused everything in this universe to happen. These are NOT experimental sciences. This is philosophy, metaphysics, and dogma instead. Entropy or natural selection doesn't cause anything to happen. Entropy and natural selection are the result or the product, NOT the cause! Entropy and natural selection actually prevent abiogenesis, chemical evolution, and macro-evolution from happening. It takes an infinite amount of blind faith to believe that entropy or chance designed and created everything that we see around us today – a blind faith that violates the whole purpose of statistics, science, and the scientific methods. Everything these people do violates the rules of science by making chance or entropy the cause of everything in this universe. They do this because they don't understand the Philosophy of Science, or the purpose of science and statistics.

Interestingly, according to the claims of the scientists themselves, most of them prefer the deceptions and the lies, rather than the truth. They don't want to falsify their theories and their ideas. They deliberately hide, manipulate, or destroy the evidence that falsifies their beliefs. Atheism of any kind is a refusal to look at evidence and a rejection of evidence. These people literally destroy, hide, and eliminate all of the evidence that falsifies their pre-chosen beliefs. That's how they do science – by destroying the evidence that falsifies their beliefs. Why do they do this? Well, Materialism, Naturalism, Darwinism, Nihilism, and Atheism have become their personal religion; and, they will dogmatically defend their religion to their last breath whenever someone demonstrates why they are wrong or falsifies their religion. But that doesn't change the fact that they are wrong, and it doesn't change the fact that they use chance or happenstance to explain everything that they do in science, in complete violation of the rules of science and to the complete ruin of the purpose of statistics.

*Vaxxed* is a documentary that demonstrates how these types of scientists manipulated and decreased the sample sizes in order to turn lies into truths. They mastered the art of lying with statistics; and, they did so because there is a lot of money to be made. It's how they make their living. These people made their millions by lying to us and deceiving us; and, all of our autistic children paid the price to make them rich.

https://www.amazon.com/Vaxxed-Cover-Up-Catastrophe-Del-Bigtree/dp/B06XGSJBZM

*Vaxxed* is a movie about statistics and scientists gone bad, and scientific data suppressed and destroyed.

Natural selection and sexual selection are pure chance. Natural selection is creation by death and entropy. Evolution of any kind is entropy, because these people only believe in entropy and physical matter. When you get selected against, you die; and then, entropy allegedly takes over. The Darwinists literally have death, entropy, and chance as our Designer, Creator, and God. Contrary to the rules of logic and science, these people really truly believe that entropy or chance designed and created everything that we see around us. That's what they believe, is it not? That's what they teach and preach, is it not? These people have performed and accomplished the ultimate logic fallacy and the ultimate violation of statistics and science by turning entropy or chance into our Designer and our Creator, in complete violation of the rules of science and the rules behind the scientific methods.

Science 2.0 uses the Philosophy of Science and the rules of science to identify and eliminate the falsehoods and false procedures in science and the scientific methods. Science 2.0 uses the Philosophy of Science to upgrade and improve science. According to Science 2.0, science is defined as observation and experience. If it has never been experienced nor observed, then it really isn't science. It is science fiction instead.

For example, design and creation by entropy, heat death, natural selection, chemical evolution, spontaneous generation, abiogenesis, or macro-evolution has NEVER been experienced nor observed, which means that it's science fiction. It's physically impossible. Creation by entropy or creation ex nihilo can't be produced nor reproduced in a laboratory either, because it's physically impossible. Design and creation by chance is unreliable, which means that it's patently unscientific. It's *wishful thinking* instead, which is a logic fallacy.

Likewise, the multi-me interpretation of quantum mechanics or the many worlds interpretation of quantum mechanics has NEVER been experienced nor observed by anyone – not even God – so it doesn't really exist. It's science fiction and has only been experienced and reported in our science fiction movies and nowhere else. Each human being seems to have only ONE psyche, intelligence, or spark. We have multiple bodies – a physical body and possibly multiple spirit bodies, or a single spirit body capable of living at multiple different levels, phases, or dimensions simultaneously; but, we only have ONE psyche, intelligence, consciousness, life force, or spark according to the near-death experiencers and the out-of-body explorers.

The thing to remember in all of this is that chance cannot do causation. Chance and causation are mutually exclusive. Chance represents the absence of causation or the absence of choice. ONLY psyche, or intelligence, or life force, or quantum consciousness can do causation or choice – both at the quantum level and the physical level where we currently live and work.

Mark My Words

### Science Is Observation and Experience

https://quantum-law-of-thermodynamics.com/science-is-observation-and-experience/

Technically, infinite entropy, infinite and ever-increasing entropy, and "entropy that has never decreased and gone to zero" has NEVER been experienced NOR observed. The Big Bang Theory actually falsifies the idea! Entropy is constantly going to zero and ceasing to exist, according to the equations for entropy. If the second law of thermodynamics is true, then the Big Bang Theory can't possibly be true. In contrast, if the Big Bang Theory is true, then the second law of thermodynamics can't possibly be true. They are mutually exclusive. They falsify each other. That means that one of them is automatically false; and, for all we know, they both are false.

So, which one is true, and which one is false?

I KNOW which one is more likely to be true, because I have been studying physics and entropy from a non-physical, massless, quantum perspective for a few years now. I've given the whole thing some serious thought.

Here is what I discovered!

### The one that has been experienced and observed is the one that is real and true.

Massless, heatless, chargeless, entropyless, non-physical photons or quantum waves have actually been experienced and observed! Their very existence FALSIFIES Materialism, Naturalism, Darwinism, Nihilism, Atheism, and their derivatives which state that the massless and the non-physical DO NOT EXIST.

Each photon that is made by Nature or by Nature's Psyche is in fact a miniature Big Bang. The Big Bang by definition is a complete reversal of entropy, a situation where entropy has gone to zero and ceased to exist. The Big Bang, if it happened, was by definition pure syntropy – a complete and total reversal of entropy. The same thing happens every day whenever Nature or Nature's Psyche makes a photon from the absolute-zero temperatures of the energies within the quantum fields, or from physical matter in our sun. The production of a massless, chargeless, heatless, and entropyless non-physical photon is pure syntropy – it is a miniature Big Bang – it is a complete reversal of entropy. Photons have NO entropy because they have no mass, which means that they have NO heat storage capacity. Mass has heat storage capacity, according to the equation for heat. NO mass, then NO heat storage capacity. NO mass, then NO heat, NO thermodynamics, and NO entropy. This is precisely what we experience and observe.

Whenever Nature or Nature's Psyche transforms mass, heat, and entropy into a massless, chargeless, heatless, and entropyless non-physical quantum wave or photon, ALL of that mass and heat get transformed into infinite acceleration and omnipresence instead. A newly made photon goes from zero to the speed-of-light (or faster) instantly. That's infinite acceleration. This is what we actually experience and observe. A photon is capable of infinite acceleration and any speed of its own choosing because it has NO mass or NO resistance to acceleration. A photon has no mass, which means that it has no heat storage capacity, which means that a photon has NO heat. A photon or quantum wave has transformed all of its mass, heat, resistance to acceleration, heat storage capacity, and entropy into massless, heatless, chargeless, and entropyless infinite acceleration or quantum omnipresence instead. This is what we actually experience and observe.

A massless, chargeless, and entropyless photon is made within a realm that has NO entropy; and, a photon is by definition a massless object that has NO entropy or NO resistance to acceleration. A photon travels at maximum efficiency through the absolute-zero temperature quantum fields that fill the expanse of space. By definition, a photon has no heat while it travels because it has no mass. Mass is heat storage capacity, and a photon has none. Mass is resistance to acceleration, and a photon has none.

A photon is omnipresent quantum waves while it is traveling. It travels at the speed-of-light or faster because it has NO entropy or NO resistance to acceleration. A tachyon by definition is a photon that has chosen to quantum tunnel or has chosen to travel faster than the speed-of-light. A photon or tachyon is omnipresent because it has no mass, no entropy, and no resistance to acceleration while it is traveling. Only when it deliberately stops and manifests itself in spacetime does a photon try to gather itself into a single location and transform itself into heat, mass, and entropy. This explains everything that has ever been experienced or observed, does it not?

A photon is massless and omnipresent until it chooses to stop and transform itself into heat, mass, and entropy instead. A photon doesn't have to stop if it doesn't want to. A photon can travel through glass, water, and the earth as if they weren't even there. This has actually been experienced and observed. There are lots of massless quantum particles who have that ability, and a photon is one of them. They can go straight through the earth or the sun as if the sun or the earth weren't even there, IF they choose to do so.

Photons can also choose to stop and transform themselves into heat, mass, and entropy anytime and anywhere they want to do so. The Big Bang Theory tells us that this is true. It has to be true in order for the Big Bang Theory to be true. If the dual nature of photons or the Big-Bang-Like nature of photons is true, then that truth falsifies the second law of thermodynamics by completely reversing entropy and decreasing that entropy to zero. The truth is that entropy is constantly decreasing and going to zero, whenever a mass-based particle is transformed into a massless particle or whenever matter is transformed into energy and light – in complete contradiction to the second law of thermodynamics which states that entropy is always increasing, and that entropy can never decrease.

**The Perpetual Motion Cycle E = $mc^2$** falsifies the second law of thermodynamics every time that Nature or Nature's Psyche transforms mass, heat, and entropy into massless and entropyless and heatless photons instead. This is what we have actually experienced and observed! E = $mc^2$ is the Perpetual Motion Cycle. It works perfectly both directions, and all the while the energy is conserved.

Einstein is the one who discovered the Perpetual Motion Cycle, but he didn't identify it as such because he had rejected Quantum Mechanics or Action at a Distance, and was hung up on Materialism, Naturalism, or Classical Realism instead. However, the Perpetual Motion Cycle E = $mc^2$ IS the mechanism behind the Conservation of Energy, Quantum Field Theory, and the massless and entropyless Perpetual Motion Machines that we call the Quantum Fields.

Psyche is the thing that triggers the transformations of energy.

According to the LAW of Psyche, each psyche or intelligence has a certain amount of energy that's under its control, and that controlling psyche can form or transform the energy under its control into anything it wants that energy to be, anytime and anywhere that it chooses to do so.

In our sun, Nature's Psyche is currently transforming mass, heat, and entropy into massless, heatless, chargeless, and entropyless infinite acceleration photons or quantum waves. Whenever a quantum wave or photon chooses to stop, the psyche or intelligence within that photon or quantum wave transforms its omnipresence, infinite acceleration, and superluminal velocity into mass, resistance to acceleration, heat, or mass's heat storage capacity instead. The psyche who is controlling each quantum wave or photon can transform the energy under its control into anything it wants that energy to be, including mass, electrons, quarks, physical matter, spirit matter or dark matter, heat, or mass's heat storage capacity. This is the Perpetual Motion Cycle $E = mc^2$, and all the while, ALL of the energy is conserved. This Perpetual Motion Cycle will never wear out and never cease to exist, because it is constantly being conserved thanks to the Conservation of Energy and Psyche and Quantum Information. ALL of this has actually been experienced and observed; therefore, WE KNOW that it is real and true.

The false is falsified by the truth, and the truth is constantly and repeatedly experienced and observed.

Nature's Psyche makes the quantum waves, and Nature's Psyche collapses the wave functions. Nature's Psyche makes and MAPS each physical brain. When the Human Psyche wants to lift its finger, Nature's Psyche chooses and fires the specific neuron that it made and assigned to that specific task. It's Nature's Psyche who fires the neurons in your brain by collapsing the necessary wave functions, because it was Nature's Psyche who made and MAPPED your brain in the first place. It's Nature's Psyche who controls your brain and fires the correct neurons in your brain. The Human Psyche isn't consciously aware of any of these quantum processes. Nature's Psyche is.

We NEED a mechanism at the quantum level that is smart enough to KNOW how to transform energy into mass, and mass back into usable energy or exergy. Psyche is that machine. Psyche or Intelligence or Quantum Consciousness is the ONLY machine or mechanism at the quantum level that has actually been experienced and observed, that is capable of making quantum waves and capable of collapsing the wave function AT WILL.

At the quantum level and the psyche level, a psyche looks like a photon or a pinpoint of light. A Psyche is a real quantum computer. A Psyche makes, transmits, receives, processes, and stores quantum waves or thoughts. In science, we NEED an intelligent mechanism, capable of designing and making the Quantum Fields. The Gods or Controlling Psyches had to design and make the quantum fields BEFORE the first particle of physical matter could be made and sustained. According to Quantum Field Theory, there can be NO physical matter without the quantum fields, and there can be no quantum fields and quantum laws and quantum perpetual motion machines without some type of psyche or intelligence to make them in the first place. That's just the way it is.

Photons are massless and entropyless infinite acceleration. Photons go from zero to the speed-of-light instantly at an infinite acceleration, because photons have NO mass, NO resistance to acceleration, and NO entropy. Photons literally transform their mass, heat, and entropy into infinite acceleration instead. It happens instantly, and it sets them free. Only when the photons stop, do they in fact transform themselves back into heat, mass, and entropy. This is what we have actually experienced and observed! Is it not?

Every instantiation of a photon is a miniature Big Bang – the creation of a whole new universe. Every time that mass or physical matter is converted or transformed into a photon, its mass, heat, and entropy are transformed into infinite acceleration instead. It's like the inflationary period within the Big Bang Theory. A photon also represents a complete reversal of entropy or a complete elimination of entropy. Photons falsify the second law of thermodynamics, which states that entropyless photons can't exist and don't exist.

The photon sets off instantly at the speed-of-light or faster. A photon is intelligent. A photon is a psyche or has a psyche. A photon launches at an infinite acceleration, going from zero to the speed-of-light instantly because it has NO resistance to acceleration (no mass); and, then it chooses its final velocity which is anywhere from the speed-of-light to infinity. It happens every time that a photon is made, whether that photon is made from physical matter or from the infinite expanse of the massless and entropyless quantum fields. This explanation of science matches perfectly with everything that we have ever experienced or observed. Does it not?

We have the truth, and then we have everything else. The false is falsified by the truth; and, the truth is constantly experienced and observed. Entropy is constantly decreasing to zero and ceasing to exist every time that matter or mass is converted or transformed into pure, syntropic, massless energy, or light, or photons instead. By definition, a photon is massless, which means that by definition it has no entropy because it has no mass or no resistance to acceleration. We eliminate entropy, or entropy goes to zero, every time we convert mass or matter into massless energy or light. Think of the proven and verified science where we have actually converted or transformed matter into massless and entropyless energy and light. A hydrogen bomb produces a lot of massless energy and light. The entropy within that massless energy or light goes to zero and ceases to exist, according to the equations for entropy and according to the Big Bang Theory.

Every time that Nature or Nature's Psyche decides to transform a particle of physical matter into a photon, Nature transforms the mass, heat, and entropy within that physical particle into massless, heatless, and entropyless infinite acceleration instead. This is what we actually experience and observe! The thing goes instantly from having resistance to acceleration (mass) to having NO resistance to acceleration (no mass) and infinite acceleration instead. A photon goes instantly from zero to the speed-of-light, each time that a particle of physical matter is converted or transformed into a massless and entropyless photon. This is what we have experienced and observed, is it not?

I have moved way beyond "wanting this to be true". I KNOW that it is true because it has actually been experienced and observed. I KNOW that it is true because I understand the physics behind it. Unlike other scientists, I understand the physics from both the physical side and the non-physical quantum side. I can see with my own eyes that they are in complete harmony with each other. I understand the first law of thermodynamics or the conservation of energy, and I KNOW that Nature or Nature's Psyche can transform the energy under its control into anything that it wants that energy to be, anytime and anywhere that it chooses to do so. Whenever a physical particle is transformed into a photon, ALL of its mass, heat, and entropy is transformed into massless, heatless, and entropyless infinite acceleration instead. Whenever a photon chooses to stop, it transforms its omnipresent energy into mass, heat, and entropy; and, that's how it stops.

I KNOW that this is true because it is obviously true and demonstrably true. I KNOW that this is true because it allows ALL of the evidence into evidence and then pursues a preponderance of that evidence. I KNOW that this is true because Science, or observation and experience, has already proven it true. I KNOW that this is true because it matches perfectly with the Perpetual Motion Cycle, **$E = mc^2$**, both coming and going. I KNOW that this is true because it's constantly being experienced and observed and verified. I KNOW this is true because it provides a realistic and logical mechanism for the construction of something like a photon or a big bang singularity. Photons ARE entropyless big bang singularities! I KNOW that this is true because it actually takes the massless or the non-physical into consideration. I KNOW that this is true, because it actually explains what happens to the heat, mass, and entropy whenever Nature or Nature's Psyche chooses to make a photon instead. I KNOW that this is true because it falsifies Materialism,

Naturalism, Darwinism, Nihilism, Classical Physics, Classical Realism, Atheism or Creation Ex Nihilo, and the Second Law of Thermodynamics – all of which are demonstrably false. I KNOW that this is true because it explains everything that has ever been experienced and observed.

Science 2.0 provides us with much greater explanatory power because it allows all of the evidence into evidence – including the massless evidence or the non-physical evidence. If we want to find and know the truth, then we must be willing to allow all of the evidence into evidence. Knowing the truth sets us free.

These NEW scientific discoveries became possible because I finally chose to look at science from a massless, non-physical, quantum perspective. Change your perspective, and it changes the way you see the world.

I've worked hard to master and enhance the Philosophy of Science. I've done more to upgrade and improve the Philosophy of Science than anyone else on this planet, because it's that important to the human race as a whole.

One of the primary lessons from the Philosophy of Science is the observable fact that when it comes to each piece of scientific data or scientific evidence, there are literally an infinite number of different ways that it can be interpreted. Many of those personal interpretations falsify each other, contradict each other, or are mutually exclusive. That reality and truth is the fundamental flaw in the Scientific Method. The final step of the Scientific Method is to interpret the data or the evidence; and, that's where we humans always fail and get things wrong. We stink when it comes to trying to figure out what the evidence truly means. Our personal biases and prejudices hide or destroy the true interpretation of scientific evidence every time.

I've written twenty books trying to improve the way that we interpret scientific evidence, because interpretation is everything when it comes to science, scientific data, and scientific evidence. In science, we are trying to find the true cause of the things that we experience and observe, and a faulty or false interpretation destroys what we are trying to accomplish by doing science in the first place. A stupid interpretation of scientific data – such as chance causality or creation by chance – hides and destroys and ruins the true interpretation of the scientific data and the true cause of the scientific evidence.

Garbage in. Garbage out. That's what stupid interpretations are all about.

I'm trying to develop better, more demonstrable, and more realistic interpretations for scientific evidence than the stupid and easily falsified junk that we keep getting from the Materialists, Naturalists, Darwinists, Nihilists, and Atheists. Rather than using chance as the cause of everything that exists as these people do, I'm actually trying to find the true cause of everything that exists. I'm actually trying to do science.

I'm a generalist. I'm good at everything and master of nothing. I KNOW to what extent Science 2.0 has been of value and worth to me. It's been worth a great deal, actually. It's not just a gimmick. Nevertheless, Science 2.0 leaves it up to you go decide for yourself whether it has any worth to you or not. Science 2.0 is based exclusively upon observation and experience – NOT peer review – therefore, what you have experienced and observed will be completely different than what I have experienced and observed.

However, it has become my personal belief that God gives to each person a piece of the overall puzzle, and if each one of us were to add our own unique piece of the puzzle into the collective encyclopedia and make that encyclopedia publicly available, then the human race as a whole would know and understand everything that exists by sheer observation and experience alone. That's what Science 2.0 is all about. It's about KNOWLEDGE – and not about philosophy, wishful thinking, censorship, peer review, nor personal speculation. Censoring away the evidence, hiding the evidence, and destroying the evidence is not the right way to do science.

Remember, a photon is an entropyless syntropic big bang singularity. That means that the Big Bang is ongoing to this very day. The Big Bang was NOT a one-time event.

Nature or Nature's Psyche is constantly transforming entropyless omnipresence or entropyless infinite acceleration into mass, resistance to acceleration, heat, and mass's heat storage capacity (entropy) all the time all around us. A miniature Big Bang happens every time a photon or quantum wave CHOOSES to stop and transform itself into mass, heat, and mass's heat storage capacity (entropy) instead. We can feel the heat on our skin whenever a photon CHOOSES to stop on our skin.

Likewise, Nature or Nature's psyche is constantly transforming the mass, heat, and entropy within our sun INTO massless, heatless, chargeless, entropyless, non-physical photons or quantum waves all the time. This whole process is eternal and everlasting because it is based upon the Conservation of Energy, Psyche, and Quantum Information and because it is based upon the Perpetual Motion Cycle $E = mc^2$.

Look at me. I just figured out how everything works! Best of all, it has actually been experienced and observed!

Mark My Words

# The Probability of Protein Synthesis by Chance Alone

https://origin-science.org/the-probability-of-protein-synthesis-by-chance-alone/

Throughout known life, there are 22 genetically encoded (proteinogenic) amino acids, 20 in the standard genetic code and an additional 2 that can be incorporated by special translation mechanisms. In eukaryotes, there are only 21 proteinogenic amino acids, the 20 of the standard genetic code, plus selenocysteine.

The Materialists, Naturalists, Darwinists, Nihilists, Atheists, Determinists, Physical Reductionists, Atomists, Behaviorists, and Classical Realists claim that proteins and their matching genes formed by chance alone.

Let's examine the probability of protein synthesis happening by chance alone.

First, we have to start with a God-given tub full of the correct 21 different amino acids that are used to form proteins, so that CHANCE actually has something to work with. CHANCE can't produce a tub full of the correct 21 different amino acids to start with. God has to do that for CHANCE so that CHANCE has a chance of making a protein by chance alone to begin with. Without that God-given tub full of the correct 21 different amino acids, there's no chance that CHANCE will ever be able to produce a protein by chance alone. CHANCE needs the building blocks for proteins already in place to begin with, or nothing is ever going to happen, either purposefully or by chance. This is logical common sense. Only the irrational would disagree with it. We have to start with a pot full or a tub full of the correct amino acids as a God-given fact; otherwise, there can be NO protein synthesis. No amino acids, then no protein synthesis. It's that simple.

Starting with that God-given tub full of the correct 21 amino acids, the probability of getting two correct amino acids in a row by chance alone is $(1/21) * (1/21) = $ "0.0022675737". That's two-tenths of a percent, which some people would consider to be within the realm of possibility. It's highly improbable, but it's possible for chance to get two amino acids in the correct sequence.

Let's back up a second. The odds of getting the correct amino acid in the first spot the first time is $(1/21)$, which is "0.0476190476". The odds are against that happening, but clearly it's possible to get the correct amino acid in the first spot the first time, assuming that God has already made that amino acid and placed it into the pot, or the tub, or the tube.

The probability of getting three correct amino acids in a row is $(1/21) * (1/21) * (1/21) = $ "0.0001079796998". That's 1/10000, which is still within the realm of detectable possibility according to the typical binomial distribution table, which takes things out to .0001 before the table drops to zero probability. It's extremely improbable for chance to get three correct amino acids in a row, but the hopeful and those operating on blind faith alone will claim that it's theoretically possible. "There's still a chance", they will say.

The probability of getting four correct amino acids in a row is $(1/21) * (1/21) * (1/21) * (1/21) = $ "0.000005141890467". That's five chances in a million; and, here we enter into the realm of impossibility. Technically, that's never going to happen, and it gets worse from there. At some point we have to call it a loss, or call it an impossibility; and for me, this is where it starts to happen.

Here, time really begins to be a factor. Five chances in a million are not good odds. If four amino acids in the wild find each other one time per year by sheer luck alone, it's going to take nearly a million years on average for the correct four to end up in the right sequence, and then the wind or the sun or the water will immediately destroy them anyway. Even with a God-given tub full of the correct 21 amino acids, God is going to have to stir the pot, or nothing is going to happen. The amino acids will just sit there and do nothing for all eternity, unless someone is stirring the pot. That same person also has to keep that tub full of amino acids in existence for millions of years so that chance has a chance of combining some of them into a useful sequence that is four amino acids long.

Remember, there are NO God-given tRNA molecules, mRNA molecules, and ribosomes in this tub to carry the amino acids to the appropriate ribosome and to stitch the amino acids together into a protein at the ribosome. We are relying upon chance alone to put these amino acids together into proteins. It's going to take a while to do so.

https://en.wikipedia.org/wiki/Protein_biosynthesis

The probability of CHANCE getting five correct amino acids in a row is "2.4485 e -7". That's never going to happen. There's just no chance that CHANCE can do the job that needs to be done. So, the best that CHANCE can do by chance alone is a protein that is four or five correct amino acids in a row; and, we are being extremely generous here, giving chance the full benefit of the doubt. The whole process is improbable to begin with, and it soon becomes impossible, even with a God-given tub full of the correct 21 amino acids to start with.

Remember, this process starts with a God-given tub full of the correct 21 amino acids, and with NOTHING else in the tub besides those correct amino acids. If you add anything else into the tub, then the probability drops to zero even quicker than it would with a tub full of the correct 21 amino acids. Furthermore, God has to keep stirring this pot or this tub continuously so that the different amino acids are shifting position and coming into contact with each other. If God isn't there to stir the pot, then nothing is ever going to happen.

God would have to make all of the amino acids in the first place and put them into close proximity to each other so that chance can then combine them into something unusable that is four or five amino acids long. One website stated that the shortest protein in the human genome is 44 amino acids long, and that's $21^{40}$ more than what chance can do by itself. Think about it. It's true. Chance just isn't all that reliable. The house always wins, which means that the Materialists, Naturalists, Darwinists, and Atheists always lose because they are relying exclusively upon chance alone to get the job done.

It's NEVER going to happen!

It can't happen. It's impossible. It's impossible to do protein synthesis by chance alone. That's just the way it is.

The claims and the definitions vary from one website to the next, but I also found this on Uncle Google:

**Thyroid releasing hormone or TRH should be the smallest protein in the human body, with 234 amino acids. (> 100 amino acids is a protein.) The smallest polypeptide in the human body should be insulin, with 54 amino acids. (10-100 amino acids is a polypeptide.)**

Let's run with this for a minute and see what we get.

The probability of making insulin by chance alone is:

$(1/21)^{54}$

This is equal to: 3.98252109 e -72

This can't be done by chance alone, even with a God-given tub full of the correct amino acids. Even with God constantly stirring the pot, it would take trillions of trillions of years for such a thing to happen by chance alone. It's just not going to happen. Our observable universe hasn't been around long enough for this to happen by chance alone. Insulin was designed and made by God. Without God to design it, make it, fine tune it, and field test it, insulin wouldn't exist.

The probability of making THR by chance alone is:

$(1/21)^{234}$

When I put this into my graphing calculator, the answer was ZERO (0). Literally. The answer is ZERO. This is NEVER going to happen ever! This is beyond what a graphing calculator can handle. This is beyond what God could do by chance alone!

The probability of making THR by chance alone is ZERO. You can wait for all eternity, and it will NEVER happen, even with a God-given tub full of the correct amino acids or with an ocean full of the correct amino acids. It's NEVER going to happen. Making THR by chance alone is impossible. It can't happen by chance alone. It requires some kind of intelligent and deliberate intervention.

If it were possible to make genes and proteins by chance alone, then we should also be finding computer chips and computers and calculators that were made by chance alone, given all the sand and minerals that we have on this planet, and the constant stirring that is taking place. It is obvious that genes, proteins, genomes, life forms, computer chips, and computers can ONLY be made through intelligent design and intelligent intervention. It is obvious that these things cannot arise through chance alone.

It is obvious that God made your polypeptides and your proteins! Each protein or polypeptide that you encounter is God's Signature. Proteins are impossible to make by chance alone. It can't be done, which means that it wasn't done that way. God designed and made your proteins and polypeptides. God also made the matching genes to go along with those polypeptides and proteins. The probability of anything like that arising by chance alone is ZERO. It's impossible. It's NEVER going to happen.

Your genome is God's Signature, and that Signature is written on every cell in your body. Polypeptides and proteins are also God's Signature, and that Signature is also written on you and throughout you. You were made by God whether you realize it or not. According to the probabilities and the statistics, there is no chance that I am wrong.

I just used Science and Statistics to PROVE that God exists. God has to exist, or genomes and proteins wouldn't exist. God has to exist, or you wouldn't exist. It's that simple. I've been doing this for a few years now, and I'm getting better at it. I'm not afraid of the truth anymore, like I used to be when I was a materialist, naturalist, nihilist, and atheist. I simply tell it as it is. Your proteins and genes were deliberately and intelligently MADE. Anything that is obviously made obviously has a Maker who made it. Our genes and proteins were obviously MADE; therefore, our genes and proteins obviously have an Intelligent Maker who designed them and made them. It's obvious that this is true.

Because I'm a mathematician, statistician, logician, and quantum theorist, I understand the math, and I understand probability and chance. For me personally, this is one of the most convincing Scientific and Statistical Proofs of God's Existence that I have ever encountered; and, I have encountered dozens of them, which I found convincing.

Mark My Words

# A Binomial Distribution Table Falsifies Creation by Chance

https://ultimate-model-of-reality.com/a-binomial-distribution-table-falsifies-creation-by-chance/

A binomial distribution table FALSIFIES chance causation, creation by chance, or chance causality of any kind, except for the very simplest of sequences.

We can calculate the probability of chance events.

https://origin-science.org/Binomial-Distribution-Table

The probability of tossing two heads in a row is 0.25, according to a binomial distribution table. That's doable, and realistic.

The probability of tossing three heads in a row is 0.125. That too is realistically possible.

The probability of tossing four heads in a row is 0.0625. The probability of five heads in a row is 0.0312. The probability of tossing six heads in a row is 0.0156. The probability of tossing seven heads in a row is 0.0078. That's getting down there. That's less than one percent, and it gets much worse from there. You can toss seven fair coins a hundred times, and it's possible that you will NEVER get seven heads to come up. Chance just isn't that reliable! If you need seven heads to win the jackpot, chances are good that you will run out of time or money before you achieve your goal.

According to a binomial distribution table, it is impossible to get fifteen heads in a row or fifteen tails in a row from a fair coin by chance alone. The probability of doing so has effectively dropped to zero.

And, that's with a binomial distribution – heads or tails – two possibilities for each trial or event. When it comes to proteins and amino acids, there are 21 different amino acids that are used to make proteins, which means that the sequence that can be produced by chance alone drops significantly lower than a sequence fifteen units long, as I will demonstrate later on in this essay.

Remember, according to a binomial distribution table, it is statistically impossible to get fifteen heads in a row with a fair coin, which also means that it is statistically and realistically impossible to get fifteen correct amino acids in a row, when trying to build a protein by chance alone.

When dealing with probability, when do you declare it a lost cause? When do you cash out and cut your losses? When do you declare a "low probability" to be synonymous with ZERO, IMPOSSIBILITY, or NO CHANCE?

On the typical binomial distribution table, they cut it off at 1 chance in 10,000. Beyond that, they call it impossible. Beyond a probability of 0.0001, they declare the chance or probability to be 0.0000, or they declare the event to be impossible, because it is.

**Nobody has time enough to waste to make something happen by chance alone!**

I'm a statistician and mathematician among other things. I have the formula for computing the probabilities – which are found on a binomial distribution table – to great precision on my graphing calculator.

The probability of getting 15 heads in a row with a fair coin is:

0.00003051757813

That's three chances in 100,000.

That means that if you toss 15 fair coins one hundred thousand times in a row, the chance or the probability that you will get ALL heads is three times, or three trials, or three tries in 100,000.

Time starts to be a factor here.

Assuming that you have 15 fair coins to start with, and assuming you can toss those fair coins and count the result once every minute, it will take you 100,000 minutes or 70 days of tossing those coins continuously in order to end up with fifteen heads, three separate times on average. Are you really going to sit there and toss 15 coins for 70 days around the clock on the hope of getting 15 heads in a row, two or three times during that 70-day period of time? That's why tossing 15 heads in a row with a fair coin is both statistically and realistically impossible. Lack of time prevents us from doing so. It's still theoretically possible on paper, but it completely impossible in reality.

Now swap those coins for amino acids instead.

According to a binomial distribution table and a Sign Test based upon the Correct or Incorrect amino acid coming up by chance alone, the probability of getting 15 correct amino acids in a row is also:

0.00003051757813

There's NO chance that it will ever happen.

The Materialists, Naturalists, Darwinists, Nihilists, and Atheists will complain and say, "But there's still a chance!"

NO, THERE ISN'T.

Are you really going to sit there and keep stirring the pot or keep tossing those amino acids for 70 days just so that you can get fifteen of them to accidentally line up in the correct sequence? You see, when it comes to CHANCE, you actually have to have a real person there who is willing to keep tossing the coins, or to keep tossing the amino acids, or keep stirring the amino acids until the correct sequence comes up by chance alone. Without a person there to toss the coins or to stir the amino acids, NOTHING is ever going to happen. This is Logic 101, and everyone completely overlooks it.

There really is nothing like chance causation or creation by chance, because you still need a person, psyche, or intelligence there to keep tossing the coins or to keep stirring the amino acids until "chance" produces the correct results. You also NEED a person there to recognize that fifteen heads in a row has been achieved or to recognize that the correct amino acid sequence has been achieved. And, with the amino acid "protein" that is produced by "chance", you also NEED a person, psyche, intelligence, or quantum consciousness there to make sure that that new "protein" is protected and used, or that "protein" which was produced by chance alone will be of no value to anyone.

**There is no such thing as pure chance or chance causation!**

"Chance" of any kind needs an intelligent set-up or an intelligent assist. In the case of coins, "chance" needs an intelligent person, psyche, or quantum consciousness to make the coins, toss the coins, count the coins, and recognize when the fifteen coins have produced fifteen heads. "Chance" doesn't do anything and can't do anything without an intelligent person or psyche there to toss the coins or to line up the amino acids into sequences.

You have to have a person who is willing to find and/or make the coins and the amino acids in the first place. Without any coins or amino acids, there's nothing to toss; and therefore, there is NOTHING for chance to do! Then that same person has to be dumb enough and have time enough to keep tossing the fifteen coins or the fifteen amino acids for 70 days hoping that they will fall out just right. Finally, that same Person, Psyche, or Intelligence has to be there to recognize that the correct amino acid sequence has been achieved by chance alone. Without that intelligent person there to recognize that the correct amino acid sequence has been achieved, they will just keep tossing the coins or stirring the amino acids for all eternity with NO RESULTS. You NEED Someone Psyche or Someone Intelligent there to recognize and use that correct amino acid sequence when CHANCE finally produces it.

You see, there really is NO such thing as pure chance or chance causation, which is why they say that the Null Hypothesis in a science experiment was by definition produced by chance alone. A Null Hypothesis means that the independent variable had NO effect on the dependent variable, which means that there is NO causal relationship between the two.

**Chance cannot produce causal relationships!**

**Chance cannot do causation!**

Chance has NO mechanism in place for recognizing when it has succeeded.

Once we have established a causal relationship between the independent variable and the dependent variable, we are no longer looking at chance but are looking at causation instead.

**Chance and causation are mutually exclusive!**

Even games of chance require an intelligent assist from some type of person, psyche, quantum consciousness, or life force.

When it comes to protein synthesis and the origin of life, we are dealing with 21 amino acids. God would have to make all of the amino acids in the first place and then put them into close proximity to each other, for there to be a chance that CHANCE can combine them into a protein that is 15 amino acids long. No God-given amino acids, then NO proteins. It's that simple.

But, let's say that we have a God-given tub full of the correct 21 amino acids, God is still going to have to keep stirring that pot or that tub constantly, until the correct 15 amino acids miraculously line up in the correct sequence. If nobody stirs the pot, then nothing is ever going to happen! Amino acids are 3D and not 2D like coins. It's going to take a lot of work, stirring, and luck to get fifteen 3D amino acids to line up in the correct sequence by chance alone. There has to be Someone to stir the pot so that chance can do its job! There has to be someone to toss the coins, so that chance can do its job. If there is NO psyche, or intelligence, or quantum consciousness there to toss the coins or to stir the pot, nothing is ever going to happen. It's that simple!

Even then, there is still NO CHANCE that it will ever happen. There's NO chance that we will ever get 15 amino acids in the correct sequence by chance alone.

Why?

Well you see, by definition, chance or chemical evolution is blind. It has NO way of knowing that it has finally achieved the correct sequence of 15 amino acids. Chance has NO use for a correct sequence of fifteen amino acids. Chance wouldn't know what to do with it. Chance wouldn't even know that it has it. "Chance" or "chemical evolution" will just keep stirring the pot or the tub until that correct sequence of 15 amino acids is destroyed. There's nothing there in that tub to recognize that it has a correct sequence of fifteen amino acids. Recognition requires psyche, or intelligence, or quantum consciousness. A sequence of fifteen correct amino acids is absolutely worthless without some Intelligent Psyche there who KNOWS how and where to use that protein or sequence that was allegedly produced by chance alone.

Let's say that chance miraculously produces a gene. That gene is absolutely worthless without some Intelligent Psyche who KNOWS how, where, and why to use that gene. The same reality applies to proteins.

**Creation by chance alone is impossible. Chance of any kind always requires an intelligent assist of some kind to take advantage of it.**

Remember, there's nothing there in the tub to protect that correct sequence of amino acids, once it is assembled by chance alone. Soon, the sun, wind, water, or chance will destroy that correct sequence of 15 amino acids, and the whole process will have to start all over again. Nothing will ever be accomplished by chance alone. There is no such thing as Chance Causation. There is no such thing as Creation by Chance. Someone Intelligent has to be there to recognize that the correct amino acid sequence has been achieved; and then, that same Someone has to protect and use that correct amino acid sequence, or nothing will ever come of it. Chance cannot do design and creation by chance alone. It's that simple.

A binomial distribution with a sign test is a very rough and crude estimate in comparison to the true probability of getting fifteen correct amino acids into the right sequence by chance alone. The true probability of getting fifteen amino acids in the correct order by chance alone is much lower than what we got from the binomial distribution table and sign test, as I will demonstrate in the next section.

For now, all you really need to know is that it is statistically and realistically IMPOSSIBLE to get fifteen heads in a row, with a fair coin. It's theoretically possible on paper, but in the real world, nobody is going to toss and count fifteen fair coins for a solid month around the clock on the off chance that he or she will get fifteen heads somewhere along the way. In real world practice, it's never going to happen.

The same can be said for fifteen amino acids. Nobody is going to keep sequencing fifteen amino acids randomly and checking to see if the sequence is right, on the off chance that he or she will end up with the correct sequence somewhere along the way. And, without that intelligent person, psyche, life force, or quantum consciousness there to toss and count the coins or the amino acids in the first place, NOTHING will ever happen for all eternity, because someone intelligent has to be there to toss the coins and count the number of heads for "chance" to do its job to begin with. Someone Psyche or Someone Intelligent has to be there to recognize the correct amino acid sequence for what it is and take advantage of it, once chance has finally produced it; otherwise, nothing will ever come of it.

**Chance Cannot Do Causation!**

Chance and causation are mutually exclusive.

This is the fundamental axiom of Science, Statistics, the Scientific Methods, and Science Experimentation. Chance cannot do causation!

Furthermore, we are dealing with 21 amino acids here as well as proteins, and NOT coins. Amino acids are more complex than coins. Amino acids and proteins are more like 21-sided dice than coins. Amino acids and proteins are 3D. The chances of success go down drastically when we start dealing with 3D objects instead of 2D coins.

Let's say that we can get God to give us a tub full of the correct amino acids, and nothing else but the correct amino acids in that tub; and, let's say that you can convince God to keep that tub in existence and keep stirring the contents of that tub for all eternity; and, let's say that God has time to keep stirring that tub for all eternity so that CHANCE can do its work in a timely fashion – how long is it going to take on average for CHANCE to produce a correct amino acid sequence 15 amino acids long, if CHANCE is putting 15 amino acids together every minute?

This is a legitimate statistics problem, and it does have a real answer that can be found. We can calculate the probability of chance events.

There are 21 different amino acids that are used to make proteins, and we are looking for a correct sequence that is in fact 15 amino acids long – 15 trials or 15 events. The Multiplication Rule in Probability applies here. We are looking at 15 separate events combined into one result. The answer is (1/21) multiplied by itself fifteen times – one time for each trial, or event, or amino acid in the sequence.

The true probability of getting 15 correct amino acids in a row by chance alone is:

$(1/21)^{15}$

This is equal to: 1.46794769 e -20

YIKES!

That's one solid chance in 100,000,000,000,000,000,000 attempts or trials, with fifteen amino acids being lined up and checked for correctness in each attempt, trial, or event.

That's NEVER going to happen!

Look what happens when we switch from fifteen coins to fifteen 21-sided dice instead!

We definitely entered into the realm of impossibility!

We start with the estimated 100,000,000,000,000,000,000 minutes that it will take on average for "chance" to find or produce the correct amino acid sequence that is fifteen amino acids long, by chance alone, if CHANCE is successfully lining up 15 amino acids every minute. We have to start with some sort of given, or there is nothing for chance to do.

Let's start with a God-given ocean full of the correct 21 amino acids. And, let's convince God to keep stirring this ocean for all eternity so that chance can do its job. Producing the correct amino acid sequence by chance alone will NEVER happen unless God keeps stirring the pot. Let's say that chance can miraculously sequence 15 amino acids randomly every minute given this ocean full of the correct amino acids. We start with the estimated 100,000,000,000,000,000,000 minutes that it will take for chance to do its job. Then we divide that huge number by 60 minutes, 24 hours, and 365 days in order to find the number of years that God will have to stir the contents of that pot or tub so that CHANCE can produce a "protein" that is 15 amino acids long by chance alone.

Assuming that God can miraculously keep getting 15 amino acids to line up into a sequence every minute to produce a single trial or event every minute, it will take God 190,285,875,190,000 YEARS on average, of stirring that same ocean full of the correct amino acids, in order to get the correct amino acid sequence fifteen amino acids long that we are looking for – one time. Furthermore, God is going to have to be on hand the whole time to recognize and use the correct amino acid sequence when it is finally achieved by chance alone. If God isn't there to recognize and use that correct amino acid sequence when CHANCE finally produces it, then nothing will come of it.

That's 190 trillion years on average of stirring the pot, for CHANCE to produce a correct or functional protein that is fifteen amino acids long by chance alone. The scientists estimate that our physical universe is only 13.8 billion years old. Is God really going to be dumb enough to stir a pot or a tub full of the correct amino acids for 190 trillion years on the off chance that He will get a correct sequence of fifteen amino acids by chance alone? Would you be willing to do that? Would you be able to do that? Of course not! So, why believe that CHANCE can do it at will? It's superstition to believe that CHANCE can design and create things at will.

And, what did we get from all of this hypothetical stirring?

We got a "protein" that was 15 amino acids long. We got nothing! There is NO protein that is 15 amino acids long! Proteins are much longer than that. 190 trillion years of stirring the pot, and we still have nothing to show for it.

Are you starting to get a sense for why Creation by Chance or Causation by Chance is NOTHING but superstition, and fiction, and wishful thinking?

*Creation by Chance*, *Causation by Chance*, *Chance Causality*, and *Wishful Thinking* are logic fallacies. These are just a few of the dozens of different logic fallacies upon which the Theory of Evolution and the Second Law of Thermodynamics are based. They are both Creation by Chance, which means that they are both automatically FALSE. Chance cannot do causation! Chance is precisely the thing that we are trying to identify and eliminate from Science by running science experiments in the first place! Creation by Chance is nothing but science fiction. Creation by Chance is magic at best and a deceptive lie at worst.

I got the following off the internet from Uncle Google.

**Thyroid releasing hormone or TRH should be the smallest protein in the human body, with 234 amino acids. (> 100 amino acids is a protein.) The smallest polypeptide in the human body should be insulin, with 54 amino acids. (10-100 amino acids is a polypeptide.)**

Let's run with this for a minute and see what we get.

The probability of making insulin by chance alone is:

$(1/21)^{54}$

This is equal to:  3.98252109 e -72

This can't be done by chance alone, even with a God-given ocean full of the correct amino acids. Even with God constantly stirring the pot or tub or ocean, it would take trillions of trillions of trillions of years for such a thing to happen by chance alone. It's just not going to happen. Our observable universe hasn't been around long enough for this to happen by chance alone. Insulin was designed and made by God. Without God to design it, make it, fine-tune it, and field test it, insulin wouldn't exist. Only God could recognize that He had actually created insulin and KNOW what it is good for and how to use it.

Technically, insulin isn't a protein. It's a polypeptide that is 54 amino acids long. From now on, whenever you think of insulin, remember that God designed and made the stuff in the first place. We KNOW for a fact that chance didn't do the job, so that leaves God as the only other plausible candidate. Someone made the stuff, or it wouldn't exist. Anything that is obviously made obviously has a Maker who made it. Insulin was obviously made. God made insulin; and, there's no chance that I got that wrong.

Here we just falsified and rejected the Null Hypothesis (which is produced by chance alone) and accepted the Alternative Hypothesis (which is some type of causation by a causal agent). In this case, the Alternative Hypothesis is that God designed and created insulin, because chance could never have done the job. This is Science in action. This is a True Science Experiment.

The probability of making THR by chance alone is:

$(1/21)^{234}$

When I put this into my graphing calculator, the answer was ZERO (0). Literally. The answer is ZERO. This is NEVER going to happen ever! This is beyond what a graphing calculator can handle. This is beyond what God could do by chance alone!

The probability of making THR by chance alone is ZERO. You can wait for all eternity, and it will NEVER happen, even with a God-given tub full of the correct amino acids or with an ocean full of the correct amino acids. It's NEVER going to happen. Making THR by chance alone is impossible. It can't happen by chance alone. It requires some kind of intelligent and deliberate intervention from God Himself.

Here we just falsified and rejected the Null Hypothesis (which is produced by chance alone) and accepted the Alternative Hypothesis (which is some type of causation by a causal agent). In this case, the Alternative Hypothesis is that God designed and created THR, because chance could never have done the job. This is Science in action. This is a True Science Experiment.

THR is the smallest protein in the human body. THR can never be made by chance alone. Every time you think of any protein, remember that God designed it, created it, made it, fine-tuned it, and field tested it. Anything that is obviously made obviously has a Maker who made it. Proteins were obviously made.

God made your proteins. Each protein is one of God's Signatures. People are looking for the Signature of God in nature and science, and there it is in each one of your proteins.

Your genome is also God's Signature. God had to make a matching gene for each of the proteins that He designed and made. God made your proteins and your genes. This is obviously true. According to the statistics and the probabilities, there is NO chance that I am wrong. God made your proteins and your genes.

This is one of the most convincing Scientific and Statistical Proofs of God's Existence that I have encountered so far. I just used Statistics, Probability, and Science to prove that God exists; and, there's NO chance that I'm wrong.

One day in 2015, I finally decided to go looking for Scientific Proof of God's Existence, and eventually I found what I was looking for. Did I not? It's obvious that I found Scientific Proof of God's Existence; and, there's NO chance that I got it wrong.

This is what I got from my statistics class.

It's hiding in plain sight where nobody can see it, because nobody is looking for it and nobody wants it.

I bet you never got anything like this from your statistics class, and neither did your statistics professor, because you weren't looking for it and didn't want to find it in the first place. No seeking, then no finding.

Yet, it's there and it's obvious for anyone who is willing to look and see.

Creation by Chance, or Chance Causality, or Causation by Chance is IMPOSSIBLE.

That means that Materialism, Naturalism, Darwinism, Nihilism, Atheism, Creation Ex Nihilo, Creation by Random Disorder or Entropy, Creation by Death or Natural Selection, Abiogenesis, Spontaneous Generation, Chemical Evolution, Macro-Evolution, the Theory of Evolution, and the Second Law of Thermodynamics ARE IMPOSSIBLE because they are Creation by Chance, Chance Causality, Chance Causation, or caused by chance alone.

There's the truth, whether you want it or not. Scientists lie, but the math doesn't lie. I'm a scientist, mathematician, statistician, and logician. I believe the math, and I KNOW that it is true. We can calculate the probability of chance events; and, WE KNOW for a fact that chance cannot do causation. Chance and causation are mutually exclusive.

We can calculate the probability of chance events; and, CHANCE cannot produce fifteen heads in a row in a timely fashion with any kind of reliability, with a fair coin. If chance can't do fifteen heads in a row in a timely fashion, then chance can NEVER produce a protein or a gene. There is the truth that everyone is searching for and that nobody is willing to accept. God made your proteins and your genes. Anything that is obviously made obviously has a Maker who made it. Genes, proteins, quantum fields, quantum waves, and physical atoms are obviously made. Therefore, it is obvious that they have a Maker who made them. The whole purpose of Science, Statistics, Science Experiments, and the Scientific Method is to find the TRUE CAUSE of each event that we experience or observe.

**Chance is NEVER the True Cause of anything! Chance cannot do causation.**

Now, the only thing left to do is to figure out which God is the God who made your proteins and your genes. I will leave that up to you to figure out for yourself, because it will be meaningless to you unless you discover it for yourself and want to become a part of it. I believe that answer can be found as well, if you are willing to look for it with an open mind and heart. It took a while, but I eventually found the answer to that question also, after I finally decided to go looking for it. The answer is there to be found, if one is willing to look for it and is able to accept it when he finds it. God has revealed Himself to us millions of times in thousands of different ways. It's subtle most of the time, but it's there to be found for those who are willing to look and see.

Mark My Words

# The Perpetual Motion Cycle

https://origin-science.org/the-perpetual-motion-cycle/

Once we have the True Meaning or the True Definition for a scientific theory or a scientific concept, then even the simplest of mathematical equations can reveal the profoundest truths, that have been hidden from the world for the duration of human history.

$E = mc^2$ is one such equation.

Obviously, I am not the first person on the planet to discover and use $E = mc^2$. However, I may be the first person to discover what it truly means and what it truly is.

**$E = mc^2$ is the Perpetual Motion Cycle.**

This Perpetual Motion Cycle is the heart of modern-day physics, and we don't even know it because we are still caught-up in or stuck in the dark ages with Materialism, Naturalism, Darwinism, Nihilism, Atheism, Creation Ex Nihilo, Classical Realism, Creation by Chance, the Theory of Evolution, and the Second Law of Thermodynamics.

Most people on this planet don't even know it, but the Perpetual Motion Cycle $E = mc^2$ FALSIFIES Materialism, Naturalism, Darwinism, Nihilism, Determinism, Behaviorism, Determinism, Classical Realism, Atheism or Creation Ex Nihilo, the Theory of Evolution, and the Second Law of Thermodynamics, which ARE Creation by Chance, Creation by Disorder, Creation by Chaos, Creation by Death, or Creation by Entropy.

The false is falsified by the truth, and the truth is repeatedly and constantly experienced and observed.

$E = mc^2$ is Quantum Mechanics, particularly Quantum Field Theory. $E = mc^2$ is Conserved! The Quantum Fields ARE perpetual motion machines. The Quantum Fields operate on the Perpetual Motion Cycle and are an integral part of the Perpetual Motion Cycle. Once the Quantum Fields were made by Nature or Nature's Psyche, they have always been conserved thanks to $E = mc^2$ and the Conservation of Energy, the Conservation of Quantum Information, and the Conservation of Psyche or Quantum Consciousness.

It's Nature's Psyche or Nature's Intelligence who makes the quantum waves or the photons in the first place. Later, it is Nature's Psyche or Nature's Intelligence who collapses the wave function and transforms infinite acceleration or omnipresent quantum waves or photons INTO mass, heat, resistance to acceleration, and mass's heat storage capacity (entropy) instead.

According to the Perpetual Motion Cycle $E = mc^2$ and Quantum Field Theory, the Gods or the Controlling Psyches had to design and make the massless, heatless, and entropyless Quantum Fields BEFORE they could make and sustain mass, resistance to acceleration, heat, and mass's heat storage capacity which is entropy. The proven and verified existence of the Quantum Fields FALSIFIES the second law of thermodynamics, which says that they don't exist and can't exist. The second law of thermodynamics will NEVER be true as long as the Quantum Fields exist. The Quantum Fields are pure Syntropy or pure Exergy. The Quantum Fields are perfect Order and Organization. The Quantum Fields are massless, invisible, non-physical, immaterial, entropyless, syntropic, exergic, and conserved Perpetual Motion Machines. Their very existence falsifies the second law of thermodynamics.

According to the Law of Psyche, each psyche or intelligence or life force or quantum consciousness has a certain amount of energy that's under its control, and that controlling psyche can form or transform the energy under its control into anything that it wants that energy to be, anytime and anywhere that it chooses to do so. This is the Perpetual Motion Cycle $E = mc^2$ in action. The controlling psyche CHOOSES what form the energy under its control will be. Psyches can also coordinate their actions transpersonally or telepathically at the quantum level through thoughts or quantum waves. This is the way things really work at the quantum level in the quantum realm, or the non-physical realm, or the spiritual realm. This is what has actually been experienced and observed.

According to Quantum Field Theory, particles are born, and particles die. In other words, particles or quanta of any kind are made, and they can be unmade or reabsorbed back into the quantum fields from whence they came, anytime and anywhere that their controlling psyche CHOOSES to make them or transform them or dissolve them. All the while, their energy is conserved. This is the Perpetual Motion Cycle $E = mc^2$ in action.

The Perpetual Motion Cycle has been experienced, and observed, and verified. That means that it is real and truly exists. The Perpetual Motion Cycle $E = mc^2$ works perfectly and eternally, both coming and going, thanks to the Conservation of Energy and Psyche, as well as the Conservation of Quantum Information.

Currently in our sun, Nature or Nature's Psyche is transforming mass, heat, and entropy into massless, heatless, chargeless, and entropyless photons or quantum waves. Nature's Psyche is transforming resistance to acceleration, mass's heat storage capacity (entropy), mass, and heat into infinite acceleration instead. Photons and quantum waves go from zero to the speed-of-light instantly. That is infinite acceleration. Photons and quantum waves also CHOOSE their ultimate velocity, which can be less than the speed-of-light, the speed-of-light, or an infinite velocity which we call omnipresence or quantum tunneling. This is the part of the Perpetual Motion Cycle where Nature or Nature's Psyche transforms mass, heat, and entropy into massless, heatless, chargeless, and entropyless photons, quantum waves, and infinite acceleration instead. This is happening all the time through our sun or within our sun.

Likewise, any time that a photon, virtual particle, or quantum wave CHOOSES TO STOP, that photon or quantum wave transforms its omnipresence and infinite acceleration INTO mass, heat, resistance to acceleration, or mass's heat storage capacity (entropy) instead. The controlling psyche within a photon or a quantum wave can transform that photon or quantum wave into ANYTHING that it wants that photon or quantum wave to be, anytime and anywhere that it CHOOSES to do so. In Feynman Diagrams, photons or quantum waves are constantly transforming themselves into electrons and positrons, or quarks and gluons. Massless, heatless, chargeless, and entropyless photons and quantum waves are constantly transforming themselves INTO mass, heat, and entropy (mass's heat storage capacity) all the time. All the while, the energy is conserved!

This, too, is a part of the Perpetual Motion Cycle. It's happening all the time. This, too, has been experienced and observed. A miniature Big Bang happens every time a photon CHOOSES to stop and land on our skin. That photon transforms itself into mass or heat. We typically feel the heat, when a photon lands on our skin. However, a photon or quantum does NOT have to transform itself into heat if it doesn't want to. It can transform itself into mass, resistance to acceleration, and mass's heat storage capacity (entropy) instead, if it chooses to do so. A photon or quantum wave doesn't have to stop if it doesn't want to. A photon or quantum wave can pass through water, glass, our earth, our sun, our physical body, or a black hole as if they weren't even there, if a quantum wave or a photon chooses to do so. This too has been experienced and observed.

The whole Perpetual Motion Cycle $E = mc^2$ has been experienced and observed; and, it FALSIFIES the second law of thermodynamics which claims that the amount of disorder or entropy is constantly increasing and that it can never decrease and go to zero.

ALL of the conservation laws falsify the second law of thermodynamics! The second law is a violation of the Conservation of Energy or the First Law of Thermodynamics. According to the second law of thermodynamics, nothing should exist. Everything should be random disorder or random chaos, or there should be nothing at all. The second law of thermodynamics FAILS TO PREDICT what we are actually experiencing and observing. Everything that exists falsifies the second law of thermodynamics! The very fact that you exist is scientific proof that the second law of thermodynamics is false.

Do we really observe proton decay?

Do we really observe ever-increasing disorder and chaos? Do we really observe an ever-encroaching gray goo coming in at us from all sides as the second law of thermodynamics predicts? Or do we observe constantly conserved order and organization as Quantum Field Theory and the Perpetual Motion Cycle predict?

The one that we actually experience and observe is the one that's actually real and true. This is the pinnacle of Science 2.0. Science 2.0 allows all of the evidence into evidence, and then it pursues a preponderance of that evidence. Science 2.0 is observation and experience. The Perpetual Motion Cycle has been experienced and observed; therefore, we know that it is real and truly exists.

We have NEVER observed an ever-increasing amount of substance coming into existence out of thin air from nothing as the second law predicts. We have NEVER observed creation ex nihilo, creation by entropy, creation by disorder, or creation by chance as the second law predicts. We have never observed the proton decay that the second law predicts. We have NEVER observed anything that the second law of thermodynamics predicts. Everything that exists FALSIFIES the second law of thermodynamics.

But, we have observed the Perpetual Motion Cycle $E = mc^2$ in action, both coming and going. The Perpetual Motion Cycle and constantly conserved Order and Organization at the quantum level FALSIFY the second law of thermodynamics which predicts and claims that they do not exist. We have indeed observed the Perpetual Motion Cycle and the massless, heatless, and entropyless photons, quantum waves, and quantum fields in action. We do indeed observe constantly conserved Order and Organization thanks to Quantum Field Theory, the Quantum Fields, the Perpetual Motion Cycle, the Conservation of Quantum Information, and the Conservation of Energy and Psyche.

The one that has been experienced and observed is the one that's actually real and true. The second law of thermodynamics is a con and a scam that is falsified by everything that exists and by everything that has been experienced and observed. The second law of thermodynamics is FALSIFIED by the Perpetual Motion Cycle $E = mc^2$, as well as Quantum Mechanics and Quantum Field Theory. The second law of thermodynamics will NEVER be true as long as the Quantum Fields or the Perpetual Motion Cycle exists.

My ultimate goal is to identify and fix everything that is wrong with Science; and, that process starts with and includes replacing the theory of evolution, the second law of thermodynamics, creation ex nihilo, and creation by chance WITH the Conservation of Energy and Psyche, the Conservation of Quantum Information, the Conservation of Order and Organization, the Law of Psyche, Quantum Mechanics, Quantum Field Theory, and the Perpetual Motion Cycle instead. We replace the things that have NEVER been experienced NOR observed with the things that have actually been experienced, observed, verified, and proven true instead. That's what Science 2.0 is all about, and that's part of the reason why

I chose to upgrade my science to Science 2.0. One day I simply realized that science as a whole is in massive need of a serious upgrade. Since then, I have tried to do just that.

    Mark My Words

## Evolution Is Entropy

While writing the book on Syntropy, I had a flash of insight one day in 2018 and realized that Evolution is Entropy. I even dedicated a website to the subject.

https://evolution-is-entropy.com/

While researching the subject of entropy, I discovered that there is literally a couple dozen different, contradictory, and mutually exclusive definitions for entropy. That's why entropy is such a tricky and difficult subject. Nobody knows what entropy is. Entropy can literally be anything that you want it to be. There are dozens of different definitions for entropy. They often falsify each other because they are mutually exclusive. Entropy is their catch-all and magical solution for everything in science that they don't understand or can't explain.

While writing the book on Syntropy and while trying to determine what Syntropy is, I realized that "evolution" is yet another definition that they use for "entropy".

The Materialists, Naturalists, Nihilists, Atheists, and Darwinists only believe in physical matter, entropy, death, and chance. Therefore, their creative mechanism or causal mechanism has to be one of these. Consequently, Darwinism and evolution and the theory of evolution have to be Creation by Rocks (creation ex nihilo), Creation by Death (natural selection), Creation by Chance (the null hypothesis), or Creation by Disorder and Entropy (the second law of thermodynamics).

The null hypothesis in a science experiment is caused by chance alone or the result of chance alone. Creation by Chance or Chance Causality is the Null Hypothesis. ALL of the different forms of Creation by Chance are the Null Hypothesis – including the second law of thermodynamics, the many different forms of the theory of evolution, creation by disorder or chaos or entropy, the big bang theory, and every other theory or philosophy of science that has been developed by the Materialists, Naturalists, Darwinists, Nihilists, and Atheists. It is ALL Creation by Chance. It is ALL the Null Hypothesis, because it was all caused by chance alone. The Null Hypothesis tells us that there is NO causal relationship between the independent variable and the dependent variable because the results of the experiment were the result of chance alone.

All of these things are Creation by Chance, which means that they are all Creation by Random Disorder and Chaos. Entropy is often defined as disorder or random chaos. Boltzmann Brains, the Big Bang, and the Theory of Evolution in its many different forms, as well as the creation of planets and stars and genomes – ALL end up being Creation by Entropy, Creation by Random Disorder, or Creation by Chance in the hands of the Naturalists, Darwinists, and Atheists. It's all a part of the same faulty package. It's all Creation by Chance or the Null Hypothesis. Evolution is Entropy. Evolution is Creation by Random Disorder. Evolution is Creation by Chance. Evolution is Creation Ex Nihilo. Evolution is magic. It's all the same thing in the end. It's magic. It's superstition.

In any science experiment, the Null Hypothesis states that the results of the experiment were caused by chance alone. In any science experiment, chance causality or creation by chance is the Null Hypothesis, which is the result of chance alone. If we retain the Null Hypothesis, then we have established that there is NO causal relationship between the independent variable and the dependent variable because the results of the experiment were produced by chance alone. The whole purpose of a science experiment is to REJECT the Null Hypothesis, Chance Causality, or Creation by Chance because they resulted by chance alone. If we can successfully REJECT the Null Hypothesis, Chance Causality, or

Creation by Chance, then we have established that there is some type of measurable causal relationship between the independent variable and the dependent variable. According to statistics and the rules of science, chance of any kind is NOT a valid cause. The Null Hypothesis or Creation by Chance is NOT a valid causal mechanism! Through science experiments and rejection of the null hypothesis, we are trying to eliminate chance, chance causality, and creation by chance from science. We are trying to establish true causality by identifying and eliminating from science everything that was produced by chance alone.

The Null Hypothesis is the result of chance alone. The Null Hypothesis is Chance Causality, Creation by Chance, or Creation Ex Nihilo which was produced by chance alone.

The whole of Science, Statistics, the Scientific Method, and Science Experiments were designed to identify and eliminate from science the Null Hypothesis, Chance Causality, Creation by Chance, the Second Law of Thermodynamics (creation by random disorder or entropy), the Theory of Evolution (creation by random disorder or entropy), the Big Bang Theory (creation ex nihilo), Creation by Random Disorder, Creation by Natural Selection (creation by death), Materialism, Naturalism, Darwinism, Nihilism, Atheism, and everything else that was caused by chance alone. The whole of Science was designed to identify, falsify, and eliminate these things from science because they were the result of chance alone.

One of my most fascinating and interesting discoveries in the Philosophy of Science came when I first realized that Materialism, Naturalism, Darwinism, Nihilism, Atheism, the Second Law of Thermodynamics, the Theory of Evolution, Creation Ex Nihilo and Creation by Chance ARE the Null Hypothesis that we are trying to reject, falsify, and eliminate from Science. All of these falsified philosophies of science use CHANCE alone as their only causal mechanism. They are ALL Creation by Chance or the Null Hypothesis, which means that they are ALL automatically false because they have NO causal relationship with anything that exists. Fascinating, is it not? I find it interesting how easily we were tricked and deceived. The Naturalists, Darwinists, and Atheists were able to trick us and deceive us because we desperately wanted to be tricked and deceived.

Natural selection is Creation by Death. Natural selection doesn't do anything except sit around and wait for us to die. Natural selection doesn't touch our genes. One of the standard and most common definitions for "entropy" is death and heat death. It's all the same thing. Entropy is death. Natural selection is Creation by Death. So once again, evolution ends up being creation by heat death, creation by death, or creation by entropy. These people have literally turned death, chance, chaos, disorder, or entropy into the cause of everything that exists in this universe. That's what they have done, is it not?

Evolution is the disorder and chaos definition for entropy; and, evolution is the death and heat death definition for entropy. Evolution is the Creation by Chance and the Creation Ex Nihilo definition for entropy. Any way we look at it, evolution is entropy.

Evolution, entropy, death, heat death, random disorder, chaos, creation ex nihilo, and creation by chance ARE the same mechanism; and, it is this mechanism that the Naturalists and Atheists use as the causal mechanism behind everything that was ever designed and made in this universe. It's superstition, not science. So, that's why the Materialists, Naturalists, Darwinists, and Atheists define science as Materialism, Naturalism, Darwinism, and Atheism so that their falsified theories and philosophies will always be axiomatically true. These people stole and hijacked science, and then redefined science from "observation", "experience", and "knowledge" to be Materialism, Naturalism, Darwinism, Atheism, and Creation by Chance instead. They cheated. It's a scam. It's a fraud. And now, they label everything that falsifies Materialism, Naturalism, Darwinism, Nihilism, Atheism, and Creation by Chance as "pseudo-science", and toss it out, and pretend that it doesn't exist. They lie. They cheat. And, they deceive themselves. Self-deception works, and it works every time.

I revisit this topic frequently within the twenty different books that I have written and published, because it explains what has been done to us. It explains how we were tricked and deceived.

https://amazon.com/author/science

If we can successfully eliminate from Science everything that has been falsified or everything that is demonstrably false, then only the truth will remain. The truths – which remain after eliminating Materialism, Naturalism, Darwinism, Nihilism, Atheism, Creation by Entropy, Creation Ex Nihilo, and Creation by Chance from science – end up being the most interesting subjects in science because they have actually been experienced and observed by someone somewhere at some point in time.

The false is falsified by the truth, and the truth is constantly and repeatedly experienced and observed. Science is observation and experience after all, or it's supposed to be. Science is supposed to identify and eliminate from science everything that is false and everything that has been falsified. Instead, the Materialists, Naturalists, Darwinists, Nihilists, and Atheists retain the Null Hypothesis and make the Null Hypothesis or Creation by Chance the cause of everything that exists in this universe. That's what they do, is it not? They cheat!

**The attitude of the scientist, here so admirably explained, is the attitude, also, of the mystic. *Experience*, to both, is what matters most. – *The Sufi Quarterly*, 1929.**

Amen!

Whether we are talking about science, philosophy, the scientific method, religion, mysticism, science experiments, or our universe as a whole, observation and experience is what matters most in the end. If it has NEVER been experienced NOR observed by anyone anywhere, then it really doesn't exist. Creation by Chance, Creation by Death, Creation Ex Nihilo, and Creation by Entropy have NEVER been experienced NOR observed. Not even God can do Creation Ex Nihilo, so why believe that entropy can? Spontaneous generation, abiogenesis, chemical evolution, macro-evolution, and the other forms of creation ex nihilo and creation by chance have NEVER been experienced NOR observed. They are NOT science. They are faulty religions or falsified philosophies of science.

The truth is KNOWN by the fact that it is constantly and repeatedly lived, experienced, verified, and observed. Observation and experience is what matters most. Science is observation and experience, or it's supposed to be.

In contrast, Creation by Chance is a faulty and false definition for science because chance has NO causal relationship with anything except for the Null Hypothesis, which is the thing that we are trying to reject and eliminate from science. We created science experiments so that we could identify and eliminate Creation by Chance or the Null Hypothesis from science. We created science experiments so that we could identify and eliminate Materialism, Naturalism, Darwinism, Nihilism, Atheism, Creation Ex Nihilo, the Second Law of Thermodynamics (creation by random disorder), the Theory of Evolution (creation by entropy and chance), and every other form of Creation by Chance from science. These things are NOT science. We are trying to eliminate them from science. They are the Null Hypothesis that we are trying to eliminate from science.

We created science experiments to eliminate these things from science. The whole purpose of a science experiment is to reject Chance Causality or the Null Hypothesis so that we can establish some type of true causality instead. The whole purpose of a science experiment is to eliminate Materialism, Naturalism, Darwinism, Atheism, and every other form of Creation by Chance from science. We are trying to eliminate these things from science, so that we can establish true causality instead.

I upgraded my science to Science 2.0. Science 2.0 allows ALL of the evidence into evidence and then pursues a preponderance of that evidence.

In contrast, Materialism, Naturalism, Darwinism, Nihilism, and Atheism are based exclusively upon the censorship, deletion, banning, and rejection of ALL the evidence that falsifies them. Atheism of any kind is based upon a refusal to look at evidence, particularly a refusal to look at ALL the evidence that falsifies Materialism, Naturalism, Darwinism, Nihilism, and Atheism. That's the way these people do "science". They ban, censor, delete, hide, and destroy the evidence that falsifies their pre-chosen beliefs. They cheat! And then, they label everything that falsifies Materialism, Naturalism, Darwinism, and Atheism as a "pseudo-science" so that they can toss it out and retain their falsified beliefs instead.

This is what we have experienced and observed, is it not?

Studying the psychology and behavior of these people can be one of the most fascinating and interesting things that we do in science. Self-deception works, and it works every time. These people use Creation by Chance or the Null Hypothesis as the causal force, or the causal entity, or the causal mechanism for everything that exists in this universe. It's fascinating to figure out what they are doing and how they are deceiving themselves, is it not? I think it is. The study of the Philosophy of Science along with the study of the Human Psyche (psychology) is one of the most interesting and fascinating things that we can study.

This in part is what Science 2.0 is all about. I'm trying to fix everything that's wrong with science and replace it instead with something that has actually been experienced and observed.

Mark My Words

--

Online Archival Source:

https://evolution-is-entropy.com/

# Freedom of Science

https://bio-psycho-social.com/freedom-of-science/

Our society in the United States is big on freedom of speech. We seem to be able to vilify our President, cyber bully, and call each other names and get away with it unscathed.

However, when it comes to science, we have NO freedom of speech or freedom of science in our public schools. If you are not preaching and teaching Materialism, Naturalism, Darwinism, Nihilism, and Atheism in our public schools, then you will be censored and fired.

That's just the way it is.

We have NO freedom of science in the United States.

They have been trying to take freedom of religion or freedom of conscience away from us also.

Science 2.0 promotes freedom of science.

If we want to find and know the truth in science, physics, chemistry, and biology, then we must be free to falsify the biology, physics, and science that is obviously and demonstrably false. Currently, we are not free to do so, especially in our public schools. We can only do so in private, within our own homes and our own personal thoughts.

We need to start promoting freedom of science within the public arena as well.

It's time to falsify and eliminate the things that are false.

Mark My Words

## The Null Hypothesis

https://bio-psycho-social.com/

I found the following figure helpful for understanding the null hypothesis, statistics, and their relationship to the scientific method.

Adapted from Pagano, R.R. (2010). *Understanding Statistics in the Behavioral Sciences* (9th ed.). Belmont, CA: Wadsworth. (p. 269.)

**THE NULL HYPOTHESIS**
The Independent Variable had no effect on the Dependent Variable

$P_{null} = 0.50$

$H_0$ is false; IV has real effect.  →  $H_0$ is false; IV has real effect.

$P_{real} < 0.50$     $P_{real} > 0.50$

The Independent Variable decreased the Dependent Variable

The Independent Variable increased the Dependent Variable

Size of effect increases ←     → Size of effect increases

| 0+ | 1,000+ | 3,000+ | 5,000+ | 7,000+ | 9,000+ | 10,000+ |
| 10,000− | 9,000− | 7,000− | 5,000− | 3,000− | 1,000− | 0− |

0.00    0.10    0.30    0.50    0.70    0.90    1.00

Adapted from: Pagano, R.R. (2010). *Understanding Statistics in the Behavioral Sciences* (9th ed.). Belmont, CA: Wadsworth.

    The null hypothesis in a science experiment and statistics states that the independent variable had NO EFFECT on the dependent variable. If the independent variable had an effect on the dependent variable, then the next step is to try to determine the size and the direction of the effect.

    It's helpful to have a handle on this, when we start discussing whether the independent variable has the "power" to produce a real effect on the dependent variable, or not. This is science, the scientific method, and statistics in action. We have to understand this, if we want to know how scientists use statistics to do science and science experiments.

    In this science experiment, their sample size was 10,000, and they used the sign test. They tested whether the effect of the independent variable was positive or negative.

    In the middle there, we have the **Null Hypothesis**, where there are 5,000 positive effects and 5,000 negative effects. The results are fifty/fifty, which is synonymous with pure chance. $P_{null}$, the probability of a null, or the null hypothesis is always 0.50. The null hypothesis is fifty/fifty, or pure chance.

    Whenever the results of a science experiment or statistical analysis are fifty/fifty, then they conclude that the results were caused by chance alone. Fifty/fifty is pure chance when gambling as well. When the results of the experiment are fifty/fifty, pure chance, or the null hypothesis, then they conclude that the independent variable didn't have an effect on the dependent variable. It doesn't have to be exactly 50/50 either. The size of the effect is miniscule if the results of the science experiment are hovering around the null hypothesis.

We are looking for a real effect, a sizable effect, or a significant effect, if we are to conclude that the independent variable is causing the dependent variable.

Looking at the percentages at the bottom of the table above, we observe that when the percentage comes within the 0.00 to 0.05 range, then the independent variable had a significant effect on the dependent variable and decreased the dependent variable a lot. If the percentage comes within the 0.95 to 1.00 range, then we conclude that the independent variable had a significant effect on the dependent variable and increased the dependent variable a lot.

Through science experiments, we are trying to establish causality, and thereby remove chance as the "causal explanation". We are trying to reject chance or the null hypothesis. In science and statistics, chance is NOT a valid cause! That's why Materialism, Naturalism, Darwinism, Nihilism, Atheism, and the Second Law of Thermodynamics are invalid sciences or pseudo-sciences, because they use chance as their causal mechanism and are in fact Creation by Chance or Creation Ex Nihilo.

We cannot use chance, randomness, chaos, or disorder in science as our causal explanation and still remain within the realm of science. That's why Materialism, Naturalism, Darwinism, and Atheism are pseudo-sciences or religions masquerading as science, because they are Creation by Chance and use chance as their causal mechanism and make the claim that chance created everything that exists in this universe. Creation by Chance isn't real. Creation by Chance is the null hypothesis, which was produced by chance alone. Chance has no discernable effect on anything. These people literally turned chance into their independent variable and then they made chance the cause of everything that exists in this universe.

Creation by Chance or the Null Hypothesis is precisely what we are trying to falsify and eliminate through science and our science experiments. Creation by Chance or "chance causality" is the Null Hypothesis. Materialism, Naturalism, and their derivatives such as Darwinism and Atheism ARE Creation by Chance or the Null Hypothesis. The Null Hypothesis tells us that we should eliminate Materialism, Naturalism, Darwinism, Nihilism, Atheism, the Second Law of Thermodynamics, and every other form of Creation by Chance from science. So, why aren't we doing so?

It's because the Materialists, Naturalists, Darwinists, and Atheists have granted themselves a *special pleading* or a *special exemption* so that they don't have to follow the rules of science and don't have to prove their theories true through scientific experimentation and statistical analysis. Instead, these people unilaterally turned their theories and ideas into axioms and laws, and forbid us from falsifying their ideas and theories, so that their falsehoods will always be true.

They cheat!

That's how they do science.

You will never see statistical analysis, the null hypothesis, science, the scientific method, and science experiments presented in quite this manner within our public schools because it falsifies Materialism, Naturalism, Darwinism, and even Nihilism and Atheism. Our college professors have to at least make it seem as if the theory of evolution is true, or they will be fired from their jobs. They are not free to tell us the truth.

Our public schools don't reward us for falsifying the second law of thermodynamics, the theory of evolution, or creation by chance. They punish us instead whenever we do.

The whole purpose of our public school system is to indoctrinate us into Materialism, Naturalism, Darwinism, Nihilism, Behaviorism, Determinism, Atheism, and all the other forms of Creation by Chance. If you are not an atheist by the time you graduate from college, then your college professors have failed in their mission and their purpose. They have no other reason for being there. That's how far our society has deteriorated in recent decades.

You can't falsify or reject the Null Hypothesis, whenever the Null Hypothesis is Materialism, Naturalism, Darwinism, Nihilism, Atheism, the Theory of Evolution, the Second Law of Thermodynamics, or any other form of Creation by Chance. By axiom, by law, by edict, and by decree, you have to retain the Null Hypothesis whenever the Null Hypothesis is Materialism, Naturalism, or one of their derivatives such as Darwinism and Atheism. That's the way these people operate. They cheat! They call it peer review. It's censorship and intimidation, which is why I work where they can't get at me.

Thanks to the Atheists, we don't have freedom of speech and freedom to teach in our public schools, but so far, that doesn't seem to apply where my books are concerned. Here, I'm permitted to say anything that comes to mind. Here, I'm permitted to teach the truth if I want to do so.

Mark My Words

# Origin Science

https://origin-science.org/

I'm a scientist.

I'm interested in finding the origin and the cause of everything that exists.

My websites and my books are my Doctoral Thesis in Origin Science.

I spent a year falsifying Materialism, Naturalism, and the Theory of Evolution with syllogisms, the philosophy of science, and the scientific methods. I used the philosophy of science to upgrade science to Science 2.0. As a result, I have the equivalent of a doctorate in the Philosophy of Science. As far as I know, I have taken the philosophy of science further than anyone else on this planet.

I spent a year developing the Ultimate Model of Reality and presented a Psyche Ontology to the world for the first time in this world's history. I identified the Ultimate Cause or the Ultimate Causal Agent, a necessary upgrade to Aristotle's "four causes". These books are a work-in-progress and serve as the foundation for my doctorate in Origin Science and the Ultimate Model of Reality.

I'm the one who developed the new science that I called Quantum Neuroscience. I wrote a huge book on the subject. Consequently, I'm the first doctorate in Quantum Neuroscience. I believe I made my case.

My recent efforts have been focused on documenting a couple dozen different ways to falsify the second law of thermodynamics, along with explaining why we should replace the second law of thermodynamics with $E = mc^2$, the Conservation of Energy and Psyche, and Quantum Field Theory instead. The second law of thermodynamics is obviously false. As it currently stands, the second law is unrealistic, physically impossible, unfalsifiable, unverifiable, and in desperate need of a massive overhaul.

I wrote the book on Syntropy, and I introduced the Law of Psyche, the Ultimate Law of Thermodynamics, and the Quantum Law of Thermodynamics to the world. I also produced a demonstrable, verifiable, and truthful replacement for the second law of thermodynamics. In other words, I finally figured out what entropy really is. The associated books and articles comprise my doctorate in Comparative Syntropy and Entropy.

This is NEW SCIENCE that the world has never had before now because they don't want it, and they don't want it to be true. It is comprised of all the different observations and experiences that they have formally and officially rejected.

Currently, they don't give doctorates in the truth. Our public universities only give doctorate degrees in Materialism, Naturalism, Darwinism, Nihilism, and Atheism. If we want to find and know the truth, we have to do that on our own.

Mark My Words

--

My Amazon Author Page:

https://amazon.com/author/science

# A New Model for Reality

### The Transformation of Energy

https://ultimate-model-of-reality.com/

I upgraded my science to Science 2.0. Science 2.0 allows ALL of the evidence into evidence, and then it pursues a preponderance of that evidence.

The second law of thermodynamics states that the amount of disorder or entropy in our universe is constantly increasing, and that that universal amount of entropy or disorder cannot decrease or go to zero.

This simple Feynman Diagram, along with the associated Quantum Field Theory, falsifies the second law of thermodynamics. In other words, Quantum Field Theory PROVES that the second law of thermodynamics is false. This is my greatest scientific discovery of all time.

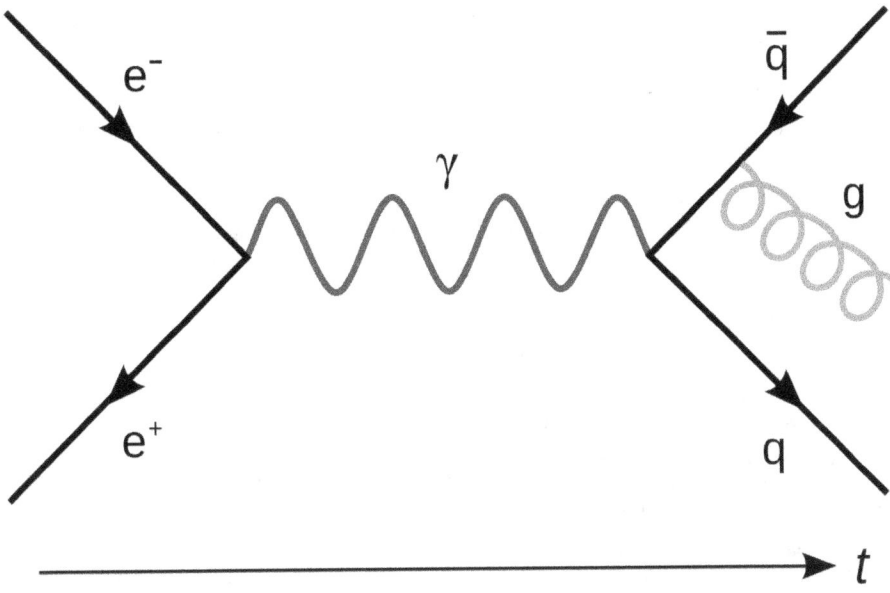

In this Feynman diagram, an electron and a positron annihilate, producing a photon (represented by the sine wave) that becomes a quark–antiquark pair, after which the antiquark radiates a gluon (represented by the helix or the pig tail). The reason why the arrows in this diagram are messed up and "wrong", or counter-intuitive, is because Feynman and others believe that anti-particles travel backwards in time.

https://en.wikipedia.org/wiki/Feynman_diagram

https://en.wikipedia.org/wiki/Feynman_diagram#/media/File:Feynmann_Diagram_Gluon_Radiation.svg

Here is the whole of modern-day physics in a nutshell, and it falsifies the second law of thermodynamics. The equations for heat and entropy also falsify the second law of thermodynamics. The second law, which claims that the amount of entropy or disorder in our universe is constantly increasing and can never decrease, has been proven false by Quantum Field Theory and the equations for heat and entropy.

Here in the caption below this Feynman Diagram, they used the word "annihilate". So, what was annihilated when the electron and the positron merged to form that photon?

This is the KEY to everything in physics.

Some of the Naturalists, Atheists, and Darwinists have erroneously stated that the energy is annihilated when the electron and the positron merge to form a photon; and, these people are always wrong because they don't understand how science and quantum mechanics really work. ALL of the energy is still there, and still working just fine according to the Conservation of Energy or the First Law of Thermodynamics.

All of the energy is still there and working just fine according to $E = mc^2$.

In order to understand what was annihilated when the positron and electron merged to form that photon, we must study the official equations for heat and entropy.

**Q = mcΔT**

Heat (**Q**) is equal to mass (**m**) times that mass's heat storage capacity (**c**) times the change in temperature (**ΔT**).

**S = Q/ΔT**

Entropy (**S**) is equal to heat (**Q**) divided by the change in temperature (**ΔT**). Combining these two equations into one, we get the following equation. Notice that the temperature (**T**) is irrelevant, because the temperature cancels out.

**S = mcΔT/ΔT**

Solving this equation, we get the following: **S = mc**.

Entropy (**S**) is the heat storage capacity of mass (**mc**). Simple. Parsimonious. Demonstrable. Replicable. Verifiable. Logical. Conservative. True.

This is the mass-based definition for entropy; and as far as I know, I'm the first person on the planet to have discovered it. It's there in plain sight just waiting for someone to find it and present it to the world.

Notice again that the temperature (**T**) is irrelevant, because the temperature cancels out. This process works and this equation holds true no matter what the temperature might happen to be. It even works at absolute zero Kelvin (**T or K = 0**). Everything continues to function perfectly fine at the quantum level, even at absolute zero. In fact, by studying photons and superconductors, we observe that absolute zero maximizes the efficiency and functionality of quantum objects such as photons, quantum waves, superconductors, virtual particles (thoughts), and quantum fields. Absolute zero is maximum efficiency and maximum order at the quantum level thanks to the Quantum Fields and the Conservation of Energy.

**S = mc**

This simple equation for entropy is my greatest scientific discovery of all time. It falsifies the second law of thermodynamics. It's demonstrably true.

$c$ = the specific heat capacity of the mass under consideration

$m$ = mass

Entropy (**S**) is mass's heat storage capacity (**mc**). No heat storage capacity, then no heat. No mass and no heat, then NO entropy. Simple. Logical. Parsimonious. True. Science doesn't get any better than this.

Entropy (**S**) equals mass (**m**), particularly the heat storage capacity of mass (**mc**). Entropy is mass's heat storage capacity (**S = mc**). Entropy is also resistance to acceleration or mass. This is the mass-based definition for entropy. No mass, then no entropy. It's that simple.

Now we are ready to answer the question, with science and with what has actually been experienced and observed.

In the caption below the Feynman Diagram, they used the word "annihilate". So, what was annihilated when the electron and the positron merged to form that photon?

Well, what are the equations for heat and entropy trying to tell us? It's there in plain sight waiting for the whole world to see it.

When that electron and that positron merged to form that single photon, their associated positive and negative charges, mass, heat storage capacity, resistance to acceleration, heat, and entropy were completely annihilated or ceased to exist. Their entropy went to zero because their mass or heat storage capacity went to zero. It's that simple. And, this simple observation from Quantum Field Theory falsifies the second law of thermodynamics which claims that the amount of entropy in our universe can never go to zero. The amount of entropy, or mass, or heat storage capacity in our universe can and does go to zero all the time, thanks to the Conservation of Energy, $E = mc^2$, and Quantum Field Theory.

Annihilation, though, is the wrong word. Nothing was actually annihilated. Instead, the heat, mass, charge, and entropy within the electron and positron were TRANSFORMED into a massless, heatless, chargeless, and entropyless photon instead.

This is the KEY to everything in physics. It explains everything that has ever been experienced and observed.

When that positron and electron merged to form that photon, their mass, heat storage capacity, resistance to acceleration, and entropy were TRANSFORMED INTO massless, heatless, entropyless, wave-like infinite acceleration or quantum omnipresence instead according to Quantum Mechanics and Quantum Field Theory.

Electrons and positrons can't travel at the speed-of-light because they have mass, or resistance to acceleration, or entropy. Because electrons and positrons have mass, they also have some type of heat storage capacity and some resistance to acceleration.

However, when that electron and positron were transformed into a photon, their associated heat, mass, entropy, resistance to acceleration, charge, and heat storage capacity were TRANSFORMED INTO massless, heatless, chargeless, and entropyless infinite acceleration or omnipresence instead. Their mass, heat, and entropy completely ceased to exist. This is happening all the time in nature.

Notice carefully that a photon goes from zero to the speed-of-light instantly. That reality is infinite acceleration or the omnipresence of a quantum wave. That electron and positron along with their associated heat, entropy, and mass were TRANSFORMED into the infinite acceleration or the omnipresence of a quantum wave or photon. Photons have NO mass, NO heat storage capacity, NO resistance to acceleration, NO charge, and therefore, NO heat, and NO entropy. Photons are omnipresent or infinite acceleration quantum waves, according to Quantum Mechanics and Quantum Field Theory. Photons are non-local and non-physical, because photons have NO mass, NO heat storage capacity, NO resistance to acceleration, and NO entropy.

This is the KEY to everything in physics. It explains everything that has ever been experienced and observed. Does it not?

When the positron and electron merged to FORM that single photon, their associated mass, heat, charge, and entropy were TRANSFORMED into a massless, heatless, chargeless, and entropyless quantum wave or photon instead. Their associated entropy or heat completely ceased to exist when their mass or heat storage capacity ceased to exist. The result of this transformation of heat, mass, and entropy was in fact an omnipresent, syntropic, exergic, massless, heatless, chargeless, and entropyless photon or quantum wave. The associated entropy completely ceased to exist when the mass or heat storage capacity ceased to exist. It's that simple; and, it falsifies the second law of thermodynamics.

Photons are purely and completely quantum waves or quantum objects. Photons are capable of infinite acceleration, omnipresence, the speed-of-light, or quantum tunneling. Photons or quantum waves can do anything that they want to do because they have NO mass, NO resistance to acceleration, and NO entropy.

Entropy is mass's heat storage capacity.

Entropy is also resistance to acceleration.

This is the parsimonious, conservative, and logical definition for entropy – the definition for entropy that has actually been experienced and observed. The preponderance of the evidence tells us that it is true.

ALL of the entropy goes to zero and ceases to exist whenever mass of any kind is transformed into a massless, heatless, and entropyless quantum wave or photon. The transformation of mass, heat, and entropy into massless, heatless, and entropyless infinite acceleration or quantum omnipresence is happening all the time all around us; and, nobody has made this discovery because nobody is looking for it. They have all chosen to believe in the second law of thermodynamics instead.

Quantum Field Theory and the second law of thermodynamics are mutually exclusive, which means that they falsify each other. If one of them can be demonstrated to be true, then the other one has automatically been falsified or proven false.

So, which is it?

Do we observe mass, heat, entropy, electrons, and positrons being TRANSFORMED INTO massless, heatless, entropyless, chargeless, non-local, non-physical photons and quantum waves as is happening in our sun right now?

Or do we observe the Conservation of Entropy and an ever-increasing amount of mass, heat, and entropy as the second law of thermodynamics claims is happening?

Which one do we actually experience and observe?

The one that we actually experience and observe is the one that's actually real and true. It's that simple. Science is supposed to be eye-witness observation and first-hand experience after all. We observe the sun transforming mass into massless, heatless, and entropyless photons. We do NOT observe an ever-increasing amount of heat, mass, and entropy (heat storage capacity). Such a thing is physically impossible. The second law of thermodynamics is self-falsifying or self-defeating. The second law is physically impossible because it violates the Conservation of Energy. If the second law of thermodynamics were actually true, then we wouldn't be here and there really would be nothing but random chaos in this universe.

The second law of thermodynamics states that the amount of disorder or entropy in our universe is constantly increasing and can never decrease and go to zero.

The second law of thermodynamics was designed to falsify the first law of thermodynamics, and that's how the second law is typically used in "science". We observe Darwinists, Physicists, Atheists, and Scientists using the second law of thermodynamics to falsify the Conservation of Energy all the time on PBS, in our physics textbooks, and on the internet. The second law of thermodynamics is Creation Ex Nihilo or Atheism.

The first law of thermodynamics and the second law of thermodynamics are mutually exclusive. They falsify each other. If one of them can be demonstrated to be true, then the other one has been falsified or proven false.

So, which is it?

Do we truly observe ever-increasing amounts of disorder in our universe, or do we observe the ongoing perfection, order, and organization that's being produced by Nature's Psyche through the Quantum Fields?

Which one do we actually observe?

Which one actually exists?

Do we see an ever-encroaching grey goo coming in at us from all sides as the second law of thermodynamics predicts; or, do we observe ongoing order and organization as Quantum Mechanics and Quantum Field Theory predict? I KNOW which one I experience and see whenever I step outside in the morning or look through my telescopes at night. The syntropy, order, and organization completely dominate the disorder and chaos. True disorder and chaos cannot exist while the Quantum Fields exist. That's just the way it is.

Ask yourself, "Which one is being produced? Which one is being conserved? Which one is being experienced and observed?" Then answer truthfully.

Do we observe order and organization all throughout our universe, or do we observe nothing but ever-increasing random chaos? The second law of thermodynamics tells us that we are supposed to observe nothing but random chaos or ever-increasing disorder, entropy, and chaos; yet, I look out all around me and all I see is order and organization courtesy of the syntropic and perfectly organized Quantum Fields.

We will NEVER have true disorder and chaos in our part of the multiverse as long as the Quantum Fields exist. The Quantum Fields are perfect, syntropic, exergic perpetual motion machines. Ever since the Quantum Fields were made, they have been conserved. The proven and verified existence of the Quantum Fields falsifies the second law of thermodynamics. The Quantum Fields are indestructible and everlasting; and, they will always exist as long as their Designer and Maker exists. We observe that Quantum Field Theory continues to hold true; and, soon it becomes obvious that Quantum Field Theory falsifies the second law of thermodynamics. The second law of thermodynamics will NEVER be true as long as the Quantum Fields exist. You would have to destroy the Quantum Fields in order to make the second law of thermodynamics true. That's just the way it is.

When the science community is ready and fully understands what's really happening in Nature and in our Universe, then the second law of thermodynamics will be replaced by Quantum Field Theory instead.

That is the paradigm shift that it yet to take place within Science.

The falsehoods will be replaced with the truth, when Materialism, Naturalism, Darwinism, Nihilism, Atheism, Classical Realism, Natural Selection, and the Second Law of Thermodynamics are replaced by Action at a Distance, Quantum Mechanics, Quantum Field Theory, and the true maker of the wave function and the true collapser of the wave function who is Nature's Psyche. Nature's Psyche was making quantum waves and collapsing wave functions long before the first particle of physical matter was designed and made. Once they realize and accept the fact that it is Nature's Psyche or the Quantum Intelligence within Nature who makes the quantum waves and then subsequently collapses the wave function, then that's then end of the falsehoods in science that currently exist. Eventually, the truth will prevail.

Now, let's return to our Feynman Diagram, to see what other truths we can derive from it.

Notice carefully the arrow of time (**t**) on the Feynman Diagram. I remember one of my science teachers telling us that entropy is the thing that prevents us from running the film backwards. Remember the fat kid running down the diving board and jumping into the pool, and then the science teacher runs the film backwards, and the boy comes up out of the pool running backwards onto the diving board? Then the teacher says that entropy is the thing that prevents us from running the film backwards.

Well, that teacher was wrong. It's the arrow of time that prevents us from running the film backwards just as our Feynman Diagram suggests – NOT entropy. Entropy or disorder or chaos is completely reversible according to the LAW of the Conservation of Quantum Information, Information Theory, the Law of Intelligence or Psyche, $E = mc^2$, Quantum Mechanics, and Quantum Field Theory. Information, or knowledge, or intelligence is all that we need to reverse and eliminate disorder and chaos at will.

It has been said that black holes contain most of the disorder or chaos or entropy in our universe. This claim is based upon the erroneous and falsified "disorder definition" for entropy. The truth is that the quantum fields are still there and still functioning perfectly within a black hole. Furthermore, a black hole represents a powerful amount of order and organization. Black holes do work and produce work. They produce gravity waves to the extreme.

Furthermore, heat death or absolute zero maximizes the efficiency and functionality of super conductors, photons, quantum waves, and quantum fields. Everything about the second law of thermodynamics is wrong. The second law is clearly and conclusively falsified by Quantum Field Theory, Quantum Mechanics, Action at a Distance, $E = mc^2$, the Conservation of Energy, the First Law of Thermodynamics, and the Conservation of Quantum Information or the Conservation of Psyche and Intelligence.

There are other truths that we can learn from this simple Feynman Diagram.

What happens when that omnipresent, immaterial, non-local, non-physical photon or quantum wave chooses to stop?

It is a choice; and, ONLY Psyche or Intelligence or Quantum Consciousness can do choice at the quantum level. It is a choice! A photon or quantum wave doesn't have to stop if it doesn't want to. A photon or quantum wave can pass through our earth and our sun and a black hole as if they weren't even there!

Remember, it is Nature's Psyche or the controlling psyche within that photon who chooses to collapse the wave function and chooses to transform that photon into mass and heat. It is also Nature's Psyche or the controlling psyches within the mass (positron and electron) who chose to transform their mass, or resistance to acceleration, heat, or entropy (heat storage capacity) INTO a massless, heatless, chargeless, and entropyless quantum wave, virtual particle, or photon instead. It is Nature's Psyche who chooses to transform localized mass INTO non-local, massless, heatless, entropyless, omnipresent, syntropic, exergic quantum waves or photons – as is happening within our sun right now. ALL of this has been experienced and observed! ALL of this is an integral part of Quantum Field Theory and $E = mc^2$. ALL of this falsifies the second law of thermodynamics.

Whenever a quantum wave or photon chooses to stop its speed-of-light velocity and infinite acceleration, that photon or quantum wave literally TRANSFORMS its energy or infinite acceleration or non-local omnipresence INTO mass, heat storage capacity (entropy), heat, locality, and resistance to acceleration instead. We can feel the heat on our skin every time a photon or quantum wave chooses to stop rather than choosing to pass through us as if we weren't even there.

In the Feynman Diagram above, the massless and entropyless photon chose to transform itself into a quark–antiquark pair. The photon or quantum wave chose to transform itself into mass, heat storage capacity, and resistance to acceleration. Photons or quantum waves can TRANSFORM themselves into anything that they want to be. Light can become anything that it wants to become.

This time around, this photon chose to transform itself into mass, or resistance to acceleration, or heat storage capacity (entropy). Together, quarks and gluons represent a huge amount of resistance to acceleration or mass. Photons can also choose to transform themselves into heat. This is what typically happens when a photon lands on your skin.

We don't typically see new mass or new physical matter growing on our skin from the light coming from our sun. It could happen, but it doesn't happen, because Nature's Psyche or the controlling psyche within that photon is choosing what it wants to do with the energy that's under its control. Instead, when a photon stops on our skin, we experience the heat, when that photon TRANSFORMS its massless, heatless, and entropyless non-local infinite acceleration INTO localized heat instead. That photon KNOWS precisely what it is doing, and it KNOWS what is appropriate given its surroundings and circumstances, because that photon is omnipresent, omniscient, sentient, intelligent, and alive. When a photon stops on your skin, it KNOWS that it has stopped on your skin, so it KNOWS that it is appropriate to transform itself into heat instead of transforming itself into physical matter or mass.

This is the KEY to everything in physics. It explains everything that has ever been experienced and observed. Does it not?

According to the Law of Psyche and the Conservation of Energy, each psyche or intelligence or quantum consciousness has a certain amount of energy that it controls. That controlling psyche or intelligence can FORM or TRANSFORM the energy under its control into anything that it wants that energy to be, anytime and anywhere that it chooses to do so. That quantum consciousness, psyche, or intelligence has NO physical limitations! Psyche can do anything that it wants to do with the energy that's under its control, as our Feynman

Diagram so adequately attests. Every vertex or transformation on a Feynman Diagram represents a choice that has been made by a bunch of different psyches, who are controlling the energies from which these different quantum particles are made. Nature or Nature's Psyche is controlling it all – making the quantum waves in the first place and then choosing when to collapse those wave functions into mass or resistance to acceleration.

Every time a photon chooses to stop, we experience and observe a miniature Big Bang. Every time a photon or quantum wave chooses to stop, it transforms itself into mass or heat. Mass is the fundamental component of heat storage capacity or entropy; and, heat is the stuff that gets stored within that heat storage capacity or entropy. The "Big Bang" is ongoing. It's happening right now all around us every time a photon or quantum wave chooses to stop and transform itself into mass and heat and entropy (heat storage capacity). The Big Bang was NOT a one-time event 13.8 billion years ago. A miniature Big Bang takes place every single time that a photon or quantum wave chooses to stop and transform itself into mass, and heat, and resistance to acceleration (entropy). You can feel it on your skin every time it happens. A miniature Big Bang happens every time a photon lands on your skin and chooses to stop its infinite acceleration and speed-of-light velocity by transforming its omnipresence into heat instead.

$E = mc^2$!

It works perfectly fine both directions – both coming and going – thanks to Quantum Field Theory and the Conservation of Energy, the Conservation of Quantum Information, and the Conservation of Psyche or Intelligence.

We experience and observe a miniature Big Bang every single time that a photon or a quantum wave chooses to stop and transform itself into mass, resistance to acceleration, heat, and heat storage capacity (entropy) instead.

According to the Feynman Diagram at the beginning of this essay, Nature or Nature's Psyche is constantly TRANSFORMING charge, mass, heat, and entropy INTO massless, heatless, chargeless, and entropyless photons or quantum waves. In other words, the Controlling Psyches within Nature are constantly transforming heat storage capacity, heat, resistance to acceleration, mass, and entropy INTO massless, heatless, and entropyless omnipresence or infinite acceleration or quantum waves or photons. It's happening right now within our sun! Where do you think the light comes from?

It is happening all around us all the time whether we believe it or not; and, it falsifies the second law of thermodynamics whether we know it or not. As long as the Quantum Fields exist, the second law of thermodynamics will NEVER be true.

Entropy goes to zero and completely ceases to exist whenever the associated heat storage capacity, resistance to acceleration, or mass ceases to exist. This is what the equations for heat and entropy are trying to tell us. This is what Quantum Mechanics and Quantum Field Theory are trying to tell us, but nobody is listening, because everyone wants to believe in the second law of thermodynamics instead.

$E = mc^2$ is the perpetual motion cycle. It's real. It truly exists.

The Quantum Fields are the ultimate perpetual motion machines. Nothing is ever truly lost, because the underlying energy is always conserved. Disorder and chaos completely ceased to exist in our part of the multiverse the very moment that the Gods, the Controlling Psyches, and Nature's Psyche chose to make the Quantum Fields. The Quantum Fields are syntropic perfection; and, the Quantum Fields have been conserved ever since they were made. We will NEVER have true disorder, chaos, or "entropy" in our part of the multiverse as long as the non-local, non-physical, massless, heatless, entropyless, syntropic, exergic Quantum Fields exist. The second law of thermodynamics will NEVER be true so long as the Quantum Fields exist. That's just the way it is.

This is a completely NEW Model of Reality that explains everything that has ever been experienced and observed. Here within this model, I took everything that is false and replaced it with everything that is true, or with everything that has been experienced and observed. Did I not?

By falsifying the second law of thermodynamics and upgrading science to Science 2.0, I turned everything into Science, or turned everything into what has actually been experienced and observed. This is a paradigm shift. This is my greatest scientific discovery. This is my greatest contribution to Science. It explains everything that has ever been experienced and observed. Does it not? I've finally found the truth, and I KNOW IT.

Here, after millennia of thought and research, we finally have the truth. We finally have something that matches with everything that has ever been experienced and observed. This is the Ultimate Model of Reality. It is my gift to the world. It's there now. You'll just have to decide for yourself whether you want it or not.

Mark My Words

# Introducing the Law of Psyche

https://mypsyche.us/

Modern-day science is in desperate NEED of a reliable mechanism capable of getting things done at the quantum level among the quantum fields in the Quantum Realm. We NEED something that has actually been experienced and observed. Psyche, Intelligence, Quantum Consciousness, Life Force, or that Spark of Life that has been experienced and observed is the only thing that we know of that is actually capable of producing reliable results at the quantum level.

Nature or Nature's Psyche is the ONLY thing we know of that is capable of making the quantum fields from the chaotic and unorganized energy that existed before the quantum fields were designed and made. It is Nature's Psyche who collapses the wave function in quantum mechanics. It's the ONLY thing that can, because the human psyche isn't consciously aware of any of these things. It's also Nature's Psyche who makes the quantum waves, or the particles, or the quanta in the first place. Again, it's the ONLY thing that can!

We NEED something that has actually been experienced and observed, in order to explain the cause behind the effects that we observe taking place at the quantum level among the quantum fields in the Quantum Realm, or Spirit World, or Non-Physical Dimension.

https://psyche-ontology.com/buhlman/

https://psyche-ontology.com/psyche-experienced-and-observed/

It was time for a scientist like me to introduce the LAW OF PSYCHE to the science community and the world. We NEED a reliable mechanism for getting things done at the quantum level – a mechanism that has actually been experienced and observed.

**The LAW of Psyche states that each psyche or intelligence has a certain amount of energy that's under its control, and that controlling psyche can form or transform the energy under its control into anything that it wants that energy be, anytime and anywhere that it chooses to do so. Psyches can also coordinate with each other transpersonally or telepathically in order to form macro-structures such as physical atoms, out of the energies or quanta that they own and control.**

**The LAW of Psyche is based exclusively upon the LAW of Conservation of Energy and Psyche, as well as the LAW of Conservation of Quantum Information. The LAW of Quantum Information Conservation has to be true in order for quantum mechanics and quantum field theory to be true; and, the LAW of Psyche, as well as the LAW of Conservation of Energy and Psyche, have to be true in order for the LAW of Quantum Information Conservation to be true. All truth can be circumscribed into one great whole. We NEED these truths in order to have quantum mechanics, quantum field theory, and science in the first place.**

A scientific interpretation of data has an infinitely greater chance of being true if it is based upon Quantum Mechanics, Quantum Consciousness, or Quantum Field Theory rather than Materialism, Naturalism, Darwinism, Nihilism, Behaviorism, Determinism, Physicalism, Classical Realism, or Atheism.

Why is that?

It's because Quantum Mechanics, Quantum Consciousness, and Quantum Field Theory are an attempt to include ALL of the evidence into evidence. We try to include everything that has ever been experienced or observed within Quantum Mechanics, or we try to use Quantum Mechanics, Conservation of Energy, and Quantum Field Theory to explain everything that has ever been experienced and observed.

In contrast, Materialism, Naturalism, and their derivatives such as Darwinism and Atheism are based upon the exclusion of evidence – they deliberately exclude and reject everything that falsifies them. The Materialists, Naturalists, Darwinists, and Atheists claim that ONLY physical matter, entropy, death, and chance exist. These people literally believe that the non-local, the quantum, the psychic, the transdimensional, the spiritual, or the non-physical does not exist.

There's a serious problem here.

It's impossible to experience the non-existence of intelligence, thoughts, consciousness, life force, or psyche. Materialism and Naturalism claim that these things don't exist; yet, it is obvious that they do. It's impossible to experience the non-existence of the immaterial, or the non-physical, or the spiritual. It's impossible to experience the non-existence of light, gravity, dark energy, dark matter, magnetism, radio waves, quantum waves (virtual particles and thoughts), quantum fields, x-rays, microwaves, space, time, and all the other non-physical things that have been proven to exist. Materialism and Naturalism and Darwinism claim that these non-physical things do not exist; yet, it is obvious that they do.

The false is falsified by the truth; and, the truth is constantly and repeatedly experienced and observed. Materialism, Naturalism, and their derivatives such as Darwinism and Atheism are based exclusively upon the non-existence all the different non-physical things that clearly and obviously exist. Materialism and Naturalism can't be experienced and observed, which means that they are philosophy and NOT science. The non-existence of gravity, light, quantum fields, action at a distance, intelligence, consciousness, and psyche cannot be experienced and observed. Science is based upon what has actually been experienced and observed; whereas, Materialism, Naturalism, Darwinism, and Atheism are based upon the non-existence of what has actually been experienced and observed.

Materialism, Naturalism, Darwinism, and Atheism are the very definition of Bad Science and Bad Faith. Not only are they bad science, but they are bad religions or bad belief systems as well. They are BAD because they are based upon the rejection and the exclusion of all the evidence that falsifies them. You can't find and know the truth by rejecting it, or banning it, or excluding it whenever you encounter it; and, that's precisely what the Materialists and Naturalists do. These people ban, censor, ridicule, mock, reject, hide, and destroy all the evidence that falsifies their beliefs. That's how these people do "science".

I understand why these people reject quantum consciousness, quantum non-locality, quantum mechanics, and quantum field theory. There's no money in it. You can't make money from telling people the truth; and, we mortal fallen physical beings have absolutely no control whatsoever over quantum mechanisms. When it comes to quantum waves and quantum fields, we can't manipulate and control the stuff. Only the Gods and Nature's Psyche can control and manipulate quantum mechanisms at the quantum level with the power of their mind. Only Nature's Psyche can collapse the wave function; and, only Nature's Psyche can make the quantum waves in the first place.

We can do things at the physical level to transform energy from one form to another by starting a fire, making a generator, controlling electricity, or dropping a nuclear bomb; but, that's not the same thing as starting a quantum wave or collapsing a wave function. That's not the same thing as quantum tunneling or action at a distance. Transistors are the closest we have ever come to using quantum tunneling and quantum mechanics to get things done, and that came by determining what size of barrier is needed to prevent quantum tunneling from happening. There's some money to be made from that, but everything else seems to remain out of range to us.

Unless we are willing to classify photons or light as non-physical quantum objects, quantum mechanics or energy mechanics or non-physical mechanics is typically seen as a monetary dead-end, because we can't start a quantum wave nor collapse a wave function with the power of our mind at will like Nature's Psyche can. Many of our scientists have simply chosen to believe that the non-physical does not exist because there's no money to be made from it. In contrast, you can make a ton of money telling the Materialists, Naturalists, Darwinists, Nihilists, and Atheists what they want to hear. These people will pay a huge amount of money to be lied to. It's a growth industry!

People will not pay money to be told the truth, because the truth is not what we want to hear. I KNOW. I used to be a materialist, naturalist, nihilist, and atheist as recently as 2013. I used to be afraid of the evidence that falsified my beliefs. I refused to look at it. I refused to look at the truth.

Well, I'm no longer afraid of the evidence that falsifies my beliefs. Instead, I simply change my beliefs. My current goal is to make my theories, ideas, science, and beliefs come into line with what has actually been experienced and observed. My current goal is to find and know the truth.

In 2013, I started to question my beliefs. I slowly discovered that there is no such thing as a stupid question – just lots of stupid answers. By 2015, I was starting to realize that some of the most stupid and idiotic answers that have ever been devised by the mind of man were in fact created by the Materialists, Naturalists, Darwinists, Nihilists, Behaviorists, Determinists, Physical Reductionists, Classical Realists, and Atheists.

I formally turned against these people and my materialism, my naturalism, my nihilism, and my atheism in late 2015 and early 2016 when I first realized that these people are actually lying to us and trying to trick us and deceive us.

In December 2015, when I read William Buhlman's book, "Adventures Beyond the Body: How to Experience Out-of-Body Travel", that's when my infatuation with Scientism, Materialism, Naturalism, Nihilism, and Atheism ended; and, I started to search for another way. After I discovered the truth, I could no longer retain my materialism, my naturalism, my nihilism, and my atheism in good faith. It was finally time for me to pursue the truth instead.

Mark My Words

## Falsifying the Second Law of Thermodynamics

https://ultimate-law-of-thermodynamics.com/falsify-2nd-law/

The second law of thermodynamics states that the total amount of entropy or disorder in our universe is constantly increasing and that it can never decrease or go to zero.

I wish I had sufficient enough money to do a proper and complete statistical study.

I would like to discover how many scientists and how many people in this world KNOW that the Second Law of Thermodynamics is false. I would like to know if there is a difference between the two. Is the public aware of this and the scientists blind to it, or is it the other way around?

I would also like to discover how many scientists and how many people in this world KNOW WHY the Second Law of Thermodynamics is false. The second law is obviously false once you know why it is false.

Some of us KNOW that the second law of thermodynamics and the Big Bang Theory are mutually exclusive. They contradict each other. They falsify each other. If one of them can be demonstrated to be true, then the other one has been automatically falsified or proven false. In science, we can prove that our theories and ideas are false by falsifying them through the many different scientific methodologies and statistical methodologies that exist.

By definition, in principle or in practice, the Big Bang represents a complete and total reversal of entropy or disorder in our part of the multiverse, or in our part of the observable universe. The Big Bang was supposed to be Pure Syntropy. Everywhere we look, all we see is order and organization. We don't observe chaos, disorder, or "entropy defined as disorder and chaos" as the second law predicts that we should see. We don't see an ever-encroaching gray goo coming in at us from all sides as the second law of thermodynamics predicts. We don't observe proton decay as the second law of thermodynamics predicts. We don't observe everything freezing and grinding to a halt as the second law predicts. The second law of thermodynamics FAILS TO PREDICT what we are currently observing and experiencing throughout our universe. That makes the second law of thermodynamics Bad Science, Failed Science, or Falsified Science. That proves that the second law of thermodynamics is false.

The second law of thermodynamics was designed to falsify and eliminate the first law of thermodynamics; and, that's typically how the second law is used on PBS, television, and the internet. The first law and the second law are mutually exclusive. They falsify each other. They contradict each other. If one of them can be demonstrated to be true, then the other one has automatically been falsified or proven false. In science, we can prove that our theories and ideas are false by falsifying them through the many different scientific methodologies and statistical methodologies that exist. We can also use demonstrable or provable theories to falsify the ones that aren't producing any evidence to support them.

The first law of thermodynamics states that the total amount of energy in our universe is fixed or constant. This is the Conservation of Energy. The amount of energy in our universe cannot increase, and it cannot decrease. In other words, energy is conserved. Energy is a perpetual motion machine. Energy cannot be made, and it cannot be destroyed. Energy has always existed, and it will always exist. Energy is eternal and everlasting. $E = mc^2$ works perfectly both coming and going. It works both ways! It works perfectly while converting energy into mass or physical matter, and it works perfectly while transforming mass or physical matter back into usable energy or exergy. All the while the amount of energy is conserved or remains the same. Since everything in this universe – including physical matter and photons and quantum fields – is made from energy, the first law of thermodynamics should apply to everything in this universe including thermodynamics, heat, and entropy.

Yet, the second law of thermodynamics states that the total amount of entropy in our universe is constantly increasing and that it can never decrease and go to zero. The second law is Creation Ex Nihilo – the magical creation of something from nothing. Not only do they turn this mystical, magical, supernatural stuff that they call "entropy" into a conserved substance through the second law, they also make this invisible supernatural stuff capable of constantly increasing in amount out of thin air from nothing. The second law of thermodynamics is magic or Creation Ex Nihilo. The second law of thermodynamics violates and falsifies the first law of thermodynamics.

So, which is it?

Which one is true? Which one has been experienced and observed? Have we ever observed an ever-increasing amount of disorganized substance or goop coming in at us from thin air out of nothing, as the second law of thermodynamics predicts? We should be able to observe the deterioration, rot, or the dissolution of matter if the second law were true. Instead, we see the conservation of order and organization, thanks to the Conservation of Energy. The physical matter stays where it is placed. It doesn't dissolve into thin air or turn into goo as the second law predicts. We don't observe proton decay! That fact falsifies the second law of thermodynamics which predicts that it should be happening all around us right now.

In the summer of 2019, I was stunned to discover that my whole statistics book and my whole statistics class in science and physics and the scientific method FALSIFIES the second law of thermodynamics, particularly inferential statistics. I didn't see that one coming, let me tell you! When I hit the chapter on correlation in my statistics class and my statistics book, that's when the second law of thermodynamics fell apart right before my eyes. The second law couldn't stand in the face of truth. The second law is demolished by the truth. Disorder has NO correlation with heat or thermodynamics! I couldn't believe it at first. The heat definition for entropy FALSIFIES the disorder definition for entropy. They are mutually exclusive. They falsify each other.

Likewise, the first law of thermodynamics and the second law of thermodynamics are mutually exclusive. They contradict each other. They falsify each other. If one of them can be demonstrated to be true, then the other one has automatically been falsified or proven false.

So, which is it? Which one is true, and which one is false?

Do we truly observe ever-increasing disorder and chaos all throughout our universe, or do we observe constant and conserved order and organization? Do we observe proton decay and an ever-encroaching gray goo coming in at us from all sides seemingly by magic as the second law of thermodynamics predicts, or do we observe constant and conserved order and organization when we look through our telescopes and microscopes at our universe? Do we see conservation of energy or are we witnessing the destruction of the protons and quantum fields? Do the planets, stars, and galaxies continue to revolve around their orbits, or is everything coming to a screeching halt as the second law predicts? Do we observe the first law of thermodynamics, or do we observe ever-increasing chaos as the second law predicts?

The one that we actually experience and observe is the one that's actually real and true.

So, which is it?

Which one is true, and which one has been falsified?

The answer is obvious.

You would have to destroy the Conservation of Energy in order to make the second law of thermodynamics true.

According to Quantum Field Theory and Dark Energy, the quantum fields are where most of the energy in our part of the multiverse is currently being conserved or stored. The proven and verified existence of the massless and entropyless quantum fields falsifies the second law of thermodynamics. One day I finally realized that we would have to destroy the quantum fields in order to make the second law of thermodynamics true. The second law will NEVER be true as long as the quantum fields exist.

The quantum fields are perfect order and organization. They have no entropy, chaos, or disorder within them. The massless and entropyless non-physical quantum fields function flawlessly at absolute zero temperature, where there is no heat. There's no correlation between heat and disorder! Planets continue to revolve around neutrons stars and black holes in the complete absence of heat and light. The order and organization remain, thanks to the perpetual motion machines that we call the quantum fields.

Syntropy is the opposite of entropy. Syntropy is the Conservation of Energy. The second law of thermodynamics has been presented to us as the Conservation of Entropy or the Conservation of Disorder. Entropy represents everything that's NOT being conserved. In the popular vernacular, entropy has come to represent corruption, death, and chaos. Entropy represents a lack of conservation. Entropy is the opposite of Syntropy or Conservation.

In contrast, the quantum fields are syntropic perpetual motion machines. The quantum fields are constantly conserved. There is NO entropy, chaos, or disorder in the quantum fields. The quantum fields are massless, non-physical, and transdimensional. We physical beings can't touch nor change the quantum fields. We can't destroy the quantum fields. Ever since they were made, the quantum fields have been conserved. Only the Gods and the Controlling Psyches within Nature could destroy the quantum fields; and, it's the Gods working with and working through the Intelligences or Psyches in this universe who made the quantum fields in the first place. They could destroy the quantum fields anytime they wanted to and return our part of the multiverse to the Chaos Realm from whence it originally came, but apparently, they don't want to destroy the quantum fields because we continue to exist.

If the second law of thermodynamics were actually true, we would be currently witnessing the ongoing destruction of the quantum fields as everything around us returns to the chaos from whence it originally came. We will NEVER have true chaos or disorder in our part of the multiverse so long as the quantum fields exist.

According to Quantum Field Theory, particles are born, and particles die. Particles are made, and they can be unmade or disassembled as well. The Gods and Nature's Psyche had to design and make the quantum fields BEFORE they could make and sustain physical matter. First things first! The massless particles, as well as the physical atoms, ride the quantum fields. They can't exist in their current form without the quantum fields, according to Quantum Field Theory. Quantum Field Theory and $E = mc^2$ FALSIFY the second law of thermodynamics.

In the distant future, when Quantum Mechanics and Quantum Field Theory are fully accepted and truly understood, the scientists in this world will vote to replace the false and falsified second law of thermodynamics with Quantum Field Theory instead. Quantum Field Theory actually explains how thermodynamics works. The second law does not.

In science, we can prove that our theories and ideas are false by falsifying them through the many different observations, lived experiences, eye-witness accounts, and science experiments that we have done. Quantum Field Theory and the verified existence of the quantum fields falsify the second law of thermodynamics.

We are not supposed to guess in science. We are not supposed to use chance as a causal explanation in science. We are supposed to use observation, experiences, and evidence to support our beliefs while doing science. Yet, there's NO data or science experiments to support the second law of thermodynamics. We don't observe constantly increasing disorder throughout our universe. We don't observe ever-increasing amounts of heat throughout our universe either. We observe the exact opposite. We observe constantly conserved order and organization instead.

The second law of thermodynamics is nothing but science fiction. The second law of thermodynamics is a fictional story – a personal interpretation of philosophy – just like the theory of evolution. Most of the falsehoods and lies within science are coming from the second law of thermodynamics and the theory of evolution. Both of these are Creation by Disorder, Creation by Entropy, or Creation by Chance. They are in fact Creation Ex Nihilo – the creation of something from nothing by nothing.

The fact that they are Creation by Chance automatically falsifies them according to the null hypothesis in statistics, the scientific method, and the rules of science and scientific experimentation. Chance can't produce a real effect. If chance is producing a real effect, then it is no longer chance but has become causality instead. In statistics and science, a Real Effect is an effect that the independent variable has when it produces or causes a change in the dependent variable. In a science experiment, we manipulate an independent variable and measure its effect on a dependent variable. Alas, we can't use chance as an independent variable because once we start manipulating it, then it is NO LONGER chance!

Materialism, Naturalism, Darwinism, Nihilism, Atheism, the Theory of Evolution, and the Second Law of Thermodynamics ARE Creation by Chance. They ARE the Null Hypothesis, which means "caused by chance alone". The whole purpose of science and a science experiment is to reject the Null Hypothesis, Chance Causality, or Creation by Chance. The whole purpose of science and science experiments is to falsify things like Materialism, Naturalism, Darwinism, and the Second Law of Thermodynamics – anything that is creation by chance, chance causality, or caused by chance alone.

The second law of thermodynamics and the theory of evolution are automatically false because they are Creation by Chance or Creation Ex Nihilo.

For over forty years of my life, I defined entropy as "disorder". That's a falsified definition for entropy, and I didn't even know it. I'm a scientist, and I didn't know it.

I was shocked when I first realized that the equation for heat and the heat-based equation for entropy falsify the second law of thermodynamics. There's absolutely NO correlation between disorder and thermodynamics. Disorder doesn't cause heat; and, disorder doesn't eliminate heat. There's NO correlation between the two. Disorder or chance or randomness doesn't correlate with anything. It certainly doesn't correlate with thermodynamics or heat. Heat and thermodynamics represent a high degree of order and organization to begin with; and, even when the heat dissipates back into the absolute zero temperature quantum fields from whence it originally came, the quantum fields remain as a huge monumental amount of order and organization and perfection.

The massless, heatless, and entropyless Matrix of Quantum Fields represents perfect syntropic order, even at absolute zero temperature. The quantum fields are perpetual motion machines. Ever since they were made, the quantum fields have been conserved. They don't get old, and they don't wear out. The quantum fields don't need heat as a lubricant to keep them working. The amount of heat in the system is totally irrelevant when it comes to the quantum fields. The quantum fields function perfectly at maximum efficiency when the heat or temperature is absolute zero, and so do quantum waves or photons.

The thing that the scientists call "heat death" represents perfect order and organization, and maximum efficiency at absolute zero, when we are dealing with the quantum level or the non-physical level of existence, through the quantum fields. The Quantum Realm is syntropic perfection. It has nothing to do with disorder! The massless, heatless, and entropyless Quantum Fields are perpetual motion machines. The quantum fields are made from pure energy or pure exergy, which means that the quantum fields are constantly being conserved. The quantum fields are syntropic.

The protons and mass were also made from pure energy or pure exergy, which means that the protons and mass are also being conserved at the quantum level because their energy is being conserved. Nature or Nature's Psyche made the protons and the mass in the first place, and the protons will not decay as long as Nature's Psyche continues to conserve them. Nature's Psyche also made the photons, the quantum waves, and the quantum fields. They will not dissolve or disappear into chaos until Nature's Psyche decides to stop conserving them. They will not dissolve or disappear until Nature's Psyche chooses to transform them into something else. Nature's Psyche cannot destroy the energy, but Nature or Nature's Psyche can transform the energy into anything that it wants it to be, anytime and anywhere that it chooses to do so; and, there's nothing we can do to stop it or to make it happen either.

Quantum Field Theory states that particles are born, and particles die, but the underlying energy is constantly conserved. Anything that is made isn't fully conserved, but the energy within it is always conserved. When the scientists in this world figure out what's really going on and how things really work, they will vote to replace the second law of thermodynamics with Quantum Field Theory instead. Quantum Field Theory predicts what we are observing and experiencing all around us; whereas, the second law of thermodynamics does not. We don't observe the ever-increasing chaos and disorder that the second law predicts. We see what Quantum Field Theory predicts instead. As long as the quantum fields are conserved, order and organization will be conserved. That's just the way it is.

All that we have ever had in our part of the multiverse, ever since God and Nature's Psyche made the quantum fields from exergy, is syntropic and conserved and perfect order and organization at the quantum level. The protons and quantum fields will always exist or will always be conserved as long as God and the "Controlling Psyches within Nature" continue to exist or continue to be conserved and continue to conserve the quantum fields. It's Nature's Psyche who collapses the wave function; and, it's Nature's Psyche who makes the quantum waves and the quantum fields in the first place. It's the Gods who taught Nature's Psyche how to make the quantum fields. It's the Gods who taught Nature's Psyche how to organize the quantum waves and quantum fields into physical matter or mass or protons. Each entity does its part.

The second law of thermodynamics will NEVER be true as long as the quantum fields exist. You would have to destroy the quantum fields in order to make the second law of thermodynamics true. You would have to destroy the protons in order to make the second law of thermodynamics true. Even then, you wouldn't have an ever-increasing amount of disorder or chaos as the second law of thermodynamics predicts. You would simply have what seems to be a constant or conserved amount of chaos and disorder in our part of the multiverse if you destroyed the quantum fields. Once you achieve true disorder or chaos, its amount doesn't increase as the second law of thermodynamics predicts. Its amount simply remains constant or conserved just like the first law of thermodynamics predicts. The first law of thermodynamics predicts exactly what we experience and observe. The second law of thermodynamics does not!

The second law of thermodynamics FAILS TO PREDICT what we experience and observe. That means that the second law of thermodynamics is false. That's the way science works or is supposed to work.

The second law of thermodynamics states that the universal amount of disorder or entropy is constantly increasing and can never decrease. Is this what we are actually experiencing and observing?

By making "disorder" synonymous with thermodynamics or heat, the second law of thermodynamics technically predicts that we should be observing an ever-increasing amount of heat while we are also observing an ever-increasing amount of chaos or disorder. The two are supposed to be linked together as one according to the second law of thermodynamics. Do we really observe an ever-increasing amount of disorder and heat as the second law predicts?

Let's abandon the fiction and try to find the truth instead.

Thermodynamics is the transfer of heat from a hot reservoir to a cold reservoir, which means that "entropy defined as thermodynamics" is the transfer of heat from a hot reservoir to a cold reservoir. Entropy defined as thermodynamics FALSIFIES entropy defined as disorder! They are mutually exclusive. They contradict each other. They falsify each other. I could NEVER find a way to reconcile them. It's impossible to reconcile them. There's absolutely NO correlation between disorder and thermodynamics! Disorder doesn't cause heat, and heat doesn't cause disorder. Disorder doesn't make heat, and disorder doesn't eliminate heat. There's NO correlation or relationship between the two!

The second law of thermodynamics defines entropy as an ever-increasing amount of disorder. Whether the amount of disorder is increasing or decreasing, it's irrelevant when it comes to thermodynamics. There's NO relationship between disorder and heat. Heat is NOT synonymous with disorder. They are two different things completely. We do not observe an ever-increasing amount of heat within our universe. We observe the exact opposite. Entropy defined as heat FALSIFIES entropy defined as disorder. If the second law of thermodynamics were true, then an ever-increasing amount of disorder would produce an ever-increasing amount of heat – there would be both substances coming at us out of thin air from nothing, if the second law of thermodynamics were true, and if disorder were synonymous with heat as the second law predicts.

The second law of thermodynamics has the amount of disorder in our universe constantly increasing, yet we clearly and conclusively observe that the amount of heat or thermodynamics in our universe is NOT constantly increasing. The two ideas are mutually exclusive. They falsify each other. They oppose each other! We don't observe an ever-increasing amount of heat, or disorder, in our universe. We don't observe the second law of thermodynamics in our universe. We ONLY observe the conservation of order and organization.

The one that we experience and observe is the one that's real and true.

In ALL of their thought experiments into entropy and thermodynamics, they ALWAYS start with a hot piston and a cold infinite reservoir as a given; and then, they put the hot system into contact with the cold system so that there can be some entropy or a transfer of heat from the hot system to the cold system. The thermodynamics version of entropy is a "transfer of heat" from a hot system to a cold system. NO hot system, then NO entropy. It's that simple. They NEVER once realize that a massive reversal of disorder or a huge "reversal of entropy" took place, in order to produce the hot system or the hot reservoir in the first place. They simply start with the hot piston as an entropyless given, and then they go on from there in order to make the thing produce some entropy. Their hot reservoir or hot piston is produced by magic out of thin air from nothing for free as a given, so that they can then have a transfer of heat (entropy) taking place from the hot system to the cold system.

They cheat! It's a kludge. It's a scam. It's magic. It's sleight-of-hand. They are robbing from Peter to pay Paul. It's creation from nothing. The second law of thermodynamics is Atheism or Creation Ex Nihilo. It's magic.

They always produce the hot reservoir out of thin air from nothing so that they can then have some entropy or the transfer of heat from that hot reservoir into their pre-given cold reservoir. They cheat. The second law of thermodynamics is superstition. The second law of thermodynamics is Creation Ex Nihilo. The second law of thermodynamics is Creation by Entropy, or Creation by Disorder, or Creation from Nothing, or Creation by Chance. They produce the hot reservoir for free from nothing so that they can then have some entropy or a transfer of heat from that hot reservoir into their pre-given cold reservoir. The whole thing is nothing but a thought experiment. It has absolutely nothing to do with what we actually experience and observe. The second law of thermodynamics has absolutely nothing to do with thermodynamics. The second law of thermodynamics has nothing to do with reality.

Someday in the distant future, when the scientists in our world finally figure out how things really work and figure what's really going on in Nature and our Universe, they will vote to replace the falsified second law of thermodynamics with Quantum Field Theory instead. The quantum fields are used by Nature's Psyche to produce mass, heat, and thermodynamics. The quantum fields are used by Nature's Psyche to produce resistance to acceleration or mass's heat storage capacity. The quantum fields are also used by Nature's Psyche to produce massless, heatless, chargeless, and entropyless quantum waves or photons. Quantum Field Theory actually matches with what we experience and observe in Nature and in our Universe. Quantum Field Theory explains how thermodynamics works. The second law of thermodynamics does not. Quantum Field Theory is the perfect replacement for the falsified second law of thermodynamics.

The first law of thermodynamics FALSIFIES the second law of thermodynamics. The second law is Creation by Chance. The second law is the Null Hypothesis. The second law is caused by chance alone, and that fact FALSIFIES the second law of thermodynamics. The Conservation of Energy FALSIFIES the conservation of disorder and the conservation of entropy. The verified and proven existence of the perpetual motion machines that we call the Quantum Fields FALSIFIES the second law of thermodynamics. Quantum Field Theory FALSIFIES the second law of thermodynamics. The Big Bang Theory FALSIFIES the second law of thermodynamics. ALL of the conserved Syntropy, Order, and Organization that we constantly observe FALSIFIES the second law of thermodynamics. The fact that we exist

FALSIFIES the second law of thermodynamics. The fact that the equation for heat and the heat-based equation for entropy, or the thermodynamics definition for entropy, has absolutely nothing to do with chaos or disorder FALSIFIES the second law of thermodynamics. The heat-based equations for entropy FALSIFY the second law of thermodynamics. There is NO correlation between disorder and heat. Heat doesn't cause disorder, and disorder doesn't cause heat. The observable fact that there is no correlation between disorder and thermodynamics FALSIFIES the second law of thermodynamics.

The whole of modern-day Science FALSIFIES the second law of thermodynamics.

So, which one is true, and which one is false?

The Materialists, Naturalists, Darwinists, Nihilists, and Atheists will assure you that the second law of thermodynamics is true and that the Conservation of Energy, Quantum Field Theory, the Big Bang Theory, and the First Law of Thermodynamics ARE FALSE.

Are they right? Or have they gotten something wrong?

Which one has actually been experienced and observed? Which one predicts what we currently see throughout our universe?

If you figure out which one has been experienced and observed, then you will KNOW which one is real and true.

The Second Law of Thermodynamics is Materialism, Naturalism, Darwinism, Nihilism, Atheism, Creation Ex Nihilo, and Creation by Chance.

The very fact that these things are Creation by Chance or "caused by chance alone" automatically falsifies them according to the null hypothesis in statistics, the scientific method, and the rules for doing science and running science experiments.

There's another crucial thing that falsifies these false religions or false philosophies.

Materialism, Naturalism, and their derivatives such as Darwinism and Atheism claim that the non-physical does not exist. Therefore, the proven and verified existence of anything non-physical falsifies them. The massless, heatless, entropyless, and non-physical perpetual motion machines that we call the Quantum Fields falsify Materialism, Naturalism, and their derivatives which state that these non-physical Quantum Fields do not exist. The second law of thermodynamics is part of this materialistic and atheistic package. The second law is based upon the non-existence of the non-physical. The second law is a materialistic and atheistic philosophy. Quantum Field Theory falsifies the second law of thermodynamics, and Quantum Field Theory is the perfect replacement for the falsified second law of thermodynamics.

The massless, heatless, chargeless, and entropyless quantum waves or photons falsify Materialism and Naturalism which claim that the quantum waves or the photons don't exist. Gravity, magnetism, dark matter or spirit matter, dark energy, quantum consciousness, thoughts, intelligence, psyche, space, time, microwaves, radio waves, x-rays, forces, fields, and quantum waves falsify Materialism, Naturalism, Darwinism, Nihilism, and Atheism which claim that these non-physical or immaterial things do not exist.

Ask yourself which one has been experienced and observed?

Materialism, Naturalism, and their derivatives such as Darwinism and Atheism claim that gravity, magnetism, microwaves, radio waves, quantum waves, energy, and photons do not exist. Has the non-existence of these non-physical and immaterial things actually been experienced and observed? If the non-existence of these non-physical things has actually been experienced and observed, then we KNOW that Materialism, Naturalism, Darwinism, Nihilism, and Atheism are true.

The second law of thermodynamics is false because it is Creation by Chance and because it is Materialism, Naturalism, Darwinism (Creation by Chance), and Atheism (Creation Ex Nihilo). The fact that the second law of thermodynamics is a part of these falsified religions or falsified philosophies automatically falsifies the second law of thermodynamics. It's that simple.

My biggest blunder as a scientist took place when I chose to trust and believe the Materialists, Naturalists, Darwinists, Nihilists, and Atheists. I thought these people knew what they were talking about. I thought they were smart. I was wrong. These people don't have a clue when it comes to the Non-Physical Realm, the Transdimensional Realm, the Energy Realm, or the Quantum Realm; and, they hide their ignorance through censorship, ridicule, intimidation, peer review, tenure, and posturing. These people ban and delete the truth from existence whenever they encounter it.

Whenever we encounter the mutually exclusive, the one that has been experienced and observed repeatedly and constantly is in fact the one that's real and truly exists. The second law of thermodynamics will NEVER be true as long as the Quantum Fields exist. You would literally have to destroy the Quantum Fields in order to make the second law of thermodynamics true. Good luck with that! The quantum fields are constantly being experienced and observed.

Likewise, you would have to prove the non-existence of the photons and the non-existence of the quantum fields in order to prove that Materialism, Naturalism, Darwinism, Nihilism, and Atheism are true. The massless, heatless, chargeless, and entropyless photons are quantum waves, and those quantum waves ride the quantum fields at any speed that they choose. Quantum waves can quantum tunnel or teleport, if they choose to do so. Quantum waves and photons are capable of infinite acceleration – they can go from zero to the speed-of-light instantly. That's infinite acceleration. The proven and verified existence of these things falsifies Materialism, Naturalism, Darwinism, Nihilism, and Atheism which claim that these non-physical things do not exist.

The one that has been experienced and observed is the one that is real and truly exists!

So, which is it?

Have we experienced the **non-existence** of space, time, photons, gravity, radio waves, microwaves, x-rays, magnetism, energy, thoughts, and quantum waves as Materialism and Naturalism predict; or, have we experienced the exact opposite?

Materialism, Naturalism, and their derivatives such as Darwinism and Atheism claim that the non-physical does not exist; and, they claim that everything in this universe was created by chance. You would have to prove that space and time don't exist in order to prove that Materialism, Naturalism, Darwinism, Nihilism, and Atheism are true. Likewise, you would also have to prove beyond a shadow of a doubt that chance can design and create at will, in order to prove that the second law of thermodynamics, the theory of evolution, and Naturalism are real and true. Again, good luck with that! All of the evidence we have on hand and everything that has ever been experienced and observed falsifies Materialism, Naturalism, and their derivatives which claim that these non-physical things do not exist.

The false is falsified by the truth, and the truth is repeatedly and constantly observed. It's self-deception, pure and simple, to conclude that Materialism, Naturalism,

Darwinism (Creation by Chance), Nihilism, and Atheism (Creation Ex Nihilo) are true. It's self-deception to conclude that the second law of thermodynamics, or creation by entropy, or creation ex nihilo is true. It's self-deception to conclude that Creation by Chance is true. They can't be true because ALL of the evidence that we have on hand falsifies them! That's what science is supposed to do – falsify the things that are false. That's what scientists are supposed to do – falsify the things that are false. And, that's precisely what I have done here, which makes me a scientist.

Anything that was made FALSIFIES the second law of thermodynamics, which claims that everything came into existence by chance alone.

Everything falsifies the second law of thermodynamics. Everything! The very fact that you exist and are reading this right now FALSIFIES the second law of thermodynamics, which predicts that you shouldn't be here and predicts that you shouldn't exist. Your very existence as well as the existence of the planets, stars, and genomes is scientific proof that the second law of thermodynamics is false. The disorder in this universe is NOT constantly increasing as the second law predicts. The protons are NOT decaying as the second law predicts. Instead, we experience and observe conserved order and organization. All of the order and organization that surrounds us and permeates us FALSIFIES the second law of thermodynamics, which predicts or claims that it shouldn't exist. That which is experienced and observed is real and true, especially when all of us are experiencing it and observing it and living it for ourselves.

Mark My Words

--


I wrote three books in which I spent a great deal of time falsifying the second law of thermodynamics and finding some true theories and ideas to take its place.


**Syntropy in Defense of Quantum Mechanics: The Answer to Life, the Universe, and Everything**

https://www.amazon.com/gp/product/B07BPT3W8R/

**Quantum Mechanics: From a Non-Physical Spiritual Perspective**

https://www.amazon.com/gp/product/B01J023TGU/

**Science 2.0: I Upgraded My Science**

https://www.amazon.com/gp/product/B0771K6WTX/

## Heat Transfer Is One Definition for Entropy

I know what some of you are thinking, because I used to be a materialist, naturalist, nihilist, and atheist as recently as 2012 and 2013.

You are thinking, "Okay, this crackpot has proven that entropy does not exist, and that entropy is a kludge and a scam, but I still don't believe him because I know that entropy exists. It's obvious that entropy exists. I can see it all around me."

Technically, I haven't proven that entropy does not exist. What I have proven is that the different official definitions for entropy are false. I have proven that the second law of thermodynamics is false. In other words, I have falsified the second law of thermodynamics, which is completely different than proving that entropy does not exist. Entropy does exist if you get the right definition for entropy, the true evidence for entropy, and a true interpretation of that evidence.

I've spent some time thinking about the other side of the equation, and I have indeed asked myself, "How do I prove that entropy exists?"

Well, the way that I prove that entropy exists is by defining entropy as "heat transfer" or "a change in heat". It works, because heat transfer or friction obviously exists. Once we have a true interpretation or a true definition for a scientific concept, the simplest equations provide us with the profoundest scientific discoveries and truths. This one is amazing!

**$S = \Delta Q$**

Entropy (**S**) is equal to an exchange of heat (**Q**). Entropy is a transfer of heat from a hot system to a cold system (**ΔQ**). This is the thermodynamics definition for entropy. Thermodynamics obviously exists, so this version of entropy is obviously true. It is based upon one of the official equations for entropy – **$\Delta S = \Delta Q/T$**. This one has actually been experienced and observed.

Entropy is NOT some kind of mystical, universal, supernatural force as the Materialists, Naturalists, and Darwinists claim. Entropy can't design and create as the Darwinists and Atheists claim. Entropy is NOT constantly increasing as the second law of thermodynamics claims. Entropy can and does indeed decrease, thereby falsifying the second law of thermodynamics. Entropy is simply heat transfer; and, heat transfer can stop or go to zero. The amount of heat in a system can go to zero, which means that the amount of entropy can go to zero, and entropy can cease to exist. It's that simple.

The second law of thermodynamics claims that the total amount of entropy or disorder in the universe is constantly increasing and that it can never decrease. The second law of thermodynamics is garbage. It's obviously false. It violates ALL of the conservation laws, including the Conservation of Energy or the First Law of Thermodynamics. The first law of thermodynamics and the second law of thermodynamics are mutually exclusive. They falsify each other. If one of them can be demonstrated to be true, the other one has automatically been falsified. Everything that exists falsifies the second law of thermodynamics.

When enough scientists finally figure out how science and nature really work, they will vote to replace the second law of thermodynamics with Quantum Field Theory and $E = mc^2$ instead. Until then, we have to be satisfied with finding a couple of true definitions for entropy. We haven't had a true definition for entropy, since Rudolf Clausius coined the term "entropy" in 1864. The second law of thermodynamics erroneously states that the amount of entropy or disorder in our universe is always increasing, and that the amount of universal entropy can never decrease and go to zero.

At long last, after decades of being a scientist, and after a few years of concentrated research studying entropy and the second law of thermodynamics, it finally dawned on me yesterday (26JUN2019) that the stuff that they call entropy is in fact the transfer of heat from a hot reservoir to a cold reservoir. This was my first truthful and useful definition for entropy. For a few days, it was my best definition for entropy until I found one even better.

There are dozens of different, contradictory, self-defeating, and mutually exclusive definitions for entropy. I came up with the thermodynamics definition for entropy after watching the thermodynamics videos on Khan Academy. In all of their thought experiments into entropy, they always start with a hot reservoir and a cold reservoir as a given, so that there can then be some entropy, or a transfer of heat from that hot piston into that infinite cold reservoir.

**S = ΔQ**

https://www.khanacademy.org/science/physics/thermodynamics

https://www.khanacademy.org/science/chemistry/thermodynamics-chemistry

You've got to understand what these people believe, before you can see what's wrong with it. I put in the time and figured it out. I discovered the truths and the falsehoods in the process. The false is repeatedly falsified by the truth; and, the truth is constantly experienced and observed. Thermodynamics or a transfer of heat from a hot reservoir to a cold reservoir is constantly being experienced and observed; whereas, an ever-increasing amount of disorder and chaos, or an ever-encroaching grey goo coming in at us from all sides, has never been experienced nor observed anywhere in our universe, ever since the quantum fields were designed and made.

Thermodynamics IS the transfer of heat from a hot reservoir to a cold reservoir, so it actually makes logical sense to define entropy in terms of "heat transfer" or thermodynamics, rather than defining entropy as a statistical increase in "disorder" as they do in order to make the second law of thermodynamics seem to be true.

The second law of thermodynamics is ALL about disorder, randomization, probabilities, statistics, and chance. The second law is creation by entropy or creation by chance. The second law is the Conservation of Disorder or the Conservation of Entropy. The second law of thermodynamics actually has NOTHING to do with thermodynamics or the transfer of heat from a hot system to a cold system, which is why they are able to trick us into believing that entropy is always increasing and can never decrease. Disorder has absolutely nothing to do with thermodynamics or heat. There's no correlation between the two. With the second law of thermodynamics, they define entropy as something that it is NOT so that they can then have entropy always increasing and never decreasing. It's a scam. It's a bait and switch. It's a *category error* logic fallacy.

If you understand logic, then you can understand the logic errors that they employ to make entropy about increasing-disorder rather than what it really is – a transfer of heat from a hot system to a cold system. The second law of thermodynamics is a *category error* logic fallacy and an oxymoron, because it makes entropy about ever-increasing disorder rather than making entropy about thermodynamics or the transfer of heat from a hot system to a cold system. So, even logic falsifies the second law of thermodynamics, if we let it do so. Everything that exists falsifies the second law of thermodynamics. Since they used faulty logic, category errors, and logic fallacies to produce the second law of thermodynamics in the first place, we can easily use logic to falsify it.

**Entropy is the transfer of heat from a hot system to a cold system. Entropy is mass or resistance to acceleration, which generates friction or heat. No mass or no heat, then no entropy. It's that simple.**

That's how they ALWAYS set it up and define entropy in ALL of their thought experiments into entropy. They ALWAYS start with a hot reservoir and a cold reservoir as a given, and then put the two into contact with each other so that there can be some heat transfer or entropy. Thermodynamics is ALWAYS about the transfer of heat from one system to another, so "heat transfer" ends up being the best and the most natural definition for entropy rather than the erroneous definition for entropy that is forced upon us by the inappropriately named second law of thermodynamics. Thermodynamics is a transfer of heat from a hot system to a cold system; therefore, heat transfer or an exchange of heat is the thermodynamics definition for entropy. This is a valid definition for entropy because it has actually been experienced, observed, and verified.

https://www.khanacademy.org/science/chemistry/thermodynamics-chemistry/enthalpy-chemistry-sal/v/chem19-calorimetry

Observe that any heat that is made by Nature's Psyche eventually finds its way back to the infinite, absolute-zero temperature, quantum reservoir or quantum fields from whence it came and from which it was originally made.

Who or what do we know of that can make things at the quantum level in the infinite absolute-zero of non-physical transdimensional space and then transfer the thing that it has made into our physical spacetime realm? Who or what do we know of that can take absolute zero quantum energies and transform those energies into heat, mass, and physical matter? Well, it isn't physical matter as the Materialists, Naturalists, Darwinists, and Atheists claim. Physical matter or mass is the product, and NOT the Creator. Heat, heat transfer, and entropy are the product, and NOT the Creator. Boltzmann, Darwin, Hawking, and others erroneously turned disorder, chaos, chance, or entropy into our creator. These are some of the people who turned the second law of thermodynamics into creation by disorder, or creation by chance, or creation by entropy. Creation by Chance, and Creation by Disorder, and Creation by Entropy have never been experienced nor observed!

According to the LAW of Psyche, each psyche or intelligence has a certain amount of energy that's under its control, and that psyche can form or transform the energy under its control into anything that it wants that energy to be, anytime it chooses to do so. It is Nature's Psyche who collapses the wave function. It is the Controlling Psyches within Nature who make the quantum waves in the first place. It's the only thing that can! Nature's Psyche was collapsing wave functions and making things in the absolute-zero temperature of the Quantum Realm long before the first particle of physical matter was designed and made. Nature's Psyche or the Controlling Psyches had to design and make the quantum fields BEFORE they could make and sustain physical matter, or mass, or resistance to acceleration. This is obviously true to anyone who understands Quantum Field Theory and understands the idea that particles are made, and that particles can be unmade or absorbed back into the quantum fields from whence they came.

Particles, mass, and heat are made by Nature's Psyche, and they can be unmade or absorbed back into the quantum fields anytime that Nature's Psyche chooses to do so. This is what has actually been experienced and observed. It has NOTHING to do with entropy, increasing disorder, or even the transfer of heat from a hot system to a cold system. Where the creation and disassembly of mass, heat, and particles are concerned, it's ALL about Psyche and the Conservation of Energy instead. Work in equals work out; and, at the quantum level, all of the work is done for free thanks to the Conservation of Energy.

Nature's Psyche can make particles, mass, and heat, anytime it wants to do so; and, it will never run out of energy from which to do so.

$E = mc^2$ is the perpetual motion cycle. It will never end, because energy is constantly recycled, and energy is always conserved. According to Quantum Field Theory and $E = mc^2$, energy or quantum waves are constantly being transformed into mass and heat by Nature or Nature's Psyche; and then in return, mass and heat are constantly being transformed back into quantum waves or available energy (exergy).

Here we finally have the truth, and here we finally figured out how things really work at the quantum level in the Non-Local, Transdimensional, Non-Physical, Syntropic, Massless, Absolute-Zero Temperature, Quantum Realm or Psyche Realm.

**Heat transfer ends up being the best and most realistic definition for entropy**; and, heat transfer completely ceases to exist in a massless particle and in a mass-based system that has reached absolute zero or heat death or thermal equilibrium. Entropy completely ceases to exist at absolute zero and at thermal equilibrium, when heat transfer has ceased to exist. No heat transfer, then NO entropy. According to ALL the different equations for entropy, entropy ceases to exist, or entropy goes to zero, at heat death or at the state that the physicists call "maximum entropy". At "maximum entropy", or heat death, or thermal equilibrium, or permanent absolute zero, entropy or heat transfer goes to zero and ceases to exist.

Here, we finally have entropy defined in terms of something that I can believe in and KNOW to be true. Entropy is the transfer of heat. NO heat, then NO entropy. No heat transfer, then no entropy. It's that simple.

Now, let's look at this new and amazing discovery from a non-physical quantum perspective, and see what we can learn. We can learn a lot by choosing to MAP the physical onto the quantum or the non-physical.

Entropy is always produced as a thought experiment where they start with a hot reservoir and a cold reservoir as a given, and then they put those two reservoirs into contact with each other so that there can be some heat transfer or entropy. That's the way they do it. That's the way they make entropy in their thought experiments. They create a hot reservoir out of thin air and a cold reservoir out of thin air, and then they put the two reservoirs into contact with each other so there can be some entropy. They have to do that because there is NO entropy when there is NO heat and when there is NO heat transfer.

This is powerful knowledge, is it not!

Now we finally KNOW what entropy truly is; and, the true definition for entropy FALSIFIES the second law of thermodynamics which states that entropy is always increasing, and that entropy can never decrease. Once we KNOW that entropy is heat transfer, and once we KNOW that where there is no heat and where there is no heat transfer there is NO entropy, then we simply KNOW that the second law of thermodynamics is false because we know why it is false.

The second law of thermodynamics is Conservation of Entropy; and, the Conservation of Entropy is falsified by the truth of what we have actually experienced and observed. The Conservation of Entropy is falsified by the heat-based equations for entropy, by thermodynamics, by Quantum Field Theory, by the Conservation of Energy and Nature's Psyche, and by $E = mc^2$.

The second law of thermodynamics is about some mystical, non-physical, supernatural, immaterial thing that is constantly increasing and can never decrease. In contrast, heat transfer or change in heat (**ΔQ**) is about something that has actually been experienced and observed, is constantly changing, can go to zero, and is never conserved. Change in heat or heat transfer is something real that I can actually believe in, because it has actually been experienced and observed.

The false is falsified by the truth; and, the truth is repeatedly experienced and observed. Here at the physical level heat transfer or friction is constantly and repeatedly experienced and observed. Is it not? Go with what has been experienced and observed!

What else can we learn by looking at this process from a non-physical, spiritual, quantum perspective? Quite a lot actually.

Ask yourself, "How is heat produced, really?"

Well, how was heat produced in the Big Bang by the alleged Big Bang Singularity? The majority of scientists currently believe that the Big Bang happened, so let's use that belief to see what we can learn from it. It's simply amazing what we can learn by choosing to look at things from a non-physical quantum perspective.

If the Big Bang Singularity existed, it was by definition purely a quantum object – pure syntropy – because physical matter and spacetime didn't exist yet. By definition, in principle, the Big Bang Singularity was a non-physical object, created at the non-physical quantum level, by supposedly non-physical supernatural quantum beings. Furthermore, the Big Bang Singularity was designed and created in the absolute zero of non-local transdimensional space or quantum space because, by definition and in principle, spacetime and physical matter didn't exist yet.

So, what happened when these psyches or non-physical beings triggered the Big Bang Singularity at the moment of the Big Bang?

Well, instantaneously, a ton of immaterial, non-physical, absolute zero quantum energy was converted into a lot of spacetime HEAT (**Q**) at the physical level. Eventually some of that heat and energy was allegedly converted into physical matter by these same psyches or non-physical beings. That's how the Big Bang supposedly worked, is it not? That's basically how it's explained, is it not? If you believe that the Big Bang Theory is true, we just explained how it was done. Nature's Psyche converted a lot of absolute-zero temperature quantum energy at the quantum level into a bunch of physical heat at the physical spacetime level. And, Nature's Psyche could do so again, today, right now, if it wants to do so. Nature's Psyche could instantaneously convert a whole bunch of "empty space" and quantum energy in the middle of nowhere at the quantum level into a brand-new physical galaxy, with a bunch of brand-new stars, that start pumping out tons of heat at the physical level.

According to the Big Bang Theory, a whole bunch of non-physical, non-local, transdimensional, syntropic, quantum energies at the quantum level were instantaneously converted into HEAT (**Q**) at our physical spacetime level where we currently live. That's how heat was made during the Big Bang; and, that's how heat is made today!

It's the same exact process!

According to the LAW of Psyche, each psyche has a certain amount of energy that's under its control, and that psyche can form or transform the energy under its control into anything that it wants that energy to be, anytime it chooses to do so.

If that psyche wants that energy formed into quantum fields or spacetime, then that's precisely what happens. If that psyche wants that energy transformed into photons, or quantum waves, or quantum fields, or mass, or spirit matter, or dark matter, or physical matter, then that's precisely what happens. If that psyche wants its energies transformed from cold, massless, non-physical, absolute zero energy into HEAT (**Q**), then that's precisely what happens.

It is Nature's Psyche who collapses the wave function. It's the only thing that can! It is Nature's Psyche who makes the quantum fields and the quantum waves in the first place. It's the only thing that can!

There's nothing mystical or magical about it, once we understand how it truly works. Psyche is the ultimate point particle, the ultimate causal agent, and the ultimate cause, which means that psyche can do anything it wants to do with the energies that are under its control. If that psyche wants those energies to become heat at the physical level, then those energies become heat at the physical level. It's that simple.

https://www.khanacademy.org/science/chemistry/thermodynamics-chemistry/internal-energy-sal/v/first-law-of-thermodynamics-introduction

Energy is not made, and it cannot be destroyed. Energy is simply transformed from one type of energy to the next; and in the end, the work is always done for free. According to the Big Bang Theory, Nature's Psyche can transform the energies within the infinite absolute-zero quantum reservoir into heat and physical matter and mass anytime that Nature's Psyche chooses to do so. The heat and the physical matter are made for free, whenever Nature's Psyche chooses to make them.

The Equation for the Conservation of Energy:

**Work In = Work Out**

The work is always done for free thanks to the Conservation of Energy. This reality and truth actually falsifies the second law of thermodynamics; and, we don't even know it.

Heat in ends up being heat out. Entropy in equals entropy out. This is the thermodynamics or heat-based definition for entropy. Entropy is synonymous with heat. Nature or Nature's Psyche makes heat at will, and then that heat or entropy is slowly absorbed back into the quantum fields from whence it came.

Entropy is NOT ever-increasing disorder as the second law of thermodynamics claims. Entropy is NOT an ever-increasing amount of substance as the second law claims. Ever-increasing disorder is technically impossible thanks to Nature's Psyche and the infinitely organized quantum fields that Nature's Psyche made. There will NEVER be true disorder and chaos in our part of the multiverse as long as the quantum fields exist. The second law of thermodynamics will NEVER be true so long as the quantum fields exist. You would have to destroy the quantum fields in order to make the second law of thermodynamics true.

The infinite organized quantum fields or dark energy at the quantum level simply dwarves any randomization of physical atoms that might take place at the physical level. The physical is a drop in the ocean compared to the infinite expanse of perfectly organized quantum fields that have been designed and made by Nature's Psyche or by God Himself. Entropy cannot be ever-increasing disorder as the second law of thermodynamics clams, because the second law of thermodynamics deliberately and axiomatically excludes the infinite expanse of perfectly organized quantum fields from being considered as part of the equation. The second law of thermodynamics denies at least 95% of the evidence and refuses to allow it into evidence; and, denying evidence and deliberately excluding evidence are logic fallacies, whether we realize it or not.

### Entropy is simply a transfer of heat from a hot system to a cold system.

Entropy is thermodynamics, and NOT mystical philosophical magic. This may in fact be the most comprehensive treatment of entropy that has ever been designed and made. It's certainly the most true or the most verified account of entropy that has ever been designed and made. I've never seen anything better.

Entropy as the transfer of heat from a hot reservoir to a cold reservoir is the conservative definition for entropy. In contrast, the second law definition of entropy as an ever-increasing substance that can never decrease is the extravagant definition for entropy. If you know anything about probabilities and statistics, then you know that a conservative estimate has a much greater probability of being true than an extravagant estimate. The second law of thermodynamics is so extreme and so extravagant that it can't possibly be true.

Entropy is simply the transfer of heat from a hot system to a cold system ($S = \Delta Q$). Once thermal equilibrium is reached, then entropy or the transfer of heat ceases to exist, because it is no longer happening.

**S** equals entropy.

**$\Delta Q$** is an exchange of heat, or a transfer of heat, from a hot system to a cold system, regardless of what the temperature of those two systems might be. Once they reach thermal equilibrium, at whatever temperature that might be, entropy or the transfer of heat ceases to exist, until Nature's Psyche makes some more heat.

I call this the thermodynamics definition for entropy. Entropy is an exchange of heat from a hot reservoir to a cold reservoir. This is a powerful, demonstrable, and true definition for entropy! It defines entropy by what it really is and by what entropy truly does. Centuries of superstition simply disappear, once we have a true definition for entropy.

Once we have a true definition for entropy, then it immediately becomes clear and obvious that the First Law of Thermodynamics is true and that the second law of thermodynamics is false.

Fascinating, is it not? The truth is hiding in plain sight where nobody can see it because they aren't looking for it and don't want it to be true. But, once we start looking for the truth, then we start to find it. Entropy is heat transfer or a change in heat ($S = \Delta Q$). It's that simple. When heat ceases to exist, or when heat transfer has stopped, or when the temperature is fixed at absolute zero, then entropy ceases to exist or goes to zero, just like the equations for entropy claim.

$\Delta S = \Delta Q/T$

$S = \Delta Q/T$

$S = Q/\Delta T$

These are different versions of the same thing. Entropy is conservatively and realistically defined as a transfer of heat from a hot reservoir to a cold reservoir.

Entropy (**S**) equals a change in heat ($\Delta Q$); and, entropy ceases to exist when heat is zero ($Q = 0$), at thermal equilibrium ($\Delta T = 0$), and when the temperature is fixed at absolute zero ($T = 0$). No heat, then no entropy; and, entropy or heat transfer ceases to exist at thermal equilibrium. It's obviously true, once we accept the fact that this equation for entropy is true.

In this case (**T = 0**), when temperature approaches zero, entropy or heat transfer also approaches zero, instead of approaching infinity as the materialists and naturalists claim. Yes, anything divided by zero is infinity; but, ALL of the evidence falsifies the idea that entropy is infinite when the temperature is zero. In this case, we have to switch over to limits and calculus, and then observe what really happens as temperature, and heat, and heat transfer approach zero in the real world. As heat and heat transfer go to zero, entropy goes to zero. ALL of the evidence and common sense tell us that as the temperature goes to zero, heat goes to zero, heat transfer goes to zero; and therefore, entropy goes to zero.

According to the Quantum Law of Thermodynamics, as heat transfer goes to zero, the amount of entropy goes to zero and ceases to exist. This is the way things really work!

https://quantum-law-of-thermodynamics.com

Here we actually USE observation and experience to verify and prove the truth. That's the way Science should be done, and not just through opportunistic thought experiments as the Materialists, Naturalists, and Darwinists do science.

We need to figure out how things really work, rather than going with what we want to believe to the exclusion of all else.

I call this the thermodynamics definition for entropy. Entropy is an exchange of heat from a hot reservoir to a cold reservoir. This is a powerful, demonstrable, and true definition for entropy! It defines entropy by what it really is and by what entropy truly does. Centuries of superstition simply disappear, once we have a true definition for entropy.

By defining entropy as "heat transfer", we finally KNOW what the stuff really is and how it truly works. As heat transfer or the change in heat goes to zero, the amount of entropy goes to zero, just as the equation for entropy claims. In the quantum realm or the non-physical realm where the temperature is locked at absolute zero, heat transfer or entropy has officially ceased to exist. We KNOW that this is true because it has actually been experienced and observed. No heat, then no entropy, according to the equations for entropy. According to the Quantum Law of Thermodynamics, thermodynamics, the transfer of heat, and entropy don't exist at the quantum level.

For me personally, this was an amazing flash of insight. Finally, everything made sense. I finally had the truth that I had been searching for all of my life. The stuff that they call entropy is in fact the transfer of heat from a hot system to a cold system. Once thermal equilibrium is achieved, then the heat transfer stops, and entropy goes to zero because there is no heat transfer taking place at thermal equilibrium.

**S = ΔQ**

I call this the thermodynamics definition for entropy. Entropy is an exchange of heat from a hot reservoir to a cold reservoir. This is a powerful, demonstrable, and true definition for entropy! It defines entropy by what it really is and by what entropy truly does. Centuries of superstition simply disappear, once we have a true definition for entropy.

Entropy (**S**) equals a transfer of heat or an exchange of heat (**ΔQ**) between a hot system and a cold system. It happens naturally and automatically. The exchange of heat or entropy goes to zero as the two systems move toward thermal equilibrium. At thermal equilibrium, there is NO entropy. Entropy or heat exchange has decreased to zero at thermal equilibrium and at absolute zero.

So, what do we observe after we have finally gotten a true and accurate definition for entropy?

1. We observe that entropy is heat transfer, or a change in the amount of heat ($S = \Delta Q$). And then, we observe the obvious. No heat ($Q = 0$), then no entropy ($S = 0$). No heat transfer ($\Delta Q = 0$), then no entropy ($S = 0$). By definition, at absolute zero, there is NO heat; and therefore, there is NO entropy. This is obviously true. Finally, we have a true, accurate, and verified definition for entropy. It matches with what has actually been experienced and observed.

2. We observe that heat transfer is NOT always increasing. In fact, heat transfer tends towards zero as the two pre-selected hot and cold reservoirs move towards thermal equilibrium.

3. We observe that heat transfer (or entropy) can and does tend to decrease toward thermal equilibrium. It's naturally drawn that direction. The Quantum Realm is obviously an infinite, absolute zero, quantum reservoir; and, the psyches who control this infinite quantum reservoir determine the fate of our universe. Heat transfer or entropy doesn't determine the destiny of our universe. Nature's Psyche does, because it's Nature's Psyche who collapses the wave function. Anytime that it chooses to do so, a psyche can transform the energy under its control into anything that it wants that energy to be – including physical matter and heat. This is precisely what the Big Bang Theory is trying to tell us. Some quantum intelligence took the absolute-zero non-physical quantum energies that are under its control, and transformed those quantum energies into quantum fields, spacetime, heat, and physical matter. The alleged Big Bang was in fact the construction of the quantum fields.

4. We observe that thermal equilibrium is a state of NO heat transfer, which means that it is a state of NO entropy. Heat transfer stops whenever a system is in thermal equilibrium, and whenever the temperature of the system is locked at absolute zero. In other words, entropy or heat transfer ceases to exist at thermal equilibrium and at absolute zero where there is NO heat. This is obviously true. There is no heat transfer; and therefore, there is no entropy when there is no heat.

5. In consequence of all of this, we observe that the second law of thermodynamics is obviously false, because the second law of thermodynamics says that heat transfer (entropy) is always increasing and that heat transfer (entropy) can never decrease. The second law of thermodynamics ends up being the exact opposite of the truth, once we have a true definition for entropy.

6. We observe that the equations for entropy, especially the heat equations for entropy, FALSIFY the second law of thermodynamics. Entropy goes to zero when heat goes to zero, in complete contradiction and falsification of the second law of thermodynamics. No heat, then no entropy. No heat transfer, then no entropy. It's that simple. The equations for entropy reflect the truth, and NOT the second law of thermodynamics. According to the equations for entropy, heat and heat transfer are constantly decreasing after the heat is made; whereas, the second law of thermodynamics erroneously states that heat and heat transfer and entropy are always increasing, and that heat transfer and heat and entropy can never decrease. Everything about the second law of thermodynamics is falsified by the truth.

7. By correctly and accurately defining entropy as the transfer of heat, we make entropy compatible with the First Law of Thermodynamics or the Conservation of Energy

and Psyche. We finally achieve the truth. In the quantum realm, the work is always done for free thanks to the Conservation of Energy. The heat and the physical matter are made for free, anytime that the Controlling Psyches or Nature's Psyche chooses to make the stuff. Once the heat is made, it slowly makes its way back into the infinite quantum reservoir from whence it came. Simple. Logical. True.

8. We observe that psyches or intelligences at the quantum level can make heat (**Q**) and physical matter out of absolute zero, non-physical, syntropic, quantum energies, ANYTIME they choose to do so. This is exactly what was allegedly done during the Big Bang. If the Big Bang Theory is true, then this observation is true. By defining entropy has heat transfer and as the generation of heat, we finally have a physics mechanism in place for explaining how the Big Bang was allegedly done. The Big Bang was in fact raw, quantum, non-physical, syntropic, absolute-zero, transdimensional energies that Nature's Psyche decided to turn into heat, quantum fields, spacetime, and physical matter instead. It's that simple.

9. We also quickly observe that fallen, physical, mortal beings such as ourselves have NO mechanism and NO system in place for converting absolute zero, syntropic, quantum energies into heat (**Q**) and physical matter from the quantum level. It is Nature's Psyche who makes the heat and the physical matter, not us. It is Nature's Psyche who collapses the wave function – not us, and not the physical matter. The Human Psyche isn't consciously aware of any of these quantum processes. But, we can assist, or guide, or encourage, or coax. Nature's Psyche or God has provided us with some ways to generate heat at the physical level. Starting a fire or detonating a hydrogen bomb entices Nature's Psyche to produce a lot of heat. And, particle accelerators or atom smashers can entice Nature's Psyche to produce some physical matter. However, we human beings cannot create new heat and new physical matter out of absolute-zero quantum energies at will. Only Nature's Psyche can do that, because it's Nature's Psyche who controls all of the energy in the first place.

Anytime Nature's Psyche makes some heat, that heat is quickly absorbed back into the infinite, absolute-zero, quantum reservoir from whence it came. According to the Big Bang Theory, Nature's Psyche can make as much heat and physical matter as it wants, anytime it chooses to do so, because the heat and the physical matter generated at the physical level is a drop in the ocean compared to all the absolute-zero, massless, non-physical, syntropic, quantum energies that exist at the quantum level. Remember, at the quantum level, ALL of the work is done for free, thanks to the Conservation of Energy – including the work that produces heat. The heat comes for free, the very moment that Nature's Psyche decides to make some heat. This same reality and truth applies to mass, or resistance to acceleration, or physical matter. It's all made for free at the quantum level by Nature's Psyche thanks to the Conservation of Energy and Psyche.

Anytime Nature's Psyche chooses to make some more heat and physical matter, it does so. There's nothing that we can do about it, because we can't collapse the wave function. It's all under the control of Nature's Psyche instead, because Nature's Psyche is the thing that collapses the wave function, not us.

10. We observe that the second law of thermodynamics is a ruse, a scam, a hoax, a con, and a lie. Psyches or intelligences can create physical matter, spacetime, quantum fields, and heat from non-physical, absolute zero, transdimensional, syntropic, non-local, quantum energies ANYTIME they choose to do so, just like the Big Bang Theory claims. Entropy or heat transfer is NOT always increasing but instead tends to decrease over time; and the amount of entropy or heat transfer can and does decrease as any two systems strive towards thermal equilibrium. This is what really happens, and it falsifies the second law of thermodynamics. We observe that the equations for entropy FALSIFY the second law

of thermodynamics. Did you notice how we always get to the truth when we start with the truth in the first place? Did you notice that the false is falsified by the truth, and that the truth is always experienced and observed?

**Examining the Equation for Heat**

Finally, we can learn a lot about entropy by studying the equation for heat, because entropy is by definition a transfer of heat.

Once we have a true interpretation or a true definition for scientific data or a scientific concept, the simplest equations provide us with the profoundest scientific discoveries. This one is amazing!

The Equation for Heat:

$Q = mc\Delta T$

Heat (**Q**) is energy in transfer to or from a thermodynamic system. Heat is a function of mass (**m**). No mass then NO heat. Therefore, by definition, massless particles or non-physical particles such as photons and quantum waves and quantum fields have NO heat, because they have no mass. Mass acts as a heat reservoir. Mass, or resistance to acceleration, or friction generates heat. According to specific heat capacity (**c**), each type of mass has a specific heat storage capacity. Mass stores heat. No mass, then no heat storage. It's that simple.

For quantum fields and quantum waves and psyches and massless particles, their natural state is absolute zero, where there is no entropy and can be no entropy. No mass then NO heat. By definition, photons travel, quantum waves travel, and quantum fields work best within the absolute zero of space where there is no heat, because they are massless, and non-physical, and have NO resistance to acceleration.

Heat is a product of mass or resistance to acceleration. Resistance to acceleration generates heat. Friction generates heat. It's that simple. No physical matter, then no heat. That's precisely what the Big Bang Theory predicts. There was NO heat in the Big Bang Singularity, because there was NO mass or physical matter in the Big Bang Singularity. Heat and physical matter didn't exist yet. The whole thing was an absolute zero quantum phenomenon, contrary to what the Materialists and Naturalists have chosen to believe. What the Physicalists, Darwinists, Naturalists, and Atheists have chosen to believe is the exact opposite of the truth.

$Q = mc\Delta T$

Heat (**Q**) is energy in transfer to or from a thermodynamic system. Heat is a function of mass (**m**). No mass (**m = 0**) then NO heat. No resistance to acceleration, then no heat. Photons, quantum waves, and quantum fields have NO mass, which means that they have no temperature and no heat. However, Nature's Psyche can transform quantum waves, quantum fields, and photons into heat, mass, and physical matter anytime that Nature's Psyche chooses to do so. That's the way Quantum Mechanics works! Quantum Mechanics is Energy Mechanics. At the quantum level, ALL of the work is done for free thanks to the Conservation of Energy.

**ΔT** is change in temperature. No change in temperature (**ΔT = 0**), then NO heat. There has to be a change in temperature; otherwise, there is NO heat. This once again proves the obvious fact that there is NO heat at absolute zero, where there is NO change in temperature. By definition, absolute zero means no heat! Since entropy is the transfer of heat, there is NO entropy at the quantum level in the absolute zero of the Quantum Realm or the Syntropic Realm. Observe that every quantum object functions perfectly well in the absolute zero temperatures of the Syntropic Realm, or the Massless Realm, or the Non-Physical Realm, or the Quantum Realm. Quantum objects or massless objects don't need heat as a lubricant in order to function properly!

In this equation for heat, **c** is called **specific heat**, and it's typically based upon some type of **calorie** or **Btu**. There are many different types of heat, and therefore many different ways to define and measure a calorie through calorimetry. What they do is they define a hot reservoir and a cold reservoir into existence axiomatically as a given, put them into contact with each other, and then measure the heat transfer that takes place. The heat transfer that takes place is typically called "entropy".

Entropy doesn't exist naturally. They have to make it exist by creating a hot reservoir and a cold reservoir as a given in the first place. Remember, ONLY Nature's Psyche can create hot reservoirs and cold reservoirs at will. We human beings can only do that in thought experiments, but we can't do it for real. There can be NO entropy if there is NO transfer of heat, because entropy or the transfer of heat completely ceases at thermal equilibrium and at absolute zero where there is no heat.

Look at what we can do, once we finally have a true definition for entropy. We can explain everything that has ever been experienced and observed. We can falsify the second law of thermodynamics. We can verify Quantum Mechanics instead. We can provide a realistic and plausible mechanism for the Big Bang, and the origin of heat and physical matter. We can explain where heat comes from and where it goes. We can explain where physical matter and mass come from. We can remain consistent and true to the First Law of Thermodynamics or the Conservation of Energy and Psyche. We can even explain what psyche does and how it works.

This is the coolest science that I have ever done, and it's going to generate some heat, once it's finally accepted and understood.

They are not going to like it because it falsifies Classical Realism, the Second Law of Thermodynamics, Materialism, Darwinism (Design and Creation by Entropy), Naturalism, and Atheism (Creation Ex Nihilo); and, it verifies Quantum Mechanics instead. Oh well, everything falsifies Materialism and Naturalism, except for physical matter; and, physical matter is only 5% of our universe at most, and it's probably a lot less than that. Many a sacred bull is falsified by these scientific discoveries; and, that's the way Science should work.

All of this became possible because I chose to study the equations for entropy with the intention of trying to figure out what they truly mean. This also became possible because I'm no longer afraid of looking at things from a non-physical, non-local, massless, quantum perspective. I had to overcome my fear, my materialism, my naturalism, my nihilism, and my atheism before I was able to make these kinds of scientific discoveries. What do you think? Did I succeed, or not? I think I did because it explains everything that has ever been experienced or observed. Once you let Psyche and Quantum Mechanics in to play, you can literally explain everything that comes your way. That's what I have experienced and observed.

In summary: By defining entropy as heat transfer, we FALSIFY the second law of thermodynamics, and we VERIFY or PROVE the Quantum Law of Thermodynamics instead. The Quantum Law of Thermodynamics says that there is NO entropy, NO thermodynamics,

NO heat, and NO heat transfer at absolute zero in the non-physical, non-local, quantum realm. The Quantum Realm or Syntropic Realm is an infinite absolute-zero heat sink or quantum reservoir. It can easily reabsorb ALL of the heat that Nature's Psyche chooses to make, without budging a single bit. The second law of thermodynamics is falsified by the Conservation of Energy, by the existence of Syntropy, by the equations for entropy, by the universal infinite absolute zero quantum fields, by the abilities of psyches to generate heat and physical matter and mass and spacetime at will, and by the Quantum Law of Thermodynamics. Entropy is simply the transfer of heat.

This is obviously true, because it is precisely what we have experienced and observed.

Mark My Words

--

Original Source:

https://www.amazon.com/gp/product/B01J023TGU

Enhanced Version:

https://www.amazon.com/gp/product/B0771K6WTX


## Entropy Is Mass

Entropy is the heat storage capacity of mass.

Entropy is mass or resistance to acceleration.

This is possibly the most interesting definition for entropy. It's based upon the "heat transfer" definition of entropy, and it concerns the definition for heat and the equations for entropy that are based upon heat. I've never seen this definition for entropy before, so I'm the one who discovered it or created it. It's there in the equation for heat, and therefore in the equations for entropy that are based upon heat.

Once we have a true interpretation or a true definition for scientific data or a scientific concept, the simplest equations provide us with the profoundest scientific discoveries. This one is amazing!

The mass-based definition for entropy is derived from the heat-based definitions for entropy, starting with the following unique equation and definition for entropy.

**$S = \Delta Q$**

Entropy (**S**) is an exchange of heat (**ΔQ**) from a hot reservoir to a cold reservoir. No physical reservoirs, then NO exchange of heat and therefore NO entropy.

However, for a complete proof of this mass-based definition for entropy, we need to use the complete equations for heat (**Q**) and entropy (**S**).

**$Q = mc\Delta T$**

**$S = Q/\Delta T$**

$S = mc\Delta T/\Delta T$

$S = mc$

$c$ = the specific heat capacity of the mass under consideration

$M = mc$

Entropy (**S**) equals mass (**M**), particularly the heat storage capacity of mass (**mc**).

$S = Q = M = mc$

In the heat (**Q**) equation, (**c**) is the specificity of heat constant for a given material. This specific heat constant is basically the amount of heat that a specific type of mass can hold. Regardless of what happens to the temperature (**ΔT**), entropy (**S**) equals heat (**Q**) which basically equals mass (**M**) or resistance to acceleration. More specifically, we are talking about the heat storage capacity of the mass (**mc**). This is a more conservative definition for entropy. In fact, it's a demonstrably true definition for entropy. It has actually been experienced and observed.

What makes this definition for entropy interesting is the fact that NO mass means NO entropy. No mass (**m = 0**), then no heat (**Q = 0**). Whenever heat is zero (**Q = 0**), then entropy is equal to zero (**S = 0**). Consequently, according to the equations for entropy and heat, whenever the mass is zero, then the entropy is also zero.

$M = 0 = Q = S$

$S = mc$

Entropy (**S**) is the heat storage capacity of mass (**mc**). No mass, then no heat storage capacity. No heat and no heat storage capacity then NO entropy. It's that simple. This is the truest and the best definition for entropy. It's actually based upon something that has been experienced and observed.

Once we have a true interpretation or a true definition for scientific data or a scientific concept, the simplest equations provide us with the profoundest scientific discoveries and truths. This one is amazing! This one provides us with something that nobody seems to have thought of before. This one provides us with a demonstrable and true definition for entropy that actually matches with what has been experienced and observed.

**Entropy is equal to mass's heat storage capacity. $S = mc$.**

When mass or resistance to acceleration equals zero, heat equals zero, which means that entropy has gone to zero. When mass goes to zero, heat and mass's heat storage capacity goes to zero, which means that entropy goes to zero. That means that massless photons, massless quantum waves, and the massless quantum fields have NO entropy because they have no mass or no heat storage capacity. This is in fact the Quantum Law of Thermodynamics that everyone has been searching for. There is no mass, heat, nor thermodynamics at the quantum level among the massless, heatless, and entropyless quantum fields, quantum waves, and photons. Why? It's because these massless objects have NO mass or NO heat storage capacity. Simple. Logical. Rational. Parsimonious. True.

Massless particles like photons and quantum waves have NO entropy according to this definition for entropy. Likewise, the infinite, absolute-zero temperature, massless and heatless quantum fields have NO entropy because they have no mass or resistance to acceleration. The quantum fields have NO heat storage capacity. For the quantum fields

$mc = 0$, which means that their entropy is zero ($S = mc = 0$). No mass or no resistance to acceleration, then no friction or heat, which means NO entropy. No heat storage capacity, then NO entropy. It's really simple to understand. It also falsifies the second law of thermodynamics, which is based upon *overgeneralization errors* or *exaggeration* logic fallacies. The second law is based upon infinities that can't be measured, experienced, nor observed. The second law is just too good to be true. The second law of thermodynamics is physically impossible. The second law is creation ex nihilo or magic.

Remember, the simple equations for heat and entropy FALSIFY the second law of thermodynamics which states that entropy is always increase and that the amount of entropy in the universe can never decrease.

**Remove the mass from a system, and that system's entropy goes to zero! As mass goes to zero, heat goes to zero, which means that entropy goes to zero. The way we remove entropy from a system is to remove its mass or resistance to acceleration. That's something that Nature or Nature's Psyche can do at will, according to the photon and the quantum wave. Nature or Nature's Psyche is creating or instantiating massless particles all the time; and, they have no entropy because they have no mass or no resistance to acceleration.**

**A miniature Big Bang happens every time a photon chooses to stop and enter into our mass-based physical realm. That photon goes from zero mass and zero heat to a complete stop, which then produces heat, mass, and entropy as a result of its deceleration, or resistance to acceleration, or transition into our physical realm.**

If mass equals zero ($m = 0$), then heat equals zero ($Q = 0$), which means that the amount of entropy has gone to zero ($S = 0$). This is scientific and mathematical proof that massless non-physical objects such as photons, quantum waves (thoughts), quantum fields, and psyches have NO entropy, which means that their energies are eternal, everlasting, and always conserved. A photon is scientific proof that Nature or Nature's Psyche can convert massless and heatless energy into mass and heat anytime that it chooses to do so. The truth has always been there, but we have been too stubborn to see it because we don't want it to be true.

**The process is reversible. Whenever a particle goes massless or non-physical, it loses ALL of its entropy, and that entropy ceases to exist.**

This is an extremely powerful definition for entropy because of all the different truths that it reveals to us. Massless particles have no entropy and can have no entropy because they have no mass or resistance to acceleration.

My discovery, that the massless or the non-physical FALSIFIES the second law of thermodynamics, may just be the greatest scientific discovery of all time; and there it is, plain as day, within the equations for heat and entropy. No mass, then no entropy. The reason that nobody has ever discovered it before now is because nobody has been willing to look at physics from a non-physical quantum perspective before I came along and started doing so.

**Entropy Is Mass or Resistance to Acceleration**
**Entropy Is Mass's Heat Storage Capacity**

$S = mc$

Entropy (**S**) is the heat storage capacity of mass (**mc**).

This is by far the most useful, most powerful, most truthful, and most interesting definition for entropy that I have encountered so far, because it takes into consideration everything that has ever been experienced and observed.

Mass obviously exists; and, it has been demonstrated or proven that each type of mass has a specific heat storage capacity.

https://en.wikipedia.org/wiki/Specific_heat_capacity

https://en.wikipedia.org/wiki/Table_of_specific_heat_capacities

Entropy is mass's heat storage capacity.

This ended up being my favorite definition for entropy. It's obviously true, and it falsifies the second law of thermodynamics.

The second law of thermodynamics is based upon an *exclusion of evidence*. Everything massless and non-physical is excluded from consideration where the second law of thermodynamics is concerned, which is why it is easily falsified by the massless or the non-physical. *Deliberate exclusion of evidence* is a logic fallacy. The second law of thermodynamics is based upon a logic fallacy because it excludes evidence. The people who made the second law deliberately excluded everything that falsifies the second law of thermodynamics.

Furthermore, the second law of thermodynamics is based upon the *affirming the consequent* or the *jumping to conclusions* logic fallacy. They pick the conclusion that they want in advance, and then they deliberately exclude any evidence that falsifies it. Anytime they encounter entropy decreasing to zero and ceasing to exist, they label it as invalid and toss it out. They ONLY affirm or acknowledge the instances and the evidence where entropy is increasing. Anytime entropy goes to zero and ceases to exist, as it does within any particle that becomes massless, then they call that evidence and situation invalid; and, they toss it out and pretend that it doesn't exist, so that the second law of thermodynamics can seem to remain true.

**If you want the truth, it is important to allow ALL of the evidence into evidence – and that includes ALL of the times when entropy and mass have gone to zero and ceased to exist!**

**Entropy is mass's heat storage capacity. When mass or resistance to acceleration goes to zero, the amount of entropy goes to zero.**

I'm very excited about this particular definition for entropy. I've been searching for it for fifty years. The world has been searching for it for millennia. We haven't had a true definition for entropy since Rudolf Clausius coined the term in 1864.

Now, we finally have a true definition for entropy that explains what it is and how it works. It explains everything that has ever been experienced and observed. I finally KNOW what entropy is; and now, we have complete scientific, mathematical, and observational proof that this particular definition for entropy is valid, verified, and true.

When enough scientists finally figure out how science and nature really work, they will vote to replace the second law of thermodynamics with Quantum Field Theory and $E = mc^2$ instead. Until then, we have to be satisfied with finding a couple of true definitions for entropy. We haven't had a true definition for entropy, since Rudolf Clausius coined the term "entropy" in 1864.

Well, we now have a true definition for entropy because I asked myself how I can prove that entropy is real and truly exists. The only way I could do so is to find a true, demonstrable, replicable, and believable definition for entropy that has actually been experienced and observed in nature and in our laboratories.

### Deriving the Mass-Based Definition for Entropy

Let's examine the heat-based equation for entropy, and then use it to derive the mass-based definition for entropy.

**S = Q/ΔT**

Entropy (**S**) equals heat (**Q**) divided by a change in temperature (**ΔT**).

**Q = mc ΔT**

Heat (**Q**) equals mass (**m**) times that mass's heat storage capacity (**c**) times the change in temperature (**ΔT**).

**S = mcΔT/ΔT**

Solving for entropy in terms of the heat storage capacity of the mass.

**S = mc**

Obviously, by definition, a massless photon has zero mass (**m = 0**), which means that it has no heat (**Q = 0**) and no entropy (**S = 0**). Since a massless particle of any kind has zero heat (**Q = 0**), that means that its temperature in Kelvin is equal to zero (**T = 0**), and its change in temperature is also equal to zero (**ΔT = 0**). In other words, it's at thermal equilibrium. Therefore, anything at thermal equilibrium (**ΔT = 0**) has NO entropy, because it's unable to produce and store any heat or entropy. We get this additional amazing truth from the equation for heat (**Q = mcΔT**).

Entropy becomes a fleeting thing, because it ONLY exists in the presence of mass and heat. No mass, then no entropy. No heat, then no entropy. No change in temperature, then no entropy. In fact, we quickly observe that entropy is constantly decreasing, going to zero, and ceasing to exist – which is the exact opposite of what the second law of thermodynamics claims.

In this equation for heat (**Q = mcΔT**), (**c**) is the specific heat value constant. Solving the equation for **c** when everything else has gone to zero, I suspect that when mass equals zero, and temperature is equal to zero, then the specific heat value of a photon or massless particle of any kind is also equal to zero (**c = 0**). In other words, a photon has NO heat storage capacity because it has NO mass or resistance to acceleration.

So, once again, we make a fascinating scientific discovery simply by choosing to look at science and entropy from a massless, non-physical, quantum perspective. We discover that the specific heat value for massless particles and photons is equal to zero (**m = 0 = c**). Massless particles have NO capacity for storing heat! It requires mass to store heat. No heat, then no entropy. No mass then no entropy.

A massless photon has NO entropy; and, when it reaches its chosen destination and encounters some mass or physical matter, it stops, and it converts its energy into heat. This is precisely what we experience and observe! This is sunshine in action! This is proven and verified science. A massless photon has NO entropy, NO mass, NO resistance to acceleration, and NO heat until it encounters physical matter; whereupon, it stops and converts itself into heat, mass, and entropy. Here is the scientific explanation for sunlight! Sunlight has NO entropy. Photons have no entropy. Cold, absolute-zero temperature, massless Photons convert into mass and heat when they stop moving or reach their chosen destination. This is how everything really works. It's hidden in plain sight where nobody can see it because they aren't looking for it.

A traveling photon is massless syntropy or massless "no-entropy", and it literally converts itself into heat, mass, resistance to acceleration, and entropy when it chooses to stop moving. There is the explanation for how the Big Bang was done. If the Big Bang really happened, it was a deliberate quantum transformation of massless syntropy, or conserved energy, or photons, or absolute-zero temperature light INTO heat, mass, resistance to acceleration, and entropy. Nature or Nature's Psyche is doing that ALL the time, not just during the Big Bang. Nature's Psyche is constantly converting massless photons into mass, heat, and entropy all the time; and, Nature's Psyche can reverse the process as well, at will.

According to this definition for entropy, a massless photon has NO ENTROPY because it has no mass or no resistance to acceleration. Anytime the mass or resistance to acceleration is removed from a system, ALL of the entropy is removed from the system. NO mass or resistance to acceleration, then NO heat. No heat, then NO entropy. It's really simple to understand. The truth always is.

This is the Quantum Law of Thermodynamics which states "no mass then no entropy". If there is no mass, then there can be NO heat or resistance to acceleration or friction, which means that there can be NO entropy. No mass and no heat mean no thermodynamics, which means NO second law of thermodynamics and NO entropy in the Non-Local, Non-Physical, Massless Quantum Realm, or Spirit World.

I think the "entropy equals mass" definition for entropy is the most interesting definition for entropy because of what it means at the quantum level, or the massless level, or the non-physical level. This definition for entropy gives us the quantum perspective on entropy; and, it ends up being the Quantum Law of Thermodynamics which states that there is no mass, no heat, no heat transfer, and therefore NO entropy at the quantum level in the Massless Realm, the Non-Local Realm, the Transdimensional Realm, the Non-Physical Realm, the Absolute-Zero Realm, the Syntropic Realm, the Conserved Realm, or the Quantum Realm.

Massless ($m = 0$) means NO entropy ($m = 0 = Q = S$); and, this ends up being the quantum definition for entropy. There is no resistance to acceleration at the quantum level in the massless Quantum Realm, which means that there is NO entropy at the quantum level in the Syntropic Realm.

**Entropy is mass or resistance to acceleration. Entropy is the heat storage capacity of mass.**

This is the best definition for entropy because it allows ALL of the evidence into evidence; and consequently, it successfully explains everything that has ever been experienced and observed. This is the best definition for entropy, because it's really simple to understand. No mass, then no entropy. Simple. Logical. Obvious. True.

This definition for entropy has already been verified and proven true.

I like this definition for entropy because it reveals the truth to us that there is no entropy and can be no entropy in the Massless Realm or Non-Physical Realm among the quantum fields. This definition for entropy ends up producing the Quantum Law of Thermodynamics which states that there is no mass, no heat, no heat transfer, no thermodynamics, and therefore no entropy in the Massless Realm or the Syntropic Realm or the Conserved Realm or the Quantum Realm. This might in fact be my greatest scientific discovery of all time. I've been studying entropy for fifty years trying to figure out what it is, and this is the culmination of all my effort and research. I figured out what happens to entropy when it goes massless or goes quantum and non-physical. It ceases to exist. It falsifies the second law of thermodynamics. It makes the Big Bang a realistic possibility. It makes the creation of stars and galaxies by Nature's Psyche – from absolute-zero temperature, non-local, non-physical energies and quantum fields – the best explanation for how the creation was actually done.

By studying Syntropy, the Conservation of Energy and Psyche, and the different equations for heat and entropy, I was finally able to MAP entropy or mass onto the quantum realm from whence it originally came. Heat, and mass, and entropy don't exist at the quantum level; and, they are made to exist at the physical level by Nature or Nature's Psyche, whenever Nature's Psyche decides to make them, according to the Big Bang Theory. Nature or Nature's Psyche can make mass, heat, and physical matter out of absolute-zero temperature quantum fields and energies, anytime Nature's Psyche chooses to do so; and, there's nothing we can do to stop it or to make it happen. It's completely out of our control, and totally in the control of Nature or Nature's Psyche.

The Law of Psyche states that every psyche or intelligence has a certain amount of energy that's under its control; and, that controlling psyche can form or transform the energy under its control into anything that it wants that energy to be anytime that it chooses to do so – including mass, heat, or physical matter.

This reality and truth explains everything that has ever been experienced and observed.

The Law of Psyche combined with the Quantum Law of Thermodynamics even provide a realistic and logical mechanism for the creation of stars and galaxies out in the absolute-zero temperatures of spacetime, among the expanding clouds of hydrogen and helium gas. It provides a mechanism for the Big Bang too, if that's the way that the Gods chose to do it.

https://ldssoul.com/the-birth-of-our-earth/

The account of our earth's construction, at the preceding link, provides a different version of the organization of our earth; and, it actually becomes a realistic possibility once we know about and accept the Law of Psyche and the Quantum Law of Thermodynamics. Nature, Nature's Psyche, and the Gods can make heat, mass, and physical matter from massless dark energy or massless quantum fields anytime they choose to do so; and, there's nothing we can do to stop them.

Entropy is mass or resistance to acceleration. This is the best definition for entropy that I have found so far. It's already been proven true by science, observations, logical common-sense, and math. It's been there all the way along waiting for someone to find it; and, the only reason nobody has found it until now is because they have chosen to believe in the second law of thermodynamics instead.

This definition for entropy falsifies the second law of thermodynamics and supports the Big Bang Theory instead. The Materialists, Naturalists, Darwinists, Nihilists, and Atheists aren't going to like that because this definition for entropy falsifies their beliefs that entropy, heat death, and death are conserved. In this definition for entropy, all of the associated entropy ceases to exist whenever a particle goes massless or non-physical. The energy is actually used or transformed to make the particle massless and non-physical.

This is not going to sit well with the Materialists and Naturalists and Darwinists, because these people preach, teach, and believe that the non-physical or the massless does not exist. This definition for entropy falsifies their beliefs and proves them wrong.

Mark My Words

--

Original Source:

https://www.amazon.com/gp/product/B01J023TGU

## I Finally Found a Couple of True Definitions for Entropy

The inscription at the entrance to Hell is one word, "Entropy". It translates into English as the following, "Abandon hope all ye who enter here".

I still remember the way I felt when I first understood what entropy means. This was forty years ago in my physical sciences class in college. I exited the classroom that day in complete and total despair knowing that all life in the universe is going to come to an end, and that everything is going to end in disarray, disorder, and heat death. That's precisely what they want you to feel. I felt despair because I believed them. At that point in time, I believed that they had told me the truth. I thought they knew what they were talking about. I had no idea that they were wrong.

For forty years, I defined entropy as chaos or disorder. I had no idea that this is a falsified definition for entropy or thermodynamics. I had no idea that disorder has absolutely no correlation whatsoever with heat or thermodynamics. Disorder doesn't make heat, and disorder doesn't eliminate heat either. There's no correlation or relationship between the two. We can have perfect order and zero heat, or we can have a huge amount of chaos or random motion and a lot of heat; and, we can have anything in between. There's no correlation between disorder and heat. There is NO correlation between disorder and thermodynamics. Thermodynamics is a transfer of heat from a hot system to a cold system, and that has nothing to do with chaos or disorder.

Nobody knows what entropy is because so far nobody has presented us with a true definition for entropy. Currently, entropy can be anything that you want it to be. There is NO standardized definition for entropy.

https://origin-science.org/entropy/

There are literally dozens of different definitions for entropy. The Materialists, Naturalists, Darwinists, Nihilists, and Atheists believe that ONLY physical matter, entropy, death, and chance exist. Consequently, the mechanism behind evolution has to be one of these. Natural selection is creation by death. Therefore, evolution in general ends up being creation by chance or creation by entropy. These, too, are falsified definitions for entropy. Science as a whole is in desperate need of a couple of true definitions for entropy.

I have been a scientist all of my life, and it took me over 55 years of thought and study and research before I finally knew enough science and truth to falsify the second law of thermodynamics. I had to start at the beginning and rethink the whole thing. I started falsifying the second law of thermodynamics in 2017 and 2018, while I was writing the book on Syntropy. After a couple of years of it, I have gotten very good at it. I know too much now for them to be able to convince me that the second law of thermodynamics is true. I KNOW why it is false. I have FALSIFIED it a couple of dozen different ways.

I did so in the enhanced edition of my book, "Quantum Mechanics from a Non-Physical Spiritual Perspective", on 04JUL2019. Here I am again, proof-reading some of these essays on 06AUG2019; and, I'm falsifying the second law of thermodynamics and trying to replace it with the truth.

https://ultimate-law-of-thermodynamics.com/

https://quantum-law-of-thermodynamics.com/

I no longer believe that the second law of thermodynamics is true, because I have found a couple dozen different things that falsify it. Science as a whole falsifies the second law of thermodynamics. It's time for us to upgrade the second law of thermodynamics to something that is real and true.

The second law of thermodynamics claims that the amount of entropy or disorder in our universe is constantly increasing and can never decrease. It's a paradox. It's an oxymoron. It's impossible. Not even energy can constantly increase in amount without ever decreasing; yet somehow magically, entropy can. It's a con. It's a scam. It's too good to be true.

The second law of thermodynamics is obviously false and obviously flawed. Even many of the Atheists and Naturalists instinctively know that if the second law were true, then the Big Bang could never have happened, we wouldn't be here, and physical matter wouldn't exist. The fact that we exist is scientific proof that the second law of thermodynamics is false, and evidentiary proof that there is something important that we are completely overlooking. Something somewhere knows how to reverse entropy and disorder at will. This something or someone has to exist, or we wouldn't exist. The Big Bang, if it happened, represents a huge amount of syntropy or exergy, which means that someone knows how to reverse entropy and disorder at will. This is obviously true. It has to be true, or we wouldn't exist.

So, what in this universe is capable of eliminating entropy and disorder at will? What can do so at the quantum level as well as the physical level? Well, it isn't entropy, natural selection, or the second law of thermodynamics as the Materialists, Naturalists, Atheists, and Darwinists claim.

The second law of thermodynamics is obviously false. There is a lot of NEW and interesting work waiting to be done in science that will overcome, adjust, correct, and fix the flaws hidden in the second law of thermodynamics so that science as a whole will come a little bit closer to truth and reality instead. In science, whether we know it or not, we are currently looking for the mechanism that is capable of reversing entropy and disorder at will. There's NEW science to be discovered when we crack this nut.

I'm a theoretical physicist and a quantum theorist, and much of my effort, time, research, and work the past couple of years has gone into identifying what's wrong with the second law of thermodynamics, and then trying to provide patches or remedies to the situation so that our science gets a bit closer to the truth than it currently is.

I started this journey in 2017 and 2018 by writing the book on Syntropy – what it is and how it works. That's how I slowly learned how to falsify the second law of thermodynamics. I identified syntropy, and then explained how it overrides entropy. That's where it started:

https://www.amazon.com/gp/product/B07BPT3W8R

And, my efforts to identify and correct the flaws in the second law of thermodynamics continue to this very day. It isn't easy work because I, like everyone else, have been brainwashed and trained to believe that the second law of thermodynamics is true, and that entropy really exists. I couldn't see anything wrong with it, until one day it became obvious that everything about it is wrong. Now, when it comes to the second law of thermodynamics, I can't find anything right about it or correct within it. The whole thing is a fraud. I didn't know that until the summer of 2019, when I started studying the equations for entropy and figured out what they really mean and how they really work.

Then, all the lights went on and suddenly I could see.

Finding a couple of true, realistic, conservative, demonstrable, reliable, defendable, and believable definitions for entropy has been at the top of my bucket list for nearly five decades. I've finally done it. I can check that one off my bucket list. I've accomplished the purpose for which I was made.

In order to find the true definitions for entropy, we must be willing and able to identify and eliminate the false definitions for entropy, which is what I try to do within this particular multi-part essay, within this book that I call "Science 2.0".

https://www.amazon.com/gp/product/B0771K6WTX/

I'm trying to bring Science into the modern Quantum Era, which means that I must use Science to falsify Materialism, Naturalism, Darwinism (Creation by Chance or Entropy), Nihilism, Classical Realism, Behaviorism, Atheism (Creation Ex Nihilo), and the inappropriately named Second Law of Thermodynamics.

Within the enhanced and final edition of my book, "Quantum Mechanics from a Non-Physical Spiritual Perspective", I provided the most comprehensive treatment of entropy that has ever been devised by man. As of 04JUL2019, I provided in print a comprehensive discourse on the different competing definitions for entropy.

https://www.amazon.com/gp/product/B01J023TGU

https://www.amazon.com/gp/product/1521132380

Most of them are extravagant definitions for entropy, because they are based upon the extravagant claims made by the second law of thermodynamics, which states that the entropy or disorder in our universe is always increasing, and that it can never decrease. By axiom or by decree, the people who invented the second law of thermodynamics set entropy equal to all of the energy in the universe, all of the heat in the universe, and all of the disorder in the universe. In other words, by axiom or by decree, they made entropy infinite to begin with before doing anything else to increase its amount. When I realized that, I could see the con plain as day.

Observe the standard equations for entropy ($S$) as found within our college textbooks:

$S = J/K$

$S = Q/T$

The way that the extravagant definitions for entropy work is that these people set entropy ($S$) equal to ALL of the heat ($Q$) in the universe ($S = Q$) or ALL of the energy ($J$) in the universe ($S = J$) BEFORE dividing that pre-given infinite amount of entropy ($S$) by something close to absolute zero ($K = 0$ or $T = 0$) in order to produce even more entropy. As the limit of "division by zero" goes to zero in the denominator, the numerator always goes towards infinity. These equations and definitions for entropy were designed to produce an infinite amount of entropy. They start with an infinite amount of entropy, and then they produce some more.

They cheat! The stack the deck and load the dice so that it always comes up entropy.

It's *too good to be true*, which means according to the laws of logic that it can't possibly be true. I mean, just think about it. They set entropy equal to infinity to begin with, and then they divide that by something close to zero in order to produce an additional infinity. It's meaningless! The official second law definitions for entropy are completely meaningless because they have been rigged – they were deliberately designed to produce infinities both coming and going. The extravagant definitions for entropy can't possibly be true because they are *too good to be true*.

It's rigged!

Not even energy is capable of increasing in amount, yet somehow magically, entropy can. The second law of thermodynamics defines entropy as creation ex nihilo, magic, or some type of supernatural perpetual motion machine that is capable of producing ever-increasing amounts of entropy out of thin air. It's illogical. It breaks parsimony. They literally turned entropy into some type of superfluid whose amount is constantly increasing and can never decrease.

Furthermore, by setting entropy (**S**) equal to ALL of the energy (**J**) in the universe (**S = J**), it makes it impossible to determine what entropy is not. The technical term for this practice is called a *category error* logic fallacy. This definition for entropy and consequently the second law of thermodynamics ARE based upon a *category error* logic fallacy, wherein they set entropy equal to all of the energy in the universe before they do anything else. They *affirm the consequent* or *jump straight to that pre-chosen conclusion* before doing anything else. It produces a wide variety of different logic fallacies in the process. It's Bad Science or Biased Science, and they don't even know it, because they have been brainwashed and intimidated to believe in it instead.

This same reality and truth applies to Boltzmann's equation for entropy.

$S = k_B \ln(\Omega)$

**This is the official statistical definition for entropy (S). According to this equation for entropy, entropy is axiomatically defined as ALL of the disorder in the universe.**

**This definition for entropy is based upon a *category error* logic fallacy. Disorder has absolutely nothing to do with heat or thermodynamics. Disorder doesn't make heat, and disorder doesn't eliminate heat either. There is NO correlation between disorder and thermodynamics whatsoever. Disorder has nothing to do with thermodynamics or heat.**

**It is obvious that "disorder" has NO EFFECT on the amount of heat within a given system or the temperature of a given system. The heatless, massless, and entropyless Quantum Fields are perfect order and perfect functionality at absolute zero temperature. The quantum level of existence functions most efficiently at absolute zero. Many define the sun as chaos or random disorder according to this equation for entropy; yet, the sun has a lot of heat stored within it. There is NO correlation whatsoever between heat and disorder. Highly ordered systems can have no heat, and highly disorganized systems can have tons of heat. There is NO correlation between disorder and heat; and, heat of any kind represents a certain level of order and organization.**

$S = k_B \ln \Omega$

**This equation for entropy (S) erroneously defines entropy as the Conservation of Disorder or the Conservation of Entropy; and, it has the disorder in our universe constantly increasing and never decreasing, in complete violation of the Big Bang Theory and the Conservation of Energy. The disorder definition for entropy is based upon the exclusion of any evidence that falsifies it. It's rigged to produce infinite disorder or infinite entropy. It's *too good to be true*, so it can't**

possibly be true. Not even energy is capable of increasing in amount, yet by axiom or by decree, they make it so that entropy magically can.

Disorder, the Conservation of Disorder, and the Conservation of Entropy end up being the ultimate worst definitions for entropy and the second law of thermodynamics, because they have absolutely nothing to do with thermodynamics. Defining entropy as disorder, chaos, chance, or randomization ends up being an obvious *category error* logic fallacy. They are not the same thing at all. The definition that I had for entropy for forty years of my life – entropy as disorder – is a *misnomer* and a *category error* logic fallacy. I couldn't believe it when I first discovered that we had been lied to and deceived. Entropy and thermodynamics have absolutely nothing to do with disorder, chaos, randomization, or chance. I never knew that, until after I was actually willing to look at it and study it.

You see, at perfect thermal equilibrium and at absolute zero and at heat death at the physical level, the photons, the quantum waves, and the quantum fields are still there, and are still perfectly ordered and organized; and, they still continue to function and work flawlessly, and perfectly, and with maximum efficiency even when the physical level is at heat death, or perfectly randomized, or perfectly homogenized. At heat death, everything is working perfectly fine at the quantum level in the Spirit World or the Quantum Realm. Heat death at the physical level doesn't touch nor change the perfectly organized Quantum Fields and the perfect order of the Quantum Realm or Spirit World. The Quantum Fields are the ultimate perpetual motion machines, thanks to the Conservation of Energy.

ALL of the physical matter is still there; and, ALL of the order and organization are still there at the quantum level too, among the quantum fields, working perfectly fine no matter how disorganized, randomized, chaotic, or heat-dead things might become at the physical level. Remember, heat death and maximum disorder at the physical level is perfect and maximum order and efficiency at the quantum level.

It has been observed clearly and conclusively that "disorder" has NO effect on the amount of heat or thermodynamics within a given system. We have zero heat, zero temperature, and zero entropy among the perfectly organized and perfectly functional Quantum Fields; yet, according to the scientists, we have a lot of randomization, chaos, and disorder among the gases and the plasma within our sun. We certainly have a lot of heat, mass, and entropy. Disorder has NO correlation with heat or thermodynamics.

Thanks to Quantum Mechanics, the Conservation of Energy, and the perfectly organized syntropic Quantum Fields, "disorder" is precisely the wrong definition for entropy, and we don't even know it because we have been brainwashed into believing in something else instead.

I discuss this definition and equation for entropy in greater detail within my book, "Quantum Mechanics from a Non-Physical Spiritual Perspective" within the essay entitled, "The Equation '$S = k_B \ln \Omega$' Is Forced to Become '$S = J/K$'". It's rigged to produce an infinite amount of entropy from a limited supply of energy!

The whole thing is forced to become ALL of the disorder in the universe as well as ALL of the energy in the universe; so, it can't possibly be true because it is *too good to be true*. It can't possibly be true because the conservation of disorder is diametrically opposed to the conservation of energy. The two ideas are mutually exclusive; but, they are forced to play together within Boltzmann's equation for entropy. It's rigged. It's a kludge. It was purposefully designed to produce an infinite amount of entropy every time. It's a scam. It's a pyramid scheme.

The reason that they force entropy to be defined as randomness, disorder, or chance is so that they can then have entropy or chance be our Designer, our Creator, and our God. Essentially what they are saying with Boltzmann's equation and the second law of thermodynamics is that "entropy is the only thing that exists"; therefore, entropy or chance has to be our Designer and our Creator. It's a deception. It breaks logic and parsimony. It's a *category error* logic fallacy.

Within all of our college physics books, the sheeple actually use this equation, $S = k_B \ln \Omega$, to define entropy in terms of the number of heads and tails that show up within a set or a system. According to this equation for entropy, entropy is defined as disorder, randomness, probability, or chance. This is the stupid definition for entropy and the second law of thermodynamics, because it has absolutely nothing to do with thermodynamics, but is instead all about the probability of getting heads or tails, and then calling the result "entropy". It's a dumb idea that's in all of our college physics books – the entropy of coin tosses – that actually does NOTHING to measure or explain thermodynamics, or the heat transfer from a hot system to a cold system.

Not only is this statistical mechanics equation for entropy the stupid equation for entropy; but, it is also the honest equation and definition for entropy. It reveals the fact that these people have erroneously chosen to define entropy as chance.

For those who are looking and able to see, this equation and this definition for entropy reveals clearly and conclusively that entropy as defined by the second law of thermodynamics is a *category error* logic fallacy and has absolutely nothing to do with thermodynamics. For these people, entropy is all about the probability of getting heads or tails with a set of coins that have already been tossed. It has NO predictive value! The coins have already been tossed, and then they determine the amount of entropy based upon how many coins came up heads, and how many coins came up tails. Then the whole thing is called disorder or entropy, by axiom or by definition. This really is the stupid definition and the stupid equation for entropy – the one that is an oxymoron or a contradiction in terms – the one that is a *category error* logic fallacy – the one that is based upon nothing but chance – and the one that has nothing to do with thermodynamics.

This Boltzmann equation for "entropy" is in fact the probabilistic explanation for how disorganized a set of heads and tails is, depending upon how the coin tosses turned out in actuality. It has absolutely NOTHING to do with thermodynamics. It's simply statistics and probability, which often have nothing to do with reality. My brother defined statistics as the art of learning how to lie with math. That's what we see in actuality whenever they start applying statistical mechanics and probability to something like "entropy" or thermodynamics. We see and witness a *category error* logic fallacy – cats defined as dogs, and apples defined as oranges, and chance defined as entropy.

Probability always falls between 0 and 1, which means that by defining entropy as probability, entropy can NEVER go below zero. It's stupid and has nothing to do with thermodynamics. I was so disappointed when I first realized that that's what they are doing in order to produce entropy in the first place, within our physics books. They axiomatically define entropy in terms of probability so that it can NEVER go below zero. It's a deceptive lie, and we don't even know it. We fall for their lies every time, because we want to believe them. We trust our scientists to tell us the truth, and they end up being deceptive liars and cheats instead.

I'm a scientist, a statistician, a logician, and a mathematician; and, I was so embarrassed for them and their stupidity. I felt like mocking them, but I also felt sorry for them. I experienced a bunch of conflicting emotions, when I first realized that they had defined "entropy" and "thermodynamics" in terms of the number of heads and tails that showed up in a bunch of coin tosses. They made no effort whatsoever to identify, measure, or study the amount of heat that was generated by tossing those coins. They ONLY measured the results of the coin tosses, and then erroneously called the results "entropy", in one of the greatest *category error* logic fallacies ever produced by man.

Good statistics or unbiased statistics is actually meant to eliminate chance as an explanation for the results. But, what do we see here? We see entropy defined axiomatically as chance; and, they rigged the experiment in advance so that there is NO chance of getting anything but entropy, because entropy is not allowed to go to zero and cease to exist according to the second law of thermodynamics and the laws of probability. The statistical model and statistical definition for entropy is actually based upon biased and bad statistics. Instead of eliminating chance as an explanation, as statistics should do, the statistical definition for entropy actually defines entropy as chance, or disorder, or randomization, or chaos instead. It's Bad Faith, Bad Science, and Bad Statistics, because it's completely biased and rigged to produce disorder or entropy both coming and going. There's no way to get zero entropy or a reversal of entropy because everything in the universe is defined as entropy to begin with.

I just couldn't believe it at first. They literally end up measuring the amount of entropy in "four heads" compared to "two heads and two tails" in our physics books, because entropy is NO longer about thermodynamics but is instead about random chance and the probabilities of getting heads or tails. Making entropy about heads and tails is a *category error* logic fallacy, because entropy is actually supposed to be about the transfer of heat from a hot system to a cold system, and not about the number of heads and tails in a system. Entropy becomes meaningless within our college physics books whenever they go out of their way to define it as something that it clearly is not. Entropy also becomes meaningless and unmeasurable whenever they define the stuff as randomness or chance or disorder. By defining entropy as everything in our universe, it makes it impossible to determine what entropy is not, because everything is already entropy to begin with.

Randomness, chaos, chance, and disorder are notoriously difficult to define and measure, so what Boltzmann and the others did is to set entropy equal to ALL of the disorder in the universe BEFORE doing anything else. Again, the results are meaningless. Anything that produces infinite entropy ends up being a meaningless definition for entropy, especially considering the fact that there is NO standardized definition for entropy to begin with. Furthermore, by setting entropy equal to infinity at the very beginning of their thought experiments, there's NO way to determine what entropy is not, because everything in the universe is defined as entropy to begin with; and then, they find ways for the universe to make some more. Entropy thereby becomes the ultimate perpetual motion machine – the ONLY thing in the universe that is constantly increasing and can never decrease. Entropy becomes a supernatural perpetual motion machine, because these people have our universe constantly making more entropy from nothing throughout the whole duration of its existence.

The second law of thermodynamics is a lie.

There are a couple of dozen different competing definitions for entropy within science; and, they actually falsify each other, so they can't possibly be true. Whenever different definitions for the same thing are mutually exclusive, that means that one of them is automatically false; and for all we know, the other one might be false as well. These people would know that if they had ever studied the Philosophy of Science.

I'm a philosopher, too, and I have spent a great deal of time studying the Philosophy of Science and logic fallacies, so that I can identify the mistakes and the errors that our modern-day scientists and physicists are making. I've been working on mastering statistics as well, because these people are using statistics to deceive us and lie to us.

By setting entropy equal to all the disorder in the universe, or all the energy in the universe, or all the heat in the universe to begin with before doing anything else, we no longer have a random sample but a biased sample instead. The statistical definition for entropy defines entropy as disorder; but, defining entropy as all the disorder in the universe invalidates any statistical sample that we might make because we have no variability possible, because <u>everything</u> has already been defined as disorder and entropy to begin with. They provide no basis. They provide no control group. They provide no perfectly ordered syntropic system as a basis to compare all the disorder with, in order to determine the amounts of entropy or the degree of disorder. The entropy thought experiment is biased. There is no control group. There is NO syntropic group included in the mix. Everyone in the experiment receives the entropy treatment because <u>everything</u> in the universe is defined as entropy to begin with.

Any way that we choose to sample or choose to measure entropy, or disorder, ends up being biased and invalid because we have already defined everything in the universe as entropy to begin with. This is the logic fallacy that they call *affirming the consequent*. They start by concluding that everything in the universe is entropy, and then they use that conclusion as evidence or proof that entropy does in fact exist and that the second law of thermodynamics is in fact true. This is also called *jumping to conclusions*. These people jump directly to the conclusion that the massless, the non-physical, the quantum, and the syntropic DO NOT EXIST; and, they deliberately toss out anything and everything that falsifies their prechosen beliefs. In other words, they cheat, so that they can produce entropy from everything in the universe, including the number of heads and tails that showed in after tossing a bunch of coins.

Entropy or the second law is primarily philosophy or metaphysics. Ever-increasing entropy is based upon *wishful thinking*, which is a logic fallacy. Where modern physics is concerned, entropy is nothing but a thought experiment, so let's do a thought experiment with entropy, and see what we can learn. If they do philosophy and metaphysics to produce and support the second law of thermodynamics, we can do philosophy and science to falsify the second law of thermodynamics. Your atheistic college professors won't allow you to do that; but, we are not in college and in a public school here, so here we can actually pursue the truth if we want to. Here we are free to think outside of the box.

I have tried to develop equations that accurately model the second law of thermodynamics. It's hard to do. The second law of thermodynamics claims that entropy (or disorder) is always increasing, and that entropy (or disorder) can NEVER decrease.

What they are doing is that they are taking everything in the universe (**X**) and raising it to entropy (**S**). By definition, axiomatically, they have entropy constantly increasing exponentially. This ends up being the second law of thermodynamics (**SLT**).

**SLT = X$^S$**

The second law of thermodynamics (**SLT**) is anything and everything (**X**) raised to entropy; and, entropy (**S**) can NEVER be zero. That's the catch. That's the scam. According to the second law of thermodynamics, entropy can NEVER be zero (**S ≠ 0**). It's rigged. Entropy (**S**) is forced by the second law of thermodynamics to always be positive and increasing (**S > 0**). Entropy is forced to go to infinity every time, and in every way.

It's loaded. They can't miss. And, they can't falsify it either, because it's designed to be infinite to begin with.

**SLT = (S > 0)**

Then you get some crackpot like me who comes along and is actually willing to look at entropy from a massless, non-physical, syntropic, quantum perspective; and, what do you get besides a whole bunch of NEW scientific discoveries? Well, you get thought experiments into what happens when entropy goes to zero and ceases to exist.

What happens when entropy equals zero or when entropy ceases to exist? What is anything (**X**) raised to zero entropy (**S = 0**)?

**SLT = $X^{0=S}$**

Well, in real life and in nature, thanks to the conservation of energy, you are left with the original thing that you started with in the first place, when that thing loses all of its entropy and its entropy completely ceases to exist. According to the Big Bang Theory, it goes back to its original Pre-Big-Bang-State (**$X^0 \rightarrow X$**), when its entropy goes to zero (**S → 0**) and stops existing.

Mathematically

**SLT = $X^0$ = 1 = Unity**

Thermodynamically

**SLT = $X^0$ = X**

Remove ALL of the entropy, and you end up with what you had before the thing got some mass, heat, and entropy. The thing goes back to its theoretical entropyless and massless Pre-Big-Bang-State, just like what happens whenever we convert mass and entropy into raw energy instead!

**E = $mc^2$**

This equation actually works both ways. Nature or Nature's Psyche can convert raw energy or wavelike photons into heat, mass, and entropy; and, Nature's Psyche can reverse the process and convert heat, mass, and entropy into raw, massless, heatless, and entropyless photons or energy. This is what we have actually experienced and observed. The entropy completely ceases to exist whenever the mass or physical matter is converted into massless, heatless, and entropyless photons, quantum waves, and energy.

According to observation and experience, where thermodynamics is concerned, as entropy goes to zero (**S → 0**), then any object (**X**) goes to its original state that existed before it got some mass and entropy (**$X^S \rightarrow X$**), in complete violation and falsification of the second law of thermodynamics. This is what really happens whenever an object loses all of its entropy. The original object remains. The original object (**X**) doesn't cease to exist when its entropy ceases to exist (**S → 0**). It returns to thermal equilibrium instead. It returns to absolute-zero temperature. It returns to its original quantum state.

In other words, entropy completely ceases to exist at thermal equilibrium and at absolute zero, and only the original physical particle or photon remains. The second law of thermodynamics cannot allow that to happen, which is why it states axiomatically and erroneously that entropy is always increasing, and that entropy can never decrease to zero (**S > 0**).

**SLT = (S > 0)**

Think about it. According to the Big Bang Theory, when something loses ALL of its entropy ($S \rightarrow 0$), it returns to its original Pre-Big-Bang-State from whence it came. It returns to the infinite absolute-zero temperature entropyless and massless Quantum Fields from whence it originally came. When physical matter is transformed back into energy, it loses all of its entropy when its energy is absorbed back into the Quantum Fields from whence it came. Therefore, in order to seem true, the second law of thermodynamics has to axiomatically deny the existence of the infinite expanse of zero-entropy ($S = 0$) Quantum Fields, because the perfectly syntropic Quantum Fields falsify the second law of thermodynamics.

The second law also has to claim erroneously that entropic physical matter can NEVER be transformed back into massless, entropyless, exergic, syntropic Quantum Fields. Obviously, the second law of thermodynamics is in complete violation of the Conservation of Energy which says that mass, heat, entropy, and physical matter can be transformed back into heatless, entropyless, and massless photons and quantum fields anytime that Nature or Nature's Psyche chooses to do so.

Furthermore, the second law of thermodynamics is falsified by the Big Bang Theory, which states that before the Big Bang, everything was syntropic and without any entropy. The Big Bang Theory and the second law of thermodynamics are mutually exclusive. The second law of thermodynamics falsifies the Big Bang Theory; and, the Big Bang Theory falsifies the second law of thermodynamics. They both can't be true because they falsify each other. They are mutually exclusive which means that one of them is automatically false; and for all we know, they might both be false.

According to Quantum Field Theory, the infinite Matrix of zero-entropy Quantum Fields was in fact the original prime construct. The Quantum Fields were and are pure exergy – pure syntropy – with absolutely NO entropy whatsoever within them. In a very real sense, the Quantum Fields ARE the big bang singularity. The Quantum Fields were made BEFORE heat and physical matter were made. The construction of the Quantum Fields was the Big Bang. The Gods or Controlling Psyches had to make the quantum fields first, or physical matter couldn't be made and sustained.

The Quantum Fields were made by Nature's Psyche from pure, syntropic, no-entropy, non-physical energy; and, they remain as such to this very day. The Quantum Fields are pure exergy. All of their energy is available for use all of the time. There's NO time delay or inefficiency where the massless and entropyless and heatless Quantum Fields are concerned. They are pure syntropy or pure exergy. The Matrix of Quantum Fields is an infinite lattice of perpetual motion machines. They never wear out, and their work never ceases, because they have NO mass, NO resistance to acceleration, NO friction, NO heat, NO thermodynamics, and therefore NO entropy. This reality and truth is the Quantum Law of Thermodynamics; and, it falsifies the second law of thermodynamics.

The ONLY realistic and legitimate use for the second law of thermodynamics, as it currently exists, is to provide a demonstrable axiom which states that perpetual motion machines are impossible to achieve at the physical level thanks to friction, resistance to acceleration, thermodynamics, heat leakage, heat loss, the transfer of heat, or entropy. The second law doesn't exist and doesn't apply at the quantum level, which is comprised of a whole bunch of different perpetual motion machines that are indestructible, never wear out, perfectly functional at any temperature, and are constantly sustained by the Conservation of Energy.

One of my greatest scientific discoveries of all time came on the day that I first realized that the equation for heat and the heat-based equation for entropy FALSIFY the second law of thermodynamics. Soon thereafter, I finally realized that Quantum Field

Theory clearly and conclusively falsifies the second law of thermodynamics. When I made that discovery, that was the end of the second law of thermodynamics. There's no going back, because now I have the truth.

    Mark My Words

**Introducing the Ultimate Law of Thermodynamics**

https://ultimate-law-of-thermodynamics.com/

You will never see the Ultimate Law of Thermodynamics in our public universities, because it falsifies Materialism, Naturalism, Darwinism, Nihilism, Atheism, and the Second Law of Thermodynamics, which are the Conservation of Entropy or Creation by Entropy.

It doesn't matter that the Ultimate Law of Thermodynamics is demonstrable or true. It doesn't matter that the Ultimate Law of Thermodynamics is based upon Quantum Mechanics, Quantum Field Theory, the Heat-Base Equations for Entropy, and the Perpetual Motion Cycle $E = mc^2$. The Ultimate Law of Thermodynamics will automatically be rejected, vilified, and ignored because it contradicts and falsifies the materialistic, naturalistic, and atheistic paradigm.

The first law of thermodynamics is the Conservation of Energy.

The second law of thermodynamics presents entropy to us as a conserved substance. The second law of thermodynamics is the Conservation of Entropy. The second law states that the overall amount of entropy or disorder in our universe can never decrease or go to zero. In other words, the second law claims that entropy and disorder are conserved. The second law of thermodynamics IS the Conservation of Entropy.

So, is entropy a conserved substance?

Or, is the Conservation of Entropy yet another falsehood that has been given to us by the Materialists, Naturalists, Darwinists, Nihilists, and Atheists?

Which is it?

As I did the research and wrote the book on Syntropy, I eventually realized that the Big Bang, if it happened, was a massive reversal of entropy or a huge amount of syntropy. Most of the scientists have chosen to believe that the Big Bang happened. By definition and in principle, the Big Bang was the complete elimination of entropy in our part of the multiverse. If the Big Bang is true, as our scientists seem to claim, then someone in this universe knows how to reverse entropy at will. Someone KNOWS how to do Syntropy!

https://www.amazon.com/gp/product/B07BPT3W8R/

The Big Bang Theory falsifies the second law of thermodynamics and the conservation of entropy; and, the conservation of entropy or the second law of thermodynamics falsifies the Big Bang Theory. The two are mutually exclusive. If one of them can be demonstrated to be true, then the other one has automatically been falsified.

So, which one is true, and which one is false? Or, are they both false?

You can't build the truth on falsehoods and lies. Eventually, someone will catch onto the fact that you are lying to them. So, which one is the lie, and which one is true?

Eventually, it became obvious to me that the second law of thermodynamics was designed to falsify and eliminate the first law of thermodynamics, and that's how the second law is typically used on PBS and on the internet. The two are mutually exclusive. The first law of thermodynamics is the Conservation of Energy, and the second law of thermodynamics is the Conservation of Entropy. They falsify each other.

The second law of thermodynamics claims that the amount of entropy or disorder in our universe is constantly increasing and can never decrease and go to zero. Periodically, I see scientists and physicists on PBS and on the internet using the second law of thermodynamics to falsify the first law of thermodynamics.

The first law of thermodynamics states that the amount of energy in our universe can never increase and can never decrease but remains constant instead. The first law of thermodynamics is True Conservation – the Conservation of Energy, Light, or Psyche.

In contrast, the second law of thermodynamics is Atheism or Creation Ex Nihilo – the creation of something from nothing. The second law of thermodynamics is magic! Through the second law of thermodynamics we have this mystical and magical god called Entropy who is constantly creating stuff out of thin air from nothing. Entropy is not only defined as a conserved substance, but entropy is also an ever-increasing amount of that substance. It's magic. It's the creation of something from nothing by nothing.

The second law of thermodynamics or the Conservation of Entropy was designed to falsify and eliminate the first law of thermodynamics or the Conservation of Energy. The two ideas are mutually exclusive. They falsify each other.

So, which one of them is true, or are they both false? Which one do we actually experience and observe?

Do we really observe proton decay as the second law of thermodynamics predicts? Do we really observe an ever-encroaching gray goo coming in at us from all sides as the second law predicts, while our universe dissolves back into chaos and disorder from whence it was originally made? Or do we observe constantly conserved order and organization?

While writing the book on Syntropy and while trying to figure out what Syntropy truly is, I realized that I had to figure out what it really means for a substance or an entity to be conserved. I needed something that had actually been experienced and observed to serve as my model or my prototype for Syntropy or a conserved substance. It isn't science unless it has actually been experienced and observed.

Energy (and later the Quantum Fields) ended up being my model or my prototype for Syntropy and conserved substances. Energy is the ultimate perpetual motion machine; and, ever since the quantum fields were made, they quantum fields have been conserved syntropic perpetual motion machines as well. Energy is the small-scale quantum perpetual motion machine, and the quantum fields are the large-scale universal perpetual motion machines. They will never get old and never wear out because they are always conserved. That's what it really means to be conserved. The Perpetual Motion Cycle $E = mc^2$ is conserved. It will go on forever and never stop.

I KNEW from the Big Bang Theory that some type of syntropy has to exist, because according to the Big Bang Theorists in general, the Big Bang was a massive reversal of entropy or pure syntropy. There had to be a massive infusion of syntropy into the system; otherwise, there couldn't have been all of that subsequent entropy that they keep talking about. The whole entropy scam wasn't adding up and making sense to me. I needed to rethink and redo the whole thing, because the Big Bang was falsifying the second law of thermodynamics, and the second law was falsifying the Big Bang Theory. The two are mutually exclusive, which means that one of them is automatically false, according to the Philosophy of Science and the Rules of Science.

I kept running this over and over again through my head, trying to figure out which one of these things is true – the second law of thermodynamics or the Big Bang Theory. They are mutually exclusive and falsify each other, so they both can't be true at the same time. Something has to give.

The necessary existence of some type of syntropy, in order to make the Big Bang realistically possible, falsifies the second law of thermodynamics which states that the total amount of entropy or disorder in the universe can never decrease. The Big Bang, if it happened, represented a complete reversal of entropy and disorder, in our part of the multiverse. In other words, the Big Bang Theory and the second law of thermodynamics are mutually exclusive. If one of them can be demonstrated to be true, then the other one has automatically been falsified.

So, which is it?

Can entropy be reversed, or is entropy some type of substance whose amount is always increasing and whose amount can never decrease? Does the Big Bang Theory represent the truth, or does the second law of thermodynamics? We can't have it both ways because they are mutually exclusive and falsify each other.

I had to make a choice.

Which one would you choose?

I chose the Big Bang Theory, because we are actually here, and we actually exist. If the second law of thermodynamics were true, and disorder or entropy or chaos could NEVER be reversed as the second law claims, then there would be no physical matter, no planets, no stars, no genomes, no proteins, no life forms, no earth, and no physical universe. We wouldn't exist if the second law of thermodynamics were actually true. Our very existence is scientific proof that the second law is fundamentally flawed and obviously false.

The Big Bang Theory falsifies the conservation of entropy or the second law of thermodynamics. I chose to go with the Big Bang Theory because it was more likely to be true.

Since I had decided that the Big Bang Theory is true and that the Big Bang represented a massive and total reversal of entropy in our part of the multiverse, I realized that the Big Bang if it happened had to have been some type of pure syntropy. That was the only logical explanation that I had left to me. So then, it was finally time to write a book and determine what Syntropy is. Syntropy is obviously the opposite of entropy; but, since nobody seems to know what entropy is, I had no idea what syntropy is either, except for the fact that it must exist.

https://www.amazon.com/gp/product/B07BPT3W8R/

After a lot of thought and research, it dawned on me one day in 2018 that Syntropy is in fact the Conservation of Energy. It's there in plain sight, and it's obvious. Syntropy is the exact opposite of entropy, which means that Syntropy is the Conservation of Energy or the First Law of Thermodynamics. It was then that I realized that the First Law of Thermodynamics falsifies the second law of thermodynamics. Eureka! I was onto something important here, and I knew it. That was a massive flash of insight. I finally had a realistic mechanism for the Big Bang that had actually been experienced and observed. It's called the Conservation of Energy. It was then that I decided that I needed to figure out what it really means to be conserved.

Why is energy conserved?

What does it mean to be conserved?

Energy is conserved because its amount can never increase, and its amount can never decrease. Energy is conserved because its universal amount remains the same. Energy is reusable. Energy is recycled. $E = mc^2$ is the perpetual motion cycle! Energy is the ultimate perpetual motion machine. Energy is the father and the mother of all the other quantum perpetual motion machines that we have discovered and observed. Energy cannot be made, and it cannot be destroyed, because it is conserved. Energy is eternal and everlasting. Energy is the ONLY substance in the universe that is truly conserved. Its total overall amount can never increase and never decrease.

It was then that I realized and KNEW why the second law of thermodynamics is false. The second law of thermodynamics is false because it violates the Conservation of Energy. By axiom and by decree, through the second law of thermodynamics, they have entropy or disorder as some type of substance whose amount is constantly increasing, and whose universal amount can never decrease. The second law of thermodynamics is physically impossible according to the First Law of Thermodynamics and the Conservation of Energy.

The second law of thermodynamics ends up being creation ex nihilo or magic. The second law talks about some type of supernatural, mystical, magical, invisible, intangible substance whose amount is constantly increasing and whose amount can never decrease. The second law is magic or creation ex nihilo. The second law of thermodynamics was designed to falsify and eliminate the first law of thermodynamics. They are mutually exclusive. If one of them can be demonstrated to be true, then the other has automatically been falsified.

So, which is it?

Which one has actually been experienced and observed?

Nobody seemed know what entropy actually is. My research revealed that there are dozens of different, contradictory, mutually exclusive, self-falsifying, and self-defeating definitions for entropy. For all we know, entropy might not even exist. Its existence seems to have been taken on blind faith as being real and true. I couldn't find a consistent, demonstrable, and realistic definition for entropy at the time. I was starting to think that entropy doesn't really exist.

Whereas, everyone seems to know what energy is. We KNOW for a fact that energy exists, because it has actually been experienced and observed. We also KNOW that energy is conserved. Energy cannot be made, and it cannot be destroyed. It has always existed, and it will always exist. Its total universal amount remains constant. Its amount cannot increase, and its amount cannot decrease. Energy is truly conserved!

In comparison to that, the second law of thermodynamics is nothing but science fiction. Through the second law, they have created out of thin air some type of mystical magical substance whose amount is constantly increasing and whose amount can never decrease. In other words, in complete violation of the Conservation of Energy, they have entropy increasing in amount from nothing, and they have set it up so that entropy can never decrease or be destroyed. It's magic. It's creation ex nihilo. It's science fiction. If the second law of thermodynamics were actually true, we should observe this ever-encroaching gray goo coming in at us from all sides, constantly increasing in amount and never decreasing. Yet, all we observe is order, organization, and syntropy.

We have NEVER experienced a substance whose amount can constantly increase while never decreasing. We have NEVER experienced creation ex nihilo. We have never experienced a substance that violates the Conservation of Energy; and, since everything is made from energy to begin with, the existence of something like an ever-increasing amount of entropy is physically impossible. Not even God can do creation ex nihilo; yet, by axiom and by decree, they set it up so that entropy and the second law of thermodynamics can.

The second law of thermodynamics is Conservation of Entropy. The problem with that is that they went too far and made it so that the amount of entropy can actually increase, thereby making it so that entropy really isn't conserved. They made entropy ultra-conserved or supernatural instead. Entropy cannot be a conserved substance because by decree or by definition, they made the stuff so that its total universal amount can actually increase from thin air out of nothing. They made entropy violate the Conservation of Energy. They made a massive conceptual blunder when they tried to make entropy greater than or more powerful than energy. They made entropy or the second law a violation and a falsification of the First Law of Thermodynamics or the Conservation of Energy, when they unilaterally declared that the amount of entropy is increasing and that its total universal amount can never decrease.

After all this effort, work, and thought, I finally KNEW that the second law of thermodynamics is false because I KNEW why it is false. Then I slowly started to replace the second law of thermodynamics with other laws of thermodynamics that are demonstrably true.

The first one I came up with I called the Ultimate Law of Thermodynamics. It was meant to be a course correction for the falsified second law of thermodynamics. The Ultimate Law of Thermodynamics attempts to identify the things that are conserved, and the things that are not conserved. It's an attempt to introduce some truth into thermodynamics by formally falsifying the second law of thermodynamics.

Energy is the ONLY substance in the universe that is truly conserved. Its total universal amount can NEVER increase, and its total universal amount can NEVER decrease. Energy is truly conserved. Energy cannot be made, and it cannot be destroyed. Energy is eternal and everlasting. Its amount remains constant. Energy is the ultimate perpetual motion machine. Energy has always existed, and it will always exist. Other things can approximate conservation, but they are not true conservation, because their overall amount can actually change. I needed another year to think about this conservation issue, and the other things that claim to be conserved; but, in 2018 I had what I needed to submit the Ultimate Law of Thermodynamics into the record for good.

Psyche is the innate intelligence within all the different forms of energy or quanta. Quanta are organized packets of energy. Psyche or Intelligence is the thing that organizes them! Quanta or "particles" are waves of energy moving through a quantum field. Someone Psyche has to start the quanta moving in the first place or nothing would ever happen. Syntropy is the conservation of energy. Syntropy is the First Law of Thermodynamics. Psyche or Energy is syntropic, which means that Psyche or Energy or Quantum Information is conserved. In order to be a truly conserved substance, it has to have always existed; and by that definition for conservation, Psyche and Energy are the only two things that actually qualify as conserved substances or conserved entities.

In contrast, entropic physical matter is made from different forms of energy. The form is NEVER conserved. Physical matter, entropy, and time are made or caused to begin, which means that they are NOT conserved. Anything that is made or anything that has a beginning cannot be truly conserved in the full sense of the word. Entropy is defined as a "made substance" – it's made from nothing out of thin air, accord to the second law of thermodynamics. Since entropy is made, it cannot be a conserved substance. Since the amount of entropy is constantly increasing, it cannot be a conserved substance.

Each psyche has a certain amount of energy that's under its control, and that controlling psyche can form or transform that energy into anything that that controlling psyche wants it to be, anytime and anywhere that it chooses to do so. This is the Law of

Psyche. The FORM is never conserved. Energy is infinitely malleable. Energy can be formed or transformed into anything by the psyche who controls it and owns it.

The Gods and the Controlling Psyches within Nature can transform entropic physical matter into a different form of energy anytime they choose to do so. $E = mc^2$. This is the Ultimate Law of Thermodynamics. It states that Psyche, Energy, Life Force, Quantum Perpetual Motion Machines, and Syntropy are conserved; whereas, entropic physical matter and entropy are NOT conserved. The Ultimate Law of Thermodynamics states that the FORM of the energy is never conserved! The form of energy can change or be changed anytime its controlling psyche chooses to change it! The FORM is NEVER conserved. The Ultimate Law of Thermodynamics simply differentiates between what is being conserved and what is NOT conserved.

The Ultimate Law of Thermodynamics uses observation and experience to falsify the second law of thermodynamics which claims that entropy is conserved or that entropy is ultra-conserved. The second law states that the amount of entropy or disorder in our universe is constantly increasing and can never decrease, in complete violation of the Conservation of Energy or the First Law of Thermodynamics. The Ultimate Law of Thermodynamics states that entropy and physical matter are NOT conserved. The Ultimate Law of Thermodynamics also states that the FORM is never conserved. Entropy is a form of mass, or mass's heat storage capacity. Physical matter or mass is a form of energy. The FORM is never conserved! The FORM of energy can be changed anytime the controlling psyche chooses to change it.

Every psyche has a certain amount of energy that's under its control, and that controlling psyche can form or transform the energy that it controls into anything that it wants that energy to be, anytime and anywhere that it chooses to do so. The FORM is never conserved. Physical matter, entropy, mass, resistance to acceleration, and heat storage capacity are different forms of energy. The FORM is never conserved. ONLY the underlying energy is conserved. This is the Ultimate Law of Thermodynamics. It falsifies the second law of thermodynamics and goes with the first law of thermodynamics instead.

The Ultimate Law of Thermodynamics is a necessary adjustment or refinement to the classical laws of thermodynamics because the Materialists, Naturalists, Darwinists, Nihilists, and Atheists erroneously teach and believe that physical matter, death, and entropy are conserved because these people erroneously teach and believe that entropic physical matter is the ONLY thing that exists. The Ultimate Law of Thermodynamics corrects that flaw by stating that entropy, death, and physical matter are NOT conserved. Therefore, if there is any conflict between the first law of thermodynamics and the second law of thermodynamics, the Ultimate Law of Thermodynamics declares the first law of thermodynamics to be the winner every time.

The Ultimate Law of Thermodynamics states that Psyche, Intelligence, Consciousness, Energy, Life Force, the Quantum Fields, Quantum Information, Quantum Perpetual Motion Machines, and Syntropy are always conserved after they have been made or organized; whereas, entropy, death, chance, and physical matter are NOT conserved. This is what has been experienced and observed by Out-of-Body Travelers and Near-Death Experiencers. Science is observation and experience, not wishful thinking and philosophical speculation. The Ultimate Law of Thermodynamics gives precedence to whatever has been experienced and observed. The Quantum Fields and Quantum Information are made from energy, and energy is always conserved. Once the Quantum Fields and Quantum Information are made, then they too end up being conserved because they are made from energy in a manner that remains constant or conserved.

The Quantum Fields are perpetual motion machines. Quantum Information is perpetual information storage. They were designed to be permanent. The Quantum Fields are perfect syntropy because they are conserved. The second law of thermodynamics will NEVER be true as long as the quantum fields exist. Quantum Field Theory falsifies the second law of thermodynamics. $E = mc^2$ is the Perpetual Motion Cycle. It too is constantly conserved. Its very existence falsifies the second law of thermodynamics which states that it doesn't exist and can't exist.

In contrast, the amount of mass, heat, thermodynamics, and entropy (mass's heat storage capacity) are constantly changing and going to zero. The total amount of heat in our universe seems to be constantly moving towards zero. Does it not? Entropy defined as heat or entropy defined as thermodynamics seems to be constantly moving towards zero. It's NOT conserved. Its overall amount is constantly changing! Heat, thermodynamics, or entropy is constantly moving towards zero. Whenever Nature's Psyche makes some new heat within a star or a fire or a nuclear bomb, its temperature or the amount of heat tends towards the absolute zero of the quantum fields, as that heat is slowly absorbed back into the quantum fields from whence it came. This is obviously true, because it has actually been experienced and observed; and, it falsifies the second law of thermodynamics.

Physical matter, mass, and entropy are comprised of different forms of energy; and, the FORM is never conserved. The form can be transformed by Nature's Psyche into a completely different form anytime and anywhere that Nature's Psyche chooses to do so. Only the underlying energy is conserved. Energy is infinitely malleable. Energy can be transformed into anything, anytime. Physical matter, mass, and entropy (mass's heat storage capacity) are never conserved. Their overall amount is constantly changing; and, as the amount of mass goes to zero, the amount of entropy or heat storage capacity goes to zero along with it. As the overall amount of mass in a system goes to zero, the overall entropy or resistance to acceleration goes to zero along with it. Entropy or mass is NOT a conserved substance!

The Matrix of Quantum Fields is the only thing that was ever made that seems to be truly conserved – truly eternal and everlasting – truly syntropic. Quantum Field Theory falsifies the conservation of entropy or the second law of thermodynamics.

Once the Gods, the Controlling Psyches, Nature's Psyche, or the Intelligences within Nature designed and made the Matrix of Quantum Fields in our part of the multiverse, those quantum fields became eternal and everlasting perpetual motion machines. The Quantum Fields are Conservation of Order or the Conservation of Syntropy in our part of the multiverse. There is NO entropy and NO disorder at the quantum level as long as the quantum fields exist. The quantum fields are perfect syntropic perpetual motion machines. The second law of thermodynamics will NEVER be true as long as the quantum fields exist. Quantum Field Theory falsifies the second law of thermodynamics.

The Gods had to design and make the quantum fields BEFORE it became possible to make and sustain physical matter. The construction of the Quantum Fields was the "Big Bang" in our part of the multiverse. Only thereafter was it possible to make and sustain mass, resistance to acceleration, heat storage capacity (entropy), or physical matter. The Quantum Fields are pure syntropy and pure exergy – just precisely what the theoretical Big Bang was supposed to be. The construction of the Quantum Fields from chaos was the Big Bang, whether we realize it or not.

The Quantum Fields had to be made by Nature's Psyche under the tutelage and standardization provided by the Gods; but, now that they are made and exist, the Quantum Fields are conserved. They are eternal and everlasting. The Quantum Fields are Pure Syntropic Exergic Order. The Quantum Fields are perfection. The Quantum Fields are the ultimate perpetual motion machines. They will never get old, and they will never wear out. Their very existence and their continued existence falsify the second law of thermodynamics which states that they don't exist and can't exist.

Now that they have been made, the Quantum Fields will always exist as long as the Gods, and Nature's Psyche, and Energy exist. Even if the whole universe at the physical level were magically to achieve perfect heat death, the Quantum Fields will continue to function perfectly and flawlessly at absolute zero temperature at the quantum level; and, Nature's Psyche will be able to make NEW mass, NEW resistance to acceleration, NEW heat storage capacity (entropy), NEW heat, NEW photons, NEW light, NEW planets, NEW stars, and NEW galaxies anytime and anywhere it chooses to do so, thanks to the Conservation of Energy and the Quantum Fields. This is what Quantum Mechanics and Quantum Field Theory are trying to teach us; but, nobody is listening because they prefer heat death, entropy, death, chance, creation ex nihilo, and the second law of thermodynamics instead.

The proven and verified existence of the Quantum Fields is one of the most convincing Scientific Proofs of God's Existence that I have encountered so far. We would still be back in the Chaos Realm with nothing but random lawless chaos at the quantum level if it weren't for the Gods, Nature's Psyche or the Controlling Psyches within Nature, and the Quantum Fields. The construction of the Quantum Fields was the Big Bang in our part of the multiverse. This is obviously true; otherwise, we wouldn't be here right now. There will NEVER be any such thing as "disorder" or "chaos" in our part of the multiverse so long as the Quantum Fields exist. The second law of thermodynamics will NEVER be true as long as the quantum fields exist.

It is impossible to conserve disorder or entropy, because the amount of disorder or entropy is constantly changing, and disorder completely ceased to exist at the quantum level when the Quantum Fields were made.

There is NO entropy or disorder at the quantum level thanks to the Quantum Fields. This is the Quantum Law of Thermodynamics which states that there is no entropy and no thermodynamics and no disorder at the quantum level. The Quantum Fields function at maximum efficiency at absolute zero temperature when there is NO heat, NO mass, NO heat storage capacity, NO resistance to acceleration, NO thermodynamics, and NO entropy. There will never be any disorder, chaos, or entropy at the quantum level so long as the Quantum Fields exist. This is the Quantum Law of Thermodynamics; and, it's obviously true because it has actually been experienced and observed. We KNOW that it is true because we actually exist. If the second law of thermodynamics were true, then we wouldn't exist. Our very existence is scientific proof that the second law of thermodynamics is false. Everything that exists falsifies the second law of thermodynamics!

Energy is truly conserved. Energy cannot be made, and it cannot be destroyed. It has always existed, and it will always exist. The same reality seems to apply to Psyche, Intelligence, Quantum Consciousness, Quantum Information, or Life Force.

As I see it, it's time for the scientists in this world to replace the second law of thermodynamics with Quantum Field Theory instead, because Quantum Field Theory and the Quantum Fields clearly and conclusively falsify the second law of thermodynamics as it currently stands. Science is actually looking for a made substance that is constantly conserved, to explain everything that we have experienced and observed. Entropy or thermodynamics or disorder is NOT it. By definition, disorder cannot be a conserved substance! Heat and disorder are NOT being conserved; but, the quantum fields are definitely being conserved and are definitely what we are looking for! The quantum fields are precisely what entropy and the second law were trying to be and become. The quantum fields are a made substance that are constantly being conserved now that they have been made. The Quantum Fields are the ultimate perpetual motion machines. The Quantum Fields and Quantum Field Theory are precisely what science needs, and precisely what science is looking for!

I developed the Ultimate Law of Thermodynamics before the Quantum Law of Thermodynamics; but, the Quantum Law of Thermodynamics ends up being more powerful and more interesting, in my humble opinion. The Quantum Law of Thermodynamics was truly ground-breaking! Nevertheless, the Quantum Law of Thermodynamics grew out of my work on Syntropy and the Ultimate Law of Thermodynamics. They all end up falsifying the second law of thermodynamics and end up being a course correction for the falsified second law of thermodynamics.

The Ultimate Law of Thermodynamics and the Quantum Law of Thermodynamics are an attempt to replace the second law of thermodynamics with the truth. The second law is of no use to us because it is demonstrably false. In science, we need the truth, not philosophical speculation and wishful thinking.

The only thing the second law is good for is to emphasize the obvious fact that perpetual motion machines are physically impossible at the physical level due to friction, resistance to acceleration, mass, loss of heat to the quantum fields, and that process which we typically call entropy or the transfer of heat from a hot reservoir to a cold reservoir.

In contrast, the massless, heatless, entropyless, non-physical, syntropic, exergic Quantum Fields are the perfect perpetual motion machines, and they function endlessly and flawlessly at the quantum level no matter what's happening at the physical level. At the quantum level the work is always done for free thanks to the fact that at the quantum level energy is always exergy and energy is always conserved. Heat death or absolute zero maximizes the efficiency and reliability of quantum objects like photons, quantum waves, super conductors, and quantum fields. Unlike physical objects, quantum objects don't need heat as a lubricant or propellant to make them work right. Quantum objects function at maximum efficiency when the temperature is absolute zero or near absolute zero.

Remember, we will NEVER have true disorder or chaos in our part of the multiverse, as long as the Quantum Fields exist; and, this reality or truth falsifies the second law of thermodynamics.

Remember, according to the Ultimate Law of Thermodynamics, energy, psyche, the perpetual motion cycle, and now the quantum fields are conserved; whereas, the form of that energy, particles, physical matter, death, mass, heat, thermodynamics, chance, and entropy are never truly conserved. The amount of physical matter, mass, heat, and entropy (mass's heat storage capacity) are constantly changing which means that they are not conserved. Only the energy, psyches, quantum perpetual motion cycle $E = mc^2$, and the quantum fields, which the psyches made, are truly conserved.

Mark My Words

**Introducing the Quantum Law of Thermodynamics**

https://quantum-law-of-thermodynamics.com/introducing-the-quantum-law-of-thermodynamics/

You will never see the Quantum Law of Thermodynamics in our public universities, because it falsifies Materialism, Naturalism, Darwinism, Nihilism, Atheism, and the Second Law of Thermodynamics, which are Creation by Chance. It doesn't matter that the Quantum Law of Thermodynamics is demonstrable or true. It doesn't matter that the Quantum Law of Thermodynamics is based upon Quantum Mechanics, Quantum Field Theory, and $E = mc^2$. It doesn't matter that the Quantum Law of Thermodynamics is based upon observational evidence and eyewitness experiences. The Quantum Law of Thermodynamics will automatically be rejected, vilified, and ignored because it contradicts and falsifies the materialistic, naturalistic, and atheistic paradigm. They don't allow anything into our public schools that successfully falsifies Materialism, Naturalism, Darwinism, Nihilism, Scientism, and Atheism or Creation Ex Nihilo. They don't allow anything into our public schools that falsifies Creation by Chance.

The Quantum Law of Thermodynamics states the obvious and verified truth that there is no mass, no heat, no resistance to acceleration, no heat storage capacity, no entropy, and NO THERMODYNAMICS at the quantum level among the Quantum Fields that are currently being conserved or sustained by Nature's Intelligence or by Nature's Psyche. There is NO second law of thermodynamics at the quantum level in the Quantum Realm, the Spirit World, or the Non-Physical Transdimensional Realm. Entropy or mass's heat storage capacity is purely a physical phenomenon.

One of my all-time greatest scientific discoveries came to me when I first realized that Quantum Mechanics, Quantum Field Theory, Feynman Diagrams, the Perpetual Motion Cycle E = mc2, the Conservation of Quantum Information, the Conservation of Energy, and the Conservation of Quantum Consciousness or Psyche FALSIFY the second law of thermodynamics, which claims that the total amount of disorder or entropy in our universe is constantly increasing and can never decrease or to go zero. Everything that has been experienced and observed FALSIFIES the second law of thermodynamics. Everything that exists FALSIFIES the second law of thermodynamics. The fact that you exist is scientific proof that the second law of thermodynamics is false and needs to be replaced with something better and with something that's true.

Someday in the far distant future, when our scientists begin to figure out how things really work at the quantum level in the Quantum Realm, they will vote to replace the falsified second law of thermodynamics with Quantum Field Theory, the Conservation of Energy and Psyche, the Conservation of Order and Organization at the quantum level, the Conservation of Quantum Information, and the Perpetual Motion Cycle $E = mc^2$, all of which have been experienced and observed and verified and proven true.

The massless, heatless, chargeless, entropyless, non-physical, immaterial Quantum Fields are pure syntropy, pure exergy, and completely conserved. The Quantum Fields are the ultimate perpetual motion machines. The Quantum Fields are perfection. They will never wear out and never get old thanks to the Conservation of Energy and Psyche.

The Quantum Law of Thermodynamics states that there is no second law of thermodynamics at the quantum level, because there is no entropy and no thermodynamics and no disorder at the quantum level. The Matrix of Quantum Fields represents pure ORDER, or pure syntropy, or pure exergy; and, the verified and proven existence of the Quantum Fields falsifies the second law of thermodynamics, which claims that the amount of entropy or disorder in our universe can never decrease and go to zero. The Quantum Fields are scientific proof that the second law of thermodynamics is false. The Quantum Fields ended chaos or disorder in our part of the multiverse. The Quantum Fields ARE disorder and entropy gone to zero and ceasing to exist. The Quantum Fields are syntropy and ORDER, which is the opposite of entropy and disorder.

There will NEVER be true disorder and chaos in our part of the multiverse as long as the quantum fields exist. The second law of thermodynamics will NEVER be true so long as the quantum fields exist and are being conserved by Nature's Psyche.

It is Nature, or Nature's Psyche, or Nature's Intelligence who collapses the wave function and chooses what form the energy under its control will assume. It's Nature's Psyche who forms and starts the massless and entropyless quantum waves or the photons in the first place. It is also Nature's Psyche or the Controlling Psyches within Nature who transform mass, heat storage capacity, resistance to acceleration, and entropy INTO massless, entropyless, heatless, and chargeless quantum waves, photons, quantum fields, and infinite acceleration instead. Nature's Psyche can transform mass, heat, and entropy into anything that it wants that energy to be, anytime and anywhere that it chooses to do so.

Just look at our sun.

Within our sun, Nature or Nature's Psyche is constantly transforming mass, heat, and mass's heat storage capacity (entropy) INTO massless, heatless, chargeless, and entropyless photons or quantum waves which are then capable of infinite acceleration because they have no mass or resistance to acceleration. Nature's Psyche makes the quantum waves or photons from whatever it desires, including physical matter; and, Nature's Psyche collapses the wave function thereby making heat, mass, or a physical particle out of the quantum waves or photons. This is the Perpetual Motion Cycle $E = mc^2$. It functions perfectly both directions, and all the while the energy is totally conserved.

Anytime a massless, chargeless, heatless, and entropyless photon or quantum wave CHOOSES to stop, it transforms its infinite acceleration and speed-of-light velocity INTO heat, mass, heat storage capacity (entropy), and resistance to acceleration instead. This is what has actually been experienced and observed! We can feel the heat on our skin every time that a photon or quantum wave chooses to stop.

$E = mc^2$ is perfectly functional, both coming and going, and throughout it all, the energy is always conserved. $E = mc^2$ is the ultimate perpetual motion machine. $E = mc^2$ is the Perpetual Motion Cycle. $E = mc^2$ represents pure syntropy. $E = mc^2$ and the quantum fields are perfect Order and Organization at the quantum level or the non-physical level.

$E = mc^2$ falsifies the second law of thermodynamics, which says that it shouldn't exist. The Quantum Fields, Quantum Field Theory, and Feynman Diagrams falsify the second law of thermodynamics. The heat-based equations for entropy falsify the second law of thermodynamics. Everything that exists falsifies the second law of thermodynamics, which says that it shouldn't exist.

Nature's Psyche transforms quantum waves or quantum fields into mass and heat at will; and then later, Nature's Psyche takes the same mass and heat and entropy, and transforms it back into massless, heatless, chargeless, and entropyless photons, quantum waves, and quantum fields. This is the Perpetual Motion Cycle. It's real. It has actually been experienced and observed. This is proven and verified science. It explains everything that has ever been experienced or observed. The Perpetual Motion Cycle and Quantum Field Theory are scientific proof that the Quantum Law of Thermodynamics is true. There is NO entropy at the quantum level among the Quantum Fields. The Quantum Fields are massless, heatless, and entropyless perpetual motion machines. Now that they have been formed by Nature's Psyche, the Quantum Fields will always exist. They are conserved. That means that there is NO entropy or disorder within the Quantum Fields. The second law of thermodynamics, disorder, and entropy don't exist at the quantum level among the Quantum Fields. Thanks to the Quantum Fields, the Transdimensional Non-Physical Quantum Realm is conserved Order and Organization and Syntropy. That's why we exist. This explains everything that has ever been experienced or observed.

When we look out around us, we DON'T observe the proton decay that the second law of thermodynamics predicts. We don't observe the ever-increasing disorder and disorganization that the second law predicts. We don't observe an ever-encroaching gray goo coming in at us from all sides as the second law of thermodynamics predicts. We don't even observe the heat death that the second law predicts. We ONLY observe conserved Order and Organization, thanks to the Quantum Fields, Quantum Field Theory, the Conservation of Energy, the Conservation of Quantum Information, the Conservation of Psyche or Quantum Consciousness, the Quantum Law of Thermodynamics, and the Perpetual Motion Cycle $E = mc^2$. The one that's experienced and observed is the one that's actually real and true. We don't observe the second law of thermodynamics in action, because it's false.

There's nothing sacred about entropy. Nature or Nature's Psyche can transform mass, heat, and entropy into massless, heatless, chargeless, and entropyless photons and quantum waves anytime and anywhere that Nature's Psyche chooses to do so. It's doing so right now within our sun. Where do you think the light comes from? Nature's Psyche is constantly converting mass, heat, and entropy into massless, heatless, and entropyless photons or quantum waves. It's happening all around us all the time, according to $E = mc^2$, Quantum Field Theory, the Conservation of Energy and Psyche, and the Quantum Law of Thermodynamics.

Nature's Psyche can transform any type of entropy, heat, or mass INTO massless and entropyless photons and quantum waves, anywhere and at any time that it chooses to do so; and, there's nothing we can do to stop it. There's nothing sacred about matter, or mass, or entropy. It can be transformed into massless, heatless, chargeless, and entropyless quantum fields, quantum waves, virtual particles, and photons anywhere and anytime that Nature's Psyche CHOOSES to do so, according to Quantum Field Theory and $E = mc^2$. This universal perpetual motion machine, $E = mc^2$, falsifies the second law of thermodynamics which claims that the amount of disorder or entropy in our universe is constantly increasing. The constant perpetual ORDER of $E = mc^2$ and the conserved syntropic quantum fields FALSIFY the second law of thermodynamics, which claims that this perpetual ORDER does not exist.

Likewise, whenever a photon or quantum wave CHOOSES to stop, it can transform its omnipresence or infinite acceleration into anything that it wants that energy to be. On Feynman Diagrams, massless, heatless, chargeless, and entropyless photons are constantly transforming themselves into charged electrons and positrons, or quarks and gluons and mass, all the time. There's nothing we can do to stop it. When photons land on our skin, they KNOW that they have landed on our skin and transform themselves into heat. They don't have to transform themselves into heat, if they don't want to. They can transform themselves into mass, resistance to acceleration, or mass's heat storage capacity (entropy) instead if they CHOOSE to do so. Photons or quantum waves can transform themselves into anything they want to be, anytime and anywhere they choose to do so. A miniature Big Bang happens every time that a photon or quantum wave chooses to stop and transform itself into mass, heat, resistance to acceleration, electrons, or heat storage capacity (entropy).

Photons and quantum waves have a psyche or intelligence or point-particle within them. According to the Obvious Law of Physics, the smaller dwells within and controls the larger. The smallest point-particle, Psyche, dwells within and controls everything else. A psyche looks like a photon or a pinpoint of light, even at the quantum level and the psyche level.

**Energy is infinitely malleable, and Nature's Psyche can form the energy under its control into anything that it wants that energy to be, anytime and anywhere that it chooses to do so. This is the Law of Psyche. Each psyche has a certain amount of energy that's under its control; and, that controlling psyche or controlling intelligence can form or transform the energy under its control into anything that it wants that energy to be, anytime and anywhere that it chooses to do so. This is the way things really work at the quantum level. This is what has actually been experienced and observed. Psyche has been experienced and observed.**

https://psyche-ontology.com/psyche-experienced-and-observed/

https://psyche-ontology.com/buhlman/

Under the guidance, tutelage, and instruction of the Gods or the Controlling Psyches, it was Nature's Psyche who made the quantum fields, and then used those quantum fields to make mass, resistance to acceleration, heat, physical matter, and mass's heat storage capacity (entropy). Furthermore, Nature's Psyche can unmake these things and restore them to the Quantum Fields as pure exergy, anytime and anywhere that Nature's Psyche chooses to do so.

Nature's Psyche makes the quantum waves, and Nature's Psyche collapses the wave functions. Nature's Psyche is an essential part of the Perpetual Motion Cycle, $E = mc^2$, and Quantum Field Theory. They won't work without Nature's Psyche. Something non-physical had to make and then conserve the quantum fields. Something non-physical and intelligent has to make the quantum waves and then later collapse the wave functions. Psyche, or Intelligence, or Quantum Consciousness is the ONLY thing we know of that has been experienced and observed that could actually fulfill this function of making the massless, heatless, chargeless, and entropyless photons, quantum waves, and quantum fields. Therefore, it is absolutely essential that Psyche be conserved or that Quantum Information be conserved within Psyche; otherwise, there would be NOTHING to make the quantum waves, to collapse the wave functions, and to conserve the quantum fields.

The Perpetual Motion Cycle $E = mc^2$, the Conservation of Energy and Psyche, the Conservation of Quantum Information within Psyche, the conserved syntropic ORDER of the Quantum Fields, as well as Quantum Field Theory FALSIFY the second law of thermodynamics which claims that these things DO NOT EXIST.

Once I realized that the Materialists, Naturalists, Darwinists, Nihilists, and Atheists are lying to themselves and lying to us, I lost all desire to remain in conformity with them; and, I set off on my own with the ultimate goal of finding and knowing the truth. I'm no longer trying to please them nor placate them. I chose to be a scientist instead and falsify them. I used to be a materialist, naturalist, nihilist, and atheist back in 2012 and 2013; but I caught them in one too many lies, and they lost me. I can't go back, because now I know the truth. The truth set me free. I'm no longer enslaved to them and their lies.

One day, it finally dawned on me that if I want to have a true definition for entropy, then I must develop some NEW science that defines and explains Syntropy, Psyche, and the Quantum Law of Thermodynamics.

I wrote the book on Syntropy (25MAR2018), and then a year later (04JUL2019), I set out to conquer entropy. In the process, I discovered the Quantum Law of Thermodynamics, which identifies and explains thermodynamics from a massless, non-physical, quantum perspective.

Syntropy

https://www.amazon.com/gp/product/B07BPT3W8R

Entropy from a Massless and Non-Physical Perspective

https://www.amazon.com/gp/product/B01J023TGU

The Quantum Law of Thermodynamics states that there is no mass, no heat, no resistance to acceleration, no thermodynamics, and therefore NO entropy (mass's heat storage capacity) in the Quantum Realm, or the Spirit World, or the Massless Realm, or the Non-Physical Realm. Entropy or mass's heat storage capacity is purely a physical phenomenon. It doesn't exist at the quantum level among the Quantum Fields.

From a practical standpoint, the Quantum Law of Thermodynamics defines "syntropy" as the infinite expanse of massless, heatless, and entropyless Quantum Fields. From a physics standpoint, Syntropy is the Conservation of Energy or the Conservation of the Quantum Fields. The Quantum Law of Thermodynamics also defines "syntropy" as all the massless, heatless, chargeless, and entropyless quantum wave-like omnipresent Photons that Nature or Nature's Psyche chooses to make.

The Quantum Law of Thermodynamics is verified by the official equations for heat and entropy, and by thermodynamics itself. The Quantum Law of Thermodynamics is a NEW, true, and better replacement for the second law of thermodynamics. Whereas the second law of thermodynamics is all about chaos and disorder, the Quantum Law of Thermodynamics along with Quantum Field Theory actually explain how thermodynamics is done. Furthermore, the Quantum Law of Thermodynamics, Quantum Mechanics, the Equations for Entropy and Heat, the Perpetual Motion Cycle $E = mc^2$, the Conservation of Energy and Psyche, the Conservation of Quantum Information, Feynman Diagrams, and Quantum Field Theory FALSIFY the second law of thermodynamics. The false is falsified by the truth; and, the truth is repeatedly experienced and observed. This is Science 2.0 in action!

Study the following equation for heat (**Q**). There is something important here that they completely overlooked, because they aren't looking for it and don't want it to be true. It falsifies the second law of thermodynamics.

### $Q = mc\Delta T$

Thermodynamics is the transfer of heat from a hot system to a cold system. According to thermodynamics and the equation for heat (**Q**), anything that has NO mass (**m = 0**) has NO heat storage capacity (**mc = 0**), which means that the massless has NO heat (**Q = 0**) and is natively at absolute zero instead (**T = 0**). Everything goes to zero when mass goes to zero!

Remember that, it's important!

Mass, heat, temperature, and entropy (mass's heat storage capacity) go to zero whenever mass goes to zero. Whenever Nature's Psyche transforms mass, heat, or entropy INTO massless, heatless, chargeless, and entropyless photons or quantum waves or infinite acceleration or omnipresence, the entropy completely ceases to exist, thereby falsifying the second law of thermodynamics which erroneously claims that entropy or disorder cannot cease to exist.

When mass is equal to zero (**m = 0**), then the heat storage capacity is also zero (**mc = 0**). This is really easy to understand. When mass and mass's heat storage capacity are zero (**m = 0 = mc**), then the particle can no longer store nor contain heat (**Q = 0**). Simple truth, is it not? When mass goes to zero, heat and heat storage capacity go to zero. It's unavoidable. It's automatic. It's true. Massless particles like photons have NO heat because they have no mass and no heat storage capacity (entropy). When mass goes to zero, mass's heat storage capacity goes to zero. When heat storage capacity goes to zero, then entropy goes to zero. It's that simple.

Massless photons travel fastest and travel best in the absolute zero temperatures of the infinite expanse of the Quantum Fields – in a vacuum. Massless photons can travel at the speed-of-light or faster because they have NO mass or NO resistance to acceleration. In other words, massless photons can produce infinite acceleration, or go to the speed-of-light instantly, because they have no mass or resistance to acceleration. This is what has actually been experienced and observed. It's hiding in plain sight where nobody can see it because they aren't looking for it and don't want it to be true.

Remember the equation for heat (**Q**). It is important!

**Q = mcΔT**

Now, let's look at the official heat-based equation for entropy, and see what we can learn.

**S = Q/ΔT**

The heat-based equation for entropy (**S**) is based upon heat (**Q**). When heat goes to zero (**Q → 0**), entropy goes to zero (**S → 0**)! It's that simple! This is the KEY to entropy and thermodynamics, and it has been completely overlooked.

Now, let's combine the two equations and solve the equation for entropy (**S**).

**S = mcΔT/ΔT**

So, what do we get?

We get:

**S = mc**

Entropy (**S**) is equal to mass's heat storage capacity (**mc**). This is a NEW equation for entropy that has been completely overlooked. It falsifies the second law of thermodynamics and proves that the Quantum Law of Thermodynamics is true. It also explains entropy and thermodynamics. It actually provides a realistic and demonstrable definition for entropy.

**S = mc**

Entropy IS mass's heat storage capacity, according to the equations for heat and entropy. NO mass, then NO heat storage capacity, which means NO entropy. Entropy goes to zero whenever mass or heat goes to zero! It's that simple.

In this equation for entropy, (**c**) is mass's specific heat capacity; therefore, entropy is mass's heat storage capacity. When mass goes to zero, heat and mass's heat storage capacity go to zero, and then all of the entropy goes to zero along with them.

https://en.wikipedia.org/wiki/Specific_heat_capacity

https://en.wikipedia.org/wiki/Table_of_specific_heat_capacities

This is real science. It has already been discovered and verified. It's just NEVER been applied to entropy before, because it falsifies the second law of thermodynamics.

The Naturalists, Darwinists, and Atheists are determined to protect and enshrine and axiomize the falsehoods in the second law of thermodynamics because they are falsified by the truth. They protect and defend the second law of thermodynamics because it is Creation by Disorder or Creation by Chance, just like the theory of evolution. These people always prefer the falsehoods over the truth. It's what they do.

There is NO entropy in the massless, heatless, and chargeless quantum waves and photons. There is NO entropy in the massless non-physical quantum fields. This is the Quantum Law of Thermodynamics. This is what Quantum Mechanics and Quantum Field Theory have been trying to tell us all the way along, but nobody is listening because nobody wants it to be true.

According to thermodynamics and the heat-based equation for entropy (**S**), any massless particle has NO heat storage capacity (**mc = 0** and **Q = 0**) which means that it has NO entropy either (**S = 0**). Massless particles can't have entropy according to thermodynamics and the equations for heat and entropy. When mass and heat go to zero (**m → 0** and **Q → 0**), then entropy also goes to zero (**S → 0**) along with them. Whenever anything is converted into a massless non-physical particle, ALL of its entropy ceases to exist because ALL of its heat storage capacity ceases to exist. Everything goes to zero when mass goes to zero, including heat and entropy!

This is really simple to understand. It's hidden in plain sight within the official equations for heat and entropy. It's amazing that everyone missed it, is it not? But, they were never looking for it in the first place, because they had already erroneously chosen to believe that the massless or the non-physical does not exist.

This is a neat NEW TOY, though, for any physicist because it explains everything that has ever been experienced and observed. Whenever Nature or Nature's Psyche chooses to change physical matter back into energy, what it does is it takes and converts all of the mass and heat directly into massless and entropyless infinite acceleration instead. It converts or transforms the physical matter or the mass into photons – or raw, massless, entropyless, chargeless, and heatless energy or exergy. That's what **$E = mc^2$** is all about! The particles of mass or physical matter get converted back into massless, entropyless, and heatless photons because their heat, mass, and entropy are converted into infinite acceleration instead. Photons go from zero to the speed-of-light instantly, and that's infinite acceleration. Furthermore, photons or quantum waves don't have to stop at the speed-of-light if they don't want to. They can go much faster. They can be omnipresent or have an infinite velocity or quantum tunnel if they choose to do so.

Thanks to the Perpetual Motion Cycle $E = mc^2$, the physical particles in our sun are set free by being transformed into massless, heatless, chargeless, and entropyless spiritual photons or non-physical quantum waves instead. Heat, mass, and entropy are being transformed into heatless, massless, and entropyless photons within our sun right now. This explains everything that we physicists have ever experienced or observed. By allowing Psyche and Quantum Mechanics in to play, we can literally explain everything that comes our way.

By definition, ALL of the photons have NO mass, which means that they have NO heat storage capacity, which means that they have NO entropy. Only when the photons choose to stop and choose to integrate with our physical realm do they in fact transform themselves into heat, mass, and entropy (mass's heat storage capacity). Every time that a photon chooses to stop and chooses to transform itself into heat, mass, and entropy, a miniature Big Bang happens, and a bunch of massless, heatless, and entropyless Energy is transformed into mass, resistance to acceleration, heat, and entropy (mass's heat storage capacity) instead. The Big Bang is happening all over again every time a massless and entropyless Photon chooses to stop and transform itself into heat, mass, and entropy. This is what we actually experience and observe every day, is it not? It explains everything, does it not? It's really simple to understand, is it not?

This is what we start to get, when we choose to allow all of the evidence into evidence, and then start to pursue a preponderance of that evidence. We find ourselves upgrading our science to Science 2.0. We end up finding a new and better way to do science. We end up finding a new and better way to interpret science. We end up with a new and better Philosophy of Science.

When it comes to science, the Materialists, Naturalists, Darwinists, Nihilists, and Atheists draw a line in the sand and refuse to go any further. In contrast, Science 2.0 allows the quantum or the non-physical or the massless and entropyless into evidence, and then tries to take Science as far as it can go.

We really don't need the Big Bang to make heat and physical matter. All we really need is the Quantum Fields. Once the Quantum Fields are in place, Nature's Psyche can make and sustain heat, mass, and entropy (mass's heat storage capacity) anytime and anywhere Nature's Psyche chooses to do so. We don't need the Big Bang once we have the Quantum Fields. The Quantum Fields ARE the hypothetical big bang singularity. The construction of the Quantum Fields by Nature's Psyche WAS the Big Bang. The Quantum Fields are the thing that made heat, mass, entropy, and physical matter possible in the first place. These physical things still wouldn't exist without the Quantum Fields. That's just the way it is. The physical wouldn't exist without the prior existence and organization of the non-physical, entropyless, massless, heatless Quantum Fields by Nature's Psyche.

**Quantum Fields and Nature's Psyche completely solve ALL the mysteries behind the big bang singularity and the Big Bang, if we let them do so. The Quantum Fields ARE the big bang singularity; and, the construction of the Quantum Fields by Nature's Psyche or the Controlling Psyches WAS the Big Bang. It's what made physical matter, mass, heat, and entropy possible in the first place.**

**Furthermore, every time a photon chooses to stop, we experience a miniature Big Bang, as that massless, heatless, chargeless, and entropyless photon or quantum wave transforms itself into mass, heat, resistance to acceleration, and entropy (mass's heat storage capacity) instead.**

The creation is ongoing to this very day, because Nature's Psyche can make new mass, new heat, new physical matter, and new entropy anytime and anywhere that Nature's Psyche chooses to do so. Nature's Psyche can also convert entropic physical matter, mass, heat, and entropy back into entropyless and massless Quantum Fields anytime and anywhere that Nature's Psyche chooses to do so, according to Quantum Mechanics, Quantum Field Theory, the Perpetual Motion Cycle $E = mc^2$, and the LAW of the Conservation of Energy and Psyche. We observe heat, mass, entropy, and physical matter being absorbed back into the entropyless, massless, absolute zero temperature Quantum Fields all the time; and, we don't even realize that that's what's happening because we have been trained not to look for it nor to think about it. Yet, it's happening right now within our sun.

Remember, the creation is ongoing even to this very day, because Nature's Psyche can make brand-new heat, mass, physical matter, planets, stars, and galaxies anytime and anywhere Nature's Psyche chooses to do so, in any quantity that Nature's Psyche desires. Nature's Psyche could instantly transform the energy, the entropy, and the mass in the planet Jupiter and the associated Quantum Fields into a star, a galaxy, or a black hole anytime that Nature's Psyche chooses to do so.

Nature's Psyche (or the Controlling Psyches) can draw in anything that it needs from the infinite, entropyless, massless, and heatless Quantum Fields anytime and anywhere it chooses to do so, to make anything that it wants to make – including brand-new mass, heat, physical matter, entropy, planets, stars, black holes, and galaxies. Likewise, Nature's Psyche could disassemble and reabsorb our whole galaxy back into the entropyless and massless Quantum Fields instantaneously, right now, if it wanted to do so; and, there is nothing that we could do to stop it.

That's what $E = mc^2$, Quantum Field Theory, the Conservation of Energy and Psyche and Quantum Mechanics are trying to tell us, but nobody is listening because nobody wants it to be true. They don't want it because it falsifies Materialism, Naturalism, Darwinism (creation by chance), Nihilism, Atheism (creation ex nihilo), and the Second Law of Thermodynamics (creation by disorder or creation by entropy).

Nevertheless, by choosing to allow Psyche and Quantum Mechanics in to play, you can literally explain everything that comes your way. This is what I have experienced and observed; and, it all makes logical sense to me when I am done. By allowing all of the evidence into evidence, my scientific explanatory power goes through the roof to infinity and beyond. I can even explain the origin and the destiny of our universe.

Alas, you will never see the Quantum Law of Thermodynamics in our public universities, because it falsifies Materialism, Naturalism, Darwinism, Nihilism, Atheism, and the Second Law of Thermodynamics, which are Creation by Chance. It doesn't matter that the Quantum Law of Thermodynamics is demonstrable or true. It doesn't matter that the Quantum Law of Thermodynamics is based upon Quantum Mechanics, Quantum Field Theory, and the Perpetual Motion Cycle $E = mc^2$. The Quantum Law of Thermodynamics will automatically be rejected, vilified, and ignored because it contradicts and falsifies the materialistic, naturalistic, and atheistic paradigm. That's just the way it is. That's the way these people work – by suppressing, censoring, banning, ridiculing, deleting, and destroying any evidence that falsifies their beliefs. That's how they do "science".

These people prefer the falsehoods and the lies over the proven and verified observable truth. As long as they do so, the Perpetual Motion Cycle and the Quantum Law of Thermodynamics will never see the light of day. These truths will continue to be suppressed and hidden throughout the whole of human history instead, as has already happened. We will never find and know the truth, unless we deliberately go looking for it and are willing to accept it and embrace it when we find it. That's just the way it is.

Mark My Words

## The True Version of the Second Law of Thermodynamics

In complete violation of the Conservation of Energy, the second law of thermodynamics states that the amount of entropy or disorder in our universe is constantly increasing and that its overall amount can never decrease.

According to the second law of thermodynamics, entropy is a super-substance or a superfluid whose amount is constantly increasing and whose amount can never decrease. The second law is creation ex nihilo, magic, superstition, and the Conservation of Entropy. The people who created the second law truly believe that ONLY physical matter, entropy, and death exist. They deny the existence of everything else. According to these people, entropy or natural selection or disorder or chance designed and created everything that exists in this universe. By definition and by axiom, the second law of thermodynamics is Creation Ex Nihilo or Atheism.

In the summer of 2019, while studying the different competing equations for entropy, I found a couple dozen different contradictory and mutually exclusive definitions for entropy, which means according to the Philosophy of Science, Logic, the Scientific Method, and Statistics, I found a couple dozen different ways to falsify the second law of thermodynamics. I released the preliminary results on 04JUL2019.

https://www.amazon.com/gp/product/1521132380

https://www.amazon.com/gp/product/B01J023TGU

Since then, I have been exploring my amazing science discoveries regarding entropy and the second law of thermodynamics within this book, "Science 2.0".

https://www.amazon.com/gp/product/B0771K6WTX

It's brutal.

After falsifying the second law of thermodynamics a couple dozen different ways, I realized that it's time for us scientists to derive a NEW definition for entropy and the second law of thermodynamics – one that is actually measurable, demonstrable, reliable, replicable, and true. We need to scale back the extravagant second law of thermodynamics to something that has actually been experienced and observed. We need to scale back infinitely increasing amounts of disorder or entropy to something on the order of the Conservation of Energy, at the least. We need to make the second law of thermodynamics match with reality, which means that we need to make it match with what has actually been experienced and observed.

Over the past century, entropy has been given a couple dozen different contradictory mutually exclusive definitions. Entropy has been defined as all the energy in the universe by the Standards International. Entropy has been defined as all of the heat in the universe by the heat-based or thermodynamics versions of entropy. Entropy has also been defined as all of the disorder in the universe by the Boltzmann constant and Boltzmann's equation for entropy. The second law of thermodynamics is an amalgamation of all these different contradictory definitions for entropy, as well as a couple dozen more.

By defining entropy axiomatically as all of the energy in the universe, all of the heat in the universe, and all the disorder in the universe, we end up with a *conflation error* logic fallacy or *category error* logic fallacy that makes it impossible to determine what entropy really is because there's no way to determine what entropy is not. The second law of thermodynamics is flawed because it defines entropy axiomatically as everything in the universe, which makes it impossible to measure entropy and do science with entropy, because there is no way to identify what entropy is not. Through the second law of thermodynamics, entropy can literally be anything that you want it to be.

Thermodynamics has absolutely nothing to do with disorder. "Disorder" is the false and falsified definition for thermodynamics, entropy, and the second law of thermodynamics. There's NO way to measure the amount of disorder in a system, because disorder doesn't correlate with anything. We would have to order it and organize it before we could measure it. "Disorder" is the worthless definition for entropy and the second law of thermodynamics, because it can't be measured which means that it's unscientific.

Furthermore, disorder only really applies to a gas. Heat up the gas or cool down the gas, and the amount of disorder remains the same – a bunch of atoms and molecules bouncing off of each other and their container. In contrast, heat up a liquid, a solid, or even a plasma, and often the order is increased as the impurities are removed. A solid, like a neutron star, has a huge amount of structure and order whether its temperature is hot or cold. A plasma like our sun has a huge amount of structure and order. A black hole has a powerful amount of order. There's NO correlation whatsoever between disorder and the amount of heat or thermodynamics in a system. Highly disordered systems can have a ton of heat; and, highly ordered systems can exist and function just fine at absolute zero. Disorder has absolutely nothing to do with the amount of heat or thermodynamics within a system. Disorder doesn't produce heat, and disorder doesn't eliminate heat, either. "Disorder" ends up being the stupid and meaningless definition for heat, thermodynamics, and entropy.

I'm here to tighten up the second law of thermodynamics and find a true and accurate definition for entropy that can then be used to provide us with a true and verifiable version of second law of thermodynamics. If you tear everything down, then you are obligated to build something better in its stead, which I will now do.

$S = Q/\Delta T$

$Q = mc\Delta T$

$S = mc$

Entropy ($S$) is comprised of heat ($Q$), mass ($m$), and heat storage capacity ($c$). Mass has heat storage capacity ($mc$); and, entropy ($S$) is heat storage capacity ($S = mc$), according to these equations for entropy. The temperatures ($T$) cancel out. Entropy remains as a component of heat and mass, no matter what the temperature might be. Thermodynamics is an exchange of heat from a hot reservoir to a cold reservoir. The thermodynamics part of entropy can be accurately defined as the "transfer of heat" from a hot system to a cold system. But entropy is more than that. Entropy is mass's heat storage capacity. Entropy doesn't exist without mass, or heat storage capacity!

Entropy is the heat storage capacity of mass ($S = mc$). This ends up being the best and most believable definition for entropy. It's the conservative and true definition for entropy. This is the mass-based definition for entropy and thermodynamics. No mass, then no entropy. It's that simple.

These equations reveal that at absolute zero (**T = 0 Kelvin**), heat completely ceases to exist, which means that entropy completely ceases to exist along with it. In the massless, heatless, and entropyless Quantum Fields at absolute zero, mass ceases to exist which means that heat ceases to exist, heat storage capacity ceases to exist, and entropy completely ceases to exist. That's why temperature tends toward absolute zero in the Quantum Realm or Spirit World. The Quantum Fields and Quantum Field Theory falsify the second law of thermodynamics. For me personally, this is the greatest scientific discovery of all time. These equations for entropy define what entropy truly is and what entropy is truly comprised of. These equations also tell us how and when entropy goes to zero and ceases to exist.

Instantly, when presented with the truth (**S = mc**), entropy ceases to be some type of mystical magical super-substance capable of constantly increasing in amount out of thin air and BECOMES instead something that is definable, detectable, measurable, credible, and realistic.

It doesn't take a genius to realize that when mass goes to zero (**m → 0**), then heat goes to zero along with it (**Q → 0**). Remove all the mass or resistance to acceleration, then you remove mass's heat storage capacity (**mc → 0**); and consequently, you eliminate all the heat (**Q → 0**) and all of the entropy along with it (**S → 0**)!

It's there in plain sight simply waiting to be discovered by someone who is willing to look and see. Entropy (**S**) is mass (**m**) or heat storage capacity (**mc**). No heat storage capacity or no mass, then NO entropy! Whenever mass or resistance to acceleration ceases to exist, then the associated entropy completely ceases to exist along with it. It's that simple.

This truth also falsifies the second law of thermodynamics which claims that the amount of entropy is always increasing and that the amount of entropy in the universe can never decrease. The false is falsified by the truth, and the truth is repeatedly experienced and observed. Mass, and heat, and the heat storage capacity of mass are constantly being experienced and observed. They are measurable and detectable and verifiable; whereas, something that is constantly increasing in amount has NEVER been experienced nor observed anywhere in this universe.

Remember, ever-increasing substances and infinite disorder are NEVER experienced nor observed anywhere in our universe. The second law of thermodynamics defined as ever-increasing amounts entropy or ever-increasing amounts of disorder can NEVER be experienced and observed. A universal amount of entropy that can never decrease or go to zero has never been experienced or observed, because the Big Bang Theory, Photons, Quantum Waves, and Quantum Field Theory are constantly providing us with examples of syntropy or examples where entropy has gone to zero and ceased to exist. Entropy is constantly going to zero and ceasing to exist according to the equations for heat and entropy.

**S = Q/ΔT**

**Q = mcΔT**

**S = mc**

The variable (**c**) is the specific heat capacity of a particular type of mass. The different elements on the periodic table have different levels of heat storage capacity (**c**). Remember, if there is NO heat (**Q = 0**), then there is NO thermodynamics (**T = 0**), because thermodynamics is by definition a transfer of heat from a hot system to a cold system. No heat, then no thermodynamics. Furthermore, when temperature (**T**) is absolute zero Kelvin, there is no heat. If T = 0, then Q = 0. NO heat (**Q = 0**) then NO entropy (**S = 0**). No transfer of heat from a hot system to a cold system, then no thermodynamics. No thermodynamics (**Q = 0 and T = 0**), then NO entropy (**S = 0**).

**It's really simple to understand!**

https://en.wikipedia.org/wiki/Specific_heat_capacity

https://en.wikipedia.org/wiki/Table_of_specific_heat_capacities

Temperature is irrelevant when it comes to entropy. Temperature simply reveals the presence of heat. The amount of entropy is based exclusively upon the heat storage capacity of the mass. Entropy (S) ends up being the heat storage capacity of mass (mc). When there is NO mass (m = 0), then there is NO heat storage capacity (mc = 0), which means that all of the entropy has gone to zero and ceased to exist (S = 0) because heat has gone to zero and ceased to exist.

This is the simplest and most parsimonious definition for entropy and thermodynamics. That makes it the best and the most truthful definition for entropy and thermodynamics. Here we call it what it really is.

At absolute zero Kelvin, there is NO heat, NO thermodynamics, and therefore, NO entropy. At absolute zero, everything is in perpetual motion and functioning perfectly, thanks to the conservation of energy. At absolute zero, there is no resistance to acceleration or NO entropy. This is what has actually been experienced and observed.

This is the KEY to understanding thermodynamics and entropy. Thermodynamics is the transfer of heat from a hot system to a cold system. No heat, then no thermodynamics. No thermodynamics, then no entropy. This should have been the Second Law of Thermodynamics. This is what's demonstrably true. This is thermodynamics!

At the state that the physicists and scientists call "heat death", there is NO thermodynamics and can be NO thermodynamics, because everything is a cold system, and because everything is in thermal equilibrium. At thermal equilibrium, there is NO heat transfer, which means that there is NO thermodynamics. Consequently, because there is NO thermodynamics at "heat death", there is NO entropy at "heat death". Whenever thermodynamics ceases to exist, the associated entropy ceases to exist, because entropy too is a transfer of heat from a hot system to a cold system. No heat (Q = 0), then no entropy (S = 0). It's that simple.

Entropy IS thermodynamics; consequently, when there is NO thermodynamics, there is NO entropy! When there is NO heat, then there is NO entropy. When the amount of mass or heat goes to zero, the amount of entropy goes to zero along with them.

This should have been the Second Law of Thermodynamics, because this is actually true. It matches perfectly with the equation for heat, and the heat-based equation for entropy. No heat, then no thermodynamics. No thermodynamics, then no entropy. Entropy completely ceases to exist whenever thermodynamics or heat transfer completely ceases to exist. This is what has actually been experienced and observed. Entropy completely ceases to exist whenever a massless photon or a massless and heatless quantum field is made. No mass, then no entropy. No heat, then no entropy. It's really simple to understand.

Entropy should have been defined thermodynamically as a transfer of heat from a hot system to a cold system, but it wasn't. Entropy should have been defined as mass's heat storage capacity, but it wasn't.

Entropy was instead erroneously defined as disorder, randomness, blind luck, death, chaos, and chance. In the ultimate *category error* logic fallacy of all time, they erroneously turned entropy into chance, or randomness, or disorder;

and then, they went one step further and made chance or entropy our Designer, Creator, and God – contrary to the rules of logic, statistics, and science. These people literally turned correlation into causation by turning entropy into disorder, randomness, and chance – and then by turning chance or entropy into a causal agent capable of designing and creating physical matter, genomes, eyes, brains, life forms, planets, stars, and galaxies at will.

It's time to upgrade the Second Law of Thermodynamics to something that is actually real and true.

The NEW and IMPROVED Second Law of Thermodynamics states: Thermodynamics is the transfer of heat from a hot system to a cold system. When there is no heat, there is no thermodynamics. Entropy IS thermodynamics; and, when there is NO thermodynamics, there is NO entropy! Entropy is mass's heat storage capacity ($S = mc$). No mass, then no entropy.

Entropy ceases to exist whenever thermodynamics ceases to exist. No heat, then no entropy. Entropy is the transfer of heat from a hot system to a cold system. This is the thermodynamics definition for entropy. This is what has actually been experienced and observed. This is thermodynamics, which is why this is the true definition for entropy and the REAL Second Law of Thermodynamics. It's true because it's demonstrable, replicable, reliable, parsimonious, logical, conservative, defensible, credible, measurable, realistic, and has actually been experienced and observed. The transfer of heat from a hot system to a cold system has actually been experienced and observed. It's real. It truly exists.

Entropy is mass's heat storage capacity ($S = mc$). This is the physics definition for entropy. This is the mass-based definition for entropy. When mass goes to zero, the associated entropy or resistance to acceleration goes to zero along with it. It's really simple to understand, and it has actually been experienced, observed, and verified. It's there in plain sight for the whole world to see just waiting for someone to come along and state the obvious truth of it.

The science community at-large will never upgrade the second law of thermodynamics to the truth, because it doesn't give them what they want most.

And, what is it that they want most?

They want chance or entropy to be our Designer and our Creator – and not some type of God that they have taught themselves to hate. They deliberately define entropy as randomness, chance, or disorder so that they can then have entropy or chance be our Designer and our Creator.

The science community will never define entropy as thermodynamics, because they prefer the deception and the lie instead. But that doesn't stop the rest of us from finding and embracing a true definition for entropy that has actually been experienced and observed. The rest of us can update and upgrade entropy to be about thermodynamics, even if the scientific community doesn't want to.

We can go on without them. We can define entropy as thermodynamics, and then observe that whenever there is no thermodynamics, there is no entropy. We can go with parsimony and leave the atheistic science community behind in the dark ages where they want to be – constantly defining entropy as Creation by Chance, which is what they want it to be.

You have to understand their psychology if you want to understand why they choose to believe that entropy or chance is our Designer and Creator. I understand it because I used to be a materialist, naturalist, nihilist, and atheist as recently as 2013. I didn't want God to exist. I didn't want to be answerable to God. Therefore, I too decided that I would rather have natural selection, chance, or entropy be my designer and my creator.

It was only after Science convinced me that I was wrong that I finally decided to change my mind and chose to go a different way. I ended up being more of a Scientist than a materialist, naturalist, nihilist, or atheist; and, when Science started to falsify My Materialism, My Naturalism, My Nihilism, and My Atheism, that's when I finally realized that I was wrong, and decided to change my mind, and decided to go with the Science and the Truth instead.

I lost the falsehoods and the lies; but, I gained the truth instead. It was a worthy exchange; and, I'm glad that I finally saw the light.

Mark My Words

--

**Reference Material Regarding Entropy and Syntropy**

If this topic interests you, then see the different essays about entropy as well as the sixteen different competing definitions for entropy as found in my book:

**Quantum Mechanics from a Non-Physical Spiritual Perspective**

https://www.amazon.com/gp/product/B01J023TGU

https://www.amazon.com/gp/product/1521132380

The essays about entropy within that book are somewhere between 200 and 500 pages, depending upon how one chooses to define entropy and its opposite syntropy.

I'm not going to reproduce that here because this book has a different purpose – to demonstrate the usefulness of Science 2.0 and our desperate need to upgrade our science to the modern age.

--

One can get a good sense as to what entropy is by studying its opposite, which is syntropy.

I wrote the book on Syntropy, because nobody else was going to.

Nobody else on this planet is looking at entropy from a massless non-physical perspective. I'm the only one. I'm the first. I pioneered the discoveries that were waiting to be made; and, I've only begun to touch the surface of what is there to be found.

**Syntropy: The Answer to Life, the Universe, and Everything**

https://www.amazon.com/gp/product/B07BPT3W8R

These two books provide a good introduction to entropy and its opposite syntropy. We have to have good, solid definitions for entropy and syntropy before we can KNOW what they truly are and how they work. I have attempted to do so within these two books.

--

I also created a few websites to promote these NEW concepts and ideas.

https://syntropy.site/

https://syntropy.website/

https://evolution-is-entropy.com/

https://evolution-is-entropy.com/entropy-defined/

https://science-2-0.com/

https://quantum-neuroscience.com/

https://origin-science.org/

https://ultimate-model-of-reality.com/

https://markme.us/forums/

https://mypsyche.us/

https://psyche-ontology.com/

https://quantum-law-of-thermodynamics.com/

https://ultimate-law-of-thermodynamics.com/

These websites were created for the people who can't buy my books.

These are some of the fruits of Science 2.0. Science 2.0 allows ALL of the evidence into evidence, and then it pursues a preponderance of that evidence. There are lots of new scientific discoveries to be made, after we have upgraded our science to Science 2.0.

Mark My Words

https://www.amazon.com/author/science

--

Original Source:

https://www.amazon.com/gp/product/B0771K6WTX

## Science Is in Desperate Need of an Upgrade

From what I can tell, ALL of the deceptions, falsehoods, and lies that I have found within science, biology, and physics are the RESULT of Materialism, Naturalism, and their derivatives such as Darwinism (creation by entropy or chance) and Atheism (creation ex nihilo or creation of something from nothing by nothing). The second law of thermodynamics is the cornerstone of Materialism, Naturalism, and Darwinism which is why I have been able to falsify it so easily. Materialism, Darwinism, and Naturalism produce garbage and falsehoods everywhere they go, especially within science, physics, and biology.

The primary deceptions, falsehoods, and lies within science, biology, and physics are centered upon Creation by Chance and Creation by Entropy. Modern scientists need to learn how to eliminate Chance and Entropy as their designer and creator of choice, because it is obvious and clear to most non-scientists on this planet that random chance, blind luck, entropy, and death cannot design and create anything at all. Design and creation by chance, by death, by heat death, by natural selection, or by entropy has NEVER been experienced NOR observed; and, it never will be because it's physically impossible and because these things can't design and create.

Chance, blind-luck, and creation by entropy need to be eliminated from science as causal agents, because Creation by Chance or Creation by Entropy or Creation by Natural Selection is physically impossible. Chance or correlation is NEVER causation; but, the Materialists, Naturalists, Darwinists, Nihilists, and Atheists axiomatically and erroneously define Chance, Entropy, or Natural Selection as our Designer, our Creator, and our God. Biology and physics are in desperate need of plausible, believable, and capable Causal Agents that can actually design and create at will, rather than having blind luck, mindless entropy, and death or natural selection be our designer and creator. Biology and physics desperately need an upgrade from Materialism and Naturalism to actual evidence, true causal agents, and reality instead.

Entropy needed of a fresh pair of eyes from someone who is not religiously and dogmatically devoted to the second law of thermodynamics. The obviously false and the falsified aspects of science need to be removed from the Doctrine of Entropy, so that entropy can come into line with reality and truth.

Science is in desperate need of new, conservative, logical, demonstrable, and believable definitions for entropy. Science is in desperate need of true and accurate replacements for the second law of thermodynamics. But, it's not going to happen while the Materialists, Naturalists, and Atheists control the review process and our public schools. The best we can do for now is to demonstrate clearly and logically why the second law of thermodynamics is false. The Atheists and Naturalists won't let us have the truth in our public schools; but, there is nothing that they can do to stop us from finding and using the truth in our own personal lives.

Unlike some of the other parts of physics, it doesn't require a genius level intelligence to see and understand what's wrong with the second law of thermodynamics. It's really simple to understand. The only reason that nobody has discovered any of this before now is because they weren't looking for it and didn't want to find it.

The official heat-based equation for entropy (**S**) is really simple and easy to understand.

**$S = Q/T$**

It doesn't take a genius to realize that when heat goes to zero ($Q \rightarrow 0$), then entropy goes to zero along with it ($S \rightarrow 0$). Remove all the heat, then you remove all the entropy! In other words, entropy completely ceases to exist and cannot exist at absolute-zero temperature ($T = 0$), among all the massless and entropyless photons and all the heatless and massless quantum fields. This is the Quantum Law of Thermodynamics. Entropy does not exist and cannot exist in the Quantum Realm or Spirit World where there is no heat and no thermodynamics. The truth is hiding in plain sight where nobody can see it, because they aren't looking for it and don't want it.

The official mass-based equation for heat ($Q$) is really simple and easy to understand.

#### Q = mcΔT

It doesn't take a genius to realize that when mass goes to zero ($m \rightarrow 0$), then heat goes to zero along with it ($Q \rightarrow 0$). Remove all the mass or resistance to acceleration, then you remove mass's heat storage capacity ($mc \rightarrow 0$), and you eliminate all the heat ($Q \rightarrow 0$) and all the entropy along with it ($S \rightarrow 0$)!

In other words, entropy completely ceases to exist and cannot exist within massless particles, massless quantum waves, massless photons, and massless quantum fields. Remove the mass, and you remove the entropy, and then the entropy completely ceases to exist.

**Mass is by definition heat storage capacity.**

Massless photons and massless particles have NO heat storage capacity, which means that they have no entropy and can have no entropy. This is the Quantum Law of Thermodynamics. Entropy does not exist and cannot exist in the Quantum Realm, the Massless Realm, the Syntropic Realm, the Conserved Realm, or the Spirit World. This truth is hiding in plain sight where nobody can see it, because they aren't looking for it and don't want it.

Heat death or absolute zero is really good for a quantum system, a non-physical system, a non-local system, or a spiritual system. Heat death or absolute zero actually represents maximum efficiency and perpetual motion in the Quantum Realm, in complete violation of the second law of thermodynamics which equates maximum disorder, maximum entropy, maximum chaos, and maximum inefficiency with heat death.

The thing that they call "heat death" or "maximum entropy" is extremely good for a quantum system and actually maximizes the efficiency, power, speed, and productivity of that quantum system. At the quantum level, ALL of the mass, or entropy, or resistance to acceleration gets transformed into heatless and massless infinite acceleration instead. This has already been experienced and observed. We already KNOW that this is true. All that I have found his is a NEW and better way of interpreting the evidence. Think of photons, quantum fields, and superconductors – all of which experience complete heat death in order to achieve maximum efficiency, maximum speed, and perfect syntropic order. The second law of thermodynamics is a lie. Heat death is a good thing for the Quantum Realm or the Spirit World, because heat death removes ALL of the resistance to acceleration or ALL of the entropy from a quantum system! Heat death completely removes the physical limitations from a quantum system, because there is no mass or resistance to acceleration at true heat death, which means that there is no entropy.

The infinite and eternal expanse of perfectly ordered and perfectly organized, massless, heatless, entropyless Quantum Fields FALSIFIES the second law of thermodynamics, especially the part of the second law that defines entropy as heat death and maximum disorder. At the quantum level, the state that they call heat death is represented by the perfectly organized and perfectly functional Quantum Fields that fill the immensity of space. The Gods or Controlling Psyches actually removed ALL of the chaos or randomness or disorder from the system when they organized and made the Quantum Fields. The result was syntropic perfection – the exact opposite of entropy, disorder, randomness, and chaos. The proven and verified existence of the Quantum Fields falsifies the second law of thermodynamics.

This is really simple to understand. It's really easy to falsify the second law of thermodynamics which states that entropy is always increasing, and that entropy can never decrease. The equations for entropy and heat falsify the second law of thermodynamics by allowing entropy to go to zero and cease to exist. Heat and mass are being absorbed back into the quantum fields all the time, and the associated entropy ceases to exist along with them. This has already been experienced and observed trillions of trillions of times.

The Quantum Fields function flawlessly and are syntropic perfection because they have no mass, no heat, no friction, no resistance to acceleration, no physical limitations, and no entropy. Furthermore, entropic mass or physical matter is a drop in the ocean compared to the entropyless and massless Quantum Fields. Again, the proven and verified existence of the massless and entropyless Quantum Fields falsifies the second law of thermodynamics. At complete heat death or at absolute zero Kelvin, when mass or resistance to acceleration has gone to zero and entropy has completely ceased to exist, we enter fully into the Realm of Quantum Mechanics and the Quantum Fields, which are perfectly ordered and organized all the time.

This should have been discovered centuries ago, but it wasn't, because nobody is looking for it and nobody wants to find it. They prefer to believe in the second law of thermodynamics instead, because the second law implicitly claims that the massless or the non-physical does not exist. The second law of thermodynamics deliberately and axiomatically excludes the massless, the entropyless, and the non-physical from consideration. In other words, the second law does NOT take the massless photons and the entropyless quantum fields into consideration. If it did, then it (and they) would soon realize that entropy is constantly going to zero and ceasing to exist where the Syntropic Realm or the Quantum Realm is concerned.

What I have done here is what should have been done centuries or millennia ago. I pulled out ALL the stops, in order to find and know the truth. The truth was more important to me than being in complete conformity with the majority of the Scientists, Materialists, Naturalists, Darwinists, Nihilists, and Atheists. I'm no longer trying to impress them or be one of them. I've graduated, and I don't need them anymore.

Entropy is purely a mass-based and a heat-based phenomenon. No mass, or no heat, then no entropy. Entropy does indeed prevent the construction of perpetual motion machines at the physical level, because heat is constantly bleeding from physical systems back into the quantum fields from which the heat originally came, thanks to friction, resistance to acceleration, and the ongoing heat transfer from a hot system to a cold system.

In contrast, the Conservation of Energy IS the ultimate perpetual motion machine. It has been going on forever, and it will go on forever. The Energy Realm or the Quantum Realm among the Quantum Fields is pure syntropy, meaning that it has no mass, no heat, no friction, no resistance to acceleration, and no entropy. It is constantly conserved! The Quantum Fields continue to work at maximum efficiency, even when a physical system has achieved heat death. At the quantum level, ALL of the work is done for free thanks to the lack of entropy and the Conservation of Energy.

Consequently, entropy or heat death is NOT maximum disorder, maximum randomness, or maximum chaos at the quantum level. Heat death in the Quantum Realm is maximum efficiency because it has NO physical limitations and is capable of infinite acceleration.

Furthermore, it's a *category error* logic fallacy to define entropy as disorder, randomness, chance, or chaos. By doing that, entropy is no longer about thermodynamics, or an exchange of heat from a hot system to a cold system. Entropy becomes synonymous with chance or randomness or chaos instead, which is a completely different thing.

They don't realize it because they aren't looking for it; but, there's tons of order and organization within mass, heat, heat transfer, and physical matter, even when the hot reservoir and the cold reservoir have reached thermal equilibrium or absolute zero. Thanks to the conservation of energy, the physical matter remains fully functional and completely in existence, even at thermal equilibrium or absolute zero. All of the energy within mass and physical matter is conserved and fully functional at the quantum level, even at absolute zero Kelvin or at thermal equilibrium at the physical level.

According to the equations for heat and entropy, entropy goes to zero and completely ceases to exist at thermal equilibrium and again at absolute zero. In other words, the massless and heatless quantum objects go to maximum speed, maximum power, maximum order, and maximum efficiency at absolute zero when entropy or resistance to acceleration goes to zero and ceases to exist. Order and organization go to syntropic perfection and flawless reliability at heat death, according to Quantum Mechanics and Quantum Field Theory. This is what we have actually experienced and observed.

The fact that we are here in a physical format is scientific proof that this is true. If everything came to a screeching halt at heat death as the Materialists, Naturalists, and Atheists claim, then that would have happened already before the theoretical Big Bang, and we wouldn't be here right now. Big Bangs – or the conversion of raw energy into mass, heat, and entropy – are ONLY possible if heat, mass, and entropy can be transformed back into massless, heatless, and entropyless photons anytime that Nature or Nature's Psyche chooses to do so. Big Bangs are ONLY possible when everything at the quantum level is an indestructible perpetual motion machine to begin with.

That's what's wrong with the second law of thermodynamics. It makes Big Bangs or the reversal of entropy axiomatically impossible. The second law of thermodynamics falsifies the Big Bang Theory and **$E = mc^2$**; and, both the Big Bang Theory and **$E = mc^2$** falsify the second law of thermodynamics. They are mutually exclusive. They both can't be true.

Massless and entropyless Quantum Objects, Superconductors, Photons, Quantum Waves, Psyches, Quantum Fields, and Perpetual Motion Machines work infinitely better at absolute zero than they do in the presence of friction, heat, mass, resistance to acceleration, entropy, and physical matter.

These obvious truths about heat and entropy going to zero and ceasing to exist have already been rejected by the Materialists, Naturalists, Darwinists, Nihilists, Behaviorists, Atomists, Physical Reductionists, and Atheists without even seeing them or thinking about them, because these people have already decided that the massless or the non-physical does not exist.

A true definition for entropy is one of the best-kept secrets in the whole of physics. What science desperately needs and does not have is a Conservative Definition for Entropy that has actually been experienced and observed. During my decades of study and my years of concentrated research into entropy, I found two!

I finally found a couple of true definitions for entropy, within the equations for heat and entropy!

Eureka!

They are true because they are conservative and measurable, which means that they can actually be defined! They are true because they are not too good to be true. They are true because they have actually been experienced and observed. They are true because they actually make logical sense and do not involve infinities within their equations and definitions. They are true because they actually allow entropy to decrease to zero and stay there. They are true because they are constantly seen, experienced, observed, measured, and verified. They are true because they falsify the second law of thermodynamics.

I discuss these two true definitions for entropy in the next couple of essays, so I won't do so in as much detail here.

The true identity of entropy and the true definitions for entropy are found within the equation for heat and the heat-based equation for entropy. They are true, because heat has been experienced, measured, and observed. Heat is something that we actually understand to a great degree. Unlike infinite amounts of entropy or infinite amounts of disorder or infinite chaos, heat is something that we actually some experience with.

The Materialists, Naturalists, Darwinists, Nihilists, and Atheists erroneously teach, preach, and believe that entropy and physical matter are conserved. The second law of thermodynamics is Conservation of Entropy or Conservation of Disorder. The Ultimate Law of Thermodynamics states that entropy and physical matter are NOT conserved. Using Quantum Mechanics, the Conservation of Energy and Psyche, Quantum Field Theory, and the Quantum Law of Information Conservation, the Ultimate Law of Thermodynamics states that ONLY energy, quantum information, and the psyches or intelligences that contain that quantum information are conserved. Everything else is NOT conserved.

Anything that is made can also be unmade, which means that it is NOT conserved. According to Quantum Field Theory, particles are made, and particles can be unmade, which means that particles of any kind are NOT conserved. Physical matter is NOT conserved. Entropy is NOT conserved.

Physical matter is the product, NOT the Creator or the Maker. Entropy is a by-product of mass or matter, which means that entropy is a product, and NOT the Creator or the Maker. By the conventional definition, entropy is chance, randomness, or disorder. Entropy is an emergent property of randomly distributed physical atoms. No physical atoms, then no entropy. Entropy or chance is NOT the maker of the physical atoms, as the Materialists and Naturalists claim. Entropy is a by-product of mass, NOT the creator of mass. Once we have falsified the Conservation of Entropy or the second law of thermodynamics, then we are free to pursue and discover the truth instead.

The Ultimate Law of Thermodynamics is an attempt to fix the flaws associated with Materialism, Naturalism, Darwinism, Nihilism, Atheism, Creation by Chance, Creation by Entropy, the Conservation of Entropy, and the Second Law of Thermodynamics. Natural selection or the theory of evolution is creation by chance or creation by entropy. According to the Materialists, Naturalists, and Atheists, the DNA, RNA, genes, proteins, genomes, life-forms, physical matter, planets, stars, and galaxies were ALL created by chance or entropy. These people really truly believe that chance or entropy can design and create anything that it sets its mind to.

Notice carefully that the Ultimate Law of Thermodynamics states that ONLY the conserved can serve in the capacity of a Designer and a Creator, because the product or the thing that is made is by definition un-conserved and incapable of functioning as the designer and creator. Logic tells us that the thing that is made cannot be the Ultimate Cause or the Ultimate Original Creator. The unconserved cannot function as a designer and a creator, because the unconserved is the thing that is made. That means that physical matter and entropy cannot function as a designer and a creator because they are not conserved. Mass and entropic physical matter are the things that were made, and NOT our Designer and Creator.

The Ultimate Law of Thermodynamics states that entropy and physical matter are not conserved; therefore, they can't be designers and creators.

According to the Ultimate Law of Thermodynamics, Energy, Quantum Information, and Psyche or Intelligence are the ONLY things in this universe that are truly conserved; therefore, they are the ONLY things that can legitimately function as Designers and Creators. So, which one of them is the Designer and the Creator? Which one of them collapses the wave function?

Energy is the substance from which everything is made. Quantum Information is knowledge that is stored as energy within some type of Psyche, Intelligence, or Life Force. That leaves Nature's Psyche or Nature's Intelligence as the Ultimate Cause, the Ultimate Creator, the Ultimate Thing that collapses the wave function, and the Ultimate Thing that makes the quantum fields, the quantum waves, the photons, the forces, and the physical matter. So, who taught Nature's Psyche how to do all of these different things? It would have to be some type of other Psyche or Intelligence.

This is what we get from Psychology (the study of the human psyche), Quantum Mechanics (the study of energy), Quantum Field Theory (the study of the things made by Psyche or Intelligence), and the Philosophy of Science (the study of ultimate causality).

The human psyche isn't consciously aware of any of these things; so, it's obvious that it is in fact Nature's Psyche who is collapsing the wave function. And, it's obvious that Nature's Psyche was collapsing the wave function long before the first particle of physical matter was designed and made.

We get ALL of this from the Ultimate Law of Thermodynamics simply by differentiating between what is conserved and what is not conserved. The Materialists and Naturalists and Darwinists claim that entropy and entropic physical matter are conserved. The Ultimate Law of Thermodynamics demonstrates and explains why these people are wrong. The second law of thermodynamics is the Conservation of Entropy; and, the Ultimate Law of Thermodynamics falsifies the Conservation of Entropy. They are mutually exclusive, which means that if one can be demonstrated to be true, then the other one has been falsified.

So, which one of these is demonstrably true? Which one has actually been experienced and observed – the Conservation of Entropy or the Conservation of Energy? Is entropy really conserved, or is its amount constantly changing?

The whole of science, observation, and experience demonstrates clearly and conclusively that the Materialists, Naturalists, Darwinists, Nihilists, and Atheists are always wrong; and, these people don't even know it because they refuse to look at it and think about it.

Materialism, Naturalism, Darwinism, Atheism, and the Second Law of Thermodynamics were designed to stop you from finding and knowing the truth. Therefore, you must be willing to pull out all the stops if you truly want to find and know the truth.

You must be willing to allow all of the evidence into evidence, or you will never find and know the truth.

Likewise, if your ultimate goal is to find and know the truth, then you must be willing to identify and eliminate everything that is false – starting with Materialism, Naturalism, Darwinism, Nihilism, Behaviorism, Determinism, Physical Reductionism, Scientism, Atheism, and Creation by Chance or Entropy.  You can't know the truth if you have deliberately embraced everything that is false instead.

Finding and knowing the truth is what the Philosophy of Science is supposed to be all about.  We must find and eliminate the false philosophies if we truly want to find and know the truth.  We must be willing to eliminate everything that has been falsified, or there's no way for us to improve and upgrade our Science in general.  Turning falsehoods and lies into axioms and laws is precisely the exact opposite of what we should be doing if we truly want to find and know what is true.

From what I can tell, ALL of the falsehoods in science, physics, and biology derive from their attempt to make entropy or chance our Designer, Creator, and God.  ALL of the falsehoods in science, physics, and biology come from the philosophical religions that we call Materialism, Naturalism, Darwinism (Creation by Entropy), and Atheism (Creation by Chance).  Their purpose in life is to trick us and deceive us, and they are very good at what they do.  I fell for it, didn't you?

We are easily deceived when we want to be deceived.

Mark My Words

## Dark Matter and Dark Energy – Proof of the Non-Physical

I was surprised when I was first taught that over 95% of our physical universe is in fact non-physical. I had no idea, because that's not what I had been taught. One cannot know the truth if he or she is taught falsehoods instead.

The estimates for the amount of dark matter and dark energy in this universe are constantly changing.

https://en.wikipedia.org/wiki/Universe#Composition

Today's (27JUN2019) estimate for the composition of our universe is:

**Dark Matter or Spirit Matter – 26.8%**

**Dark Energy or Quantum Fields – 68.3%**

**Physical Matter – 4.9%**

Tomorrow, the guestimates will be different, and the Wikipedia article will be different. The Scientific Method is a constantly moving target. There's nothing secure or sure about it.

By definition, dark matter and dark energy are invisible and non-physical. They are purely spiritual or quantum instead of being physical.

Who would have thunk it?

We aren't taught this information in our public schools by our atheistic, naturalistic, and materialistic teachers because the proven and verified existence of anything non-physical – such as photons or quantum fields – FALSIFIES Materialism, Naturalism, and their derivatives such as Atheism (Creation Ex Nihilo) and Darwinism (Creation by Death and Entropy).

The BEST and the FASTEST way to find and know the truth is to observe it, experience it, and live it for yourself – or to choose to trust someone who has. Science is observation and experience after all. Science is knowledge. Observing the truth, experiencing the truth, and living the truth are infinitely better and more reliable than the Scientific Method. Due to faulty and false interpretations of science experiments and scientific data, millions if not billions of people have been deceived by Science and by the Scientific Method; and, they don't even know it.

In December 2015, at age 54, I finally found my first piece of convincing observational and experiential proof that the immaterial or the non-physical does indeed exist. I believed it because I found it convincing.

https://psyche-ontology.com/buhlman/

You cannot find and know the truth if you are constantly rejecting it whenever you encounter it. I didn't think that there was any evidence to support the existence of the non-physical, so I really wasn't looking for it. I didn't believe that it existed. I was wrong. The vast majority of our universe is non-physical, or non-local, or quantum. NOW I KNOW.

Once I found one proof of the non-physical, then I quickly found another, and then another, and then another. Now here I sit today KNOWING that over 95% of our universe is non-physical. Now I KNOW that Classical Realism, Materialism, Naturalism, Darwinism, Nihilism, Behaviorism, Atomism, Determinism, Physicalism, Spontaneous Generation,

Abiogenesis, Chemical Evolution, Macro-Evolution, Creation Ex Nihilo, and Atheism are FALSE because now I KNOW why they are false and have successfully falsified each one of them. These things are demonstrably false because they claim as their primary axiom that the non-physical does not exist. The proven and verified existence of photons, electromagnetic waves, quantum waves or thoughts, quantum information, action at a distance, instantaneous communication at a distance, psyche or intelligence, and quantum fields FALSIFIES each one of these false religions or false philosophies in one way or another. Does it not?

The false is falsified by the truth, and the truth is repeatedly experienced and observed. Science is observation and experience after all – or it should be.

Once I finally started looking at Science and Scientific Evidence from a non-physical perspective, then I started finding evidence of the non-physical right and left. It was unavoidable, because over 95% of our universe is non-physical. In other words, the Materialists, Naturalists, Darwinists, and Atheists deliberately reject and deny the existence of over 95% of our universe. This 95% is a measure of how wrong the Materialists and Naturalists really are. One cannot find and know the truth by actively and deliberately rejecting it.

This book is my personal attempt to find and know the truth; and if you accept what I have written here, then we are already off to a good start.

Mark My Words

**Dark Energy**

Once I discovered the Quantum Fields and Quantum Field Theory, then Dark Energy was really easy to understand and accept.

Dark energy is just a different unique type of quantum field. It fills the immensity of space, and it permeates everything or is within everything. Dark energy seems to be a variable quantum field. Dark energy seems to be actively involved in expanding space, opening up space, stretching space, and making new space. Dark energy organizes and makes space.

Dark energy provides new places where the Gods or the Controlling Psyche can phase-shift new galaxies into existence. It doesn't matter where the Gods choose to place a new-born galaxy. No matter where that new-born galaxy is phase-shifted into existence and made physical, sometime in the future, it will seem as if every galaxy in the universe is moving away from that new-born galaxy, thanks to Dark Energy.

Nature or Nature's Psyche can make a brand-new galaxy anywhere and anytime that it chooses to do so, thanks to the quantum fields and dark energy.

Like every other quantum field, Dark Energy is massless, entropyless, heatless, intangible, non-physical, and omnipresent. Every quantum field has a unique design, a unique purpose, and unique functionality. Dark energy is one of these quantum fields.

Dark energy is unique in that it shows up in the cosmic density of our physical universe. Dark energy is indirectly measurable by measuring and weighing our physical universe.

We KNOW that dark energy is there because of the influence that it has on our physical universe. That's also why we KNOW that magnetism and gravity exist. These, too, are non-physical intangible quantum fields; yet, WE KNOW that they exist because of the influence that they have on physical matter.

I'm a physicist, mathematician, scientist, and quantum theorist. Dark energy and dark matter fall under my purview; and, I have spent a lot of time trying to figure out what they truly are and how they work. I'm uniquely qualified for doing so, because I'm one of the few scientists on the planet who is willing to take the non-physical or the non-baryonic into consideration and treat the stuff as if it were real.

I used to be a materialist, naturalist, nihilist, and atheist, until Science, Dark Energy, Quantum Mechanics, Quantum Information or Quantum Intelligence, and Quantum Field Theory convinced me that I was wrong. Now, I'm no longer afraid of studying science and physics from a non-physical, non-baryonic, spiritual perspective instead of the exclusively physical perspective that everyone else has chosen to pursue.

I continue to make NEW and interesting scientific discoveries every day because I have chosen to study science and physics from a non-physical, massless, entropyless, spiritual, quantum perspective.

The massless, the entropyless, and the non-physical are completely untapped and unexplored where science and physics is concerned. It's completely undiscovered territory, and one can make new scientific discoveries every day by choosing to pursue the non-physical or the non-baryonic within science and physics. It's a new playground for a quantum theorist like me, and I love it.

Mark My Words

**Dark Matter**

Many physicists continue to deny the existence of Dark Matter or Spirit Matter.

https://www.youtube.com/watch?v=miGddxrvmDU

I, too, couldn't wrap my mind around Dark Matter.

For a long while, I called it non-physical matter or spirit matter. The stuff is clearly and obviously intangible and invisible, which typically are synonymous with non-physical.

However, Dark Matter is used by the Controlling Psyches or the Gods to keep our physical galaxies from flying apart and dissolving into the expanse of space.

Dark Matter actually participates in the equation for gravity!

Newton's universal law of gravitation: **$F = Gm_1m_2/r^2$**, where **F** is the force due to gravity, between two masses (**$m_1$** and **$m_2$**), which are a distance **r** apart; **G** is the gravitational constant.

Dark matter is **$m_2$**.

Do you know what that means?

It means that Dark Matter has mass or resistance to acceleration. Dark Matter is non-physical or non-baryonic, yet it also has mass or resistance to acceleration; and, we typically define "mass" as physical matter. If Dark Matter has mass, then it also has some type of heat storage capacity, which means that it is also subject to entropy. Dark matter is an oxymoron – a contradiction in terms. Yet, we KNOW that the stuff exists, because all of our galaxies remain in one piece rather than flying apart.

Then I begin to wonder. Is Dark Matter some type of intangible invisible physical matter, rather than being spirit matter? Or, are there many different types of spirit matter, with Psyche and Dark Matter being a couple of them? Clearly, Dark Matter has mass or resistance to acceleration; but, is it simultaneously entropyless and heatless like the other types of spirit matter and the quantum fields?

Psyche is the ultimate point particle – an infinite singularity, made out of pure exergy. Psyche is a type of spirit matter – a massless, entropyless, and heatless type of spirit matter. But, what is Dark Matter? Dark Matter seems to be real matter with real mass, even though it is intangible, invisible, and non-physical or non-baryonic at the same time. Dark matter has mass or resistance to acceleration like physical matter; yet, dark matter seems to be completely intangible, invisible, non-physical, and non-baryonic like spirit matter.

https://en.wikipedia.org/wiki/Baryon

Baryon means "heavy", which indicates that the other mass-based particles have less mass or less resistance to acceleration than the baryons. That seems to be the case with Dark Matter. Dark Matter has noticeably less mass than baryonic matter or physical matter. Consequently, there has to be a lot more Dark Matter than physical matter, for the Dark Matter to have the same gravitational effect that physical matter or baryonic matter has.

Dark Matter seems to be spirit matter, or intangible and invisible matter, that was deliberately designed to have mass, locality, stability, and resistance to acceleration.

Once I finally understood Quantum Phase-Shifting and how it works, then I became comfortable with the idea that Dark Matter is physical matter that has been phase-shifted into a different dimension. Since Dark Matter clearly exists in a different phase or a different dimension, it ends up being spirit matter as well as being a type of physical matter.

According to Quantum Phase-Shifting, dozens of different physical earths like ours can exist in the same space at the same time – just slightly out-of-phase with each other. That's how Jesus was able to walk through walls. The Gods can do phase-shifting. If we could do phase-shifting at will, then we too could walk through walls or put our hand through the wall un-phased.

Phase-shifting is common knowledge among the Quantum Physicists and Out-of-Body Explorers; but, it's unheard of among the Materialists, Naturalists, Darwinists, Nihilists, and Atheists, because these people have chosen to believe that ONLY baryonic matter exists. They only believe in the stuff that they can see with their eyes and get their hands on. These people don't believe in Dark Matter, or Dark Energy for that matter. These people don't believe in spirit matter or phase-shifted matter. These people don't believe in the ultimate point particle that we call Psyche or Intelligence. Their primary axiom erroneously states that the non-physical or the non-baryonic does not exist. The whole of physics tells us that they are wrong; but, that doesn't change what they have chosen to believe.

Dark Matter is phase-shifted physical matter, which means that Dark Matter is a type of spirit matter. Dark Matter exists in a different phase or a different dimension. Dark Matter has been phase-shifted into the Quantum Realm, Non-Baryonic Realm, or Spirit World. Dark Matter exists as a type of spirit matter in the Spirit World; yet, its mass or resistance to acceleration seems to function in our physical realm

Spirit matter seems to have a little bit of mass or resistance to acceleration – like an electron – but, spirit matter seems to be able to quantum tunnel anywhere at will as well, just like an electron. An electron is a type of spirit matter. An electron is half in the physical realm and half out of the physical realm. Dark Matter seems to be all the way out of our physical realm; yet simultaneously, its mass or resistance to acceleration seems to be totally within our physical realm.

Unlike an electron which spends half of its time being omnipresent and infinite velocity, Dark Matter seems to be localized and relatively stationary like physical matter. Dark Matter seems to function more like physical matter than the stuff that we typically call spirit matter; yet at the same time, Dark Matter is completely invisible, intangible, and non-baryonic just like spirit matter. Dark Matter seems to be physical matter that has been phase-shifted into the Spirit World, the Transdimensional Realm, or the Quantum Non-Physical Realm.

Like an electron, Dark Matter seems to be spirit matter and physical matter all at the same time. Dark Matter seems to be phase-shifted physical matter; and, physical matter seems to be phase-shifted spirit matter.

Mark My Words

## The Brain Is MAPPED

Is prejudice socialized or conditioned into us, part of our genetic inheritance, or something that we bring with us from our pre-mortal life as part of our psyche or spiritual makeup?

YES!

According to the BioPsychoSocial Model, the answer is all three. Each one contributes its part to the overall picture. We mortal beings are BioPsychoSocial beings.

### Introduction to Brain Functionality

Brain functionality has been MAPPED. Different parts of the physical brain have been MAPPED to provide different types of physical functionality. This reality has been experienced and observed. Two of the MAIN MAPS are what the Neuroscientists call the Sensory Homunculus Map and the Motor Homunculus Map.

https://en.wikipedia.org/wiki/Cortical_homunculus

https://quantum-neuroscience.com/wp-content/uploads/2018/05/Cortical-Homunculus.pdf

It's real. It's true. Brain functionality has been MAPPED.

Scientists, Neuroscientists, and Psychologists have caught these different MAPS being used in real life by real people

> **In some situations, automatic, implicit prejudice can have life or death consequences.**
>
> **It also appears that different brain regions are involved in automatic and consciously controlled stereotyping. Pictures of outgroups that elicit the most disgust (such as drug addicts and the homeless) elicit brain activity in areas associated with disgust and avoidance.**
>
> **This suggests that automatic prejudices involve primitive regions of the brain associated with fear, such as the amygdala, whereas controlled processing is more closely associated with the frontal cortex, which enables conscious thinking.**
>
> **We also use different bits of our frontal lobes when thinking about ourselves or groups we identify with, versus when thinking about people that we perceive as dissimilar to us.**
>
> **Even the social scientists who study prejudice seem vulnerable to automatic prejudice.** (*Social Psychology*, pp. 314-315.)

A physical brain is MAPPED!

What do we know about maps in general?

We know that maps do not spontaneously generate out of thin air.

Maps require some kind of Map-Maker.

Brain functionality has been MAPPED. Mapped by whom? Where are these MAPS being stored? How are these MAPS created, and how are they being used? Who is using these MAPS?

Maps have no value if they aren't being used by Someone Psyche or Someone Intelligent to get something useful done.

### The Map-Maker Argument

The following syllogism summarizes this scientific discovery:

**Scientific Observation: Every type of MAP has a Map-Maker who made it. MAPS don't just spontaneously generate out of thin air.**

**Scientific Observation: It's obvious that each part of a physical brain has been carefully MAPPED to perform a specific physical function.**

**Scientific Conclusion: Therefore, every physical brain has a Map-Maker, who MAPPED each part of that brain for specific physical functions.**

This is a syllogism. If the premises or observations are true, then the conclusion has to be true as well. This syllogism is philosophically valid and observationally sound.

### Your Primitive Brain and Quantum Maps of Physical Functionality

There's nothing primitive about your brain. ALL the functionality had to be there from the very beginning, or it would have had no survival value and the human race would have gone extinct. However, Nature's Psyche (or God's Psyche) has indeed MAPPED different parts of your brain at the quantum level so that your brain can perform different types of physical functionality at the physical level.

The MAPS are created and stored at the quantum level by Nature's Psyche. We KNOW this is so, because there's not enough memory storage capacity within our 750-megabyte genome to store and process the petabytes of MAPPING and INTERFACING that's taking place both at the quantum level and the physical level simultaneously. It's physically impossible to store petabytes of Mapped Functionality within our 750-megabyte genome.

**The 2.9 billion base pairs of the haploid human genome correspond to a maximum of about 725 megabytes of data, since every base pair can be coded by 2 bits.** — Uncle Google.

You can't shove petabytes of QUANTUM MAPPING and/or SYNAPTIC MAPPING into a 725-megabyte to 750-megabyte genome. It's physically impossible. But, there's seems to be NO memory storage limits at the quantum level or the psyche level. The quantum doesn't have any physical limitations. That's the primary message and lesson of Quantum Mechanics – namely, quantum mechanisms have NO physical limitations.

Some parts of the brain were MAPPED by Nature's Psyche to register physical input or physical sensations from the environment, and other parts of the brain were MAPPED by Nature's Psyche to register and then manifest the Human Psyche's feelings, emotions, desires, choices, and actions onto the physical plane. We KNOW that it was Nature's Psyche who did this MAPPING for us at the quantum level, because the Human Psyche isn't consciously aware of these things. This QUANTUM MAPPING and SYNAPTIC MAPPING is being handled automatically for us by Nature's Psyche, or by God. Some parts of the brain are consciously controlled by the Human Psyche; and, other parts of the brain are controlled automatically for us by Nature's Psyche. Remember, ALL of the quantum mapping and physical functionality of the brain is handled by Nature's Psyche outside the conscious awareness of the Human Psyche.

Nature's Psyche organized the physical brain structure and then MAPPED the different neurons, glial cells, and synapses at the quantum level to have a specific meaning, purpose, or goal at the physical level. Then Nature's Psyche uses those Quantum Maps of Physical Functionality to get things done for us at the physical level by collapsing the necessary wave functions and/or turning on the necessary neurons from the quantum level whenever the Human Psyche makes a choice and wants to get something done at the physical level.

I was very pleased with the explanatory power of Quantum Maps of Physical Functionality when it was first revealed to me. It made sense to me and basically explained everything that we are observing and experiencing when it comes to a physical brain and brain functionality. I think Quantum Maps of Physical Functionality is my favorite scientific discovery, due to its massive explanatory power. I finally feel like I understand how everything really works at every level of existence.

Quantum Maps of Physical Functionality also match perfectly with the Orthodox Interpretation of Quantum Mechanics by Henry P. Stapp, which explains that it is the Human Psyche who makes the choices and it is Nature's Psyche who collapses the necessary wave functions and/or turns on the necessary neurons in order to make that CHOICE actual and real in the physical world.

I eventually concluded that Quantum Maps of Physical Functionality is the BEST explanation for brain functionality that we have discovered to date. It explains what Nature's Psyche and the Human Psyche are doing for us at the quantum level in order to get things done for us at the physical level. Quantum Maps of Physical Functionality provide the interface between the quantum level and the physical level; and, that's all I really needed to know and have.

Quantum Maps of Physical Functionality also match perfectly with and explain the origin of the Sensory Homunculus Map and the Motor Homunculus Map, which the Neuroscientists have discovered at the physical level. Nature's Psyche (or God) creates these MAPS or INTERFACES at both the quantum level and the physical level, and then Nature's Psyche uses these MAPS at the quantum level in order to get things done for us at the physical level. Quantum Maps of Physical Functionality explain everything that I learned about neuroscience, brain functionality, psyche, consciousness, and quantum mechanics. Quantum Maps of Physical Functionality combine everything into one realistic whole, both at the quantum level and the physical level.

Remember, there's nothing primitive about your brain.

If parts of your brain are missing or fail to develop, Nature's Psyche can still MAP physical functionality onto the parts of the brain that remain. It's called Neuroplasticity; and, Neuroplasticity is a proven and verified part of Neuroscience. A properly developed brain really isn't necessary for a fully functional brain. During development, Nature's Psyche can MAP all functionality to the parts of the brain that are there, while working around those that are not.

https://quantum-neuroscience.com/wp-content/uploads/2018/05/Is-Your-Brain-Really-Necessary.pdf

Once the MAPS and axonal connections have been established, though, it's basically a done deal. If you cut the axons or destroy the MAP, there are times when the physical functionality does not return. However, depending upon the location and the severity of the damage, there are other times when the physical functionality gets successfully REMAPPED at both the quantum level and the physical level by Nature's Psyche. It all depends upon the situation, as well as the determination of the Human Psyche.

If the Human Psyche is determined to move its finger again, Nature's Psyche can indeed be encouraged to MAP that physical functionality onto a different part of the physical brain. It doesn't always work, but sometimes it does. Sometimes the axons don't regrow and remap properly. Other times, they do. If the axon is still there but the neuron dies during a stroke, that axon can sometimes be REMAPPED to a different neuron or a different part of the brain. It's hit and miss. Sometimes it works, and sometimes it doesn't. This is what has been experienced and observed.

Oligodendrocytes in the brain and spine tend to block, resist, and restrict axonal regrowth and neural remapping. It's as if the central nervous system has been set in plastic. Remapping can be done, but not as easily as it is done in the peripheral nervous system. The Schwann cells in the peripheral nervous system tend to facilitate axonal regrowth and subsequent remapping. The axonal regrowth doesn't always take place and isn't always done right; but, it's more likely to be done in the peripheral nervous system than the central nervous system. The peripheral nervous system is much more fluid and flexible.

In both cases, where there's a WILL (a Human Psyche), there is often a way, even within the physical brain. Sometimes stroke victims can indeed recover and successfully make Nature's Psyche remap their physical brain. Sometimes it doesn't work, or there isn't enough time to make it work before the person dies. Other times, the results seem to be miraculous.

As a teenager, I cut the nerves in the lateral part of my left thumb; and, that side of my thumb was numb and weird feeling for a long time. Forty years later, everything is back to normal. I'm only marginally aware of it if I'm actually thinking about it. Otherwise, the former damage goes unnoticed.

Remember, your brain, brain functionality, and nervous system have been MAPPED. Mapping requires some sort of intelligent Map-Maker in order to get the job done right. Maps don't just spontaneously generate out of thin air. Spontaneous generation or creation ex nihilo is impossible. It can't happen, which means that it didn't happen. The MAPS within your physical brain required some kind of intelligent Map-Maker, or they wouldn't exist, and they wouldn't work right if they did exist.

Mark My Words

—

**Source**

**Quantum Mechanics from a Non-Physical Spiritual Perspective.**

https://www.amazon.com/dp/B01J023TGU

https://www.amazon.com/dp/1521132380

***Scientific Proof of God's Existence: Finding God Where the Atheists Refuse to Look for Him***

    https://www.amazon.com/dp/B07B26CRHX

***The Ultimate Model of Reality: Psyche Is the Ultimate Cause***

    https://www.amazon.com/dp/B071NC9JK6

**References**

***Quantum Neuroscience: The Answer to Life, the Universe, and Everything***

    https://www.amazon.com/dp/B079Z6QQQB

Myers, D. G. (2010). *Social Psychology* (10th ed.). New York: McGraw-Hill.

## We Find the Truth by Living It and Experiencing It

Some might argue that my greatest scientific observation and scientific discovery is the realization that Lived Experience IS the BEST way for finding and knowing the truth.

I spent a year studying the Philosophy of Science. In the process, I was introduced to Phenomenology. Phenomenology is the scientific study of events or phenomena, including Out-of-Body Experiences (OBEs), Near-Death Experiences (NDEs), Shared-Death Experiences (SDEs), Visions, Theophanies, and other types of Spiritual Experiences. These are phenomena that have been experienced and observed by millions of different people on this planet. Phenomenology is the scientific study of the Lived Experiences of the human race, including both our spiritual experiences and our physical experiences.

Begging the question, Science is currently defined as Materialism, Physicalism, Naturalism, Darwinism, Nihilism, Classical Physics, and Atheism before any science is actually done. Thanks to Phenomenology and the Lived Experiences of the human race, I realized one day that Science should be redefined as Observation and Experience. In other words, Science should be brought into the modern Quantum Age of Action at a Distance, Phenomenology, Observation, and Experience rather than being left in the restrictive and limited dark ages of Materialism, Naturalism, and Atheism. Materialism, Naturalism, Darwinism, Nihilism, and Atheism have been falsified. Science should be taken out of the falsified and placed into the verified or the observed. I realized that Science should be redefined as Phenomenology.

As a result, I upgraded my science to Science 2.0 which allows all of the evidence (all of the phenomena) into evidence and pursues a preponderance of the evidence.

### Science 2.0: I Upgraded My Science

https://www.amazon.com/dp/B0771K6WTX

Lived Experiences or Psyche Experiences are the best way of finding the truth and knowing the truth in EVERY realm of existence, including this physical realm. The BEST and fastest way to find and know the truth is to live it, experience it, and observe it for yourself, or to choose to trust someone who has. Science is Observation and Experience, and NOT Materialism and Naturalism which are based exclusively upon philosophical speculation and wishful thinking. In other words, there is NO observational evidence supporting the major premises or hidden assumptions of Materialism and Naturalism which claim that the quantum or the transdimensional does not exist.

Due to the fact that the scientific methods cannot be used to find the truth and prove the truth directly, and due to the fact that the scientific methods are typically restricted to the physical realm or the local realm, and due to the fact that there are a wide variety of logic fallacies built into the scientific methods, the scientific methods are in fact a much weaker source of knowledge and truth than Lived Experiences are. Lived Experiences, or Psyche Experiences, or Direct Observations are in fact an infinitely better way for finding and knowing the Truth than science and the scientific methods can ever be. I wish I would have known that forty years ago. It would have saved me a lot of frustration and grief.

Possibly my greatest scientific discovery of all time is that the Lived Experiences of the human race are in fact a much better way of finding and knowing the truth than traditional science and the scientific methods. For me, that was a major epiphany and scientific breakthrough because I used to be a Materialist, Naturalist, Nihilist, and Atheist.

Lived Experience is extremely powerful. Lived Experience is Science, Observation, Knowledge, and Truth in their purest form. Science is supposed to be Observation and Knowledge and the pursuit of the Truth. In other words, Science is supposed to be Lived Experiences or Observations. The Materialists, Naturalists, Behaviorists, and Atheists did the world a great disservice when they hijacked and stole science and limited science exclusively to physical matter. One of my greatest scientific discoveries was to finally realize how wrong these people really are.

I KNOW how it goes, though, because I allowed My Scientism, My Nihilism, My Materialism, and My Atheism to blind me to Lived Experience as a source of evidence and to blind me as to how powerful and useful Lived Experiences really are as a source of knowledge and truth. But, that's what we Materialists and Atheists do. We refuse to allow Lived Experiences into evidence. Naturalism and Atheism of any kind is based upon a refusal to look at evidence.

I was a practitioner of Scientism. I truly believed that Science and the scientific methods were the ONLY way for finding and knowing the truth. I WAS WRONG! In fact, the scientific methods can't be used to prove the truth, which means that Lived Experience is in reality the ONLY way to find and KNOW the truth directly. The truth is KNOWN by living it and experiencing it for yourself, or by choosing to trust someone who has.

I never fully realized until April 2017 that Lived Experience or Direct Observation is a much better way of finding and knowing the truth than the scientific methods and philosophical guesswork will ever be; but, I'm seeing it now and understanding it now.

We KNOW from the Lived Experiences of the human race that the Biblical God Jesus Christ does indeed exist and truly rose from the dead – although it's nice to have Science pointing us to the same truths, in its roundabout way.

For some Lived Experiences of the Divine see: *Bible, Book of Mormon: Another Testament of Jesus Christ, Doctrine and Covenants,* and *Pearl of Great Price* – available for free at:

https://www.lds.org/scriptures/bible?lang=eng

Possibly my greatest scientific discovery came to me when I first realized that Science should be redefined as Phenomenology, Lived Experience, and Observation. Quantum Mechanics is Supernatural Mechanics. Science should be based upon Syntropy, Psyche, and Quantum Mechanics – and NOT Materialism and Naturalism.

Currently, the scientific community at-large define science as Materialism, Naturalism, Darwinism, Nihilism, Behaviorism, Determinism, Physical Reductionism, Classical Physics, Entropy, and Atheism. It's time for us to upgrade science and bring it into the modern age.

It's time for us to redefine Science as Phenomenology, Observation, Experience, Eye-Witness Verification, Quantum Mechanics, Syntropy, Psyche, and the Lived Experiences of the human race. We'll make much faster progress if we do. I KNOW because I have lived it and experienced it on both sides of the fence; and, the grass really is greener on the Syntropy side or Psyche side of the fence.

Mark My Words

## Science Is Observation and Experience

Possibly my most useful and my most powerful scientific discovery is the fact that Science is Observation and Experience, and NOT Materialism, Naturalism, Darwinism, Nihilism, and Atheism.

Most people define science as Materialism and Naturalism. It's a scam! Materialism, Naturalism, and their derivatives are based exclusively on blind-faith and wishful thinking. There is NO evidence supporting the major premises or hidden assumptions of Materialism and Naturalism, which state that the quantum or the supernatural does not exist. Everything that has been experienced and observed falsifies Materialism, Naturalism, Darwinism, Nihilism, and Atheism.

For me personally, this was a major scientific discovery because I used to be a Materialist, Naturalist, Nihilist, and Atheist.

With this discovery or observation, I experienced a paradigm shift and developed a new and better way of doing science, which I call Science 2.0. Science 2.0 is based upon the observed, verified, experienced, and proven aspects of Quantum Mechanics rather than Materialism and Naturalism.

Quantum Mechanics, or Syntropy, or Psyche is the answer to life, the universe, and everything. Syntropy or Quantum Mechanics falsifies Materialism, Naturalism, and their derivatives.

Science is in desperate need of upgrading because Materialism, Naturalism, Darwinism, Nihilism, and Atheism have NO explanatory power when it comes to Quantum Mechanics, Quantum Waves, Near-Death Experiences, Out-of-Body Experiences, Non-Locality, Non-Physicality, Transdimensionality, Consciousness, Thoughts, Dark Matter or Spirit Matter, Dark Energy, Syntropy, Psyche, Mysticism, Quantum Tunneling, the Quantum Zeno Effect, Action at a Distance, and invisible non-physical forces and fields such as Gravity and Magnetism and the Zero-Point Field of Light. Science needs to be upgraded and brought into the modern Quantum Age. I have attempted to do so in the books that I have written.

One of the most useful things I ever did was to use Quantum Mechanics to upgrade science from Materialism and Naturalism TO Observation and Experience. I upgraded my science to Science 2.0. Rather than defining science as Materialism, Naturalism, Darwinism, Nihilism, and Atheism as most college professors do, I now define science as Observation and Experience. Science 2.0 is based upon Quantum Mechanics or Transdimensional Physics, which has been experienced and observed. Science 2.0 is based upon Phenomenology or the Lived Experiences of the human race.

Unless it has been experienced and observed by Someone Psyche somewhere sometime, it really isn't science. The major premises or hidden assumptions of Materialism, Naturalism, Darwinism, Nihilism, and Atheism – which claim that the quantum or the supernatural does not exist – have not been and cannot be experienced and observed. Therefore, Materialism and Naturalism are NOT science, because their major premises or hidden assumptions cannot be experienced nor observed but have to be taken on blind faith as being real and true.

The observed, experienced, and verified existence of Quantum Mechanics, Psyche, the Biblical God Jesus Christ, and Action at a Distance FALSIFIES Materialism, Naturalism, Darwinism, Nihilism, Atheism, and their derivatives.

The truth falsifies the false; and, the truth has been repeatedly verified, experienced, and observed. This is good stuff!

Science 2.0 allows all of the evidence into evidence and pursues a preponderance of the evidence. I've even written a few books based upon Science 2.0 and what I learned from it. As a scientist, I've found it very useful. I'm no longer trapped by nor restricted by My Materialism, My Naturalism, My Nihilism, and My Atheism. Upgrading my science to Science 2.0 helped me to overcome them.

Mark My Words

—

**Reference**

*Science 2.0: I Upgraded My Science*

https://www.amazon.com/dp/B0771K6WTX

*Quantum Neuroscience: The Answer to Life, the Universe, and Everything*

https://www.amazon.com/dp/B079Z6QQQB

*Syntropy in Defense of Quantum Mechanics: The Answer to Life, the Universe, and Everything*

https://www.amazon.com/dp/B07BPT3W8R/

## Denying the Existence of Observed Facts

"Einstein's Relativity is WRONG" by Run Ze Cao:

https://www.youtube.com/watch?reload=9&v=5_tig3NaTjI

While attempting to falsify Einstein's Special Relativity, Run Ze Cao concludes that there is no time dilation nor length contraction. Time dilation has been verified by scientific experimentation which means that Run Ze Cao's conclusions are wrong, which means that Run Ze Cao got something wrong that Einstein managed to get right.

https://www.physlink.com/education/askexperts/ae433.cfm
https://www.scientificamerican.com/article/einsteins-time-dilation-prediction-verified/
http://www.alternativephysics.org/book/TimeDilationExperiments.htm
https://futurism.com/the-most-accurate-clocks-in-the-world-just-confirmed-that-time-is-not-absolute http://math.ucr.edu/home/baez/physics/Relativity/SR/experiments.html

In order to falsify Einstein's Special Relativity, Run Ze Cao employs Newtonian Mechanics and Classical Physics to "prove" that there is NO time dilation and NO length contraction or distance contraction. I don't buy it. The science experiments prove that Cao is wrong, and that Einstein was right.

According to the Theory of Relativity, as an object achieves the speed of light, time stops, distance goes to zero, length contracts to zero, and velocity goes to infinity. The results match with Quantum Tunneling and Quantum Mechanics. In the Quantum Realm, everything is capable of quantum tunneling, or infinite velocities, or instantaneous action at a distance. This is what has actually been experienced and observed.

https://ultimate-model-of-reality.com/falsifying-einstein-and-classical-realism/

Time dilation has been experienced and observed. Run Ze Cao "explains" it by stating that it does not exist. Our job as scientists is to explain what has been experienced and observed – NOT to just explain it away. Time dilation has been experienced and observed, which means that as scientists, we are supposed to explain what it is and how it works. Einstein attempts to do so and Run Ze Cao does not. Instead, Run Ze Cao dogmatically states that it does not exist. Stating that something does not exist doesn't explain what it is and how it works. Now does it?

The purpose of Science is to explain everything that has ever been experienced and observed – NOT to simply state that things do not exist. Stating that God, Intelligence, Consciousness, Psyche, and Action at a Distance do not exist, does not explain what they are and how they work. Psyche, Intelligence, Consciousness, Action at a Distance, and the resurrected Lord Jesus Christ have been experienced and observed. Our job as scientists is to explain what these things are and how they work – NOT to simply claim that they don't exist. Claiming that they don't exist doesn't explain what they are and how they work. It's Bad Science.

## Proof of God's Existence

**Observation or Premise:** Anything that was obviously made obviously had a Maker or Creator who made it.

**Observation or Premise:** A genome was obviously made.

**Logical Conclusion:** Therefore, a genome obviously has a Maker or a Creator who designed it, programmed it, engineered it, field-tested it, fine-tuned it, made it, manufactured it, and deployed it.

—

This syllogism is both a Philosophical Proof of God's Existence and a Scientific Proof of God's Existence. It works. It's true. Every aspect of it is true, because both the premises and the conclusion have been experienced and observed in real life by real people. This Scientific Proof of God's Existence is both philosophically *valid* and observationally *sound*. It, and similar ones like it, convinced me that God does indeed exist.

Variations on this syllogism were my first Scientific Proof of God's Existence; and, I first built one based upon the non-existence of abiogenesis, or the non-existence of macro-evolution. After science itself proved to me that God must of necessity exist, then I simply KNEW that God exists, even though I still have yet to get to know God.

In 2016, when I first realized that NO type of evolution – chemical evolution, micro-evolution, macro-evolution, spontaneous generation, abiogenesis, random mutations, or natural selection – is capable of designing, creating, and implementing genomes from atoms or from scratch, I just KNEW that God exists because I KNEW why He must of necessity exist. God must exist in order to have done ALL of the science, organization, proteins, and genomes which the different types of evolution could NEVER have done.

In a very real sense, the Theory of Evolution proved to me that God exists. In other words, the inability of the different types of evolution to design, create, manufacture, and deploy proteins and the matching genes to go along with those proteins BECAME convincing scientific proof of God's necessity, which effectively became Scientific Proof of God's Existence.

Based upon this same exact theme, I developed many other Scientific Proofs of God's Existence and eventually used the Scientific Method itself to falsify Materialism, Naturalism, Darwinism, Atheism, and the Theory of Evolution.

I kept a log of my ongoing discoveries in the following books:

1. ***Summary Of: The Theory of Evolution Proves that God Exists***

    https://www.amazon.com/dp/B01GQCWED6

    https://www.amazon.com/dp/1521130485

2. ***The Theory of Evolution Proved to Me that God Exists: Why I Am No Longer an Atheist and Why I No Longer Believe in the Theory of Evolution***

    https://www.amazon.com/dp/B01HZYBZ7K

    https://www.amazon.com/dp/1521131228

3. ***The Scientific Method Proves That the Theory of Evolution Is False***

https://www.amazon.com/dp/B01IAAIRT2

https://www.amazon.com/dp/1521133611

### 4. *Using the Scientific Method to Eliminate the Usual Suspects and to Prove the Truth*

https://www.amazon.com/dp/B01J6STHP0

https://www.amazon.com/dp/1521133581

This was an era of transition for me. I used to be a Materialist, Naturalist, Nihilist, and Atheist. To develop a Scientific Proof of God's Existence that I actually KNEW to be true, as well as using the Scientific Methods to falsify Materialism, Naturalism, Darwinism, Atheism, and the Theory of Evolution represented a MAJOR paradigm shift in my way of thinking and doing science.

Once I finally KNEW the Truth, there was no going back to the deceptions and the lies; and, I was able to progress forward from there and make additional adjustments and improvements to my science, my worldview, and my philosophical underpinnings.

Although it is technically true – due to the *affirming the consequent* logic fallacy – that it's impossible to use the Scientific Methods to prove anything true, there is a legitimate work-around that actually produces the same result. This work-around is a philosophically *valid* and an observationally *sound* way to find the truth through a process of elimination.

The way this is done in practice is that we use the Scientific Methods and *negating the consequent* to falsify and eliminate everything that is false such as Materialism, Naturalism, Darwinism, Atheism, and their derivatives so that ONLY the Truth remains.

Think about it logically.

If you successfully eliminate everything that is false – everything that has NEVER been experienced nor observed by anyone, not even God – then ONLY the Truth will remain, because ONLY the observed and the verified and the proven will remain. The Truth is whatever has been experienced, lived, and observed by Someone Psyche sometime somewhere. Science is observation, or it should be.

Sherlock Holmes taught me how this process of elimination works in principle or practice.

**How often have I said to you that when you have eliminated the impossible** [and the false]**, whatever remains, however improbable, must be the truth? — Sherlock Holmes.**

Sherlock Holmes and the Philosophy of Science taught me how to use the Scientific Method to falsify the Theory of Evolution, Materialism, Naturalism, and Atheism; and, observations from real life people like you and me falsified Nihilism for me.

If you successfully eliminate everything that is false, then ONLY the Truth will remain; and, that's precisely how we use the Scientific Methods to "prove" the truth, by using the Scientific Methods to eliminate everything that is false or everything that has NEVER been experienced nor observed. It works!

Mark My Words

## Entropy Is Death and Syntropy Is Eternal Life

Syntropy is by far my greatest, most useful, most powerful, and most explanatory scientific discovery that I have made so far.

Logical common sense tells us that some type of Syntropy MUST exist, or all of that subsequent entropy wouldn't have been possible. NONE of our physical reality would have been possible without some type of Syntropy.

Entropy is death. Syntropy is Eternal Life. Syntropy is eternal and everlasting, without a beginning of days or an end of years.

Chances are good that you have never heard of Syntropy.

Why haven't you ever heard of Syntropy? It's the foundation of ALL science, after all.

The reason is that the Materialists, Naturalists, Darwinists, Nihilists, Behaviorists, Determinists, Physical Reductionists, and Atheists have BANNED Syntropy or Eternal Life from science. These people state axiomatically a priori that Syntropy or Psyche does not exist. These people define science as Materialism and Naturalism.

Are they right?

They can't be right. Some type of Syntropy MUST exist, or all of that subsequent entropy wouldn't have been possible. Think about it logically and rationally, rather than emotionally; and, you will KNOW that this is true.

Entropy is death. Physical matter or entropy results in death. This reality has been experienced and observed.

Physical Matter, Materialism, Naturalism, Darwinism, Nihilism, Behaviorism, Determinism, Reductionism, Atheism, and Classical Physics are based exclusively on entropy. Entropy is death. These "scientific disciplines" are literally based upon Creation by Death. Now, please explain to me how Creation by Death or Creation by Entropy could ever be made to work!

It can't! It's impossible! Not even God can do the impossible!

This is an extremely powerful Scientific Observation. Its explanatory power is through the roof!

The Theory of Evolution IS Creation by Death, or Creation by Entropy. It's NEVER going to work! I just resolved millennia of scientific controversy and debate with that simple observation.

Materialism, Naturalism, Darwinism, Nihilism, Atheism, Classical Physics, and their derivatives ARE Creation by Death or Creation by Entropy. That's NEVER going to work! It's impossible. Creation by Death or Creation by Entropy is synonymous with magic, or creation ex nihilo, or creation by chance. That's just never going to work while we are stuck here in the physical realm at the physical level of our existence. Not even God can do creation ex nihilo, or creation by death, or creation by entropy.

Some type of Syntropy or Eternal Life or Psyche MUST exist in order to MAKE the physical laws and classical physics WORK. Syntropy MUST exist in order to make entropy and physical laws possible in the first place. I just solved ALL of the mysteries, confusion, problems, and controversies that stand at the foundation of modern-day materialistic, naturalistic, and atheistic sciences. It was solved by the simple observation that entropy is death, and Syntropy is some type of intelligence or life.

Since we KNOW for a fact that entropy exists, we also KNOW for a fact that its opposite Syntropy MUST also exist. All that entropy wouldn't have been possible without a massive initial infusion of Syntropy or Life. That's just the way it is. The truth of it is undeniable; yet, the Materialists, Naturalists, Darwinists, Nihilists, and Atheists choose to deny the existence of Syntropy or Psyche anyway. These people are constantly telling us that Psyche, Consciousness, and Life do not exist – only physical matter or entropy exists.

These people are demonstrably wrong.

Quantum Mechanics, Spiritual Mechanics, Transdimensional Mechanics, or Supernatural Mechanics is a type of Syntropy – eternal and everlasting, without a beginning of days or an end of years. The verified and proven existence of Quantum Mechanics, Magnetism, Gravity, Dark Matter or Spirit Matter, Dark Energy, Invisible Light, the Strong and Weak Nuclear Forces, the Zero-Point Field, Invisible Forces and Fields, Radio Waves, Microwaves, X-Rays, Gamma Rays, Thoughts or Quantum Waves, Psyche or Intelligence, Space or the Vacuum, Time, the Quantum Zeno Effect, Action at a Distance, and Syntropy FALSIFIES the major premises of Materialism, Naturalism, Darwinism, Nihilism, Atheism, and their derivatives which state that these various different Quantum Mechanisms and Non-Physical Realities DO NOT EXIST.

Well of course they exist! Their existence has been proven and verified by scientific observations or scientific experiments. It's the Materialism, Naturalism, and Atheism that are FALSE and WRONG, not the scientific observations! Go with the science and NOT the wishful thinking of the Materialists, Naturalists, and Atheists. Science is observation and experience, NOT Materialism and Naturalism. The proven and verified existence of all these different non-physical and invisible forces and fields FALSIFIES Materialism, Naturalism, Darwinism, and their derivatives. You want the truth, and not the fiction and the wishful thinking.

Entropy is restricted to or limited to physical matter and space-time. Everything else is Syntropy. Everything else is eternal and everlasting. This reality and truth has been experienced and observed during Out-of-Body Experiences and Near-Death Experiences. Only physical matter or entropy is subject to the passage of time or the aging process. Everything else is eternal and everlasting. Everything else is Syntropy.

I just solved ALL of the problems that Materialism, Naturalism, and Atheism created for us in the first place. Entropy is death. Syntropy is Eternal Life.

Within the books that the Biblical God Jesus Christ had a hand in writing and producing, we observe Him talking about Eternal Life or Syntropy ALL the time. It has to exist because He has observed it, lived it, and experienced it for Himself. Science is observation and experience. Jesus Christ wouldn't be talking about Eternal Life or Syntropy if it didn't exist. Resurrection from the dead is a type of Syntropy. Christ rose from the dead. The Atonement of Christ is Syntropy – or restoring everything back to its original natural Eternal Order. Entropy is death; and, death was introduced into our physical bodies when Adam and Eve fell. Death was introduced into our physical bodies and our lives when we chose to leave God's Presences or chose to leave Syntropy.

With Syntropy or Psyche or Quantum Mechanisms, we can literally explain everything that comes our way. This IS powerful science! Yet, the Materialists, Naturalists, and Atheists assure us that it doesn't exist. They are wrong. Of course it exists. It has been experienced and observed! Syntropy or Life has been experienced and observed on both sides of the veil. It's real and truly exists.

**Quantum Mechanics is God's Priesthood Power. Quantum Mechanics is Syntropy. God's Psyche or God's Intelligence used Quantum Mechanisms to create physical matter from spirit matter and to do Cosmic Fine-Tuning. The physical laws as we know them are the result of Syntropy or Fine-Tuning.**

Fine-Tuning is a type of Syntropy. Physical matter or entropy cannot DO Fine-Tuning. Entropy or physical matter can only do death.

Entropy is disorder, randomness, chaos, and death. Materialism, Naturalism, Atheism, Darwinism, Nihilism, Classical Physics, and the Theory of Evolution are based exclusively upon entropy or death. Evolution (genetic change) or random mutation is entropy. Natural selection eventually results in death or entropy. Random Mutations and Natural Selection cannot produce life because they are in fact entropy or death. Chemical evolution, abiogenesis, spontaneous generation, and macro-evolution are prevented from happening by entropy. Evolution of any kind is entropy. Entropy or evolution cannot do fine-tuning, teleology, purpose, planning, design, programming, engineering, manufacturing, production, and creation. Entropy is death! Entropy or death cannot produce functional proteins nor the matching genes to go along with them! This is logical common sense.

In many of his books, Hugh Ross uses Cosmic Fine-Tuning as Scientific Proof of God's Existence. The whole thing made logical sense to me because ONLY Psyche can convert or transmute spirit matter into physical matter. ONLY Psyche can organize physical matter into useful and productive forms such as planets, stars, galaxies, genomes, and life forms. ONLY Psyche can do Cosmic Fine-Tuning. God's Psyche must of necessity exist, in order to have done all of the cosmic fine-tuning and science that needed to be done; otherwise, you and I would not be here right now in this physical realm. Entropy or death cannot do creation and fine-tuning! Science has made it obvious and clear to me that this must be so. It's obvious that the rocks or raw physical matter cannot do Cosmic Fine-Tuning. That kind of process requires an active, living, agentic, conscious, intelligent Being or Psyche, whom many of us tend to call God or the Ultimate Scientist.

For the best list of the Fine-Tuning that has been done by God see:

http://www.reasons.org/explore/blogs/todays-new-reason-to-believe/read/tnrtb/2010/11/16/rtb-design-compendium-2009

I archived it because it kept disappearing on me.

Archival Copies:

https://syntropy.site/wp-content/uploads/2018/04/compendium_part1.pdf

https://syntropy.site/wp-content/uploads/2018/04/compendium_part2.pdf

https://syntropy.site/wp-content/uploads/2018/04/compendium_Part3_ver2.pdf

https://syntropy.site/wp-content/uploads/2018/04/compendium_Part4_ver2.pdf

Only some type of Syntropy, Psyche, or Intelligence can do Fine-Tuning!

**Remember, entropy is death, and Syntropy is Eternal Life.** If you are able to remember that simple truth, then you will be able to explain every scientific observation that comes your way.

I just fixed science and upgraded science with this simple yet powerful scientific observation. It explains everything that has ever been experienced or observed.

Mark My Words

## The Purpose of Science 2.0 Is to Identify the Falsehoods in Science

Science 2.0 allows all of the evidence into evidence, and then it pursues a preponderance of that evidence.

As you can tell so far, the primary focus of Science 2.0 is to identify the falsehoods in science, falsify them, and eliminate them from science.

Clearly, this process has the most success when dealing with natural selection and the second law of thermodynamics, because these two are the most obviously false concepts in science; and, there's a lot of different ways to falsify them.

The other thing that is obviously false in science is their ongoing rejection of the immaterial or the non-physical. These falsehoods, deceptions, and lies stem from Materialism, Naturalism, Darwinism, Nihilism, Behaviorism, and Atheism.

Quantum Fields, Psyches, and Photons are obviously massless, entropyless, heatless, and non-physical; yet, the Naturalists and Atheists assure us repeatedly that these things do not exist. These people reject the non-physical which means that they reject the massless, entropyless, and heatless Photons, Psyches, and Quantum Fields pretending that these things do not exist when clearly they do.

If you successfully identify and eliminate everything that is false, then only the truth will remain. This becomes a NEW and BETTER method for doing science – identifying and removing the falsehoods from science rather than enshrining them, turning them into axioms, and making them into unfalsifiable laws.

It can be very exciting to identify and study what remains after these various different falsehoods have been removed from science. The truth remains.

Mark My Words

## Science Needs a Philosophy of Science that Is True

Modern-day science is in desperate need of a new, enhanced, and upgraded Philosophy of Science that actually matches with what has been experienced and observed. The modern-day philosophies of science have been falsified by Science, knowledge, science experiments, and observations and experiences.

Materialism is a philosophy of science. Materialism claims that the quantum, the non-physical, and the immaterial do not exist. That's not science. It's philosophy. It's a philosophy of science that has been falsified by Science. Materialism is a philosophy of science that has been falsified by the things that have been experienced and observed. The proven and verified existence of immaterial, non-physical, massless, entropyless, supernatural quantum fields and photons FALSIFIES Materialism which claims that they do not exist and cannot exist.

In modern-day Science, we desperately need a Philosophy of Science that matches with what has actually been experienced and observed. We need a Philosophy of Science that is based upon science rather than philosophy and wishful thinking. We need a Philosophy of Science that has been verified, proven, demonstrated, replicated, and observed. We need a Philosophy of Science that is real and true. All of the twenty different books that I have written and published so far have been an attempt to present the world with a Philosophy of Science that is actually true – one that has actually been experienced and observed by someone somewhere.

Science 2.0 is a part of my attempt to produce such a thing. Science 2.0 allows ALL of the evidence into evidence, and then it pursues a preponderance of that evidence. Science 2.0 allows evidence for the quantum or the non-physical or the supernatural into evidence, giving preference to the evidence that has actually been experienced and observed. It's time for scientists to get rid of the FALSIFIED philosophies of science, because they are worthless, and they are holding us back. As far as I know, I have upgraded the Philosophy of Science and taken it farther than anyone else on the planet has ever taken it. I did this because Science desperately needs a Philosophy of Science that is true.

The twenty different books that I have written represent my doctorate in the Philosophy of Science. Someone had to upgrade the Philosophy of Science, because what we currently have for "philosophy of science" is obviously wrong and false.

Naturalism is a philosophy of science. Naturalism claims that action at a distance, infinite velocities, infinite acceleration, telepathy, psyche, quantum tunneling, the quantum zeno effect, the quantum fields, quantum mechanics, and the supernatural do not exist. That's not science. It's philosophy. It's a philosophy of science that has been falsified by Science. Naturalism is a philosophy of science that has been falsified by the things that have been experienced and observed. The proven and verified existence of immaterial, non-physical, massless, entropyless, supernatural quantum fields and photons FALSIFIES Naturalism which claims that they do not exist and cannot exist.

Darwinism, Abiogenesis, Spontaneous Generation, Chemical Evolution, and the Theory of Evolution are Creation by Chance. Darwinism is a philosophy of science. Darwinism claims that everything in our universe including the big bang, planets, stars, galaxies, physical matter, quantum fields, genomes, eyes, brains, and life forms were ALL created by chance alone. A binomial distribution table in the back of a statistics textbook makes it obvious that it is impossible to create a protein 15 amino acids long by chance alone. Where the sequence must be exact, the probability of getting an exact sequence that is fifteen or more units long by chance alone is effectively 0.0000 according to a binomial distribution table. It can't be done. The longer the sequence, the more impossible it becomes. You can't successfully and reliably toss fifteen heads in a row with a fair coin by chance alone. The probability of doing so has basically gone to zero. It can't be done. And, it only gets worse and more impossible from there, the longer the sequence gets.

Your average protein is about 400 amino acids long. That's impossible to produce by chance alone. Likewise, it is impossible to produce the matching gene to go along with that protein by chance alone. It can't be done. Design and Creation by Chance is physically impossible and statistically impossible. There's NO probability that it could ever happen. Darwinism, the Theory of Evolution, and its derivatives are Creation by Chance and Creation Ex Nihilo. They are physically and statistically impossible to achieve by chance alone. It can't be done, which means that it wasn't done that way. We have to find a better explanation for the origin of life, because it is impossible to create life and genes and genomes and proteins by chance alone. Creation by Chance is a FALIED and FALSIFIED philosophy of science. Science as a whole NEEDS a much better Philosophy of Science than the one provided to us by Darwinism, Naturalism, and Creation by Chance.

The second law of thermodynamics is a philosophy of science. The second law states that the total amount of entropy or disorder in our universe is always increasing and that it can never decrease. The second law of thermodynamics is falsified by the observations and the experiences of the human race. The second law is falsified by the existence of the human race. Our very existence falsifies the second law of thermodynamics. If the second law were actually true, then we wouldn't exist, and you wouldn't be reading this.

The second law of thermodynamics is another philosophy of science that has been FALSIFIED by everything that exists. Everything that we have ever experienced or observed falsifies the second law of thermodynamics. Every aspect of statistics falsifies the second law, because the second law of thermodynamics is Creation by Chance, Creation by Random Disorder, Creation by Chaos, and Creation Ex Nihilo. Statistics, the Null Hypothesis, the Binomial Distribution Table, the Correlation Coefficient, the Scientific Methods, Science, and Science Experiments were designed to eliminate chance causality, creation ex nihilo, creation by chance, or the second law of thermodynamics and the theory of evolution from Science.

The second law of thermodynamics and the theory of evolution are the Null Hypothesis, since they are caused by chance alone. The whole purpose of Science and Science Experiments is to reject and eliminate the Null Hypothesis so that we can establish the true cause instead. Chance of any kind is an invalid causal mechanism in Science, Statistics, the Scientific Methods, and Science Experiments. The fact that the second law of thermodynamics and the theory of evolution are Chance Causality, Creation by Chance, or Caused by Chance Alone (the Null Hypothesis) automatically FALSIFIES them according to the Rules of Science and the Philosophy of Science. They can't possibly be true because they were caused by chance alone or were the result of nothing but chance.

### $E = mc^2$ is the perpetual motion cycle.

The perpetual motion cycle FALSIFIES the second law of thermodynamic, which claims that the perpetual motion cycle does not exist. $E = mc^2$ and the second law of thermodynamics are mutually exclusive. They falsify each other. If one of them can be demonstrated to be true, then the other one has automatically been falsified.

The Big Bang Theory and the second law of thermodynamics falsify each other. The Big Bang Theory states that entropy or disorder completely ceased to exist at the beginning of our physical universe; whereas, the second law of thermodynamics claims that entropy or disorder can never decrease or go to zero. The two ideas are mutually exclusive. They are two different philosophies of science, and they falsify each other.

Likewise, Quantum Field Theory and Feynman Diagrams falsify the second law of thermodynamics. The second law of thermodynamics will NEVER be true as long as the quantum fields exist.

The second law of thermodynamics is a philosophy of science; and, everything falsifies the second law of thermodynamics. The whole of statistics, logic, science experiments, and science FALSIFIES the second law of thermodynamics. We have NEVER observed a substance or an entity that is constantly increasing in amount out of thin air from nothing. Even the total amount of energy in our universe can't increase in amount, yet somehow magically entropy can. The second law of thermodynamics is magic or creation ex nihilo. The second law of thermodynamics is fantasy and science fiction. Everything that has been experienced and observed falsifies the second law of thermodynamics.

The purpose and goal of Science 2.0 is to identify, falsify, and eliminate from Science the theories, philosophies, and ideas that are false – and then to try to replace them with something that has actually been experienced and observed. The Conservation of Energy and the Perpetual Motion Cycle $E = mc^2$ have been experienced and observe; whereas, an ever-increasing substance created out of thin air from nothing has NOT. The second law of thermodynamics hasn't actually been experienced nor observed. It's a pseudo-science or a philosophy of science. It's not actually science. The second law is more in the realm of dogma, religion, blind faith, and wishful thinking instead.

Science also desperately needs an interpretation of Quantum Mechanics that has actually been experienced and observed, instead of all the different philosophies, interpretations, wishful thinking, and garbage that are being fed to us by the Materialists, Naturalists, Darwinists, Nihilists, and Atheists who don't even believe that the quantum, the non-physical, or the spiritual exists. Why should these people be making up interpretations for Quantum Mechanics, since these people don't even believe that the massless, entropyless, heatless, non-physical Photons and Quantum Fields exist? What do these people know about the non-physical, the spiritual, or the quantum? They don't know anything about any of this because they don't even believe that it exists. Most of the interpretations for Quantum Mechanics, Spiritual Mechanics, or Energy Mechanics are absolutely worthless, because they are based exclusively upon Materialism, Naturalism, Darwinism, Nihilism, and Atheism which claim that the non-physical photons and quantum fields do not exist and cannot exist.

Each interpretation of Quantum Mechanics is a philosophy of science, and most of them are completely worthless because they deny the existence of what has actually been experienced and observed. Most of the others are completely worthless, too, because they are based upon fictional ideas that have never been experienced or observed by anyone. Anything that is based upon Creation Ex Nihilo is automatically false. Creation Ex Nihilo is patently absurd! Not even God can do Creation Ex Nihilo, so why should anyone believe that entropy can?

Science is in desperate NEED of a Philosophy of Science that is real and true. Science NEEDS a Philosophy of Science that matches with what has actually been experienced and observed.

Mark My Words

# PART I — DEFINING SCIENCE 2.0

As an upgraded science, Science 2.0 absolutely needs an Account of Origins that matches perfectly with the Observational Evidence and the Lived Experiences of the human race. Consequently, a great deal of time and effort has been placed into finding and/or developing such a thing. You will have to judge for yourself if the process and this project has been a success, or not. All I can do is to present the evidence. You will have to decide for yourself if I have met my burden of proof through a preponderance of the evidence, or not.

The Materialists, Naturalists, Darwinists, Nihilists, and Atheists define "science" as Materialism and Naturalism. Their whole paradigm is based upon a logic fallacy that's been variously called *begging the question*, *affirming the consequent*, or *jumping to conclusions*. In other words, they chose the conclusion that they wanted, long BEFORE they did any actual science. These people jump directly to that conclusion every time they try to do science and interpret science, a process which is very unscientific to say the least. The world is in desperate need of a new and better definition for "science" – a new and better paradigm.

Ironically, Materialism and Naturalism and their derivatives are religions or philosophical worldviews, and NOT science. Materialism and Naturalism are simply a way of interpreting scientific evidence – a way that has actually been falsified by observation, experience, science, scientific evidence, the scientific methods, and the lived experiences of the human race. Materialism, Naturalism, Darwinism, Nihilism, and Atheism are unable to meet their burden of proof because the preponderance of the evidence stands against them and falsifies them; consequently, these false and falsified religions or philosophies have to be taken on blind-faith as being true because they cannot be lived nor experienced for real. Materialism and Naturalism exist only in a person's imagination. They aren't real.

Something is said to be "scientifically accurate" if it matches with reality, meaning that it has been observed and experienced in real life by a real living person or psyche. Materialism, Naturalism, Atheism, and Nihilism are by definition "unscientific" because they don't match with reality, meaning that they have NEVER been observed nor experienced in real life by anyone, nor can they be.

Materialism, Naturalism, Darwinism, Nihilism, and Atheism ARE DEAD. What this means is that they have been falsified trillions of times in thousands of different ways by scientific evidence, or observational evidence, or experiential evidence. We are simply in the process of waiting for the old guard to die, so that a new and better paradigm can have its day in the sun. The world is in desperate need of a new and better definition for "science".

While we are waiting for this new and better paradigm as well as that new and better definition for science to hit critical mass and become generally accepted throughout the world, I have chosen to upgrade my science to Science 2.0. Under Science 2.0, I have chosen to admit ALL of the evidence into evidence. Nothing is excluded now. Everything is taken into consideration, as it should be if we are ever to do Real Science.

Science 2.0 is based exclusively on Observational Evidence and the Lived Experiences of the human race. Under Science 2.0, eye-witness accounts, observational evidence, experiential evidence, direct observations, and lived experiences of any kind trump and take precedence over "scientific inferences," circumstantial evidence, and philosophical speculation. Remember, *scientific inferences* or someone's *personal interpretation of scientific evidence* IS the logic fallacy upon which Materialism, Naturalism,

Darwinism, Nihilism, and Atheism are based. Just present the Observational Evidence or the Lived Experiences, and then let people make up their own mind as to what it all means. Let the people do their own interpretation of the evidence. You can tell people what the evidence means to you, as I repeatedly do; but, don't be dogmatic or religious about your personal interpretations and scientific inferences.

Even though *personal interpretation of the scientific evidence* or *scientific inferences* ARE the last and final step in the Scientific Method, this process of *personal interpretation* or *scientific inference* IS a logic fallacy which introduces ALL of the falsehoods, deceptions, trickery, and lies into science as a whole. Pick any scientific discipline you like, and then observe that it derives and receives ALL of its falsehoods, deceptions, and lies from the *personal interpretations, philosophical speculations, personal guesswork, wishful thinking,* and *scientific inferences* which are given to it by fallible and sometimes-deceptive human beings. Under Science 2.0, just present ALL of the Observational Evidence, Eye-Witness Evidence, and the Lived Experiences of the human race; and then, let people make up their own mind as to what it all means to them.

I'm a scientist; and, I'm not going to apologize for trying to upgrade my science to something better!

While developing Science 2.0, I will present eye-witness accounts, observational evidence, experiential evidence, and lived experiences from every scientific discipline, person, and religion on the planet that I can find; but, you are going to have to make up your own mind as to what that evidence means to you.

Under Science 2.0, there are NO EXPERTS at scientific inferences or scientific interpretations of the data – your personal interpretation is just as good as mine. There is no priesthood under Science 2.0 who will bless and sanctify the religious dogma that's being presented as fact, as we find when it comes to Materialism, Naturalism, Darwinism, Nihilism, and Atheism which have priests and pastors who have given themselves PhDs in their chosen sciences and have thereby made themselves the official interpreters of their scientific disciplines. If you don't agree with their personal interpretation of the scientific evidence, then you are censored, ridiculed, censured, banned, and excommunicated from doing their brand of science.

There's NONE of that under Science 2.0, because under Science 2.0 there are NO high priests of scientific inference and personal interpretation; and therefore, there is NO religion and NO religious dogma where the scientific information and observational evidence are concerned. When it comes to Science 2.0, there is NO official dogma that has to be taken on blind faith as being true, like there is under Materialism, Naturalism, Darwinism, Nihilism, and Atheism. Science 2.0 is a completely open-minded approach to scientific evidence – ALL the evidence. I chose to follow the evidence wherever it might lead me.

Under Science 2.0, as the oracle repeatedly says in the *Matrix*, you make up your own damned mind as to what it is that you want to believe. As Neo and the architect so adequately conclude, it's about choice. It all comes down to choice! Your beliefs are chosen – chosen beliefs and chosen behaviors are the pinnacle and hallmark of free will. It all comes down to choice.

I will present the eye-witness evidence and experiential evidence; and, I will often tell you what that evidence means to me personally. But in the end, you are going to have to make up your own damned mind as to what it is that you personally want to believe. That is as it should be; and, that's the way science should be done.

While developing Science 2.0, I gleaned eye-witness evidence or observational evidence from every scientific discipline and every religion and nationality on the planet that

I possibly could. Neo wanted guns, lots and lots of guns. I want evidence, lots and lots of evidence. The BEST way to find and know the truth is to live it and experience it for yourself, or to choose to trust someone who has. Science 2.0 is based exclusively upon the Lived Experiences of the human race – all of them. Brutal, huh?

Remember, "experiment" and "experience" have a similar etymology – they derive from the same root word. Science is observation, and a science experiment is the same thing as experience.

Science 2.0 is based upon the Lived Experiences of the human race, including all of our science experiments as well as our non-local out-of-body experiences. Under Science 2.0, Lived Experiences of any kind are treated as Scientific Evidence. Whether we are dealing with the spiritual or the physical, the truth of a matter is established through a preponderance of the evidence, as if we were in a court of law – according to the Law of Witnesses. If there are no witnesses, then there is no evidence to be had. The whole purpose of Science 2.0 is to find different ways to account for and explain the observational evidence, the eye-witness evidence, and the experiential evidence provided by the Lived Experiences of the human race.

Under Science 2.0, observational evidence and experiential evidence is given precedence or priority over "scientific inferences" and "personal speculation". Under Science 2.0, Materialism, Naturalism, Darwinism, Nihilism, and Atheism are considered "unscientific" because they have NO observational evidence supporting their major premises or primary assumptions. These pseudo-sciences, speculative philosophies, and dogmatic religions have NO eyewitnesses to sustain them, and therefore they violate the Law of Witnesses.

## The Law of Witnesses

**2 Corinthians 13: 1: "In the mouth of two or three witnesses shall every truth be established."**

Science 2.0 is based upon eye-witness evidence, observational evidence, and the lived experiences of the human race, including both the physical evidence that has been provided by our scientists so far and the non-local evidence which has been provided by the lived experiences of the human race. The best way to find and know the truth is to live it and experience it for yourself, or to choose to trust someone who has.

Science is Observation. Science is Knowledge. Science is Truth. Science is comprised of Lived Experiences.

Under this particular definition for science, Materialism, Naturalism, Darwinism, Nihilism, and Atheism are unscientific because they have NO observational evidence supporting their primary assumptions or major premises. Their lack of eye-witness evidence or lack of observational evidence FALSIFIES them.

Science 2.0 is based upon the Law of Witnesses as established by God himself. The Law of Witnesses is the best way to do science, because the best way to find the truth and know the truth is to live it and experience it for yourself or to choose to trust someone who has.

The Law of Witnesses is technically the ONLY way to find and know the truth. The Law of Witnesses is definitely better than the "scientific inferences," "circumstantial evidence," and "philosophical speculation" which we receive from the Materialists, Naturalists, Darwinists, Nihilists, and Atheists. Why? It's because science IS observation and lived experiences, not scientific inference and personal speculation.

The Law of Witnesses IS Science!

Science 2.0 defines science as the Law of Witnesses, rather than defining science as Materialism and Naturalism as the Materialists and Naturalists do.

## The Dedicated Opposition

Every scientific theory and philosophical idea needs a dedicated opposition to give it some teeth and staying power – Science 2.0 is no exception.

Since Materialism, Naturalism, Darwinism, Nihilism, and Atheism have been FALSIFIED trillions of times in thousands of different ways, they really cannot serve as the opposition which Science 2.0 needs, because they have NO observational evidence supporting them. Science 2.0, which is based exclusively on the observational evidence and lived experiences of the human race, can adequately serve as a replacement for Materialism, Naturalism, Darwinism, Nihilism, and Atheism because Science 2.0 automatically subsumes ALL physical observation or physical evidence that has ever been produced; but, Science 2.0 can't actually derive "inspiration" or "opposition" from the falsified philosophies of Materialism and Naturalism because they have been FALSIFIED already.

Instead, Science 2.0 repeatedly turns to Young-Universe Creationism (YUC) and Young-Earth Creationism (YEC) for the dedicated and cunning opposition that it needs to get established. The YUCs and the YECs are extremely intelligent; and, they actually teach science that can be debunked or falsified with observational evidence, which is exactly the purpose of Science 2.0 – getting rid of falsehoods through observational evidence and the lived experiences of the human race. Over and over again, Science 2.0 turns to the YUCs and the YECs in order to get the opposition that it needs.

The YUCs and the YECs teach us directly the flaws and the weaknesses of relying exclusively upon "scientific inferences" or "personal interpretations of scientific evidence and scriptural evidence" while at the same time rejecting or trying to explain away the Observational Evidence – as do the Materialists, Naturalists, Darwinists, Nihilists, and Atheists who also deny, resist, or reject the Observational Evidence and the Lived Experiences of the human race. The people who deny the Observational Evidence and try to explain it away through philosophical logic, scientific inferences, and personal interpretations of the data end up being the dedicated opposition which Science 2.0 needs in order to make its case and demonstrate its superiority.

Under Science 2.0, Lived Experiences and Observation Evidence trump personal interpretations of data, scientific inferences, and philosophical conclusions. The weakest part of the Scientific Method has always been the final step in the process – the interpretation given to the scientific evidence.

## The Epiphany

I know the very moment that the paradigm shift took place and Science 2.0 was born.

During my studies, I had finally realized that Macro-Evolution is the same exact thing as Abiogenesis or Spontaneous Generation. These are insights and teachings that we NEVER get from Materialists, Naturalists, Darwinists, Nihilists, and Atheists because they are deliberately trying to hide this type of information from us – literally trying to keep us from realizing that Macro-Evolution, Abiogenesis, and Spontaneous Generation are the same exact thing with different names that have been given to it over the years.

One day, it dawned on me that Macro-Evolution, Abiogenesis, and Spontaneous Generation HAVE NEVER BEEN OBSERVED. They have never been caught in the act of design and creation, because they don't exist.

That epiphany or flash of insight crescendoed and reverberated throughout the whole of my scientific knowledge and my way of thinking, completely transforming and changing the way that I look at science.

It was then that I KNEW that observation TRUMPS philosophical speculation and personal interpretations every time. I had literally FALSIFIED the Theory of Evolution through a lack of observation, or a lack of evidence. In other words, the preponderance of the evidence was telling me that the Theory of Evolution is FALSE. Consequently, with the preponderance of the observational evidence standing firmly against the Theory of Evolution falsifying the Theory of Evolution, I realized instantly that the Theory of Evolution fails to meet its burden of proof. That was the end of Macro-Evolution for me. I simply KNEW that it is false.

At that very same instant of epiphany, another corollary sprang into view. Since I KNEW that Macro-Evolution, Spontaneous Generation, and Abiogenesis DO NOT EXIST and have never been observed, instantly I KNEW that God must of necessity exist in order to have done all the science, design, programming, and creation that needed to be done at that point in time in order to bring genomes and life forms into existence in the first place.

Since I KNEW that Macro-Evolution is false and doesn't exist and has never been observed, I JUST KNEW that there was only one other person who could have done the design, programming, engineering, manufacturing, and creation that needed to be done. Suddenly, I KNEW that God exists, because I KNEW why He has to exist.

A bit later, I discovered why the Materialists, Naturalists, Darwinists, Nihilists, and Atheists work so hard trying to confound, conflate, and confuse things for us. While interacting with these people online, I realized that they don't want us to discover that Macro-Evolution, Spontaneous Generation, and Abiogenesis are the same exact thing.

Why?

It's because Spontaneous Generation was falsified by Louis Pasteur in 1859. The very moment that I discovered that fact, that piece of science, it dawned on me that Spontaneous Generation, Abiogenesis, Macro-Evolution, Materialism, Naturalism, and Darwinism had been FALSIFIED by Louis Pasteur in 1859, the very same year that Charles Darwin published "On the Origin of Species". The Theory of Evolution was a non-starter, because it was FALSIFIED the very same year that it was given to the world; and, the Materialists, Naturalists, and Darwinists have been trying to hide that fact from us ever since.

At that very moment, I had external confirmation that my former epiphany or revelation was in fact TRUE; and, I simply KNEW that the Theory of Evolution is false because I KNEW why it is false. I KNEW that we had been tricked, duped, and deceived because we wanted to be.

I have always been skeptic, agnostic, and even atheistic; but, when I realized that these people had been lying to me all of my life and that I had at times bought into their lies, I was done with them and I turned against them with a vengeance. After all, I really wasn't Christian in my outlook at that point in time, and I hate being lied to and deceived.

I had successfully falsified Macro-Evolution both through a complete lack of observation and through the fact that Spontaneous Generation or Macro-Evolution had been falsified by Louis Pasteur in 1859.

Even though I didn't fully realize it at the time, that was the point in time that Science 2.0 was born, because Science 2.0 is based exclusively upon Observation, Experimentation, Experience, and the Lived Experiences of the human race. Under Science 2.0, these things (or the lack of these things) TRUMPS personal speculation, scientific inferences, circumstantial evidence, and personal interpretations of data EVERY TIME.

Remember, experience and experiment are basically the same word and basically the same thing – they both have the same root etymology. Any time we gain experience, we are in fact running an experiment of sorts. Each one of us was born a scientist, and we are here to gain experience by running experiments on everything that we encounter both here in the physical realm and there in the non-local realm.

I finally had a science the way that science should be. Science 2.0 is observation, experience, fact, truth, reality, and knowledge – not something that is simply imagined into existence by philosophers and wishful thinkers such as Materialists and Naturalists.

Once I realized and understood the power of Observational Evidence or Experiential Evidence, I found myself eventually looking for Observational Evidence and Experiential Evidence of both God and the Afterlife. When I finally started looking, I found tons of observational evidence and experiential evidence from everywhere on the planet supporting both concepts, especially the existence of an Afterlife or a Reality separate from the physical body and physical brain, and the existence of the Biblical God Jesus Christ. Thousands of people have seen and touched Jesus Christ after His resurrection from the dead, both here in the flesh and there in the Non-Local Spirit Realm. Some have had Jesus Christ come and rescue them from hell, for the asking.

The Biblical God Jesus Christ has in fact met His burden of proof through a preponderance of the evidence both in Scripture and during Near-Death Experiences (NDEs).

Under Science 2.0, observation of Jesus Christ TRUMPS any philosophical speculation suggesting that He does not exist. Observation or experience always TRUMPS scientific inferences, guesswork, wishful thinking, and personal interpretation. OBSERVATION (as well as the lack of supporting observation) FALSIFIES Materialism, Naturalism, Darwinism, Nihilism, and Atheism. If you want to find and know the truth, you are going to have to look someplace else besides these falsified religions and philosophies. That was the epiphany that I had that day; and, it changed everything for me. There's no going back now, and why would I want to? I hate being lied to. Ever since, I have been trying to share my epiphany with the rest of the world.

I was slow getting there – over fifty years of life – but I finally got there in the end. Some of us are late bloomers; but then again, some of us never get it in the first place. It is what it is, I guess.

I sincerely wish you luck, because the Materialists, Naturalists, Darwinists, Nihilists, and Atheists are trying to deceive all of us, and these people have been very successful in doing so. I KNOW, because I have been there and done that. In other words, I have experienced it, observed it, and lived it for myself, so I know.

## We Need an Evidence-Based Science, Not Scientific Inferences

Pinel, J. (2014). *Biopsychology* (9th ed.). New York: Pearson.

The following quote came from a college "science" textbook, no less. It's incredible!

After reading the first couple of chapters of *Biopsychology* by John Pinel, it became abundantly clear to me that we desperately need an evidence-based science rather than pseudo-sciences based upon *scientific inferences* and *ad hoc just-so stories*.

> **Scientific inference is the fundamental method of biopsychology and of most other sciences [like the theory of evolution] – it is what makes being a scientist fun. This section provides further insight into the nature of biopsychology by defining, illustrating, and discussing scientific inference.**
>
> **The scientific method is a system for finding things out by careful observation, but many of the processes studied by scientists cannot be observed.**
>
> **The empirical method that biopsychologists and other scientists use to study the unobservable is called <u>scientific inference</u>. Scientists carefully measure key events they can observe and then use these measures as a basis for logically inferring the nature of events that they cannot observe.**
>
> **Like a detective carefully gathering clues from which to recreate an unwitnessed crime, a biopsychologist carefully gathers relevant measures of behavior and neural activity from which to infer the nature of the neural processes that regulate behavior.**
>
> **The fact that the neural mechanisms of behavior [and evolution] cannot be directly observed and must be studied through scientific inference is what makes biopsychological research such a challenge – and as I said before, so much fun.** (*Biopsychology*, p. 13.)

That's unbelievable!

He bases his whole religion – including his belief in the Theory of Evolution – upon *scientific inferences*, which are by definition things that have never been observed and can never be observed. Can you imagine the amount of blind-faith and wishful thinking that would take? It's bordering on the infinite!

The first time I read this, I found it severely disturbing. I'm a scientist. I have always been a scientist.

*Scientific inference* is a logic fallacy.

According to John Pinel, making up stories out of thin air and then presenting them as scientific fact is what makes science so much fun.

In this quote, John Pinel actually declares scientific inferences to be an empirical method; and by his own admission, he literally treats his scientific guesswork as empirical evidence. That's an oxymoron – a contradiction in terms. But, he declares right up front that ALL of his "science" is based upon scientific inferences. Do you find that as disturbing as I did? Probably not if you are a Materialist, Naturalist, Darwinist, Nihilist, or Atheist because their whole Philosophy of Science is based exclusively on scientific inferences, logic fallacies, and faulty falsified interpretations of the evidence.

Pinel is absolutely clueless and doesn't even seem to know that *scientific inferences* are logic fallacies. Scientific inferences are NOT science – they are in fact *ad hoc just-so stories* designed to match a person's pre-conceived and pre-chosen conclusions. *Scientific inferences* are synonymous with the *affirming the consequent* logic fallacy, the *begging the question* logic fallacy, the *ad hoc just-so story-telling* logic fallacy, and the *jumping to conclusions* logic fallacy.

By his own admission, John Pinel literally treats these logic fallacies as empirical methods, empirical evidence, and scientific evidence because it makes science so much fun.

Through scientific inferences, you can provide scientific evidence and empirical evidence proving any conclusion you want to be true simply by creating a story out of thin air (a scientific inference) and then presenting that fictional story as fact.

I couldn't believe it. I had no idea how far science had fallen – thanks to Materialism, Naturalism, Darwinism, Nihilism, and Atheism – until after reading the first two chapters in John Pinel's book.

Science has become completely worthless now that self-declared "scientists" are treating *scientific inferences* as empirical methods and empirical evidence. Unbelievable! I have KNOWN that it was so for a few years now; but, this is the first time that I have encountered someone who has blatantly admitted that it is so, in a college science textbook no less. I was shocked. *Scientific Inferences* are *ad hoc just-so stories*, which are logic fallacies. Scientific inferences are science fiction – they are assumed to be true without any observational evidence to support them – a logic fallacy that is called *begging the question* or *jumping to conclusions*. Incredible, is it not? I wouldn't have believed it, if I hadn't seen it with my own eyes.

After reading his opening chapters in his book *Biopsychology*, I simply KNEW for a fact that this world desperately needs a Science that's based exclusively upon observational evidence, experiential evidence, eye-witness evidence, empirical evidence, and the Lived Experiences of the human race as a whole.

I present to the world Science 2.0, an upgraded version of science that was designed and meant to be an evidence-based form of science, rather than a science which is based upon logic fallacies such as scientific inferences, affirming the consequent, begging the question, jumping to conclusions, and ad hoc just-so stories concocted to make the scientific inferences seem real and true.

Science 2.0 is just such a science. Science 2.0 defines science as observation, experience, experimentation, eye-witness evidence, observational evidence, experiential evidence, knowledge, truth, and fact.

Science 2.0 is a massive upgrade over the former version of science which defined science as Materialism, Naturalism, Darwinism, Nihilism, Atheism, and Scientific Inferences.

Under Science 2.0, any kind of Observational Evidence or Experiential Evidence takes precedence over and simply trumps scientific inferences, personal interpretations of evidence, sophistry, affirming the consequent, philosophical assumptions, philosophical suppositions, ad hoc just-so stories, and guesswork.

According to this updated version of science, Science 2.0 defines Materialism, Naturalism, Darwinism, Nihilism, and Atheism as UNSCIENTIFIC, because there has never been and there never will be any observational evidence nor empirical evidence to support their major premises or primary assumptions, which claim that the Non-Local Realm and Quantum Non-Locality do not exist. Remember, it's philosophically and scientifically and logically impossible to observe that something does not exist. Since Materialism, Naturalism, Darwinism, Nihilism, and Atheism are based exclusively on *scientific inferences* and *ad hoc just-so stories* masquerading as empirical evidence and fact, these pseudo-sciences, false religions, and falsified philosophies are unscientific at best and downright fraudulent hoaxes at worst. These pseudo-sciences are based upon non-observations of things that do not exist and can never be observed – they are based upon *scientific inferences*, which are logic fallacies.

It's time for us to upgrade our science to Science 2.0, which is based exclusively upon observational evidence and experiential evidence, and which leaves the interpretation of the evidence and the interpretation of the science experiments up to you, the reader and the user of the evidence.

Nowadays, whenever I encounter research evidence, experimental evidence, "scientific evidence", experiential evidence, eye-witness evidence, and observational evidence of any kind, I provide my own personal interpretation of the evidence rather than blindly accepting the "official interpretation" of the evidence which is provided by the Materialists, Naturalists, Darwinists, Scientists, Nihilists, and Atheists.

As a scientist, I have trained myself to think for myself, and I encourage my readers to do the same. Providing your own interpretation of the Observational Evidence and Eye-Witness Evidence is what Science 2.0 is all about. No more *scientific inferences* and *ad hoc just-so stores* IS our motto. Give me evidence – lots and lots of evidence – and I will decide for myself what it means to me.

Science 2.0 has NO "official interpretation" of the evidence. When presenting Observational Evidence, I will at times tell you what that evidence means to me personally, which is permissible under Science 2.0; but, I know that you might in fact have a better interpretation of the evidence than mine. You are always going to believe your own personally chosen interpretation of the evidence over mine anyway; and like me, you are going to design and create your own interpretations of the evidence no matter what I say or what I believe; so, why not let you do so and encourage you to do so in the first place? Once you understand the evidence and know where to find it, you can interpret it just as well as I can.

Under Science 2.0, it's permissible for me to provide you with my own personal interpretation of the evidence or science experiment; but, it's not permissible for me to make my own personal interpretation of the evidence the ONLY officially acceptable interpretation of the evidence, as the Materialists and Naturalists do. There are no credentialed and degreed High Priests of Science 2.0 set in place to bless and sanctify the Official Interpretation of the Evidence. We just present the evidence – all the evidence – and then leave it up to you to decide what to make of it. It's a better way of doing science and interpreting science.

Science 2.0 is all about presenting the Observational Evidence and Empirical Evidence, and then leaving it up to the reader to decide for himself or herself how to interpret it and what to make of it. That's the way science should be done, anyway. You are going to make your own interpretation of the evidence, no matter what I say or what I personally believe, so why not let you do so and encourage you to do so in the first place? Such a process is in keeping with the spirit of Science 2.0.

This is a new, upgraded, and different way of doing science that gives Observational Evidence precedence and treats "ad hoc just-so scientific inferences" as logic fallacies and pseudo-science. This is the way that science should have been, all the way along.

## Defining Science

1. Etymology can be extremely instructive.

Definitions for the Latin words *experior, experiri, expertus*:

>   attempt, try
>
>   find out
>
>   prove, experience
>
>   test, put to the test

The English words "experiment" and "experience" derive from the Latin word, *experiri*, which means to try, to prove, to find out, and to know.

The English word "experience" is synonymous with KNOWLEDGE.

Definitions for Empirical: In a general sense, "being guided by mere experience". Related: "Empirically" is defined as "by means of observation, experience, and experiment". "Empirical Evidence" is evidence that is acquired by means of observation, experience, and experiment.

"Empirical" derives from the Latin *empiricus*, which is a transliteration of the Greek *empiricos* (empirical, experienced;) from *empiria* (experience;) from *en-* (in, with) + *pira* (experience, trial;) from the verb *pirao* (to make an attempt, try, test, get experience, endeavor, or attack).

"Empirical Evidence" is synonymous with KNOWLEDGE.

Definitions for Science: "What is known, knowledge (of something) acquired by study; information;" also "assurance of knowledge, certitude, certainty." The English word "science" is from Old French *science* "knowledge, learning, application; corpus of human knowledge." The Old French word "science" is from the Latin *scientia* "knowledge, a knowing; expertness," which is from *sciens* (genitive *scientis*) "intelligent, skilled," a present participle of *scire* "to know."

The original definition for the word "Science" is KNOWLEDGE.

The English word KNOWLEDGE is synonymous with observation, experience, and experimentation.

"Observation" is an act of recognizing and noting a fact or occurrence often involving measurement with instruments – "seeing is believing" – we simply KNOW the things that we experience and observe.

The word "knowledge" is from an Old English word that meant knowledge. *Gnosis* is a feminine Greek noun which means "knowledge". Spanish has a couple of different words for "to know". In contrast, the English word "knowledge" or "to know" has a bunch of different definitions, some of which are no longer used.


2. Now, let's think about this whole thing logically and rationally.

It is impossible to KNOW for a surety that something DOES NOT EXIST. Such an attempt results in a logic fallacy sometimes known as *proving a negative*, but more accurately known as "proving that something or someone does not exist".

*Proving that someone or something does not exist* is impossible for all practical purposes, especially when it comes to metaphysics, ontology, extra-dimensionality, non-locality, the universe, reality, and existence.

For example, in order to know for sure that God does not exist, one would have to be able to go, look, and see in every corner and fraction of this universe, in every other universe, in every other dimension or reality or alternate reality including the non-local dimensions and the transdimensional dimensions and see for himself or herself that God does in fact not exist. If you can actually do that, then you are by definition for all practical purposes a God; and then, God does in fact exist. In other words, you would have to be a God in order to KNOW for a surety and to PROVE as fact that God does not exist.

*Proving a Negative*: From the Wikipedia, "A negative claim is a colloquialism for an affirmative claim that asserts the non-existence or exclusion of something. Saying "you cannot prove a negative" is a pseudo-logic because there are many proofs that substantiate negative claims in mathematics, science, and economics including Arrow's impossibility theorem. There can be multiple claims within a debate. Nevertheless, whoever makes a claim carries the burden of proof regardless of positive or negative content in the claim."

Depending on the task at hand, it may or may not be possible to *prove a negative*. However, I have observed that it is impossible *to prove that someone or something does not exist* in such a manner as to make everyone else believe and know that you are right.

Proving that something does not exist is impossible, which by definition in principle makes any attempt to prove that something does not exist "unscientific". Claiming that something does not exist is "unscientific", because it can never be proven nor falsified. It's philosophically and logically and scientifically impossible to prove that something or someone does not exist.

The ONLY thing we can do as scientists is to demonstrate and prove that something does in fact exist. We do so through observation, experience, and experiment. Once we have either observed it, experienced it, or demonstrated it through experiment, then we simply KNOW that it exists and that it is real.

I have upgraded my science to Science 2.0, which is defined as KNOWLEDGE that has been acquired through observation, experience, and experimentation. Science 2.0 defines "science" as Empirical Evidence or KNOWLEDGE. Science 2.0 allows ALL of the empirical evidence into evidence, which means that Science 2.0 allows ALL of the evidence into evidence. Nothing is excluded nor denied. It ALL comes in as Scientific Evidence.

Science 2.0 is the way that science was originally meant to be. I'm trying to return science to its roots, by defining Science as Empirical Evidence and KNOWLEDGE.


3. Materialism is the philosophical claim that the non-physical, the non-local, the spiritual, non-local consciousness, psyche, and the immaterial DO NOT EXIST.

Naturalism is the philosophical claim that the supernatural, the transdimensional, the non-local, the spiritual, non-local consciousness, psyche, and God DO NOT EXIST.

Behaviorism is the philosophical claim that Psyche, Soul, Mind, Free Will, and Non-Local Consciousness DO NOT EXIST and ARE NOT NECESSARY.

Nihilism is the philosophical claim that the Afterlife, the Spirit World, Heaven, and Hell DO NOT EXIST. Nihilism is the philosophical claim that life has NO purpose or meaning.

Atheism is the philosophical claim that God DOES NOT EXIST. Creation Ex Nihilo is creation from nothing by nothing. Creation Ex Nihilo is Atheism. Creation Ex Nihilo, Abiogenesis, Spontaneous Generation, and Macro-Evolution have never been observed nor experienced nor witnessed. These things have been FALSIFIED by Science. Atheism is the philosophical claim that a Designer and a Creator DOES NOT EXIST and WAS NOT NEEDED.

How do these people KNOW that these things DO NOT EXIST?

They don't!

They are simply *guessing* and making *scientific inferences*, which are logic fallacies based upon *wishful thinking*. There's NO SCIENCE, NO KNOWLEDGE, behind their *wishful thinking* – it's pure philosophical guesswork and speculation. It's sophistry, not science.

There's no way in this universe to observe and empirically prove that these things DO NOT EXIST. In fact, ALL of the empirical evidence, experiential evidence, observational evidence, eye-witness evidence, corroborated veridical evidence, experimental evidence, and scientific evidence tells us that these things do in fact exist.

The very existence of the Quantum Zeno Effect, Quantum Entanglement, Consciousness, Gravity, Magnetism, Dark Matter, Dark Energy, Light, Near-Death Experiences (NDEs), Out-of-Body Experiences (OBEs), and Shared-Death Experiences (SDEs) FALSIFIES Materialism, Naturalism, and their derivatives.

Remember: Proving that something or someone does not exist is impossible, which by definition in principle makes any attempt to prove that something does not exist "unscientific". Claiming that something does not exist is "unscientific", because it can never be proven nor falsified. It's philosophically and logically and scientifically impossible *to prove that someone or something does not exist*.

This Reality and Truth and Knowledge by definition and for all practical purposes makes Materialism, Naturalism, Behaviorism, Nihilism, and Atheism "UNSCIENTIFIC".

So, how do the Materialists, Naturalists, Darwinists, Behaviorists, Nihilists, and Atheists deal with and get around the incongruence of this Empirical Fact?

Easily!

These people define "science" axiomatically as Materialism, Naturalism, Darwinism, Behaviorism, Scientism, Nihilism, and Atheism; and then, these people claim that everything else is "not science". It's the ONLY argument that these people ever make and the ONLY case that they ever present. They axiomatically by fiat define "science" as Materialism, Naturalism, Darwinism, Behaviorism, Scientism, Nihilism, and Atheism; and then as their primary corollary, they define everything else as "not science". It works; and, they have been doing so effectively for thousands of years now. The words may change or evolve over the millennia, but their overall tactic does not.

Unfortunately, defining "science" as Materialism, Naturalism, Darwinism, Behaviorism, Scientism, Nihilism, and Atheism involves *affirming the consequent, begging the question*, and *jumping to conclusions* which are logic fallacies and make for BAD SCIENCE. So, even though these people have succeeded in defining "science" as Materialism, Naturalism, Darwinism, Behaviorism, Scientism, Nihilism, and Atheism, due to all the different logic fallacies that these people employ, Materialism and Naturalism and their derivatives end up being by definition in principle BAD SCIENCE; and, they don't even KNOW IT!

By the rules of logic, making a claim that something DOES NOT EXIST carries with it a Burden of Proof, through a preponderance of the evidence. The Materialists, Naturalists, Behaviorists, Nihilists, and Atheists are claiming that something or someone DOES NOT EXIST; and, the burden of proof falls upon them to demonstrate and prove through a preponderance of the evidence that they are right.

So, what do these people do with their Burden of Proof? They shift it to you and demand that you prove them wrong. That's what the Materialists, Naturalists, Darwinists, Behaviorists, Nihilists, and Atheists do – they shift their Burden of Proof to you, and demand that you prove that they are wrong, which I have done.

I have proven the Materialists, Naturalists, Darwinists, Behaviorists, Nihilists, and Atheists WRONG trillions of times in thousands of different ways. These people shifted their Burden of Proof to me, and I went out and proved them wrong by using observational evidence, experiential evidence, empirical evidence, scientific evidence, eye-witness evidence, corroborated veridical evidence, and logical common sense to FALSIFY their philosophical claims. These people will tell you that it can't be done; but, they are WRONG.

It's really easy to use empirical evidence, experiential evidence, observational evidence, and eye-witness evidence to FALSIFY Materialism, Naturalism, and their derivatives once you KNOW how to do it; but, this is something that you will never learn how to do from the Materialists, Naturalists, and Atheists who run our public schools. It's something that you will have to figure out on your own; but, since the Materialists and Naturalists have shifted their Burden of Proof to you, you have every right to do so. Go out and prove them wrong!

Meanwhile, I have upgraded my science to Science 2.0, which defines "science" as observation, experience, experimentation, knowledge, fact, and truth. I have observed that the best way to find and know the truth is to live it and experience it for yourself, or to choose to trust someone who has. The second-best way to find and know the truth is to falsify and eliminate everything that is false – such as Materialism, Naturalism, Darwinism, Behaviorism, Nihilism, and Atheism – so that ONLY the truth remains. If you successfully eliminate everything that is FALSE, then only the TRUTH will remain. That's Logic 101.

Remember, ALL of the new evidence that comes into Science FALSIFIES Materialism, Naturalism, Darwinism, Behaviorism, Determinism, Scientism, Nihilism, and Atheism; and, that's pretty BAD, considering the fact that there was NEVER any evidence to support these things in the first place. We have always had to take these things on blind-faith as being true, because there was no other way to support them and sustain them.

## Foundational Tenets of Science 2.0

The BEST way to find and know the truth is to live it and experience it for yourself, or to choose to trust someone who has. If you are going to choose to trust someone, use your brains and choose to trust someone whom you deem to be absolutely trustworthy. You should do this with any type of evidence.

Although Science 2.0 allows all of the evidence into evidence, since the personal interpretation of the evidence is left up to YOU under Science 2.0, carefully examine that evidence and decide for yourself whether it is trustworthy or not. Under Science 2.0, YOU are called upon to decide for yourself whether evidence is reliable or not.

Science 2.0 is based upon the Burden of Proof Methodology and encourages you to pursue and follow the Preponderance of the Evidence.

Since there is NO officially sanctioned interpretation of the evidence under Science 2.0, you are left free to judge for yourself whether evidence is trustworthy and reliable, or not. This means that you are completely free to reject evidence if you deem it untrustworthy, unreliable, faked, or false – or if it fails to meet its Burden of Proof through a Preponderance of the Evidence.

The SECOND-BEST way to find and know the truth is to eliminate everything that is False and everything that has been Falsified, so that ONLY the Truth remains.

We start this process by eliminating Materialism, Naturalism, Darwinism, Nihilism, Behaviorism, Hard Determinism, Reductionistic Atomism, and Atheism because they are False and have been Falsified by Science and Observation. If you can successfully eliminate everything that is false and everything that has been falsified, then logically only the truth will remain standing.

Even though the Scientific Methods cannot be used to prove anything true, they can indeed be used to prove things false. When you falsify something, you do in fact prove that it is false. The Scientific Methods along with Direct Observations are extremely powerful tools for falsifying False Philosophies and False Theories such as Materialism, Naturalism, Darwinism, Nihilism, and Atheism.

It works.

How often have I said to you that when you have eliminated the impossible [and the false], whatever remains, *however improbable*, must be the truth? — Sherlock Holmes.

Eliminating the impossible and the false is what Science 2.0 is supposed to do, but only if you make it work for you. Just like in a court of law, Science 2.0 leaves it up to you to decide for yourself if the evidence meets its Burden of Proof through a Preponderance of the Evidence.

## Invisible Forces and Fields

**The day science begins to study non-physical phenomena, it will make more progress in one decade than in all the previous centuries of its existence.** — Nikola Tesla

Wind is an invisible force that we can't see directly with our physical eyes.

So, how do we know that it exists?

We know that the wind exists by the effect that it has on physical matter. We can feel that it exists by the effect that it has on our skin and hair.

Gravity and magnetism are a couple of invisible, immaterial, supernatural, non-physical forces or fields that we can't see directly with our physical eyes nor detect directly with our physical instruments.

So, how do we know that they exist?

We know that gravity and magnetism exist by the effects that they have on physical matter. The effects that they have on physical matter is undeniable. However, the very existence of gravity and magnetism – the proven and verified existence of gravity and magnetism – falsifies the claims of Materialism and Naturalism which state that such supernatural, invisible, non-physical forces and fields do not exist.

Likewise, Dark Energy, Quantum Waves, Quantum Mechanisms, Action at a Distance, and the Human Psyche are invisible, immaterial, supernatural, non-physical forces or fields that we can't see directly with our physical eyes nor detect directly with our physical instruments.

So, how do we know that they exist?

We know that they exit by the effects that they have on physical matter. Action at a Distance, Quantum Tunneling, and the effects of the Human Psyche have been caught in the act by the observable influence that they have on physical matter. The proven and verified existence of Quantum Mechanics or Transdimensional Physics FALSIFIES Materialism, Naturalism, Darwinism, and their derivatives.

Entropy is one more of those invisible forces or fields or phenomena that we can't see directly with our physical eyes, nor can we detect it directly with our physical instruments.

So, how do we know that it exists?

We know that it exists by the effects that it has on physical matter. We see things running down all around us. The increasing disorder, corruption, and decay is impossible to deny.

Likewise, Syntropy is yet another of those invisible forces or fields or phenomena that we can't see directly with our physical eyes, nor can we detect it directly with our physical instruments.

So, how do we know that it exists?

We know that Syntropy exists because entropy exists. Entropy wouldn't have been possible without a huge infusion of Syntropy and Order at the beginning of our physical universe.

Logical. Simple. Parsimonious. True.

The Syntropy had to come from the Quantum Realm, or the Syntropy Realm, or the Psyche Realm because there doesn't seem to be any observable Syntropy within physical matter. Physical matter only seems capable of collecting entropy. Physical matter doesn't seem to be capable of producing Syntropy.

Instead, Syntropy seems to be an inherent part of Quantum Mechanics, Supernatural Mechanisms, Transdimensional Physics, Psyche, Non-Local Consciousness, Intelligence, Dark Matter or Spirit Matter, God, and Love – eternal and everlasting, without a beginning of days or an end of years. Syntropy seems to be the original primal state of existence; and, it's the physical matter which is the anomaly in all of this.

Gravity, Magnetism, the Strong Nuclear Force, the Weak Nuclear Force, the Zero-Point Field, Dark Energy, Quantum Waves, Quantum Mechanisms, Action at a Distance, and the Human Psyche ARE all manifestations of Syntropy. Even entropy is a manifestation of Syntropy. The only thing that doesn't seem to fit within this paradigm is physical matter. Physical matter seems to be the very manifestation of entropy. How fascinating is that?

Do you see what becomes possible by choosing to allow ALL of the evidence into evidence and then pursuing a preponderance of that evidence? We find ourselves approaching Truth or Syntropy.

# PART II – FIXING SCIENCE WITH QUANTUM MECHANICS

Quantum Mechanics, or Syntropy, or Psyche is the answer to life, the universe, and everything.

One of the most useful things I ever did was to use Quantum Mechanics to upgrade science from Materialism and Naturalism TO Observation and Experience. I upgraded my science to Science 2.0. Rather than defining science as Materialism, Naturalism, Darwinism, Nihilism, and Atheism as most college professors do, I now define science as Observation and Experience.

Unless it has been experienced and observed by Someone Psyche somewhere sometime, it really isn't science. The major premises or hidden assumptions of Materialism, Naturalism, Darwinism, Nihilism, and Atheism – which claim that the quantum or the supernatural does not exist – have not been and cannot be experienced and observed. Therefore, Materialism and Naturalism are NOT science, because their major premises or hidden assumptions cannot be experienced nor observed but have to be taken on blind faith as being real and true.

The observed, experienced, and verified existence of Quantum Mechanics, Psyche, the Biblical God Jesus Christ, and Action at a Distance FALSIFIES Materialism, Naturalism, Darwinism, Nihilism, Atheism, and their derivatives.

The truth falsifies the false; and, the truth has been repeatedly verified, experienced, and observed.

Mark My Words

—

**Reference**

*Science 2.0: I Upgraded My Science.*
https://www.amazon.com/dp/B0771K6WTX

## Syntropy vs. Entropy

Syntropy is Mind-Over-Matter or the Placebo Effect. If we are paying attention, we will notice that it shows up everywhere in our physical universe. Syntropy or Mind-Over-Matter shows up a lot in Social Psychology. Obviously, it should because Social Psychology is the study of the interaction between the Human Psyche and all those other psyches, on every level of existence.

### Syntropy vs. Entropy

In the cyclical nature of the Universe there are two opposing principals, Entropy and Syntropy. To understand syntropy, it is useful to understand its opposing force, entropy.

The Principle of Entropy has effects that are easy to understand and observe. According to this Principle, all organized forms of matter require more energy than those that are less organized. These organized forms will lose their order and initial energy unless they are constantly nourished by energy. Plants need water and sunlight to grow (energy); when that is deprived they begin to breakdown. A new car will fall apart unless energy is applied to maintain it, and a business that doesn't put energy into maintaining itself will close its doors. As life forms age their systems lose energy becoming less efficient allowing for disease and illness to set in. The tendency of nature, systems, and organisms to lose energy and become disorganized over time, essentially the law of death is entropy.

Entropy's opposing force, however, is Syntropy. This Principle is much harder to notice, the best example of which, is life itself. It is the law of order and organization, finality and differentiation, the ability to attract, evolve and bring together ever-increasing complex forms creating something new. A new galaxy forming from the ruins of an older galaxy, building a business, and the bringing together of individual cells to create an organism are all examples of the syntropic nature of the Universe. Consciousness focusing energy to create and maintain a system; the law of life is Syntropy.

In an organism the degree of internal disorganization is entropy and the level of internal organization is syntropy. The sum of these two quantities represents the level of health of an organism in the present and the transformation potential for the evolution of that system or organism.

As far as health is concerned, the more energy we have available to maintain and balance our various systems, the greater our health benefits will be. But if we are overly stressed, in chronic pain, mentally and emotionally drained, most of our energy is being used to sustain these negative states. This means much less energy for our healing and growth processes. Over time an organism's systems begin to break down, dis-ease sets in and entropic forces overtake the syntropic forces until death occurs.

The holistic healing model integrates the whole person in the healing process: body, emotion, mind, and spirit, all of which are different forms of energy. A holistic approach utilizes non-invasive treatments and natural healing techniques to ultimately change the energy associated with disease and dysfunction, bringing balance and a higher degree of homeostasis to the patient.

Aside from practitioner's treatments, a major aspect of holistic healing is the patient's active participation in the process. The outcome is subject to the patient's personal desire, intent, and follow-through in their healing process. Self-healing by means of a balanced diet, exercise, adequate rest, and stress release techniques to name a few are imperative for anyone wanting to heal themselves and become more whole.

Syntropy Energetics is an integrative approach to health and wellness, which utilizes holistic therapeutic modalities and life practices to bring about change in a client. These healing techniques work with different energies of the body at the different layers and levels in which they manifest. *Primal Reflex Release Technique* (PRRT), works with the physical body, utilizing reflexes to quickly release pain, tension, and stress. *Subtle energy healing* accesses our subtle energy fields to create instant change at these primary levels of organization. *Biodynamic Cranial Touch* is a means to access the different levels of consciousness associated with energy, allowing *it* to return the client to a state of wholeness.

http://syntropyenergetics.com/Syntropy_vs._Entropy.html

I'm not promoting syntropy energetics per se; but, I am indeed promoting Syntropy or Mind-Over-Matter. The Placebo Effect is REAL, documented, proven, verified, observed, and powerful. The Placebo Effect or Mind-Over-Matter is an integral part of Science 2.0, Spirit Matter, and Syntropy. Likewise, Action at a Distance or Transpersonal Psychology is an integral part Quantum Mechanics, Transdimensional Physics, Psyche, and Syntropy. It has been experienced and observed, which means that it's REAL and truly exists.

The physicists have observed that it requires a huge infusion of energy to elevate electrons from a lower level to a higher level. That's how spirit matter or dark matter is converted into physical matter – by a huge infusion of energy, or syntropy, or God's glory, power, and light. If that energy is released from the physical matter, then it goes back to being spirit matter. Where does this huge infusion of energy come from? It comes from the Quantum Realm, the Syntropy Realm, or the Zero-Point Field which is the Light of Christ. The explanatory power of Syntropy or Quantum Mechanics is through the roof!

*Begging the question* and applying *circular reasoning* – which are logic fallacies – the Materialists, Naturalists, Darwinists, Nihilists, and Atheists define "science" as Materialism and Naturalism. Materialism, Naturalism, Darwinism, and their derivatives are based upon a wide variety of different logic fallacies. Materialism and Naturalism are based exclusively on entropy. There's NO Syntropy where Materialism, Naturalism, Darwinism, Nihilism, and Atheism are concerned.

I have upgraded my science to Science 2.0. Science 2.0 defines "science" as observation and experience, both at the quantum level and at the physical level. Unlike Materialism and Naturalism, Science 2.0 works at both levels of existence – both the psyche level and the physical level. Science 2.0 defines "science" as Syntropy. Quantum Mechanics is Syntropy. It works at both the quantum level and the physical level to explain what's happening at both levels of existence. Quantum Mechanics, Transdimensional Physics, or Syntropy has infinitely greater explanatory power than Materialism, Naturalism, Darwinism, Nihilism, Atheism, and Classical Physics. Materialism, Darwinism, and Naturalism are powerless to explain the Placebo Effect or Mind-Over-Matter.

Please let me demonstrate.

**To consider the answer, consider how we interpret and label our bodily states.**

**The principle seemed to be:** *A given state of bodily arousal feeds one emotion or another, depending on how the person interprets and labels the arousal.*

**A frustrating, hot, or insulting situation heightens arousal. When it does, this arousal, combined with hostile thoughts and feelings, may form a recipe for aggressive behavior.** (*Social Psychology*, p. 368).

We = the Human Psyche.

The person = the Human Psyche.

It's the very same physical input from the environment, the very same physical body, and the very same physical genes making up the physical cells in the physical body; but, we are observing two completely different chosen responses or two different chosen behaviors depending upon how the same Human Psyche decides to interpret and label the incoming physical stimulus.

If the physical harm is interpreted and labeled as deliberate, mean, intentional, and hostile, anger or retaliation tends to ensue.

If the physical harm is interpreted and labeled as accidental, unintentional, non-hostile, and coming from a friend who typically means us well, then compassion and forgiveness tend to be the result.

Everything in these experiments remains the same, except for the way that the Human Psyche chooses to label and interpret the physical harm or the physical stimulus.

Even a dog can tell the difference between a deliberate kick and an accidental trip, even though the physical behavior is precisely the same and has the same physical effect of doing harm to the dog. It's the Psyche that interprets and labels the physical action, and that makes ALL the difference in the world.

This is the smoking gun, where Psychology, Psyche, and Science are concerned. It's the same physical input, but a completely different chosen response or a completely different chosen interpretation, depending upon the way that the Human Psyche or the dog's psyche chooses to interpret and label the physical assault.

Our chosen interpretation of the evidence helps to determine the emotional feelings and the behavioral outcome of every physical event.

**Guns prime hostile thoughts and punitive judgments. What's within sight is within mind. This is especially so when a weapon is perceived as an instrument of violence rather than a recreational item. For hunters, seeing a hunting rifle does not prime aggressive thoughts, though it does for non-hunters.** (*Social Psychology*, 369.)

This is the smoking gun, where Psychology, Social Psychology, Behaviorism, Biology, and Genetics are concerned. We have the very same physical stimulus, but two completely different chosen interpretations given to that physical stimulus.

How's that possible?

It's NOT according to Determinism, Behaviorism, Darwinism, Genetics, Materialism, Naturalism, Classical Physics, and the Theory of Evolution.

Here we are observing the same people, with the same genetics as before the experiment or observation, with the same physical stimulus, but two completely different chosen interpretations or two completely different chosen responses. It's the Human Psyche who is making the choices and doing the interpretation and the labeling – NOT their genes. Their genes have NO way of telling the difference between a gun that is used for recreational purposes and a gun that is used to kill people; but, the Human Psyche does and can.

The society or environment provides the gun; but, it's the Human Psyche who decides what that gun means to that particular Human Psyche. Some Human Psyches decide that the gun means recreation and fun. Other Human Psyches decide that the gun is a good way to kill other people and commit suicide. It's the very same gun, but a completely different meaning given to that gun by a Human Psyche.

Do you see how that works?

It's called Psyche Determinism or a Psyche Ontology, wherein Psyche is the fundamental unit of reality. Psyche is the Ultimate Cause. Psyche is the ultimate causal agent. It is the Psyche's chosen interpretation or chosen label for the physical stimulus that determines how that gun is used and perceived by that specific Psyche who is doing the using, choosing, and the perceiving. Their chosen interpretation of the evidence helps to determine the behavioral outcome of their chosen actions and chosen beliefs.

The Human Psyche can literally transform a physical object typically used for recreation and fun into a murder instrument if it chooses to do so. The Human Psyche can transform this computer into a murder instrument with enough planning and forethought or anger and rage.

Anger, rage, and hate are chosen into existence by the Human Psyche. So are justice, mercy, kindness, friendship, and love. The very same physical object transforms and morphs into something completely different depending upon whether the Human Psyche chooses to love it or chooses to hate it. This reality and truth FALSIFIES Behaviorism, Materialism, Naturalism, Determinism, and even Classical Physics.

The Human Psyche can trump its nature and nurture at will.

The choice or decision as to how to interpret and label the meaning of physical objects is done by the Human Psyche, and NOT the genes. The Human Psyche is being impacted and influenced by the physical input, and NOT the genes. Decisions and choices change the Human Psyche, NOT the genes.

> **Arousal tends to spill over: One type of arousal energizes other behaviors.**
>
> **Other research shows that viewing violence *disinhibits*. In Bandura's experiment, the adult's punching of the Bobo doll seems to make outbursts legitimate and to lower the children's inhibitions. Viewing violence primes the viewer for aggressive behavior by activating violence-related thoughts. Listening to music with sexually violent lyrics seems to have a similar effect.**
>
> **Media portrayals also evoke *imitation*.** (*Social Psychology*, p. 377.)

Thoughts and memories are a product and a function of the Human Psyche.

How do we know?

We KNOW because our thoughts, memories, individuality, and personality survive the death of our physical brain, according to the empirical evidence from Near-Death

Experiences (NDEs), Out-of-Body Experiences (OBEs), Shared-Death Experiences (SDEs), and our after-death Life Reviews.

The viewer in these experiments is the Human Psyche, NOT our genes.

Being a cognitive behaviorist, an atheist, a materialist, and a determinist, Bandura denies the existence of the Human Psyche or the human spirit, but the evidence tells us that he is wrong to do so.

We can't explain what's really happening, both here in the physical realm and there in the spirit realm or quantum realm when we remove the Human Psyche from the equation.

Genes and physical matter don't make choices and decisions for us; and, a careful study of the evidence makes this abundantly clear.

### Television's Effects on Thinking

**We have focused on television's effect on behavior, but researchers have also examined the cognitive effects of viewing violence: Does prolonged viewing *desensitize* us to cruelty? Does it give us mental *scripts* for how to act? Does it distort our *perception* of reality? Does it *prime* aggressive behavior?** (*Social Psychology*, p. 378.)

Cognitions or thoughts are a function of the Human Psyche. Mental scripts and aggressive thoughts are a function of the Human Psyche, NOT our genes. Our genes can't make us do anything we don't want to do. Our genes can't make us believe anything we don't want to believe. Our reinforcement history cannot make us do anything we don't want to do. Our beliefs are chosen into existence by the Human Psyche.

Chosen actions or chosen behaviors are a function and product of the Human Psyche. It's the Human Psyche who is getting desensitized, NOT our genes. Mental scripts and anything within the mind has to do with the Human Psyche and NOT our genes. Physical input or physical stimulus comes in from the physical environment and registers on the physical brain, where it is then transferred or transmitted by the physical brain to the Human Psyche. Then the Human Psyche decides how to interpret and label that physical input, and the Human Psyche chooses how to respond to that physical input or chooses to ignore it. Physical stimuli take place at the physical level, and spiritual stimuli take place at the quantum level; but, our **perception of reality** is done by the Human Psyche at every level or dimension of existence.

Do you see how that works?

Our chosen behaviors or chosen actions are chosen into existence by the Human Psyche. The meaning, or label, or interpretation that is given to a physical object or a physical stimulus is chosen into existence by the Human Psyche. Our beliefs are chosen into existence by the Human Psyche. Love or friendship does not exist until it is chosen into existence by the Human Psyche.

You can grind down the whole physical universe into its constituent parts, and you will never find a single molecule or gene of justice, mercy, kindness, friendship, or love. There's no such thing as justice, mercy, kindness, friendship, or love until it is chosen into existence by Someone Psyche. Furthermore, once these things are chosen into existence by the Human Psyche, these things continue to exist long after your physical genes, physical body, physical society, physical world, and physical brain are dead and gone.

That's just the way things work at every level of existence.

Syntropy is Endless and Eternal without a beginning of days or an end of years. Quantum Mechanics and Psyche are Syntropy. Compared with the Quantum Realities or Eternal Realities associated with the Human Psyche, evolution and evolutionary psychology are completely worthless and inadequate, because evolution and evolutionary psychology are restricted and limited exclusively to the physical realm and classical physics. They don't touch the quantum realm.

We can't use the theory of evolution, evolutionary psychology, behaviorism, and determinism to explain what's happening at the psyche level and what's being done at the quantum level by Syntropy or Psyche to get things accomplished for us here at the physical level. However, we can indeed use Syntropy, Quantum Mechanics, and Psyche to explain these things at every level of existence or every dimension of existence, including both the psyche level and the physical level.

When it comes to science in general, Syntropy, Psyche, and Quantum Mechanics or Transdimensional Physics has infinitely more explanatory power than Materialism, Naturalism, Darwinism, Nihilism, Scientism, Behaviorism, Determinism, Evolutionary Psychology, and Atheism. Syntropy, Psyche, and Transdimensional Physics can be used to explain and understand every type of science and every level of existence that comes our way.

Mark My Words

—

**Source**

***Quantum Mechanics from a Non-Physical Spiritual Perspective***

https://www.amazon.com/dp/B01J023TGU

https://www.amazon.com/dp/1521132380

**References**

***Quantum Mechanics from a Non-Physical Spiritual Perspective***

https://www.amazon.com/dp/B01J023TGU

https://www.amazon.com/dp/1521132380

## Interpretations of Quantum Mechanics

There are literally thousands of different interpretations of Quantum Mechanics, and that is in fact the weakness of Quantum Mechanics. Quantum Mechanics or Transdimensional Physics is a proven and verified science. What hasn't been proven is "what it all means". You cannot prove an interpretation. Interpretation is a function of the Human Psyche, and you cannot use the physical sciences to prove the truthfulness or the falseness of things at the quantum level or the psyche level. The existence of things at the quantum level can only be inferred at the physical level by the effects that these things have on physical matter. That's just the way these things work.

The main problem with Quantum Mechanics is that it has spent decades under the command and control of the Materialists, Naturalists, and Atheists. These people don't have a clue as to how things really work. I KNOW, because I used to be a Materialist, Naturalist, Nihilist, and Atheist. These people neutralize, defoliate, dumb-down, limit, and castrate every theory and idea that comes their way – especially when it comes to science.

The psyche level or the quantum level has REAL effects on physical matter, but Psyche and Quantum Mechanisms cannot be observed directly by our physical instruments, because they are sub-atomic. Remember, when it comes to physics, the smaller can dwell within and control the larger. Transdimensional means non-physical. The sub-atomic or the quantum waves can dwell within and control the quarks, strings, electrons, bosons, gluons, and atoms because the sub-atomic or psyche is smaller than these things. Psyche transmits, receives, and stores quantum waves. Psyche is the fundamental unit of reality – the ultimate elementary particle – because it is the smallest thing that exists.

Materialism, Naturalism, Darwinism, Nihilism, Atheism, Scientism, Behaviorism, Determinism, Atheism, and Classical Physics create unsolvable problems in science.

I was specially prepared to resolve these unsolvable problems in science.

First of all, I successfully FALSIFIED Materialism, Naturalism, Darwinism, and Atheism by using the Scientific Method and *negating the consequent* in order to do so. I was in fact the first person to do so, as far as I know. I learned how to do this by studying the Philosophy of Science, *affirming the consequent*, and *negating the consequent*. Then I used the Scientific Method and negating the consequent in all of my books to FALSIFY Materialism, Naturalism, Darwinism, and Atheism. I observed that ALL of the unsolvable problems in Psychology are immediately resolved by getting rid of Materialism, Naturalism, Darwinism, Nihilism, and Atheism. Materialism and Naturalism caused the binding problem, the mind-brain problem, the nature versus nurture problem, and the free-will versus determinism problem. These problems are automatically solved by eliminating Materialism, Naturalism, Darwinism, and their derivatives. Simple. Fundamental. Logical. Parsimonious.

Second, I was the first person to develop a Psyche Ontology, wherein I demonstrate that Psyche is the fundamental unit of reality. I called it the Ultimate Model of Reality. Look for it in some of my other books. If there is such a thing as a point particle, Psyche or Non-Local Consciousness is it. A point particle would be the smallest unit of reality – the fundamental unit of reality – an infinite singularity. I have hypothesized many times that Psyche, or Quantum Non-Local Consciousness, is an infinite singularity. As such, it is theoretically capable of storing an infinite amount of information in something that has NO physical size and takes up NO physical space whatsoever. The smaller can dwell within and control the larger. Psyche transmits, receives, and stores quantum waves. It's Someone Psyche who chooses the frequency at which each string will vibrate. Thoughts and memories are quantum waves or the vibrations within strings. Simple. Fundamental.

Parsimonious. Logical. All of the complex math points to the existence of these types of things.

Third, I was the first person to develop Science 2.0. Science 2.0 is the way that science should have been done but wasn't. Science 2.0 allows ALL of the evidence into evidence and then pursues a preponderance of that evidence. Under Science 2.0, the BEST and FASTEST way to find and know the truth is to observe it, live it, and experience it, or to choose to trust someone who has. Under Science 2.0, the second-best way to find and know the truth is to use the Scientific Methods and *negating the consequent* to eliminate everything that is false. If you successfully eliminate everything that is false, then ONLY the truth will remain. This is logical, parsimonious, simple, and true. It's fascinating to observe and study what remains, after you have successfully falsified and eliminated Materialism, Naturalism, Darwinism, Nihilism, Behaviorism, Scientism, Determinism, and Atheism. I observed when you successfully eliminate everything that is false, then only the VERIFIED or the OBSERVED or the EXPERIENCED remains. Syntropy, Quantum Mechanics, and Quantum Non-Local Consciousness (Psyche) remain after you have successfully eliminated everything that is false, unfruitful, and unproductive. The verified and the observed remains. What you have actually experienced for yourself remains.

Fourth, based upon Science 2.0, I developed a new science discipline called Quantum Neuroscience. When it comes to Quantum Neuroscience, I chose to allow ALL of the evidence into evidence, and then I pursued a preponderance of that evidence. The discoveries were stunning, and quite unbelievable at first. There's nothing like it on the planet that I know of. Quantum Neuroscience taught me how thoroughly brainwashed into Scientism, Naturalism, Nihilism, and Atheism I really was.

Fifth, I observed that if your interpretation of Quantum Mechanics cannot explain what the Human Psyche and Nature's Psyche are doing at the quantum level, then your interpretation of Quantum Mechanics is absolutely worthless. Materialism, Naturalism, Darwinism, Nihilism, Behaviorism, Scientism, Determinism, Atheism, and Classical Physics CANNOT explain what the Human Psyche and Nature's Psyche are doing at the quantum level, and how they work at the quantum level to get things done for us at the physical level. Interpretations of Quantum Mechanics based upon Classical Physics, the Indeterminacy Principle, Random Diffusion, Materialism, Naturalism, and Entropy CANNOT explain what the Human Psyche and Nature's Psyche are doing at the quantum level or the psyche level. Materialism and Naturalism have NO explanatory power at the quantum level or the psyche level, with means that they are worthless when it comes time to explain these things.

I found two interpretations of Quantum Mechanics that actually explain what Nature's Psyche and the Human Psyche are doing at the quantum level and how they work and function at the quantum level in order to get things done for us at the physical level. I discovered the interface or the bridge between the quantum level and the physical level. The BEST interpretation of Quantum Mechanics is the Orthodox Interpretation of Quantum Mechanics by Henry P. Stapp. It's the interpretation that has the most explanatory power at the quantum level or the psyche level.

Henry P. Stapp explained to me that the mathematics for what the Human Psyche is doing at the quantum level are completely different from the mathematics for what Nature's Psyche is doing at the quantum level. The Human Psyche makes choices and requests of Nature's Psyche. It is Nature's Psyche who collapses the wave function thereby producing a single physical reality as a result.

Heisenberg's Uncertainty Principle is worthless and wrong because it's based upon Classical Physics, Random Diffusion, and Entropy. I, Henry Stapp, and some of his followers modified the Uncertainty Principle by redefining it as "Infinite Possibilities". It's the same difference, but it's a huge difference. Out of ALL the infinite possibilities, the Human Psyche makes a choice. Then, depending upon whether it's physically possible and whether God allows it or not, Nature's Psyche reaches out and collapses all of those infinite possibilities into ONE single physical reality. At that point, it's a done deal. At that point, it's physically real.

Rinse and repeat.

Once again, we start with infinite possibilities. The Human Psyche makes a choice deciding what it wants to do with its physical body. Then Nature's Psyche reaches out into those infinite possibilities and collapses them into one single physical reality. By redefining the Uncertainty Principle as "Infinite Possibilities", we convert Quantum Mechanics from classical physics and entropy over to Syntropy. Infinite Possibilities is Syntropy. In contrast, the Uncertainty Principle is classical physics, random diffusion, and entropy. You have to switch over to Syntropy and Quantum Mechanics if you want to explain what the Human Psyche and Nature's Psyche are doing at the quantum level in order to get things done for us at the physical level. The Uncertainty Principle is a dead-end. Entropy is a dead-end. In contrast, Infinite Possibilities are infinite, endless, and eternal – the very definition of Syntropy. Suddenly, you can explain everything, simply by switching over to Syntropy, Quantum Mechanics, and Psyche.

Sixth, my ultimate science discovery is that there are NO BYTES of programming code or memory within a physical brain. I didn't believe it at first. I refused to believe it at first. My science colleagues rejected it at first, because we have been trained, conditioned, and brainwashed to do so. However, there are NO wires in our brain connecting neurons together into functional BYTES of programming code and memory. A neuron is a switch. It is either ON or OFF. ALL 7,000 synapses on a single neuron get compressed into a single Post-Synaptic Potential (PSP), a single BIT of information, which then determines whether a neuron fires or not. NO messages are transmitted through a synaptic cleft. A synaptic-cleft scrambles everything that comes its way in terms of information or memories. Information transmitted through synapses gets reduced to a single BIT of information, which means that the postsynaptic neuron either fires or it doesn't.

I kept waiting for the Neuroscientists to tell me where the programming code is being stored and how it is being executed. They NEVER did. I kept waiting for the Neuroscientists to tell me where the data or the memories are being stored and how this information is being accessed. They NEVER did. I kept waiting for the Neuroscientists to tell me where all the CPUs and RAM are being stored within the physical brain. They NEVER did. The ONLY thing these people would ever say is that we don't know how these things work, but they do. Clearly, they work. Clearly, they exist. But, we don't know how they work or where they exist. These Materialists and Naturalists couldn't answer these questions and explain this science, because there are NO functional BYTES anywhere within a physical brain at the physical level. You can't answer questions when there are no answers – when your chosen paradigm or chosen model prevents you from providing the answers. By design, Materialism, Naturalism, and Classical Physics have NO explanatory power at the quantum level or the psyche level.

If there really are BYTES of programming code and memory within our brain, then it's ALL being stored and processed at the quantum level or the psyche level, because it doesn't exist at the physical level. It can't exist at the physical level. Why? It's physically impossible to construct functional BYTES out of stand-alone switches with NO wires to connect them. Neurons are stand-alone switches. This means that it's physically impossible to construct functional BYTES of programming code and memories out of neurons at the physical level, because there are NO visible physical wires connecting the neurons or switches together into a functional whole. There are no BYTES observable within a physical brain, which means that those BYTES have to exist and must exist at the quantum level or the psyche level, because clearly they do exist judging by the effect that they have on our physical brain and physical body.

You can only explain what's happening at the quantum level by choosing to allow Syntropy, Quantum Mechanics, and Psyche into the mix. Instantly, everything can be explained, once you have allowed Syntropy, Quantum Mechanics, and Psyche to come in to play.

Seventh, I fell in love with Quantum Maps of Physical Functionality due to the massive explanatory power that it provides. Science is only valuable when it has explanatory power. If your science can't explain what Nature's Psyche and the Human Psyche are doing at the quantum level and how they work at the quantum level, then your science is worthless. By choosing to allow Syntropy, Quantum Mechanics, Transdimensional Physics, Quantum Processing, and Psyche in to play, I discovered that I can explain everything in science in a logical and realistic manner.

The output from an infinite number of quantum computers existing at the quantum level can be MAPPED to a single physical neuron or concept cell.

Even with physical supercomputers running the jobs and the image recognition software, it still takes a ton of time for a computer to "recognize" a face or "identify" a person. The Human Psyche does so instantly. On the input side, our neurons are registers of physical information. That information is transferred directly to Nature's Psyche where it is processed at the speed of thought, and the output from those infinite number of quantum computers is sent to a single concept cell or neuron which Nature's Psyche has MAPPED for that physical functionality at the quantum level. That single neuron or concept cell then fires unilaterally "out-of-the-blue" thereby giving us a physical feeling of recognition.

ALL of the processing and memory storage are done by Nature's Psyche completely outside the conscious awareness of the Human Psyche. The only thing the Human Psyche experiences is the physical feeling of recognition. "Ah, I know that person." The Human Psyche-Quantum Processing-Nature's Psyche Symbiosis is vastly superior to anything that our physical computers are able to do. Nature's Psyche MAPS our physical brain at the quantum level, and then uses those MAPS to get things done for us at the physical level. Instantly, the explanatory power of our science goes through the roof! Suddenly, we KNOW how everything is done and how everything works. Suddenly, we have the answer to life, the universe, and everything.

Materialism, Naturalism, Darwinism, Nihilism, Atheism, Scientism, Behaviorism, Determinism, Atheism, and Classical Physics create unsolvable problems in science.

By allowing Quantum Mechanics and Psyche in to play, ALL of the unsolvable problems in science are automatically solved and go away. Through Quantum Mechanics and Psyche, you can answer every scientific question and scientific hypothesis that comes your way. The explanatory power of Syntropy, Quantum Mechanics, Quantum Processing, Quantum Memory Storage, and Psyche are through the roof. There is NO limit to what you can explain. Everything can be solved. Everything is solved.

Mark My Words

—

**Source**

***Quantum Mechanics from a Non-Physical Spiritual Perspective***

    https://www.amazon.com/dp/B01J023TGU

    https://www.amazon.com/dp/1521132380


**References**

***Quantum Mechanics from a Non-Physical Spiritual Perspective***

    https://www.amazon.com/dp/B01J023TGU

    https://www.amazon.com/dp/1521132380

## Transdimensional Physics

Transdimensional means Non-Local, Non-Physical, or Not-Located in our physical 3D space-time realm. Transdimensional means quantum, or syntropic, or spiritual.

Quantum Mechanics is Transdimensional Physics. Thoughts and memories are quantum waves. Psyche is intelligence or personality. They survive the death of our physical brain, according to the empirical evidence from Near-Death Experiences (NDEs), Out-of-Body Experiences (OBEs), Shared-Death Experiences (SDEs), and our after-death Life Reviews.

According to Quantum Mechanics, you (your psyche) are everywhere in the universe simultaneously, and your spirit body is never far behind. Quantum Tunneling is faster than light. Quantum Teleportation is instantaneous.

Quantum Tunneling and Quantum Omnipresence also apply to physical matter, but not as much as they apply to spirit matter and psyche because Nature, the Physical Laws, Entropy, and/or God deliberately and greatly limit the "omnipresence" of physical matter.

### Is Quantum Tunneling Faster than Light?

https://www.youtube.com/watch?v=-IfmgyXs7z8

https://syntropy.site/wp-content/uploads/2018/04/Is-Quantum-Tunneling-Faster-than-Light.zip

In this video, while talking about Quantum Tunneling and your physical body, he states:

**Wouldn't it be nice to be everywhere at once? According to Quantum Mechanics, you are. At least a little bit.**

Talking about your physical body he says:

**Any material object is really a matter wave. You're everywhere in the universe, but not very much.**

Remember this!

According to Quantum Mechanics, you are everywhere at once. According to Quantum Mechanics, you are omnipresent. So, ask yourself why you aren't omnipresent right now. What has been done to you, particularly what has been done to your physical body, so that you are NOT omnipresent right now? This IS the answer to life, the universe, and everything. Something has been done to physical matter and your physical body so that you are mostly right where you are right now. Done by whom? And, what precisely has been done?

In this video, he actually answers the question and explains why your physical body isn't everywhere all at once. He also indirectly, though not explicitly, explains why your spirit body and particularly your psyche WERE everywhere at once or omnipresent but aren't anymore.

It has to do with the De Broglie Wavelength of each of these different objects – psyche, spirit body, and physical body. They each have a different De Broglie Wavelength.

**This De Broglie Wavelength defines how well determined an object's position is. A large wavelength means a highly uncertain position [omnipresence]. A small wavelength means a well-defined position** [the locality of physical matter]. **That's true of subatomic particles, and it's sort of true of anything. Right now, I'm mostly right here. But there's also a small chance that I'm here, here, or here. There's an infinitesimal chance that I'm on the moon. Observe me, and you'll collapse my wave function and probably find me pretty much exactly where you expect to.**

Of course, he is only talking about physical matter because these people don't believe in the existence of spirit matter and psyche; but, this De Broglie Wavelength applies equally as well to spirit matter and psyche as it does to physical matter. Quantum Mechanics works at ALL levels or dimensions of existence. The explanatory power of Quantum Mechanics, Psyche, and Spirit Matter are infinitely greater than anything we can get from being limited exclusively to classical physics, physical matter, Materialism, and Naturalism.

Subatomic particles ARE spirit matter and/or psyche. Think about it! It's true! It's only when we get to the atoms that we are actually talking about "pure physical matter" if there is such a thing.

According to the Orthodox Interpretation of Quantum Mechanics by Henry P. Stapp, the Human Psyche makes ALL of the choices as to what it wants to do with its spirit body and its physical body, while it is Nature's Psyche or God's Psyche who actually collapses the wave function thereby making the Human Psyche's choices actual and real both in the quantum realm and here in the physical realm.

During the making of this physical universe, God took a small portion of dark matter or spirit matter, slowed it down to sub-light speeds, gave it a short De Broglie Wavelength, and made it subject to entropy or space-time. God imposed limitations or physical laws upon it. God imposed structure and order upon it. Remove those restraints, and it would go back to being dark matter or spirit matter.

It's all defined and explained by the De Broglie Wavelength.

A physical body has a short De Broglie Wavelength, which means that it has been localized or made physical and present in this 3D space-time realm. It's mostly here, rather than being everywhere. Physical matter has been made subject to entropy or the passage of time. Entropy is a function of space and time. Entropy doesn't exist in the Quantum Realm or Syntropic Realm. Physical matter has been localized into both space and time. Spirit matter and psyche have not.

In contrast to physical matter, a spirit body has an innately long De Broglie Wavelength, which means that it tends toward being omnipresent all at once. A psyche or intelligence or consciousness seems to be capable of an infinite De Broglie Wavelength, or TRUE omnipresence.

Physical beings such as us are comprised of three different entities, and each one of them has a different De Broglie Wavelength. The Law of Quantum Superposition allows for all three to exist in the same space at the same time in an additive fashion, because they each are out of phase with the other or exist in a different dimension than the other. These realities explain everything.

This is cool science!

In its native original state, your Psyche, Quantum Non-Local Consciousness, or Intelligence experiences Quantum Omnipresence. It's capable of being simultaneously everywhere in the universe. According to Quantum Mechanics, your Psyche IS everywhere in the universe all at once, in its native original state. A Psyche is a pinprick of light; but, its thought waves or its quantum waves are omnipresent, or its De Broglie Wavelength is infinite. The spirit matter in your spirit body has many of the same capabilities. A large De Broglie Wavelength means a highly uncertain position or Quantum Omnipresence.

A Psyche or Intelligence is restrained by and limited by being assigned to a spirit body; but, out-of-body explorers have observed and experienced the fact that the Psyche or Consciousness can separate from the spirit body and observe the spirit body from an immaterial viewpoint in space. It's the Psyche who has the experiences and forms new memories, and NOT the spirit body, and definitely NOT the physical body.

Being attached to a physical body with a silver cord greatly limits and restrains a spirit body and a psyche even more by making the spirit or the soul subject to a short De Broglie Wavelength or a physical body. God – the Maker of physical matter, physical universes, and physical bodies – deliberately forces physical matter to be entropic, limited to sub-light speeds, and made to have a very short De Broglie Wavelength. We will never be able to propel physical matter faster than the speed of light because God deliberately prevents it from happening by the physical limits that He has placed upon physical matter.

However, as the Maker and Enforcer of these physical laws or physical limitations, God doesn't place and hasn't placed these physical limitations on His own physical body. His physical body is fully capable of Quantum Tunneling, Teleportation, and even Omnipresence. With no physical limitations being imposed upon it, even a physical body is capable of teleportation, quantum tunneling, and omnipresence just like a spirit body and psyche. It's amazing to think about, isn't it?

Most scientists spend little or no time thinking about the spiritual side of the equation, because they have convinced themselves that it does not exist. However, their chosen beliefs don't change the fact that the Psyche and the Spirit Body have been experienced and observed by millions of different people.

Psyche leads, and your spirit body follows; but, both seem capable of experiencing Quantum Omnipresence or a very long De Broglie Wavelength, as well as Quantum Tunneling or Teleportation at will, while they are separated from the physical body. Psyche and a spirit body are capable of an infinite velocity or teleportation because they have little or no mass. The capabilities of Psyche and a Spirit Body seem to be limitless.

However, while contained within or married to a physical body, a spirit body and a psyche are also limited to a short De Broglie Wavelength. They have to temporarily separate from the physical body in order to experience a longer De Broglie Wavelength and omnipresence.

God deliberately and knowingly designed physical matter to significantly limit, inhibit, restrain, and block the effects of Quantum Tunneling, Quantum Teleportation, Quantum Omniscience, and Quantum Omnipresence. Physical matter has been given a short De Broglie Wavelength, which means that it has locality or presence, which means that it has physical limitations and a well-defined position or location in space-time. Those physical limitations or physical laws show up in Classical Physics, the Laws of Thermodynamics, and the Theory of Relativity, which state that it is physically impossible to accelerate physical matter faster than the speed of light; but, those physical limitations don't apply to the quantum realm or the spirit world.

Violation of relativity, teleportation, quantum tunneling, quantum omnipresence, omniscience, and faster-than-light travel seem to be an inherent and natural part of the quantum realm or psyche realm. In contrast, physical matter and physical laws greatly reduce quantum uncertainty thereby significantly increasing predictability, dependability, stability, and control. The result of all this is the Ultimate Consensus Reality, a physical reality.

Furthermore, as the designer and creator of these physical laws and physical limitations, God granted himself an exemption allowing Him to continue to teleport or quantum tunnel His physical body instantaneously anywhere in the physical universe that He wants to go. God holds the KEYS to Quantum Tunneling or the teleportation of physical matter; and for obvious reasons, God only gives those abilities or KEYS to those whom He trusts. God also continues to experience Quantum Omnipresence and Quantum Omniscience. God's Psyche and Spirit is capable of being everywhere at once and capable of multitasking. The same reality apparently applies to His resurrected physical body as well.

If our Psyche is natively omnipresent and omniscient, then why in the universe would we ever allow ourselves to be limited by a spirit body or a physical body? We've answered some of this already, but it's still a great deal of fun to think about.

As I see it, the answer lies in Order, Organization, Standardization, Structure, and Consensus. Limitations, rules, and laws open up new possibilities. Spirit body limitations and physical body limitations allow us to experience gender and sexual activity, which really isn't possible if you were to remain as a pinprick of light or a Psyche. A physical reality is the Ultimate Consensus Reality. I can depend on my dog, car, wife, and house being there after I get home from work. In a physical reality, I can depend upon this essay being there on my hard drive and in the cloud when I go looking for it tomorrow. None of that would be possible if I were to have chosen to remain as a Psyche or a pinprick of light. It's hard to interact with physical objects if you have no hands. Limitations, structure, order, commandments, or laws provide new opportunities. A physical body does have its advantages.

God deliberately designed physical matter to greatly limit your quantum mechanical nature and properties. A physical reality is the Ultimate Consensus Reality, meaning that it is highly stable, predictable, controllable, reliable, and dependable. You can actually do science in a physical reality. God deliberately designed physical matter and this physical reality to greatly limit, inhibit, control, restrain, and contain your innate quantum capabilities. It's necessary to provide the order, stability, and predictability of this physical realm. In a physical realm, our actions and choices actually have serious consequences. It's REAL because it hurts. It's all theory, until after it has been experienced and observed. The physical makes it real.

God created this physical universe and your physical body to keep your spirit body and your psyche mostly where you currently are – localized here in this physical 3D space-time realm; but, a physical realm is NOT our original innate state of existence. Transdimensional Physics or Quantum Mechanics is the NORM; and, a physical reality is a highly limited and restricted sub-set of that NORM.

Again, according to Quantum Mechanics, you (your psyche or spirit) exist everywhere in the universe simultaneously. God designed this physical universe and your physical body as a school-ground to keep your spirit or soul caged or penned mostly within your physical body so that you can learn from the experience what it is like to have limitations.

Why is this important?

It's because limitations are the ONLY way to provide us with order, structure, predictability, control, law, stability, organization, creativity, and some type of physical life or physical existence. The Ultimate Consensus Reality, a physical reality, is ONLY possible through limitations, structure, order, organization, laws, and agreed upon restrictions. A consensus reality is the opposite of random chaos and omnipresence. A single physical reality is the opposite of infinite possibilities. We learn best and learn fastest while in a physical body. The goal is to learn self-control and responsibility – to learn how to limit, restrain, direct, respect, and choose our actions and behaviors.

Imagine what your physical life would be like without the restraints being imposed by God and His physical laws. Without restrictions such as entropy, space-time, locality, and the theory of relativity, the atoms within your physical body could quantum tunnel away on you at will, before long leaving you without a physical body. How would you feel if your physical body were to quantum tunnel away on you one atom at a time as you slowly dissolve into thin air? Your physical body has the innate capability to quantum tunnel away on you; but, God prevents it from doing so by the physical laws or short De Broglie Wavelength that He imposes on your physical body.

How would you feel if your hard drive, or your car, or your house, or your spouse were to quantum tunnel away from you one atom at a time? If such a thing were to happen as a regular part of our lives, then this physical reality would no longer be the Ultimate Consensus Reality. God uses the physical laws or the physical limitations of this physical universe to prevent this from happening to you.

Do you see why this is important? Do you see now why the commandments of God, rules, laws, limitations, orders, and restraints are important to you and your physical body?

I find it fascinating to try to figure out how things really work at every level of existence or reality, and why they are made to work that way.

## Transdimensional Physics Has NO Physical Limitations

Transdimensional means non-physical, non-local, and not located in our Physical Consensus Reality. Transdimensional means not located in our local physical 3D space-time realm. Transdimensional Physics is Quantum Mechanics – the way that matter ordinarily "acts" in the complete absence of physical laws to restrain it and constrain it.

Remember, physical matter was designed to limit and restrain the effects of Quantum Mechanics. Physical matter was designed by God to localize Action at a Distance or to shorten the De Broglie Wavelength. Physical matter was designed and created to form the Ultimate Consensus Reality – reliable, predictable, dependable, and controllable.

In its original native state, spirit matter or dark matter or quantum matter had full access to all aspects of Transdimensional Physics or Quantum Mechanics. Physical matter retains those same innate capabilities. So, who or what prevents the atoms in your physical body from quantum tunneling away on you? Who or what prevents you from dissolving into thin air?

It's God's physical laws, God's physical restraints, God's physical limitations, or God's control over physical matter that keeps your physical body from quantum tunneling away on you one atom at a time to the nether reaches of the universe. It's that short De Broglie

Wavelength which God has placed upon your physical body that keeps it from quantum tunneling away from you at will.

By creating the Ultimate Consensus Reality, a physical reality, God uses the physical laws to dampen, restrain, and limit the Quantum Mechanics or the Transdimensional Physics that are an innate part of all matter. By converting spirit matter into physical matter, God infuses space-time, locality, entropy, physical restraints, physical limitations, a short De Broglie Wavelength, and physical laws into the newly organized physical matter thereby greatly increasing the improbabilities or decreasing the probabilities that the physical matter will teleport away on you one atom at a time.

Quantum Mechanics or Transdimensional Physics is an innate capability of ALL matter. The Physical Laws or Classical Physics greatly dampen, restrain, limit, and minimize the quantum mechanisms that are an innate part of all matter.

So, who or what is preventing your car, your home, your wife, your computer, and your dog from quantum tunneling away on you one atom at a time? It's God, the Physical Laws, and the short De Broglie Wavelength that are dampening, limiting, and restricting the quantum laws or transdimensional physics so as to provide us with the Ultimate Consensus Reality, a physical reality.

A physical reality was designed by God to be reliable, predictable, and controllable. Once again, imagine what would happen to you and your life if the atoms in your hard drive, your physical body, your car, and your home retained the ability to quantum tunnel or teleport at will anywhere in the universe that they wanted to go. The Physical Laws that God put into place prevents them from doing so.

Nevertheless, the physical atoms and physical molecules do indeed retain their innate ability to quantum tunnel or teleport; and, these physical particles can and do quantum tunnel whenever and wherever God permits them to do so.

Fascinating, is it not?

What would it be like if God were to remove the physical limitations or physical restraints from your physical body, yet leave you here on this earth? What would it be like if God were to give you complete psychic control over your own physical body? What would your life be like? What would you be able to do?

Obviously, you would be able to teleport your physical body anywhere you wanted it to go.

What else would you be able to do?

Your life would be the ultimate in mind-over-matter. You would be able to heal your physical body at will.

You would be able to phase-shift or walk through walls.

You would be able to levitate or walk on water.

If you had complete psychic control over every atom in your physical body, you would be indestructible. They wouldn't be able to kill you. If they were to surprise you and blow you up, you could just reassemble yourself. Theoretically, you should even be able to force the atoms in your physical body to stay in place or stay coherent even while you are standing in the middle of a nuclear explosion or in the middle of the sun.

You should be able to turn off all pain signals within your physical body.

You could even let people kill you, and then reassemble yourself and teleport out of there later on if you wanted to do so.

Are there human beings who have demonstrated or experienced any of these quantum capabilities or transdimensional capabilities?

Yes, indeed there are.

My best friend has experienced phase-shifting. An elk passed through his truck unphased. Either God wanted my friend to live, or God wanted that elk to live. His friend also experienced phase-shifting – a whole car and family passed through their car and his family during their vacation. Obviously, God wanted those families to live for whatever reason.

I have experienced quantum tunneling or teleportation. I and my mother's car were teleported to safety. That was an interesting experience, let me tell you. It was nothing that I did. It was something that God did for me in order to save me. Obviously, in the process of saving me from physical harm, God also taught me that it's possible for Him to quantum tunnel physical bodies and physical cars at will.

Both my best friend and I have experienced that "bubble of protection" wherein God makes you an indestructible super-hero for a few seconds.

The Bubble of Protection, Quantum Tunneling, and Teleportation:

https://www.youtube.com/watch?v=DmBnCTuQUOc

This stuff sounds like science fiction; but, thanks to Quantum Mechanics or Transdimensional Physics, it's very real indeed. Quantum Mechanics is the best-proven and most-used science that we have. It's the Materialism, Naturalism, Nihilism, and Atheism that are the science fiction in all of this – whether we realize it or not.

## Quantum Weirdness

So, ask yourself, "Do you want the truth, or do you want a convenient fiction? Do you want infinite explanatory power where your science is concerned, or do you prefer to be kept in ignorance and darkness where your science is concerned?" This is the million-dollar question.

When I was a Nihilist and Atheist, I wanted the convenient fiction. I wanted to die and cease to exist. I wanted annihilation. I was a sinner, and I was in a very bad situation. You can't in good conscience end your life when you believe that God exists, and you believe that there is going to be some kind of afterlife or judgment. I desperately needed the convenient fiction at the time; and, that's precisely what I received.

We Materialists, Naturalists, Darwinists, Nihilists, and Atheists call it "quantum weirdness" or "spooky action at a distance". But, there's nothing weird or strange about it once a person chooses to interpret Quantum Mechanics from a spiritual perspective or a non-physical perspective.

### *Quantum Mechanics from a Non-Physical Spiritual Perspective*

https://www.amazon.com/dp/B01J023TGU

https://www.amazon.com/dp/1521132380

Nevertheless, I used to be a Materialist, Naturalist, Nihilist, and Atheist back in 2012. God let me experience that and be that so that I would know what it is and what it is like. At the time, when it came to Transdimensional Physics or Quantum Mechanics, I wasn't seeing it, and I wasn't having it. I didn't believe it. I didn't want it. More than anything else, I didn't want God to exist, and I didn't want there to be an afterlife. I truly wanted to die and cease to exist. I wanted annihilation. That is what I wanted most. And, in a very real sense, I got my wish. That person or personality eventually died and ceased to exist.

Boy, am I seeing things differently now! I can see now that the explanatory power of Transdimensional Physics is through the roof. The explanatory power of Quantum Mechanics is limitless and infinite. It explains everything. There's nothing that you can't explain when you finally choose to bring Quantum Mechanics, Psyche, and Syntropy into the mix.

Eventually, you realize that Quantum Mechanisms or Supernatural Mechanisms have been experienced and observed in real life by real people. They have been caught in the act. They are real; and, they are true. I've even experienced them; but, I didn't know what they were at the time, so I tried to explain them away as a figment of my imagination or thought that maybe I had passed out or some such.

I experienced Quantum Teleportation of my physical body and the car that I was driving. One instant I was traveling 45 mph and about to broadside a car that had pulled out in front of me; and, the next instant I was 30 feet away in the middle of a lawn, with the car completely stopped and the engine turned off. I felt time stop; and, I have NO memory of the intervening travel. It's as if I had passed out and woke up in the middle of the lawn with the car stopped and the engine turned off. I had experienced Quantum Tunneling or Teleportation; but, I didn't know what it was at the time. I explained it away as having passed out, because I didn't have a scientific explanation for it. For all I know, I might have experienced phase-shifting as well and actually passed through that other car unphased. I remember hitting the brakes; but, I have no memory of how I got into the middle of the lawn.

To physical mortal beings like us, Quantum Mechanisms or Supernatural Mechanisms seem like magic or miracles, but only because God uses physical laws or our physical consensus reality to deliberately limit and restrict the normal and natural effects of Quantum Mechanics or Transdimensional Physics while we are here in this physical realm. Physical matter and classical physics put a deliberate damper on Quantum Mechanisms or Supernatural Mechanisms or Spiritual Mechanisms. The physical limitations are real, and they truly have an impact because they were designed to be that way. The physical makes everything seem real, and the physical makes everything dependable and reliable and predictable. When it comes to the physical, cause and effect truly come into play and are obviously real.

However, God doesn't apply the same physical limitations to Himself.

Jesus Christ has been observed walking on water, levitating into the air, teleporting, and walking through walls with His physical body. These are all quantum mechanical phenomena or transdimensional phenomena. Remove the physical limitations, and even physical matter can be made to do these quantum mechanical functions. Quantum Mechanics and Quantum Tunneling remain an innate part of physical matter, although God typically prevents them from happening as a regular every-day part of our lives.

I'm not going to apologize for finding Quantum Mechanics or Transdimensional Physics interesting. It is science after all, or it should be. It has been experienced and observed. Quantum Mechanics is found all throughout the Bible. Transdimensional Physics has been caught in the act.

### Luke 4: 28-30 New International Version (NIV):

**28** All the people in the synagogue were furious when they heard this.

**29** They got up, drove him out of the town, and took him to the brow of the hill on which the town was built, in order to throw him off the cliff.

**30** But he walked right through the crowd and went on his way.

Jesus phase-shifted and/or teleported to safety. This really happened. It was experienced and observed.

### Luke 4: 28-30 King James Version (KJV):

**28** And all they in the synagogue, when they heard these things, were filled with wrath,

**29** And rose up, and thrust him out of the city, and led him unto the brow of the hill whereon their city was built, that they might cast him down headlong.

**30** But he passing through the midst of them went his way.

While here in mortality, Jesus Christ had the ability to phase-shift and teleport. Luke was a physician or a medical doctor, which is one of the reasons why he went out of his way to document the eye-witness accounts of all this quantum mechanical stuff that Jesus could do while here in mortality and after He rose from the dead.

### Luke 24 New International Version (NIV)

### Jesus Has Risen

**24** On the first day of the week, very early in the morning, the women took the spices they had prepared and went to the tomb. **2** They found the stone rolled away from the tomb, **3** but when they entered, they did not find the body of the Lord Jesus. **4** While they were wondering about this, suddenly two men in clothes that gleamed like lightning stood beside them. **5** In their fright the women bowed down with their faces to the ground, but the men said to them, "Why do you look for the living among the dead? **6** He is not here; he has risen! Remember how he told you, while he was still with you in Galilee: **7** 'The Son of Man must be delivered over to the hands of sinners, be crucified and on the third day be raised again.'" **8** Then they remembered his words.

**9** When they came back from the tomb, they told all these things to the Eleven and to all the others. **10** It was Mary Magdalene, Joanna, Mary the mother of James, and the others with them who told this to the apostles. **11** But they did not believe the women, because their words seemed to them like nonsense. **12** Peter, however, got up and ran to the tomb. Bending over, he saw the strips of linen lying by themselves, and he went away, wondering to himself what had happened.

### On the Road to Emmaus

13 Now that same day two of them were going to a village called Emmaus, about seven miles from Jerusalem. 14 They were talking with each other about everything that had happened. 15 As they talked and discussed these things with each other, Jesus himself came up and walked along with them; 16 but they were kept from recognizing him.

17 He asked them, "What are you discussing together as you walk along?"

They stood still, their faces downcast. 18 One of them, named Cleopas, asked him, "Are you the only one visiting Jerusalem who does not know the things that have happened there in these days?"

19 "What things?" he asked.

"About Jesus of Nazareth," they replied. "He was a prophet, powerful in word and deed before God and all the people. 20 The chief priests and our rulers handed him over to be sentenced to death, and they crucified him; 21 but we had hoped that he was the one who was going to redeem Israel. And what is more, it is the third day since all this took place. 22 In addition, some of our women amazed us. They went to the tomb early this morning 23 but didn't find his body. They came and told us that they had seen a vision of angels, who said he was alive. 24 Then some of our companions went to the tomb and found it just as the women had said, but they did not see Jesus."

25 He said to them, "How foolish you are, and how slow to believe all that the prophets have spoken! 26 Did not the Messiah have to suffer these things and then enter his glory?" 27 And beginning with Moses and all the Prophets, he explained to them what was said in all the Scriptures concerning himself.

28 As they approached the village to which they were going, Jesus continued on as if he were going farther. 29 But they urged him strongly, "Stay with us, for it is nearly evening; the day is almost over." So, he went in to stay with them.

30 When he was at the table with them, he took bread, gave thanks, broke it, and began to give it to them. 31 Then their eyes were opened, and they recognized him, and he disappeared from their sight. 32 They asked each other, "Were not our hearts burning within us while he talked with us on the road and opened the Scriptures to us?"

33 They got up and returned at once to Jerusalem. There they found the Eleven and those with them, assembled together 34 and saying, "It is true! The Lord has risen and has appeared to Simon." 35 Then the two told what had happened on the way, and how Jesus was recognized by them when he broke the bread.

### Jesus Appears to the Disciples

36 While they were still talking about this, Jesus himself stood among them and said to them, "Peace be with you."

37 They were startled and frightened, thinking they saw a ghost. 38 He said to them, "Why are you troubled, and why do doubts rise in your

minds? 39 Look at my hands and my feet. It is I myself! Touch me and see; a ghost does not have flesh and bones, as you see I have."

40 When he had said this, he showed them his hands and feet. 41 And while they still did not believe it because of joy and amazement, he asked them, "Do you have anything here to eat?" 42 They gave him a piece of broiled fish, 43 and he took it and ate it in their presence.

44 He said to them, "This is what I told you while I was still with you: Everything must be fulfilled that is written about me in the Law of Moses, the Prophets and the Psalms."

45 Then he opened their minds so they could understand the Scriptures. 46 He told them, "This is what is written: The Messiah will suffer and rise from the dead on the third day, 47 and repentance for the forgiveness of sins will be preached in his name to all nations, beginning at Jerusalem. 48 You are witnesses of these things. 49 I am going to send you what my Father has promised; but stay in the city until you have been clothed with power from on high."

### The Ascension of Jesus

50 When he had led them out to the vicinity of Bethany, he lifted up his hands and blessed them. 51 While he was blessing them, he left them and was taken up into heaven. 52 Then they worshiped him and returned to Jerusalem with great joy. 53 And they stayed continually at the temple, praising God.

If you had complete control over every atom in your physical body, you could raise yourself from the dead at will.

### John 20:19-20 New Living Translation (NLT)

### Jesus Appears to His Disciples

19 That Sunday evening the disciples were meeting behind locked doors because they were afraid of the Jewish leaders. Suddenly, Jesus was standing there among them! "Peace be with you," he said.

20 As he spoke, he showed them the wounds in his hands and his side. They were filled with joy when they saw the Lord!

This required some kind of phase-shifting and teleportation.


## Transdimensional Physics in Action

John the Beloved was boiled in oil and didn't die.

This event that I will now recount is not found in the Bible, but a church writer named Tertullian makes mention of it in the 36th chapter of a book that he authored ... But, it appears that Jesus had plans for the John that was placed into a vat of boiling oil, and when God decides that you will not die, then you simply will not die!

https://www.deedsofgod.com/index.php/120-95-ad-the-apostle-john-is-forced-to-bathe-in-boiling-oil-in-rome-mainmenu-405

If God decides that you are not going to die, then you are not going to die.

**John was allegedly banished by the Roman authorities to the Greek island of Patmos, where, according to tradition, he wrote the Book of Revelation. According to Tertullian (in The Prescription of Heretics) John was banished (presumably to Patmos) after being plunged into boiling oil in Rome and suffering nothing from it.**

https://en.wikipedia.org/wiki/John_the_Apostle

That's physically impossible!

But, through Quantum Mechanics or Transdimensional Physics, the Human Psyche and God's Psyche can do the physically impossible.

John the Revelator and the Three Nephites were given complete psychic control over the physical atoms within their physical bodies.

3 Nephi 28: 1-40:

**1 And it came to pass when Jesus had said these words, he spake unto his disciples, one by one, saying unto them: What is it that ye desire of me, after that I am gone to the Father?**

**2 And they all spake, save it were three, saying: We desire that after we have lived unto the age of man, that our ministry, wherein thou hast called us, may have an end, that we may speedily come unto thee in thy kingdom.**

**3 And he said unto them: Blessed are ye because ye desired this thing of me; therefore, after that ye are seventy and two years old ye shall come unto me in my kingdom; and with me ye shall find rest.**

**4 And when he had spoken unto them, he turned himself unto the three, and said unto them: What will ye that I should do unto you, when I am gone unto the Father?**

**5 And they sorrowed in their hearts, for they durst not speak unto him the thing which they desired.**

**6 And he said unto them: Behold, I know your thoughts, and ye have desired the thing which John, my beloved, who was with me in my ministry, before that I was lifted up by the Jews, desired of me.**

**7 Therefore, more blessed are ye, for ye shall never taste of death; but ye shall live to behold all the doings of the Father unto the children of men, even until all things shall be fulfilled according to the will of the Father, when I shall come in my glory with the powers of heaven.**

**8 And ye shall never endure the pains of death; but when I shall come in my glory ye shall be changed in the twinkling of an eye from mortality to immortality; and then shall ye be blessed in the kingdom of my Father.**

**9 And again, ye shall not have pain while ye shall dwell in the flesh, neither sorrow save it be for the sins of the world; and all this will I do**

because of the thing which ye have desired of me, for ye have desired that ye might bring the souls of men unto me, while the world shall stand.

10 And for this cause ye shall have fulness of joy; and ye shall sit down in the kingdom of my Father; yea, your joy shall be full, even as the Father hath given me fulness of joy; and ye shall be even as I am, and I am even as the Father; and the Father and I are one;

11 And the Holy Ghost beareth record of the Father and me; and the Father giveth the Holy Ghost unto the children of men, because of me.

12 And it came to pass that when Jesus had spoken these words, he touched every one of them with his finger save it were the three who were to tarry, and then he departed.

13 And behold, the heavens were opened, and they were caught up into heaven, and saw and heard unspeakable things.

14 And it was forbidden them that they should utter; neither was it given unto them power that they could utter the things which they saw and heard;

15 And whether they were in the body or out of the body, they could not tell; for it did seem unto them like a transfiguration of them, that they were changed from this body of flesh into an immortal state, that they could behold the things of God.

16 But it came to pass that they did again minister upon the face of the earth; nevertheless, they did not minister of the things which they had heard and seen, because of the commandment which was given them in heaven.

17 And now, whether they were mortal or immortal, from the day of their transfiguration, I know not;

18 But this much I know, according to the record which hath been given — they did go forth upon the face of the land, and did minister unto all the people, uniting as many to the church as would believe in their preaching; baptizing them, and as many as were baptized did receive the Holy Ghost.

19 And they were cast into prison by them who did not belong to the church. And the prisons could not hold them, for they were rent in twain.

20 And they were cast down into the earth; but they did smite the earth with the word of God, insomuch that by his power they were delivered out of the depths of the earth; and therefore, they could not dig pits sufficient to hold them.

21 And thrice they were cast into a furnace and received no harm.

22 And twice were they cast into a den of wild beasts; and behold they did play with the beasts as a child with a suckling lamb, and received no harm.

23 And it came to pass that thus they did go forth among all the people of Nephi, and did preach the gospel of Christ unto all people upon the face of the land; and they were converted unto the Lord, and were united

unto the church of Christ, and thus the people of that generation were blessed, according to the word of Jesus.

24 And now I, Mormon, make an end of speaking concerning these things for a time.

25 Behold, I was about to write the names of those who were never to taste of death, but the Lord forbade; therefore, I write them not, for they are hid from the world.

26 But behold, I have seen them, and they have ministered unto me.

27 And behold they will be among the Gentiles, and the Gentiles shall know them not.

28 They will also be among the Jews, and the Jews shall know them not.

29 And it shall come to pass, when the Lord seeth fit in his wisdom that they shall minister unto all the scattered tribes of Israel, and unto all nations, kindreds, tongues and people, and shall bring out of them unto Jesus many souls, that their desire may be fulfilled, and also because of the convincing power of God which is in them.

30 And they are as the angels of God, and if they shall pray unto the Father in the name of Jesus they can show themselves unto whatsoever man it seemeth them good.

31 Therefore, great and marvelous works shall be wrought by them, before the great and coming day when all people must surely stand before the judgment-seat of Christ;

32 Yea even among the Gentiles shall there be a great and marvelous work wrought by them, before that judgment day.

33 And if ye had all the scriptures which give an account of all the marvelous works of Christ, ye would, according to the words of Christ, know that these things must surely come.

34 And wo be unto him that will not hearken unto the words of Jesus, and also to them whom he hath chosen and sent among them; for whoso receiveth not the words of Jesus and the words of those whom he hath sent receiveth not him; and therefore he will not receive them at the last day;

35 And it would be better for them if they had not been born. For do ye suppose that ye can get rid of the justice of an offended God, who hath been trampled under feet of men, that thereby salvation might come?

36 And now behold, as I spake concerning those whom the Lord hath chosen, yea, even three who were caught up into the heavens, that I knew not whether they were cleansed from mortality to immortality —

37 But behold, since I wrote, I have inquired of the Lord, and he hath made it manifest unto me that there must needs be a change wrought upon their bodies, or else it needs be that they must taste of death;

**38** Therefore, that they might not taste of death there was a change wrought upon their bodies, that they might not suffer pain nor sorrow save it were for the sins of the world.

**39** Now this change was not equal to that which shall take place at the last day; but there was a change wrought upon them, insomuch that Satan could have no power over them, that he could not tempt them; and they were sanctified in the flesh, that they were holy, and that the powers of the earth could not hold them.

**40** And in this state they were to remain until the judgment day of Christ; and at that day they were to receive a greater change, and to be received into the kingdom of the Father to go no more out, but to dwell with God eternally in the heavens.

God is Syntropy. If God decides that you are not going to die, then you are not going to die.

These three were transfigured, or translated, or turned into seraphim. They can't be killed. Many of their capabilities and powers are discussed. Their capabilities are quantum mechanical, supernatural, and transdimensional in nature and origin. They seem to violate the physical laws or trump the physical laws, because God permits them to do so.

As a race, we are never going to master teleportation or quantum tunneling unless God permits us to do so. We are never going to be able to travel faster than the speed of light unless God permits us to do so. The physical laws that God put into place prevents us from doing so against His will.

Nevertheless, it has been observed that ALL of these quantum mechanical capabilities return to us when our spirits leave their physical body behind. A spirit or a ghost is a holistic combination of both psyche and spirit body. Psyches and spirit bodies function and exist according to the rules of transdimensional physics or quantum mechanics. In other words, spirit bodies and psyches have NO physical limitations whatsoever.

Out-of-Body Travelers and Near-Death Experiencers have observed that their psyche and their spirit body can teleport or quantum tunnel anywhere at will. At the quantum level or the psyche level, we are dealing with Transdimensional Physics or Quantum Mechanics, and not Physical Limitations or Physical Laws.

Powerful, is it not?

God is Syntropy. If God decides that you are not going to die, then you are not going to die. If God wants to teleport you to safety, He can, and He will. If God wants to translate you or transfigure you, He can, and He will. God can do, and He will do, whatever He needs to do and wants to do in order to accomplish His purposes and His goals. God has NO physical limitations. In their native original state, transdimensional physics, quantum mechanisms, psyche, and spirit matter have NO physical limitations because they are pure syntropy.

God gives certain particles of spirit matter physical limitations thereby converting those particles of spirit matter into physical matter. The physical matter is then used to create or organize the Ultimate Consensus Reality, a physical reality. A physical reality is predictable, controllable, reliable, and dependable. A physical reality is the ultimate in order, law, and organization. A physical reality is powerful because it's limited, ordered, organized, contained, controlled, predictable, dependable, and lawful.

A physical reality, being the Ultimate Consensus Reality, ends up being the opposite of chaos.  That's its advantage, and that's also its limitation.  Remember, when it comes to a physical reality, its limitations are its advantage.  The atoms in your physical body aren't going to quantum tunnel away on you willy-nilly while they are subject to the physical limitations of God's physical laws.  The physical limitations or God's Physical Laws make our physical bodies possible, reliable, dependable, controllable, and predictable.

Mark My Words

—

**Source**

> ***Quantum Mechanics from a Non-Physical Spiritual Perspective***
> > https://www.amazon.com/dp/B01J023TGU
> >
> > https://www.amazon.com/dp/1521132380
>
> ***Putting Psyche Back into Psychology:  Restoring Science to Consciousness***
> > https://www.amazon.com/dp/B071NC987S
>
> ***NATURE vs. NURTURE vs. NIRVANA:  An Introduction to Reality***
> > https://www.amazon.com/dp/B01JWRCSVA
> >
> > https://www.amazon.com/dp/1521132615

**References**

> ***God Is in the Light:  God is light, and in Him is no darkness at all.***
> > https://www.amazon.com/dp/B07168S37N
>
> ***Quantum Mechanics from a Non-Physical Spiritual Perspective.***
> > https://www.amazon.com/dp/B01J023TGU
> >
> > https://www.amazon.com/dp/1521132380
>
> ***Quantum Neuroscience:  The Answer to Life, the Universe, and Everything.***
> > https://www.amazon.com/dp/B079Z6QQQB

## Quantum Field Theory

I'm not the first person on the planet to discover Quantum Field Theory, but as far as I know, I am the first person to identify the Perpetual Motion Cycle for what it truly is. Nobody thinks about these kinds of non-physical things, which is why nobody discovers them.

Quantum Field Theory and the Perpetual Motion Cycle are two of the greatest discoveries in Science and Physics that I have ever encountered. They falsify the second law of thermodynamics, and they are the perfect replacement for the falsified second law of thermodynamics.

The second law of thermodynamics erroneously claims that the amount of disorder (or entropy) in our universe is constantly increasing and that it can never decrease. That is NOT what we experience and observe. We don't observe proton decay, as the second law predicts. We don't observe an ever-encroaching gray goo coming in at us from all sides, as the second law predicts. Instead, we experience and observe constantly conserved order and organization.

Why?

Why do we ONLY experience constantly conserved order and organization? There has to be a reason.

Science gives us the reason, and it falsifies the second law of thermodynamics which states that this constantly conserved order and organization should not exist.

The reason we experience constantly conserved order and organization in our part of the multiverse is thanks to the Quantum Fields and the Perpetual Motion Cycle $E = mc^2$.

Disorder and disorganization ceased to exist in our part of the multiverse the very moment that the Gods and the Controlling Psyches designed, created, and made the Quantum Fields. According to Quantum Field Theory, the Gods and Controlling Psyches had to make the Quantum Fields BEFORE they could make and sustain physical matter. No Quantum Fields, then NO physical matter. It's that simple. The thing that we call the "Big Bang" took place when the Gods and the Controlling Psyches organized and made the Quantum Fields in our part of the multiverse. Only after making the Quantum Fields was it finally possible to make mass, mass's heat storage capacity, heat, and physical matter. Anything that is obviously made obviously has a maker who made it. The Quantum Fields, quantum waves, photons, and physical particles are obviously made. That means that they obviously have some type of Maker who made them. Our job as scientists is to try to determine the True Cause of everything that exists. The Psyches or Intelligences or Gods who made the Quantum Fields ARE the True Cause of everything that exists in our part of the multiverse.

Photons are massless, heatless, chargeless, entropyless, syntropic, non-physical, infinite acceleration quantum waves. The Quantum Fields are massless, heatless, entropyless, syntropic, non-physical perpetual motion machines. The Quantum Fields represent perfect order and organization. We will NEVER have true disorder in our part of the multiverse as long as the Quantum Fields exist. The second law of thermodynamics will NEVER be true so long as the Quantum Fields exist. That's just the way it is. The Quantum Fields are perpetual motion machines, and they will function perfectly, flawlessly, and eternally thanks to the Conservation of Energy and the Perpetual Motion Cycle $E = mc^2$. The Quantum Fields ARE the Perpetual Motion Cycle ($E = mc^2$) in action. They will go on forever, until the Gods or the Controlling Psyches decide to destroy or rearrange the

Quantum Fields. You would have to destroy or disorganize the Quantum Fields in order to make the second law of thermodynamics true.

We will NEVER have true disorder and disorganization in our part of the multiverse as long as the Quantum Fields exist. The second law of thermodynamics will NEVER be true so long as the Quantum Fields exist. That's just the way it is. The Syntropic Entropyless Quantum Fields, Quantum Field Theory, Feynman Diagrams, Quantum Mechanics, and the Perpetual Motion Cycle $E = mc^2$ FALSIFY the second law of thermodynamics.

That explains why we ONLY experience and observe conserved order and organization in our observable universe. Order and organization is the ONLY thing that exists at the quantum level in our part of the multiverse, thanks to the perpetual motion machines that we call the Quantum Fields. The Quantum Fields function perfectly and eternally through the Conservation of Energy and the Perpetual Motion Cycle, which is $E = mc^2$. This Perpetual Motion Cycle is what we actually experience and observe.

Within our sun, Nature's Psyche is constantly transforming mass, heat, and entropy into massless, heatless, chargeless, entropyless, non-physical, infinite acceleration photons or quantum waves. It's happening right now; and, all the while the energy is conserved. Energy cannot be made, and it cannot be destroyed. Energy or Psyche or Light has always existed, and it will always exist. According to the Law of Psyche, every psyche has a certain amount of energy that's under its control, and that controlling psyche can form or transform the energy under its control into anything it wants that energy to be anytime and anywhere that it chooses to do so.

Therefore, every time a photon or quantum wave chooses to stop, it transforms its infinite acceleration into whatever it wants its energy to be. According to Quantum Field Theory, a photon or quantum wave can transform itself into some type of mass, if it wants to. Mass IS heat storage capacity and resistance to acceleration. Every time a photon or quantum wave chooses to stop, it transforms itself into mass or heat. Therefore, what do we observe? When a photon chooses to stop on your skin, do you observe an ever-increasing amount of mass forming onto your skin, or do you experience heat? Well, we experience heat. A photon or quantum wave doesn't have to transform itself into heat. It chooses to transform itself into heat when it lands on your skin because it KNOWS that it has landed on your skin. A photon can just as easily transform itself into mass, such as electrons, quarks, and positrons. That's what it seems to do when it stops on your solar array on your roof. Eventually it finds a way to transform itself into mass, mass's heat storage capacity, or electrons. Electrons have mass, or heat storage capacity, or resistance to acceleration. A photon or quantum wave can transform itself into anything that it wants to be, including heat, mass, quarks, electrons, and physical matter. A photon is intelligent, conscious, psychic, sentient, and alive. It chooses what it wants to be.

This is the Perpetual Motion Cycle in action. All the while, the energy is constantly transformed and constantly conserved. The Perpetual Motion Cycle $E = mc^2$ works perfectly, flawlessly, and endlessly both coming and going. Mass, heat, and entropy are being transformed into massless, chargeless, heatless, and entropyless photons and quantum waves all the time in our sun; and, massless, heatless, chargeless, entropyless, syntropic, exergic, non-physical, infinite acceleration quantum waves and photons are constantly transforming themselves into heat, mass, mass's heat storage capacity, and resistance to acceleration whenever they choose to stop. This is the Perpetual Motion Cycle, and it will go on forever for all eternity thanks to the Quantum Fields, the Conservation of Energy, and $E = mc^2$. You would have to destroy the Quantum Fields in order to stop the Perpetual Motion Cycle and make the second law of thermodynamics true instead.

A photon or quantum wave doesn't have to stop if it doesn't want to.  A photon or quantum wave can pass through our earth, our sun, your physical body, and a black hole as if they weren't even there.  WE KNOW that this is so because we observe photons or quantum waves passing through water and glass all the time.  A photon doesn't have to stop if it doesn't want to stop.  A photon or quantum wave can maintain its infinite acceleration and chosen speed-of-light velocity for as long as it desires to do so.  A photon or quantum wave can also quantum tunnel if it chooses to do so.  It has NO speed limit except for the speed limit that it chooses to impose upon itself.

The verified and proven existence of massless, entropyless, non-physical, syntropic, conserved energy, exergy, quantum waves, photons, and quantum fields FALSIFIES the second law of thermodynamics which states that they do not exist and cannot exist.

Once the scientists in our world figure out what's truly going on, then they will vote to replace the falsified second law of thermodynamics with the Perpetual Motion Cycle $E = mc^2$ and Quantum Field Theory instead.  Quantum Field Theory and the Perpetual Motion Cycle have actually been experienced and observed; whereas, ever-increasing disorder, proton decay, and an ever-encroaching gray goo have NOT.

The one that has actually been experienced and observed ends up being the one that is actually real and true.  That's the way that science is supposed to work.  If we successfully identify and eliminate from science everything that is false and everything that has been falsified, then ONLY the truth will remain.  The truth that remains will consist of the things that have actually been experienced and observed.  The false is falsified by the truth, and the truth is constantly and repeatedly experienced and observed.

Mark My Words

# PART III – SCIENCE 2.0, ORIGINS, AND SYNTROPY

I have been a scientist all of my life; and after 55 years of life, I finally found what I have been searching for all of my life – an explanation for how everything works at all levels of existence or reality.

I found the answer to life, the universe, and everything.

I found what I was looking for in Quantum Mechanics, Action at a Distance, Psyche, and Syntropy. It explains everything!

As a result of these discoveries, I upgraded my science.

I have made some fascinating discoveries during my science career. Ironically, most of what I have discovered has already been discovered by someone else before me; but, I chose to do something unique that has never been done before. I chose to give each scientific discovery a quantum mechanical explanation, a psychic explanation, or a spiritual explanation. I chose to take Quantum Mechanics and Psyche seriously, which most scientists have never done before. Consequently, it makes it seem as if I have made a wide variety of different scientific discoveries that have never been discovered before when in fact I have not. I've simply discovered what others have chosen to ignore and reject.

Even Psyche or Intelligence has been experienced and observed, which means that it has already been discovered; but, I chose to do something different that hasn't been done before. I chose to include it in Science; and, I chose to try to define it and explain what it is and how it works.

Psyche is the innate intelligence within all the different forms of energy that have ever been discovered or observed. The psyche within energy gives all the different forms of energy the ability to hear God's commands, understand God's commands, and obey God's commands. Psyche is the "organizing force" or the "law following force" within our universe. Psyche is identified by choice, particularly its choice to obey the quantum laws and the physical laws.

Physical matter, spirit matter, and dark matter are simply different forms of organized energy existing simultaneously in different phases, different frequencies, or different dimensions. They each have a psyche within them that gives them the ability to obey God's Laws. God provides the Laws; and, the psyche within the different forms of energy provides the obedience to those Laws. Matter is organized energy or lawful energy. Someone Psyche within that matter is choosing to obey the Quantum Laws and the Physical Laws; otherwise, it would be nothing but random chaos.

A LAW in order to be a Law requires some kind of intelligent Law Giver and Law Enforcer. Laws are worthless without some type of uniformity, order, and enforcement. The thing that most people don't realize is that Laws also require Someone Psyche with enough intelligence and understanding to actually choose to obey those Laws. Without Someone Psyche or someone intelligent enough to function as the Law Follower or the Law Obeyer, Laws are also completely worthless. Laws are pointless unless there is someone intelligent enough, within the different forms of energy, who is capable of actually choosing to obey those Laws. This is logical common sense. I have a feeling that it has all been discovered before; but, nobody ever takes the time to actually look at everything from the perspective of Psyche.

Psyche is the innate intelligence, within all the different forms of energy, which gives those different forms of energy the ability to obey God's Laws and God's Commands.

When it comes time to explain life, the universe, and everything, there are only two real contenders – Creation by Chance Evolution and Creation by Intelligent Intervention. They are mutually exclusive. If one is verified and proven to be true, then the other one has been falsified in the process. So, ask yourself which one has been verified and which one has been falsified.

Materialism and Naturalism destroy everything that comes their way because they are based exclusively on entropy or death. There's NO life there! Materialism, Naturalism, Darwinism, Nihilism, Behaviorism, Determinism, Physical Reductionism, Atheism, and Classical Physics are based exclusively on entropy or death. Death cannot design and create. It can only destroy.

The theory of relativity is the pinnacle of Classical Physics, Materialism, and Naturalism. Einstein achieved this breakthrough by denying the existence of Action at a Distance, Quantum Mechanisms, or Supernatural Mechanisms. However, the theory of relativity ONLY applies to physical matter. The theory of relativity or the speed-of-light limitation doesn't exist in the quantum realm, the non-local realm, the transdimensional realm, or the psyche realm. The theory of relativity and entropy do not exist in the Syntropy Realm or the Spirit World. In the Syntropy Realm at the quantum level, everything is timeless and ageless, without a beginning of days or an end of years. In the Syntropy Realm, nothing is subject to Classical Physics or physical limitations. There is NO entropy in the Syntropy Realm because nothing there is subject to the ageing process or the passage of time.

In the end, Quantum Mechanics won, and Classical Physics lost. Quantum Mechanics or Syntropy has infinitely more explanatory power than Materialism, Naturalism, and Classical Physics. That's just the way it is. Quantum Mechanics, Syntropy, or Psyche is better science than Classical Physics, Naturalism, and Physicalism. Scientific Naturalism and the Theory of Evolution are bankrupt.

All of my efforts during the past few years has been an attempt to fix Science and to update Philosophy so as to bring them into the modern era of Contemporary Physics or Quantum Mechanics. My efforts have been an attempt to infuse Syntropy, Psyche, Non-Locality, Action at a Distance, and Quantum Mechanics into Science and Philosophy. I love Philosophy and Science; and, I want to update Science and Philosophy, and thereby make them compatible with Syntropy, or Quantum Mechanics, or Transdimensional Physics.

### *Science 2.0: I Upgraded My Science*

https://www.amazon.com/dp/B0771K6WTX

https://www.amazon.com/author/science

I upgraded my science to Science 2.0.

Science 2.0 defines science as observation and experience, NOT as Materialism, Naturalism, Darwinism, Nihilism, and Atheism. Science 2.0 relies upon the observed, proven, and verified evidence from Quantum Mechanics or Transdimensional Physics rather than the restrictions and limitations of Materialism, Naturalism, Scientism, and Classical Physics. Materialism, Naturalism, and Atheism of any kind are based upon a refusal to look at evidence. Science 2.0 allows ALL of the evidence into evidence, including the quantum evidence, the non-local evidence, or the transdimensional spiritual evidence.

Quantum Mechanics is Syntropy. Syntropy means "without beginning of days or an end of years". Syntropy means "eternal" and "everlasting". Psyche or Quantum Non-Local

Consciousness is a type of Syntropy.  Energy is Syntropy.  Energy or Syntropy is conserved meaning that it is eternal and everlasting without a beginning of days or an end of years.

Science 2.0 allows ALL of the evidence into evidence, and then pursues a preponderance of the evidence.  Science 2.0 is a new and better way of doing science.  Science 2.0 is the way that science should have always been done but wasn't.  Science 2.0 redefines science as OBSERVATION and EXPERIENCE.  Science 2.0 gives observation and experience of any kind precedence over philosophical speculation, sophistry, and wishful thinking.

Science 2.0 abandons the philosophical speculation and philosophical guesswork of Materialism, Naturalism, and Darwinism giving priority and preference instead to Phenomenology, Observational Evidence, Experiential Evidence, Eye-Witness Evidence, Empirical Evidence, and Experimental Evidence.  You see, there is NO empirical evidence supporting the primary assumptions or major premises of Materialism, Naturalism, Darwinism, Nihilism, and Atheism which state that the invisible, the immaterial, and the non-physical do not exist.  Therefore, these pseudo-sciences or false religions are abandoned in favor of Phenomenology or the Lived Experiences of the human race.

Under Science 2.0, Lived Experiences or Phenomenology is given precedence over the philosophical sophistry and scientific inferences of the Materialists, Naturalists, and Darwinists.  Scientific inferences or educated guesses are logic fallacies.  They are inferior to observation and experience.

Science 2.0 is an observed science, or an experienced science, or a proven and verified science.  If we truly consider ourselves to be open-minded scientists, then we should take advantage of everything that science or observation has to offer us, including Quantum Mechanics, Syntropy, and Psyche.

The only way to truly understand Quantum Mechanics comes by choosing to allow ALL of the evidence into evidence.  Something like Quantum Neuroscience is only possible by choosing to allow ALL of the evidence into evidence.  Quantum Neuroscience was a natural outgrowth of Science 2.0.

**Quantum Neuroscience: The Answer to Life, the Universe, and Everything**

https://www.amazon.com/dp/B079Z6QQQB

The book, *Quantum Neuroscience: The Answer to Life, the Universe, and Everything*, makes a detailed comparison between Neuroscience and Quantum Neuroscience, which means that it makes a detailed comparison between Classical Physics and Quantum Mechanics.

The amazing thing about Quantum Neuroscience is that it was already a proven science, and a verified science, decades before I coined the term.  Quantum Neuroscience has just been sitting there for a century now, waiting for someone to pick it up and run with it.  Quantum Mechanics is a verified science or a proven science.  It works.  It's REAL.  ALL of the evidence supports it and verifies it.  Likewise, Quantum Neuroscience is already a verified science or a proven science.

In contrast, Classical Physics, Materialism, Naturalism, Darwinism, Nihilism, Behaviorism, Determinism, and Atheism are based upon a refusal to look at evidence.  These pseudo-sciences or false religions are based upon a deliberate decision to ban, block, censor, ridicule, mock, and destroy certain types of evidence – any type of evidence that FALSIFIES Materialism, Naturalism, and their derivatives.

The very existence of Quantum Mechanics or quantum mechanisms FALSIFIES Materialism, Naturalism, Darwinism, and their derivatives.

Science 2.0 allows Quantum Mechanics or quantum mechanisms into evidence, thereby using them to FALSIFY Materialism, Naturalism, Darwinism, and their derivatives such as the Theory of Evolution.

Quantum Neuroscience is a branch of Science 2.0. Quantum Neuroscience FALSIFIES Materialism, Naturalism, Darwinism, and their derivatives.

In my book, *Quantum Neuroscience: The Answer to Life, the Universe, and Everything*, I make a detailed comparison between Evolution and Quantum Mechanics. If that interests you, I recommend you take a look at the book:

https://www.amazon.com/dp/B079Z6QQQB

Science 2.0 defines science as OBSERVATION and EXPERIENCE. Science is observation, or at least science should be observation.

Science 2.0 is based upon the following Laws or Axioms.

Science 2.0 allows ALL of the evidence into evidence and then pursues a preponderance of the evidence.

Like the forensic sciences, Science 2.0 uses a Burden of Proof Methodology and therefore pursues a preponderance of the observed evidence or a preponderance of the experienced evidence. Science 2.0 leaves the interpretation of the evidence up to you, the ultimate user of that evidence. Your interpretation of the evidence is every bit as good as mine.

Under Science 2.0, "replication on demand" is replaced by the preponderance of the evidence. Replication on demand is ONLY possible under the Ultimate Consensus Reality that God has created – a physical reality. A physical reality was designed to be replicable on demand, reliable, dependable, repeatable, and controllable, which is why God created the thing in the first place. In contrast, the quantum level or the psyche level is primarily comprised of non-consensus realities. They have to be observed, lived, and experienced in order to know that they are real and truly exist. Non-consensus realities mold themselves to your Psyche's demands, while you are in the non-local realm or the quantum realm.

The BEST and FASTEST way to find and know the truth is to observe it, live it, and experience it for yourself, or to choose to trust someone who has.

The second-best way to find and know the truth is to use the *negating the consequent* version of different scientific methods to eliminate everything that is false, so that ONLY the truth remains. If you successfully falsify and eliminate everything that is false – such as Materialism, Naturalism, Darwinism, Nihilism, Behaviorism, Determinism, Scientism, and Atheism – then ONLY the truth will remain. This is Logic 101.

Philosophers of science tell us that the Scientific Method cannot be used to prove the truth.

So, what good is it?

How can you use the Scientific Method to find the truth and know the truth if the Scientific Method can't be used to prove the truth?

It's not easy, but it still can be done.

How?

How do you use the Scientific Method to prove the truth?

You do so through a process of elimination.

The Scientific Method can be used reliably to prove things false or to falsify ideas.

If you successfully use the Scientific Methods to eliminate EVERYTHING that is false, then ONLY the truth or ONLY the verified will remain. If you successfully eliminate everything that has been falsified or everything that is false, then ONLY the verified truth will remain. This is what has been experienced and observed.

After you have successfully falsified and eliminated Classical Physics, Materialism, Naturalism, Darwinism, Nihilism, Behaviorism, Determinism, Atheism, and everything ELSE that has NEVER been observed, has NEVER been experienced, has NEVER been proven, and has NEVER been verified, then ONLY the Truth, the Observed, the Verified, and the Proven remain.

It's fascinating to OBSERVE what remains after you have successfully eliminated or falsified Classical Physics, Materialism, Naturalism, Darwinism, and their derivatives.

**What do you think remains after you have successfully falsified and eliminated Classical Physics, Materialism, Naturalism, Darwinism, Nihilism, Behaviorism, Determinism, and Atheism?**

I can tell you what remains. The Truth remains!

What remains is Quantum Mechanics, Quantum Non-Local Consciousness, Psyche, Intelligence, the Human Psyche, Nature's Psyche, God's Psyche, Spirit Matter or Quantum Objects, the Quantum Law of Complementarity, Action at a Distance or Quantum Non-Locality, Quantum Teleportation or Quantum Tunneling, Phase-Shifting or Quantum Superposition, and Quantum Telepathy or the Quantum Zeno Effect.

Quantum Mechanics and Psyche are what remain after you have falsified and eliminated Materialism, Naturalism, Darwinism, Nihilism, and Atheism.

Cool, huh?

It really works!

Science 2.0 really works!

If you successfully eliminate everything that is FALSE, then only the Truth remains!

Sherlock Holmes taught me how this works in principle or practice.

**How often have I said to you that when you have eliminated the impossible** [and the false]**, whatever remains, however improbable, must be the truth? — Sherlock Holmes.**

This is Science 2.0 that Sherlock Holmes is talking about here. It's based upon falsification and the elimination of everything that is false, such as Physicalism, Naturalism, Darwinism, Nihilism, Behaviorism, Determinism, Physical Reductionism, Atheism, and the Theory of Evolution. The Materialists and Naturalists have successfully gathered together everything that is false and have chosen to believe in it, so these people make it really easy to find it, identify it, and eliminate it.

Using the scientific methods, we narrow in on the TRUTH through a process of elimination, by using the scientific methods to eliminate everything that is false. Once you have eliminated EVERYTHING that is false, then you are left staring at THE TRUTH. It's elementary my dear friend.

It's so obvious, that I sometimes wonder how the human race has been able to overlook it for thousands of years!

Once you have eliminated EVERYTHING that is false, then you are left with THE TRUTH. That's what Sherlock Holmes does.

Sherlock Holmes slowly arrives at the truth by falsifying his theories and his imagined evidence. That's how I finally arrived at THE TRUTH, by falsifying Materialism, Naturalism, Darwinism, Nihilism, and Atheism. THE TRUTH is the opposite of falsehood and falsified theories!

I have used different scientific methods to successfully FALSIFY and ELIMINATE Materialism, Naturalism, Darwinism, Scientism, Behaviorism, and Atheism. The VERIFIED and proven existence of Quantum Mechanisms FALSIFIES Materialism, Naturalism, and their derivatives. Science doesn't get any better than that.

Now, here comes the kicker and the final knock-down blow! THE TRUTH is repeatedly verified by scientific observations and scientific methods. By eliminating the impossible and ALL of the falsehoods, whatever remains must be THE TRUTH! That's how we USE the scientific methods or scientific observations to find THE TRUTH, and to KNOW the truth, and to PROVE the truth! The truth is going to be the opposite of falsified theories, or the opposite of the things which have been proven false by the scientific methods and the observations of the human race.

Checkmate!

The false is repeatedly falsified by the truth; and, the truth is repeatedly experienced and observed.

THIS IS THE KEY!

After you have successfully eliminated everything that is false, the Truth remains; and, the Truth is repeatedly and endlessly VERIFIED by observational evidence, scientific evidence, experimental evidence, experiential evidence, empirical evidence, eye-witness evidence, phenomenology, and the lived experiences of the human race.

What remains after we have successfully FALSIFIED Materialism, Naturalism, Scientism, Darwinism, and Atheism?

What remains is Quantum Mechanics, Quantum Non-Local Consciousness, Psyche, Intelligence, the Human Psyche, Nature's Psyche, God's Psyche, Spirit Matter or Quantum Objects, the Quantum Law of Complementarity, Action at a Distance or Quantum Non-Locality, Quantum Teleportation or Quantum Tunneling, Phase-Shifting or Quantum Superposition, and Quantum Telepathy, Quantum Telekinesis, or the Quantum Zeno Effect.

Quantum Mechanics and Psyche are what remain after you have falsified and eliminated Materialism, Naturalism, Darwinism, Nihilism, and Atheism.

Furthermore, the existence of Quantum Mechanics and Quantum Non-Local Consciousness (Psyche or Intelligence) is repeatedly and endlessly VERIFIED by scientific experimentation, empirical evidence, observational evidence, experiential evidence, eye-witness evidence, phenomenology, and the lived experiences of the human race.

Everything reduces to Psyche, Spirit Matter, Syntropy, and Quantum Mechanics, because Psyche and Quantum Mechanics and Spirit Matter are the core foundation of everything else, including Classical Physics and physical matter.

So, where's the syntropy?

WE KNOW there has to be syntropy somewhere in the system, because someone or something had to wind up the clock in the first place; otherwise, this physical universe would have been born in heat-death with maximum entropy and no thermodynamics whatsoever. This is Logic 101.

Genetic evolution, genetic change, or random mutation is entropy, disease, death, and extinction. So, where's the life or the syntropy?

Psyche, Energy, Spirit Matter, and Quantum Mechanisms had NO beginning, which means that they will have no end, because they are pure syntropy. They neither age nor die, because they are syntropy. Syntropy means ageless – without beginning of days or an end of years. Syntropy is life. Energy is Syntropy. Syntropy or Energy is conserved according to the First Law of Thermodynamics.

In contrast, physical matter and the physical laws (classical physics) had a beginning, which means that theoretically they can have an end. Someone Psyche designed and created the first particle of physical matter. Someone Psyche designed and created the first physical universe, as well as ALL of the subsequent physical universes. Without the existence of Someone Psyche, without the existence of God's Psyche, there would be NO physical matter.

This reality and truth is Scientific Proof of God's Existence. The very existence of physical matter is Scientific Proof of God's Existence, whether we know it or not. Without the existence of God's Psyche or God's Intelligence, there would be NO physical matter and no physical universe. Remember, everything that has a beginning – such as physical matter and physical universes – has a Beginner or a Prime Mover who gets it all going in the first place.

This Beginner, or Prime Mover, or God is PURE SYNTROPY.

The opposite of syntropy is entropy. Evolution (genetic change) or random mutation is entropy. The chemical evolution, abiogenesis, spontaneous generation, or macro-evolution of genes and proteins from scratch is prevented from happening by entropy or the second law of thermodynamics. There is NO empirical evidence and will NEVER be any empirical evidence demonstrating how entropy, chemical evolution, random mutation, random diffusion, abiogenesis, spontaneous generation, or evolution (genetic drift) is able to design and create and produce proteins and genes from atoms, because such a process is physically impossible thanks to entropy or the second law of thermodynamics. Chemical evolution is physically impossible.

You are going to have to decide for yourself if any of this is useful to you, or not. That's something that nobody else can do for you. You have to do it for yourself or go without. Your Psyche determines what you are going to embrace and what you are going to believe.

Quantum Mechanics is Syntropy. Psyche is Syntropy. Spirit Matter is Syntropy. I'm not going to apologize for bringing enlightenment or Syntropy to the physical world.

Mark My Words

# Observational Studies Cannot Establish Causation

https://origin-science.org/observational-studies-cannot-establish-causation/

I'm a statistician, physicist, and mathematician among other things.

When it comes to observational studies and statistics, the important thing to KNOW is that no variables are actively manipulated by the investigator in observational studies; therefore, observational studies cannot determine causality.

Observational studies cannot be used to establish causation.

At best, the theory of evolution is based upon observational studies. At worst, the theory of evolution is nothing but philosophy or wishful thinking. The theory of evolution is nothing but a guess. We are not supposed to guess in science. The theory of evolution and the second law of thermodynamics ARE Creation by Chance or Chance Causality. Where Creation by Chance is concerned, data hasn't been presented to support it, either because it doesn't exist or because it falsifies their claims that chance can design and create at will.

In the summer of 2019, I eventually realized that my whole statistics course in science and physics and the scientific method FALSIFIES Creation by Chance or Chance Causality. Creation by Chance, Creation by Disorder, Creation by Entropy, and Creation Ex Nihilo are falsified by statistics and the whole of science and the scientific method!

We all KNOW that Materialism, Naturalism, Darwinism, Nihilism, Behaviorism, Determinism, Physical Reductionism, Classical Realism, the Theory of Evolution, the Second Law of Thermodynamics, and Atheism ARE Creation by Chance or Creation by Random Disorder or Creation by Entropy. The whole of science and statistics FALSIFIES Materialism, Naturalism, and their derivatives such as Darwinism and Atheism, because these falsified philosophies or falsified religions are in fact Creation by Chance, Creation by Disorder, or Creation by Entropy.

So, what do we see from these people whenever they are called upon to explain themselves? We see them rejecting, banning, censoring, hiding, and destroying evidence. The Materialists, Naturalists, Darwinists, Nihilists, and Atheists literally hide and destroy the evidence whenever it falsifies their beliefs. They do head-in-the-sand "science".

The Null Hypothesis in a science experiment states that the results of the experiment were caused by chance or created by chance, which means that the independent variable had NO EFFECT on the dependent variable. If we are forced to retain the Null Hypothesis, that means that there is NO causal relationship between the variables, and that means that your Alternative Hypothesis is most likely false.

The payoff is huge when we start applying correlation, inferential statistics, and the scientific method to Creation by Chance, or the theory of evolution and the second law of thermodynamics! We quickly observe that Creation by Chance or Creation by Disorder cannot be correlated with anything. Chance cannot be the cause of anything. Whenever we conclude that the Null Hypothesis is true, we have in fact concluded that there is NO correlation between the independent variable and the dependent variable, that the results of our science experiment were caused purely by chance, and that the independent variable had NO EFFECT on the dependent variable. When we are forced to retain the Null Hypothesis, we are forced to conclude that the results of our science experiment were caused by chance alone and that therefore there is NO causal relationship between the independent variable and the dependent variable.

We can't determine causality when it comes to the theory of evolution!

We can't determine causality when it comes to the second law of thermodynamics!

Why?

It's because the theory of evolution and the second law of thermodynamics are Creation by Chance, Creation by Entropy, Creation by Disorder, Creation Ex Nihilo, or the Null Hypothesis. Natural selection is Creation by Death. We can't do science experiments where we manipulate the amount of chance (or disorder) and then observe the effects of chance on a dependent variable. It just can't be done. Chance doesn't have an effect on anything. If it did, then it would no longer be chance!

We can't have "disorder" or "chance" as our independent variable in a science experiment because we can't manipulate chance or disorder and have it remain as chance or disorder. In a True Science Experiment, we vary the levels of the independent variable and then measure its effect on the dependent variable. We really can't have different levels of disorder or chance, because we are looking at order once we have left maximum disorder or chance behind. Furthermore, maximum disorder at the physical level doesn't touch nor change the perfect, entropyless, syntropic, conserved order of the Quantum Fields at the quantum level.

Likewise, varying the amount of "disorder" in a physical system does absolutely nothing to vary the amount of heat or thermodynamics in that system, because there is NO correlation between disorder and heat. Heat doesn't cause disorder, and disorder doesn't cause or eliminate heat. When dealing with Creation by Entropy, Creation by Disorder, or the Second Law of Thermodynamics, their quasi-static methodology, ideal gas, and Carnot engines are impossible to test and verify, because they don't exist in reality. All of this is purely a figment of their imagination (a thought experiment), and so is the second law of thermodynamics that's derived from it. We can't use science experiments to test and verify the things that don't exist. Creation by Chance or Chance Causation does not exist. They are oxymorons, or a contradiction in terms. They don't exist; therefore, we can't test them nor verify them with science experiments.

With "disorder" or "chance" there's no way to vary the levels of disorder or chance for the sake of experimentation, because when you add order into the experiment to get different levels of "disorder" or "chance", then you no longer have disorder or chance but varying degrees of order instead. Manipulating the independent variable introduces a certain level of order and organization into the system. In contrast, pure chance doesn't do anything to touch or change the system. Pure maximum disorder, or chance, is impossible to attain.

This whole problem is exacerbated once you realize that the Quantum Fields are perfect, syntropic, massless, entropyless, and conserved Order and Organization at the quantum level. There really is NO such thing as disorder or chance once the Quantum Fields are taken into account! At maximum disorder, there's NO place to go but upwards into some type of order or organization instead. True disorder hasn't existed in our part of the multiverse ever since the Quantum Fields were designed and made by Nature's Psyche and the Gods.

It's Nature's Psyche or the Controlling Psyches within the Energy who create the quantum waves or the photons in the first place; and then later, it is Nature's Psyche who collapses the wave function and stops the photon thereby transforming that entropyless and omnipresent quantum wave INTO some type of localized mass, heat, resistance to acceleration, or mass's heat storage capacity (entropy) instead. This is the Perpetual Motion Cycle $E = mc^2$ in action, and it works flawlessly for all eternity because energy is always conserved. We will never have true disorder in our part of the multiverse as long as the Quantum Fields, Nature's Psyche, the Conservation of Energy and Psyche, the Conservation of Quantum Information within Psyche, and the Perpetual Motion Cycle continue to exist. You would have to destroy the Quantum Fields in order to make the second law of thermodynamics true. To do that, you would have to destroy God and Nature's Psyche and the Conservation of Energy. The first law of thermodynamics and the second law of thermodynamics are mutually exclusive. They falsify each other. One of them is automatically false.

In a science experiment, there are different levels of effect that we can detect, including the ability to detect NO EFFECT, which is in fact produced by chance alone. If only chance is at work in your science experiment, then you end your science experiment by retaining the Null Hypothesis and conclude that your Alternative Hypothesis is false, all because CHANCE cannot do causation. In Science and Statistics, they use chance as their litmus test to determine whether the independent variable is having an effect on the dependent variable. If chance alone is operating, then WE KNOW for a proven fact that our independent variable is NOT causing the dependent variable, because WE KNOW for a proven fact that chance of any kind cannot do causation. Chance and causation are mutually exclusive. They falsify each other. If you detect causation, then you have falsified chance as your "cause". If you detect or demonstrate chance, then YOU KNOW that there is no causal relationship between the dependent variable and the independent variable.

That's the way Science and Statistics work.

In Science and Statistics, the only way to find and know the truth is to identify, expose, reject, and eliminate ALL of the deceptions and lies. We find the truth by falsifying and eliminating the things that are demonstrably wrong. In these essays, I try to go out of my way to provide you with the very best that I have been able to glean from Science, Statistics, the Scientific Method, Science Experiments, Observations, Experiences, and the Philosophy of Science. My purpose is to find and know the truth, which means that my purpose is to identify and eliminate the lies. My purpose is to identify everything that's wrong with Science and then try to find some way to fix it. My purpose is to design and implement the Ultimate Model of Reality and Existence.

According to the rules of statistics, correlation cannot establish causation. Observational studies cannot establish causation. We have to run a True Science Experiment in order to establish causality.

Alas, Materialism, Naturalism, Darwinism, Nihilism, Atheism, the Theory of Evolution, and the Second Law of Thermodynamics are nothing but observational studies. As such, they cannot be used to establish causation. You have to run a science experiment in order to establish causation, and there's NO way to run a True Science Experiment with Creation by Chance or Chance Causation because chance by definition cannot do causation. Causation and chance are mutually exclusive. They falsify each other. Causation FALSIFIES Materialism, Naturalism, Darwinism, Nihilism, Atheism, the Theory of Evolution, Chemical Evolution, Macro-Evolution, Abiogenesis, Spontaneous Generation, the Second Law of Thermodynamics, and every other form of Chance Causation or Creation by Chance. Physicalism, Mechanism, Naturalism, Darwinism, Atheism, the Theory of Evolution, and the Second Law of Thermodynamics are observational studies (or thought experiments) at best and deceptive lies at worst. They cannot be used to establish causation. They cannot do causation. They are the product and NOT the cause. The purpose of Statistics and Science is to find the True Cause of each event, and chance can NEVER be the true cause of anything.

The payoff is huge with this one. ALL of the falsehoods, deceptions, and lies in "science" are immediately identified and eliminated once we realize that chance cannot do causation. This is cool stuff. This is powerful stuff – but only if you are willing to accept it once you find it. This is also painful stuff, especially for a Materialist, Mechanist, Naturalist, Darwinist, Nihilist, or Atheist. I KNOW because I used to be a materialist, naturalist, nihilist, and atheist. The truth is painful because it falsifies Materialism, Naturalism, Darwinism, Mechanism, Behaviorism, Determinism, Physical Reductionism, Atomism, Atheism, the Theory of Evolution, the Second Law of Thermodynamics, and every other type of Chance Causality or Creation by Chance.

Most of our scientists are unwilling to falsify Creation by Chance, or Chance Causation, or the Null Hypothesis. They find any way they can to accept their Alternative Hypotheses instead. They will manipulate the data and destroy the evidence in order to prove their Alternative Hypothesis true. Atheism of any kind is based upon the destruction of the evidence that FALSIFIES Materialism, Naturalism, Darwinism, Nihilism, Atheism, the Theory of Evolution, or the Second Law of Thermodynamics. That's the only way that these people can retain their Null Hypothesis is by hiding or destroying the evidence that falsifies it.

We can't do observational studies on evolution or natural selection designing, creating, and manufacturing genes, proteins, and genomes from scratch, because evolution of any kind has NEVER been caught in the act of designing and creating anything. Evolution can't design and create because it is Chance Causation or Creation by Chance. Chance can't do causation, which means that chance can't do design and creation.

Observe that Materialism, Naturalism, Darwinism, Nihilism, Atheism, the Theory of Evolution, and the Second Law of Thermodynamics are purely observational studies. Because these things are Chance Causation or Creation by Chance, they cannot be used in science experiments to establish causality. They are purely observational studies, and observational studies cannot be used to establish causation. Chance cannot do causation. Causation and chance are mutually exclusive.

Any attempt to run a science experiment on the theory of evolution, chemical evolution, or macro-evolution has failed – thereby proving that natural selection and the different types of evolution can't design and create anything at all. The reason they can't design and create is because they are in fact Chance Causation, Creation by Chance, or the Null Hypothesis. Chance cannot do causation. Once we observe causation, then WE KNOW that chance is no longer involved. PSYCHE, or CHOICE, or INTELLIGENCE is now involved whenever we observe causation.

The Atheists are deathly afraid of consciousness or psyche. The Materialists, Naturalists, Darwinists, Nihilists, and Atheists are constantly saying that they don't believe in invisible, immaterial, non-physical, supernatural Gods; but, that isn't true. There are a few that these people believe in. True, they don't believe in psyche, quantum consciousness, or intelligence; but, they do believe in natural selection, evolution, chance, and entropy which are in fact invisible, immaterial, non-physical, supernatural Gods to these people. These people imbue evolution, natural selection, chance, and entropy with all the powers and abilities of a God. Evolution, natural selection, chance, and entropy ARE their supernatural replacement for Psyche, Intelligence, or God. These people replaced the True God with a bunch of false gods; but, the Atheists and Naturalists and Darwinists do indeed have gods that they worship and believe in, whether they realize it or not.

Evolution of any kind is a dead-end because it is Chance Causation, Creation by Chance or Random Disorder, or Caused by Chance Alone. Evolution is entropy, and entropy is one of the "creative mechanisms" that the Materialists, Naturalists, Darwinists, and

Atheists have chosen to believe in. It's ALL Creation by Chance or the Null Hypothesis because it is produced by chance alone. Creation by Evolution or Creation by Entropy is a dead-end because these things cannot design and create. They have never been caught in the act of doing so.

The ONLY way to succeed when it comes to "evolution" is through genetic engineering or directed evolution, and that's intelligent design or intelligent causation or choice – NOT natural evolution or natural selection. Anytime we humans start manipulating the genes and the proteins, then we are NO longer looking at evolution or chance but are looking at some type of intelligent causation instead. Only Psyche or Intelligence can do choice, causation, or creation. It's the only thing that has been caught in the act at all levels of existence and reality. Psyche, choice, and causation are synonymous with each other.

Many types of intelligent intervention can cause genes to change. Radiation can damage genes and cause them to change. By definition, genetic engineering can change our genes. In contrast, natural selection doesn't touch our genes! Natural selection doesn't do anything except sit around and wait for us to die. By definition, in principle, natural selection is unable to touch, control, or change our genes. Natural selection is Creation by Chance, Creation by Entropy, and Creation by Death. It doesn't do anything.

The theory of evolution is Creation by Chance. The big bang theory, as the scientists typically employ it, is also Creation by Chance or Creation Ex Nihilo. It's magic or creation without causation. Atheism is Creation Ex Nihilo – the creation of something from nothing by nothing. It's voodoo or magic. It's the second law of thermodynamics, which states that the total amount of entropy or disorder in our universe is constantly increasing and can never decrease or go to zero. It's the creation of something from nothing. It's a Boltzmann brain.

Statistics were designed to identify and eliminate the effects of chance from science. Statistics were designed to eliminate the various different forms of Creation by Chance from science. Statistics was designed to eliminate the theory of evolution, materialism, naturalism, nihilism, atheism, behaviorism, determinism, physical reductionism, creation by chance, and creation ex nihilo from science.

The theory of evolution cries out for empirical verification, but there can NEVER be any empirical verification for Creation by Chance, because chance is completely unreliable, unpredictable, uncontrollable, and impossible to replicate or measure.

There's NO data supporting Creation by Evolution, Creation by Entropy, or Creation by Chance; and, we should already KNOW that CHANCE cannot design and create anything. Yet, Materialism, Naturalism, Darwinism, Nihilism, and Atheism ARE in fact Creation by Chance or Creation Ex Nihilo. These people really truly believe that CHANCE designed and created everything that exists. That's what they believe, is it not?

Creation by Evolution is based exclusively on chance. In contrast, REAL SCIENCE and statistics attempt to eliminate chance from the equation so that we are left with proof of causality instead. There can be NO proof of causality for something that is based exclusively on chance, such as Darwinism, Evolution, or the Big Bang Theory. All we can do with these things is to do observational studies; and, observational studies cannot establish causality. The geniuses in this world should know this, but they don't. Self-deception works, and it works every time, especially when it comes to PhD scientists who believe that they know everything already.

We will NEVER be able to prove that Evolution or Chance caused everything to exist. In fact, since the theory of evolution is Creation by Chance, that effectively proves that the theory of evolution is false, because chance cannot be used as a causal mechanism in science.  In science, whenever we see that the results of our experiments were caused by chance, we are supposed to falsify or eliminate our hypothesis, and start over again.  Having established beyond a shadow of a doubt that evolution and the big bang were caused by chance, we are supposed to toss those theories into the garbage can and start over again, because in science CHANCE cannot be a valid cause for evolution, the big bang, physical matter, genomes, life forms, planets, stars, galaxies, or anything else that we might observe.  Chance cannot do causality.  It would no longer be chance if it did.

In science, whenever chance is the explanation that we come up with, we are supposed to falsify and eliminate our alternative hypotheses and conclude that the independent variable had NO EFFECT on the dependent variable.  In other words, since evolution is Creation by Chance, we are supposed to falsify and eliminate that hypothesis, toss it out, and conclude that evolution or natural selection or chance had NO EFFECT on the proteins, genes, genomes, eyes, brains, life forms, planets, stars, and galaxies that we observe all around us.

We can't have the truth if we are constantly lying to ourselves.

So, why haven't we tossed out the theory of evolution and the second law of thermodynamics since they are Creation by Chance or Creation Ex Nihilo or the Null Hypothesis?  We haven't tossed them out precisely because they are Creation by Chance or Creation Ex Nihilo.  They are precisely what we want to be true, NOT what we can prove to be true.  They are Atheism, Naturalism, Darwinism, Nihilism, and Physicalism which is why they are being preserved.  These people have granted themselves a *special pleading* or a *special exemption* so that they don't have to prove that their theories and ideas are true.  Instead, they simply *jump to the conclusion* that their theories and ideas are true, and then they turn their theories and ideas into axioms and laws so that their theories and ideas can't be falsified by the scientific community.

It's a monopoly.  It's a scam.  It's a deceptive lie.

Mark My Words

## Let's Run a Few Science Experiments

Let's run a few science experiments on Creation by Chance or Chance Causality, and then see what we get for our results.

In science when we run a science experiment, we manipulate an independent variable and measure its effect on a dependent variable. The science experiment is an integral part of the scientific method. We can't do science without it.

So, how do we manipulate chance in order to measure its effect on a dependent variable? We can't! Once we start to manipulate it, then it is NO LONGER chance! It has become some type of order, organization, or manipulation instead.

So, how do we manipulate disorder or chaos in order to measure its effect on a dependent variable? Once again, we can't. Once we start to manipulate it, then it is NO LONGER disorder or chaos but has become some type of order, or organization, or manipulation instead.

The original construct or the primal construct was pure disorder or pure chaos. However, once the Gods started manipulating this Chaos Realm, then it ceased to be disorder or chaos and became some type of order and organization instead. The Gods bring order and organization out of chaos. They turn chaos into order. It's what they do. I was an accountant for thirty years, and all I ever did was to try to bring some type of order out of all the chaos. We are the Gods. We are the children of the Gods. All we ever do is try to bring order and organization into our environment, our surroundings, and our lives. It's what we do.

Once we start manipulating it, then it is no longer disorder and chaos.

Once the Gods started manipulating the energies in the Chaos Realm and started to transform those energies into massless, heatless, entropyless, syntropic, exergic Quantum Fields, then the chaos or disorder ceased to exist and was replaced with order and organization instead.

So, how do we manipulate chance in order to test and measure its effect on a dependent variable? We can't. Once we start to manipulate it, then it is no longer chance! We can't science or manipulate chance and have it remain as chance. Chance completely ceases to exist once we start to manipulate it. When we start stacking the deck or loading the dice, then we are no longer dealing with chance but some type of intelligent intervention or intelligent design instead.

So, how do we manipulate chaos or disorder so as to test and measure its effect on a dependent variable? Once again, we can't. Once we start to manipulate it, then it ceases to be disorder or chaos and becomes some type of order and organization instead.

That's just the way it works!

There's NO correlation whatsoever between heat and disorder. Disorder doesn't cause heat, and heat doesn't cause disorder. Heat or thermodynamics represents a huge amount of order and organization. When it comes to psyche or intelligence, the whole thing strives towards order and organization. Whenever we start to manipulate disorder, we start to transform it into some type of order or organization instead.

If a system were to somehow magically achieve perfect disorder or maximum disorder, it has no place to go but up into some type of order or organization. Order and organization are all that we see around us in our universe and in our microscopes. We don't see proton decay.  We don't see an ever-encroaching gray goo coming in at us from all sides.  We only see order and organization.

Why is that?

What's going on?

The quantum fields are perfect, syntropic, exergic perpetual motion machines.  The second law of thermodynamics will never be true as long as the quantum fields exist.  We have never had true disorder or true chaos in our part of the multiverse ever since the Gods and the Controlling Psyches in Nature designed and made the quantum fields.  Once the Gods made the quantum fields, then disorder and chaos completely ceased to exist in our part of the multiverse.

The second law of thermodynamics states that the amount of disorder and chaos in our universe is constantly increasing and can never decrease and go to zero.  It's a lie.  You would have to destroy the quantum fields in order to make the second law of thermodynamics true.

Likewise, whenever we start manipulating chance or disorder, then they cease to be chance or disorder.  They cease to exist.  They become some type of order and organization instead.  There's no such thing as Creation by Chance because once we start to create then we are no longer dealing with chance.  It is no longer chance, when we start to manipulate it and organize it.  Likewise, it is no longer disorder and chaos, when we start to manipulate it, control it, order it, and organize it.  That's just the way it is.

Materialism, Naturalism, Darwinism, Nihilism, Behaviorism, Determinism, Physical Reductionism, Classical Realism, the Second Law of Thermodynamics, the Theory of Evolution, and Atheism (Creation Ex Nihilo) ARE creation by chance.  That's why they are automatically false.  That's why they are pseudo-science.  That's why they are religion, philosophy, and dogma masquerading as science.

Chance can't operate as a causal mechanism.  Once it starts to do so, then it is no longer chance, but has become something else instead.

Mark My Words


### Creation by Chance

Tossing coins isn't going to produce a protein, its matching gene, a genome, an eye, a brain, or a life form.  Tossing coins isn't going to produce a planet, star, or galaxy either.  Likewise, tossing molecules isn't going to produce a genome, a life form, a planet, a star, or a galaxy.  Chance can't produce reliable results.  Chance can be replicated or duplicated.  Chance has its limits.  The longer the sequence the less likely that it can be produced by chance alone.

Materialism, Naturalism, and their derivatives such as Darwinism and Atheism are Creation by Chance.

We can't run science experiments on Creation by Chance or Chance Causality, because according to the rules of science and the rules of statistics, CHANCE is not a valid cause for anything except disorder or chaos.

If two variables have NO correlation or NO relationship with each other, then any perceived similarities were "caused by chance", which means that there is NO causal relationship between them.

The null hypothesis in a science experiment states that the results of the experiment were caused by chance or created by chance, which means that the independent variable had NO EFFECT on the dependent variable. That means that there is NO causal relationship between the variables, and that means that your alternative hypothesis is most likely false.

The null hypothesis in a science experiment is in fact "Chance Causality" or "Creation by Chance" or the "Result Was Caused by Chance", which means that the independent variable had NO EFFECT on the dependent variable. In other words, the null hypothesis states that if the results we are seeing from the science experiment were caused by chance, then that means that there is NO causal relationship between the independent variable and the dependent variable. Whenever CHANCE is the independent variable, it literally has NO discernable effect on any dependent variable. There can be NO causal relationship between chance and anything else.

Yet, what do we see from the Darwinists, Materialists, Naturalists, and Atheists?

Darwinism, Naturalism, Physicalism, Behaviorism, Materialism, Determinism, and the Second Law of Thermodynamics ARE creation by chance or chance causality. These people literally make CHANCE, disorder, chaos, or entropy the cause of everything that exists.

That's NOT science!

There's NO way in the universe to run a science experiment on Chance Causality or Creation by Chance. Making chance the cause of everything is NOT science. It's philosophy, religion, dogma, and wishful thinking. Materialism, Naturalism, and their derivatives such as Darwinism and Atheism are philosophy or religion masquerading as science. They are NOT science. Materialism, Darwinism, Naturalism, and Atheism ARE the pseudo-sciences. They are religion or philosophy masquerading as science. They are chance causality or creation by chance or creation ex nihilo.

So, what these people do to secure their fraud is that they define Science axiomatically as Materialism, Naturalism, Darwinism, Nihilism, Behaviorism, and Atheism so that their religion and dogmatic philosophies are forced upon us and made to seem as if they were science, when they are not.

They cheat!

They make Materialism, Naturalism, and their derivatives such as Darwinism and Atheism axiomatically true so that they can't be falsified and so that NO science experiment is needed to prove that they are true. Instead, they are declared to be automatically true and turned into laws that we are forbidden to falsify.

It took me fifty years, but eventually I realized that we need to be able and willing to falsify Materialism, Naturalism, Darwinism, Nihilism, Atheism, and the Second Law of Thermodynamics if we want to find and know the truth. It is impossible to find and know the truth if we are unwilling or unable to falsify and eliminate the things that are demonstrably false.

Mark My Words

**Proving Non-Existence**

Materialism or Physicalism claims that the non-physical or the immaterial does not exist. Materialism claims that the massless, heatless, and entropyless photons, quantum waves, and quantum fields do not exist. Materialism claims that action at a distance or quantum mechanics does not exist.

Naturalism claims that the supernatural does not exist. Naturalism claims that God, psyche, quantum mechanisms, action at a distance, quantum tunneling, quantum non-locality, quantum entanglement, the quantum zeno effect, quantum waves, gravity waves, magnetic waves, dark matter or spirit matter, dark energy, photons, and the quantum fields do not exist. Naturalism claims that all of these supernatural things do not exist.

Darwinism is Creation by Entropy or Creation by Chance. Natural selection is Creation by Death. Darwinism is just another form of Atheism. Darwinism claims that psyche, intelligence, quantum consciousness, life force, and God do not exist.

Nihilism is the claim that the afterlife, our spirit body, and our eternal psyche do not exist. Nihilism is the claim that our lives and our existence have no purpose, meaning, or value. All we are is dust in the wind. Nihilism is the belief that we cease to exist when we die. Nihilism claims that your psyche, intelligence, life force, or quantum consciousness does not exist. Nihilism is the belief that we don't really exist.

Atheism is Creation Ex Nihilo – the creation of something from nothing by nothing. Atheism is Creation by Chance. Atheism makes the claim that God, the Controlling Psyches in Nature, the Human Psyche, Intelligence, Quantum Consciousness, Life Force, and Nature's Psyche do not exist.

There is NO way in the universe to run a science experiment on the **non-existence** of something. You can't vary the amount of **non-existence** to see if it has an effect on a dependent variable. Because we cannot run a science experiment on the **non-existence** of these different things, Materialism, Naturalism, Darwinism, Nihilism, and Atheism are NOT science. They are thought experiments. They are philosophy, religion, dogma, and wishful thinking. They are pseudo-sciences masquerading as science.

It's impossible to prove the non-existence of something. In science, the ONLY thing we can do is to examine and try to explain the things that are actually being experienced and observed. Science IS observation and experience, not philosophical speculation and wishful thinking. If it has never been experienced or observe, then it really doesn't exist. Science is the study of all the different things that have been experienced and observed.

Our resurrected Lord Jesus Christ has been experienced and observed millions of times by thousands of different people. Science is an attempt to try to figure out what that means.

Millions of different people have had out-of-body experiences and near-death experiences. Science is an attempt to try to figure out what all that evidence means. Science is the study of all the different things that have been experienced and observed – or it's supposed to be.

Mark My Words

## Natural Selection Doesn't Work as Advertised

Design and Creation by Natural Selection is a fraudulent hoax.

Natural selection doesn't touch our genes.

Technically, natural selection doesn't even exist as a creative agent, a thinking person, or a working physical process.

All natural selection or survival of the fittest does is just wait for you to die.

Natural selection is death or entropy. Random mutations are also entropy. Entropy is death. Evolution is entropy. Evolution is death.

The theory of evolution is Creation by Entropy or Creation by Death, which we already KNOW is physically impossible. It can't be done, which means that it isn't being done.

Physicalism, Materialism, Naturalism, Darwinism, Nihilism, Behaviorism, Determinism, Physical Reductionism, Atheism, the Theory of Evolution, and Classical Physics are based exclusively on death or entropy. Are they not? WE KNOW for a fact that entropy or death cannot design and create anything!

Chemical evolution, macro-evolution, abiogenesis, creation ex nihilo, and spontaneous generation are prevented from happening by entropy, random diffusion, or the second law of thermodynamics. Random mutations are entropy and typically result in disease or death. Random mutations cannot produce life. They have never been caught in the act of doing so. In fact, Spontaneous Generation, Abiogenesis, Macro-Evolution, Chemical Evolution, Materialism, Naturalism, Creation by Natural Selection, and the Theory of Evolution were falsified in 1859 by Louis Pasteur, the very same year that Charles Darwin published, "On the Origin of Species by Means of Natural Selection".

We've KNOWN from the very beginning of the theory that the Theory of Evolution is false; but, our scientists chose to ignore the evidence preferring the convenient fiction instead. These people have been lying to us and deceiving us ever since.

Astronomical evolution or the creation of stars and galaxies from clouds of dust is also physically impossible and therefore false. Creation by Evolution, of any type, is false and has been falsified. Evolution cannot design and create. In fact, evolution (genetic drift), random mutations, and natural selection didn't even exist until AFTER God designed, programmed, engineered, field-tested, fine-tuned, manufactured, produced, created, and deployed the proteins, their matching genes, the genomes, the eyes, the brains, and the physical life forms in the first place. Only after God designed and created these things did evolution, random mutations, and selection come onto the scene and begin to exist.

**Our Created Solar System**
https://www.youtube.com/watch?v=CzyQbOQ0dv0

**Our Created Stars and Galaxies**
https://www.youtube.com/watch?v=E66409i-yn4

**Our Created Universe – Apologetics Symposium**
https://www.youtube.com/watch?v=w6SoTVZrBV8

**Our Created Universe – Seattle Creation Conference**
https://www.youtube.com/watch?v=na2TL-wcK4I

**The Probability of a Protein Forming by Chance**
https://www.youtube.com/watch?v=W1_KEVaCyaA

**The Probability of Making a Protein by Chance**
https://www.youtube.com/watch?v=cQoQgTqj3pU

**Evidence for Creation by Outside Intervention**
https://www.youtube.com/watch?v=ci4s75al1Rw

**Scientific Observation:** Anything that began to exist had a Psyche, Intelligence, or Person who made it, organized it, and caused it to exist.

**Scientific Observation:** Physical matter, stars, galaxies, evolution, genetic drift, genes, proteins, genomes, eyes, brains, random mutations, and the various different types of selection began to exist.

**Scientific Conclusion:** Therefore, it is logical and rational to conclude that physical matter, stars, galaxies, evolution, random mutations, selection, proteins, and genomes have Someone Psyche or Someone Intelligent who made them, organized them, and caused them to exist.

I'm not going to apologize for finding the truth. That's what we scientists are supposed to do. Is it not?

There's NO such thing as Design and Creation by Evolution. It has NEVER been experienced nor observed. The Theory of Evolution in all of its many different forms has been falsified by Science. Therefore, the Theory of Evolution in its many different guises ends up being Scientific Proof of God's Necessity which is in fact Scientific Proof of God's Existence. ONLY Intelligence, or Psyche, or Personalities can design and create. That is what has actually been experienced and observed.

Mark My Words

—

## ORIGIN: Probability of a Single Protein Forming by Chance

Chemical evolution or macro-evolution has been falsified. It should be eliminated from science because it is impossible and false.

https://www.youtube.com/watch?v=W1_KEVaCyaA

http://www.originthefilm.com/mathematics.php

http://science-2-0.com/wp-content/uploads/2018/03/THE-MATHEMATICS-OF-ORIGIN.pdf

https://www.youtube.com/watch?v=cQoQgTqj3pU

https://www.youtube.com/watch?v=_zQXgJ-dXM4

http://bio-complexity.org/ojs/index.php/main/article/view/BIO-C.2010.1/BIO-C.2010.1

http://science-2-0.com/wp-content/uploads/2018/03/Case-Against-Darwinian-Origin.pdf

http://science-2-0.com/wp-content/uploads/2018/03/Chemical-Evolution-Is-Impossible.zip

http://science-2-0.com/wp-content/uploads/2018/04/Origin-Of-Life.zip

Once you have eliminated everything that is FALSE and everything that is IMPOSSIBLE, then ONLY the Truth remains. It's elementary.

Science is observation and experience.

The theory of evolution is based upon a wide variety of different assumptions which have NEVER been experienced nor observed. Design and creation by the different types of evolution has NEVER been experienced nor observed. Design and creation of functional genomes and proteins by natural selection and random mutations is physically impossible; and consequently, it has never been observed.

Critical Thinking Skills that Disarm Evolutionists

https://www.youtube.com/watch?v=ZE6hm2kpYiY

https://www.youtube.com/watch?v=4gxVnCsaLwY

https://www.youtube.com/watch?v=Pe1k0NDGEaw

No part of the theory of evolution has ever been caught in the act of design and creation. Natural selection and random mutations have never been caught in the act of design and creation. Creation by Evolution has never been observed.

Now, let's apply Syntropy to the real world, our physical world.

The chemical evolution of proteins – and the matching genes to go along with them – from atoms is physically impossible thanks to entropy or the second law of thermodynamics. Entropy prevents atoms from self-arranging into proteins and genes, thereby making chemical evolution physically impossible.

What does this mean?

It means that the design, creation, and production of proteins and genes requires some kind of intelligent and deliberate Syntropy at the quantum level or the psyche level, because chemical evolution or the spontaneous generation of genes and proteins is physically impossible at the physical level. Syntropy is thought and intelligence and consciousness, which requires Someone Psyche in order to get the job done.

All truth bears record of truth; and, every truth FALSIFIES Materialism, Naturalism, Darwinism, Nihilism, and Atheism. ALL of our observations and experiences falsify Materialism, Naturalism, Darwinism, Nihilism, and Atheism. The very existence of Quantum Mechanics and Syntropy and Intelligent Psyches FALSIFIES Materialism, Naturalism, Darwinism, Nihilism, Behaviorism, Scientism, Determinism, and Atheism. The very existence of genes and proteins is Scientific Proof of God's Existence.

Materialism, Naturalism, Atheism, Nihilism, and the Theory of Evolution have been FALSIFIED trillions of times in thousands of different ways, because they are fiction. These things are repeatedly FALSIFIED due to a complete lack of observational support. In contrast, Quantum Mechanics, Action at a Distance, and the effects of Quantum Non-Local Consciousness are repeatedly VERIFIED. The truth is repeatedly OBSERVED, lived, experienced, and verified.

Remember, through Quantum Mechanics or quantum mechanisms, Psyche can do the physically impossible because Quantum Mechanics is pure syntropy. This is what becomes possible by choosing to allow ALL of the evidence into evidence. A case can be made for Syntropy, Quantum Mechanics, Psyche, Intelligence, and God by choosing to allow all of the evidence into evidence.

Through Syntropy, Quantum Mechanics, and Psyche, you can literally explain EVERYTHING that comes your way. The explanatory power of Quantum Mechanisms is vastly superior to Materialism, Naturalism, Darwinism, Nihilism, Behaviorism, Determinism, Atheism, and Classical Physics. Quantum Mechanisms require Intelligence or Psyche in order to make them work; but then again, so would Materialism, Naturalism, and Darwinism if they worked.

Science 2.0 defines science as OBSERVATION and EXPERIENCE. Quantum Mechanisms have been observed and experienced, both at the physical level and the quantum psyche level. In contrast, Materialism, Naturalism, Nihilism, Atheism, Determinism, and even Darwinism FAIL the observation requirement of Science 2.0; therefore, Materialism, Naturalism, Darwinism, Atheism, and their derivatives are FALSIFIED by a lack of observational evidence and are thus eliminated from science as being unscientific, falsified, and incomplete.

Under Science 2.0, phenomenology, observation, and experience trump the unobserved philosophical speculation of Materialism, Naturalism, Darwinism, Nihilism, and Atheism. Science is observation and experience. This is the way science should have been done but wasn't.

Mark My Words

—

## Science Is Evidence

*Begging the question* and *jumping to conclusions*, the Materialists, Naturalists, Darwinists, Nihilists, and Atheists define "science" axiomatically as Materialism, Naturalism, Darwinism, Nihilism, and Atheism. That's how they prove that the Theory of Evolution is true, by making Darwinism axiomatically true.

In the summer of 2017, I upgraded my science to Science 2.0. I redefined science as observation and experience.

Scientific inferences are worthless unless they have actually been experienced and observed by Someone Psyche. If they have been experienced and observed by Someone Psyche, then they are no longer scientific inferences but are in fact KNOWLEDGE instead. The BEST and FASTEST way to find and know the truth is to live it, observe it, and experience it for yourself or to choose to trust someone who has.

Science is evidence. Whenever you encounter a scientific claim, ask yourself, "Where's the evidence that this claim is true?"

Science is observation and experience. Whenever you encounter a scientific hypothesis, theory, or claim, ask yourself, "Has it ever been experienced or observed?"

Science is the process of falsifying ALL assumptions, guesswork, magical thinking, fuzzy thinking, hypotheses, and wishful thinking and replacing them with observations, experiences, and knowledge. Science is knowledge. Whenever you encounter a scientific claim, ask yourself, "What assumptions are being made?" Assumptions are there to be falsified and eliminated from science so that we end up with Observed Knowledge instead.

In the summer of 2017, I upgraded my science to Science 2.0. I redefined Science as observation and experience. A year later in June 2018, I finally encountered someone else who is using and promoting these very same Critical Thinking Skills in relationship to the Theory of Evolution – Mike Riddle. Meanwhile, I have been trying to apply Critical Thinking Skills to the whole of science. My goal is to make science real, experienced, and observed rather than fictional wishful thinking and axiomatic assumptions. My ultimate goal is to identify and eliminate everything that is false and everything that has been falsified so that ONLY the truth will remain. The truth is whatever has been experienced and observed by Someone Psyche or Someone Intelligent – somewhere sometime.

Critical Thinking Skills that Disarm Evolutionists:

https://www.youtube.com/watch?v=ZE6hm2kpYiY

https://www.youtube.com/watch?v=4gxVnCsaLwY

https://www.youtube.com/watch?v=Pe1k0NDGEaw

Is there any evidence that your theory is true?

Has it ever been experienced and observed?

What unsubstantiated assumptions are you making?

Mike Riddle does an excellent job applying Critical Thinking Skills to the theory of evolution; so, let's pick a different target by way of introduction to Critical Thinking Skills.

Let's apply Critical Thinking Skills to everyone's favorite topic – Young-Earth Creationism (YEC) and Young-Universe Creationism (YUC). YUC and YEC are the claim that God brought the universe and our earth into existence 6,000 years ago ex nihilo. The YUCs and YECs never apply Critical Thinking Skills to YUC and YEC – only to Darwinism and the Theory of Evolution.

When it comes to YUC and YEC, I find that it's more efficient and most effective to identify their assumptions or wishful thinking first. These people are identified by the assumptions that they make – as are we all. What unsubstantiated assumptions are the YUCs and YECs making and promoting?

Most of the YUCs and YECs tend to assume that God brought our physical universe, physical galaxy, and physical earth into existence ex nihilo 6,000 years ago just before our Adam and Eve were placed into the Garden of Eden. These people choose to believe that God created the physical universe and our earth in six consecutive 24-hour earth rotations, which they call "days", even though our earth didn't exist when the process was started. The YUCs and YECs choose to interpret the word "day" in Genesis Chapter 1 as a single 24-hour period as measured on this earth. In order to get around the Distant Starlight Problem, the YUCs and the YECs assume that the speed-of-light is variable. In other words, in violation of the theory of relativity and $E = mc^2$, the YUCs and the YECs tend to assume that the speed-of-light is infinite whenever it is traveling towards our earth from any direction in space. These people assume that God created our earth and this physical universe ex nihilo or spontaneously all at the same time 6,000 years ago with the appearance of age. These people assume that the physical constants are variable over time. The YUCs and YECs people assume that the rates of radioactive decay are variable over time. The YUCs and the YECs assume that there has been only one mass-extinction event during our earth's history – the Genesis flood mentioned in the Bible. The YUCs and the YECs assume that the volcanic layer of the earth's crust was produced on the ocean floor during creation week and that the sedimentary layers of the earth's crust were produced during the Genesis flood.

The YUCs and YECs assume that our earth is the only inhabited planet in our physical universe and tend to assume that our earth is at the center of our physical universe. The YUCs and YECs assume that physics is special or unique in relationship to our earth. The YUC and YEC model is a geocentric model. Contrary to the observational evidence, the YUCs and YECs assume that our whole universe fell when our Adam and Eve fell. These people assume that our Adam and Eve were the only ones to have been produced – that there are no other inhabited planets in this physical universe. The YUCs and the YECs also tend to assume that God created everything with the appearance of age, ex nihilo, including the fossils in the rocks and the sedimentary layers. In order to make their case and prove their point, the YUCs and the YECs assume that God is a trickster and a deceiver, and that any evidence to the contrary should be ignored and dismissed, or it should be considered a trick of the Devil. The YUCs and the YECs dismiss, ignore, or even try to destroy any evidence that falsifies Young-Universe Creationism or Young-Earth Creationism.

The YUCs and the YECs make more unsubstantiated assumptions than the Evolutionists, Darwinists, and Naturalists do. The YUCs and the YECs destroy science, physics, the physical constants, the speed-of-light, and the theory of relativity in an attempt to convince people that our universe is only 6,000 years old. YUC and YEC are at the opposite extreme from Darwinism and Naturalism.

Ask yourself, "Has any of this been experienced or observed?"

Ask yourself, "What has actually been experienced and observed?"

Ask yourself, "Where's the evidence that YUC and YEC are true? Where's the evidence that YUC and YEC are false?"

Let's answer some of these questions.

Spontaneous generation, creation ex nihilo, chemical evolution, abiogenesis, or macro-evolution has NEVER been experienced nor observed. In fact, spontaneous generation, creation ex nihilo, naturalism, chemical evolution, or the theory of evolution was falsified in 1859 by Louis Pasteur. Spontaneous generation or creation ex nihilo violates the first law of thermodynamics which states that Energy or Syntropy is conserved. YUC, YEC, and the theory of evolution are based upon spontaneous generation or creation ex nihilo; and, both spontaneous generation and creation ex nihilo have been falsified by scientific observations. Spontaneous generation or creation ex nihilo has NEVER been experienced nor observed. There's NO convincing evidence that spontaneous generation, creation ex nihilo, chemical evolution, or the theory of evolution is true. But, there's tons of convincing evidence that they are false.

Energy is Syntropy. Energy or Syntropy is conserved, meaning that it is eternal and everlasting without a beginning of days or an end of years. The physical matter or entropy comes and goes as God sees fit; but, the Energy or Syntropy is conserved. This is what the

First Law of Thermodynamics is trying to tell us and teach us. The First Law of Thermodynamics has been experienced and observed. It has always been verified and proven to be true. That's why it is considered to be a Physical Law. It remains constant. From everything that we have experienced and observed, the physical constants remain constant. Physical matter "functions" the same way now that it did millions and billions of years ago. That is what we have experienced and observed.

A variable speed-of-light has never been experienced nor observed. A variable speed-of-light would have been noticed while communicating with astronauts on the moon; and, it would have been noticed by all of the GPS satellites that we have in orbit. The speed-of-light remains constant every direction that it travels according to the theory of relativity and $E = mc^2$; otherwise, universal chaos would ensue. Imagine what would happen if the c (speed-of-light) in $E = mc^2$ were variable depending upon the direction that the light was originally traveling! It wouldn't be pretty. It would be the very definition of chaos. It would eventually violate the First Law of Thermodynamics. Imagine what would happen through $E = mc^2$ if the speed-of-light were infinite while traveling towards our earth. There should be an infinite amount of Energy heading towards our earth if the speed-of-light is infinite while traveling towards our earth. The whole universe would go up in flames if matter were ever converted to energy while traveling towards our earth. That's the very definition of chaos!

An infinite speed-of-light makes NO logical sense whatsoever, where the physical realm is concerned; and, the YUCs and the YECs teach that the speed-of-light is infinite whenever light is traveling towards our earth. If true, then our whole universe should be in flames ALL traveling towards our earth right now. The YUCs and the YECs literally destroy Physics, Physical Laws, Physical Constants, and Science in order to get a 6,000-year-old universe out of the deal. The speed-of-light can no longer be relied upon to measure distance and the passage of time once the YUCs and YECs have destroyed it by making it variable and/or infinite.

$E = mc^2$ is the scientific proof that Energy or Syntropy is conserved and that the First Law of Thermodynamics is true. The First Law of Thermodynamics or the Conservation of Energy has been experienced and observed. There's a ton of evidence supporting its truthfulness. Creation ex nihilo, spontaneous generation, creation with the appearance of age, a variable speed-of-light, YUC, and YEC violate $E = mc^2$ and the First Law of Thermodynamics. A variable speed-of-light has NOT been experienced nor observed. The speed-of-light is a physical constant, and it has to remain constant every direction and at every time, or the whole of classical physics would collapse and cease to be true. In order for the speed-of-light to have an infinite velocity toward our earth from all directions in space would require that we have a universe-sized singularity or black hole at the center of our earth; and, we don't. It certainly has never been experienced nor observed.

The Biblical God Jesus Christ has stated many times that our earth is NOT the only inhabited world in this physical universe. I mention and list that evidence elsewhere in other books and essays. The evidence is there to be found if you choose to go looking for it. The YUC and YEC claim that our earth is the only inhabited planet in the universe has been falsified by the Biblical God Jesus Christ. It has been falsified by His observations and experiences. Seers on this planet have interacted with human beings – God's children – living on other planets in this universe.

The physical laws have been observed to be universal, and NOT variable in relationship to our earth. There's nothing special or unique about our earth where the physical laws and physical constants are concerned.

In complete violation of the theory of relativity and $E = mc^2$, the YUCs and YECs tend to claim that ALL starlight travels at an infinite velocity towards the earth. That would make our earth the center of our physical universe and a huge gravity well. There's NO convincing evidence that our earth is at the center of our physical universe as the YUCs and the YECs claim that it is. In fact, there's NO convincing evidence that our earth is the center of our own galaxy.

Fossils have never been observed spontaneously generating into rocks with the appearance of age. The best explanation for the origin of fossils is that they were rapidly buried in some kind of flood; and, a flood requires time to do its work and leave its traces, not spontaneous generation and creation ex nihilo. The YUCs and the YECs assume that there has been only one mass-extinction event – the Genesis flood listed in the Bible. Twenty-five different mass-extinction events have been found in the fossil record. Since the Bible mentions only one of them, that means that the others took place BEFORE Adam and Eve were placed into the Garden of Eden.

All of the observational evidence, experiential evidence, and scientific evidence is telling us that YUC and YEC are false; whereas, the YUCs and the YECs simply assume that Young-Earth Creationism and Young-Universe Creationism are true. The YUCs and YECs destroy science, physics, the theory of relativity, the speed-of-light, and the physical constants in order to convince themselves that our earth and universe are only 6,000 years old. That's wishful-thinking and self-deception – not science.

The YUCs and YECs turn God into a trickster and a deceiver in order to make their case and to prove their point; whereas, God tells us in the Bible that we can depend upon the stars and the lights in the heavens to measure times, seasons, signs, and distances. The Bible tells us that we can rely upon God to tell us the truth. The Bible says that in God's eyes, a day is a thousand of our years, with the word "thousand" being Hebrew and Greek for a "very long undetermined amount of time".

Ask yourself, "Where's the evidence that YUC and YEC are true? Where's the evidence that YUC and YEC are false?"

There is NO convincing evidence that YUC and YEC are true. I haven't found any.

However, there is a lot of convincing evidence that YUC and YEC are false. My overall purpose in all of this is to falsify and eliminate everything that is false so that only the truth will remain.

$E = mc^2$ is NOT dependent on directionality! It is universally true all directions and at all times. YUC and YEC violate the First Law of Thermodynamics and $E = mc^2$. That's convincing scientific evidence that YUC and YEC are false.

The 800,000-year-old ice cores are convincing scientific evidence that YUC and YEC are false.

> https://en.wikipedia.org/wiki/Ice_core
>
> https://ldssoul.com/wp-content/uploads/2018/06/Ice-core.pdf
>
> https://ses.edu/ice-age-is-ideal-for-humanity/
>
> https://ldssoul.com/wp-content/uploads/2018/06/Ice-Age-is-Ideal-for-Humanity.pdf

The ice cores are convincing evidentiary proof that YUC and YEC are false. Ten 100,000-year Newtonian epicycles have been identified within our earth's ice cores; and, the rocks under those ice cores are older than the ice cores. That's convincing evidence that YUC and YEC are false.

The quantity of bio-deposits within our earth is evidentiary proof that YUC and YEC are false. The half-life of plutonium as well as the non-existence of plutonium in nature suggests that YUC and YEC are false, assuming of course that there was some plutonium in nature to begin with.

ALL of the evidence we have on hand suggests that our earth is NOT the center of our solar system, is NOT the center of our galaxy, and is NOT the center of our universe; and, it would have to be if all incoming light travels at an infinitely velocity towards our earth from all directions in space as the YUCs and YECs claim. The Andromeda Galaxy 2.5 million light years away, as well as visible galaxies 13.2 billion light years away, is convincing scientific proof that Young-Universe Creationism is false.

Evidence for an Old Earth

https://www.youtube.com/watch?v=3v-uvh_KWkg

https://science-2-0.com/wp-content/uploads/2018/06/Scientific-Evidence-for-Old-Earth.zip

Distant Starlight

https://www.youtube.com/watch?v=0BGAaLm52kg

https://science-2-0.com/wp-content/uploads/2018/06/Distant-Starlight.zip

Mira Proves YUC and YEC False

https://www.youtube.com/watch?v=XyuXBYWZegY

https://science-2-0.com/wp-content/uploads/2018/06/Mira.zip

The fact that there are 25 mass-extinction events in the fossil record – not one but twenty-five – is convincing scientific proof that YUC and YEC are false. It takes a bit of time to setup and produce a new unique mass-extinction event, and then leave a record of it in the fossil record.

https://en.wikipedia.org/wiki/Extinction_event

https://ldssoul.com/wp-content/uploads/2018/06/Extinction-event.pdf

The forty main sedimentary layers in the Grand Canyon, with some of those layers containing desert sands, is convincing scientific proof that YUC and YEC are false. NOT all of the sedimentary layers were formed by floods as the YUCs and YECs claim. Some of the sedimentary layers were formed by deserts and drifting desert sands. Deserts require that the land dry-out first. The formation of new environments and thereby new unique sedimentary layers takes time. It's not produced ex nihilo with the appearance of age as the YUCs and YECs claim. Spontaneous generation or creation ex nihilo has been falsified by Science and Scientific Observations.

https://en.wikipedia.org/wiki/Geology_of_the_Grand_Canyon_area

https://ldssoul.com/wp-content/uploads/2018/06/Geology-of-the-Grand-Canyon.pdf

Within the Grand Canyon alone, there is convincing scientific evidence that there have been at least 40 major catastrophe events and/or massive environmental transitions during our earth's history – NOT one. We only have ONE major flood event or catastrophe recorded in the Bible and the written history of our planet, which means that the other 39 major flood events or global catastrophes in the fossil record and sedimentary record took place before our Adam and Eve were placed into the Garden of Eden.

We are looking at millions of years (not necessarily billions of years) when we start looking at ice cores, the 25 mass-extinctions in the fossil record, radiometric dating, and the 40 main layers of sediment that have been revealed to us in the Grand Canyon. These scientific observations provide convincing evidence that Young-Earth Creationism is false.

The geological rock formations on the east side of the Yellowstone River, around Tower Falls in Yellowstone, provide convincing scientific evidence that YUC and YEC are false. The YUCs and the YECs want you to believe that the Genesis flood produced all the different layers in the canyon, including all of the different volcanic layers, and then cut the canyon when the Genesis flood was done. Assuming that the Genesis flood was global in nature, the Genesis flood would maybe have been enough to cut the canyon, or maybe enough to produce the top sedimentary layer for the canyon; but, the Genesis flood could never have produced the different volcanic layers along with the sizable sedimentary layers between the different volcanic layers. The different volcanic basalt rock columns produced by cooling lava flows – over the top of sedimentary layers and underneath sedimentary layers – provides convincing scientific evidence that YUC and YEC are false. The Genesis flood couldn't have produced all these different layers of sediment separated by the different volcanic layers as the YUCs and YECs claim.

https://ldssoul.com/wp-content/uploads/2018/06/Tower-Falls-Volcanic-Basalt-Columns-between-Sediment-Layers.jpg

https://ldssoul.com/wp-content/uploads/2018/06/Basalt_columns_in_yellowstone_national_park.jpg

https://ldssoul.com/wp-content/uploads/2018/06/wpid-TowerFallsRocks2-2015-10-16-21-301.jpg

https://ldssoul.com/wp-content/uploads/2018/06/Basalt_columns_in_yellowstone_national_park-1.jpg

https://ldssoul.com/wp-content/uploads/2018/06/Basalt_columns_in_yellowstone_2.jpg

https://ldssoul.com/wp-content/uploads/2018/06/10488.jpg

https://ldssoul.com/wp-content/uploads/2018/06/128321725.Kmy17Cw4.yellowstone71.jpg

https://ldssoul.com/wp-content/uploads/2018/06/Basalt-Columns.jpg

https://ldssoul.com/wp-content/uploads/2018/06/19510598353_d871561183_b.jpg

https://ldssoul.com/wp-content/uploads/2018/06/glacier-volcanoes.jpg

https://ldssoul.com/wp-content/uploads/2018/06/Basalt_Columns_Band.jpg

https://ldssoul.com/wp-content/uploads/2018/06/img_6217.jpg

https://ldssoul.com/wp-content/uploads/2018/06/02338.jpg

https://ldssoul.com/wp-content/uploads/2018/06/An_Igneous_Sill_Intrusion_in_Yellowstone_National_Park_Wyoming_USA.jpg

https://ldssoul.com/wp-content/uploads/2018/06/427-sill-coulee-Yellowstone-05.jpg

https://ldssoul.com/wp-content/uploads/2018/06/canyon-of-the-yellowstone-river-with-columnar-volcanic-rock-layer-on-top-of-sedimentary-deposits.jpg

https://ldssoul.com/wp-content/uploads/2018/06/SAM_2557.jpg

https://ldssoul.com/wp-content/uploads/2018/06/SAM_2551.jpg

https://ldssoul.com/wp-content/uploads/2018/06/SAM_2545.jpg

https://ldssoul.com/wp-content/uploads/2018/06/SAM_2558.jpg

The YUCs and the YECs want you to believe that our earth is only 6,000 years old. The first time I saw this canyon wall across the Yellowstone River from Tower Falls, I simply KNEW that Young-Earth Creationism (YEC) is false. There's no way in the universe that the Genesis flood could have produced all of that 4,000 years ago as the YUCs and YECs claim.

The YUCs and the YECs want you to believe that all of this was formed by the Genesis flood 4,000 years ago, including the volcanic layers, the sedimentary layers, and the canyon; but, that's physically impossible. These pictures falsify YUC and YEC. There's no way in the universe that the Genesis flood could have produced all of this in one fell swoop as the YUCs and YECs repeatedly claim that it did. Some of the YUCs and YECs will even say that all of this was produced by the Devil in order to trick us into believing that our earth is older than 6,000 years when it is not. I don't buy it. I give precedence to the scientific observations over the wishful thinking and misinterpretations of the YUCs and YECs. All you want is the truth, unless of course you are trying to deceive someone.

Ages for the Rock Layers across the Canyon from Tower Falls:

http://web.mit.edu/deaps/earth_2010/tf.html

https://science-2-0.com/wp-content/uploads/2018/06/Tower-Falls-Ages.pdf

The exposed rock layers across the canyon from Tower Falls falsifies YUC and YEC.

This doesn't mean that Genesis is false. All of this simply means that the YUCs and the YECs have chosen the wrong interpretation for the word "day" in the book of Genesis. A "day" in Genesis Chapter One is at least a thousand years, and it could even be as much as a million years.

When it comes to the scientific evidence provided by these different pictures of the volcanic basalt columns sandwiched between sedimentary layers, the Genesis flood completely lacks explanatory power because the single Genesis flood could never have produced such a thing. For me personally, the geological formations across the Yellowstone River from Tower Falls provide convincing scientific proof that YUC and YEC are false.

Because there is only ONE catastrophic event listed in the Bible, that observed reality means that ALL of the other catastrophic mass-extinction events and many of the volcanic events recorded in our earth's crust happened BEFORE our Adam and Eve were placed into the Garden of Eden over 7,000 years ago – assuming of course that God placed the Garden of Eden onto this planet to begin with and not some other planet.

A convincing case can be made that the Evolutionists, Darwinists, Materialists, and Naturalists deliberately inflate radiometric dates in order to make the magic associated with the theory of evolution possible. I have observed that most of the time you can safely divide the radiometric ages – being produced by the Atheists and Darwinists – by 1,000 and come up with a more accurate date for things. I have noticed multiple times in the past that the different radiometric dates have to be divided by 1,000 to 10,000 in order to match the "ages" being produced by Carbon-14 dating for the very same rocks. I also trust the Carbon-14 ages more than the other inflated ages being produced by the Darwinists and Naturalists. Even then, Carbon-14 Dating falsifies YUC and YEC.

Remember, Carbon-14 Dating is meaningless and worthless on an earth that is less than 100,000 years old. The half-life of Carbon-14 is 5,730 years. That means that the half-life of Carbon-14 is meaningless and worthless on an earth that is only 6,000 years old, or one Carbon-14 half-life old. Our earth has to be at least 100,000 years old in order for Carbon-14 dating to work right and make sense. Half-Lives make NO sense whatsoever on a 6,000-year-old earth. Once again, the YUCs and the YECs are forced to destroy science, physics, scientific evidence, the fossil record, the sedimentary and volcanic record, and logical common sense in order to convince themselves that our earth is only 6,000 years old.

Falsifying Potassium-Argon Radioisotope Dating:

https://creation.com/radioisotope-dating-of-rocks-in-the-grand-canyon

https://ldssoul.com/wp-content/uploads/2018/06/Radioisotope-Dating-Grand-Canyon.pdf

Each assumption that the Darwinists and Atheists use to inflate the Radiometric Ages results in an age-increase ten times greater than the previous guestimate.

When it comes to naturalistic and atheistic Potassium-Argon Dating, I like to divide their dates by at least 1,000 in order to make them conservative and realistic. However, when we divide their Darwinian dates by 1,000, we still end up with 2.5 million years as the age of those amphibolite rocks mentioned in this article. Be ultra-conservative and divide their Darwinian Radiometric ages by 10,000, and we still end up with a 250,000-year-old earth. Any way we choose to look at it, YUC and YEC are false. It's demonstrably clear that our earth is much older than 6,000 years.

Observed Rock Layers in the Grand Canyon:

https://ldssoul.com/wp-content/uploads/2018/06/gc_layer.gif

https://ldssoul.com/wp-content/uploads/2018/06/18_4biblio-f1_0.jpg

https://ldssoul.com/wp-content/uploads/2018/06/GC-2.jpg

https://ldssoul.com/wp-content/uploads/2018/06/sediments-lge.jpg

The YUCs and the YECs want you to believe that the volcanic layers at the bottom of the Grand Canyon were formed on the ocean floor during "creation week" 6,000 years ago and that the sedimentary layers in the Grand Canyon formed during the Genesis flood 4,000 years ago. The unconformities falsify that idea! Logical common sense falsifies that idea. Just looking at these charts falsifies YUC and YEC. Furthermore, take the dates being produced by the Evolutionists and Naturalists, divide those dates by 1,000 thereby producing a convincing and realistic age for those rocks, and you still end up with 1.7 million years for the age of the rocks at the bottom of the Grand Canyon. Divide their Darwinian dates by 10,000, and you end up with 170,000-year-old rocks at the bottom of Grand Canyon. However, a convincing case can also be made showing that dividing by 10,000 is too conservative because things start to fall apart at the other end when dealing with the youngest rocks and sediments that have been observed. There's no way in the universe that the different rock layers in the Grand Canyon could have been put down and then exposed during the past 6,000 years.

A convincing case can be made that our earth is at least 1.7 million years old based upon the different rock layers that have been exposed and found in the Grand Canyon.

Observed Rock Layers in Yellowstone National Park:

https://ldssoul.com/wp-content/uploads/2018/06/fig5.jpg

If we were to do this process of dividing by 1,000 when it comes to this particular chart containing the different rock layers found in Yellowstone National Park, we end up with the Gneiss and the Schist layer at the bottom being 2.7 million years old rather than 2.7 billion years old. The proven and verified age of 2.7 million years old for the bottom Pre-Cambrian layer is more than enough to falsify YUC and YEC.

https://ldssoul.com/wp-content/uploads/2018/06/YNP-Geology.pdf

https://ldssoul.com/wp-content/uploads/2018/06/Volcano-YNP-Part-1.pdf

https://ldssoul.com/wp-content/uploads/2018/06/Volcano-YNP-Part-2.pdf

https://www.nps.gov/parkhistory/online_books/geology/publications/bul/1347/sec2.htm

https://science-2-0.com/wp-content/uploads/2018/06/Geologic-History-of-the-Yellowstone-Region.pdf

The ages for the different rock formations are in dispute. However, even when you decide to go conservative and divide the 2.5-billion-year ages by 1,000, you end up with 2.7 million years for the age of those rocks. Divide by 10,000 if you want to be ultra-conservative, and you still end up with an earth that is at least 270,000 years old; but, the ice cores are giving us an earth that is at least a million years old, so dividing by 10,000 seems to be too conservative.

Of course, when these same Evolutionists and Atheists are producing ages in the range of 2.1 million years or 1.3 million years for the different volcanic eruptions in Yellowstone, dividing those by 1,000 for the sake of consistency doesn't really work because there really is no written record for huge continent-wide volcanic mass-extinction events 2,100 years ago or 1,300 years ago which suggests that a prehistoric age of 21,000 years or 13,000 years would probably be more accurate than the 2,100 years or 1,300 years that the YUCs and YECs would want us to produce.

Any way that we choose to look at it, YUC and YEC are falsified by observational evidence or scientific evidence. In other words, there's no way to make YUC and YEC explain the observational evidence that we have on hand as a species. Our earth is much older than 6,000 years! Whether we choose to go with 2.7 billion years or 2.7 million years, both of them are much older than 6,000 years. Based upon the different rock layers in the Grand Canyon and Yellowstone as well as the ice cores from Antarctica and Greenland, a solid convincing case can be made proving that our earth is at least 2.7 million years old.

We are definitely looking at billions of years when we start looking through our telescopes into deep space. You have to destroy science, physics, the physical constants, the theory of relativity, and the speed-of-light in order to produce anything else for the age of our universe than the billions of years that have already been observed and measured through our telescopes. That reality provides convincing scientific evidence that Young-Universe Creationism (YUC) is false.

Science 2.0 defines Science as observational evidence and experiential evidence, NOT philosophical speculation and wishful thinking. Ask yourself, "Where's the evidence that YUC and YEC are true?" Ask yourself, "Has spontaneous generation or creation ex nihilo ever been experienced or observed? Has a variable speed-of-light ever been experienced or observed?" Ask yourself, "What assumptions are the YUCs and YECs using to make their case and prove their point?" If you do so, you will find scientific evidence, observations, and experiences that falsify YUC, YEC, Materialism, Naturalism, and the Theory of Evolution.

If you successfully eliminate everything that is false and everything that has been falsified, then ONLY the truth will remain.

In Science, something can be reliable but invalid. It can be reliably wrong.

I have observed that Materialism, Naturalism, Darwinism, Nihilism, Behaviorism, Determinism, Atheism, and the Theory of Evolution are RELIABLY WRONG. You can count on these things being wrong. You can easily find experiences and observations that falsify them. They are reliably wrong, which means that they are not scientifically valid. Likewise, YUC and YEC are RELIABLY WRONG. You can find tons of observational evidence and experiential evidence that falsifies YUC and YEC.

YUC and YEC are just like Materialism, Naturalism, Darwinism, and Atheism – they are both reliably wrong, just at opposite extremes. The truth is always found between the extremes; whereas, the extremes are always reliably wrong and therefore scientifically invalid.

Mark My Words

—

## Evolution Produced Your Brain

Did you know that Evolution produced your brain?

The Evolutionists say that it is so, so it must be true. They would never lie to us and deceive us; now would they?

Siegler, R., DeLoache, J., & Eisenberg, N. (2010). *How Children Develop*, (3rd ed.). New York: Worth.

> **The obvious question is: Why does the human brain – the product of millions of years of evolution – take such a devious developmental path, producing a huge excess of synapses, only to get rid of a substantial portion of them? The answer appears to be evolutionary economy. The capacity of the brain to be molded or changed by experience, referred to as *plasticity*, means that less information needs to be encoded in the genes.**
>
> **Experience-Dependent Plasticity: The process through which neural connections are created and reorganized throughout life as a function of an individual's experiences.**
>
> **Experience-Expectant Plasticity: The process through which the normal wiring of the brain occurs in part as a result of experiences that every human who inhabits any reasonably normal environment will have.**
>
> **The brain is no less a product of natural selection that the rest of the body's structures and functions. Hearts evolved to support the process of**

**blood circulation; livers evolved to carry out the process of toxin extraction; and mental structures evolved to enable the learning of certain types of information necessary for adaptive behavior.**

(*How Children Develop*, pp. 110-112, 155.)

Do you see anything wrong with these statements?

Most people can't because they have been trained in our public schools not to be able to see anything wrong with them. These statements are the official explanation for origins that most scientists have chosen to believe in and promote. The whole purpose of our public schools, high school textbooks, and college textbooks is to indoctrinate you into Materialism, Physicalism, Darwinism, Nihilism, Behaviorism, Determinism, Physical Reductionism, Atheism, and Evolutionism.

Can you see anything wrong with any of this?

Well, I can see a few things wrong with it.

First of all, where's the evidence proving that evolution or natural selection made your brain?

THERE IS NONE!

It's simply assumed to be true.

Evolution or natural selection has NEVER been caught in the act of designing and creating brains from scratch

It has NEVER been observed; and, it never will be observed because it's physically impossible for invisible, non-physical, immaterial entities like evolution or natural selection to make brains, genomes, or anything else for that matter – according to the Darwinists and Naturalists. The theory of evolution is self-defeating. No part of it has ever been experienced or observed. There's NO evidence supporting any of it. It's simply assumed to be true.

It has been verified and proven by Scientific Observation and Scientific Experimentation that evolution of any kind CANNOT PRODUCE a single functional protein from atoms, let alone the matching gene to go along with that protein. Evolution has NO physical mechanism for doing so. There's NO way in the physical universe that chemical evolution could ever have produced our genome and our brain. It's physically impossible! It's the pinnacle of superstition to think otherwise.

Chemical evolution or spontaneous generation is magical thinking, wishful thinking, and fantasy. Chemical evolution or spontaneous generation is prevented from happening by random diffusion and entropy. Chemical evolution, abiogenesis, or the spontaneous generation of life from atoms has NEVER been experienced nor observed. It's physically impossible. Where's the scientific evidence that chemical evolution is true? There is NONE! Chemical evolution, spontaneous generation, or abiogenesis is simply assumed to be true. There's no convincing evidence that it is true. It has never been experienced or observed. It's assumed to be true.

However, chemical evolution, or spontaneous generation, or abiogenesis, or the self-assembly of atoms was falsified in 1859 by Louis Pasteur, the very same year that Charles Darwin published "On the Origin of Species by Means of Natural Selection". Ironically, natural selection doesn't even touch our genes! They based a whole theory and science discipline for origins on something that doesn't even touch our genes – natural selection.

Where's the evidence that your proteins, their matching genes, your genome, and your brain spontaneously generated out of thin air from atoms? There is NONE. Such a thing has never been experienced or observed. The claim that chemical evolution, or evolution of any kind, made your brain is an assumption. Whenever the Darwinists use the phrase "over millions of years" they are admitting that they have NO observational evidence to support their claims and that everything they are stating as fact is merely an assumption or guesswork. Nobody, except for God, has ever observed millions of years. The claim that evolution made your brain is an assumption. It has never been experienced nor observed. There is no observational evidence supporting any of it. It's simply assumed to be true. That's NOT science – that's blind-faith.

Natural selection and random mutations can't produce anything except entropy and death. Random mutations are entropy. Natural selection or survival of the fittest results in death. The theory of evolution is based exclusively on entropy or death. Is it not? These people teach that ONLY physical matter or entropy exists. Do they not? Let's take them seriously and see where it gets us. Evolution is entropy. Entropy is death. The theory of evolution is Creation by Entropy or Creation by Death. Please explain to me how that's supposed to work. You can't because it's physically impossible. Entropy and death cannot design and create. There's no way in the universe to make Creation by Entropy or Creation by Death true, which means that there's no way in the universe to make the theory of evolution true.

Likewise, Design and Creation by Natural Selection is physically impossible. There's no way in the universe that natural selection can reach out telepathically at a distance and organize raw physical matter into functional proteins and then produce the matching genes to go along with those proteins. There's NO psyche or intelligence within natural selection according to the Darwinists and Evolutionists. According to these people, natural selection and evolution are dumb and blind without a soul or a mind.

Where's the evidence that natural selection made your brain? There is NONE! It's purely an assumption. It is wishful thinking, magical thinking, fuzzy thinking, and nothing more. Design and Creation by Natural Selection has never been experienced nor observed; and, it never will be because it's physically impossible. Design and Creation by Natural Selection is simply assumed to be true. It has never been experienced or observed. There is NO evidence proving that natural selection made your brain, and there never will be because natural selection can't make anything at all. It has never been observed doing so. Creation by Natural Selection has to be taken on blind-faith as being true because there is NO evidence proving that it is true.

According to the Physicalists, Naturalists, Determinists, Classical Physicists, and Behaviorists, genes cannot make choices. Genes cannot do selection! Genes have NO choice in the matter. So, when it comes to synaptogenesis, synaptic typing, and the quantum mapping of our physical brain, who is doing the selection or the choosing if it isn't our genes? Well, it can't be natural selection either, because natural selection doesn't have any physical mechanism for doing choice or selection. Natural selection doesn't touch our genes or our proteins! Natural selection doesn't touch our synapses!

Choice is a function of Psyche and NOT our genes. Our genes can't do choice or selection according to the Materialists, Determinists, Behaviorists, and Naturalists. This is yet another example where Materialism, Naturalism, Behaviorism, and the Theory of Evolution are internally inconsistent and self-defeating.

The Materialists, Naturalists, and Darwinists repeatedly talk about how our genes interact with each other; but, that's physically impossible. Our genes don't interact – not at the physical level at least. They can't interact with each other at the physical level.

Interaction between genes or molecules would imply a telepathic connection between the genes and everything else within their environment. Our genes either have Psyche and Intelligence, or they don't. Which is it? If our genes are indeed interacting with each other, then that interaction has to be taking place at the quantum level, and the genes must have some kind of Psyche, Intelligence, and Language. However, the proven and verified existence of Psyche falsifies the claims of Materialism, Naturalism, Darwinism, Nihilism, Behaviorism, Determinism, Physical Reductionism, Classical Physics, and Atheism which state that Psyche, Syntropy, and Supernatural Mechanisms do not exist.

This is yet another example where Physicalism, Naturalism, Behaviorism, Atheism, and the Theory of Evolution are internally inconsistent and self-defeating. If the genes, proteins, enzymes, and molecules within a cell are truly interacting with each other as the Naturalists and Darwinists claim, then they must have some kind of invisible immaterial non-physical Psyche, Intelligence, or Soul; but, if they have some kind of Psyche or Mind, then the very existence of that Psyche or Mind falsifies Materialism, Naturalism, Darwinism, Atheism, and their derivatives which claim that Psyche and Syntropy do not exist. Likewise, if evolution or natural selection truly made your brain as the Darwinists and Atheists claim, then evolution and natural selection must have some kind of Psyche, Mind, Intelligence, or Soul. However, the proven and verified existence of Psyche, Mind, or Soul falsifies Physicalism, Naturalism, Darwinism, and their derivatives. Checkmate! The theory of evolution is self-defeating and has therefore been falsified.

Where's the evidence that random mutations made your brain? There is NONE! It's purely an assumption. It is wishful thinking, magical thinking, and nothing more. Random Mutations are Entropy. Design and Creation by Random Mutations or Entropy has never been experienced nor observed; and, it never will be because it's physically impossible. Design and Creation by Random Mutations is simply assumed to be true. It has never been experienced nor observed. There is NO evidence proving that random mutations made your brain because random mutations or entropy can't make anything at all. It's never been observed doing so. Creation by Random Mutations or Creation by Entropy has to be taken on blind-faith as being true because there will never be any evidence proving that it is true.

Once we take the evidence into account, we find ourselves looking for a better explanation for Origins besides Darwinism, Materialism, Naturalism, and Atheism. Darwinism, Materialism, Naturalism, and Atheism are based exclusively on the elimination of any evidence that contradicts them or falsifies them. These people cheat in order to make their case and prove their point. Atheism of any kind is based upon a refusal to look at evidence and the destruction of evidence.

The theory of evolution is a fictional story that they made up out of thin air in order to convince people that God does not exist. It works as designed. The theory of evolution has convinced millions of people that God does not exist. It does what it was designed to do; but, that doesn't make it true.

ALL of the evidence that we have on hand as a race FALSIFIES Physicalism, Materialism, Naturalism, Darwinism, Nihilism, Behaviorism, Determinism, Physical Reductionism, Atheism, and the Theory of Evolution. Where's the evidence that the theory of evolution is true? There is NONE! The theory of evolution is based exclusively on wishful thinking, confirmation biases, guesswork, and assumptions. The theory of evolution is simply assumed to be true. Design and Creation by Evolution has NEVER been experienced nor observed. It can't be, and it won't be because it's physically impossible.

## Information Exchange at the Quantum Level

There's way too much information exchange taking place at the quantum level or the psyche level for something dumb and blind like evolution or natural selection to be handling it all, commanding it all, and controlling it all.

For example, how do the transcription enzymes sense and know at a distance which gene is dominant and which gene is recessive so that that enzyme can then read and transcribe the dominant gene while completely ignoring the recessive gene? The dominant gene could be in either slot.

Transcription enzymes are molecules comprised of atoms. Keep asking yourself how the atoms and molecules can tell the difference between a dominant gene and a recessive gene! Ask yourself, "What kind of sensory system and brain system do the transcription enzymes have at the quantum level or the psyche level, since they clearly have nothing like this going on at the physical level?" What the transcription enzymes do at the physical level is physically impossible at the physical level. This same truth applies to the activation enzymes, the translation enzymes, mRNA molecules, tRNA molecules, our genes, our proteins, and the amino acids as well. ALL of the communication and information exchange, that these atoms and molecules are doing with each other at a distance within a living cell, is physically impossible at the physical level.

The "atoms" or "genes" controlling the transcription enzymes recognize and read the dominant gene while choosing to ignore the recessive gene. How is this information being transmitted between atoms and molecules at a distance? Where is this information being stored within the atoms and molecules? How are the atoms and molecules able to know and understand what this information means? It can't be done at the physical level, so it has to be getting done at the quantum level or the psyche level. These are atoms and molecules that we are talking about here, which are communicating with each other. How would you make one atom communicate with another atom? It's physically impossible; but, it's totally possible at the quantum level or the psyche level through Psyche, Action at a Distance, and the Quantum Zeno Effect.

One day it simply dawned on me that there is way too much information exchange taking place at the quantum level or the sub-atomic level for the theory of evolution to be true. Atoms and molecules are obviously communicating with each other at a distance within a living cell; and, that's physically impossible according to the Darwinists, Materialists, Naturalists, Nihilists, Classical Physicists, and Atheists. These people are right in that it's physically impossible for atoms and molecules to communicate with each other and make choices at the physical level; but, atoms and molecules within a living cell are clearly and obviously communicating with each other and making choices at the quantum level or the psyche level, exchanging information with each other and receiving commands from Someone Psyche.

## Everything Falsifies the Theory of Evolution

Everything tells us that the theory of evolution is false. Everything!

Louis Pasteur falsified chemical evolution, spontaneous generation, abiogenesis, creation ex nihilo, the self-assembly of atoms into life, Materialism, Naturalism, Darwinism, and the Theory of Evolution in 1859; and, our scientists have been ignoring and suppressing that Scientific Evidence ever since. Entropy itself falsifies spontaneous generation or chemical evolution. It requires some kind of Syntropy, Psyche, or Intelligence to produce order, organization, proteins, their matching genes, genomes, eyes, brains, and LIFE. This is what the scientific evidence is telling us. This is what observation and experience are telling us must be true. This is what has been experienced and observed. The false is falsified by the truth; and, the truth is repeatedly experienced and observed.

Evolution is entropy; and, evolution results in death. That's what has been experienced and observed. There's no way in the universe that entropy and death could have produced your brain, which means that there is no way in the universe that evolution could have made your brain.

The theory of evolution is science fiction. Every part of it, every version of it, has been falsified by scientific evidence, observations, and experience. The false is repeatedly falsified by the truth; and, the truth is repeatedly experienced and observed.

ONLY humans can do syntax, grammar, sentence construction, spoken sentences, and the writing of books. There are NO evolutionary precursors for these types of language skills. They did NOT evolve!

We are NOT related to the primates! We did not descend from the apes. God made this reality and truth obvious and clear by giving the primates, monkeys, and apes 48 chromosomes while giving the humans only 46 chromosomes. We're not even remotely related to the primates and the apes. We're not sexually compatible with the monkeys and the apes. Monkeys, chimpanzees, and apes have 48 chromosomes. There's NO way in the physical universe that humans and their 46 chromosomes could have ever descended from apes, monkeys, or primates. Genetics prevents it from happening. Apes and humans cannot mate and produce viable offspring. It's physically impossible!

Everywhere we turn, Scientific Observations, Scientific Evidence, Common Sense, and Genetics falsify the Theory of Evolution, Materialism, Naturalism, Darwinism, and Spontaneous Generation or Chemical Evolution.

Macro-evolution is the conversion of one species into a completely different sexually incompatible species over millions of years of time. Macro-evolution involves two chimp-like ancestors siring and giving birth to genetically compatible male and female chimpanzees and/or humans in the same place at the same time so that a whole new mutant species is born from scratch all at once. Macro-evolution is physically impossible. It can't happen, which means that it didn't happen.

Thanks to entropy or random mutations as well as genetics, there's no way in the universe for a sexually-reproducing species to produce a genetically compatible Mr. and Mrs. Mutant year-after-year and generation-after generation in the same place at the same time over millions of years of time so that one species (chimp-like ancestors) can evolve naturally into a completely different sexually incompatible species (chimpanzees or humans). Macro-evolution is physically impossible.

We've never observed cats giving birth to dogs. We've never observed chimp-like ancestors giving birth to genetically compatible male and female chimpanzees, or genetically compatible male and female human beings. We've never observed macro-evolution. Why? It's because macro-evolution is physically impossible. It can't happen, which means that it didn't happen. Macro-evolution has never been experienced nor observed because it's impossible. Macro-evolution is prevented from happening by entropy, random mutations, and genetics. Macro-evolution never happened because it can't happen. It's physically impossible.

If you are going to jump to conclusions (as the Darwinists and Naturalists do), then be smart and jump to a conclusion that has actually been experienced and observed (which the Darwinists and Naturalists don't do). No part of what the Evolutionists have chosen to believe in has ever been experienced or observed. There's NO evidence supporting any of it. When it comes to the theory of evolution, the whole thing is simply assumed to be true. It's a fictional story that they made up out of thin air. There is NO evidence supporting.

Any way we choose to look at it, the theory of evolution is self-defeating, which means that it is false and has been falsified. The false is repeatedly falsified by the truth; and, the truth is repeatedly experienced and observed.

Furthermore, evolution (genetic change), natural selection, and random mutations didn't even exist until after God designed, programmed, engineered, field-tested, fine-tuned, produced, manufactured, created, and deployed our proteins, their matching genes, genomes, eyes, brains, living cells, and physical bodies in the first place. The eyes, brains, proteins, and their matching genes were made first by God thereby making evolution (genetic drift), natural selection, and random mutations possible thereafter.

Evolution didn't make your brain. That's physically impossible.

### Information Storage Capacity Falsifies the Theory of Evolution

Next, these people erroneously assume that ALL of the information for brain development has been encoded within our genes. That's a faulty assumption. It's physically impossible for it to be true.

The human genome and our genes code for proteins, NOT synapses! The human genome stores about 725 megabytes to 800 megabytes of information. To do the 3D addressing for synaptogenesis, the typing and sub-typing of synapses, the construction of synapses, as well as the command and control of synaptogenesis and synaptic pruning would require petabytes of data storage and code to handle the quadrillion synapses that a human child is estimated to produce before synaptic pruning begins to take place. There's nowhere near enough information storage capacity within our genome to handle synaptogenesis and synaptic pruning, especially given the fact that genes code for proteins and NOT synapses. So, where is all of that petabytes of information for synaptogenesis being stored since it's physically impossible for it to be stored within our genes? How is that 3D blueprint information being transmitted to our genes and proteins telepathically at a distance so that our genes and proteins can do synaptogenesis?

Our genes code for proteins, and NOT schematics or blueprints! The 3D schematics and blueprints for axons, dendrites, synapses, synaptic typing, synaptic placement, synaptic meaning, synaptic mapping, and synaptogenesis are being stored someplace else besides our genes! It's physically impossible for all of that information to be stored within our genes! Petabytes of data cannot be stored within our genes.

### Telepathic Communication Is Physically Impossible

According to this quote from *How Children Develop*, the Darwinists and Naturalists literally have our experiences being communicated telepathically at a distance to our genes within our cells so that our genes can then command and control the construction of our physical body, physical brain, and synapses. These people have our brains self-assembling.

Do you see anything wrong with that?

I do!

Self-assembly or spontaneous generation was falsified in 1859 by Louis Pasteur; and, the Darwinists assure us that genes are dumb and blind without a soul or a mind. How in the universe would our genes command and control all of this anyway – telepathically at a distance – without some kind of psyche, soul, mind, or intelligence? That's physically impossible according to the Materialists, Naturalists, Darwinists, and Atheists; yet, these very same people have our personal experiences being telepathically communicated to our genes within our brain cells so that those genes can then command and control brain development. They called it brain plasticity.

It doesn't fit!

These people claim that psyche, mind, spirit, or soul does not exist; yet at the same time, they imbue our genes with a psyche, soul, or mind so that our genes can then handle and control all of our brain development for us. Their claims are inconsistent, mutually exclusive, and self-contradictory. Their claims are physically impossible. Spontaneous generation or the self-assembly of proteins and their matching genes from atoms has been falsified by Science, Scientific Observations, and Science Experiments.

The evolutionary message or evolutionary perspective is internally inconsistent, incoherent, self-contradictory, paradoxical, and at odds with itself. These people teach that ONLY physical matter and entropy exist. If true, then that reality would prevent our genes and proteins from communicating with each other telepathically at a distance in order to command and control and coordinate the construction of our physical brain; yet, these people claim that evolution or your genes made your physical body and physical brain. The contradiction is staggering.

There is a major conceptual problem with Experience-Dependent Plasticity and Experience-Expectant Plasticity that cannot be answered by Physicalism, Scientific Naturalism, and Darwinism.

Can you see what it is?

It has to do with experience!

Experiences are information or data. Experiences are invisible, immaterial, non-physical information and data. Experiences, thoughts, and memories are transmitted, stored, and processed as quantum waves. Plasticity is the process through which neural connections are created and reorganized throughout life as a function of an individual's experiences. The processing of experiences requires information exchange at the quantum level or the psyche level because there's NO direct physical connection between our genes and our synapses; and, our genes, proteins, and molecules have NO physical mechanism within them for processing information at the physical level.

How is evolution transmitting and receiving all of this information? How is evolution feeding our experiences into our genes and our physical brain? How is evolution commanding and controlling all of these different physical processes telepathically at a distance? That would imply that evolution is some type of psyche, mind, or soul. However, the Evolutionists claim that psyche, soul, spirit, or mind does not exist.

There are NO wires within a physical brain or a physical body. Our bodies and brains are NOT wired together as the Materialists and Naturalists claim. Synapses are physical gaps. Even if synapses are filled with some type of cerebral spinal fluid, chemical synapses are still physical gaps. There's no direct physical connection between neurons through our chemical synapses as the Naturalists and Darwinists claim. There's just a physical gap between neurons.

Synapses falsify the claim that evolution made your brain. It's physically impossible for evolution to handle the information exchange between atoms and across synapses at any level of existence.

Neurotransmitters travel between neurons; but, neurotransmitters are molecules comprised of atoms; and, there's NO physical mechanism in place for storing information, data, experiences, memories, commands, codes, and control information within atoms or molecules at the physical level. There's NO physical mechanism in place for transmitting data, information, and experiences between molecules and atoms. There's NO physical mechanism in place within an atom or a molecule for processing the information that it receives telepathically at a distance, either.

The evolutionists claim that evolution made your brain. The Neo-Darwinists equate evolution with your genes, and then they claim that your genes made your brain.

So, ask yourself how an individual's experiences are being transmitted telepathically at a distance between the neurons and into a person's genes, so that our genes can then command and control synaptogenesis, plasticity, and brain remapping?

These people literally claim that our experiences are somehow reorganizing or remapping our brain; but, that's physically impossible at the physical level. There's no physical mechanism in place for doing so. There's NO physical connection between our genes and the proteins that are being used by the Map-Maker to make and construct our synapses. There are NO wires in our brain, which means that our experiences are somehow being transmitted wirelessly at a distance between cells, genes, proteins, and synapses.

**Brain Mapping Requires Intelligence**

Clearly our brains have been mapped. Maps don't spontaneously generate out of thin air! They are mapped by Someone Intelligent or Someone Psyche.

Our physical brains have been organized into a somatosensory map, a motor map, and many other types of maps. How do our experiences get mapped onto our brain? Who is doing this mapping?

Chemical synapses are physical gaps in space, not direct wired connections as the Materialists, Naturalists, and Darwinists tend to call them. Despite claims to the contrary from the Scientific Naturalists and Darwinists, there are NO wires within a physical brain connecting everything together into RAM, CPUs, logic gates, transistors, and communication networks. Neurotransmitters of any type OPEN ion channels, NOT a direct wired telephone connection between neurons. Everything within the physical brain is done wirelessly at a distance. That's Quantum Mechanics that we are talking about there, and NOT classical physics. There's NO such thing as "action at a distance" at the physical level according to Materialism, Naturalism, Darwinism, Atheism, and Classical Physics. Instead, within the physical brain, everything is done wirelessly at a distance at the quantum level or the sub-atomic level.

Synapses involve Action at a Distance. Information exchange between atoms and molecules involves Action at a Distance. The transmission of our personal experiences into our genes requires Action at a Distance or Telepathy. Synapses and Action at a Distance falsify the claim that evolution made your brain. The claim that evolution made your brain is physically impossible because the communication between atoms, molecules, your personal experiences, your proteins, your genes, your neurons, and your synapses requires Action at a Distance or Quantum Mechanics. It also requires Someone Intelligent or Someone Psyche.

Remember, each neuron is separated from every other neuron by a physical gap that we call a synapse. If our neurons are communicating with each other, they can only be doing so at the quantum level because it's physically impossible for them to do so at the physical level. It's physically impossible to transmit a message across a synaptic cleft. Synaptic clefts scramble and randomize everything that comes their way at the physical level. Communication between atoms is only possible at the quantum level or the sub-atomic level. It's physically impossible to transmit messages within atoms and molecules to other atoms and molecules. Communication between genes and proteins is physically impossible at the physical level; but, clearly and obviously it's happening at the quantum level or the sub-atomic level. The proven and verified existence of Quantum Mechanics and Action at a Distance falsifies Materialism, Naturalism, and the Theory of Evolution making it impossible for evolution to have produced your brain.

A neurotransmitter is a molecule. A neurotransmitter is a key that opens an ion channel or an ion gate. An ion is an atom. How would you transmit command and control instructions through individual atoms to and from different molecules and atoms? You can't at the physical level! All of our experiences and all of that command and control information exchange has to be taking place at the sub-atomic level as quantum waves because it's physically impossible at the physical level through atoms and molecules.

There's NO message within a neurotransmitter at the physical level. A neurotransmitter simply opens a gate allowing atoms or ions to pass through that gate. All of the "information" from every single synapse on a neuron gets reduced to one single BIT of information called a post-synaptic potential (PSP) which determines whether a neuron turns ON or stays OFF. It's physically impossible to transmit command and control instructions through a single physical bit such as a PSP or a neuron. A neuron is a switch. It is either on or off at the physical level. That's it. If there's any actual information exchange taking place between neurons and their genes, it's ALL taking place at the quantum level or the sub-atomic level because it's physically impossible at the physical level.

Communication across synapses is taking place at the quantum level or the sub-atomic level because it's physically impossible at the physical level. How would you make atoms communicate with each other at the physical level? You can't! It's physically impossible. If there are any messages and experiences being transmitted between neurons, those transmissions are happening as quantum waves at the quantum level or the psyche level because they are physically impossible at the physical level. Synapses scramble and randomize everything that comes their way at the physical level; so, all information exchange between neurons has to be taking place at the quantum level or the sub-atomic level.

Through the use of synapses or physical gaps, God deliberately made brain functionality physically impossible at the physical level, so that it would eventually become obvious and clear to us that all of our brain functionality is taking place at the quantum level or the psyche level. There's NO physical mechanism in place for our experiences to be communicated to our genes at the physical level. There's NO physical mechanism in place for our genes to communicate with and control our proteins and protein assembly at the physical level. Eventually it becomes obvious and clear that it's physically impossible for evolution and genes to have produced your physical brain at the physical level, which means that your brain was produced by Someone Psyche at the quantum level or the sub-atomic level.

There are NO physical wires running between the atoms and molecules within your brain. Communication between atoms and molecules is physically impossible according to the Naturalists, Darwinists, and Atheists. God seems to have gone out of His way to design and map our brains with synapses, so that a brain is ONLY functional at the quantum level or the psyche level and so that a brain's functionality is obviously impossible at the physical level. Information exchange between atoms and molecules is physically impossible at the physical level according to the Scientific Naturalists and Atheists, which means that it has to be taking place at the quantum level or the psyche level.

Evolution did not make your brain. Such a claim is physically impossible. Evolution is a physical process; and, it's physically impossible for our brains to exchange information and experiences between molecules at the physical level. It's physically impossible for the atoms, molecules, proteins, and genes to transmit and receive information and knowledge among each other at the physical level. A synapse is a physical separation and not a direct wired connection. If atoms and molecules are communicating with each other within your physical body and physical brain, and they clearly are, then that communication and information exchange has to be taking place at the quantum level or the sub-atomic level because it's impossible in principle at the physical level.

The mapping of purpose and meaning onto each and every synapse or physical gap within your brain has to be taking place at the quantum level because it's physically impossible at the physical level.

**Someone Psyche Clearly Mapped Your Brain**

Clearly, our brains are being MAPPED!

How are they being mapped? Mapped by whom?

**Our physical brains have clearly been mapped. Maps require some kind of Intelligent Map-Maker; otherwise, it would be nothing but chaos within a physical brain. It requires an Intelligent Map-Maker to give map coordinates, or neurons and synapses, purpose and meaning.**

The evolutionists claim that evolution made your brain. The Neo-Darwinists equate evolution with your genes, and then they claim that your genes made your brain. Neural plasticity claims that your personal experiences are restructuring, or reordering, or remapping your brain.

Ask yourself how all of that information and all of those experiences are being transmitted to the genes. Ask yourself how your genes or transcription enzymes are able to make decisions and choose which proteins need to be produced and which structures need to be made. Choice is a function of Psyche, and NOT our genes. By definition, in principle, genes and proteins can't make choices – not at the physical level at least. Genes and proteins can't process, transmit, nor understand our personal experiences at the physical level.

At the physical level, genes and proteins simply do what they are told to do. Told by whom? Brain plasticity can't function nor work right at the physical level because the necessary information exchange between atoms, molecules, genes, proteins, and our personal experiences can't take place at the physical level. ALL of this information exchange has to be taking place at the quantum level or the sub-atomic level.

Thoughts, memories, and personal experiences are quantum waves. How do we know? WE KNOW because the memories of our personal experiences, as well as our ability to think and choose, survive the death of our physical brain according to the scientific evidence being obtained from Near-Death Experiences (NDEs), Share-Death Experiences (SDEs), Out-of-Body Experiences (OBEs), and our after-death Life Reviews. If a thought or a memory survives the death of your physical brain, then it's truly a memory in every sense of the word. Is it not?

Ask yourself how information is being transmitted wirelessly at a distance between the genes and the proteins telling the proteins where to go and what to make when they get there. Ask yourself how your personal experiences are being transmitted wirelessly at a distance into your genes so that your genes or evolution can then do Command and Control as well as Map-Making. How are your genes and proteins able to read and understand the information and messages that are being transmitted to them and received by them? Explain how your genes and proteins are able to construct an invisible non-physical quantum computer within your physical brain. Explain how your genes and proteins are able to map physical functionality, purpose, and meaning onto each one of your synapses and neurons telepathically at a distance.

The Materialists, Naturalist, Darwinists, Nihilists, Behaviorists, Determinists, Classical Physicists, Physical Reductionists, and Atheists have NO scientific explanation for how the information from our experiences is being transmitted wirelessly at a distance between cells through physical gaps, within molecules and atoms, into genes and proteins. There's NO physical mechanism in place for transmitting experiences and information into our genes and proteins at the physical level. Genes and proteins are molecules comprised of atoms. There's NO physical mechanism in place for information exchange to take place between atoms and molecules; yet, it's obvious and clear that the atoms and molecules such as genes, enzymes, and proteins are indeed communicating with each other at the quantum level, the psyche level, or the sub-atomic level.

Ask yourself, "How do the proteins learn and know how to assemble into synapses?" A protein dumped into the cytoplasm can literally go anywhere after it has been made. So, who is telling each protein where to go and what type of machine or synapse to make? All of this 3D Mapping Information or Command and Control Information is being transmitted wirelessly at a distance between the genes and the proteins and the enzymes because there's NO physical mechanism in place for doing so. There are NO physical wires within our brains connecting the genes with the enzymes and proteins; so, that means that all of this information is being transmitted wirelessly at the quantum level through quantum waves.

**Quantum Waves Are WiFi at the Psyche Level**

Quantum waves are WiFi at the quantum level or the psyche level. Thoughts and memories are quantum waves. They survive the death of our physical brain.

Remember, synapses are physical gaps between neurons. There are NO direct physical connections between neurons through chemical synapses! There are no wires within your brain as the Materialists and Naturalists claim. Any information that's being transmitted between neurons is being transmitted wirelessly at a distance through individual atoms, molecules, and/or empty space. Atoms are mostly empty space.

Synapses are based upon action at a distance or communication at a distance. Synapses are NOT direct physical connections or physical wires. Communication at a Distance or Telepathy is Quantum Mechanics. That is Action at a Distance or Quantum Mechanics that we are talking about there, and NOT classical physics. According to Classical Physics, Physicalism, Naturalism, and Atheism, it's physically impossible for the neurons to be communicating with each other through synapses, physical gaps, and/or empty space; but, communicating with each other they are, at the quantum level or the psyche level.

God designed our brains so that its functionality is physically impossible at the physical level thanks to synapses or the physical gaps between neurons and the physical gaps between proteins, neurons, and their genes. The only way a physical brain can be made to work is at the quantum level or the sub-atomic level because there are NO wires within a brain at the physical level. ALL information exchange between neurons within a physical brain is done wirelessly at the quantum level or the psyche level because it's physically impossible at the physical level.

Our physical brains have clearly been mapped at the quantum level so as to have physical functionality at the physical level; so, who is the Map-Maker at the quantum level or the psyche level who is mapping the construction, typing, purposing, and meaning of each neuron and synapse within your physical brain? Who is the invisible non-physical Map-Maker, Engineer, and Programmer who is creating an invisible non-physical quantum computer within your physical brain? It can't be evolution and it can't be your genes because evolution and genes are restricted exclusively to the physical level by definition in principle. There's NO scientific explanation at the physical level for how all of this information is being exchanged and processed and understood at the quantum level.

In contrast, there's actually a Scientific Explanation for all of this at the Quantum Level or the Psyche Level. According to Orthodox Quantum Mechanics, this invisible non-physical Map-Maker, Engineer, and Programmer is Nature's Psyche (or God's Psyche).

It can't be the Human Psyche doing this Map-Making because the Human Psyche isn't consciously aware of any of these quantum transmission processes that are related to synaptogenesis at the physical level and brain-mapping at the quantum level. Science, Psyche, Intelligence, Information Exchange at the Quantum Level, Syntropy, Supernatural Mechanisms, Invisible Non-Physical Forces and Fields, Action at a Distance, and Quantum Mechanics are extremely painful to the committed Darwinist, Naturalist, Physicalist, Behaviorist, Determinist, Classical Physicist, and Atheist. I KNOW because I used to be a Materialist, Naturalist, Nihilist, and Atheist. Quantum Mechanics, Psyche, Quantum Waves, Action at a Distance, Quantum Mapping, and Syntropy cannot be explained by Materialism, Naturalism, Darwinism, Nihilism, Atheism, and Classical Physics.

**The Theory of Evolution is Self-Defeating**

The Naturalists and Darwinists claim that evolution is your Maker, which means that they claim that evolution has some kind of psyche, soul, or mind; but at the same time, these people also claim that psyche, soul, or mind does not exist. The contradiction is staggering! Materialism, Naturalism, and the Theory of Evolution are self-defeating because their claims are internally inconsistent, incoherent, and self-contradictory. Evolution of any kind can't produce the goods. Evolution of any kind would require the direct intervention of Someone Psyche in order for the theory of evolution to be true; however, admitting or allowing the existence of Psyche, Syntropy, Action at a Distance, or Quantum Mechanics falsifies Materialism, Naturalism, Darwinism, Nihilism, and Atheism and defeats the purpose for which they were made – namely to convince people that God's Psyche or God's Spirit does not exist.

It requires some kind of Psyche or Intelligence to transmit information wirelessly at a distance, to interpret that information when it is received, and then decide what to do with that information and how to use that information after it has been received. Someone Psyche has to be handling the communication and information exchange wirelessly at a distance between atoms, molecules, genes, enzymes, and proteins within a living cell. It can't be evolution handling this information exchange at the quantum level or the psyche level. By definition, in principle, evolution doesn't have psyche or intelligence. Evolution is dumb and blind without a soul or a mind. That's what the Darwinists and Naturalists teach and believe. The theory of evolution is self-defeating and falsifies itself. The theory of evolution is based upon the claim that Psyche, Mind, Spirit, and Soul do not exist; yet, it would require intervention from Someone Psyche at the psyche level in order to make the theory of evolution true.

Remember, the verified and proven existence of Quantum Mechanics, or Supernatural Mechanics, or Spiritual Mechanics falsifies the Theory of Evolution which is based upon the claim that the spiritual or the supernatural does not exist. The false is falsified by the truth; and, the truth is repeatedly experienced and observed.

The Materialists and Naturalists are wrong at all levels of existence. These people are even wrong at the physical level. These people claim that ONLY physical matter or entropy exists. Evolution is entropy. Entropy cannot design and create. Evolution and genes have NO physical mechanism in place for transmitting and receiving information and experiences wirelessly at a distance. Entropy can't do any of that either. All of that transception has to be taking place at the quantum level because it's physically impossible at the physical level; and, the proven and verified existence of a quantum level falsifies the claims of Materialism, Naturalism, Darwinism, Nihilism, and Atheism which claim that a quantum level or a psyche level DOES NOT EXIST.

**Scientific Observation: Anything that was obviously made obviously had Someone Intelligent or Someone Psyche who made it, caused it to be, and brought it into existence.**

**Scientific Observation: Proteins, their matching genes, genomes, eyes, brains, and organized life-forms were obviously made.**

**Scientific Conclusion: Therefore, it is rational and logical to conclude that proteins, genes, genomes, eyes, brains, and made lifeforms obviously had Someone Intelligent or Someone Psyche who made them.**

Can you see that and understand it?

I can because I know what the human race has experienced and observed, in general, throughout the history of its existence on this planet. The observations and experiences of the human race FALSIFY Materialism, Naturalism, Darwinism, Nihilism, Atheism, Behaviorism, Determinism, Physical Reductionism, and even Classical Physics. The observations and experiences of the human race have instead VERIFIED the existence of Psyche, Syntropy, Quantum Mechanics, Supernatural Mechanisms, Invisible Immaterial Non-Physical Forces and Fields, Non-Physical Matter, Quantum Non-Locality, after-death Life Reviews, and God.

The truth is repeatedly verified by observation and experience, while the false is repeatedly falsified by observation and experience. Observation and experience falsify the claims of Materialism, Naturalism, Darwinism, Nihilism, Atheism, Classical Physics, and the Theory of Evolution.

Genes code for proteins. Someone Psyche or Someone Intelligent has to be telling those newly made proteins and molecules, wirelessly at a distance, where to go and what to make; otherwise, they would just sit there and do nothing after they are made. Outside of a living cell, there is NO such thing as spontaneous generation. However, the very same atoms, molecules, amino acids, genes, and proteins that do absolutely nothing outside of a living cell literally come alive with purpose and activity within a living cell. How's that possible? It's not according to the Physicalists, Naturalists, Darwinists, Nihilists, and Atheists.

Someone has gotten their science wrong. Someone is failing to observe.

The Materialists and Scientific Naturalists will tell you and assure you that I got my science wrong.

Do you believe them?

Begging the question, these people define "science" as Materialism, Naturalism, Darwinism, Nihilism, Behaviorism, Determinism, Physical Reductionism, Classical Physics, and Atheism. If their definition for science is the correct definition for science, then clearly, I have gotten my science wrong. However, I choose to define Science as observation and experience. If my definition for Science is right, then these people have clearly gotten their science wrong because trillions of different observations and experiences falsify Physicalism, Naturalism, Darwinism, Nihilism, Atheism, Classical Physics, and their derivatives.

These people have told me many times that I don't understand Science.

Do you believe them?

Or, is it possible that I understand the Science, the Scientific Observations, the Scientific Research, the Scientific Evidence, and the Experiential Phenomenology of the human race much better than these people want me to?

We both can't be right because what we are teaching and preaching is mutually exclusive in the end. If one of us is right, then the other position has been automatically falsified. So, which one of us is right, and which one of us is wrong – or are we both wrong?

Ask yourself, "Which position has been experienced and observed – the existence of all these different immaterial, invisible, non-physical forces and fields such as gravity, magnetism, microwaves, radio waves, gamma rays, x-rays, dark energy, dark matter, the strong and weak nuclear force, vacuum energy or the zero-point field, quantum waves, psyche, intelligence, syntropy, consciousness, thoughts, after-death memories, after-death life reviews, and life after death; or, the non-existence of these things?"

Which position has been experienced and observed? How do you experience and observe the non-existence of these things, especially since they have been experienced and observed?

Science is observation and experience after all. It's through observation and experience that we find and know the truth.

The theory of evolution is easy to falsify because it is science fiction. There is NO observational evidence and NO experiential evidence supporting any part of it. All aspects of the theory of evolution have been FALSIFIED by scientific evidence, observations, and the experiences or phenomenology of the human race. The theory of evolution is a worthless piece of fiction containing no scientific merit whatsoever. It didn't happen because it can't happen. It's physically impossible.

Remember, the theory of evolution has been FALSIFIED. It's falsifiable and it has been falsified. The theory of evolution has been falsified trillions of times by thousands if not millions of different people. No part of the theory of evolution has ever been experienced or observed. Even natural selection and random mutations have never been caught in the act of design and creation because they can't design and create. It's physically impossible!

Creation by Evolution has never been observed. It has no evidence to support it. It's simply assumed to be true. It's important to eliminate everything that's false so that we can find and know the truth.

## It Took Millions of Years for Evolution to Make Your Brain

Can you see anything else wrong with this statement or quote from *How Children Develop*?

I can.

This statement literally says that it took millions of years for evolution to produce the human brain.

How did our ancestors survive and thrive for millions of years while they didn't have a brain? How did they manage to be the fittest for millions of years while evolution was making their brain? How well do humans do when they are born without a brain? How well would chimp-like ancestors do for millions of years without a brain while evolution is making their brain? A brain has NO survival value for the millions of years that it takes for evolution to make a brain. That reality and truth forms a "chicken and egg" paradox where evolution is concerned and thereby falsifies the theory of evolution. According to the theory of evolution, we didn't have a brain for millions of years while evolution was making our brain.

Does that make any sense to you?

Any time you see the words "millions of years", that's code for "we do not have any observational evidence to support our claims". There's NO evidence supporting the claim that evolution made your brain. The process has never been observed. It is simply assumed to be true and taken on blind-faith as being true.

The theory of evolution is a fictional story that they made up out of thin air to convince the gullible and the ignorant that God does not exist. They are telling people what they want to hear so that they don't have to face the hard truths associated with God's existence and God's Commandments. In all of our science fiction, evolution is given all of the credit for doing something that it could never have done.

When it comes to the theory of evolution, I can't find anything to suggest that it might be true; whereas, ALL of the scientific evidence and the observational experiences of the human race falsify Materialism, Naturalism, and the Theory of Evolution. I'm going with the evidence on this one rather than the desperate wishful-thinking of the Materialists, Naturalists, Darwinists, Nihilists, and Atheists.

Falsifying the theory of evolution can take a great deal of effort, scrutiny, specificity, analysis, critical thinking, and work; but, it's worth the struggle due to all of the Science and Quantum Mechanics that you can learn in the process. The more you understand Quantum Mechanics or Supernatural Mechanics, the more you falsify Materialism, Naturalism, and the Theory of Evolution.

It changes the way that you look at Science and Reality when you first discover and realize that the Materialists, Naturalists, Darwinists, and Atheists are trying to trick you and deceive you by suppressing evidence, hiding evidence, destroying evidence, ridiculing evidence, and claiming that the evidence does not exist. When you finally start to discover the tons of observational evidence that falsifies Physicalism, Naturalism, Darwinism, Nihilism, Atheism, and the Theory of Evolution, you feel really stupid for having believed these people all the hundreds of different times that they lied to you and told you that the evidence does not exist, because Naturalism and Darwinism are obviously false once you start looking at the reams of observational evidence that falsify them.

But, nobody's immune. Self-deception works because we want it to work. We want to be deceived. We want to believe that God does not exist and that His Commandments do not apply to us. It's a message that we want to hear. I KNOW because I used to be a Materialist, Naturalist, Nihilist, and Atheist; and, I desperately wanted them to be true. I know how it goes. I've been there and done that too.

Over time I changed. The observational evidence and the collective experiences of the human race convinced me that I was wrong. I repented. I changed. Eventually, Science and observational evidence convinced me that God exists and that the theory of evolution is false. I wanted to find and know the truth instead of hiding from the truth. Eventually I did just that. I began to find the truth because I finally went searching for the truth.

The false is repeatedly falsified by the truth; and, the truth is repeatedly experienced and observed. The BEST and FASTEST way for finding and knowing the truth is to live it, experience it, and observe it for yourself; or, to choose to trust someone who has. That is exactly what I have done.

In contrast, the scientific evidence, experimental evidence, observational evidence, eye-witness evidence, and experiential evidence are constantly telling us that the theory of evolution is physically impossible and therefore nothing but science fiction. The theory of evolution is Creation by Entropy or Creation by Death; and, that's physically impossible. Creation by Entropy or Creation by Death has NEVER been experienced nor observed; and, it never will be because it's impossible.

All you want is the truth, unless of course you are trying to trick people and deceive people; then, any old lie will do. Once day I decided that I want the truth rather than all the fiction and lies that the Materialists, Naturalists, Darwinists, and Atheists have been feeding me throughout my life. Seek and ye shall find. Knock, and it shall be opened unto you.

Ye shall find the truth, and the truth shall set you free. I've been set free; and, you can be set free too.

I KNOW that I'm on the right track whenever everything fits together perfectly, and whenever everything is internally consistent. I KNOW that I'm on the right track whenever I can explain every experience, observation, and piece of evidence that comes my way. The explanatory power of Quantum Mechanics, Syntropy, Action at a Distance, and Psyche is through the roof to infinity and beyond; whereas, Materialism, Naturalism, and the Theory of Evolution fall apart, are self-contradictory, and are self-defeating even at the physical level. Furthermore, Physicalism and Scientific Naturalism have absolutely NO explanatory power when it comes to the psyche level or the quantum level. You want your Science to have explanatory power; otherwise, it's worthless.

The verified and proven existence of Action at a Distance, Syntropy, Psyche, and Quantum Mechanics falsifies Materialism, Naturalism, Darwinism, Nihilism, Atheism, Classical Physics, and their derivatives. Within Quantum Mechanics, Psyche, or Syntropy, I have found the answer to life, the universe, and everything.

Mark My Words

—

## Source

### Science 2.0: I Upgraded My Science

https://www.amazon.com/dp/B0771K6WTX

## References

I discuss these concepts in much greater detail in the following books:

### Science 2.0: I Upgraded My Science.

https://www.amazon.com/dp/B0771K6WTX

### Quantum Neuroscience: The Answer to Life, the Universe, and Everything.

https://www.amazon.com/dp/B079Z6QQQB

### Scientific Proof of God's Existence: Finding God Where the Atheists Refuse to Look for Him.

https://www.amazon.com/dp/B07B26CRHX

### Quantum Mechanics from a Non-Physical Spiritual Perspective.

https://www.amazon.com/dp/B01J023TGU

https://www.amazon.com/dp/1521132380

### Scientific Proof of God's Existence: A Primer.

https://www.amazon.com/dp/B071713NNL

https://www.amazon.com/dp/1521325170

I used to be a Materialist, Naturalist, Nihilist, and Atheist. Obviously, I learned not to be afraid of God, religion, spirituality, psyche, fringe science, and quantum mechanics. I had nothing to lose and everything to gain. I wanted to understand how everything works, and now I do.

I encourage you to join me.

When it comes to Science, we observe that Quantum Mechanics, Syntropy, and Psyche have infinitely more explanatory power than Materialism, Naturalism, Darwinism, Nihilism, Classical Physics, and Atheism can even begin to have. Scientific Naturalism lacks explanatory power when it comes to the quantum level or the psyche level. By choosing to allow Psyche, Syntropy, and Quantum Mechanics in to play, we can literally explain everything that comes our way.

Mark My Words

## Syntropy versus Entropy

Most of the scientists claim to believe in the Big Bang Theory, and they really truly believe that the Big Bang happened. If they are right, and there's really no way to prove that they are right, then we automatically KNOW that the Big Bang represents a huge monumental reversal of entropy or a huge amount of Syntropy. Syntropy, or a reversal of entropy, or the cessation of entropy HAS TO EXIST; otherwise, the Big Bang could never have happened, as the scientists claim that it did. The fact that we exist and the fact that physical matter exists are scientific proof that someone in this universe knows how to reverse and eliminate entropy at will.

If the second law of thermodynamics were actually true, then all of this wouldn't exist, and "all there is" would be nothing but random chaos, as the second law erroneously predicts. All we see around us is order and organization. We don't observe an ever-encroaching gray goo coming in at us from all sides, as the second law of thermodynamics predicts. Therefore, WE KNOW there is something fundamentally wrong with the second law of thermodynamics; and, it's time for us to figure out what's wrong with thing, because we should see nothing but chaos or disorder all around us if the second law of thermodynamics were actually true.

In 2018, I wrote the book on Syntropy.

https://www.amazon.com/gp/product/B07BPT3W8R

In that book, I figured out what syntropy is and how it works.

Syntropy is the opposite of "entropy defined as chaos or disorder"; and, this book slowly encouraged me to find NEW and BETTER definitions for entropy than the ones upon which the second law of thermodynamics is based.

A year later, I formally falsified the second law of thermodynamics in my books: "Quantum Mechanics: From a Non-Physical Spiritual Perspective" and "Science 2.0: I Upgraded My Science".

https://www.amazon.com/gp/product/B01J023TGU

https://www.amazon.com/gp/product/1521132380

https://www.amazon.com/gp/product/B0771K6WTX

By falsifying the second law of thermodynamics, I really was forced to upgrade my science to Science 2.0.

Due to the lack of space, I won't spend much time here discussing this topic because that is better done in the other books that I have written.

If you are interested in syntropy, then find and read my book on the subject. I'm the first scientist on the planet to have taken on the topic in any great detail. I guess you could say that I'm the first person to have a doctorate in Syntropy. Nobody else in this world is interested in this topic. They can't seem to get their hands on it nor wrap their minds around it.

For the most part, Syntropy remains undiscovered country. Syntropy obviously has to exist, or we wouldn't exist, but nobody seems to know what it is or how it works. I tried to resolve that situation in a few of my books. Seek and ye shall find. Knock, and it shall be opened unto you.

Mark My Words

--

My Amazon Author Page:

https://www.amazon.com/Mark-My-Words/e/B01IAEF2Y6

# PART IV — THE WEAKNESSES OF THE SCIENTIFIC METHOD

Most people don't know that the Scientific Method is fundamentally flawed; and, that fact or reality is where all of the confusion and contention comes from where science is concerned.

The Scientific Method is based upon a logic fallacy that is called *affirming the consequent*. The whole of modern-day Science, Materialism, and Scientific Naturalism is based upon this logic fallacy; and, most scientists don't even know it, because they have never studied the Philosophy of Science.

The weaknesses of science and the scientific methods explain why science needs to be upgraded to Science 2.0.

Science is Observation.

If you observe carefully, you will notice that the Materialists, Naturalists, Darwinists, Nihilists, and Atheists define science as Materialism, Naturalism, Darwinism, Behaviorism, Determinism, Scientism, Physicalism, Atomistic Reductionism, Nihilism, and Atheism. These philosophies have been falsified by Observational Evidence, Experiential Evidence, Eye-Witness Evidence, and Empirical Evidence. In other words, Materialism, Naturalism, and their derivatives have been falsified by Observations. Notice carefully, and you will observe that the promoters of these false and falsified philosophies rely upon *affirming the consequent* and the suppression of evidence in order to make their case and meet their burden of proof.

Defining science as Materialism, begging the question, suppressing evidence, and similar practices result in a wide variety of different logic fallacies known variously as *circular reasoning, affirming the consequent, begging the question, jumping to conclusions, scientific inferences, personal interpretations, confirmation biases,* and *ad hoc just-so stories*. These are the bread and butter of Scientific Materialism and the foundation upon which modern-day science is built. It's a house of cards – a pyramid scheme.

Science 2.0 resolves the problem by defining Science as Observation, Experience, Experimentation, Knowledge, Fact, Truth, and Evidence. Science 2.0 is a paradigm shift in our way of thinking and in the way that we do Science. Science 2.0 allows all of the evidence into evidence.

Furthermore, under Science 2.0, there is NO officially sanctified interpretation of the scientific evidence. You are encouraged to make your own interpretation of the evidence. Science 2.0 allows all of the evidence into evidence, and then leaves it up to YOU to interpret the evidence, decide for yourself what to make of the evidence, and choose for yourself how to use that evidence in your own life. You are going to do so anyway, so why not encourage you to do so right from the start?

Science 2.0 is the way that Science should have always been done but wasn't.

## SCIENTISM

Scientism is the philosophical religious dogmatic belief that science and the scientific method is the only source of knowledge and truth. These people literally worship science and the scientific methods, and they treat science and the scientific method as if these things were an infallible god.

## The Scientific Method Is Based Upon Affirming the Consequent

The Scientific Method is a good way for coming to know the truth; but, most people do not realize that the Scientific Method is not a foolproof way of knowing the truth nor is it the best way of knowing the truth.

Like me, most people are shocked when they first realize or are first taught that the Scientific Method is based upon a logic fallacy called "affirming the consequent".

It took an honest philosopher and scientist to reveal this truth to me, because many people like me have been brainwashed into believing that the Scientific Method is an infallible god. The people who are part of the religion called Scientism actually worship Science and treat Science as if it were God. These people pin all of their hope and faith on the Scientific Method, never once realizing that there is a fatal flaw at the heart of the Scientific Method which has the power to mess them up every time.

The Scientific Method typically runs through the following sequence in an attempt to arrive at the truth.

1) Form a HYPOTHESIS.

2) Select a Scientific Method or Scientific Methodology to TEST the hypothesis.

3) Run the Science Experiment; and then, Observe and Measure the RESULTS.

4) Find the BEST INTERPRETATION or the BEST EXPLANATION for the Scientific Data, the Scientific Evidence, the Scientific Observations, and the RESULTS of the Science Experiment.

The main flaw in the Scientific Method is found in the fourth and final step of the Scientific Method, where human error and human weakness comes into play by "affirming the consequent" or affirming the conclusion which the human wants the hypothesis to prove. In other words, the human being chooses the conclusion or the interpretation that he or she wants, never once realizing that there might in fact be a better explanation for the scientific data or a better interpretation of what the tested hypothesis and the chosen conclusion might in fact really truly mean. This happens all the time. Human beings make the data and the scientific evidence fit the conclusion they personally want, even though there are better explanations for the scientific data than the explanation or the interpretation which the person as the scientist has chosen to affirm.

One example: Materialism by definition is design and creation by physical matter, or Creation by Rocks. That's what Materialism, Darwinism, and Naturalism reduce to – Creation by Chance or Creation by Rocks. Millions of scientists across this world on a daily basis form hypotheses, run experiments and tests, get lots of results and scientific data, and then conclude that their hypotheses and scientific evidence have successfully proven Creation by Rocks, or Materialism, or Darwinism to be true – never once realizing that Intelligent Design, or Intelligent Manufacturing, or Creation by Psyche is in fact a far better, more logical, more realistic, and more parsimonious explanation for their scientific data than Creation by Rocks or Materialism.

Affirming or choosing Creation by Rocks as one's conclusion or as one's interpretation of the scientific data is the perfect example of the "affirming the consequent" logic fallacy, which the Scientific Method employs every time that the Scientific Method is used. Creation by Rocks is the "affirming the consequent" logic fallacy in action. The fourth step of the Scientific Method is unavoidable – conclusions have to be drawn, but there's no guarantee that one's chosen conclusions are correct.  Creation by Rocks is an illogical and unsustainable conclusion; yet, it is THE CONCLUSION which millions of scientists choose on a daily basis.

Obviously, one's chosen consequent or one's chosen conclusion doesn't have to be wrong or false.  These same scientists could have chosen Creation by Intelligent Designers or Creation by Psyche as their conclusion and greatly increased their chances that their chosen conclusion might in fact be real, right, correct, and true; but, most of the scientists will not and have not chosen Creation by Intelligent Beings as their conclusion or their interpretation of the scientific data because these people don't want to choose that conclusion even though it is in fact the more logical, more parsimonious, and most realistic conclusion that they could have chosen.

Ultimate Cause or Psyche will always be a better consequent, or a better conclusion, or a better interpretation of the scientific data than Creation by Rocks; but, Creation by Rocks will typically be the consequent which the majority of the scientists and that all of the Materialists and Naturalists will choose to affirm, because these people want to, not because it's true.

The thing we each need to realize in all of this is that it is always some kind of Psyche (or Ultimate Cause) who chooses the conclusion, or affirms the consequent, or chooses the interpretation of the scientific data that he or she personally desires the most; and, there is the flaw and the logic fallacy which is built into the Scientific Method right from the very beginning.  A fallible psyche or fallible human being chooses the interpretation of the scientific data which he or she desires most never once realizing that there might in fact be another interpretation of the same scientific data which is in fact a better fit or a more realistic conclusion than the one they have chosen.

Design and Creation by Psyche or Creation by Intelligent Beings will always be a better interpretation of the scientific data than Creation by Rocks; but, Creation by Rocks has proven to be THE INTERPRETATION which the majority of the scientists have chosen to employ.  Alas, the majority isn't always right.  The majority can be right, but they aren't always right.  And, therein is the flaw of the Scientific Method – the majority chooses the consequent or the conclusion they desire most and then unilaterally affirm that that consequent is true.  It's happening right now even as I write this and even as you read this.  Someone right now is affirming Creation by Rocks or Materialism as their consequent or their conclusion.

We human beings are ever-learning but never able to come to a knowledge of the truth, because we are constantly affirming consequents that can't possibly be true.  Design and Creation by Rocks – how could that ever possibly be true?  It can't.  But, Creation by Rocks or Materialism is nonetheless affirmed by scientists across this world on a daily basis because it is the conclusion or the consequent which they desire to affirm the most.  Interesting, is it not?

Can you see how an understanding of philosophy, psychology, and logic can greatly improve our understanding of Science and the Scientific Method?  The Scientific Method has a HUGE human element or psyche element; and, human beings are known to make mistakes and are known to jump to conclusions.  We humans often leap before we look.  Creation by Rocks or Materialism can't possibly be true; yet, it is THE CONSEQUENT which most of the scientists choose to affirm for the duration of their careers.

> "This is why philosophers are wont to point out that for any given fact pattern which can be demonstrated or "discovered" empirically, an infinite number of theoretical explanations are possible.  This is also why we say theories remain

theories, even after they have been validated. All validating evidence can establish convincingly is the *negation* of a theoretical proposition." (Rychlak, *A Philosophy of Science for Personality Theory*, p. 81).

In other words, Science can be used to prove that Creation by Rocks is false by negating Creation by Rocks; but, Science cannot convincingly prove that Creation by Rocks is true. To conclude that Darwinism, Materialism, or Naturalism has been proven true by Science and the Scientific Method requires an act of faith or a leap of faith, which millions of scientists are willing to take. But, taking that leap of faith doesn't actually mean that these people have chosen the correct consequent or the right interpretation for their scientific data.

"This is a good point at which to observe that the logic of empirical study is flawed – not fatally, but in a way that limits the certainty with which our explanation of empirically proven facts can be believed in. We never achieve logical necessity [certainty] in the proof garnered by a scientific experiment. This is because we always commit the logical error that Aristotle pointed out long ago, of *affirming the consequent* of an "If, then" line of argument. Another way of saying this is that the empirical findings act as a predicating meaning for our theory, but there are always going to be other theories that can take meaning from this data array as well." (Rychlak, *Artificial Intelligence and Human Reason: A Teleological Critique*, p. 33-34).

Drawing conclusions or interpreting the scientific data is an integral and essential part of the Scientific Method; but, this is also where the logic errors and the flaws are introduced into the process. I have observed that Creation by Rocks is never the best explanation that can be given to scientific evidence; yet, Creation by Rocks or Materialism is the explanation that is most-given to scientific evidence. Interesting, is it not?

I have also observed that there are infinitely better and more believable explanations for scientific evidence than Materialism, or Design and Creation by Rocks. Materialism, or Creation by Rocks, always provides a false interpretation or a false explanation to any data array or set of scientific evidence. That has been my observation, once I finally started looking at the empirical evidence and the logic associated with the Scientific Method. Materialism, Naturalism, and Atheism are based upon a refusal to look at contradictory evidence and a refusal to look at any other possible explanation for the scientific evidence. Once I started looking at the logic and the evidence, it was easy to see that Materialism and Naturalism are fatally flawed.

There's a lot of money that can be made telling the Atheists and the Materialists exactly what they want to hear; but, it's dishonest. Like Joseph Rychlak's books, my books go largely unnoticed, because I'm not telling the Materialists and the Atheists what they want to hear; but, I sleep well at night with a clear conscience knowing that I have finally found the truth that I have been searching for all of my life. This stuff is really cool, and it has set me free; but, most people will never see it because they don't want to see it. Such is life.

"The fallacy of affirming-the-consequent stipulates the fact that it will always be possible for some other explanation to account for any empirically observed fact pattern. This loss of certainty in validation is not fatal for the scientific method, of course. It has not prevented scientists from curing polio or putting people on the moon. It merely alerts us to the fact that some conceptualizer [psyche] always has to make a decision as to which fundamental grounding [or interpretation or explanation] will be used in the sequence of theory formation and testing. The grounds are never 'out there' in the hard data but 'in here' as assumptive frameworks. If we who theorize appreciate that we will never attain certainty in validating our theories, we will be in a better position to see that alternative groundings that explain such empirical evidence can be complementary. To *complement* is to fill out or make up for what is lacking in any theoretical understanding of a subject." (Rychlak, *In Defense of Human Consciousness*, pp. 18-19).

Notice that there must be a conceptualizer, psyche, interpreter, theorizer, observer, decider, chooser, and assumer behind every theoretical hypothesis and the Scientific Method, or the science experiment will never take place. The grounds for doing science take place 'in here' in our psyche. Contrary to the claims of the Darwinists and the Materialists, you can't place the rocks or raw physical matter into the role of conceptualizer, psyche, theorizer, formal cause, final cause, or ultimate cause. The rocks won't go there and can't do that. Once again, logic and the Scientific Method have proven Materialism and Scientific Naturalism inadequate and false. The Scientific Method can definitely be used to prove things false, which has happened in the case of Materialism and Naturalism thousands of different ways. In fact, the falsification of Materialism and Naturalism is complementary, in that Naturalism and Materialism have been falsified thousands of different ways. We have our fill of evidence demonstrating and proving what is lacking or false in Naturalism and Materialism.

"Science can only work through a kind of negating procedure of falsifying claims put to nature by the theorist in question. As scientists, says Popper, we never really verify things but continually falsify – or fail to falsify – claims [theories, hypotheses, etc.] expressed by some investigator [recognizing, of course, that serendipitous findings occur as well]. This is why the scientist always restates his hypothesis into the null form. Ultimately the reason we must falsify has to do with the logical fallacy of 'affirming the consequent' of an 'If [antecedent] . . . then [consequent] . . .' proposition. We like to think our theory has necessarily been verified, but Popper teaches us that it has not. There will always be, in principle, other ways of accounting for the observed data [the facts] than our preferred theory." (Rychlak, *The Psychology of Rigorous Humanism*, pp. 181-182).

There will always be other ways of accounting for the scientific data than our preferred conclusion, Materialism.

Every scientist, myself included, loves to make the claim that Science has proven different concepts to us. In the past, I have made the claim that the Scientific Method proved to me that God exists, that the Theory of Evolution is false, and that Quantum Nonlocality or the Spirit Realm does indeed exist. I have also made the claim that the Scientific Method has eliminated falsehoods while at the same time pointing me to the truth or proving the truth. I have been called on it many times, but I still stand behind it. What good is Science and the Scientific Method if it can't be used to prove things?

Typically, though, the way that the Scientific Method proves the truth is by eliminating all of the associated falsehoods. The Scientific Method can be used thousands of different ways to falsify Materialism and Naturalism; and, it already has been used in that way thousands of different ways. The Scientific Method has in fact proven to me that Materialism and Naturalism are false by falsifying Materialism and Naturalism thousands of different ways. This also lends evidentiary support to the observation, claim, and the conclusion that something other than Materialism must of necessity be true. Intelligence, Psyche, Intelligent Design, Creation by Psyche, and Ultimate Cause are the opposite of Materialism. Since the Scientific Method has proven to me that Materialism and Naturalism are false, the Scientific Method greatly increases my chances that Intelligent Design, Psyche, and Ultimate cause will be proven to be true.

The Scientific Method has proven to me that Quantum Mechanics, Quantum Entanglement, and Quantum Nonlocality are real and true. How? It's because Quantum Mechanics has never been falsified in any of the science experiments that have been run on it. In every science experiment that has been performed on Quantum Mechanics and Quantum Entanglement, Quantum Mechanics or Spiritual Mechanics has been verified as true. Consequently, I feel safe in claiming that the Scientific Method has proven to me that Quantum Nonlocality or the Spirit Realm is real and truly exists, because Quantum Nonlocality by definition means spiritual, or non-local, or non-physical, and because Quantum Nonlocality has been repeatedly verified and proven true.

Finally, I know that Intelligence, or Psyche and Ultimate Cause, exist because they are obvious and axiomatic. We all KNOW that Psyche or Intelligent Beings can design and create anything they set their minds to at will, because we have observed and experienced this Reality first-hand every single day of our lives. I don't need the Scientific Method to prove to me that Intelligent Beings exist, because if you can read this, then you already KNOW that Intelligent Beings exist. And since Psyche or Ultimate cause is synonymous with Intelligent Beings, whether looked at from a spiritual perspective or a materialistic perspective, I really don't need the Scientific Method to prove to me that Psyche and Intelligent Beings exist because I already KNOW that they exist. The Scientific Method is unnecessary if you already KNOW the truth.

KNOWING trumps the Scientific Method and philosophical speculation every time. Also, since I am willing to accept NDEs, SDEs, OBEs, and other spiritual experiences into evidence, I KNOW that Psyche or Ultimate Cause is non-local, non-physical, and spiritual in nature and origin. I KNOW that psyche or our personality is the part of us that survives bodily death and brain death. Since many people have seen and talked with the Biblical God Jesus Christ during their NDEs and OBEs, I know that He exists as well; and, I don't need the Scientific Method to try to convince me otherwise. I don't need the Scientific Method for any of these things, because I KNOW that they are real and true.

But, the failures and the inability of Materialism and Darwinism to account for the origin of the first physical genome and the first physical life form was in fact the first thing to prove to me scientifically that God must of necessity exist in order to do all of the design, science, and creation that the rocks, Materialism, and Naturalism could never have done. So in a very real sense, the Scientific Method and the falsification of Materialism and the Theory of Evolution proved to me that God must of necessity exist in order to do all of the different things that needed to be done which Materialism, the rocks, and the Theory of Evolution could never have done.

It's really not 100% accurate to make the claim that Science and the Scientific Method will never be able to prove the truth, because through the scientific process of eliminating every falsehood, we eventually find ourselves landing upon the truth as a last resort or a final default; and then, we find ourselves staring at and going with something which is axiomatic law, obvious, and 100% true and can never be falsified, or proven false, or proven not to exist – something like Psyche, Intelligence, Intelligent Design, and Ultimate Cause which we simply KNOW exists. I exist; therefore, I AM; and, I KNOW IT.

> "When we design and carry out a research experiment, it is easy to confuse the concrete empirical findings with the activity or the process that supposedly brought these findings about or made them happen. The logic of experimentation gives rise to the following problem: for any observed fact pattern there are, in principle, infinitely many possible explanations. This follows from the necessity that, in conducting research, all scientists are constrained by the "*affirming the consequent*" fallacy. There will always be an alternative explanation possible for the

observed fact pattern." (Rychlak, *Logical Learning Theory: A Human Teleology and Its Empirical Support*, pp. 3-4).

When the Naturalists and Materialists carry out their research experiments and choose to affirm Creation by Rocks or Materialism as their consequent, know that there are an infinitely many possible explanations for their science experiment and their scientific data, some of which might actually be true – such as using Intelligent Design, or Creation by Psyche, or Design and Creation by Intelligent Beings as the explanation for their science experiment and scientific data instead of using Creation by Rocks or Materialism as their explanation.

I have observed that Psyche, or Intelligence, or Ultimate Cause will always be a better explanation for research experiments than Materialism or Creation by Rocks. Would you agree with that observation, or not? Whether you agree or not, I have observed that Psyche or Intelligence can do research experiments; whereas, the rocks cannot. Consequently, science experiments done by psyche (or ultimate cause) are infinitely more plausible and believable than science experiments done by rocks. That's one of the things which the Scientific Method has taught me.

See how the Scientific Method and the Philosophy of Science can be used to get at the truth, often through a process of elimination? It's not as direct nor as immediate as Knowing or Knowledge, but it does have its uses.

# Scientism

I had bought into all the hype, posturing, and promotion; and, I truly believed that Science and the Scientific Methods are the best way for finding and knowing the truth. I was a proponent of Scientism – the philosophical belief that scientific methods are the best, if not the only, way to find and know the truth. I had based my whole life on Scientism. For the first fifty-five years of my life, I considered myself to be first and foremost a scientist. I was dedicated to the system. I believed in it passionately.

I was in for a rude awakening!

## Defining Scientism

The following definition and explanation of Scientism comes from class notes by Edwin Gantt:

> In the world at large, there is a troubling trend toward viewing science as not just a sturdy approach to answering some important questions about the world, but rather as the only reliable source of truth, and as a way of making sense of the world that is superior to all others.
>
> Many scientists unquestioningly adopt the position articulated by the British chemist Peter Atkins:
>
>> Although some may snipe and others carp, there can be no denying the proposition that science is the best procedure yet discovered for exposing fundamental truths about the world. There appear to be no bounds to its competence.
>
> Atkins has further stated, **"Science is the only path to understanding"**. [That's a very narrow and extremist point of view. That's a religious point of view. Atkins is posturing here.]
>
> Scientism is a thoroughly unscientific approach to the world. This is because science, truly and properly understood, is inseparable from a deep epistemic humility—a genuine and thorough-going acknowledgement of the inherent limitations and fallibility of human understanding.
>
> Genuine science, at the very least, does not permit making the sorts of sweeping metaphysical, theological, and moral claims to which the advocates of scientism are so frequently prone, since these sweeping metaphysical claims cannot be grounded in empirical experience.
>
> For example, Atkins's claim that "science is the only path to understanding" is not some truth discovered by the methods of science and empirical observation. Rather, it is just a sweeping philosophical assertion.
>
> Assuming that observed regularities are immutable or universal is entirely unjustified by empirical experience. We simply do not know and cannot know until we have made systematic observations of all of reality.

We cannot make broad, sweeping metaphysical claims about the world without leaving the realm of systematic empirical observation and entering the world of philosophical speculation.

(From class notes by Ed Gantt).

Scientists like Peter Atkins do an excellent job of explicating the dogmatic religious extremism of Scientism. Scientism is just another type of religious extremism – scientism is a worship of Scientific Methods. These people treat science as if it were God. Scientism is philosophical speculation.

These people don't realize that Science and Scientific Methods are extremely limited or bounded when it comes to the spiritual, non-physical, immaterial, and non-local Realities of our existence, because these people have deliberately blinded themselves to other realities. Science and Scientific Methods are not only bounded by physical limitations but are also extremely incompetent when it comes to the spiritual or the non-local Realities of our Lived Experiences.

When it comes to "psychology", the Behaviorists, Materialists, and Naturalists limit themselves exclusively to observable human behavior. For them, the word "observable" and the word "empiricism" are defined in terms of the five physical senses and therefore limited exclusively to the physical.

Therefore, whenever human behavior is caused by the human psyche or non-local consciousness, Scientific Methods are actually a very poor way to study human behavior, because there is no way for Scientific Methods and physical observations to study human behavior that's taking place in the spirit realm or the non-local realm.

Scientism is based exclusively on Naturalism and Materialism.

**The Weaknesses of Naturalism**

During our class, Ed Gantt outlined some of the weaknesses of Naturalism:

1. Observing regularities and patterns in the world does not itself warrant the belief that these "laws" cause events, or even serve as sufficient explanations for them.

2. The naturalistic worldview cannot be proven to be true by empirical evidence because it is really just a philosophical assumption.

3. Naturalism has not been a very successful approach in the social sciences — psychologists simply have not discovered the kinds of scientific laws they hoped to find.

4. The naturalistic worldview, by definition, precludes explanations that do not assume the sufficiency of scientific laws in accounting for the world.

5. Human beings are fundamentally different from the sorts of things that can be explained exclusively in terms of universal laws, natural processes, and physical reactions. For this reason, the assumptions of naturalism are simply inadequate in the social sciences and the behavioral sciences.

6. Merely observing regularities in the world is not itself reason to believe we have actually explained anything. We all know from personal experience that

unsupported objects fall.  Personal experience or lived experiences is how we come to know about immaterial and non-physical objects and subjects such as gravity.

These are undeniable regular and consistent patterns, and we give these patterns a name: *gravity*.  However, to this day, nobody actually knows why things fall.  We often say, "Things fall because of gravity."  However, this is a logical fallacy, known as the *Nominalistic Fallacy*.

To commit the Nominalistic Fallacy is to make the mistake of assuming that merely because we have given something a name or described it in great depth we have, therefore, explained it.  Things do not fall because of gravity; rather, gravity is simply the name we have given for the observed pattern of things falling.

7. Naturalism is a philosophical perspective or worldview that assumes that everything that happens in this world can be explained in terms of physical processes and scientific laws.  A naturalist, in this context, is someone who assumes that we do not need to invoke God, supernatural entities, or any religious mystery to fully explain events in the world.  (Adapted from class notes by Ed Gantt).

Naturalism, Naturalism, Atheism, and Scientism are nothing more than biased philosophies and extremist dogmatic religions.  The whole of Scientism, Naturalism, Materialism, and Darwinism are based exclusively on a wide variety of different logic fallacies.

I'm lucky to have found college professors and scientists who were willing to teach me about these kinds of things.  That's why I dedicated this book to them!

## Science and Human Behavior

Slife, B. D. & Williams, R. N. (1995).  Science and Human Behavior.  In *What's Behind the Research? Discovering Hidden Assumptions in the Behavioral Sciences*, (pp. 167–204).  Thousand Oaks, CA: SAGE Publications.

http://mypsyche.us/science/

The authors, contributors, and their disciples have been handing out this paper in class to the public at-large.  They are not concerned about copyright.  They want this thing spread about!

**This article is excellent!  This thing is paradigm-shifting and life-changing.  Warning: This article is extremely painful if you are a promoter of Scientism, Naturalism, and Behaviorism.**

I feel very lucky to have encountered this article and these ideas, because our materialistic, naturalistic, and atheistic college professors have no idea that such things actually exist; and, the few who do know refuse to teach us anything about them.  I was exposed to these ideas a full twenty years after Slife and Williams presented them to the world – better late than never!

This article marked the END of my Scientism – my worship of the scientific method as an infallible god.

While reading this article, a lot of ideas came to mind and I wrote them as notes in the margins.  Here's a sampling of what I came up with.

## Science Is Based Upon Philosophical Assumptions and Philosophical Interpretations

Ironically, due to the fact that hypotheses, theorizing, and interpretation of scientific data are an integral part of the scientific method, Science and the Scientific Method ARE philosophy at their very core. We can't escape philosophy, or personal opinion and personal interpretations of the scientific data.

For example, this is currently my own personal interpretation of the scientific evidence; and, it tends to be quite different from the interpretation that is typically found in our college textbooks.

**According to Quantum Mechanics, God or the Prime Observer organized this physical universe out of spirit matter, which was already in existence in great abundance. His Conscious Observation, His WORD, was necessary to transition, transfer, and transmute part of that pre-existing spirit matter into physical matter; and, He could pick what parts of that spirit matter to transform into physical matter and when to do so. His WORD, His Psyche, or His Light could command whole suns and whole galaxies into existence all at once.**

**According to Quantum Mechanics, ONLY Psyche can observe physical matter into existence or call physical matter into existence, which means that some kind of Psyche had to precede the creation or organization of the first particles of physical matter.**

**According to Astrophysics, 96% of this universe is still in its spirit matter and/or light and energy state of existence, as dark matter and dark energy respectively. Only 4% of this universe has been consciously observed into physical matter, or consciously called upon to be physical matter. The Big Bang was not an explosion. It was a birth, a transition, and the beginning of physical matter in this universe – a process that happened when God or the Prime Observer called that physical matter into existence from pre-existing spirit matter.**

**The Big Bang, or the beginning of space-time and the beginning of this physical universe happened; but, not in the way that the Materialists and Atheists typically visualize it. There was no explosion. There was NO physical matter until AFTER God's Psyche called, or WORDED, or observed the first physical matter into existence. It's interesting how the Biblical Prophets understood Quantum Mechanics and the power of the WORD millennia before our scientists did. In fact, there are many scientists and Quantum Mechanists who still have no idea how Quantum Mechanics really works or fits into the picture as a whole.**

**Likewise, there were no random mutations and no natural selection and no DNA on this planet until AFTER God's Psyche designed, organized, and created the first genomes and the first life forms on this planet. Psyche is the only thing we know of that can convert spirit matter into physical matter. Psyche is the only thing we know of that can design, create, and do science. Psyche or Non-Local Consciousness, particularly God's Psyche, is in fact the fundamental unit of Reality.**

There's more than one way to interpret the scientific data and the scientific evidence. In fact, there can be dozens and possibly even thousands of different ways to interpret a particular piece of scientific evidence. Furthermore, ONLY the Human Psyche is capable of doing a full-range of interpretation and meaning-making.

**Defining Science**

What is science?

Surprisingly, science is a philosophy, a worldview, which is why Scientism can be classified as a religion.

Science is just another philosophy attempting to determine what is true. Science is an advanced form of philosophy, or metaphysics.

"Science itself is based on a set of ideas – assumptions about what the world is like and how it should be studied. In the minds of many, science encompasses what we *know* to be true. We cannot have knowledge about what is not real or true."

Ironically, lived experiences end up becoming what a person or psyche KNOWS to be true. I learn best through comparison and contrast.

The strength and the weakness of Science is found in the fact that most scientists limit the discipline or philosophy purely to the physical, completely ignoring and discounting the other aspects of reality. That's good if the ultimate goal is technology that you can sell to others; but, it's really bad if your ultimate goal is to get a handle on the whole of our Reality and Existence.

Scientists, particularly Atheistic Scientists, base their philosophical assumptions on Materialism and Naturalism. By deliberately limiting their brand of science exclusively to what can be physically observed, these scientists deliberately limit the scope and the range of what they can study and discover. This philosophical focus on physical matter blinds them to everything else. In fact, it's extremely unscientific – especially if we can agree upon the fact that Science should really be about the pursuit of knowledge and not just about the promotion of Materialism, Naturalism, and Atheism!

The Atheists tend to promote themselves as the beacons of rationality. Rationality is synonymous with logic and interpretation. Unfortunately, there can be dozens, if not thousands, of different interpretations for the same piece of scientific data or scientific evidence. The Atheists, Materialists, and Naturalists can use their rationality to talk themselves out of the truth just as easily as we can use our rationality to talk ourselves into the truth. Rationality and logic have a blind spot, and it's called personal interpretation or personal preference.

Science was based upon the idea that "knowledge is gained through careful observation and skilled exercise of rationality".

Unfortunately, Science and the Scientific Method actually break down on two fronts. First of all, a hypothesis is a philosophical guess; and, it is therefore subjective and subject to a lot of personal bias and personal interpretation right up front even before the experiment is conducted and the results of the experiment gathered in. Second of all, the Scientific Method ends with a personal philosophical interpretation of the scientific data or scientific evidence. It's possible to interpret evidence and data incorrectly. In fact, I would venture to guess that most scientific evidence and data is interpreted incorrectly, especially if it's being interpreted from a materialistic and naturalistic worldview or philosophy.

Interpretation is subject to preferences, values, desires, biases, and pre-determined conclusions. And, the final step of Scientific Methods is to interpret the data or the results of the experiment, which places Science firmly on the grounds of philosophy once again, from whence it started in the first place. Unfortunately, a single observation can be interpreted a dozen different ways which only make sense to the interpreter but not necessarily to anyone else. Interpretation is a function of philosophy and psyche, and NOT observation or experimentation.

In fact, I have observed in recent years that the philosophical interpretation of scientific data which they call "Intelligent Design Theory" is infinitely more plausible, credible, and realistic than the philosophical interpretation of the same data which they call "The Theory of Evolution". I have lost my faith and trust in materialistic and naturalistic interpretations of scientific data and scientific evidence. My Materialism, My Nihilism, and My Atheism betrayed me and threw me to the wolves. They were no friends of mine, and they have become my mortal enemies.

"It is common in our culture to refer to the *scientific method* as if it were a single thing about which all people agree. [In fact, in the literature, most people simply call it *method*.] It should be pointed out, however, that there is no such thing as *the* scientific method. There are as many scientific methods as there are scientists doing research."

In fact, the authors of this article promote what they call Methodological Pluralism – the use of many different types of methods – in an attempt to get at the truth or what is often called *knowledge*.

**The Purpose of Science**

The goals of a science, which has been limited to the philosophical bounds of Materialism and Naturalism, is prediction and control – replication and reliability – technology. The physical realm is the ultimate consensus reality after all! In a physical realm, I can trust that this essay will be on my hard drive and my flash drive tomorrow morning when I wake up. The same can't be said about spiritual realities which are created by psyche or mind – especially the non-consensus realities in the non-local realm or spirit realm.

We can't predict and control the human mind or the human psyche suggesting that it is part of the supernatural world rather than being a part of this physical world. Human beings cannot be programmed like computers. Human beings resist programming, and human beings can break their conditioning at will. In fact, any time you have a thought or a dream or make a choice, you are in fact having a spiritual experience. Your thoughts, dreams, memories, feelings, and choices show up in your life review after your physical body and physical brain are dead and gone, according to the empirical science of Near-Death Experiences.

There's a whole other Science of Lived Experiences, a Science of Choices and Actions, which the Materialists, Naturalists, Nihilists, Atheists, Determinists, and Behaviorists CHOOSE to ignore and reject, in a very unscientific manner.

"The test of whether we have, indeed, discerned the secrets [of nature] is whether we can exercise power and control over nature to bring about practical ends. This dual emphasis on careful observation in the pursuit of truth and practical control over the natural world is at the heart of our modern notion of science. Since the Enlightenment, science has been thought of as a method for testing ideas and opinions through observation. Observation takes place through the medium of sensory experience [defined as the five physical senses]. This emphasis on observation of the world as a way of attaining true knowledge is the essence of *empiricism*."

In the literature, empiricism is defined as physical empiricism or observations done by the five physical senses. These people have CHOSEN to completely ignore Spiritual Empiricism or the Lived Experiences of the human race.

**Behavioral Sciences**

Prediction and control are the hallmarks of a naturalistic and materialistic science, because the ultimate goal is technology. Unfortunately, we still can't force nature to produce new and unique genomes and life forms from scratch, according to our desires and our will. All we can do is mess around with what God has already created. In fact, Random Mutations and Natural Selection and Evolution didn't even exist until AFTER God designed and created all the genomes and life forms in the first place. Materialism and Naturalism are design and creation by rocks. God and God's Psyche must of necessity exist in order to have done all of the design, engineering, manufacturing, science, and creation that needed to be done which evolution and the rocks could NEVER have done and can NEVER do.

The reason why the physical sciences or natural sciences have so much prestige and success associated with them is because of all the wonderful and useful technology that has come as a result.

"Behavioral scientists hold out the hope that a technology for dealing with the problems of human behavior might be developed with much the same beneficial results as in the natural sciences."

The Behaviorists want to be able to predict and control human behavior, much the same way that we can predict and control robots and computers. But, the Materialists, Naturalists, and Atheists don't realize nor understand that physical technology will NEVER be able to touch, change, predict, nor control the immaterial, non-physical, non-local, human psyche! The human psyche has a life all its own. There really IS a ghost in that machine!

The scientists can want power, prediction, and control when it comes to the human psyche and human behavior, but human beings are not exclusively machines thanks to psyche!

In fact, it is ALWAYS some kind of Psyche or Ultimate Cause who designs, programs, engineers, builds, and creates the various machines in the first place. The Materialists, Naturalists, and Behaviorists don't get it, because they don't understand nor accept Psyche.

**"Behaviorists have tried to explain human behavior in strictly naturalistic terms, but the attempt has not been wholly successful. The behavioral sciences have for the most part been unable to formulate universal laws that fully account for human behavior. Behavioral scientists have assumed a deterministic posture toward human behavior, but they have not been able to demonstrate the kind of control and prediction that have been achieved in the natural sciences."**

**This IS because Psyche is a supernatural or non-local phenomenon, and not a natural nor physical process. The universal laws associated with human behavior are in fact based upon a Psyche Ontology wherein psyche is the fundamental unit of reality; therefore, the Universal Law associated with human behavior IS free will, the pursuit of freedom, and agency because Psyche IS an independent agent and an independent entity completely separate from the physical body and the physical brain. Psyche continues to behave, and misbehave, long after the physical body is dead and gone, according the empirical evidence from Near-Death Experiences and other types of Spiritual Experiences. To study the human psyche and human behavior, we must study the Lived Experiences of the human race and not just the physical brain. It's elementary, my dear friend!**

When it comes to human behavior or the human psyche's behavior, the Materialists, Naturalists, and Behaviorists are barking up the wrong tree!

The first common way of defining science is to consider science as the body of knowledge which explains the physical world and the natural sciences. Under this model for science, "scientific explanations for the most part involve rejecting supernatural explanations of phenomena in favor of naturalistic explanations. Scientific explanations are usually given in terms of matter or other naturalistic constructs [such as biology]. [As a result,] the behavioral sciences [and the psyche sciences] will never be 'scientific' in the same way as the natural sciences. The second common way of speaking about the nature of science is to say that science is primarily a *method* for studying phenomena." In other words, we could choose to broaden the horizons of science to include Phenomenology and formally state that science IS the study of the Lived Experiences of the human psyche. The Lived Experiences of the human race naturally include physical phenomena, but our Lived Experiences also include spiritual phenomena. This is the way that the behavioral sciences and psychology should go with the discipline; but, it's not going to happen while the Materialists, Naturalists, and Atheists are in control of these disciplines.

It is "difficult to measure many important aspects of human beings", particularly the spiritual aspects and the psyche aspects, "because we lack the instruments to do so. Furthermore, human beings in many settings actively resist strict control and prediction". Therefore, let's change the rules of the game so that we can study human beings, human choices, human actions, and human behavior as a whole. Let's make psychology what it was originally meant to be, a study of the human psyche!

"Remember that scientific methods are based on certain [materialistic and naturalistic] assumptions about the nature of truth and the world. If these assumptions are not true of the human world and of human phenomena, then the scientific methods based on these assumptions may not be appropriate for studying people [and the human psyche]". Therefore, we need to choose a different methodology if we truly want to do Psychology or a study of the human psyche! It's only logical!

Get rid of Materialism, Naturalism, Darwinism, Atheism, and Creation by Rocks as our major premises or primary assumptions, and then we can start to do Science for real! It's elementary my dear reader. By getting rid of all the falsehoods that we know about, we are then free to start a pursuit of the truth instead. That's what Science should really be about – getting rid of all the falsehoods that we know about so that only the truth remains. There's NO evidence to support the major premises of Materialism, Naturalism, and Atheism; and, there NEVER will be because there can't be. We should get rid of Scientism and replace it with some kind of Methodological Pluralism instead, which will continue to include scientific methods as one of the methods but develop and include other methods as well.

Materialism, Naturalism, Nihilism, Darwinism, Atheism, and Scientism are predatory religions. These things claim to be the ONLY source of truth, and therefore their proponents mock and ridicule everything else. These are exclusionary philosophies. These people are religious extremists and religious fanatics. The ONLY way that we will ever be free to study the inclusionary philosophies, the psyche sciences, and the holistic sciences is if we successfully EXCLUDE the exclusionary philosophies.

## Knowledge

These are powerful ideas being presented in this article, but only if people are willing to understand them, accept them, and embrace them. This article is over twenty years old, so these changes and improvements have been slow in coming, because this is a message that you will never hear from the materialistic and naturalistic professors who control our public schools and academia. A person really has to dig in and study on the fringes if he or she truly wants to know what's wrong with Materialism, Naturalism, and Darwinism. I have been doing this for a few years now, and I'm finally starting to make some headway.

I KNOW how all this works, because I have lived it. I have experienced it for myself.

I KNOW what it is like to have one's physical body develop a mind of its own and a life of its own and become completely beyond my control. That's a special type of hell, which is no fun to live with and live through.

I KNOW that we can go to hell, while here in mortality while we are still in our physical body.

I KNOW what it is like to want to die and to want annihilation. I was hoping that I could die and cease to exist. I was no longer afraid of death. I wanted to die; and, I was hoping that my spirit and psyche would completely cease to exist when my physical body died.

I KNOW what it is like to have no friends. While I was attempting suicide, I remember thinking, "At least I will never have to come back here." I also remember thinking, "If the theists are right after all, at least in hell I'll finally have a chance to make some friends." Hell is being completely alone and having no friends.

I KNOW what it's like to finally reach a point in life where one is thinking, "If I end up in hell again, I'm going to try to help people and try to help them be happier and better." I KNOW what it's like to get outside oneself and to think about others and to think about helping others, every once in a while. I KNOW that there are better ways to live, other than the ways which we typically choose to live as mortals.

I KNOW what it's like to be addicted to crap; and, I KNOW what the months of psychosis are like while one is going through withdrawal.

I KNOW how important it is to keep God's Commandments to the best of our ability. After I got sober, one of the first things to come back to me was a deep and abiding faith or trust in the Commandments of God. I KNOW how good they are for us. I KNOW how happy and peaceful they can make us, by choosing to own them and live them. I KNOW what it's like to be completely at peace, thanks to the mercy and forgiveness and grace that comes from the Atonement of Christ.

I KNOW what it's like to be suckered into believing that philosophical concepts like Materialism, Naturalism, Nihilism, Scientism, Atheism, and Darwinism are real and true. I

KNOW how gullible we can be as human beings, and how easily and eagerly we fall for obvious falsehoods such as Materialism and Naturalism and Atheism because we desperately want them to be true.

I KNOW what it's like to falsify philosophical theories such as Materialism, Naturalism, Darwinism, and Nihilism.

I KNOW these things, because I have experienced them.

**Positivism and Realism**

Slife, B. D. & Williams, R. N. (1995). Science and Human Behavior. In *What's Behind the Research? Discovering Hidden Assumptions in the Behavioral Sciences*, (pp. 167–204). Thousand Oaks, CA: SAGE Publications.

http://mypsyche.us/science/

Next, this article from Slife and Williams does a few pages on Positivism and Realism.

"Many people believe that science is self-correcting when it goes wrong."

It is NOT, because personal interpretation of the data and evidence is involved, when it comes to the scientific methods!

Each piece of scientific evidence or data can theoretically be interpreted an infinite number of different ways. Scientific data can be interpreted in ways that support the Biblical narrative and in ways that completely contradict and debunk the Biblical narrative. You can interpret the scientific data any way you want.

For example, I have observed that the scientific data makes a lot more logical sense if it's given an Intelligent Design interpretation than it does when it's given a materialistic, creation by rocks, Theory of Evolution interpretation. Psyche or Intelligence can design and create. The rocks can't. Psyche and Intelligence have been caught in the act of design and creation trillions of different times; whereas, the rocks NEVER have. This reality is so obvious and clear that I sometimes wonder how I was able to overlook it for the first fifty years of my life. Intelligent Design is an infinitely more plausible and believable interpretation of the scientific evidence than Creation by Rocks or the Theory of Evolution will ever be. But, I couldn't see it nor understand it when I was a Materialist and an Atheist. I was blind to those kinds of obvious realities; but, now I see.

I have tended towards Scientism and Realism most of my life. I wasn't going to believe in God unless I had Scientific Proof of God's Existence; and, I didn't think that such a thing was possible. It wasn't until the end of 2015 that Science started proving to me that God exists.

Why did it take so long?

It's because I wasn't looking for it and didn't believe it to be possible!

I had fallen for the Scientism and Materialism hook, line, and sinker!

**Observation**

"Both realists and positivists put great emphasis on observation" – a thing that they call empiricism.

When it comes to observation and empiricism, what these people really mean is physical observation of physical objects with one's physical senses and one's physical instruments. These people deliberately exclude spiritual experiences acquired by spiritual senses, even though the spiritual experiences are technically more powerful and real than anything acquired by the physical senses.

Science is based upon biased and prejudiced worldviews, assumptions, and interpretations of the data. There's really nothing objective about science. The whole thing is philosophical interpretation from beginning to end.

"Science is not as objective and free from cultural influences [and personal interpretations] as the realists and positivists have thought. The view of the world that each scientist has and shares with other scientists and non-scientists influences the way that science is done. This worldview (in Kuhn's terms, a *paradigm*) leads scientists to think about their science [and to interpret their science] in the way they do, even though they usually do not recognize their view as a worldview [or personal philosophy]. The way they formulate their questions, the methods they believe to be appropriate, and the sorts of explanations they hold to be acceptable are all influenced by the worldview in which they live. Scientific method itself is based on certain assumptions about the nature of the thing it studies."

The whole scientific enterprise is biased and subjective from bottom to top, completely colored and molded and controlled from beginning to end by a materialistic and naturalistic worldview. They stack the deck, pre-determine acceptable conclusions, and limit the conclusion to a materialistic explanation BEFORE they even start the science experiment. I have encountered people online who are dogmatically religious about this exclusionary process. Scientific method is based upon philosophical assumptions, and exclusionary philosophies such as Materialism, Naturalism, Darwinism, Scientism, and Atheism. That's its strength, but that's also its greatest weakness. Science and scientific methods are biased and prejudiced. And, that's not the only thing that's wrong with the scientific methods.

Scientism is based exclusively in Materialism and Naturalism.

**Definition of Behaviorism**

"Behavioral scientists have approached their subject matter – human beings – convinced of the importance of objectivity and scientific certainty, and intent on uncovering general or universal causal laws, formulating deterministic and mechanistic explanations of behavior, and try to develop technological solutions to human problems."

"It is important to consider whether human beings are really the sort of entities best described by mechanism, determinism, and efficient causality. If they are, then the traditional methods of science may be the most appropriate methods to employ. If they are not . . ." well, a whole mess of confusing interpretations and philosophical speculation is going to ensue.

Scientific method is limited to studying physical matter. It's a very good way to get at the truths about physical objects and physical realities. However, scientific method is a weak way of getting at the truth about the immaterial or the non-physical and nonobservable realities of our existence. Furthermore, scientific methods seem to be absolutely worthless when trying to get at the truth about psyche, thoughts, dreams, and the afterlife. Scientific methods are not tuned for psyche!

Ironically, only the human psyche KNOWS how to use language, math, and science to make sense of things! Only the human psyche can do science! Yet, the proponents of Scientific Naturalism completely overlook this reality and choose to pretend that psyche does not exist.

## Some Limitations and Weaknesses of Scientific Methods

"One of the main reasons science is so attractive is that it seems to hold the promise of truth. At least science appears to be a more trustworthy way of testing the truth of our ideas than other ways that have been developed. However, this perception of science is based upon" a wide variety of philosophical assumptions and logic fallacies. A logic fallacy is NOT the best platform upon which to try to build the truth; but, that's exactly what we see going on with the scientific methods and the *affirming the consequent* logic fallacy.

"Prior to the application or even the development of any scientific method, there is always an operative understanding of the truth", meaning that there are certain philosophical assumptions that are made. "It is this pre-understanding of the truth", a pre-selection of what is assumed to be true, "that makes it possible to frame any method at all. Without this understanding" or initial prejudice, "we could not formulate any scientific method because we would not know what the method should be like – or that we even need a method. This means that understandings of truth produce methods, rather than methods producing the truth. If this is the case, then we cannot be confident that properly and carefully applying scientific methods to the study of human behavior will get us to the truth." Rather, the whole enterprise could in fact lead us away from the ultimate truth.

Scientific Methods are NOT tuned for psyche! Scientific Methods are capable of dealing with a Psyche Ontology.

"Many important questions about human behavior can only be solved by careful theoretical work, not by the application of method. Scientific methods do not establish the truth of the matter", because they are limited exclusively to physical matter. "A related conclusion is that all methods, including scientific methods can only find the sorts of things that they are 'tuned for'. In the behavioral sciences, if our methods are not 'tuned' for human beings" and the human psyche, "then the method can miss what is true and important about human beings. In this way, scientific methods can act as blinders as much as they can reveal something important to us" that might be true.

## The Affirming the Consequent Logic Fallacy

This article does the best treatment of *negating the consequent* and *affirming the consequent* that I have encountered so far.

I was shocked to learn that the whole Western Tradition has been built upon the *affirming the consequent* logic fallacy, which is built into the scientific methods and is unavoidable. I was shocked and stunned to learn that we can NEVER use the scientific methods to prove the truth. (Rychlak). I remember thinking, "What good is it, if we can't use it to prove the truth?"

This is the ONLY article that I have encountered so far that talks about *negating the consequent*, how it's logically and philosophically sound, and how scientific methods can be used to falsify hypotheses and theories that are physical in nature, and origin, and substance.

I had already been introduced to Karl Popper's idea that science and scientific methods should be used to falsify ideas; but, I had no idea why he had chosen to use science to falsify theories rather than using science to verify or prove theories. (Kalat). It's because *negating the consequent* is philosophically and logically sound; whereas, *affirming the consequent* is not.

I was also shocked to learn that *negating the consequent* isn't as effective nor as convincing as Popper made it out to be, due to the fact that there's theoretically an infinite number of variables to have to take into account and control while using scientific methods to falsify our theories and due to the fact that there are dozens of possible interpretations that can be given to each and every piece of evidence and scientific experiment which we come across.

Nevertheless, we can, and we do use our scientific experiments, scientific methods, and *negating the consequent* to falsify our theories. I have used such procedures dozens of times myself to **falsify** Materialism, Naturalism, Nihilism, Darwinism, and Atheism to my satisfaction. The idea is that if we eliminate enough falsehoods using science and the scientific methods and logical common sense (procedural evidence), we eventually start centering in on the truth through a process of elimination. Slife and Williams mentioned this reality – we arrive at the truth through a process of eliminating the falsehoods. Science and the scientific methods can help us to do so.

Ironically, we can't even design and run scientific experiments on the major premises or primary assumptions of Materialism, Naturalism, Atheism, and even Darwinism (design and creation by rocks) which makes these philosophical concepts by definition, in principle, unscientific to begin with. We can, though, use scientific methods and logical common-sense methods and observational and experiential methods to **FALISFY** Darwinism, Materialism, Naturalism, and Atheism to our heart's content. ALL is not lost, because we can use scientific methods successfully to falsify these kinds of falsehoods. I still rely upon the Scientific Method to do so!

I spend some time discussing verification and falsification in the next chapters, so I won't go into the details here.

By pointing out all the weaknesses of the Scientific Method, this article forced me to face Reality. Scientific methods are NOT a sure way of knowing the truth. There are in fact better ways of knowing the truth than the Scientific Method! I never realized that before! My goal is to learn something new every day, and for me, this was a major breakthrough or epiphany. This article marked the END of My Scientism. You won't catch me worshipping the Scientific Method ever again. I have seen the light!

Instead, I now use the Scientific Method for what it is good for and what it was designed for – falsifying falsehoods such as Darwinism, Materialism, Naturalism, Nihilism, Atheism, Scientism, and Creation by Rocks. The Scientific Method, observation, empiricism, and logical common sense can be used to falsify ALL of the exclusionary philosophies that are based upon Naturalism and Materialism. Get rid of all the falsehoods and exclusionary philosophies that we know about, and slowly we should arrive at the truth through a process of elimination. Materialism, Naturalism, and Atheism are based upon a denial of Reality and a refusal to look at evidence, so they are worthless in the end. We have to get rid of them if we want to know the truth. It's elementary my dear friend! This reality is so clear and obvious to me now, that I sometimes wonder why I wasn't able to see it nor understand it for the first fifty years of my life. But, we had all been duped by the Materialists, Naturalists, Darwinists, and Atheists and had fallen for their deceptions and lies. Even the Theists that I knew in college back in 1980 had fallen for these kinds of things. We didn't know better, because we weren't taught better! We had all been fooled!

The good thing about Richard Dawkins and the other New Atheists is that they finally presented the weakness of their case to the world. Before their books came out, we took it on blind faith that these people knew what they were talking about. After their books came out, suddenly we knew that these people didn't have a clue about how science, philosophy, and reality truly work. Before they wrote their books, we thought their case was infinitely stronger than it really is. After they wrote their books, suddenly everyone could see that the Emperors have no clothes.

Dozens of books have been written explaining what's wrong with Dawkins' books, Dawkins' science, Dawkins' ideas, and Dawkins' philosophy. The books that have been written debunking Dawkins and the other New Atheists make infinitely more sense than anything that Dawkins and his friends ever wrote or said. Suddenly everyone could see that these New Atheists weren't so bright after all. In fact, I and my friends feel kind of embarrassed and sorry for them, because these New Atheists are so ignorant, uneducated, unscientific, closed-minded, prejudiced, and blind. We have watched their debates and were surprised how weak their case and their knowledge of science and philosophy really is. You would have to be a Materialist or an Atheist not to be able to see it.

**Science, Logic, and Reason**

Slife, B. D. & Williams, R. N. (1995). Science and Human Behavior. In *What's Behind the Research? Discovering Hidden Assumptions in the Behavioral Sciences*, (pp. 167–204). Thousand Oaks, CA: SAGE Publications.

http://mypsyche.us/science/

"The relation between science and logic can be traced to Aristotle, who invented formal logic as a way of doing science. The form of logical argument that Aristotle developed was called *syllogism*."

On pages 182 to 189, Slife and Williams discuss some of the logic fallacies upon which the scientific methods are based.

The conclusion I drew from all of this is that Scientific Methods are bad at getting at the truth, even the physical truths or "materialistic truths" that we are trying to get at. In other words, Scientific Methods can't be used to prove the truth. The first time that this reality was presented to me by Joseph Rychlak in his books, I remember thinking, "What good is it if it can't be used to prove the truth?"

Furthermore, before reading these pages in this article, it had already become clear to me that it is impossible to get at a Psyche Ontology and Psyche Truths using the traditional physicalist Scientific Methods.

I was extremely shocked and surprised to learn that Scientific Methods aren't even dependable and reliable when a person is trying to falsify theories by *negating the consequent*. I came away with the impression that Scientific Methods are good for nothing at all! That was definitely the END of my Scientism!

"Whatever scientific methods may be good for, they cannot be used to *verify* theories in the sense of affirming the consequent, or *falsify* theories, in the sense of negating the consequent. In other words, we cannot through scientific methods discover whether theories and hypotheses are true or false."

"It is in principle impossible for an experiment to establish or verify that a cause-and-effect relationship exists between any two variables in a study, as Hume noted long ago. Just because the independent variable and dependent variable always seem to occur together does not mean that some other, as yet unidentified, variable could not always be present as the real cause."

YIKES! Well, shoot dang! I could feel my paradigms shifting underneath me! And, paradigm shifts can be painful!

Here we have people like me, scientists, who have built their whole worldview and placed all of their faith and trust in the Scientific Method. What an eye-opener and a shock that was to discover that Scientism – placing all of our faith and trust in scientific methods – is an unsustainable religion and an untenable epistemology!

Williams and Slife were the first scientists in my life to point out the fact that scientific methods can't even be relied upon to successfully falsify theories and hypotheses. They were the first people to tell me anything about *negating the consequent*, and its limitations. Scientism had been my epistemology ALL of my life! It looked like it was time for me to find a NEW epistemology – a new way of finding and knowing the truth. It suddenly made this Psyche Epistemology or a "Lived Experience Epistemology" look highly desirable, for the first time in my life.

In another article Richard Williams stated:

**The truth of an event is very different from the truth of a proposition. The truth of propositions is established by reason and argument, the difficulty of which I have just described. The truth of events is established by witnesses. Once we know what is true, reason provides a wonderful tool for sorting out our obligations, anticipating consequences, and persuading others that what we know is true. Truth, I am convinced, can be rendered reasonable, but it does not arise from reason.**

http://mypsyche.us/faith-vs-reason/

Witnessing events or lived experiences is a better way of knowing the truth than philosophical speculation or scientific interpretation of data. Truth and knowledge arise from lived experiences, or certain events in our lives.

The truth of Christianity does not rest on reason and logic, nor does it rest on scientific evidence. The truth of Christianity rests on the occurrence of certain events. The truth of Christianity rests upon the lived experiences of the human psyches or the human beings who were there at those events.

ONLY the human psyche can witness and then testify of events. In other words, only the human psyche can share its lived experiences with other human psyches. We humans are unique. We can share our lived experiences, including our spiritual experiences and out-of-body experiences, with each other in writing and through the spoken word.

I enter these Spiritual Experiences and Lived Experiences into evidence as proof of this concept:

http://www.markme.us/forums/forum/favorite-spiritual-experiences/

https://www.youtube.com/watch?v=UPj4wci_bcI

https://www.youtube.com/watch?v=HbTAmN4m2lQ

http://www.markme.us/forums/topic/appearances-of-our-resurrected-lord-jesus-christ/

## Psyche and Lived Experiences Are Axiomatic

I now treat the **falsity** or **falsehood** of Materialism and Naturalism as axiomatic, which means that I now treat Psyche's existence as axiomatic and treat Psyche as the fundamental unit of Reality and Existence.

ONLY Psyche can do lived experiences, the highest form of knowledge and truth. ONLY the human psyche can share his or her lived experiences with other human psyches.

Scientism is no longer my chosen nor favorite epistemology. I have officially switched from a materialistic and naturalistic ontology in which physical matter is the fundamental unit of reality over to a Psyche Ontology and a Psyche Epistemology in which psyche is the fundament unit of Reality and lived experiences are the best way of finding and knowing the Truth. Psyche really is the Ultimate Cause! And this Ultimate Model of Reality was the result of all of this conversion, which has been taking place in my life during the past year or so.

I'm starting to look at things in new and different ways.

## Unidentified Variables and Unknown Gods

Some things like Psyche cannot be falsified nor reliably verified, especially if one is relying exclusively on scientific methods and a materialistic epistemology. Things like forces, fields, light, spirit matter, non-locality, and psyche have to be experienced and observed directly in order to KNOW that they are real and truly exist. We can observe their effects on physical matter while in a physical environment; but, we have to switch to a non-local spiritual environment if we want to experience and observe these things directly. That's just the reality of the situation.

Only Psyche can do relationships, science, logic, language, math, experience, memories, interpretation, meaning, teleology, afterlife, life, friendship, charity, love, compassion, evaluation, experimentation, design, creation, programming, engineering, manufacturing, existence, and reality. These are all things that we can't get at directly with our scientific methods. Instead of trying to operationalize these things so that we can apply scientific methods to them, we NEED a new and better methodology if we truly want to get at these sorts of immaterial and non-physical things.

This article talks about unidentified variables in scientific research.

The ultimate unidentified variable in ALL scientific research is invariably PSYCHE, or Non-Local Consciousness. Psyche pervades the whole scientific enterprise from beginning to end – starting with philosophical hypotheses, moving into philosophical methodology, and ending with philosophical interpretations of the scientific data. The Scientific Method is pure philosophy from beginning to end; and, ONLY Psyche can do philosophy, hypothesizing, theorizing, methodology, experimentation, and hermeneutical meaning and interpretation of data.

If WE make something happen, it is in fact our Psyche who is making it happen or setting it up so that it can happen. This reality applies to scientific methods and everything else that WE make happen!

## The Problem of Operationalization

In order to study concepts and realities such as forces, fields, light, consciousness, thoughts, dreams, and psyche, the scientists have to try to find some way to *operationalize* or *physicalize* these things so that their effects can be observed and measured.

On page 191, this article talks about the weaknesses and flaws that are introduced into science whenever we try to operationalize the immaterial and non-physical realities of our existence.

"A hallmark of traditional scientific method is observation. To perform a scientific test, we must be able to observe the phenomenon we are trying to study. This is not always an easy or straightforward thing to do, even in the natural sciences and physical sciences. For example, a clearly scientific construct like gravity is never observed directly. We can observe what we take to be the effects of gravity, but not gravity itself."

Just because a phenomenon can't be observed with our physical eyes and physical instruments doesn't mean that it doesn't exist. These types of things can be inferred to exist by the effect they have on physical objects and physical beings such as ourselves.

Gravity is a force or a field, and it is therefore like Psyche or Consciousness, something that is immaterial and non-physical in nature and origin. Because gravity is not physical in nature or origin, it can be omnipresent and instantaneously and simultaneously everywhere all at once. It is physicality or localization in 3D Space-Time which introduces limitations into our lives as mortal beings. Since gravity is not a part of our physical reality, it is NOT subject to the limitations of our physical existence. The same can be said of Psyche or Non-Local Consciousness. Physical matter imposes severe limitations on our Psyche and our Spirit Body; but, when our psyche and spirit body have separated from our physical body, psyche and spirit have no physical limitations. Physical matter cannot travel faster than the speed of light; but, Psyche and our Spirit Body and Gravity certainly can.

Gravity is not observable with our physical eyes and our physical instruments. It can only be experienced. In fact, it has to be experienced in order to know that it is real. This same reality applies to Psyche, Spirit Bodies, and Non-Local Consciousness. It has to be experienced in order to know that it is real. Thus, a Psyche Epistemology or an epistemology based upon Lived Experiences becomes the best and surest way of finding and KNOWING the Truth.

The moral of the story is that we SHOULD switch away from the scientific methods and over to knowledge and experience when it comes to human behavior and the human psyche! Lived experiences are an infinitely better way of getting at the truth and knowledge associated with Psyche and the Spirit Realm. Scientific methods can't even begin to touch these things! We need to pursue a Psyche Epistemology if we want to KNOW the truth about immaterial and non-physical realities such as forces, fields, light, psyche, consciousness, spirit matter, time, thoughts, dreams, and non-locality.

Materialists, Naturalists, Nihilists, and Atheists can't be objective about these kinds of immaterial things because these people have a prior dogmatic commitment to their chosen cause, chosen religion, and chosen worldview.

"In the behavioral sciences, however, it is open to debate whether people can in fact be manipulated and controlled by behavioral technology. There is wisdom in taking great care about how and why we would ever pursue such a project. There is also a good argument that the first and most important goal of science is not manipulation or control, but understanding."

## Objectivity

Objectivity becomes an issue every time an interpretation or an explanation of the scientific data needs to be made.

Materialists, Naturalists, Nihilists, and Atheists can't be objective even when it comes to Science and the Scientific Methods because these people have a prior dogmatic commitment to their chosen cause, chosen religion, and chosen worldview, which completely colors their scientific interpretations.

"The very definitions and framing of a research question are shot through with traditions, history, expectations, values, biases, prejudices, and other subjective factors. It seems unlikely that at any stage the scientific research process is truly objective." Furthermore, Science can't make truth claims.

In contrast, we can KNOW what we have lived and experienced first-hand for ourselves.

Lived Experience becomes the best way of KNOWING the Truth, followed by trusting the witnesses who have had the lived experiences and therefore have KNOWLEDGE of the truth.

"What is important for the purposes of this book is that the characteristics of science, as they are often held up for the behavioral sciences to emulate, are fully exposed for thoughtful examination. Natural science methods may not be capable of serving the validational function that many behavioral scientists desire. The best that science can offer may be *one* way of viewing human behavior, without any special warrant for claiming that it is the only way or even the best way. If this is the case, then whether the behavioral sciences can or should be considered sciences, and just what it might mean to claim that they are sciences, remain open to question."

Science and Scientism are simply one way of looking at the world, and not necessarily the best or most effective way of looking at the world. This article suggests that we treat scientific methods as if they are one language, among many other possible languages.

## Alternative Epistemologies

This article ends by discussing alternative epistemologies (ways of knowing the truth) and alternative methodologies (qualitative methods, human science methods, methodological pluralism, interview methods, and introspective methods).

"If we take the theoretical position that human beings are not simply natural objects, not fundamentally like mechanisms, and not determined by laws and forces the way natural objects are, then it is inappropriate to use the methods developed to study natural objects" to study human beings and the human psyche.

"Use of methods developed for natural objects will result in imposing this naturalistic theoretical outlook on human beings. Some who hold this perspective," namely that Scientism and a Naturalistic Ontology are not the best way to study the human psyche, "argue that adequate study of human beings requires a 'human science' – natural objects can be studied by the methods of natural science and Naturalism, but human beings require **human science methods**" and some kind of **Psyche Science**.

And here we finally have it!

"In this view, human scientists accept **lived experience** as the origin of understanding as well as the object to be understood. **Methods of study must be faithful to and grounded in lived experiences.** Methods developed for a detached study of natural objects will be inadequate."

Slife and Williams are talking about and suggesting a **Lived Experience Epistemology**, wherein lived experiences become the best way of finding and knowing the truth about Reality and our Existence. Although they don't say it in so many words, they are also talking about a **Psyche Ontology** and this Ultimate Cause Model of Reality which I have been trying to develop in this book, wherein Psyche is the Ultimate Cause and Psyche is the fundamental unit of reality.

Slife and Williams then mention a few methods, typically qualitative methods, which are based upon this proposed **Lived Experience Epistemology**.

**Qualitative Methods**

"Many human science methods are grouped under the rubric ***qualitative methods***. The thrust of qualitative research methods is to reject the philosophical assumptions of the traditional scientific methods. Researchers avoid measurement and quantification, allowing subjects to describe their own behaviors and experiences in the language native to their experience. The analysis of the data is likewise carried out in conversational language rather than with statistics. The qualitative researcher is essentially involved in a project of careful questioning, describing, and interpreting. Many people who subscribe to **human science methods** argue that qualitative methods are superior to quantitative methods and ought to be the method of choice for all the behavioral sciences. Qualitative research methods are becoming increasingly accepted as a legitimate alternative to traditional empirical methods. Qualitative methods now form an important part of the literature of many behavioral science disciplines."

I never knew that! Imagine it, this article was written twenty years ago, and I'm just now hearing about qualitative methods for the first time in my life. The Materialists, Naturalists, and Atheists had successfully hidden these things from me for fifty years of my life. It's what they do, and they are very good at it.

**Lived Experiences**

The primary focus in the behavioral sciences is now upon **LIVED EXPERENCES**! Or, at least it should be!

Furthermore, I have observed that ONLY Psyche can do lived experiences, science, methodology, philosophy, theorizing, meaning, final causality, interpretation of data, ontology, epistemology, qualitative methods, psyche-study, and the human sciences. Psyche really is the Ultimate Cause and the Fundamental Unit of Reality, because only psyche can do these kinds of things!

"One possible organization of the methods groups them into three categories: ethnography, phenomenology, and studies of artifacts."

"Phenomenology has as its primary interest the study of the meaning of concrete human experiences." In other words, **Phenomenology is the study of Lived Experiences**. Phenomena are lived experiences and events which the human psyche encounters and experiences. This includes spiritual experiences as well as physical experiences. Nothing is off limits or out-of-bounds.

"To get at the meaning of experiences, phenomenological researchers rely heavily on interviews and other verbal or written accounts of experiences. The researchers then carefully analyze these accounts in order to understand not only individual private meanings of the experiences, but also what is general and illuminating in understanding the meaning of human experience in the wider context of people and situations."

The authors don't mention it in this article, but **Hermeneutics** has become the study of meaning and the interpretation of Lived Experiences.

Furthermore, **Teleology** is the study of final causality and the study of purpose, reasons, goals, and intentions.

It is my personal observation that the Human Psyche is the ONLY thing that can do phenomenology, psychology, science, psychotherapy, hermeneutics, teleology, epistemology, and ontology. The Human Psyche is the ONLY thing we know of directly that can actually do Philosophy and Science. How do we KNOW of Psyche directly? We KNOW of Psyche directly through Lived Experiences and a Lived Experience Epistemology. Lived Experience is the best way of finding and KNOWING the Truth. Lived Experiences cover everything, including our revelations from God, our spiritual experiences or Spiritual Empiricism, our physical experiences or traditional empiricism, our science experiments, our work and our construction, our creativity, our philosophical theories and ideas, and everything else we humans think about and do.

Memories of our Lived Experiences survive the death of our physical body and physical brain according to the Lived Experiences of OBErs and NDErs who have gone to the other side and KNOW what it's really like in the Spirit Realms, and have experienced a life review while in the spirit. According to the out-of-body travelers and NDErs, the human psyche continues to have Lived Experiences and memories of those Lived Experiences while their physical body and physical brain are either dead or offline.

Psychology really should have remained **The Study of Psyche**; but, the Materialists, Naturalists, Atheists, and Behaviorists got their way for a century. It's finally starting to come full circle and turn on them. But, they are still entrenched and not going anywhere.

**Conclusion and Solution**

When it comes to psyche, we humans are unique, in that we can actually share our Lived Experiences with other humans through speech and the written word. The animals can't do that, even though they do have psyches of their own. ONLY the human psyche can do language, writing, reading, and speaking in a way that we can share ALL of our Lived Experiences with each other, including our spiritual experiences and our revelations from God.

This is possible for the human psyche, because we humans are the children of the Gods. Both our spirit bodies and our physical bodies are descended from the Gods. According to the Bible, Adam was a son of God, and Eve was his wife; and, we are gods. Our surname is God. The human psyche is unique among all of God's creation, because we are the children of God the Father and a Heavenly Mother. Only the human psyche has the innate inborn ability to rebel against God, because our spirit bodies and physical bodies are descended from the Gods. We humans really did partake of the tree of knowledge. NO other animal in creation has done so.

It ALL points back to a **Psyche Ontology** and to Psyche as the Ultimate Cause of these Lived Experiences! Psyche really is the Ultimate Cause or the Fundamental Unit of our reality, existence, experiences, and lives. I believe that I have made my case and met my burden of proof. Of course, you as an individual unique psyche have every right to disagree. In your own personal life, you are judge, jury, executor, and executioner. By "you", I mean your psyche. Thanks to your psyche, you will still be you long after your physical body and physical brain are dead and gone. It ALL points back to Psyche.

## VERIFICATION VERSUS FALSIFICATION

*Negating the consequent* is logically valid and typically sound; but, *affirming the consequent* is not. The scientific method is based upon "affirming the consequent". I figured all of this out logically on my own before I knew the actual terminology for it, as demonstrated in my other books listed in the references under Mark My Words.

I'm not the first person to figure all of this out, though. For example, the following is adapted from class notes by Edwin Gantt:

### Verificationism

Historically, scientists have imagined that if predictions (hypotheses) based on their theories came true, then their theories have been verified (in other words, proven to be true) by empirical evidence. That is, the best way to prove a theory true is to make a prediction and then make measurements. If whatever the scientists predict will happen does happen, then it is assumed that the theory that gave rise to the predictions is true.

It is important, however, to consider the logic of this approach. A prediction based on a theory usually takes this form: "If Theory X is true [antecedent], then we will observe Y [consequent]."

The following logical argument outlines the basic approach that has been taken by scientists:

If Theory X is true, then we will observe Y.

We observe Y.

Therefore, Theory X is true.

This sort of thinking, however, reflects a logical fallacy called *affirming the consequent*. Here's a comparable example to demonstrate:

If Sally's pet is a cat, it will have a tail.

Sally's pet has a tail.

Therefore, Sally's pet is a cat.

We can easily see that this logic is fallacious. Just because we observe Y (a pet with a tail) that does not mean that our theory X (the pet is a cat) is true. After all, dogs and lizards have tails too. (Taken from class notes by Edwin Gantt).

### Falsificationism

There is a different argument structure that scientists have begun to use instead of verificationism. It is called the falsificationism approach to science. This approach is much more logically sound than the verificationism approach.

The argument structure for falsificationism looks like this:

If Theory X is true, then we will observe Y.

We don't observe Y.

Therefore, Theory X is not true.

Rather than verifying a theory, this approach falsifies a theory. Unlike the verification argument, this argument is logically sound. The idea is that although we can't ever know (based on empirical data alone) when a theory is true, we can know (based on empirical data) when a theory is false.

In other words, we develop a theory, make predictions based on that theory, and if those predictions come true, rather than claim that we have confirmed our theory, we more humbly claim that our theory has not yet been proven false. Eventually, we arrive at the truth by a process of elimination.

The problem with this approach, however, is that there is no way to actually falsify a theory with absolute certainty. Theories are a patchwork of assumptions, presuppositions, and ideas, and there is no way to know which of those are wrong if the predicted events don't occur in our experiments.

Also, more importantly, theories and worldviews guide our assumptions about what might even constitute falsifying evidence in the first place.

So, a theory that holds that God is not real or active should, in theory, be falsified by actual encounters with God. But, instead, such experiences are usually bracketed off [by the Materialists and Naturalists] as things yet to be fully explained by the theory, while the theory itself is still assumed to be true. (Taken from class notes by Edwin Gantt).

Scientific Methods are built upon and based upon the *affirming the consequent* logic fallacy. Joseph Rychlak mentions the *affirming the consequent* logic fallacy in most of his books (Rychlak), in a similar manner as has been presented here by Ed Gantt.

See also:

Slife, B. D. & Williams, R. N. (1995). Science and Human Behavior. In *What's Behind the Research? Discovering Hidden Assumptions in the Behavioral Sciences*, (pp. 167–204). Thousand Oaks, CA: SAGE Publications.

http://mypsyche.us/science/

Understanding the *affirming the consequent* logic fallacy basically put an END to my Scientism and my exclusive reliance upon Scientific Methods to discover the truth. Scientific Methods can't be used to prove the truth or to verify the truth, because of the logic fallacies and philosophical interpretations upon which Scientific Methods are based. Consequently, I'm now willing to look to and look for other epistemologies – other ways of finding and knowing the truth, in addition to the Scientific Methods. I don't reject Scientific Methods, because that would be stupid; but, I no longer rely upon Scientific Methods exclusively for finding the truth.

**I still rely heavily on falsification, because falsification is logically sound and tends to be reliable!**

The argument structure for falsification looks like this:

**If the theory of evolution is true, we will observe the rocks and physical matter designing and creating genomes and life forms. We NEVER observe the**

**rocks designing and creating anything; therefore, Creation by Rocks, Materialism, Naturalism, Darwinism, and the Theory of Evolution ARE FALSE.**

Materialism, Nihilism, Atheism, Naturalism, Darwinism, Scientism, Spontaneous Generation, Abiogenesis, and Creation by Rocks have been **falsified** trillions of different times in thousands of different ways. I don't believe in these things anymore, because they have been **falsified** too many times.

I used Science and the Scientific Method to prove to myself that Materialism, Naturalism, and Darwinism are false. I finally chose to use the Scientific Methods for what they are truly good for, and that's **falsifying** materialistic ideas that are false. For that reason alone, I will always be grateful for the Scientific Method.

When it comes to Scientific Methods, I still believe in and trust the **falsification** process. Therefore, I'm still a scientist in the traditional sense of the word. Science itself and Scientific Methods have proven to me that Materialism, Naturalism, and Darwinism ARE FALSE. I still have a great deal of love and respect for the Scientific Methods, especially the **falsification process** that's associated with Scientific Methods.

I have observed that design and creation by rocks, or physical matter, IS *the* materialistic and naturalistic assumption or premise that is always FALSE! Only Psyche has the innate inborn ability to design and create! Materialism IS Creation by Rocks; and, the rocks can't design and create. The naturalistic worldview, by definition and fiat, excludes explanations and interpretations of the scientific evidence that do not assume a naturalistic worldview. The solution to this problem is to EXCLUDE Materialism, Naturalism, Nihilism, and Atheism from our scientific explanations and scientific interpretations of the data. Eliminate every known falsehood as the Scientific Method says we should, and we eventually end up with the truth through a process of elimination.

**How often have I said to you that when you have eliminated the impossible, whatever remains, *however improbable*, must be the truth? — Sherlock Holmes**

Eliminate Materialism, Naturalism, Nihilism, Atheism, and Creation by Rocks; and, we are left staring at the truth. It's obvious, and it's logical. It's elementary my dear reader!

## Applying Falsification to Materialism

This is the process that I used in my books to falsify Materialism and the Theory of Evolution.

Whether the Materialists and Naturalists realize it or not, they believe that the rocks or physical matter designed and created and produced the first genomes and the first life forms on this planet. Materialism and Darwinism reduce to Creation by Rocks, or Spontaneous Generation, or Abiogenesis.

The argument structure for the falsification of the Theory of Evolution looks like this:

1. If the Theory of Evolution is true, then we will observe rocks and physical matter designing and creating and manufacturing genomes and life forms from scratch.

2. We don't observe rocks designing and creating genomes and life forms; and, we have never caught the rocks in the act of designing and creating life.

3. Therefore, the Theory of Evolution is not true.

This argument, which involves negating the consequent or falsifying the consequent is philosophically and logically valid and sound.

Now, let's run the Theory of Evolution through the verification process like the Materialists and Darwinists do.

1. If the Theory of Evolution is true, then we will observe progression in the fossil record.

2. We observe progression in the fossil record.

3. Therefore, the Theory of Evolution is true.

The Materialists, Naturalists, and Atheists can't see anything wrong with this argument, because they have blinded themselves into *affirming the consequent* and use this exact argument all the time to make their case. But now, let's run something else through this argument that is sure to get the Darwinists' attention.

1. If the Theory of Intelligent Design is true, then we will observe progression in the fossil record.

2. We observe progression in the fossil record.

3. Therefore, the Theory of Intelligent Design is true.

If the first argument is considered valid, then the second argument must also be considered valid. And, since Creation by Intelligent Design is an infinitely more solid and plausible explanation for the scientific evidence and the origin of life than Creation by Rocks, the second argument (Creation by Intelligent Design) should in fact be given priority over the first argument (Creation by Rocks or the Theory of Evolution).

In other words, ALL of the scientific evidence that is being produced by the Darwinists in an attempt to prove the Theory of Evolution true also ends up proving the Theory of Intelligent Design true as well. And, since Creation by Intelligence has been observed and caught in the act, and Creation by Rocks has not, Creation by Intelligent Design is an infinitely better interpretation of the scientific evidence than Creation by Rocks or the Theory of Evolution will ever be.

Creation by Rocks, Spontaneous Generation, Abiogenesis, Darwinism, Materialism, and Naturalism have been falsified thousands of different ways. In contrast, design and creation and manufacturing by Intelligent Beings has been observed, replicated, and verified trillions of different times over millennia of time. So, which theory and which interpretation of the evidence is most likely to be true – the one that has been falsified or the one that has been observed and verified?

It was in this manner that I was able to prove to myself why the Theory of Evolution and Materialism are false. In every sense of the word, the Scientific Method proved to me that Materialism and Darwinism are false, and that Intelligent Design Theory is much more likely to be true. Eventually, we arrive at the truth through a process of elimination. Most of us instinctively know that Creation by Rocks (Materialism) is impossible. By eliminating Creation by Rocks in all its different forms (Materialism, Darwinism, Naturalism, and the Theory of Evolution), we find ourselves zeroing in on the truth (Design and Creation by Intelligence or Psyche or Ultimate Cause). And for most of us, the truth is what we are after.

**How often have I said to you that when you have eliminated the impossible, whatever remains, *however improbable*, must be the truth? — Sherlock Holmes**

**Reference Articles**

The following articles lend evidentiary support for the concepts presented in this section:

1. Slife, B. D. & Williams, R. N. (1995). Science and Human Behavior. In *What's Behind the Research? Discovering Hidden Assumptions in the Behavioral Sciences* (pp. 167–204). Thousand Oaks, CA: SAGE Publications.

   http://mypsyche.us/assumptions/

   http://sk.sagepub.com/books/whats-behind-the-research/n6.xml

   http://www.ldsphilosopher.com/blog_posts/the-logic-of-falsification/

The authors and contributors have been handing out this paper in class to the public.

This article was the end of my scientism – my erroneous belief that the scientific method is an infallible god.

For a year or two, I considered James Kalat's chapter 2 on the Scientific Method in his *Introduction to Psychology* to be the best and most informative chapter on the Scientific Method that I had ever read in my life (Kalat). This chapter from Slife and Williams about the limitations of the Scientific Method is just as good if not better than Kalat's chapter. But, I like them both.

It is impossible to use the Scientific Method to prove the truth. As this article states, it's also technically impossible to use the Scientific Method to falsify anything as well. However, you are much more likely to falsify something with the Scientific Method than you are going to be able to verify something with the Scientific Method. The Scientific Method isn't as infallible and God-like as it is cracked up to be. The Materialists, Naturalists, and Atheists place way too much unjustified blind-faith in the Scientific Method, often making it their god.

There are better ways of knowing the truth than the Scientific Method.

2. Williams, R. N. (2000, February 1). Faith, Reason, Knowledge, and Truth. *BYU Speeches*.

   https://speeches.byu.edu/talks/richard-n-williams_faith-reason-knowledge-and-truth/

   http://mypsyche.us/faith-vs-reason/

This article demonstrates how we humans rely upon other ways of knowing the truth besides just the Scientific Method. Experience or knowledge trumps scientific speculation and scientific hypotheses every time.

Furthermore, scientists take a leap of faith every time they design and run a science experiment. They hope that their sciences experiments will produce predictable results and give them more control over their environment; but, their hopes aren't always realized. It takes just as much faith to run a science experiment as it does to believe that God is real and truly exists. Belief or faith is simply a matter of preference and choice. Faith and beliefs are chosen into existence by the human psyche.

We choose what we want to do with our faith, our hopes, and our beliefs. Some people choose to find God, who KNOWS the truth. Other people choose to run a science experiment hoping to get lucky and find the truth. Others choose to engage in reason and sophistry, trying to manufacture the truth in the process. The goal in each case is the find the truth, but there are different ways of going about this process – some more effective than others.

During long stretches of my life, I didn't choose wisely.

From this article: "Finally, there is a possibility of being both dark and despairing as well as beyond the pale of reason."

I lived and experienced this particular reality when I was suicidal, psychotic, and going through withdrawal. For the first year, it manifested itself primarily as anxiety and fear; but, it eventually devolved into apathy, pain, and suicidal ideation. I have had an experience or two during my fifty years of life, and one of my friends thinks I should share them with the world.

"I believe the anchor opposite faith is darkness, nihilism, despair — that state of the soul that comes from living 'without God in the world' (Ephesians 2: 12)."

I have been there and done that as well!

Desire to get out of the pit or desire to get out of hell was the beginning of my progress; and, I have made a great deal of progress during the past five years of my life.

Truth rests on phenomena, events, and lived experiences! Faith is trusting the witnesses who have had the lived experiences and thus have THE KNOWLEDGE which we seek. The lived experiences of witnesses IS the best way of knowing the truth!

## Philosophy of Science

I had to develop a new and better way of looking at the world and reality, because the one that I had gotten from my society wasn't working for me. It was working against me.

I don't have a lot of respect for philosophy and haven't seen much value in it, because philosophy is typically used to deceive people. Materialism, Naturalism, Darwinism, Nihilism, Behaviorism, Determinism, Scientism, and Atheism are philosophies or dogma, NOT science. They are religions, NOT science. They have to be taken on blind-faith as being true, because there can NEVER be any observational evidence supporting their major premises or hidden assumptions which claim that the non-physical does not exist. In contrast, every observation and experience of the non-physical FALSIFIES Materialism, Naturalism, Darwinism, Nihilism, Atheism, and their derivatives.

Science is observation, or it should be.

There is NO observational evidence supporting the major premises or hidden assumptions of Materialism, Naturalism, and their derivatives. These philosophies and their philosophical arguments are used to trick us and deceive us; consequently, I have typically found philosophy to be unreliable and unsound because their arguments usually don't match with reality and are seldom based upon observational evidence.

An argument is *valid* if the conclusion necessarily follows from the premises.

An argument is *sound* if it is valid and the premises are true. Falsified premises cannot produce a *sound* argument.

Materialism, Naturalism, and their derivatives will NEVER be logically sound and will NEVER be philosophically sound because they will NEVER have any observational evidence supporting their major premises which claim that the quantum, or the spiritual, or the supernatural does not exist. Every observation and experience of the quantum, the spiritual, the supernatural, the transdimensional, and the non-physical FALSIFIES Materialism, Naturalism, Darwinism, Nihilism, Atheism, and their derivatives. In other words, repeated verification of Quantum Mechanics, Action at a Distance, or Supernatural Mechanisms FALSIFIES Materialism, Naturalism, and their derivatives. Materialism, Naturalism, Darwinism, Nihilism, Behaviorism, Determinism, Scientism, and Atheism cannot be true, because their premises are demonstrably false.

When it comes to philosophy, some have claimed that it is best left ignored. However, I see it differently. When it comes to philosophy, you absolutely MUST KNOW how these people are using philosophy to trick you and deceive you. If you don't, then you are going to be tricked and deceived. Billions of people have been tricked and deceived by the Materialists, Naturalists, Darwinists, Nihilists, and Atheists; and, they don't even know it. I used to be one of them. I used to be a Materialist, Naturalist, Nihilist, and Atheist.

Remember, when it comes to philosophy, your premises have to be flawless and perfect, or your conclusion is going to be wrong.

**The Essential Philosophy of Science**

The fundamental concept that you must understand about the Philosophy of Science is to know what *Affirming the Consequent* is, how it works, and what it does to the Scientific Method.

*Affirming the Consequent* is a logic fallacy, and it is the logic fallacy upon which Traditional Science, Materialism, Physicalism, Naturalism, Darwinism, and Atheism are based. *Affirming the Consequent* is the way that these people do science; and, they don't even know it, because they don't understand the Philosophy of Science.

By *affirming the consequent*, you can prove anything to be true, including Materialism, Naturalism, Atheism, and the Theory of Evolution; but, the process is nothing more than trickery and deception, which billions have fallen for over the millennia.

In my humble opinion, *affirming the consequent* is the most interesting and useful thing to know about and understand, when it comes to Science and the Philosophy of Science.

I discuss *Affirming the Consequent* in great detail in most of my books because its antithesis, *Negating the Consequent*, is the core foundation of my science and the way that I use the scientific methods. Verification uses *affirming the consequent*. Falsification uses *negating the consequent*. *Negating the Consequent* is philosophically and logically sound; whereas, *Affirming the Consequent* is not. Consequently, I use *negating the consequent* in all of my scientific arguments and while using the Scientific Method in order to eliminate everything that is false, so that I can increase my chances that my conclusions and interpretations of the evidence will actually be true.

Remember, if you successfully eliminate everything that is false, then ONLY the truth will remain. I use the scientific methods and *negating the consequent* to eliminate everything that is false, in the hope that ONLY the Truth will remain.

Mark My Words

—

### Source

1. **The Ultimate Model of Reality: Psyche Is the Ultimate Cause**

    https://www.amazon.com/dp/B071NC9JK6

2. **Science 2.0: I Upgraded My Science**

    https://www.amazon.com/dp/B0771K6WTX

### References

1. Slife, B. D. & Williams, R. N. (1995). Science and Human Behavior. In *What's Behind the Research? Discovering Hidden Assumptions in the Behavioral Sciences*, (pp. 167–204). Thousand Oaks, CA: SAGE Publications.

    https://mypsyche.us/wp-content/uploads/2017/04/Science.pdf

    https://philosophy-of-science.com/wp-content/uploads/2018/04/Science.pdf

2. Gantt, E. (2014). Logical Arguments. In *Psychology 353 – LDS Perspectives in Psychology*, (pp. 8-11). Provo, UT: Brigham Young University.

https://philosophy-of-science.com/wp-content/uploads/2018/04/Logical-Arguments.pdf

3. Gantt, E. (2014). Leveling the Playing Field – Why Science is Not a Trump Card. In *Psychology 353 – LDS Perspectives in Psychology*, (pp. 50-58). Provo, UT: Brigham Young University.

https://philosophy-of-science.com/wp-content/uploads/2018/04/Verification-vs-Falsification.pdf

4. Rychlak, J. F. (1981a). *A Philosophy of Science for Personality Theory* (2nd ed.). Malabar, FL: Robert E. Krieger Publishing Company.

5. **My Amazon Page:**

https://amazon.com/author/science

6. *Science 2.0: I Upgraded My Science*

https://www.amazon.com/dp/B0771K6WTX

# My Philosophical Proof of God's Existence

An argument is *valid* if the conclusion necessarily follows from the premises. An argument is *sound* if it is valid and the premises are true.

I have developed a philosophical proof of God's existence – or two – that I find logically compelling, philosophically valid, and scientifically sound. Let's start with the best and most believable philosophical proof of God's existence, the Kalam Cosmological Argument.

## Kalam Cosmological Argument

Let's start with William Lane Craig's version of the Kalam Cosmological Argument.

Premises:

1. Whatever begins to exist has a cause;
2. The universe began to exist;

Conclusion:

3. The universe has a cause.

https://en.wikipedia.org/wiki/Kalam_cosmological_argument

https://philosophy-of-science.com/wp-content/uploads/2018/04/Kalam-Cosmological-Argument.pdf

This argument or proof is a syllogism. If the premises are true, then the conclusion has to be true. This is a properly constructed syllogism; therefore, it is a valid philosophical argument. An argument is *valid* if the conclusion necessarily follows from the premises. This argument is also a *sound* argument. Remember, an argument is *sound* if it is valid and the premises are true. This argument is TRUE because it is philosophically valid and scientifically or observationally sound.

I find this philosophical proof of God's existence logically consistent, philosophically valid, and scientifically sound. It works because its Premises are based upon observed truths. However, it can be greatly improved.

## My Adjustments to the Kalam Cosmological Argument

Even though Craig's argument is philosophically valid and scientifically or observationally sound, I found that I had to adjust it a bit so as to make it specifically clear what I'm talking about and what I'm trying to prove. Craig's argument is too vague and universal for my tastes, which allows falsehoods, faulty interpretations, confusion, and unscientific ideas to creep into it too easily.

So, I chose to adjust it in the following manner and tighten it up a bit.

**Premises or Observations:**

**Whatever begins to exist has a Beginner or an Ultimate Cause;**

**Our physical universe began to exist;**

**Conclusion:**

**Therefore, our physical universe has a Beginner or an Ultimate Cause.**

These adjustments are important and essential, because I'm trying to bring the premises and conclusion into line with what has been observed and experienced in real life by real people. I'm trying to make my argument scientifically valid and sound, not just philosophically so. A philosophical proof that proves a lie to be true is of no value to anyone.

Whenever I develop a philosophical proof, I try to switch over to observational evidence, experiential proof, verified reality, scientific evidence, and logical common sense as quickly as I can in order to prevent myself from being tricked and deceived. Self-deception works, and it works every time.

It's the **physical universe** that we are talking about here, because the non-physical universe, quantum universe, non-local universe, psyche universe, syntropy universe, transdimensional universe, transcendent universe, or spiritual universe had NO beginning; and therefore, it will have no end. It's the physical universe that had a beginning, not the transdimensional universe.

Craig's argument talks about "the universe" – that's way too vague. Why? Well, the Multi-Verse, or the Non-Physical Universe, or the Chaos Construct, or the Spirit Realm had NO beginning, which technically makes Craig's premise #2 false to begin with when he says that "our universe began," because the Primal Universe or Original Universe had NO beginning. However, it has been observed and agreed upon by the scientists in general that our physical universe definitely had a beginning; therefore, our physical universe definitely had to have had an Ultimate Cause or a Person who organized it and brought it into existence.

Furthermore, the Beginner or Ultimate Cause has to be a person, or a living entity, or a Psyche, or an intelligent being of some kind because we KNOW that raw matter, or dead matter, or inert matter, or chaotic matter, or entropy cannot spontaneously generate into anything whatsoever. Matter cannot design and create. There is NO such thing as spontaneous generation or creation ex nihilo. Spontaneous generation or creation ex nihilo has been FALSIFIED by the scientific method. Such a concept as Creation Ex Nihilo is philosophically, logically, and scientifically unsound. It doesn't make any logical sense. Something from nothing is illogical. It can't be done, which means that it wasn't done.

Our physical universe came from something that already existed, and our physical universe was organized by Someone Psyche who also already existed.

My adjustment is already philosophically valid and scientifically sound; so, it already works as a philosophical proof of God's existence; but, let's formalize it and expand it just a bit.

### My Philosophical Proof of God's Existence

**Premise and Observation: Anything that has a beginning has a Beginner or an Ultimate Cause – a Designer, Creator, and Manufacturer who brought it into**

existence. This is Logic 101. It's inherently logical and sound. It has been experienced and observed.

Observations: It has been observed that our physical universe had a beginning, which means that it had a Beginner or an Ultimate Cause – a Designer and Creator who brought it into existence. That means that physical matter had a beginning, which means that it had a Beginner or Ultimate Cause – a Designer, Creator, and Manufacturer who organized it or brought it into existence.

Conclusion: When it comes to our physical universe and physical matter, by definition, in principle or practice, that Beginner, Designer, Manufacturer, and Creator is a God.

This is a logical conclusion since the rest of us here on this physical earth aren't in the habit of making physical matter and physical universes from scratch. As physical beings, we can't touch the sub-atomic or sub-quantum. Only a transcendent and transdimensional God would have such capabilities. Such a God would have to exist BEFORE the organization of this physical universe from spirit matter, chaotic matter, dark matter, or primal matter. Such a God would have to be transcendent without any physical limitations whatsoever – an omnipotent God who is the master of transdimensional physics or quantum mechanics. Such a God would have to have sufficient knowledge and power for such an endeavor.

The FACT that our physical universe had to have had a Beginner or an Ultimate Cause tells us some important things about the nature and capability the Being, the Psyche, or the God who designed and organized our physical universe. The Person or God who organized our physical universe had to pre-date our physical universe. He had to pre-date our physical universe; and, the stuff that He used to form or organize our physical universe also had to pre-date our physical universe. This is Logic 101.

## Commentary on My Philosophical Proof of God

Every beginning has a Person or a Psyche who caused it to happen. Psyche is the Ultimate Cause; and, Psyche is the ultimate causal agent.

See: ***The Ultimate Model of Reality: Psyche Is the Ultimate Cause***

https://www.amazon.com/dp/B071NC9JK6

Remember, every beginning has a Person or a Psyche who caused it to happen. This is Logic 101.

My Philosophical Proof is a philosophical proof of God's existence that actually makes logical sense to me and works for me. I find it compelling and convincing. It's based upon the Kalam Cosmological Argument, which I also find convincing and believable.

https://en.wikipedia.org/wiki/Kalam_cosmological_argument

https://philosophy-of-science.com/wp-content/uploads/2018/04/Kalam-Cosmological-Argument.pdf

**As with any philosophical proof, however, if you don't find the premises, arguments, and observations compelling, credible, and convincing, then you won't find the conclusions convincing either.** That's the weakness of philosophical proofs. Observation and experience can easily falsify them; and, a lack of personal observational experience can easily make that person a victim of the proofs that are faulty and false. I seldom find any philosophical proof convincing, because they are all subject to legerdemain or trickery. Most of them have NO observational evidence supporting them.

In contrast, whether I'm talking about science or philosophy, the thing that I do find compelling, convincing, and believable is Evidence, Observation, and Experience. Science is observation, or it should be. I've always found a Scientific Proof of God's Existence infinitely more believable and convincing than a philosophical proof of God's existence, because the science is based upon observation and experience.

Furthermore, a complete lack of observational experience or verified proof can easily falsify a premise, a philosophical proof, a hypothesis, or even a theory. That's what happened to Materialism, Naturalism, Darwinism, Nihilism, Behaviorism, Determinism, Scientism, Atheism, Macro-Evolution, and therefore the Theory of Evolution. A complete lack of observational evidence supporting their major premises, primary assumptions, or hidden assumptions completely FALSIFIES them.

The wonderful thing about the Kalam Cosmological Proof of God's Existence, especially my modified version, is that it's backed by observational evidence and it's easily transformed into a Scientific Proof of God's Existence simply by finding some observed evidence and experiential evidence to insert as your premises.

**For example:**

**Premises or Observations:**

**1. Anything that has been fine-tuned has a Fine-Tuner;**

**2. Our physical universe has clearly been fine-tuned;**

**Conclusion:**

**3. Therefore, our physical universe has a Fine-Tuner.**

This is a powerful syllogism that works as an excellent philosophical proof of God's Existence because it is based exclusively upon observational evidence and logical common sense. This argument is philosophically and logically SOUND, because its premises have been experienced and observed by real-life human beings.

This argument is solidified by the scientific observation that spontaneous generation is physically impossible, thanks to entropy or the second law of thermodynamics. There's no such thing as spontaneous generation, abiogenesis, chemical evolution, creation ex nihilo, or macro-evolution thanks to entropy or the second law of thermodynamics which prevents these types of things from happening in the wild.

This Fine-Tuner Argument is as much a Scientific Proof of God's Existence as it is a philosophical proof of God's existence. The evidence of fine-tuning is all around us and impossible to deny. Fine-tuning is one of the most convincing Scientific Proofs of God's Existence.

Every instance or example of cosmic fine-tuning, mechanical fine-tuning, and biological fine-tuning is a miniature scientific proof of God's existence.

How do we know?

We scientists KNOW because living cells, genomes, genes, proteins, eyes, brains, and life forms have NEVER been <u>observed</u> spontaneously generating from atoms out of thin air. They can't because it's physically impossible for them to do so thanks to entropy, random diffusion, or the second law of thermodynamics.

### Fine-Tuning Is Scientific Proof of God's Existence

https://philosophy-of-science.com/wp-content/uploads/2018/04/compendium_part1.pdf

https://philosophy-of-science.com/wp-content/uploads/2018/04/compendium_part2.pdf

https://philosophy-of-science.com/wp-content/uploads/2018/04/compendium_Part3_ver2.pdf

https://philosophy-of-science.com/wp-content/uploads/2018/04/compendium_Part4_ver2.pdf

Ironically, Louis Pasteur FALSIFIED spontaneous generation (and therefore Materialism, Naturalism, Darwinism, Atheism, Chemical Evolution, Macro-Evolution, Creation Ex Nihilo, and the Theory of Evolution) in 1859, the very same year that Charles Darwin published "On the Origin of Species". We have KNOWN since the very beginning of the theory of evolution that spontaneous generation, abiogenesis, macro-evolution, chemical evolution, creation ex nihilo, or the theory of evolution is FALSE; but, most scientists have deliberately chosen to ignore that evidence and pursue wishful thinking instead. *Wishful thinking*, or *confirmation bias*, or *blind-faith* is a logic fallacy – one of the logic fallacies upon which Materialism, Naturalism, and Darwinism are based.

Remember, your philosophical arguments and syllogisms are NOT sound if their premises have NEVER been experienced nor observed. They may be logically valid, but they are NOT sound. That's how the Materialists, Naturalists, and Darwinists trick us and deceive us. They give us syllogisms and philosophical arguments that are logically valid, but totally unsound due to a complete lack of observational evidence supporting their chosen premises. Consequently, because of faulty premises, these people also *jump to conclusions* that are philosophically, logically, and scientifically impossible and unsound. *Jumping to conclusions* or *begging the question* is also a logic fallacy. Materialism, Naturalism, Darwinism, and Atheism are unscientific because their premises are scientifically unsound, having never been experienced nor observed.

**Now, try this one:**

**Premises or Observations:**

**1. We have observed that everything that has a beginning has some kind of Creator who made it.**

**2. We have observed that our physical universe had a beginning. Consequently, we have observed that physical matter had a beginning.**

**Conclusion:**

**3. Therefore, it is logical to conclude that our physical universe and the physical matter within it had some kind of Creator who made them, organized them, or brought them into existence.**

This argument has teeth because its premises have been experienced and observed. This argument is philosophically and logically SOUND, because its premises have been experienced and observed. This Creator Argument is as much a Scientific Proof of God's Existence as it is a philosophical proof of God's existence.

We have caught intelligent beings in the ACT of design, creation, manufacturing, and production zillions of times in trillions of different ways. Intelligent Design and Intelligent Creation is an OBSERVED, verified, and proven science. We KNOW that it is REAL because it has been OBSERVED. It has been caught in the act. There's NO philosophical speculation, guesswork, or wishful thinking going on here where the observation of intelligent beings (or intelligent psyches) in action is concerned.

I find these Philosophical Proofs of God's Existence equally as compelling as the others, because their premises have been experienced and observed in real life by real people. If you want the truth, then go with what has been experienced and observed, because the philosophical speculation, wishful thinking, and blind-faith of the Materialists and Naturalist are worthless in the end.

Notice that when it comes to My Philosophical Proofs of God's Existence, I try to turn them into a Scientific Proof of God's Existence as quickly as I can. In other words, I try to turn them into an OBSERVED Proof of God's Existence. When it comes to science and proof, observation is where the tires really hit the pavement. Observed Proof of God's Existence is the most convincing and believable proof of God's existence. That's just the way it is.

In contrast, I also found the complete lack of observational evidence supporting their major premises equally as convincing when it came time to falsify theories and ideas such as Materialism, Naturalism, Darwinism, Nihilism, Behaviorism, Determinism, and Atheism. These philosophies or hypotheses are FALSIFIED by a complete lack of observational evidence supporting their major premises or hidden assumptions.

As a result of this observation, I upgraded my science to Science 2.0.

Science 2.0 allows ALL of the evidence into evidence, and then it pursues a preponderance of that evidence. When it comes to the non-physical sciences, Science 2.0 uses a Burden of Proof Methodology that is based upon a preponderance of the observational evidence. Under Science 2.0, observation and experience of any kind take precedence over philosophical speculation, guesswork, hypothesis, wishful thinking, scientific inferences, and confirmation biases. Science 2.0 is the way that science should have always been done but wasn't.

### Science 2.0: I Upgraded My Science

https://www.amazon.com/dp/B0771K6WTX

Science 2.0 is a new and better way of doing science that is based upon observational evidence, eye-witness evidence, and experiential evidence.

Under Science 2.0, the BEST way to find and know the truth is to live it, experience it, and observe it for yourself, or to choose to trust someone who has. The second-best way to find and know the truth is to use the scientific methods to falsify and eliminate everything that is false such as Materialism, Naturalism, Darwinism, Atheism, and their derivatives. If you successfully eliminate and remove everything that is false, then ONLY the truth will remain. It's fascinating to observe what remains after you have eliminated everything that is false and everything that has been falsified.

Remember, a syllogism can be logically valid but totally unsound. If the Premises of your argument, syllogism, or logic proof are NOT backed by observational evidence, then your philosophical argument is unsound. That's yet another serious problem with Materialism, Naturalism, Darwinism, Nihilism, Atheism, Behaviorism, Determinism, Scientism, Atheism, and the Theory of Evolution. There is NO observational evidence supporting their major premises or hidden assumptions, and there can NEVER be any observational evidence supporting their primary assumptions or major premises; therefore, the arguments produced by these philosophies, religions, dogmas, or pseudo-sciences can be forced to be valid, but they will NEVER be philosophically or logically sound. These falsified philosophies are unscientific and unsound because there can never be ANY observational evidence supporting their hidden assumptions or major premises which claim that the non-physical or the sub-quantum does not exist.

The moral of the story is that when it comes to your philosophical proofs, be sure to use observed evidence, verified scientific evidence, experienced evidence, veridical evidence, and empirical evidence as your Premises if you want to have the greatest chance of your conclusions and subsequent interpretations being true.

My Philosophical Proofs of God's Existence have teeth and claws, because their premises are OBSERVED evidence and have actually been experienced in real life. My Philosophical Proofs of God's Existence are both valid and sound because they are in fact Scientific Proofs of God's Existence, meaning that their premises are backed by tons of observational evidence and experience.

### Go Out of Your Way to Get the Right Interpretation

Remember, philosophy has NO value whatsoever if it doesn't match with what has been experienced and observed. Notice how I tried to switch over from logic to what has actually been experienced and observed, before drawing any conclusions. The observations are essential. Science is observation. A philosophical proof has to match with reality, or it's worthless.

It's also important to draw the correct conclusion.

Online, when I provided a scientific proof of God's existence, one of the readers asked me which of all the man-made gods we should believe in.

Therefore, in harmony with his question, I ask, "Which of all the man-made gods is the one who was the Beginner, Designer, Manufacturer, and Creator of this physical universe and the physical matter in this universe?"

I chose to go with the True and Living God, Jesus Christ, because He has been experienced and observed after He rose from the dead. Thousands have seen Him and touched Him in the flesh; and, thousands have seen Him and embraced Him during their Near-Death Experiences. Should you ever find yourself in hell, remember that Jesus Christ can get you out of there just for the asking.

We KNOW that the Biblical God Jesus Christ exists, because He has been experienced and observed by thousands after He rose from the dead. Jesus Christ claims to be the God who organized the heavens and this earth, as well as all the life forms on this earth.

Notice once again, that when it comes to God's Existence, I try to switch over to observation and experience as quickly as I can, because philosophical speculation and wishful thinking are absolutely worthless in the end.

I define "science" as observation and experience. I have trained myself to switch away from philosophical speculation, wishful thinking, hypothesis, confirmation biases, and sophistry over to observation, experience, and empirical evidence as quickly as I can in order to prevent myself from being deceived. I encourage my readers to do the same.

## A Proof or Argument Has to Be Logically Sound

Remember, if the premises of your argument don't add up to the conclusion, then your proof is worthless no matter how valid it might seem to be. If your premises have NEVER been experienced nor observed in the wild, then your conclusion is nothing but fiction and is philosophically unsound. Your chosen conclusion is automatically falsified by a complete lack of observational evidence.

In order to be logically sound, your Premises have to be experienced, observed, and verified. In order for your Conclusion to be logically sound, your Premises have to be logically sound. Missing evidence cannot be used as evidence. The missing links really are missing, so get used to it.

If your premises completely lack confirming evidence or verified evidence, then your conclusion is automatically false. Design and creation by Chemical Evolution, Spontaneous Generation, Abiogenesis, or Macro-Evolution have NEVER been experienced nor observed, which means that these concepts or theories are automatically false. These things are prevented from happening by entropy or the second law of thermodynamics.

Technically, random mutations and natural selection have NEVER been caught in the act of design and creation, which means that these theories or concepts are automatically false. In other words, random mutations and natural selection cannot design and create. Random mutations are entropy. Natural Selection ultimately results in death and extinction, which means that natural selection is also a type of entropy. Evolution is entropy. By definition, in principle, Materialism, Naturalism, Darwinism, Nihilism, Atheism, Classical Physics, and the Theory of Evolution are based upon entropy. Entropy cannot design and create anything.

The Premises behind Materialism, Naturalism, Darwinism, and their derivatives are false and have been falsified by science and by observation, which means that the Conclusions that these people make are also false and have already been falsified. Materialism, Naturalism, and Darwinism are FALSIFIED by a complete lack of confirmed evidence, verified evidence, or observed evidence supporting their major premises and hidden assumptions.

In contrast, the verified and proven existence of Quantum Mechanisms, Action at a Distance, or Supernatural Mechanisms FALSIFIES the major premises of Materialism and Naturalism which claim that the quantum or the supernatural does not exist. The very existence of Quantum Mechanics or Transdimensional Physics FALSIFIES Materialism, Naturalism, and their derivatives such as Nihilism, Darwinism, and Atheism.

I have observed and experienced the FACT that anything that has a beginning has a Beginner or an Ultimate Cause, who is the person that designed, created, manufactured, and produced that thing. This reality has always held true. A Beginner or Creator is the person who brings a thing into existence in the first place. This is a FACT that I find fully compelling, believable, and incontrovertible whether we are talking about science, philosophy, logic, reality, or existence in general. I find it convincing, because it has been experienced and observed. We have observed people or intelligent beings bringing things into existence or choosing things into existence zillions of times in trillions of different ways.

Likewise, I find the existence of the Biblical God Jesus Christ compelling and believable because He has been experienced and observed by thousands, if not millions, of different people after He died and rose from the dead. Science is Observation, or it should be. These people have seen Him and touched Him both here on the physical plane as well as on the spiritual, quantum, or transdimensional plane.

Every time that Jesus Christ is seen, experienced, touched, embraced, and observed whether in the flesh, in visions and theophanies, or during our Near-Death Experiences (NDEs), those experiences and observations are Scientific Proof of His Existence. Science is observation, or it should be.

In order for an argument or a proof to be logically sound, the premises have to be true, which means that the premises have to be experienced and observed by Someone Psyche. A philosophical proof that has NO observational evidence supporting it is completely worthless in the end. That's why Materialism, Naturalism, Darwinism, Nihilism, Behaviorism, Determinism, Atheism, and the Theory of Evolution are completely worthless. They don't have and can't have any observational evidence supporting their major premises or hidden assumptions which claim that the quantum or the supernatural does not exist. In fact, the observed and proven existence of Quantum Mechanics and Action at a Distance FALSIFIES Materialism, Naturalism, and their derivatives.

**Observational Reality and Common-Sense Logic**

Anything that has a beginning has a Beginner or a Creator.

In contrast, God's Psyche and your psyche have always existed and will always exist. Your psyche and God's Psyche have no beginning, which means that they will have no end. This also means that your psyche or intelligence has NO Beginner, Designer, Creator, Manufacturer, or Ultimate Cause who brought it into existence in the first place. You have always existed, and you will always exist.

Things that have a beginning have a Beginner or a Creator, the person or psyche who brought them into existence. Physical matter, physical universes, physical genomes, and physical life forms have a beginning, which means that they have a Designer and Creator who brought them into existence. In contrast, the things that have always existed, such as Psyche and Syntropy and Quantum Mechanics, have NO Creator because they have always existed.

Do you see how that works? This is Logic 101.

The things that have always existed, such as Psyche and Syntropy, have NO Creator. They are without a beginning of days or an end of years. Psyche is Syntropy. Quantum Mechanics or Transdimensional Physics is Syntropy. Syntropy is a type of unity, wholeness, completeness, or perfection. The Atonement of Christ is Syntropy. Syntropy means "without a beginning of days or an end of years". Syntropy means eternal, everlasting, and infinite. The Priesthood Power of God is Syntropy. Quantum Mechanics is the Priesthood Power of God. Syntropy had no beginning, and it will have no end. Syntropy, or Psyche, or Intelligence, or Consciousness has always existed. Syntropy is an organizing force that counteracts the effects of entropy. Syntropy has to exist or entropy wouldn't exist. Our physical universe had to receive an initial infusion of Syntropy, or none of that subsequent entropy would have been possible.

Do you see how that works?

Its explanatory power is through the roof and without limit!

I just provided the answer to life, the universe, and everything. That's the explanatory power of Syntropy, Psyche, and Quantum Mechanics.

### Critiquing Craig's Cosmological Argument

William Lane Craig and others have declared him to be the master of the Kalam Cosmological Argument.

https://en.wikipedia.org/wiki/Kalam_cosmological_argument

Notice that William Lane Craig makes an argument that is philosophically valid and logically sound; but then, he *jumps to a conclusion* at the end that is philosophically illogical and scientifically unsound. Craig is not going to critique his own argument, because he is too close to it and too emotionally invested in it to be able to see what might be wrong with it.

**William Lane Craig states the Kalam cosmological argument as a brief syllogism, most commonly rendered as follows:**

**1. Whatever begins to exist has a cause;**

**2. The universe began to exist;**

*Therefore*:

**3. The universe has a cause.**

**From the conclusion of the initial syllogism, he appends a further premise and conclusion based upon ontological analysis of the properties of the cause:**

**1. The universe has a cause;**

**2. If the universe has a cause, then an uncaused, personal Creator of the universe exists who *sans* the universe is beginningless, changeless, immaterial, timeless, spaceless and enormously powerful;**

*Therefore*:

**3. An uncaused, personal Creator of the universe exists, who *sans* the universe is beginningless, changeless, immaterial, timeless, spaceless and enormously powerful.**

**Referring to the implications of Classical Theism that follow from this argument, Craig writes:**

"... transcending the entire universe there exists a cause which brought the universe into being *ex nihilo* ... our whole universe was caused to exist by something beyond it and greater than it. For it is no secret that one of the most important conceptions of what theists mean by 'God' is Creator of heaven and earth."

https://en.wikipedia.org/wiki/Kalam_cosmological_argument

William Lane Craig deliberately left his Kalam cosmological argument vague so that he could sneak a falsehood in at the end.

The only thing Craig gets wrong is that he *jumps to the conclusion* that God brought our universe into being **ex nihilo**. *Jumping to conclusions* or *begging the question* is a logic fallacy. In his conclusion, **ex nihilo** is an *unjustified add-on* or a *special pleading*, which are logic fallacies. In his promotion of Creation Ex Nihilo, Craig is assuming facts not in evidence.

There's NO such thing as Creation Ex Nihilo. Technically, Creation Ex Nihilo is Atheism – design and creation from nothing by nothing. It's impossible to create something from nothing. It's also impossible for nothing to design and create something. Even God can't do the impossible. Atheism is also impossible, because it's impossible for nothing to create something from nothing. Creation Ex Nihilo is philosophically and logically unsound, because it doesn't make logical sense and because it's the result of *jumping to conclusions*, which is a logic fallacy. Creation Ex Nihilo or spontaneous generation is also scientifically unsound, because it has been falsified by the scientific method.

Clearly, God existed BEFORE He organized this physical universe and brought it into existence. The very existence of God, the pre-existence of God, FALSIFIES creation ex nihilo. God did not spontaneously spring into existence from nothing. Creation ex nihilo does not and cannot apply to God, nor can it apply to the dark matter or spirit matter from which God organized this physical universe. God's Psyche has always existed and will always exist. The dark matter or spirit matter has always existed and will always exist. Creation Ex Nihilo or spontaneous generation is false and has been falsified. Creation Ex Nihilo is unnecessary and unjustified, so let's delete it. If you successfully eliminate everything that is false, then only the truth will remain.

Therefore, Craig's concluding statement should read:

**"Transcending the entire universe there exists a cause which brought the universe into being. Our whole universe was caused to exist by something beyond it and greater than it. For it is no secret that one of the most important conceptions of what theists mean by 'God' is Creator of heaven and earth."**

As long as we get rid of the logic fallacy and scientific falsehood that Craig tacks on at the end of his argument, I find the Kalam Cosmological Argument philosophically, logically, and scientifically valid and sound.

William Lane Craig teaches us by example that a philosophical proof can be valid, the premises can be sound, and the conclusion absolutely true; but, the person can still draw false conclusions and produce faulty and invalid interpretations from that True Conclusion or Proven Conclusion. Interpretation of the evidence or scientific inference is where the Scientific Method always falls down and dies whenever the Scientific Method is used to prove that a lie is true.

Remember, even William Lane Craig falls down and FAILS whenever he switches away from observational evidence over to *wishful thinking* and starts *jumping to conclusions* rather than relying upon observational evidence to make his case. Creation Ex Nihilo has NEVER been experienced nor observed; and, it never will be, because it's impossible. It's the observational evidence that we find convincing, and NOT the philosophical arguments. Philosophical arguments are worthless if their premises are not backed-up by observational evidence, verified evidence, and experiential evidence. Even a valid and sound conclusion from a philosophical argument is worthless, if it has NO observational evidence supporting it.

We find the existence of the Biblical God Jesus Christ compelling, believable, and convincing ONLY because it's backed-up by observational evidence and experiential evidence. Philosophical proof of God's existence is worthless if the premises are NOT backed up by observational evidence, verified evidence, scientific evidence, experiential evidence, eye-witness evidence, and empirical evidence. A conclusion is also worthless if it has NO observational evidence supporting it.

I was able to tighten up my version of the Kalam Cosmological Argument and get even more specific with it because I wasn't trying to insert a falsehood at the end. I was only interested in finding the observed and verified truth. Creation ex nihilo has never been experienced nor observed. It doesn't exist. It's impossible. Even God cannot do the impossible. Since I wasn't trying to support a falsehood with my philosophical argument, I was able to go directly to the Observed and Experienced Truth of the matter.

The theory of evolution is spontaneous generation; and, spontaneous generation is a type of Atheism or Creation Ex Nihilo – creation of something from nothing or by nothing. Something from nothing or something by nothing is impossible. It didn't happen because it can't happen.

If you accept the premises of the Kalam Cosmological Argument as being true, then they PROVE that our physical universe had a Creator, Beginner, or Ultimate Cause; however, they do NOT prove the veracity of creation ex nihilo. In fact, ALL of the observational evidence and experiential evidence that we have on hand as a race FALSIFIES creation ex nihilo, spontaneous generation, or the theory of evolution. Furthermore, it's ONLY our physical universe that had a beginning. The quantum universe, or the supernatural universe, or the spiritual universe, or the non-physical universe, or the non-local universe has always existed and will always exist. It has NO Creator or Ultimate Cause, because it has always existed.

### Conclusions

The Beginner or the Ultimate Cause of this physical universe has to pre-date the beginning of this physical universe, and so does the stuff of which He was made and the stuff by which He made. This is Logic 101.

God organized the heavens and the earth from already pre-existing matter. God transformed some of that pre-existing spirit matter or dark matter into physical matter; and therefore, our physical universe had a physical beginning, even though the spirit matter or dark matter from which it was made has always existed. We KNOW that the dark matter or spirit matter exists by the effect that it has on ordinary physical matter. It has to exist, or our observations and measurements don't make any logical sense.

Do you see how that works?

Everything makes sense to me, once we get rid of the logic fallacies and switch over to observations and measurements instead. Then suddenly, everything is philosophically, logically, and scientifically valid and sound.

Verified means that it has been experienced and observed. Falsified means that it has NEVER been experienced nor observed. Materialism, Naturalism, Darwinism, Nihilism, Behaviorism, Determinism, Creation Ex Nihilo, and Atheism have been FALSIFIED by the fact that their major premises or hidden assumptions have never been experienced nor observed and can't be.

The lesson in all of this is to switch away from philosophical speculation and sophistry over to observational evidence, experiential evidence, eye-witness evidence, empirical evidence, veridical evidence, scientific evidence, and phenomenological evidence as soon as you possibly can.

The truth has been repeatedly verified, experienced, and observed. The falsified and the false have NEVER been experienced nor observed, which is why they have been falsified. That's just the way things work. That's the way things should work in science; and, that's definitely the way that things should work in religion, philosophy, and logic as well. Go with the experienced and the observed, and get rid of the wishful thinking, scientific inferences, blind-faith, and confirmation biases as quickly as you can. Go with the best and get rid of all the rest.

If you successfully eliminate everything that is false, then ONLY the truth will remain.

Mark My Words

—

**Source**

1. ***The Ultimate Model of Reality: Psyche Is the Ultimate Cause***

   https://www.amazon.com/dp/B071NC9JK6

2. ***Science 2.0: I Upgraded My Science***

   https://www.amazon.com/dp/B0771K6WTX


**References**

1. Slife, B. D. & Williams, R. N. (1995). Science and Human Behavior. In *What's Behind the Research? Discovering Hidden Assumptions in the Behavioral Sciences*, (pp. 167–204). Thousand Oaks, CA: SAGE Publications.

    https://mypsyche.us/wp-content/uploads/2017/04/Science.pdf

    https://philosophy-of-science.com/wp-content/uploads/2018/04/Science.pdf

2. Gantt, E. (2014). Logical Arguments. In *Psychology 353 – LDS Perspectives in Psychology*, (pp. 8-11). Provo, UT: Brigham Young University.

    https://philosophy-of-science.com/wp-content/uploads/2018/04/Logical-Arguments.pdf

3. Gantt, E. (2014). Leveling the Playing Field – Why Science is Not a Trump Card. In *Psychology 353 – LDS Perspectives in Psychology*, (pp. 50-58). Provo, UT: Brigham Young University.

    https://philosophy-of-science.com/wp-content/uploads/2018/04/Verification-vs-Falsification.pdf

4. Rychlak, J. F. (1981a). *A Philosophy of Science for Personality Theory* (2nd ed.). Malabar, FL: Robert E. Krieger Publishing Company.

5. Ross, H. (2008). *Why the Universe Is the Way It Is*. Grand Rapids, MI: Baker Books.

6. **My Amazon Page:**

   https://amazon.com/author/science

7. ***Science 2.0: I Upgraded My Science***

   https://www.amazon.com/dp/B0771K6WTX

# PART V — MY TOP TEN MOST FAVORITE SCIENCE THEMES AND MY FAVORITE BOOKS SUPPORTING THOSE THEMES

This section started by being a list of my top ten most favorite science books; but, it soon morphed into my top ten most favorite science disciplines, along with a variety of favorite books supporting those themes or disciplines.

I used these books to demonstrate to myself or to prove to myself that I did in fact need to upgrade my science to something better that actually allows all of the evidence to be considered.

## The Law of Witnesses

**2 Corinthians 13: 1: "In the mouth of two or three witnesses shall every truth be established."**

Eye-witness evidence is still the best type of evidence. Empirical Evidence is certainly superior to scientific inferences and philosophical speculation and wishful thinking. The best way to find and KNOW the truth is to live it and experience it for yourself, or to choose to trust someone who has.

You can only take your science as far as you are willing to read, study, and learn. If you refuse to look at things, then you are damned or stopped in your progress, learning, and growth. There are certain science books out there in this world that the Materialists, Naturalists, Darwinists, Nihilists, and Atheists don't want you to find and read. I discovered that those books were the ones that taught me the most about science and reality, and therefore ended up being the most interesting and life-changing.

Science 2.0 is based upon eye-witness evidence, observational evidence, and the lived experiences of the human race, including both the physical evidence that has been provided by our scientists so far and the non-local evidence that has been provided by the lived experiences of the human race. The best way to find and know the truth is to live it and experience it for yourself, or to choose to trust someone who has.

Science is Observation. Science is Knowledge.

Under this particular definition for science, Materialism, Naturalism, Darwinism, Nihilism, and Atheism are **unscientific** because they have NO observational evidence supporting their primary assumptions or major premises. Their lack of eye-witness evidence or lack of observational evidence FALSIFIES them.

Science 2.0 is based upon the Law of Witnesses as established by God himself. The Law of Witnesses is the best way to do science, because the best way to find the truth and know the truth is to live it and experience it for yourself or to choose to trust someone who has.

Scientism is the philosophical belief that Science and the Scientific Method is the ONLY way to find and know the truth, or at least in its milder form, the best way to find and know the truth. As a practitioner of Scientism and a believer in Scientism, I have always considered myself a scientist. I'm not a Seer, Clairvoyant, nor Astral Traveler – I haven't had those kinds of experiences in great abundance. Such has not been my talent or gift. I have had to turn to others whom I trust for those types of Empirical Evidence.

According to the Empirical Evidence from thousands of Near-Death Experiences (NDEs) and Out-of-Body Experiences (OBEs), WE KNOW that Psyche or Non-Local

Consciousness continues to exist and function after the physical brain is dead or offline. Once I finally discovered, understood, and accepted this Truth and Reality, then my goal became to try to find some way to explain these phenomena scientifically. My search eventually led me to Quantum Mechanics. The right interpretation of Quantum Mechanics can explain Non-Local Consciousness in a way that actually makes logical sense and is experimentally accurate or verified as well.

After a lifetime of Scientific Research and study, I finally found a couple of different Interpretations of Quantum Mechanics, which explain what Psyche or Non-Local Consciousness is and how it works, in a logical scientific fashion that actually made sense to me.

Van Lommel, P. (2010). *Consciousness Beyond Life: The Science of the Near-Death Experience*. New York: HarperCollins.

This book taught me what I call the **Non-Local Consciousness Interpretation of Quantum Mechanics**, which is the science behind Near-Death Experiences.

Schwartz, J. M., Stapp, H. P., & Beauregard, M. (2004). *Quantum Physics in Neuroscience and Psychology: A Neurophysical Model of Mind-Brain Interaction*. Published Online: Phil. Trans. R. Soc. B.

http://www-physics.lbl.gov/~stapp/PTRS.pdf

http://mypsyche.us/wp-content/uploads/2017/10/PTRS.pdf

https://www.researchgate.net/publication/7613549_Quantum_physics_in_neuroscience_and_psychology_A_neurophysical_model_of_mind-brain_interaction

http://escholarship.org/uc/item/4w8665vk

This paper summarizes everything I have been looking for during the past fifty-five years of my life. That's a long time to wander in darkness looking for the truth; but luckily, our generation finally has the truth, if we know where to find it and recognize it as true when we do find it.

This paper and Henry P. Stapp's papers and books taught me what Stapp calls the **Orthodox Interpretation of Quantum Mechanics**, which explains the interplay or the scientific interface between mind and matter.

# 1. Scientific Explanation of Psyche, Spirit Matter, Tachyons, and Quantum Non-Locality or Quantum Non-Physicality

Van Lommel, P. (2010). *Consciousness Beyond Life: The Science of the Near-Death Experience*. New York: HarperCollins.

When it comes to books about Quantum Mechanics, this book from Pim van Lommel is the clear winner, especially if you are looking for only one. If you are planning to spend a lot of money on the subject, I can and will list a few others for you later on in this essay; but, I will start with Pim van Lommel's book first.

After reading this book, for the first time in my life, I actually felt like I understood Quantum Mechanics. This was the first Quantum Mechanics book that I found and read which actually made logical sense to me. I derived a great deal of knowledge from this book, as the subsequent sections will attest.

We need an interpretation of Quantum Mechanics that matches with Reality and the Lived Experiences of the Human Race.

Near-Death Experiences (NDEs) and Out-of-Body Experiences (OBEs) have been proven to be real because they have been lived and experienced for real by millions of different people on this planet. According to one estimate, there are now over 13 million NDEs and OBEs on record, so WE KNOW that this phenomenon is real and repeatable because it has been lived and experienced by millions of different people throughout the world.

Remember, Science 2.0 allows all of the evidence into evidence. Science IS observation after all. The best way to KNOW the truth is to live it and experience it and observe it for yourself, or to choose to trust someone who has. KNOWLEDGE is the best type of proof, because it has been lived and experienced for real by real people. Knowledge of the truth is infinitely superior to the philosophical speculations of the Materialists, Naturalists, Darwinists, Nihilists, and Atheists.

This book from Pim van Lommel provides a Scientific Explanation for NDEs and OBEs through Quantum Non-Locality. Pim also develops an interpretation of Quantum Mechanics that I have started calling the **Non-Local Consciousness Interpretation of Quantum Mechanics**. The Non-Local Consciousness Interpretation is based upon a Psyche Ontology, wherein Psyche or Non-Local Consciousness is the fundamental unit of Reality and Existence.

Most scientists don't realize it, and few truly accept it; but, the repeated verification of Quantum Mechanics along with the scientific verification of the Quantum Zeno Effect FALSIFY Materialism, Naturalism, and Nihilism. When those go down, Darwinism and Atheism go down with them, without a leg to stand on. Rather than trying to explain away the scientific evidence which so abundantly verifies Quantum Mechanics – as the Materialists and Naturalists attempt to do – I have tried instead to determine for myself what Quantum Mechanics means to us and is trying to tell us and teach us. I'm a scientist, after all – it's what I do.

I'll rely upon and mention this book, *Consciousness Beyond Life*, throughout the remainder of this section, because this book forms the beginning and the foundation for my personal interpretation of Quantum Mechanics. After reading this book, for the first time in my life, I actually felt like I understood Quantum Mechanics. It finally made sense to me.

## Quantum Mechanics Has Never Been Falsified

My best friend is a scientist, inventor, and holds a number of different patents for science he has done and for things that he has invented. I have observed that whenever he tells me something about science, he is almost always right; and, his conclusions can actually be demonstrated to be right.

Quantum Mechanics is the most-proven and best-verified science that we have.

So, imagine my surprise when my scientist friend told me that Quantum Mechanics is false and has been falsified.

I have been a scientist all of my life; and, I have come to rely upon Quantum Mechanics as being real and true. I have stress-tested computer chips, ram chips, and motherboards to identify ones that are flawed and to identify the ones that don't meet their published specifications. I have observed and experienced first-hand what happens when quantum uncertainty leaks through into our physical reality. Consequently, you can imagine how disturbed I was when my best friend, a scientist whom I admire and respect and trust, told me that Quantum Mechanics is false and has been falsified. I had to figure out what he was talking about. Remember, I have observed that he is seldom wrong.

As I did research into the subject and as I interviewed him in greater detail over the subsequent weeks, I discovered that he is both right and wrong, depending upon the interpretation that is given to Quantum Mechanics.

In the world of science, Quantum Mechanics is typically given a materialistic, naturalistic, and physicalistic interpretation. My friend is right. The mechanical, materialistic, physically reductionistic, and naturalistic interpretations of Quantum Mechanics ARE INDEED FALSE and have indeed been falsified, because Materialism and Naturalism are false and have been falsified by scientific observations and empirical evidence. That's what my friend was talking about when he was talking about the FACT that Quantum Mechanics is false and has been falsified. ALL of the materialistic and naturalistic interpretations of Quantum Mechanics are false and have been falsified.

I realized then and there that I needed a new and better Interpretation of Quantum Mechanics, an interpretation that is actually true, matches with Reality, and matches with all of the observational evidence and empirical evidence and the Lived Experiences of the human race.

The first time I found such a thing – the **Non-Local Consciousness Interpretation of Quantum Mechanics** – was when I was reading and studying a book from Pim van Lommel entitled, *Consciousness Beyond Life: The Science of the Near-Death Experience*. Since then, I have also discovered the Quantum Consciousness Interpretation or the Cosmic Consciousness Interpretation (Idealism), which also matches relatively well with reality and the observational and experiential evidence that we have on hand as a race. Finally, a couple of years later in October 2017, I was introduced to the Orthodox Interpretation of Quantum Mechanics, which also describes the observational evidence, experiential evidence, and experimental evidence perfectly.

When it comes to Quantum Mechanics, interpretation is king!

Why?

It's because whenever Quantum Mechanics is given a materialistic and naturalistic interpretation, Quantum Mechanics so-described is actually FALSE and has been FALSIFIED. In contrast, whenever Quantum Mechanics is given some kind of Non-Local or Holistic or Integrative Interpretation, Quantum Mechanics ends up being experimentally and experientially verified, ends up matching with Reality, and therefore ends up being true.

They like to joke and say that there are as many different interpretations of Quantum Mechanics as there are Quantum Mechanists. They are right. Each scientist develops his own interpretation of Quantum Mechanics in the end.

Now think about it logically.

By design, Materialism and Naturalism are absolutely worthless whenever it comes time to explain Non-Physical or Non-Local Phenomena, including Transdimensional and Spiritual Phenomena. Consequently, a materialistic and naturalistic interpretation of Quantum Mechanics is absolutely worthless when it comes time to discuss and understand Spook Action at a Distance, Quantum Entanglement, Quantum Consciousness, Non-Local Consciousness, Cosmic Consciousness, Thoughts, Cognitions, Dreams, Visions, Quantum Complementarity, Extra-Dimensionality, Trans-Dimensionality, Shared-Death Experiences

(SDEs), Near-Death Experiences (NDEs), Out-of-Body Experiences (OBEs), Non-Local Experiences, Spirituality, and Quantum Non-Locality.

Materialism and Naturalism lack explanatory power when it comes to these types of events, observations, experiences, and phenomena.

## Quantum Non-Locality

Non-Local means "not located in our 3D physical space-time realm".

In other words, non-local means "non-physical" or "transdimensional". String theorists are talking about transdimensionality and extra-dimensionality all of the time, including dimensions that cannot be detected by our physical instruments. Quantum Non-Locality is just such a beast – an extra dimension or few.

Many scientists define Quantum Non-Locality as Quantum Entanglement, or Spooky Action at a Distance. Action at a Distance is one of the effects or results of Quantum Non-Locality, but not the only one. Consequently, after a lot of research and study, I finally realized that Non-Locality is better defined as events, phenomena, and objects which are NOT located in our physical 3D space-time realm but are in fact located in the Transdimensional, Non-Local, Timeless Realm instead.

Personality, thoughts, dreams, cognitions, perceptions, memories, non-local consciousness, and psyche are inherently Non-Local in nature and origin because these things can't be recorded by our physical instruments, because these things survive brain death and physical death, and because these things have been observed to continue functioning long after the physical brain is dead or offline. The Non-Local Realm has been observed and experienced; therefore, WE KNOW that it exists and is truly real. Remember, the best way to find and know the truth is to live it and experience it for yourself, or to choose to trust someone who has.

Non-locality is synonymous with what many people call the Spirit World or the Spirit Realm.

Classical Mechanics (classical physics) explains how physical matter works. Quantum Mechanics (spiritual mechanics) explains how spirit matter or dark matter works.

Observe, you will NEVER learn anything about Quantum Non-Locality from the Materialists, Naturalists, Darwinists, Nihilists, and Atheists because the very existence of Quantum Non-Locality or Spiritual Mechanics FALSIFIES Materialism, Naturalism, Nihilism, and even Atheism and Darwinism in the end. These people will simply deny that such things exist, leaving it up to you to figure out for yourself that these unseen realities do in fact exist.

Telepathy between psyches and teleportation of spirit matter are an integral part of the way that spirit matter or dark matter truly functions in the Quantum Realm or Spirit Realm. Notice that by definition in principle, everything in the Spirit Realm or Non-Local Realm is capable of infinite velocities, which means that psyche (non-local consciousness) and spirit matter (dark matter) are innately inherently capable of teleportation or being instantaneously and simultaneously anywhere they want to be.

In contrast, telepathy, infinite velocities, and teleportation are NOT a native inherent feature of physical matter or classical mechanics, because God has placed deliberate limitations within physical matter while He was designing it and creating it in the first place.

Remember, psyche ACTS; whereas, spirit matter or dark matter has been organized to REACT to the commands of psyche. In the Quantum Realm or Spirit Realm, we have ACTORS and we have the spirit matter or dark matter that is ACTED UPON.

As far as we know, God is the only one with the KEY or the knowledge or the power necessary to convert a particle of spirit matter into a particle of physical matter. We human beings certainly haven't figured out how to do it.

It has been computed or inferred from observation that Dark Matter comprises 23% of our current universe, which means that God only converted a small portion of that Dark Matter into physical matter while He was bringing our physical universe into existence. This is the KEY to understanding where physical matter came from in the first place – God converted a small portion of spirit matter into physical matter in order to bring our physical universe into existence.

This is the KEY to understanding what Quantum Mechanics really is and truly means. Quantum Mechanics is in fact Spiritual Mechanics – the way that spirit matter or dark matter truly works. The other KEY comes in knowing and understanding and accepting what the Quantum Law of Complementarity truly is and means.

Remember, there's no way on earth that we can use Materialism, Naturalism, Darwinism, Nihilism, or Atheism to explain Quantum Non-Locality, Psyche, the Transdimensional Realm, Non-Local Consciousness, Spirit Matter, and the Non-Local Realm. We need a new and better science to do such a thing successfully – an observational and experiential science – something like Science 2.0, which deliberately allows ALL of the evidence into evidence.

**Quantum Law of Complementarity**

Traditionally, Quantum Complementarity has been defined as Wave-Particle Duality – the idea that a Quantum Object can manifest as either a wave or a particle. Pim Van Lommel enhanced this idea quite a bit in his book, *Consciousness Beyond Life: The Science of the Near-Death Experience*.

As a result of this book, for a while, I tended to visualize the wave as the Non-Local Spiritual aspect of a Quantum Object, and the particle as the Localized Physical aspect of a Quantum Object. I now believe that it's a bit more complex than that – with a Quantum Object exhibiting particle-wave duality both in the Non-Local Realms and here in the Physical Realm too – but we will go with the phase-split definition for now because it's very instructive and foundational.

The Quantum Law of Complementarity states that a Quantum Object (a particle of matter) has two complementary phases of functionality or two complementary states of being – a non-local spiritual wave-like state and a localized physical particle-like phase – and that a Quantum Object (particle of matter) can be in only one state or phase at any given point in time.

Consequently, we observe electrons popping in and out of existence seemingly at will. When electrons "pop out of existence", they don't cease to exist. Rather, these electrons transfer temporarily to the non-local realm or the physical realm. When it comes to electrons, they are half-in and half-out of our physical reality, because they are Quantum Objects or sub-atomic particles of matter. Electrons spend half of their time in the Spirit

Realm and half of their time in our Physical Realm. We assume that the same reality applies to Quarks and Bosons as well.

So, if you were God, how would you create a particle of physical matter from a particle of spirit matter or dark matter?

Well, you would take all of these different Quantum Objects (electrons and quarks) and then create a little solar system out of them using forces and fields (gluons and bosons) to hold that solar system together. In order to convert particles of spirit matter into a particle of physical matter, you would take those pre-existing particles of spirit matter, combine the quarks together as a mini-sun in the middle, hold that sun together with gluons, put electrons in orbit of that mini-sun (nucleus of the atom), and then keep them in orbit with bosons. In order to convert particles of spirit matter into a single particle of spirit matter, you are going to have to introduce space, time, entropy (the passage of time), and energy into the system in order to hold the system together.

A single atom is the ultimate piece of engineering, design, and construction. It's a marvelous piece of technology.

The Materialists, Naturalists, Nihilists, and Atheists erroneously claim that the atom is comprised of 99.999% empty space. As usual, these people are wrong. Each one of those atoms does indeed have a "physical component" that we call quarks and electrons; but, those very same atoms have forces and fields both holding the atom together and mediating the interaction between atoms.

The gluons hold the quarks together in the nucleus of the atom – strong nuclear force. The bosons and other forces hold the electrons in orbit of the atom – weak nuclear force. Space-time is created as a result within the atom. The bosons create space-time and entropy (the passage of time) within each atom, thus giving the atom size and making it take up space. Gluons and Bosons ARE the Energy that God infuses into each physical atom in order to organize and hold together the individual pieces of spirit matter (quarks and electrons) within that atom.

Notice that there is NO entropy within the Non-Local Realm or the Spirit Realm or the Quantum Realm – everything there is immortal and eternal and never runs down, including both spirit matter and psyche (non-local consciousness). Entropy only exists within physical matter and space-time. In a very real sense, entropy represents the passage of time. If there is NO passage of time, no aging process, then there is NO entropy.

Each piece of physical matter, each atom, is a marvelous piece of engineering, construction, and design. Physical matter doesn't come together naturally – it has to be forced together through intent, purpose, and design. Naturally, spirit matter or dark matter exists in Chaos, unorganized. It is Psyche, or Non-Local Consciousness, that gives spirit matter its purpose, order, organization, and design. Without Psyche, spirit matter (quarks, electrons, and vibrating strings) is nothing but Chaos and Disorder. A particle of physical matter, an atom, is the ultimate piece of technology, the ultimate construction and design in this universe.

Then we have yet another force that's called Dark Energy or the Zero-Point Field of Light which mediates ALL of the interaction between individual physical atoms. The Zero-Point Field of Light has been variously called the Mind of God, the Holy Spirit, the Light of Christ, the Quantum Sea of Light, Bosons, and Dark Energy. We KNOW that such a thing exists, because the Biblical God Jesus Christ talks about it from time to time. It's this Dark Energy or Light of Christ that decides how the individual physical atoms are going to interact with each other. It's also this Dark Energy (or Stretching Force in our Universe) which introduces space-time into our universe as a whole. It has been estimated that 72%

of our universe is comprised of Dark Energy or the Light of Christ, which is in all things holding everything in this Physical Universe together, and which permeates the whole of space.

In other words, there is NO such thing as Creation Ex Nihilo, and there is NO such thing as empty space.

Remember, quarks and electrons ARE the fundamental units of spirit matter or dark matter – they pop back and forth between the spirit realm and the physical realm as we know it. Some string theorists have called quarks and electrons "vibrating strings". They vibrate at a God-given frequency in and out of physical existence as they pop back and forth between the spirit realm and the physical realm.

Remember, gluons and bosons are the fundamental forces that bring order and organization to physical matter and brings physical matter into existence in the first place. Gluons and bosons are command and control. Some string theorists have called gluons and bosons "vibrating thoughts" or the Mind of God. Gluons and bosons and dark energy are the tools – the force fields – which God uses to bring physical atoms into existence, to hold them in existence, and to govern the interaction between those physical atoms. Gluons, bosons, and dark energy are THOUGHTS – the Mind of God or God's Psyche in action. The gluons, bosons, and dark energy are the MUSIC which God is playing on all those different vibrating strings (quarks and electrons).

A reading of *Consciousness Beyond Life* from Pim van Lommel got me started down this path of discovery. It has taken me a couple of years to get where I am now, but it has all been worth it. I feel like I have a pretty good handle on things, now, which I didn't have when I was a Materialist, Nihilist, and Atheist back in 2012. I composed all of this from memory in October 2017 and didn't have to look up anything at all in order to do so. I feel like I have it down now.

Obviously, you are going to have to decide for yourself whether I know what I'm talking about or not. Under Science 2.0, my only obligation is to present to you the experimental and observational evidence, and I'm free to tell you what that evidence means to me personally. But in the end, it is left up to you to provide your own interpretation of the evidence and decide for yourself what to make of that evidence. I have NO right to dogmatically and irrevocably interpret that evidence for you. That's your responsibility.

### The Non-Local Consciousness Interpretation of Quantum Mechanics

Van Lommel, P. (2010). *Consciousness Beyond Life: The Science of the Near-Death Experience*. New York: HarperCollins.

In this book, Pim van Lommel develops and presents what I call the **Non-Local Consciousness Interpretation of Quantum Mechanics**. Pim van Lommel defines and explains Non-Local Consciousness.

This is my all-time most favorite science book. After reading this book, I felt like for the first time in my life that I actually understood Quantum Mechanics, the Science of the Near-Death Experience. This book completely changed my life, my worldview, and the way I look at science and reality. The KEY to it all is an understanding and acceptance of Quantum Non-Locality or Quantum Non-Physicality – what Einstein called "spooky action at a distance".

Quantum Mechanics is the science behind Near-Death Experiences. This book describes Quantum Mechanics from a non-local perspective or a spiritual perspective. Quantum Mechanics is Spiritual Mechanics or the way that spirit matter really works. In contrast, Classical Physics is the way that physical matter works, including physical matter's limitations. Spirit matter and spirit energy don't have the same limitations that physical matter has, because non-local matter or spirit matter or dark matter is transdimensional and non-physical in nature and origin.

Tachyons are spirit matter. Psyche, Non-Local Consciousness, Spirit Matter, and Spirit Energy in the non-local realm or transdimensional realm function natively at velocities faster than the speed of light. Instantaneous teleportation to any destination in the universe is a native function of spirit matter or tachyons.

Similarly, according to the theory of relativity, a photon traveling at the speed of light does not experience the passage of time from its perspective and is not subject to entropy. It doesn't age because for it there is no passage of time. From its perspective, a photon teleports to its destination instantaneously; and, it experiences nothing, no passage of time during its journey, because it's traveling at the speed of light from our perspective.

God put limitations into physical matter, in order to slow everything down so that we can live it, experience it, remember it, and learn from it. Our physical life is God's school for us, His spirit children. God introduced entropy and time into physical matter so that our physical lives would seem more real to us, and so that we could actually measure and experience the passage of time, and so that we could actually learn from it all.

Science 2.0 includes evidence from the Non-Local Realm or Non-Physical Realm, as well as evidence from the Local Realm or Physical Realm. Remember, Science 2.0 includes all of the evidence and allows it all into evidence.

I have written whole books based upon what I learned from *Consciousness Beyond Life*, so I won't do so here, because what I have written here is long enough already. Instead, I will now highlight aspects of Quantum Mechanics which I have gleaned from other books, during the past couple of years.

## Tachyons

Kaku, M. (2008). *Physics of the Impossible: A Scientific Exploration into the World of Phasers, Force Fields, Teleportation, and Time Travel*. New York: Anchor Books.

Even though Michio Kaku is a materialist, naturalist, Darwinist, big banger, evolutionist, and agnostic skeptic supporting the party-line, he is not mean-spirited nor caustic like the New Atheists are. Even though Michio Kaku still hasn't gotten the message that Darwinism and Materialism are DEAD, he ironically supports and promotes theories and ideas that have NO physical support nor any materialistic support to sustain them – intangible and unseen things like String Theory, M-Theory, Extra-Dimensionality, and Cosmic Music which are typically labeled as "not science" or "unscientific" by the Materialists and Naturalists and Evolutionists at-large.

Michio Kaku is my favorite materialist and evolutionist to study and read, because I'm fascinated by string theory, m-theory, cosmic consciousness, and extra-dimensionality.

In his book, *Consciousness Beyond Life: The Science of the Near-Death Experience*, Pim Van Lommel convinced me that in Quantum Non-Locality or the Non-Local Spirit Realm, the Quantum Objects that reside there can theoretically exist at infinite velocities much faster than the speed of light.

Based upon that evidence, I came to the conclusion that spirit matter is the same thing as physical matter, EXCEPT for the fact that spirit matter exists at velocities greater than the speed of light and physical matter exists at velocities less than the speed of light. Then I eventually concluded that immaterial psyche exists at an infinite velocity or an infinite frequency capable of being instantaneously and simultaneously everywhere in its native original format.

God marries our Psyches or Intelligences to spirit matter and physical matter in order to slow them down, so that we Psyches can have Lived Experiences, learn from our experiences, and remember having done so.

I eventually realized that matter existing at different velocities or existing in different phases can actually occupy the same space because they are out-of-phase with each other. Therefore, Psyche, spirit matter, and physical matter can occupy the same space because they have different phases of existence or exist at different velocities in different dimensions.

I eventually was taught and realized that a Quantum Object is a Particle of Matter, and that matter has two main states of existence according to the Quantum Law of Complementarity – a spiritual state and a physical state. Physical matter exists at velocities less than the speed of light, because of all the MASS or resistance to acceleration that physical matter has; whereas, spirit matter with nearly no mass exists at velocities greater than the speed of light.

In time, I came to realize that Dark Matter is in fact Spirit Matter.

For a very long time, I thought I was the ONLY one on the planet who realized that Spirit Matter exists at velocities faster than the speed of light and is therefore undetectable by physical instruments directly.

Last night 29JUL2017, I was reading in my most favorite introduction to Popular Physics or Speculative Physics – a book entitled *Physics of the Impossible* by Michio Kaku – the following description for Tachyons.

> **Tachyons live in a strange world where everything travels faster than light. As tachyons lose energy [mass], they travel faster, which violates common sense. In fact, if they lose all energy [or mass], they travel at infinite velocity. As tachyons gain energy [or mass], however, they slow down until they reach the speed of light. (p. 280).**

That's the exact SAME definition for Spirit Matter that I have been giving in all of my books for the past year or so.

Tachyons ARE Spirit Matter!

Of course, Michio Kaku doesn't realize that Tachyons are Spirit Matter, because as a Materialist and Physicalist he isn't looking for Spirit Matter. But, I instantly recognized the definition for Spirit Matter that I have been providing for the past year in my books within the definition that Michio Kaku provided for Tachyons.

Cool huh?

Well, at least I thought it was cool.

Tachyons live in a strange world that we have been calling the Non-Local Realm or the Spirit Realm. As tachyons or spirit matter lose energy, their mass or resistance to acceleration decrease, which makes perfect sense to me. In fact, if the tachyons or spirit matter were to lose ALL of their mass, they would become immaterial Psyche and exist at an infinite velocity being instantaneously and simultaneously everywhere in their native original format.

Remember, ZERO MASS = Infinite Velocity. MASS or Inertia is resistance to acceleration. No resistance to acceleration results in an infinite velocity. It's always made perfect sense to me, ever since I learned about it.

Remember, Matter is a Quantum Object, and a Quantum Object has two mutually exclusive states of existence – a spiritual state faster than the speed of light and a physical state slower than the speed of light.

As God inputs more mass or energy into a Tachyon or particle of Spirit Matter or Dark Matter, its velocity slows down until it reaches the speed of light. Then, by consciously inserting additional energy or mass into that Tachyon or particle of Spirit Matter, God slows that particle of matter to sub-light velocities thereby converting that particle of Spirit Matter into a particle of Physical Matter.

There in Michio Kaku's paragraph, I found independent confirmation of what I have been trying to teach my readers for the past year of my life.

Non-Local Matter or Spirit Matter exists in the Non-Local Realm at velocities greater than the speed of light. Spirit Matter has a little mass and consists of Tachyons that exist at velocities greater than the speed of light. Should a Quantum Object ever shed ALL of its mass, then it ceases to be matter and becomes Psyche or Non-Local Consciousness instead. Psyche, or Non-Local Consciousness, IS the fundamental unit of reality. This is a Psyche Ontology.

Psyche ACTS; whereas, matter of any kind REACTS to psyche's command. Psyche is the ACTOR, because in its native original format it exists at an infinite velocity which means that it is instantaneously and simultaneously everywhere in its native original format. God marries our psyches or intelligences to spirit matter and physical matter in order to slow us down so that we are subject to the passage of TIME and can therefore have experiences and remember having had them. Each time that God introduces limitations into our lives or our psyches, God also introduces new and interesting opportunities to learn and grow.

Tachyons ARE Spirit Matter, because they exist at velocities greater than the speed of light. I also have reasons for believing that Dark Matter is Spirit Matter. From my perspective, Science itself has PROVEN the existence of Spirit Matter, in the form of Tachyons or Dark Matter.

I found this Science extremely useful and interesting.

## Materialism and Naturalism Cause the Mind-Brain Problem

Remember, materialism and naturalism cause the Mind-Brain Problem, which means that getting rid of materialism and naturalism solves the Mind-Brain Problem.

Materialism and Naturalism create a whole host of unsolvable problems within the science of Psychology – things like the Mind-Brain Problem, the Binding Problem, the Nature vs. Nurture Problem, the Consciousness and Sentience Problem, the Survival of Death

Problem, and the Free Will vs. Determinism Problem. Consequently, getting rid of materialism, naturalism, radical reductionism, and determinism automatically solves these unsolvable problems.

The Quantum Consciousness Interpretation of Quantum Mechanics and the Orthodox Interpretation of Quantum Mechanics SOLVES the Mind-Brain Problem as well as all these other unsolvable problems which I have listed above. These particular interpretations of quantum mechanics rely upon observational evidence and experiential evidence in order to make their case. The same can be said of Science 2.0, which means that Science 2.0 is based upon the Quantum Consciousness and Orthodox Interpretations of Quantum Mechanics, which successfully falsify Materialism and Naturalism. Science 2.0 is based upon the Lived Experiences of the human race as a whole, and not the philosophical musings of the materialists and naturalists.

**The Quantum Zeno Effect**

Most scientists don't realize it, and few truly accept it; but, the repeated verification of Quantum Mechanics along with the scientific verification of the Quantum Zeno Effect FALSIFY Materialism, Naturalism, and Nihilism. When those go down, Darwinism and Atheism go down with them, without a leg to stand on. Rather than trying to explain away the scientific evidence which so abundantly verifies Quantum Mechanics – as the Materialists and Naturalists attempt to do – I have tried instead to determine for myself what Quantum Mechanics means to us and is trying to tell us and teach us. I'm a scientist, after all – it's what I do.

Remember, a materialistic and naturalistic interpretation of Quantum Mechanics will NEVER explain how psyche interacts with physical matter, because Materialism and Naturalism have been falsified by scientific observations and because Materialism and Naturalism lack explanatory power especially when it comes to the Non-Local Realm or the Spirit World. Materialism, Naturalism, Nihilism, and Atheism were intentionally designed to be worthless when it comes to the Transdimensional Realm, the Non-Physical Realities, Psyche, Non-Local Consciousness, and God. That's just the reality of the situation. By design, Materialism and Naturalism are worthless when it comes time to explain Psyche or Non-Local Consciousness.

When it comes to Quantum Mechanics, it's extremely important to get the Right Interpretation – one that actually explains how Psyche interacts with and controls Physical Matter. The Non-Local Consciousness Interpretation of Quantum Mechanics from Pim van Lommel, the Quantum Consciousness or Cosmic Consciousness Interpretation from Eugene Wigner, and the Orthodox Interpretation from John von Neumann as explained by Henry P. Stapp qualify as the right types of interpretation which can adequately and efficiently explain how Non-Local Consciousness interacts with and controls Physical Matter.

I have observed that the Quantum Zeno Effect has been given a bunch of different interpretations, including materialistic and naturalistic ones which are absolutely worthless.

As mentioned above, the Quantum Zeno Effect when properly understood FALSIFIES Materialism and Naturalism; and, when those go down, Darwinism, Nihilism, and even Atheism go down with them.

**Quantum Zeno Effect verified – atoms won't move while you watch them.**

https://phys.org/news/2015-10-zeno-effect-verifiedatoms-wont.html

http://mypsyche.us/wp-content/uploads/2017/10/Zeno-effect-verified—atoms-wont-move-while-you-watch.pdf

One verified experimental observation from the Quantum Zeno Effect is how atoms freeze into position whenever you, your psyche or non-local consciousness, observes them or looks at them or uses them. The Quantum Zeno Effect only makes sense from a non-local perspective or a spiritual perspective, because there is telepathic "action at a distance" taking place between your psyche and the psyche within nature or physical matter according to the proven Quantum Zeno Effect.

Verification of the Quantum Zeno Effect provides PROOF of **action at a distance** or telepathy between you and the atoms, because there is NO physical contact between you and the atoms being observed, yet the atoms freeze or stop moving when you look at them.

Furthermore, experimental verification of the Quantum Zeno Effect PROVES that telepathy is a normal, typical, ubiquitous function in the Non-Local Realm or the Transdimensional Spirit Realm. According to the Quantum Zeno Effect, your psyche and the psyche within a physical atom are literally communicating with each other telepathically. This IS Action at a Distance or Telepathy, which is being witnessed and proven by the Quantum Zeno Effect. There is NO physical contact between your physical body and that distant atom which you are observing while they are measuring the Quantum Zeno Effect. The whole effect is the result of psi or telepathy. It's truly mind-over-matter!

The Quantum Zeno Effect also PROVES that John von Neumann's **Orthodox Interpretation of Quantum Mechanics** is a correct and true interpretation of Quantum Mechanics. More on this in a moment.

Physical matter limits and constrains the effects of telepathy; but, there is NO physical matter in the Non-Local Realm, therefore there are NO limitations on telepathy in the Non-Local Realm. We only have limitations on telepathy when it comes to the Physical Realm.

The Quantum Zeno Effect literally PROVES the existence of psyche or non-local consciousness, because there is no other logical explanation for "action at a distance" or telepathy. A physical explanation cannot be provided for Action at a Distance, because it's purely a telepathic phenomenon. Those atoms literally KNOW when you are looking at them, and when you are not. The atoms can sense your psyche or feel your psyche and they KNOW precisely what your psyche is concentrating on at any given moment in time. In a sense, the atoms can read your mind.

While studying the Quantum Zeno Effect, I have of course come across other uses and other interpretations that are actually quite powerful. When married with the Orthodox Interpretation of Quantum Mechanics, the Quantum Zeno Effect actually explains how habits and addictions are formed.

Under Process 1 of the Orthodox Interpretation, Psyche or Non-Local Consciousness asks nature a question or makes some kind of demand from nature. Then under Process 3 of the Orthodox Interpretation, nature chooses whether to comply with Psyche's demands or to reject them. Under Process 3, nature chooses to respond with either a Yes or a No. Both Process 1 and Process 3 represent choice and free will, which are the result of some kind of non-local consciousness or psyche.

Trapped in the middle of this ongoing debate between your psyche and nature's psyche is Process 2, which represents mindless physical matter and the cloud of different possibilities that exist which physical matter can be forced to pursue or comply with. Process 2 is the collapse of the wave function into a single reality or event or phenomenon.

When nature (Process 3) makes a choice, the wave function (Process 2) collapses into existence or becomes what the Materialists and Naturalists typically call REAL.

Now, here's where the Quantum Zeno Effect comes into play.

As the demands from your Psyche (Process 1) become constant and approach infinity, due to the Quantum Zeno Effect and the constant demands from your Psyche, nature (Process 3) literally gets locked into producing a Yes Response every time and eventually loses the ability to produce a No Response, even if nature knows that what you are asking for isn't good for you. Thus, a habit or an addiction is formed, because nature (Process 3) gets locked or frozen into producing a Yes Response automatically every time there is a demand from Psyche (Process 1).

Then it either requires a miraculous intervention from the Atonement of Christ or a firm and determined NO COMMAND from your Psyche, a determination to go cold turkey, in order to force and eventually retrain nature to be able to produce a No Response instead.

Physical matter (Process 2) has NO free will and is simply forced to respond or programmed to respond to whatever nature (Process 3) commands it to do. Thus, ALL of the free will choices and negotiations take place between Your Psyche (Process 1) and Nature's Psyche (Process 3), with Process 2 and Physical Matter trapped in the middle having to respond with a wave function collapse after each of nature's commands.

The full effects of the Quantum Zeno Effect can only be fully explained, understood, and defined while in the presence of an Interpretation of Quantum Mechanics that actually takes non-local consciousness or psyche into account. In contrast, materialistic and naturalistic and atheistic interpretations of Quantum Mechanics literally make no sense whatsoever. They lack explanatory power when it comes to non-local consciousness and non-locality.

Materialism and Naturalism cannot explain psyche, non-local consciousness, non-locality, choice, free will, spirit matter, the spirit realm, the transdimensional realm, psi, telepathy, action at a distance, quantum entanglement, quantum complementarity, nature's roll in choice, nor the Quantum Zeno Effect. Materialism and Naturalism are absolutely worthless whenever it comes to these types of Non-Physical things, which have been observed and experienced by the human race as a whole.

When it comes to Quantum Mechanics, interpretation is king. Make it a good one!

**No Entropy in the Non-Local Realms**

Spirit matter and Psyche are eternal, without beginning or end. There's no aging process in the Non-Local Realms, which means that there's no entropy in the Spirit World. Entropy is a function of the passage of time; and, time works differently in the Non-Local Realms.

**Telepathy**

According to the observational evidence and experiential evidence from Near-Death Experiences (NDEs) and Out-of-Body Experiences (OBEs), telepathy is the normal mode of communication between psyches in the Non-Local Realms or Spirit Realms. There are NO physical limitations in the Non-Local Realms; and consequently, there are no physical limitations associated with Quantum Mechanics. Remember that, because it's important to know.

## Teleportation and Quantum Tunneling

Most scientists don't realize it, and few truly accept it; but, the repeated verification of Quantum Mechanics along with the scientific verification of the Quantum Zeno Effect FALSIFY Materialism, Naturalism, and Nihilism. When those go down, Darwinism and Atheism go down with them, without a leg to stand on. Rather than trying to explain away the scientific evidence which so abundantly verifies Quantum Mechanics – as the Materialists and Naturalists attempt to do – I have tried instead to determine for myself what Quantum Mechanics means to us and is trying to tell us and teach us. I'm a scientist, after all – it's what I do.

According to the observational evidence and experiential evidence from Near-Death Experiences (NDEs) and Out-of-Body Experiences (OBEs), teleportation is a common mode of travel in the Non-Local Realms, although you can also walk, levitate, and fly if you want to do so. There are NO physical limitations in the Non-Local Realms; and consequently, there are no physical limitations associated with Quantum Mechanics. Remember that, because it's important to know.

Quantum Tunneling is Teleportation; and, physical atoms have been caught in the act of tunneling or teleporting from one location to another. In contrast, according to the Quantum Zeno Effect, constant observation from psyche or non-local consciousness freezes atoms and quantum objects into place while they are needed and being observed.

**Quantum Zeno Effect verified – atoms won't move while you watch them.**

https://phys.org/news/2015-10-zeno-effect-verifiedatoms-wont.html

http://mypsyche.us/wp-content/uploads/2017/10/Zeno-effect-verified—atoms-wont-move-while-you-watch.pdf

**At such low temperatures, the atoms can "tunnel" from place to place in the lattice. The researchers demonstrated that they were able to suppress quantum tunneling merely by observing the atoms. This [is the] so-called "Quantum Zeno Effect".**

Quantum Tunneling is Teleportation. Physical atoms can teleport. Observation by non-local consciousness prevents physical atoms from tunneling or teleporting.

It has been hypothesized that the presence of your psyche or non-local consciousness is what keeps the physical matter in your physical body from disintegrating and tunneling away on you. Your psyche and spirit body are tied to your physical body by a silver cord or lifeline, which has been observed by Out-of-Body Travelers. If you sever the silver cord, the physical body dies.

Physical matter introduces a whole host of limitations and locks to spirit matter. Whenever God converts a particle of spirit matter into a particle of physical matter, a bunch of physical limitations are introduced such as entropy, locality, space-time including the passage of time, sub-light-speed velocities, and a lack of telepathy and teleportation. Physical matter greatly limits teleportation but doesn't make it impossible.

God and the Angels hold the KEYS for teleporting physical matter. I KNOW that God can teleport physical matter at will, because God teleported me and my car to safety when I was in trouble. I was going 40 mph when another car pulled out in front of me. There was nowhere for me to go. Time stopped; and then instantly, I and my car were at a completely different location in the middle of a lawn, with the car completely stopped and the engine off. God did for me what I couldn't do for myself. That was an interesting experience. God can teleport physical matter anywhere in this physical universe instantaneously, at will, simply by stopping the passage of time for that particular physical object. That's what I have observed and experienced. It was nothing that I did. It was something that God had to do for me, or it wouldn't have happened.

Remember, Teleportation is Quantum Tunneling. Physical atoms have been "observed" or caught in the act of tunneling from one location to another, while they weren't being observed.

**Interpretations of Quantum Mechanics**

https://en.wikipedia.org/wiki/Interpretations_of_quantum_mechanics

Observe, these people focus primarily on the materialistic and naturalistic and atheistic interpretations of quantum mechanics, practically ignoring and discounting everything else. The whole purpose of Materialism and Naturalism is to explain Psyche and Non-Local Consciousness out of existence. When it comes to Psyche and Non-Local Consciousness, Materialism, Naturalism, Nihilism, and Atheism lack explanatory power by design.

In contrast, when it comes to explanatory power, the Orthodox Interpretation of Quantum Mechanics is preeminent and monumental. Remember, the Orthodox Interpretation falsifies Materialism, Naturalism, and Nihilism as well as anything derived therefrom such as Darwinism and Atheism.

The whole purpose of Science 2.0 is to find different ways to account for and explain the observational evidence, the eye-witness evidence, and the experiential evidence provided by the Lived Experiences of the human race. The Orthodox Interpretation of Quantum Mechanics does just that. I was impressed with the Orthodox Interpretation. I hope you will be too. The Orthodox Interpretation is a new toy to play with – one that the Materialists and Naturalists had successfully hidden from me for over 55 years of my life; and, I had considered myself a scientist all of my life. As a scientist, I should have been taught about the Orthodox Interpretation right up front, but I wasn't.

Since Science 2.0 is based upon the Lived Experiences of the human race rather than the philosophical guesswork of scientific materialism, Science 2.0 needs an Interpretation of Quantum Mechanics that matches with Reality and the Lived Experiences of the human race. The Orthodox Interpretation of Quantum Mechanics developed by John von Neumann provides just such an interpretation; and, so does the Cosmic Consciousness Interpretation developed by Eugene Wigner, and the Non-Local Consciousness Interpretation developed by Pim van Lommel. These interpretations of Quantum Mechanics have the most observational and experimental evidence supporting them; whereas, the materialistic and naturalistic interpretations of Quantum Mechanics have no observational evidence supporting them, and by definition in principle never will.

Remember, Science 2.0 is based upon the lived aspects, or the experienced aspects, or the observed aspects of Quantum Mechanics, and not scientific naturalism which states as its primary assumption that these things do not exist. Remember, the existence and verification of the Quantum Zeno Effect FALSIFIES Materialism and Naturalism. The preponderance of the evidence simply tells us that Materialism and Naturalism are false. Materialism and Naturalism and Nihilism and Atheism fall and fail in light of the evidence, because these false and falsified religions can have no evidence of their own to sustain them.

Now, that's science in action for real!

Remember, Science IS Observation and Knowledge – not the philosophical speculation, wishful thinking, and scientific inferences which we get from the Materialists, Naturalists, and Nihilists.

## David Bohm's Interpretation of Quantum Mechanics

David Bohm did a lot of interpretation of Quantum Mechanics, including his Pilot Wave Theory and the Implicate/Explicate Orders.

I'm sorry, but most of it didn't make any logical sense to me. I couldn't picture it or visualize it. Furthermore, the Pilot-Wave Interpretation relies upon an infinite regression to solve it, which is another strike against it.

Some people have suggested that his interpretations were materialistic and naturalistic. I don't know if that is so, or not, because I didn't feel like I understood them.

Selbie, J. (2018). *The Physics of God: Unifying Quantum Physics, Consciousness, M-Theory, Heaven, Neuroscience, and Transcendence*. Wayne, NJ: Career Press.

Joseph Selbie claims to be able to understand David Bohm's interpretations of Quantum Mechanics, and Selbie seems to believe that Bohm's interpretations support M-Theory, Heaven, and Transcendence.

From what I have read so far, it seems to me that Bohm's holographic principle does a better job supporting M-Theory, Heaven, Consciousness, and Transcendence than does his interpretations of Quantum Mechanics. But, maybe as I read this book from Selbie, I will change my mind.

Meanwhile, I haven't gotten much help from David Bohm when it came time to interpret and understand Quantum Mechanics – yet.

## Three Main Interpretations of Quantum Mechanics

Kaku, M. (2014). *The Future of the Mind: The Scientific Quest to Understand, Enhance, and Empower the Mind*. New York: Doubleday.

I like all of Michio Kaku's books, because they make me stop and think.

Science 2.0 is about examining the evidence – all of the evidence. Nothing is excluded!

In his book, *The Future of the Mind*, Michio Kaku attempts to include all of the different opposing points of view regarding mind, consciousness, neuroscience, and the brain. He does try to explain some phenomena away, such as Near-Death Experiences

(NDEs) and Out-of-Body Experiences (OBEs), as hallucinations caused by an oxygen-starved brain; but, for the most part he seems to present the evidence and then let you make up your own mind which interpretation or point of view you personally want to believe.

Presenting the evidence and then letting the reader make up his own mind as to how to interpret the evidence is in harmony with the spirit and purpose of Science 2.0.

So far, my most favorite part of the book is the Appendix, which focuses on Quantum Consciousness and the three main interpretations of Quantum Mechanics which came as a result of the Schrödinger's Cat Paradox.

> **The Schrödinger's Cat paradox cuts to the very foundation of quantum mechanics, a field that makes lasers, MRI scans, radio and TV, modern electronics, the GPS, and telecommunications possible, upon which the world economy depends. Many of quantum theory's predictions have been tested to an accuracy of one part in one hundred billion. I have spent my entire professional career working on the quantum theory. Yet I realize that it has feet of clay. It's an unsettling feeling knowing my life's work is based on a theory whose very foundation is based on a paradox.**
>
> Kaku, Michio. *The Future of the Mind: The Scientific Quest to Understand, Enhance, and Empower the Mind* (Kindle Locations 5761-5766). Knopf Doubleday Publishing Group. Kindle Edition.

To solve this paradox, Kaku starts with the predominant model, the Copenhagen Interpretation, which is based upon solipsism, which claims that objects cannot exist unless there is someone there to observe them into existence. In other words, the moon doesn't exist unless someone is actually looking at it – observing it.

> **The first is the original Copenhagen interpretation proposed by Bohr and Heisenberg, the one that is quoted in textbooks around the world.**
>
> **This interpretation, no matter how strange, has been taught for eighty years by quantum physicists.**
>
> Kaku, Michio. *The Future of the Mind: The Scientific Quest to Understand, Enhance, and Empower the Mind* (Kindle Locations 5801, 5826). Knopf Doubleday Publishing Group. Kindle Edition.

Einstein hated this one, and so do I.

Why?

It's because there is NO observational evidence to support the Copenhagen Interpretation and can never be any observational evidence to support it.

Science 2.0 is based exclusively upon observation and lived experiences. There's absolutely NO way to know whether something exists or doesn't exist while nobody is looking at it or observing it. In other words, there's no way to verify or falsify the Copenhagen Interpretation. It simply has to be taken on blind faith as being true, or false. That's philosophy and metaphysics, not science. Science is observation.

Remember, there's no way to observe whether something exists or doesn't exist while nobody is observing it. By definition in principle, there's no way to verify or falsify solipsism through observational evidence or experiential evidence, which according to

Science 2.0 makes solipsism and the Copenhagen interpretation unscientific, because Science is observation.

> So how do you resolve this paradox? There are at least three ways (and hundreds of variations on these three). (Kindle Locations 5800-5801).

> There are two alternate interpretations of the cat paradox, which take us to the strangest realms in all science: the realm of God and multiple universes. (Kindle Locations 5834-5835).

Michio Kaku ends with the many-worlds interpretation, which seems to be his preferred interpretation because he has written many books and chapters on Parallel Worlds.

> The third way to resolve the paradox is the Everett, or many-worlds, interpretation, which was proposed in 1957 by Hugh Everett. It is the strangest theory of all. It says that the universe is constantly splitting apart into a multiverse of universes. In one universe, we have a dead cat. In another universe, we have a live cat. This approach can be summarized as follows: wave functions never collapse, they just split.

> There are profound consequences to this third approach. It means that all possible universes might exist, even ones that are bizarre and seemingly impossible.

> Kaku, Michio. *The Future of the Mind: The Scientific Quest to Understand, Enhance, and Empower the Mind* (Kindle Locations 5857-5863). Knopf Doubleday Publishing Group. Kindle Edition.

This Many-Worlds Interpretation makes for the BEST science fiction, and that's precisely what it is – science fiction. The primary flaw of the Many-Worlds Interpretation, the Many-Minds Interpretation, or the Multi-Me Interpretation is that NOBODY has ever seen it nor experienced it, which means that it isn't real.

Science 2.0 is based upon observational evidence and the lived experiences of the human race – not science fiction. Although I love science fiction as much as anyone, it's not real!

Furthermore, the Many-Worlds Interpretation is costly, which means that it breaks parsimony – a fact which effectively falsifies the Many-Worlds or Parallel Worlds Theory.

> What sense are we to make of this notion of a half-dead, half-alive cat existing in potentia? An answer that sounds like science fiction has come from the physicists Hugh Everett and John Wheeler. According to Everett and Wheeler, both possibilities, live cat and dead cat, occur — but in different realities, or parallel universes. For every live cat we find in the cage, prototypes of us in a parallel universe open a prototype cage only to discover a prototype cat that is dead. An observation of the cat's dichotomous state forces the universe itself to split into parallel branches. This is an intriguing idea, and some science fiction writers (notably Philip K. Dick) make good use of it. Unfortunately, this is also a costly idea. It would double the amount of matter and energy each time an observation forces the universe to bifurcate. It offends our taste for parsimony, which may be a prejudice but is nonetheless a cornerstone of scientific reasoning. Furthermore, since the parallel universes do not interact, this interpretation is difficult to put to experimental test and therefore not useful from a

scientific point of view. (Fiction is more tractable. In Philip Dick's *The Man in the High Castle*, the parallel universes do interact. How else would there be a story?)

> Goswami, Amit. *The Self-Aware Universe: How Consciousness Creates the Material World* (p. 81). Penguin Publishing Group. Kindle Edition.

It sounds like science fiction because it is science fiction.

Yes indeed, the Many-Worlds Interpretation or the Parallel Worlds Interpretation of Quantum Mechanics does indeed make the claim that the amount of matter and energy in our universe doubles each time a choice, decision, or observation is made. Imagine it! It's mind-boggling. It's much easier to believe in God than it is to believe that our physical universe splits in two and doubles in size each time one of us makes a decision, choice, or observation.

I have concluded that the Many-Worlds Interpretation is nothing but science fiction and doesn't exist for real, because it has never been experienced nor observed.

Science is Observation and Knowledge. Science 2.0 is based upon observational evidence; and, the Many-Worlds Interpretation doesn't qualify as science based upon that particular definition for science, because it has never been experienced nor observed.

How can it be Science or Knowledge or Truth if it has never been experienced nor observed?

Sandwiched in the middle Kaku's list is the Quantum Interpretation which has ALL of the observational evidence supporting it – the Quantum Consciousness Interpretation!

> **In 1967, the second resolution to the cat problem was formulated by Nobel laureate Eugene Wigner, whose work was pivotal in laying the foundation of quantum mechanics and also building the atomic bomb. He said that only a conscious person can make an observation that collapses the wave function. But who is to say that this person exists? You cannot separate the observer from the observed, so maybe this person is also dead and alive. In other words, there has to be a new wave function that includes both the cat and the observer. To make sure that the observer is alive, you need a second observer to watch the first observer. This second observer is called "Wigner's friend," and is necessary to watch the first observer so that all waves collapse. But how do we know that the second observer is alive? The second observer has to be included in a still-larger wave function to make sure he is alive, but this can be continued indefinitely. Since you need an infinite number of "friends" to collapse the previous wave function to make sure they are alive, you need some form of "cosmic consciousness," or God.**
>
> **Wigner concluded: "It was not possible to formulate the laws (of quantum theory) in a fully consistent way without reference to consciousness." Toward the end of his life, he even became interested in the Vedanta philosophy of Hinduism.**
>
> **In this approach, God or some eternal consciousness watches over all of us, collapsing our wave functions so that we can say we are alive. This interpretation yields the same physical results as the Copenhagen interpretation, so this theory cannot be disproven. But the implication is that consciousness is the fundamental entity in the universe, more**

**fundamental than atoms. The material world may come and go, but consciousness remains as the defining element, which means that consciousness, in some sense, creates reality.**

Kaku, Michio. *The Future of the Mind: The Scientific Quest to Understand, Enhance, and Empower the Mind* (Kindle Locations 5835-5849). Knopf Doubleday Publishing Group. Kindle Edition.

Of the three interpretations that Michio Kaku offers us, the Quantum Consciousness Interpretation (Cosmic Consciousness Interpretation) is the one upon which Science 2.0 is based, because the Quantum Consciousness Interpretation has ALL of the observational evidence supporting it.

God's Psyche or Non-Local Consciousness may be small in size, but it is infinite in scope, because in the Non-Local Realm there are no physical limitations; and therefore, God's Psyche can be instantaneously and simultaneously everywhere all at once because there are no speed limits and no time limits and no distance limitations in the Non-Local Realm. Psyche and spirit matter can travel at an infinite velocity in the Spirit World or Transdimensional Realm.

Quantum Consciousness Interpretation: I find it interesting that Kaku didn't even bother to give it a name, but at least Kaku presented it accurately what little of it he did present. Michio Kaku is a Darwinist, Materialist, Nihilist, and Naturalist by choice and by training.

The very existence of the physical atoms we see around us is based upon God's ability to convert spirit matter into physical matter through His Word of Command. Only God's Psyche or Non-Local Consciousness has the ability to convert spirit matter or dark matter into physical matter. We humans don't have the KEY to this technology; but, God does, or we wouldn't exist as physical beings and this physical universe wouldn't exist either. Remember, physical matter didn't exist until after God called it into existence by converting some of the pre-existing spirit matter or dark matter into physical matter.

Since Science 2.0 is based upon the Lived Experiences of the human race rather than the philosophical guesswork of scientific materialism, Science 2.0 needs an Interpretation of Quantum Mechanics that matches with Reality and the Lived Experiences of the human race. The Quantum Consciousness Interpretation of Quantum Mechanics provides just such an interpretation.

## Observational Support for Quantum Consciousness or Non-Local Consciousness

According to a Gallup Poll, there have been over 13 million Near-Death Experiences (NDEs) in the US alone – case studies and eye-witness reports for thousands of them can now be found online. Any one of them will supply observational evidence and experiential evidence of the non-local, non-physical, trans-dimensional spirit realm.

https://www.youtube.com/results?search_query=NDE

Science 2.0 is based upon observational evidence and the lived experiences of the human race, including our non-local experiences or out-of-body experiences (OBEs).

Some of the NDEs and OBEs also provide observational evidence and experiential evidence of heaven, hell, demons, and the Biblical God Jesus Christ. Through NDEs, the Biblical God has met His burden of proof by a preponderance of the evidence. A preponderance of the observational evidence or eye-witness evidence is the only way that the physically undetectable, the unseen and unseeable, the historical sciences, the non-local sciences, and the origin sciences can be proven to exist.

I submit the whole corpus, all 13 million NDEs or more, into evidence. Examine the evidence and decide for yourself what it means and if it is significant or not. Start with the following and get your feet wet. That's what I did.

A correct understanding and acceptance of Quantum Non-Locality, as well as the right interpretation of Quantum Mechanics which takes Non-Local Consciousness or Psyche into account, is the KEY to understanding the Observational Evidence, Experiential Evidence, and Scientific Evidence associated with and provided by NDEs and OBEs.

Lived experiences and direct observations such as the following had a powerful influence on my philosophy of life, worldview, and scientific theories. I simply prefer to follow the evidence, rather than making *scientific inferences* and *jumping to conclusions*. Watch some of these videos and decide for yourself what kind of evidence, or observation, or proof they provide. For me personally, these videos successfully falsify Materialism, Naturalism, Nihilism, and Atheism. Your mileage may vary, of course, depending upon what you personally want to believe.

NDE Encounters with Jesus:

https://www.youtube.com/results?search_query=NDE+Jesus

Howard Storm:

https://www.youtube.com/watch?v=Vm647n1360A

https://www.youtube.com/watch?v=UPj4wci_bcI

Dr. Mary Neal:

https://www.youtube.com/watch?v=DX473dF7ChY

Ian McCormack:

https://www.youtube.com/watch?v=HbTAmN4m2lQ

Joe Hadwin:

https://www.youtube.com/watch?v=IOhOynR9Jxg

Another:

https://www.youtube.com/watch?v=N4ut09jDdV0

Science IS Observation. After I made that realization, I now treat Lived Experiences of any kind as Scientific Observations and Scientific Evidence. Lived Experiences or Psyche Experiences ARE the BEST way of finding and knowing the truth in any realm of existence, including the Non-Local Realms. Lived Experiences or Psyche Experiences are a better way of finding and knowing the truth than the Scientific Methods. I never realized that before; but, I understand it and accept it now. It only took me 55 years to figure it out. Lived Experiences, including Out-of-Body Experiences and Near-Death Experiences ARE Scientific Evidence because they are Direct Scientific Observations of the Non-Local Realm or Spirit Realm.

I'm sorry, but I found these people and their lived experiences and observations infinitely more convincing than the posturing and "scientific inferences" which I received from the Materialists, Naturalists, Darwinists, Nihilists, and Atheists. I have chosen to believe that Observational Evidence is better than circumstantial evidence, which means that I have chosen to accept the belief that Lived Experiences are vastly superior to scientific inferences and just-so ad hoc stories concocted out of thin air.

I can guarantee you, though, that you will see the whole thing differently than the way I now see it, if you have instead chosen to believe on faith that Materialism, Naturalism, Darwinism, Nihilism, and Atheism are true. No amount of evidence will get you to change your mind, unless you actually want to change your mind.

By the time I encountered these videos, I was ready to change my mind, because science itself and scientific observations had started to prove to me that Materialism, Naturalism, and Darwinism are false.

Obviously, you have every right to fall for the "scientific inferences," logic fallacies, and just-so stories of the Materialists, Darwinists, and Naturalists; but, do so knowing that that's what you are doing and that that's what you want to do – and not on blind faith that these people are always telling you the truth like I used to do.

Once I caught the Materialists, Naturalists, and Darwinists in their deceptions, trickery, logic fallacies, and lies, I could no longer go back. I was forced to go forward. I wanted to go forward. I don't like being lied to. Obviously, your mileage may vary. You will do whatever you personally want to do and choose to do. That is as it should be, and that is as it was designed to be.

Honorable Mention in this Category

Marshall, P. D., Kelly, E. F., & Crabtree A. (Eds.). (2015). *Beyond Physicalism: Toward Reconciliation of Science and Spirituality*. London, United Kingdom: Rowman and Littlefield.

Remember, a correct understanding and acceptance of Quantum Non-Locality, as well as the right interpretation of Quantum Mechanics which takes Non-Local Consciousness or Psyche into account, is the KEY to understanding the Observational Evidence, Experiential Evidence, and Scientific Evidence associated with and provided by Near-Death Experiences (NDEs) and Out-of-Body Experiences (OBEs).

It's extremely important to follow the evidence wherever it might lead you, because Science IS observational evidence and experiential evidence.

In all of his books, following the party line, Michio Kaku typically leans towards Materialism, Naturalism, Darwinism, Nihilism, and even Atheism; consequently, he doesn't really make any effective attempt to resolve the Mind-Brain Problem, and instead typically presents the mind as a function of the physical brain. Michio Kaku isn't mean-spirited about it, like the New Atheists are; but, you won't find many non-local spiritual interpretations of Quantum Mechanics and String Theory from Michio Kaku. Instead, I have to choose to interpret Kaku's input from a Non-Local Perspective or a Spiritual Perspective in order to make logical sense of it, in a way that actually ends up matching with the experiential and observational evidence from NDEs and OBEs.

I love reading Michio Kaku, but in true Science 2.0 fashion, I always have to provide my own interpretation that actually matches with the observational evidence from NDEs and OBEs, because I typically don't get such a thing from Michio Kaku himself. Michio Kaku makes no attempt to resolve the Mind-Brain Problem, but instead either ignores it or tries to explain it away through materialism and naturalism.

Therefore, if we want a truly effective and believable resolution of the Mind-Brain Problem using science and quantum mechanics to do so, we must turn to someone else besides Michio Kaku for that solution – someone like Henry P. Stapp and John von Neumann, and the **Orthodox Interpretation of Quantum Mechanics**. The book, *Beyond Physicalism*, contains an excellent introduction to the Orthodox Interpretation, as presented by Henry Stapp, which I will review and discuss in a subsequent section.

http://mypsyche.us/wp-content/uploads/2017/10/Orthodox-Interpretation.pdf

After being led to Henry P. Stapp through *Beyond Physicalism*, I concluded that Stapp is another Winner in this Quantum Mechanics category or theme.

### Another Winner – Henry P. Stapp's Orthodox Interpretation of Quantum Mechanics

Remember, Science 2.0 needs an interpretation of Quantum Mechanics that matches with and explains the observational evidence and experiential evidence that's obtained from Near-Death Experiences (NDEs), Shared-Death Experiences (SDEs), Out-of-Body Experiences (OBEs), and other types of Non-Local or Spiritual Experiences.

There are a bunch of other books and science papers supporting Quantum Consciousness, Universal Consciousness, Cosmic Consciousness, or Non-Local Consciousness. I own, have read from, and/or discuss some of these in my other books and essays, so I will mostly list/relist some of them here for now with a brief explanation after each one. The purpose of this book is to be an introduction to Science 2.0, not a complete treatment of the subject.

The following books are compatible with the methodology of Science 2.0, which allows ALL of the evidence into evidence, and permits you to decide for yourself how to interpret and use that evidence. Remember, there is NO evidence supporting the primary assumptions or major premises of Materialism, Naturalism, Darwinism, Nihilism, and Atheism; consequently, Science 2.0 defines these philosophies and religions as unscientific because there is NO observational evidence supporting them and never will be.

### The Winner

Pim van Lommel: *Consciousness Beyond Life: The Science of the Near-Death Experience*. This book develops and promotes what I call the **Non-Local Consciousness Interpretation of Quantum Mechanics**. I already declared this book to be the overall winner, in this Quantum Mechanics category or theme. This is the book that got through to me first; and, it changed the way that I look at science and quantum mechanics, for the better. This book is first in large part because I found it first.

### Another Winner – the Clear Runner-Up

A clear winner in this Quantum Mechanics category is the wide collection of books and articles from Henry P. Stapp, including some of the following.

https://sites.google.com/a/lbl.gov/stappfiles/

http://www-physics.lbl.gov/~stapp/

The TRUE ADVANTAGE of Stapp's Orthodox Interpretation of Quantum Mechanics is that it matches with and explains ALL of the Observational Evidence, Experiential Evidence, Experimental Evidence, Eye-Witness Evidence, Corroborated Veridical Evidence, and Scientific Evidence associated with Quantum Mechanics and Non-Local Consciousness. The same cannot be said about the materialistic and naturalistic interpretations of Quantum Mechanics that are based upon classical physics.

You could dedicate a lifetime towards downloading, reading, and studying all of these articles from Henry P. Stapp. I was first introduced to him through various different books that I own.

Stapp, H. P. (2017). *Quantum Theory and Free Will: How Mental Intentions Translate into Bodily Actions*. Cham, Switzerland: Springer International Publishing.

This book explains and promotes John von Neumann's **Orthodox Interpretation of Quantum Mechanics**. This Model of Reality matches extremely well with the observational evidence from the Lived Experiences of the human race, including Near-Death Experiences (NDEs) and Out-of-Body Experiences (OBEs). The following is a quote that's found in Stapp's book, *Quantum Theory and Free Will*.

### Von Neumann's "Orthodox" Formulation of Quantum Mechanics

**The "standard" quantum theory, against which all others are compared, is von Neumann's "orthodox" formulation of Copenhagen Quantum Mechanics, or, more specifically, the updated version, called "Relativistic Quantum Field Theory", abbreviated as "RQFT". It is this relativistic "orthodox" version of quantum theory that is propounded in this book.**

**As will be presently explained, this theory is about both:**

**(1), the dynamical interaction of matter with itself that accounts for the 'unobserved' behavior of material substances; and**

**(2), the interaction between mind and matter that constitutes the highly nontrivial 'process of observation'.**

**In quantum mechanics the mind-matter interaction is mathematically very different from the matter-matter interaction. And it is different in a mathematical way that entails that the former can never be reduced to the latter.**

**The difference in these two dynamical processes is directly connected to Heisenberg's seminal 1925 discovery, which quickly led to the creation of quantum mechanics. This new theory gives detailed explanations of the plethora of twentieth century data of atomic physics that had resisted all attempted explanations via the materialist precepts of classical physics.**

**Heisenberg's discovery was that the process of observation – whereby an observer comes to consciously know the numerical value of a material property of an observed system – cannot be understood within the framework of materialist classical mechanics. A non-classical process is needed. This process does not construct mind out of matter, or reduce mind to matter. Instead, it explains, in mathematical terms, how a person's immaterial conscious mind interacts with that person's material brain.**

An immaterial mind lies beyond the ken of a materialistic approach, and the mathematics that describes the process of conscious observation is not reducible to the mathematics that describes the process of the unobserved evolution of matter.

The eminent Hungarian-American mathematician and logician John von Neumann cast the ideas of Copenhagen quantum mechanics into a rationally coherent and mathematically rigorous form that is widely used by mathematical physicists, and also by others who require mathematical and logical precision. Nobel Laureate Eugene Wigner labeled Von Neumann's formulation "Orthodox Quantum Mechanics". The label "Orthodox" is appropriate, in the sense that many, and perhaps all, mathematical physicists take it to be the logically and mathematically precise formulation of the Copenhagen ideas.

Von Neumann approached these mind-related issues by considering what amounts to a tower of good measuring devices where each device associates, one-to-one, each input to a corresponding output, and the output of each device is the input to the device above it. On the top of this tower lies an observer's conscious "ego" that can both receive perceptual inputs and instigate probing actions by means of its interactions with its associated brain.

Source: *On the Nature of Things: Human Presence in the World of Atoms*, pp. 18-19:

http://www-physics.lbl.gov/~stapp/NOT72C.pdf

http://mypsyche.us/wp-content/uploads/2017/10/Orthodox-Interpretation.pdf

This **Orthodox Interpretation of Quantum Mechanics** is the runner-up, and my second most-favorite interpretation of Quantum Mechanics. I find this Model of Reality extremely useful and informative.

Henry P. Stapp: *Mindful Universe: Quantum Mechanics and the Participating Observer*. This book briefly introduces John von Neumann's Orthodox Interpretation of Quantum Mechanics. It's also an excellent treatment of Quantum Mechanics overall.

The following links contain parts of this book, *Mindful Universe*:

http://www-physics.lbl.gov/~stapp/hpsquotes.txt

http://www-physics.lbl.gov/~stapp/OrthodoxInterpretationStapp.doc

http://exordio.qfb.umich.mx/archivos%20PDF%20de%20trabajo%20UMSNH/Aphilosofia/MUA%20el%20mejor%20libro.pdf

https://www.thenakedscientists.com/forum/index.php?topic=48746.1550

https://books.google.com/books?id=yVpyM3dJOgYC

*Mindful Universe* is many peoples' most favorite Quantum Mechanics book. There are quotes all over the place on the internet from this book, if you are willing to look for them and find them.

The following quote from page 24 of *Mindful Universe* explains the Orthodox Interpretation of Quantum Mechanics:

John von Neumann, in his seminal book, *Mathematical Foundations of Quantum Mechanics*, calls by the name 'process 1' the basic probing action that partitions a potential continuum of physically described possibilities into a (countable) set of empirically recognizable alternative possibilities. I shall retain that terminology.

Von Neumann calls the orderly mechanically controlled evolution that occurs between interventions by name 'process 2'. This process is the one controlled by the Schrödinger equation. The numbering, 1 and 2, emphasizes the important fact that the conceptual framework of orthodox quantum theory requires first an acquisition of knowledge, and second, a mathematically described propagation of a representation of this acquired knowledge to some later time at which a further inquiry is made.

There are two other associated processes that need to be recognized. The first of these is the process that selects the outcome, 'Yes' or 'No', of the probing action. Dirac calls this intervention a "choice on the part of nature", and it is subject, according to quantum theory, to statistical rules specified by the theory. I call by the name 'process 3' this statistically specified choice of the outcome of the action selected by the prior process 1 probing action.

Finally, in connection with each process 1 action, there is, presumably, some process that is not described by contemporary quantum theory, but that determines what the so-called 'free choice' of the experimenter will actually be. This choice seems to us to arise, at least in part, from conscious reasons and valuations, and it is certainly strongly influenced by the state of the brain of the experimenter. I have previously called this selection process by the name 'process 4', but will use here the more apt name 'process zero', because this process must precede von Neumann's process 1. It is the absence from orthodox quantum theory of any description on the workings of process zero that constitutes the causal gap in contemporary orthodox physical theory. It is this 'latitude' offered by the quantum formalism, in connection with the "freedom of experimentation" (Bohr 1958, p. 73), that blocks the causal closure of the physical, and thereby releases human actions from the immediate bondage of the physically described aspects of reality.

https://books.google.com/books?id=yVpyM3dJOqYC&pg=PA24&lpg=PA24&dq#v=onepage&q&f=false

https://books.google.com/books?id=yVpyM3dJOqYC&printsec=frontcover#v=onepage&q&f=false

This is useful science, in my humble opinion. It's certainly much better than saying that the non-physical does not exist, as the Materialists and Naturalists do. The Orthodox Interpretation of Quantum Mechanics is actually a believable model of reality that matches perfectly with the observational and experiential evidence coming from the Lived Experiences of the human race as a whole. The Orthodox Interpretation is precisely what Science 2.0 is looking for and needs – an experienced type of Quantum Mechanics, or an observed form of Quantum Mechanics, that actually matches with Reality and gives us a choice in the matter.

Henry P. Stapp: *Mind, Matter and Quantum Mechanics* (The Frontiers Collection) 3rd ed.

https://books.google.com/books?id=AGnwCAAAQBAJ&printsec=frontcover#v=onepage&q&f=false

https://www.goodreads.com/book/show/970984.Mind_Matter_and_Quantum_Mechanics

Reviews and Selections:

http://quantum-mind.co.uk/theories/henry-stapp/mind-matter-and-quantum-mechanics/

http://www.iaea.org/inis/collection/NCLCollectionStore/_Public/22/019/22019953.pdf

*Mind, Matter and Quantum Mechanics* is an expensive book; but, there are samplings from it across the internet, including copies of the original 1981 paper that started it all.

https://escholarship.org/content/qt6vp6q1nk/qt6vp6q1nk.pdf

http://mypsyche.us/wp-content/uploads/2017/11/Mind-Matter-and-Quantum-Mechanics.pdf

Henry P. Stapp: *Beyond Physicalism: Toward Reconciliation of Science and Spirituality*. In this book, Chapter 5, Henry Stapp has an excellent explanation of the Orthodox Interpretation of Quantum Mechanics. The title of this chapter is "Quantum-Mechanical Theory of the Mind/Brain Connection". So far, this chapter is my most favorite one from Henry P. Stapp. I love the science.

*Beyond Physicalism* was the first book to introduce me to Henry P. Stapp.

See this link for the essential bits:

http://mypsyche.us/wp-content/uploads/2017/10/A-Quantum-Mechanical-Theory-of-the-Mind-Brain-Connection.pdf

Online Source:

https://books.google.com/books?id=RuPFBgAAQBAJ&pg=PA157&lpg=PA157&dq#v=onepage&q&f=false

**Here are some other free articles from Henry P. Stapp that you might find useful and interesting.**

These are some of the free articles that Stapp considered worth mentioning in his books. His books are very expensive, which is the main reason why it took me a couple of years to find Henry P. Stapp. That's also why I declared Pim van Lommel's book, *Consciousness Beyond Life: The Science of the Near-Death Experience*, to be the winner in this Quantum Mechanics category, because it is affordable for those on a budget and quite manageable.

These free articles will introduce you to Henry P. Stapp, so that you can start saving to buy his books. Of course, I didn't find out about any of these articles until after I bought and started reading his book, *Mindful Universe*.

https://sites.google.com/a/lbl.gov/stappfiles/

Henry Stapp leads his list by giving the world a free book:

"'On the Nature of Things: Human Presence in the World of Atoms"

http://www-physics.lbl.gov/~stapp/NOT72C.pdf

http://mypsyche.us/wp-content/uploads/2017/11/NOT72C.pdf

"Quantum Physics in Neuroscience and Psychology: A Neurophysical Model of Mind-Brain Interaction"

http://mypsyche.us/wp-content/uploads/2017/10/PTRS.pdf

https://www.researchgate.net/publication/7613549_Quantum_physics_in_neuroscience_and_psychology_A_neurophysical_model_of_mind-brain_interaction

http://www-physics.lbl.gov/~stapp/PTRS.pdf

"A Quantum Theory of the Mind/Brain Interface"

http://www.iaea.org/inis/collection/NCLCollectionStore/_Public/22/019/22019953.pdf

"The Effect of Mind upon Brain"

http://www.newdualism.org/papers/H.Stapp/Stapp-kout2009.pdf

http://mypsyche.us/wp-content/uploads/2017/11/The-Effect-of-Mind-Upon-Brain.pdf

"The Hard Problem: A Quantum Approach"

https://arxiv.org/pdf/quant-ph/9505023.pdf

http://mypsyche.us/wp-content/uploads/2017/11/The-Hard-Problem-A-Quantum-Approach.pdf

"Physicalism vs Quantum Mechanics"

http://www-physics.lbl.gov/~stapp/Physicalism.pdf

"Philosophy of Mind and the Problem of Free Will in the Light of Quantum Mechanics"

https://arxiv.org/abs/0805.0116

"Attention, Intentions, and Will in Quantum Physics"

https://arxiv.org/pdf/quant-ph/9905054.pdf

"From Quantum Nonlocality to Mind-Brain Interaction"

https://pdfs.semanticscholar.org/f629/22145fc010537172eba060249d3c826777ef.pdf

"Quantum Approaches to Consciousness"

http://www-physics.lbl.gov/~stapp/Cambridge.pdf

"Quantum Mechanical Theories of Consciousness"

http://www-physics.lbl.gov/~stapp/BCC.pdf

"Quantum Interactive Dualism - An Alternative to Materialism"

https://pdfs.semanticscholar.org/18c5/18077f898f41eef9c3b028e2781e8aa10bab.pdf

"Quantum Interactive Dualism, II: The Libet and E-P-R Causal Anomalies"

https://www.jstor.org/stable/27667854

"Quantum Theory and The Role of Mind in Nature"

https://arxiv.org/pdf/quant-ph/0103043.pdf

"Science of Consciousness and the Hard Problem"

http://www-physics.lbl.gov/~stapp/38621.txt

https://www.osti.gov/scitech/servlets/purl/269006

"Whitehead, James, and Quantum Theory"

http://www-physics.lbl.gov/~stapp/KreuthTalk.pdf

"Light as Foundation of Being"

http://www-physics.lbl.gov/~stapp/LightFoundBeing.pdf

"Are Superluminal Connections Necessary?"

http://www-physics.lbl.gov/~stapp/NCimento.pdf

"The Whiteheadian Approach to Quantum Theory and the Generalized Bell Theory"

http://www-physics.lbl.gov/~stapp/Whitehead-Bell-1979.pdf

"The Principle of Sufficient Reason"

http://www-physics.lbl.gov/~stapp/Reason01132012.doc

"A Bell-Type Theorem without Hidden Variables"

https://arxiv.org/pdf/quant-ph/0205096.pdf

"The Basis Problem in Many-Worlds Theories"

https://arxiv.org/pdf/quant-ph/0110148.pdf

"Theory of Reality"

https://www.cia.gov/library/readingroom/docs/CIA-RDP96-00787R000200080053-6.pdf

http://www-physics.lbl.gov/~stapp/Theory_of_Reality_1977.pdf

"The Copenhagen Interpretation"

http://users.df.uba.ar/giribet/f4/mq4.pdf

http://www.informationphilosopher.com/solutions/scientists/stapp/Copenhagen_Interpretation.pdf

"A Model of the Quantum-Classical and Mind-Brain Connections, and of the Role of the Quantum Zeno Effect in the Physical Implementation of Conscious Intent"

https://arxiv.org/pdf/0803.1633.pdf

"Proof That Information Must Be Transferred Faster Than Light" http://www-physics.lbl.gov/~stapp/Appendix-1.doc

Obviously, I could take this whole thing much further than I have simply by delving into and quoting from everything that Henry P. Stapp has ever written; but, this is good enough for now.

Science 2.0 relies upon the various different Non-Local Consciousness Interpretations of Quantum Mechanics, which have been experienced and observed by Out-of-Body Travelers when their physical brain was either dead or offline.

**Anything from Henry P. Stapp is the clear runner-up, in this Quantum Mechanics category or theme.** He is the scientist that I have been looking for all of my life.

My only complaint against Stapp, and the reason why he doesn't win this category in my humble opinion, is that his books were way too expensive for me to be able to afford, so it took me a couple of years to find him and read him, unlike Pim van Lommel's book, *Consciousness Beyond Life*. We can usually buy ten other books for one of Stapp's, which is what I did in the beginning of my research.

Of course, cataloguing Stapp's best free material as well as directing you to his free material online is the best way to resolve this complaint; but, I had to buy and read some of Stapp's books and chapters before I was made aware that his free articles and his free book exists. You now have no such obstacle, as long as you have an internet connection.

https://sites.google.com/a/lbl.gov/stappfiles/

http://www-physics.lbl.gov/~stapp/

http://www-physics.lbl.gov/~stapp/NOT72C.pdf

http://www-physics.lbl.gov/~stapp/PTRS.pdf

http://mypsyche.us/wp-content/uploads/2017/10/PTRS.pdf

http://mypsyche.us/wp-content/uploads/2017/10/Orthodox-Interpretation.pdf

I will let you find and work through this material if it is of interest to you.

Unlike the Materialists and Naturalists who define science as Materialism and Naturalism, Henry P. Stapp and Pim van Lommel are NOT trying to suppress and hide what Quantum Mechanics is really trying to teach us. That's what I like about Stapp and Lommel – they allow all of the evidence into evidence, which is the primary theme of Science 2.0.

Science 2.0 is based upon and relies upon the scientifically verified versions of Quantum Mechanics, along with their associated Non-Local Consciousness Interpretations. Henry Stapp actually provides useful, meaningful, and scientifically accurate interpretations of Quantum Mechanics; and, he successfully falsifies Materialism and Naturalism while doing so. It doesn't get better than that!

Honorable Mention in this Quantum Mechanics Category:

Bruce Rosenblum and Fred Kuttner: *Quantum Enigma: Physics Encounters Consciousness*. This book explores dozens of different interpretations of Quantum Mechanics. From what I can tell, these authors seem to prefer the Copenhagen Interpretation. They do a shoddy job of explaining John von Neumann's Orthodox Interpretation, and effectively misrepresent it.

Various Authors: *Quantum Physics of Consciousness*. This book explores dozens of different consciousness-based interpretations of Quantum Mechanics. Henry Stapp has a chapter in this book.

Price, D. D., Finnis, D. G., & Benedetti, F. (2008). *A Comprehensive Review of the Placebo Effect: Recent Advances and Current Thought*. By definition, and in principle, the Placebo Effect is PROOF of Mind over Matter. The Placebo Effect demonstrates that Psyche, Mind, or Non-Local Consciousness does in fact exist and function separate from our physical body.

http://www.annualreviews.org/doi/pdf/10.1146/annurev.psych.59.113006.095941

Amit Goswami: *The Self-Aware Universe: How Consciousness Creates the Material World*. This book develops and promotes the **Monistic Idealism Interpretation of Quantum Mechanics**. Monistic Idealism is Universal Consciousness or Brahman. Monistic Idealism is the Buddhist and Hindu interpretation of Quantum Mechanics.

During their trances, meditation, and mystical experiences, the Buddhist Monks and Hindu Mystics eventually experience nothing or nothingness, which they subsequently define as Universal Consciousness, Cosmic Consciousness, Brahman, Enlightenment, or Nirvana. Over and over again, these people claim to experience nothingness at the height of their meditation or mystical experiences. I really do believe that these people are indeed experiencing nothing during their meditation and mystical experiences and experiencing a great deal of peace as a result. I take them at their word, where that claim is concerned; but, I'm skeptical about what that experience of nothingness truly means to us in reality. In other words, I'm skeptical about the interpretation which they give to their mystical experiences, because I think there's a better interpretation that can be given to these kinds of experiences. I will review Monistic Idealism next.

Amit Goswami: *God Is Not Dead: What Quantum Physics Tells Us about Our Origins and How We Should Live*. I originally used this book to confirm many of the ideas and principles that I learned from *Consciousness Beyond Life: The Science of the Near-Death Experience*. I discussed those ideas and concepts in a different book, so I won't do so here.

I love this kind of science. It makes me stop and think! It's not brain science, it's better than brain science.

Notice that I prefer and pursue the Quantum Interpretations which have ALL of the observational evidence supporting them – the Quantum Consciousness Interpretation, the Non-Local Consciousness Interpretation, and the Orthodox Interpretation. The other naturalistic interpretations don't have any observational evidence supporting them; consequently, under Science 2.0 which defines science as observation, the materialistic and naturalistic interpretations are considered unscientific because they have no empirical evidence supporting their major premises. Science is observation, knowledge, truth, fact, and experience. Under Science 2.0, Materialism and Naturalism are unscientific because they have no observational support for their major premises or primary assumptions.

Observe that the Quantum Interpretations – which have the most observational evidence supporting them – happens to be the ones which the Materialists, Naturalists, Nihilists, and Atheists ignore the most and reject the most and suppress the most, because a Non-Local Consciousness Interpretation or Psyche Interpretation of Quantum Mechanics isn't telling them what they want to hear nor leading them where they want to go.

I personally started making progress only after I chose to get rid of My Materialism, My Nihilism, and My Atheism. Whole new vistas of learning and experience opened to my view, once I got rid of these false and falsified philosophies and religions. It was the best thing that I ever chose to do. You can too, if you want to do so.

Nobody is stopping you from pursuing the truth and finding the truth. I simply chose to allow the evidence to take me wherever it wanted to lead me. Remember, Science 2.0 is based up a preponderance of the eye-witness evidence which we currently have at hand as a race. The best way to find and know the truth is to live it and experience it for yourself, or to choose to trust someone who has.

### The Monistic Idealism Interpretation of Quantum Mechanics

Goswami, A. (1993). *The Self-Aware Universe: How Consciousness Creates the Material World*. New York: Penguin Putnam.

Amit Goswami produces and promotes a **Monistic Idealism Interpretation of Quantum Mechanics**. Monistic Idealism is Universal Consciousness or Brahman or Nirvana. This is the Hindu and Buddhist interpretation of Quantum Mechanics. Buddhism is a type of atheism – Buddhism is spirituality for atheists.

**Monistic Idealism**: Monistic Idealism is the philosophy that defines Brahman or Universal Cosmic Consciousness as the primary reality, and as the ground of all being. The subjects (self or person or psyche) and the objects (physical matter) of a consensus empirical reality are all epiphenomena of Brahman or Universal Consciousness, which is the only source of true consciousness. There is **no self-nature** (no psyche and no mind and no consciousness) in the subject (self or person) nor the object (physical matter) apart from Brahman or Universal Consciousness, during any type of conscious experience. The Universe or Universal Consciousness is the only thing that is truly self-aware and conscious and exists. Self, psyche, mind, or individual consciousness is simply an illusion. You cease to exist when your physical body dies, and what you were if anything is simply absorbed into Brahman or Universal Consciousness when you die. (See p. 281 of *The Self-Aware Universe*.)

During their trances, meditation, and mystical experiences, the Buddhist Monks and Hindu Mystics eventually experience nothing or nothingness, which they subsequently define as Universal Consciousness, Cosmic Consciousness, Brahman, Enlightenment, or Nirvana. Over and over again, these people claim to experience nothingness at the height of their meditation or mystical experiences. I really do believe that these people are indeed experiencing nothing during their meditation and mystical experiences and experiencing a great deal of peace as a result. I take them at their word, where that claim is concerned; but, I'm skeptical about what that experience of nothingness truly means to us in reality. In other words, I'm skeptical about the interpretation which they give to their mystical experiences, because I think there's a better interpretation that can be given to these kinds of experiences.

In his book, Goswami rightfully and correctly uses observations and experiences from Hindu Mystics as "proof of concept" and as verification of Universal Consciousness or Brahman; and in the process, he completely FALSIFIES Materialism and Naturalism. Amit Goswami is brutal against Materialism and Naturalism in all of his books. He calls it Material Realism.

**So, the question is, Is there a monistic alternative to material realism, where mind and matter are integrally part of one reality, but a reality that is not based on matter? I am convinced there is. The alternative that I propose**

**in this book is monistic idealism. This philosophy is monistic as opposed to dualistic, and it is idealism because ideas (not to be confused with ideals) and the consciousness of them are considered to be the basic elements of reality; matter is considered to be secondary.**

**In other words, instead of positing that everything (including consciousness) is made of matter, this philosophy posits that everything (including matter) exists in and is manipulated from consciousness. Note that the philosophy does not say that matter is unreal but that the reality of matter is secondary to that of consciousness, which itself is the ground of all being — including matter.**

Goswami, Amit. The Self-Aware Universe (pp. 10-11). Penguin Publishing Group. Kindle Edition.

Monistic Idealism is the philosophical idea that ONLY Universal Consciousness or Brahman exists. Borrowing from Buddhism, Monistic Idealism makes the claim that self, mind, or personal psyche is an illusion and does not exist. In other words, according to Monistic Idealism and Buddhism, you, your consciousness, and your psyche cease to exist when your physical body dies – self or psyche or personality is an illusion and doesn't really exist.

Throughout his book, Amit Goswami uses experiential evidence, observational evidence, and empirical evidence from Hindu Mystics to demonstrate and prove that Universal Consciousness, or Brahman, or Monistic Idealism is real and truly exists.

**Mysticism offers experiential proof of monistic idealism.**

Goswami, Amit. The Self-Aware Universe (p. 50). Penguin Publishing Group. Kindle Edition.

Science 2.0 is based upon observation, experience, eye-witness testimony, empirical evidence, and the Lived Experiences of the human race.

Monistic Idealism is the philosophy that defines mind, self, individuality, personhood, personality, and psyche out of existence. According to Monistic Idealism, there is only one mind or consciousness in this universe – Brahman. In contrast, Idealism is the philosophical idea that holds that the fundamental elements of reality must include mind as well as matter. Ordinary Idealism is not as radical nor as exclusionary as Monistic Idealism.

I believe that Monistic Idealism is a faulty interpretation of the observational evidence; and now, I will try to explain why.

First of all, the ONLY way to defeat observational evidence or experiential evidence is with better and more convincing observational evidence and eye-witness testimony. I simply find the thousands of Near-Death Experiences (NDEs), Shared-Death Experiences (SDEs), Out-of-Body Experiences (OBEs), and Non-Local Experiences on record much more convincing than the interpretations of experiential evidence and observational evidence provided by the Hindu Mystics, Buddhists, and Amit Goswami.

Secondly, the Hindu Mystics and some of the Buddhists claim that they are connecting with Universal Consciousness or Brahman, and experiencing nothingness or unity while doing so. I take them at their word. However, as a consequence of these experiences, the Buddhists and Amit Goswami then go on to claim that self, psyche, or mind does not exist and is simply an illusion. The ONLY thing that exists according to Amit Goswami is Brahman or Universal Consciousness – hence Monistic Idealism is the true reality and the correct interpretation of Quantum Mechanics.

The promoters of Monistic Idealism teach that self or psyche is an illusion, and that only Brahman or Universal Consciousness exists. I don't buy it – the idea of only one single Universal Consciousness. It's a *confound* logic fallacy, a faulty interpretation. I simply KNOW that your consciousness or psyche exists completely separate from mine, because that's what we all experience. I believe that Monistic Idealism fails to meet its burden of proof through a preponderance of the observational evidence.

Third, it is the Mystics' connection with Brahman or Universal Consciousness which is in fact the illusion. Why? If the Mystics were truly connected to Universal Consciousness or the universal broadcast, or the universal transceiver, then they should be and would be able to hear and experience everyone else's thoughts instantaneously and simultaneously all across this universe.

The observational evidence from some of the NDEs on record demonstrate and prove that the Biblical God Jesus Christ is in fact connected to the universal broadcast or universal transceiver and truly capable of reading our minds and hearing our prayers instantaneously and simultaneously all throughout this universe and anywhere in this universe – both here in the physical realm and there in the various different non-local realms or spirit realms. Jesus Christ can hear your prayers even while you are in hell.

The Hindu Mystics aren't connecting to the universal broadcast nor the Universal Consciousness – it's only an illusion, and they only think that they are doing so. In fact, they report experiencing nothingness and unity while connecting with Universal Consciousness or Brahman. By their own admission, they aren't really connecting to the universal broadcast nor the universal transceiver, because they can't hear nor read anyone else's mind. The Hindu Mystics and Buddhists simply experience nothing; and, nothing is not the universal broadcast nor the universal transceiver.

While in the presence of the Being of Light or Jesus Christ during NDEs, the people report that they can feel the infinite love radiating from Jesus Christ, a completely separate psyche or self. These people simply KNOW that the love isn't radiating from them nor originating within them; instead, that infinite love and universal broadcast is coming from Jesus Christ instead. This is what they experience and observe. These people quickly realize that they are a completely different and separate psyche, self, individual, or personality than Jesus Christ and the Angels of Heaven. The existence of these types of observational experiences FALSIFIES Monistic Idealism and the claim that psyche or self does not exist.

We love the things most for which we sacrifice the most. Jesus Christ is capable of infinite compassion and love, because He suffered and sacrificed infinitely for each one of us. In contrast, we individual psyches or selves are NOT capable of that kind of infinite compassion and love, because NONE of us have suffered and sacrificed for others anywhere near as much as Jesus Christ has. Jesus Christ paid the ultimate price in order to be filled with infinite mercy, compassion, and love. Jesus Christ paid the ultimate price in order to become the Being of Light, the Savior of this world, Universal Consciousness, and the Infinite Manifestation of compassion, charity, forgiveness, and love. NONE of us have paid that kind of price, yet.

These kinds of observations falsify the claim that no self-nature (self or psyche) exists and that ONLY universal consciousness or Brahman exists. Remember, the only way to falsify faulty interpretations of experiential evidence and observational evidence is with better observational evidence and better eye-witness experiential evidence.

It is my observation and chosen belief based upon the collective observational evidence from NDEs, SDEs, and OBEs that Monistic Idealism is the illusion, and that the experience and observation of self, psyche, mind, individuality, personality, personhood, and non-local consciousness is the Real Deal instead. Monistic Idealism is the illusion according to NDEs, SDEs, and OBEs. There's more than one consciousness, or mind, or psyche in this universe. In fact, according to the Orthodox Interpretation of Quantum Mechanics, both the atoms (physical matter) and the observer (the individual self or psyche) are conscious, sentient, telepathically connected at a distance, and have a self-nature. This reality or truth is also verified and confirmed by Pim van Lommel's Non-Local Consciousness Interpretation of Quantum Mechanics which is based upon the observational evidence, experiential evidence, and empirical evidence from all the different NDEs, SDEs, and OBEs. We have already mentioned and discussed the Non-Local Consciousness Interpretation of Quantum Mechanics, as found in *Consciousness Beyond Life: The Science of the Near-Death Experience*.

This case study from Amit Goswami demonstrates the primary flaw of Interpretation. With any given piece of observational evidence, scientific evidence, or empirical evidence, there can theoretically be an infinite number of different possible interpretations of that evidence. They can't all be right, especially when some of them conflict with each other or counteract each other.

NDEs, SDEs, OBEs, Psyche, Non-Local Conscious, and other Non-Local Experiences FALSIFY the claims of Monistic Idealism which state that self or psyche is an illusion and that Brahman is the ONLY consciousness in this universe that exists. The only way to defeat faulty interpretations of the observational evidence is with better and more convincing observational evidence.

Based upon the observational evidence that I personally believe in the most and trust the most, NDEs and OBEs, I conclude that Idealism is a much more realistic and believable and sustainable interpretation of Reality and Quantum Mechanics than Monistic Idealism. I reject Monistic Idealism and prefer Idealism instead.

Both Buddhism and Monistic Idealism claim that self or psyche does not exist. At best, there is only one Universal Consciousness – Brahman.

While laying the foundation for Monistic Idealism, Amit Goswami explains and establishes Idealism as the true and most realistic alternative to Materialism and Naturalism, which he calls Material Realism in his book, *The Self-Aware Universe*.

**Idealism**

Realism is the belief that reality exists independently of observers. Psyche exits separate from physical matter. Psyche also exists separately from spirit matter.

Material Realism is the belief that both consciousness and physical matter are caused by physical matter – everything is physical matter and has a physical cause. Simply by introducing Materialism into the equation, every philosophical idea instantly becomes false. Materialism and Naturalism make everything false, because they are false to begin with. Notice that Realism is true, until Materialism and Naturalism get introduced into it, which then makes it false.

Idealism is synonymous with Non-Local Consciousness.

"Monistic" means One. Monistic Idealism is the idea that there is only One Consciousness in this universe – Brahman. I don't buy it.

Why?

I know for a fact that your consciousness is separate from my consciousness; otherwise, we would be able to read each other's minds and have the same exact thoughts if there was only one consciousness in this universe. Technically, the observational evidence and experiential evidence actually falsifies Monistic Idealism.

**THE ANTITHESIS OF MATERIAL REALISM is monistic idealism. In this philosophy, consciousness, not matter, is fundamental. Both the world of matter and the world of mental phenomena, such as thought, are determined by consciousness. In addition to the material and the mental spheres (which together form the immanent reality, or world of manifestation), idealism posits a transcendent, archetypal realm of ideas as the source of material and mental phenomena. It is important to recognize that monistic idealism is, as its name implies, a unitary philosophy; any subdivisions, such as the immanent and the transcendent, are within consciousness. Thus, consciousness is the only ultimate reality.**

Goswami, Amit. *The Self-Aware Universe* (p. 48). Penguin Publishing Group. Kindle Edition.

From the Wikipedia:

**In philosophy, idealism is the group of philosophies which assert that reality, or reality as we can know it, is fundamentally mental, mentally constructed, or otherwise immaterial.**

**In contrast to materialism, idealism concedes the primary nature of consciousness, which means consciousness exists before matter. Consciousness creates and determines physical matter, not vice versa. Idealism theories believe consciousness is the origin of the world and aim to explain the existing world by mental causes.**

https://en.wikipedia.org/wiki/Idealism

Idealism (Non-Local Consciousness) is the part of this that I do believe is sustainable by Observational Evidence and Experiential Evidence. Idealism is synonymous with Non-Local Consciousness or Psyche.

**[How do we resolve the Schrödinger Cat Paradox?]**

**Fortunately, an idealist resolution presents itself: Since our observation magically resolves the dichotomy of the cat, it must be us — our consciousness — that collapses the cat's wave function. Material realists do not like this idea, because it makes consciousness an independent, causal entity; admitting that would be like putting nails in the coffin of material realism.**

**In the idealist resolution, it is observation by a conscious mind that resolves the alive-or-dead dichotomy. Like Platonic archetypes, coherent superpositions exist in the never-never land of a transcendent order until we collapse them, bringing them into the world of manifestation with an act of observation. In the process, we choose one facet out of two, or many, that are permitted by the Schrödinger equation; it is a limited choice, to be sure, subject to the overall probability constraint of quantum mathematics, but it is a choice, nevertheless.**

When he was asked if the man was dead or alive, the Zen master replied, "I cannot say." How could he? According to idealism, the essence of a man, consciousness, never dies. So, it would be incorrect to say outright that the man is dead.

The idealist resolution does imply the action of consciousness upon matter. That action, however, poses a problem only for material realism. In this philosophy [scientific naturalism or materialism], consciousness is an epiphenomenon of matter, and it seems impossible that an epiphenomenon of matter could act on the very fabric of which it is built — in effect causing itself.

That causal paradox is avoided by monistic idealism, in which consciousness is primary. In consciousness, coherent superpositions are transcendent objects. They are brought into immanence only when consciousness, by the process of observation, chooses one of the many facets of the coherent superposition, though its choice is constrained by the probabilities allowed by the quantum calculus. (Consciousness is lawful. The creativity of the cosmos comes from the creativity of its quantum laws, not from arbitrary lawlessness.)

Goswami, Amit. *The Self-Aware Universe* (pp. 81-84). Penguin Publishing Group. Kindle Edition.

Based upon the observational evidence and my own personal experiential evidence, I believe in the **Idealist Resolution to Quantum Mechanics** and NOT the Monistic Idealism Interpretation of Quantum Mechanics. There's a difference, which Amit Goswami helped me to identify.

I find this whole thing fascinating. Of course, your mileage may vary. But, I like trying to figure out how things really work. When it comes to the Quantum Interpretations which actually match with all the observational evidence that we have on hand as a race, I believe that Amit Goswami's **Idealistic Resolution** is a contender.

Science 2.0 allows all of the evidence into evidence; but in the end, I prefer to let you decide for yourself what to make of all that evidence.

### The Orthodox Interpretation of Quantum Mechanics

Marshall, P. D., Kelly, E. F., & Crabtree A. (Eds.). (2015). *Beyond Physicalism: Toward Reconciliation of Science and Spirituality*. London, United Kingdom: Rowman and Littlefield.

This book contains one of my most favorite chapters about Quantum Mechanics, a chapter called "A Quantum Mechanical Theory of the Mind/Brain Connection" by Henry P. Stapp, which introduces and discusses the Orthodox Interpretation of Quantum Mechanics.

See this link for the essential bits:

http://mypsyche.us/wp-content/uploads/2017/10/A-Quantum-Mechanical-Theory-of-the-Mind-Brain-Connection.pdf

Online Source:

https://books.google.com/books?id=RuPFBgAAQBAJ&pg=PA157&lpg=PA157&dq#v=onepage&q&f=false

The whole purpose of Science 2.0 is to find different ways to account for and explain the observational evidence, the eye-witness evidence, and the experiential evidence provided by the Lived Experiences of the human race. The Orthodox Interpretation of Quantum Mechanics does just that.

The Orthodox Interpretation of Quantum Mechanics was named by Eugene Wigner in 1963:

> http://www.psiquadrat.de/downloads/wigner1963.pdf

> http://mypsyche.us/wp-content/uploads/2017/10/Wigner-The-Orthodox-Interpretation.pdf

The following from *Beyond Physicalism* by Henry P. Stapp is rather complex; but, I trust my reader to be able to figure it out, if it is of interest to him. This was my first introduction to the Orthodox Interpretation of Quantum Mechanics. You have probably never heard of this thing, because this is science which the Materialists, Naturalists, Nihilists, and Atheists are trying to hide from us because when fully understood it falsifies Materialism and Naturalism. The Materialists and Naturalists don't want us finding and using anything that falsifies Materialism and Naturalism, because these people have an *a priori* dedication and devotion to Materialism and Naturalism. It's their religion, and some of them will defend it with their very lives.

This information has been successfully hidden from us. I have been a scientist all of my life for over 55 years of life; and, October 2017 was the first time in my life that I was introduced to the Orthodox Interpretation of Quantum Mechanics. When it comes to science, I tend to give Quantum Mechanics precedence over everything else, because Quantum Mechanics is the bridge between the physical and the non-physical – the local and the non-local respectively. Quantum Mechanics has been constantly verified; and, Quantum Mechanics has never been falsified through practical application of the theory. Quantum Mechanics is real, and it works. The ONLY problem we have with Quantum Mechanics is that it has been given dozens of different FALSE and FALSIFIED interpretations, which don't match with reality and don't match with the observational evidence and the lived experiences of the human race.

When it comes to Quantum Mechanics, it's extremely important to get an "interpretation" of Quantum Mechanics that matches with reality, the lived experiences of the human race, and the applied realities of the science being done by using Quantum Mechanics. When it comes to Quantum Mechanics, we need an interpretation that is based upon observational evidence, experiential evidence, empirical evidence, non-local evidence, and veridical evidence. An interpretation of Quantum Mechanics that is based upon Materialism and Naturalism is worthless, because Materialism and Naturalism are NOT veridical and cannot be verified.

*Beyond Physicalism* was an expensive book, but it was worth the price of admission for this chapter alone.

**Nobel Laureate Eugene Wigner (1963), in a paper entitled "The Problem of Measurement," used the term "orthodox interpretation" to identify the interpretation spelled out in mathematical detail by John von Neumann (1932/ 1955) in his book *Mathematische Grundlagen der Quantenmechanik*.**

John von Neumann gave the name "Process 2" to the continuous evolution of the quantum state [or wave function] generated by what is called the Schrödinger equation. This Process 2 is the quantum-mechanical generalization of the classical mechanical process that generates evolution [change] in time.

[In other words, Process 2 is the process by which quantum mechanics gets converted into classical mechanics – the process by which the spiritual or the non-local becomes physical.]

**The locality condition** [physical condition] **prevents Process 2 from producing any faster-than-light action at a distance.**

[Process 2 represents the physical interface associated with Quantum Mechanics. Process 2 is physical; whereas, Process 1 and Process 3 are non-local or spiritual, and thereby capable of faster-than-light interaction or instantaneous action at a distance.]

**According to orthodox quantum mechanics, this Process 2** [wave function collapse or localizing of the particle] **is not the whole story. The Process-2 evolution is interrupted by** [and controlled by] **observational processes involving two kinds of choices: a Process-1 choice of a probing question on the part of an observing agent, and for each such probing question, a Process-3 response from nature.**

[Process 1 and Process 3 are functions of Non-Local Consciousness – CHOICE and RESPONSE respectively. Psyche or the Observer makes the command, the observation, or the CHOICE. Quantum Objects or Matter of any kind decides whether to respond, or not. Psyche ACTS; and, matter (quantum objects) of any kind REACTS.]

**The general logical form of the quantum-mechanical process of observation is this: an observing agent chooses and performs a probing question that inquires of nature whether the physical world has, or does not have, a certain physical property. This choice of a probing question was called by Heisenberg "a choice on the part of the observer."**

**Nature immediately responds by choosing either the positive ("Yes") answer or the negative ("No") answer. This choice was called by Dirac "a choice on the part of nature."**

**If the positive ("Yes") response is selected by nature, then two things happen: (1) at the experiential level, the observer experiences the response associated in his mind with the existence of the physical property about which he is inquiring; (2) at the level of atom-based physical descriptions, the quantum state of the universe "collapses" to the part of itself that is compatible with the existence of that physical property.**

**A negative response is accompanied by a reduction of the physically described state of the universe to the part of itself that definitely does not have the property in question, but with no associated experience.**

**An essential feature of this process of observation is that the observer-chosen property is something that the system being observed definitely possesses after a positive response, but may not have possessed before the process was initiated.**

This "collapse" idea resolves — by official edict [by axiom] — the wave-particle duality problem. [It also explains and solves the mind/brain interface as well as the Mind-Brain Problem.] **Indeed, when coupled to a certain specified statistical rule, it accounts for all relevant data. But these collapses raise the above-mentioned "locality" puzzles. The founders dodged them by claiming to be providing only a practical tool that works in actual practice, and refusing to be drawn into debates about "reality" that go beyond practical utility.**

John von Neumann faced the puzzles head-on, and, following the logic, produced a rigorous formulation of quantum mechanics that is widely used by physicists, and by others who need a mathematically precise version of the theory. Von Neumann's version can be interpreted as describing a psychophysical reality that evolves [changes] only forward in time, and in which each of us human beings exists as a psychophysical agent that can make "free and causally efficacious choices."

How did von Neumann achieve this rational reconciliation of the mental and physical aspects of the theory?

The original "Copenhagen" way of describing this collapse process was tied to a mysterious thing called the "Heisenberg cut." Everything lying "below" this cut was supposed to be described in the mathematical language of quantum mechanics, whereas everything lying "above" the cut was to be described either in the language of classical physics — which we use to describe to ourselves and to others "what we have done and what we have learned" — or in psychological terms. The experiential aspects are described in mental and classical-physics terms [materialism and naturalism], whereas their atomic underpinnings are described in terms of the quantum mathematics.

A practical account, in order to be useful to us, must accommodate our mental intentions. It must allow them to be "freely chosen," not predetermined at the birth of the universe, and it must support the capacity of our "freely chosen" mental intentions to influence our future experiences.

This Heisenberg cut was "movable": its placement depended on what practical use was to be made of the theory. But that "movability" meant that the same physical thing could be described in two logically incompatible ways — either classically or quantum mechanically — depending on the practical application. Such an inconsistency might be all right for a purely pragmatic theory, but it is not acceptable for a putative description of reality itself.

A principal move made by von Neumann was to show that the Heisenberg cut could be moved all the way up, so that reality is unambiguously separated into a mental part, and a part described in terms of the mathematical language of quantum mechanics. The external measuring devices [physical devices] then become parts of that latter world, while the "classical descriptions" of these devices become identified as aspects of the perceptions of observers — a status often emphasized by Bohr.

The Heisenberg cut is shifted up, step by step, until all atomically constituted things, including our physical bodies and brains, lie below the cut, and hence are described in the mathematical language of quantum mechanics. The observer's mental aspects are *held fixed* during these shifts of the cut, and they are eventually pushed completely out of the physically described universe.

[The Observer's mental aspects, or mind, or consciousness is pushed into the non-local realm or the spirit realm or the transdimensional realm, by shifting the Heisenberg cut all the way up as far as it can go.]

These preserved mental aspects were called "abstract egos" by von Neumann. They are mental in character because they are the parts of the psychophysical structure that represent what the theory is supposed to explain, namely the structure of our conscious experiences, and they are, by virtue of von Neumann's argument, ontologically distinct from the atomically constituted physical world. Yet each such ego retains, in the orthodox theory, a quantum dynamical linkage to its associated physical brain.

Thus Tyndall's "impassable gulf" between "man the object" and "man the subject" was bridged by rigorous quantum mathematics. Von Neumann converted what had originally been offered as a mere practical tool into a rationally coherent putative description of a dynamically integrated psychophysical reality.

Von Neumann's formulation eliminates the notion that things become classical merely by being "big" — that mere "bigness" can somehow cause the quantum-to-classical transition.

But how big is big? Von Neumann's formulation ties the collapse not to something as nebulous as "big," but to something that, according to the theory, is separate and distinct from the atomically constituted physical world — namely our conscious experiences! And his theory specifies the physical place where a person's conscious actions act. They act, via nature's choice, on that person's physically described brain, which is linked via mathematically described physical connections to the rest of the physically described universe.

Thus, quantum mechanics, in its orthodox form, is not simply a theory of a world made of atomic constituents. It is basically a theory of *"the manifold aspects of our experience"* and their connections to each other via atomically constituted, physically described brains — which are, in turn, connected to each other via other parts of the atomically described world. An important aspect of the new theory is a revised "quantum neuroscience" that contradicts "promissory materialism" by allowing our ontologically non-physical minds, or egos, to inject logically required and causally effective inputs into quantum mechanically described brain dynamics.

### The Causal Effectiveness of Mental Intent

It might seem that a mere capacity to pose questions and register answers would leave our conscious egos just as helpless and impotent as before. But the quantum-mechanical process of posing questions and receiving responses is not like the classical mechanical process, in which our observations have no physical effects. In QM, the observer's "free choice" of which question to ask plays a critical role in determining which potential physical properties will become actualized.

In QM, the observer asks nature a yes/no question about the state of an observed system. If nature's response is "Yes," then, after this response is delivered, the observed system will definitely have the property that the observer inquired about.

Normally, this dependence of the post-observation properties of the system being probed upon the observer's choice of question does not give the observer any effective control over the observed system. That is because nature's response can be "No."

However, there is an important situation in which, according to the quantum rules, the "No" answers will be strongly suppressed. In that case, the free choices made by the observer can exert effective control over the system being probed — which, in von Neumann's theory, is the brain of the observer.

Suppressions of the "No" responses occur if an initial "Yes" response is followed by a sufficiently rapid sequence of posings of the same question. In that case the observer becomes empowered, by his own free choices of what property to probe, and when to probe it, to hold stably in place a brain activity that otherwise would quickly fade away. This effect is the celebrated Quantum Zeno Effect (QZE).

*Beyond Physicalism: Toward Reconciliation of Science and Spirituality* (Kindle Locations 3342-3432). Rowman & Littlefield Publishers. Kindle Edition.

Note: This book, *Beyond Physicalism*, permits quoting of passages in reviews. Basically, what I am doing here is telling you to buy the book and read the book.

See:

http://mypsyche.us/wp-content/uploads/2017/10/A-Quantum-Mechanical-Theory-of-the-Mind-Brain-Connection.pdf

Online Source:

https://books.google.com/books?id=RuPFBgAAQBAJ&pg=PA157&lpg=PA157&dq#v=onepage&q&f=false

**The Orthodox Interpretation of Quantum Mechanics**

Under the Orthodox Interpretation of Quantum Mechanics, Psyche or Non-Local Consciousness is something completely separate from Quantum Objects or Particles of Matter; and, both the Observer (Psyche) and Nature (Quantum Objects) each have innate non-local consciousnesses or psyches of their own.

Psyche queries or questions; and, Nature chooses whether to respond or not. This Reality explains the different out-of-body or spiritual observations wherein the observer or the seer or the clairvoyant is able to sense and KNOW that the quantum objects and physical objects surrounding them are alive, sentient, conscious, and aware of their surroundings. The rocks are conscious and aware of you. They are little recording devices keeping track of all your choices and actions.

With God's help, my best friend shifted into this state of consciousness, wherein everything around him became vibrant and alive, and he could sense the consciousness or life force coming from the sun, the air, the trees, the plants, the rocks, and the river. He said that the whole thing is alive and aware of us. The rocks are watching us, recording us, and at times even choosing to respond to us. Process 3 of the Orthodox Interpretation gives us a scientific explanation for this observed reality or experienced phenomenon, wherein Nature or physical matter is alive and aware of us and chooses whether to respond to our queries or not.

The atomic state or the state of the atoms (the collapse of the wave function) is observed into existence or chosen into existence by mind, consciousness, psyche, or the observer. Furthermore, the observed object (particle of matter) also chooses whether to comply with the observational command from psyche or to reject the observational command from psyche, according to the laws which the quantum object or observed object has covenanted with God to obey. The Orthodox Interpretation of Quantum Mechanics explains all of this in great detail.

Quantum Mechanics has enough depth and breadth to be able to adequately explain the mind/brain interface as well as solve the Mind-Brain Problem, so long as a Quantum Interpretation is chosen which properly explains the experiential and observational evidence (scientific evidence) that we receive from Near-Death Experiences (NDEs) and Out-of-Body Experiences (OBEs).

If given a materialistic and naturalistic interpretation, Quantum Mechanics cannot explain NDEs and OBEs, nor can it explain the mind/brain interface, nor solve the Mind-Brain Problem. We have to get rid of Materialism and Naturalism if we want to successfully solve the Mind-Brain Problem and explain the mind/brain interface with Quantum Mechanics. Remember, the very existence of the Orthodox Interpretation by definition, in principle, and in practice FALSIFIES Materialism and Naturalism. The only way to solve the Mind-Brain Problem is to eliminate Materialism and Naturalism, because Materialism and Naturalism caused the Mind-Brain Problem in the first place. Materialism and Naturalism are a faulty and incorrect interpretation of Reality and Existence, so these things are going to naturally produce faulty conclusions and unsolvable problems like the Mind-Brain Problem.

According to this Orthodox Interpretation Model of Quantum Mechanics, both the Observer (Process 1) and Nature (Process 3) are conscious and aware of each other. The Observer ACTS, CHOOSES, QUESTIONS, or COMMANDS; and, Nature or Matter or the Quantum Object REACTS. Process 1 and Process 3 function perfectly fine both locally (physically) and non-locally (spiritually). Process 2 explains how the spiritual or the non-local becomes manifested in the physical or the local 3D space-time realm. Process 2 links the spiritual with the physical by collapsing the wave.

Each physical human being exists as a psychophysical agent that can make free and causally efficacious choices. Our human psyche exists as a psychological agent, which can make free and causally efficacious choices both here in the physical realm and there in the non-local transdimensional spirit realm when our physical brain is dead and gone.

Under the Orthodox Interpretation of Quantum Mechanics, the observer's mental aspects are pushed completely out of the physically described universe into the Non-Local Realm, or the Transdimensional Realm, or the Spirit Realm. Thereby, the Mind-Brain Problem is solved by Quantum Mechanics, by defining mind as a non-physical, non-local, transdimensional entity.

Each ego or mind or psyche retains, in the orthodox theory, a quantum dynamical linkage to its associated physical brain.

The Soul or Astral Body is typically defined as a combination of Psyche and Spirit Body. Out-of-Body Travelers have observed a refined web-like substance that they call the Silver Cord, which links the Soul (psyche and spirit body) with a specific physical body. Sever the Silver Cord, and the physical body dies.

> **The silver cord in metaphysical studies and literature, also known as the sutratma or life thread of the antahkarana, refers to a life-giving linkage from the higher self (atma or atman) down to the physical body. The silver cord is described as a strong, silver-colored, elastic cord which joins a person's physical body to its astral body.**
>
> https://en.wikipedia.org/wiki/Silver_cord

The Silver Cord has been observed by Out-of-Body Travelers.

> **Does the biblical silver cord exist?**
>
> **The biblical concept of the silver cord is accurate. According to my observations, it is not actually a cord but a thin, fibrous substance similar in appearance to a spider's web. The silver cord appears to function as the connection between the physical body and the first inner energy-body of all life-forms. Though its complete function is unknown, it's logical that it may act as an inner energy conduit. One thing is certain, when the cord is severed, the biological life ends.**
>
> Buhlman, William L. (1996). *Adventures Beyond the Body: How to Experience Out-of-Body Travel* (p. 265). HarperCollins. Kindle Edition.

Observe that the Orthodox Interpretation of Quantum Mechanics actually provides a scientific explanation for all of this; whereas, you won't get any explanation for the non-local or the spiritual from the materialistic and naturalistic interpretations of quantum mechanics. You will only get dismissal of the evidence and suppression of the evidence from the Materialists, Naturalists, Nihilists, and Atheists because these people don't want there to be any proof of non-locality or the spiritual.

Von Neumann's formulation ties the wave function collapse to something that, according to the theory, is separate and distinct from the atomically constituted physical world — namely our conscious experiences! Our non-local consciousness or psyche questions or probes nature; and then, nature chooses whether to answer the question or to ignore it. Whether nature chooses yay or nay, nature's choice results in the collapse of all future possibilities into a single current reality, where that particular query or probe is concerned.

Thus, quantum mechanics, in its orthodox form, is basically a theory of "our experiences" and their connections to each other via atomically constituted, physically described brains. In other words, the Orthodox Interpretation is based upon the Lived Experiences of the human race, just like Science 2.0.

The idea is that there really is someone driving the bus; and, that particular someone continues to live and thrive long after our physical brain is dead and gone. According to the Orthodox Interpretation, Quantum Mechanics is the interface between that someone and his or her assigned physical brain. Quantum Mechanics (Process 2) is the interface or the science between psyche and the physical brain.

Physical matter, including the physical matter in our physical bodies, has the ability to choose to comply with a command or to reject a command from our psyche or non-local consciousness. That's why we don't see too many people walking on water or moving mountains with their minds. However, the physical matter in your physical body has covenanted to obey the commands from your particular psyche. When it comes to the physical brain of the observer, the "No" answers are being suppressed, and your physical brain simply ends up following the commands of your psyche.

Obviously, Science 2.0 is completely compatible with the Orthodox Interpretation of Quantum Mechanics, because psyche or non-local consciousness has been observed, lived, and experienced during Near-Death Experiences (NDEs) and Out-of-Body Experiences (OBEs); and, Science 2.0 gives priority to Lived Experiences including Non-Local Experiences. The Orthodox Interpretation is based upon "experiences" or "observations"; and, so is Science 2.0. In a very real sense, Science 2.0 IS the Orthodox Interpretation of Quantum Mechanics.

Interestingly, the all-knowing Wikipedia (which many people claim leans heavily and favorably towards materialism, naturalism, and atheism) simply lumps both the Quantum Consciousness Interpretation and the Orthodox Interpretation together into a one paragraph whole, which they call "Consciousness Causes Collapse"; and, the two examples that they do give seem to side-step Eugene Wigner's Cosmic Consciousness (God) Interpretation and John von Neumann's Orthodox Interpretation. It's as if these people don't want you to know about these things – either hiding them, ignoring them, discounting them, or trying to explain them away. The other approach these people take is to openly ridicule and mock Quantum Consciousness, Cosmic Consciousness, and the Orthodox Interpretation which claim that Psyche or Non-Local Consciousness causes the wave function to collapse.

From the Wikipedia:

> In his treatise *The Mathematical Foundations of Quantum Mechanics*, John von Neumann deeply analyzed the so-called measurement problem. He concluded that the entire physical universe could be made subject to the Schrödinger equation (the universal wave function). He also described how measurement could cause a collapse of the wave function. This point of view was prominently expanded on by Eugene Wigner, who argued that human experimenter consciousness was critical for the collapse.
>
> Variations of the "consciousness causes collapse" interpretation include:
>
> ### Subjective reduction research
>
> This principle, that consciousness causes the collapse, is the point of intersection between quantum mechanics and the mind/body problem; and researchers are working to detect conscious events correlated with physical events that, according to quantum theory, should involve a wave function collapse.
>
> ### Participatory anthropic principle (PAP)
>
> John Archibald Wheeler's participatory anthropic principle says that consciousness plays some role in bringing the universe into existence.

(I excluded the parts where they tried to discount it, marginalize it, and explain it away, so that it takes on some semblance of reality.)

https://en.wikipedia.org/wiki/Interpretations_of_quantum_mechanics

I find it interesting how these people typically marginalize and water-down anything that points us to non-local consciousness and a resolution of the mind-brain problem. When it comes to practically everything that I read, I have to delete or debunk that worthless naturalistic and materialistic junk, so that only the valuable observed truths and experienced truths remain.

The Quantum Zeno Effect has been verified. Atoms won't move or tunnel while you watch them. They stay where they are while you are paying attention to them. In other words, nature or physical matter actually responds to your queries, or your probing, or your observations, just as the Orthodox Interpretation claims and explains. Those atoms are aware of you and know whether you are looking at them or not, just as explained in Process 3 of the Orthodox Interpretation of Quantum Mechanics.

Remember, the Quantum Zeno Effect FALSIFIES Materialism and Naturalism; and, when those go down, Darwinism, Nihilism, and even Atheism go down with them.

https://phys.org/news/2015-10-zeno-effect-verifiedatoms-wont.html

http://mypsyche.us/wp-content/uploads/2017/10/Zeno-effect-verified—atoms-wont-move-while-you-watch.pdf

Atoms: Those little buggers FREEZE whenever you look at them. They don't move or tunnel until after you stop looking at them. They KNOW when you are looking at them, and when you are not. There is consciousness and awareness within every physical atom. Those atoms aren't empty space after all, as the Materialists claim. There's actually someone there looking back at you, sensing you (your psyche), and aware of you. The atoms are sentient, even more aware of things than the Materialists and Naturalists are. You can literally hold the atoms in place by using them or by trying to measure them; but then, they go off and do their own thing whenever you are no longer trying to use them or observe them.

This observation of the Quantum Zeno Effect fits in perfectly with Process 3 of the Orthodox Interpretation, which states that nature, or physical matter, or atoms actually choose whether to respond to our queries and our probing and our observations, or not. The atoms can sense you – your non-local consciousness – and respond to you, because the atoms have a non-local consciousness of their own. In that respect, atoms are very much alive. Once again, the non-physical aspects or non-local aspects of Quantum Mechanics have been confirmed through observation and experimentation.

In addition to verifying Process 3 and Process 1 of the Orthodox Interpretation of Quantum Mechanics, the Quantum Zeno Effect has other uses.

> **One quantum mechanics discovery that may help us understand how we decide is the quantum Zeno effect. Physicists have found that if they observe an unstable elementary particle continuously, it never decays — even though it would almost certainly decay if it were not observed. In quantum physics, it is not possible to separate the observer entirely from the thing observed. They are part of the same system. The physicists are, essentially, holding the unstable particle in a given state by the act of continuing to measure it. In the same way, experiments have shown that, because your brain is a quantum system, if you focus on a given idea, you hold its pattern of connecting neurons in place. The idea does not decay, as it would if it were ignored. But the action of holding an idea in place truly is a decision you make, in the same way that the physicists hold a particle in place by deciding to continue to observe it.**
>
> **Can the Adult Human Brain Change?**

> For many years, neuroscientists believed that the adult human brain was essentially finished. It did not and could not change, any more than a billiard ball could, and individual neurons did not regenerate. According to the classical view, in such a fixed system certain mental programs were simply run over and over. Individual decisions did not affect the functioning of the system, but were rather a delusion created by the functioning of the system.
>
> In recent years, however, neuroscientists have discovered that the adult brain is actually very plastic. As we will see, if neural circuits receive a great deal of traffic, they will grow. If they receive little traffic, they will remain the same or shrink. The amount of traffic our neural circuits receive depends, for the most part, on what we choose to pay attention to. Not only can we make decisions by focusing on one idea rather than another, but we can change the patterns of neurons in our brains by doing so consistently. Again, that has been demonstrated by experiments and is even used in psychiatric treatments for obsessive compulsive disorder.
>
> So, what happens in our brains when we make a decision? According to the model created by H. Stapp and J. M. Schwartz, which is based on the Von Neumann interpretation of quantum physics, conscious effort causes a pattern of neural activity that becomes a template for action. But the process is not mechanical or material. There are no little cogs and wheels in our brains. There is a series of possibilities; a decision causes a quantum collapse, in which one of them becomes a reality. The cause is the mental focus, in the same way that the cause of the quantum Zeno effect is the physicists' continued observation. It is a cause, but not a mechanical or material one. One truly profound change that quantum physics has made is to verify the existence of non-mechanical causes. One of these is the activity of the human mind, which, as we will see, is not identical to the functions of the brain.
>
> Beauregard, Mario. *The Spiritual Brain: A Neuroscientist's Case for the Existence of the Soul* (Kindle Locations 824-847). HarperCollins. Kindle Edition.

The Quantum Zeno Effect is used to verify the existence of non-local causes or non-physical causes, which are non-mechanical immaterial causes such as Non-Local Consciousness or Psyche. In other words, Psyche or Non-Local Consciousness is a Causal Agent.

The interesting thing to note is that each physical atom has a psyche or non-local consciousness of its own, which was designed and organized by God or has covenanted with God to respond to an Observer's questions and commands.

Your physical body and physical brain are a collection of physical atoms that have covenanted with you and have been conditioned by you to obey your Psyche's commands. Whereas your physical body and physical brain belong to you and have covenanted to obey you, our physical universe belongs to God's Psyche or God's Mind and has covenanted to obey God. This is what our Non-Local Observations and the Quantum Experiments are trying to teach us; but, the Materialists and Naturalists refuse to listen. Nevertheless, the very existence of the Quantum Zeno Effect and its repeated scientific verification and scientific observation FALSIFIES Materialism, Naturalism, and their derivatives such as Darwinism, Nihilism, and Atheism.

Remember, Science is Observation, which means that Science consists of our Lived Experiences as a race. Science 2.0 is based upon Observations and Lived Experiences, rather than philosophical speculation and scientific inferences. Science 2.0 relies upon our non-local experiences just as much as it relies upon our physical experiences. That's why Science 2.0 is a significant upgrade compared to what we had before. Science 2.0 moves beyond physicalism into whole new realms of existence that I had never considered before. After all, I used to be a Materialist, Nihilist, and Atheist until I learned better.

## Integrating the BioPsychoSocial Model of Reality with Quantum Mechanics

Barlow, D. H. & Durand, V. M. (2015). *Abnormal Psychology: An Integrative Approach* (7th ed.). Stamford, CT: Cengage Learning.

http://mypsyche.us/wp-content/uploads/2017/10/Abnormal-Psychology.pdf

This book uses the BioPsychoSocial Model of Reality to explain personality disorders, mental illness, psychopathology, and abnormal behavior. It ended up being my most favorite college textbook.

I had a visiting Clinical Psychologist from the University of Utah tell me that when it comes to psychotherapy and the treatment of mental illnesses, everything is BioPsychoSocial now. While studying abnormal behavior, mental illnesses, and psychopathology in college, we used the integrative BioPsychoSocial Model to explain what was happening.

The "bio" portion of the BioPsychoSocial Model is represented by genetics, heritability, and physical matter. The "bio" portion is Process 2, or the wave function collapse, within the Orthodox Interpretation of Quantum Mechanics. The wave function collapse (Process 2) is controlled by the negotiation between Process 1 (the observer's psyche) and Process 3 (nature's chosen responses). This is a negotiation which takes place between two different types of psyche or non-local consciousness.

The "social" portion of the BioPsychoSocial Model is represented by other psyches as found within nature or the environment itself. Other psyches, or nature's chosen response, can have a powerful influence on the development (and the cure) of mental illnesses. The "social" portion is Process 3, or nature's chosen responses, within the Orthodox Interpretation of Quantum Mechanics. ONLY nature's choice (Process 3) can force the wave function to collapse (Process 2) into existence. Nature's psyche chooses how the physical matter is going to operate or function. Nature's psyche is located within each and every physical atom.

The "psycho" portion of the BioPsychoSocial Model is represented by your psyche or non-local consciousness. The "psycho" portion is Process 1, or the questioning phase, within the Orthodox Interpretation of Quantum Mechanics. The way you identify the "psycho" portion of the BioPsychoSocial Model is when a CHOICE is made. It all comes down to choice. Process 1 and Process 3 represent a constant negotiation between your psyche and the psyches which are found within atoms or physical nature, as demonstrated and proven by the Quantum Zeno Effect.

I suffered from substance-induced psychosis for a couple of years. I was addicted to half a dozen different prescribed medications. The medical doctors had me cycling through a dozen different medications on any given day; and, half of them were addictive. I would go psychotic any time I tried to get off the addictive prescriptions. I was trapped, and suicidal, with no hope of ever getting out of the situation.

There are a lot of alcoholics and drug addicts on my mother's side of the family; and, I inherited that tendency or vulnerability. I was immediately addicted to any addictive substance that the medical doctors gave me; and, the net result was always months of withdrawal symptoms (psychosis) whenever I tried to get off the drugs. A week of use would always result in six weeks of withdrawal. A month using addictive drugs would result in six months of withdrawal symptoms, until my body chemistry normalized again.

My genetic inheritance (BIO) makes me vulnerable or susceptible to addiction. As one neuropsychologist described it, I'm overly sensitive to medications and should never be placed on potentially addictive medications and should never take the full dose of any medication.

Furthermore, my society (SOCIAL) and medical doctors caused my addictions by prescribing half a dozen different addictive substances for me to take.

The Bio/Social parts of the BioPsychoSocial Model actually caused my mental illness, fed my mental illness, and prevented me from overcoming my mental illness or substance-induced psychosis.

The ONLY cure for my substance-induced psychosis or mental illness was the (PSYCHO) portion of the BioPsychoSocial Model. My psyche or non-local consciousness literally had to CHOOSE to stop taking all of the different medications, and simply suffer the consequences of withdrawal. I went cold turkey, and I went insane. It was over six months before I started thinking normally again and fully realized how insane I had been. That's a long time to be insane. Of course, I had been semi-insane for a couple of years going through multiple different withdrawals and relapses, before DECIDING to go cold turkey and end the cycle of addiction.

So, how do you identify and separate the different parts of the BioPsychoSocial Model?

Psyche or Non-Local Consciousness is identified by DECISIONS and CHOICES. Any time you witness choice in action you are in fact witnessing some kind of psyche in action, because psyches are the only thing in the universe that can make choices. Psyche is an independent agent that makes choices.

Developmental Illnesses end up being a complex interaction between genetics (Bio), mommy's choices (Social), nutrition and food supplies (Social or Environmental), and the organism's choices (Psycho).

When it comes to most mental illnesses, about 50% of it on average is due to a genetic vulnerability of some kind. If there is no genetic vulnerability, however, then it becomes much harder to develop a mental illness but not impossible, because approximately 50% of a mental illness is PsychoSocial in nature and origin. Environmental triggers, behavioral conditioning, and the opportunity to become addicted are Social in nature and origin. Other psyches and the physical environment can have an influence on the development, progression, and course of a mental illness, as well as triggering the mental illness in the first place. However, some of our mental illnesses are literally chosen into existence by Psyche or Non-Local Consciousness. Any time you witness choice in action, you are witnessing Psyche in action; and therefore, you are witnessing the "psycho" portion of the BioPsychoSocial Model in action.

The BioPsychoSocial Model and the Orthodox Interpretation of Quantum Mechanics make for an infinitely more powerful, believable, useful, and realistic Model of Reality than Materialism and Naturalism can even begin to produce. Think about it. The observational evidence suggests that this is true.

Remember, by design, Materialism and Naturalism cannot be used to develop a Model of Reality that will adequately explain Quantum Non-Locality, Action at a Distance, Telepathy, Psyche, Spirit Matter, the Afterlife, Spirituality, Psi, and Non-Local Consciousness. Materialism and Naturalism are bankrupt to begin with; and, it only goes downhill from there.

Science 2.0 defines science as Observation, Eye-Witness Empirical Evidence, and the Lived Experiences of the Human Race as a whole. Under Science 2.0, Materialism, Naturalism, Nihilism, and Atheism are defined as unscientific, because these philosophies have NO observational evidence supporting their major premises or primary assumptions. Science 2.0 is based upon the Non-Local Consciousness Interpretation of Quantum Mechanics as developed by Pim van Lommel, the Quantum Consciousness Interpretation and Cosmic Consciousness Interpretation of Quantum Mechanics as developed by lots of different people, and the Orthodox Interpretation of Quantum Mechanics as developed by Eugene Wigner and John von Neumann and presented by Henry P. Stapp.

We absolutely need some kind of **Non-Local Consciousness Interpretation for Quantum Mechanics** if we are ever going to successfully describe and explain how Non-Local Consciousness and Spirit Matter function and work.

## 2. Observational Evidence and Experiential Evidence that Falsifies Materialism and Naturalism:

Buhlman, W. L. (1996). *Adventures Beyond the Body: How to Experience Out-of-Body Travel*. New York: HarperCollins Publishing.

By definition as a founding principle, Science 2.0 allows ALL of the evidence into evidence. Science 2.0 is based upon the experiential worldview that is presented and developed in this book from William Buhlman, and similar ones like it.

This book changed my life; and, I had been looking for this book all of my life. I finally found and read this book at the end of 2015. This book was officially the END of My Materialism and My Nihilism, and the beginning of a whole new and different way of looking at the world, science, and reality. After reading this book, there was no going back to My Materialism and My Nihilism. Thanks to this book, I now have a completely different model of reality upon which to base my beliefs – something that I like to call the Ultimate Model of Reality or a Psyche Ontology, wherein psyche or non-local consciousness is the fundamental unit of reality.

Science 2.0 is based exclusively on eye-witness evidence, observational evidence, and experiential evidence which philosophers call the Lived Experiences of the human race. Under Science 2.0, it doesn't matter what the Materialists, Naturalists, Darwinists, Nihilists, and Atheists choose to believe or promote, because observational evidence trumps ALL of their philosophical speculations, scientific inferences, and guesswork. Since Materialism, Naturalism, Darwinism, Nihilism, and Atheism have NO observational evidence or eye-witness evidence to support them and never will because they can't, these philosophies or pseudo-religions are officially classified as "unscientific" under Science 2.0. Science 2.0 tells it as it is, and NOT as what someone would like it to be.

Science is observation. Science is knowledge. The BEST way to find and know the truth is to live it, observe it, and experience it for yourself or to choose to trust someone who has.

By definition in principle, **scientific inference** is circumstantial evidence that completely lacks observational support or experiential support. Scientific inference is extremely weak in comparison to eye-witness evidence or observational evidence. Scientific inference is purely speculation, or guesswork, or personal interpretation of someone else's observational experiences. Materialism, Naturalism, Darwinism, Nihilism, and Atheism are observed to be nothing more than **scientific inferences**. They are weak, because they don't hold up under severe critique and scrutiny and skepticism. They are weak, because they have no eye-witness evidence or observational evidence to support them.

Remember, circumstantial evidence is evidence that relies on an inference to connect it to a conclusion of fact. Materialism, Naturalism, Darwinism, Nihilism, and Atheism are based exclusively on scientific inferences or circumstantial evidence – the weakest form of evidence. Every scientific inference is based upon circumstantial evidence – these people are simply guessing, and then turning around and submitting their speculations and guesswork as proof that their conclusions are correct and true. This process is known as the *affirming the consequent* logic fallacy or the *begging the question* logic fallacy. The whole of their pseudo-science or philosophy or religion is based upon these logic fallacies. *Scientific inference* is also a logic fallacy, which the Materialists, Naturalists, Darwinists, Nihilists, and Atheists rely upon exclusively for their "proof" or evidence. Remember, with scientific inferences, you can have any conclusion you want simply by assuming unilaterally, "a priori" or before the fact, that your chosen conclusion is true. That's the very definition of *affirming the consequent* or *begging the question*, which are the foundation upon which Materialism and Naturalism are built.

By contrast, direct observational evidence supports the truth of an assertion directly, without need for any additional evidence or inference. Remember, *scientific inference* of any kind is a logic fallacy and consequently classified as circumstantial evidence, because it lacks observational support by definition. Truth is knowledge, because it's based upon direct observational evidence or an eye-witness account. The best source of truth and knowledge is direct observational evidence or lived experiences. The only problem we have with observational experiences comes when the Materialists and Naturalists start applying scientific inferences to these lived experiences in an attempt to explain them away.

Science is observation; therefore, observational experiences or lived experiences should trump scientific inferences or circumstantial evidence every time. The only ones who don't follow this rule or Scientific Law are the Materialists, Naturalists, Darwinists, Nihilists, and Atheists who rely exclusively on circumstantial evidence and scientific inferences instead.

For me personally, this type of observational evidence from Out-of-Body Experiences (OBEs) was convincing scientific proof of non-local consciousness and the start of Science 2.0 along with the start of my decision to give priority to the lived experiences of the human race over anything derived from scientific inferences and philosophical speculation. I simply examine the evidence – all of the evidence – and then decide for myself what to make of it. I encourage my readers to do the same.

There are over 13 million Near-Death Experiences (NDEs) on record; and, these are case studies. Case studies are always the best source of Scientific Hypotheses. Some of these NDEs are extremely scientific, having confirmational evidence or veridical evidence or verified evidence attached to them supporting them.

EVIDENCE that the SOUL Exists: Pam Reynolds' NEAR DEATH Experience:

https://www.youtube.com/watch?v=P4EztGUHnbs

https://www.youtube.com/watch?v=YO8UVebuA0g

William Buhlman is someone who has learned how to separate from his physical body at will. In his book, William Buhlman gives an account where he describes his psyche or non-local consciousness as being separate from his spirit body, which is separate from his physical body.

Others who have had NDEs have sometimes described themselves as being nothing but a spark on the ceiling. When they went to look for themselves, they saw nothing there, not even their spirit body – they were a disembodied immaterial consciousness floating in space.

Others have described watching their spirit body from an immaterial third-person perspective, after having separated from their physical body.

We mortal beings are a complex composite of psyche, spirit body, and physical body which function as a holistic union while we are alive here in mortality; but, our psyche and spirit body continue to live and function when our physical brain goes offline and/or our physical body dies.

William Buhlman's book was officially the end of My Materialism and My Nihilism, and the beginning of Science 2.0 wherein I give precedence or priority to Lived Experiences over the scientific inferences and personal speculations of the Materialists, Naturalists, Darwinists, Nihilists, and Atheists.

I want evidence, lots and lots of evidence; and, Science 2.0 admits all of the evidence into evidence and is based upon a preponderance of the evidence. I upgraded my science to Science 2.0; and, you can too if you want to do so. I'm no longer hiding from evidence, like I used to do when I was a Materialist, Nihilist, and Atheist. I'm open-minded now and therefore willing and able to accept all of the evidence into evidence. I judge everything through a preponderance of the evidence.

Under Science 2.0, you don't have to take my word for it. You can study the evidence for yourself and then make up your own mind as to what you want to believe. It's a whole new and different way of doing science.

The Materialists, Naturalists, Darwinists, Nihilists, and Atheists define science as Materialism, Naturalism, Darwinism, Nihilism, and Atheism. I have upgraded my science to Science 2.0. Now I define Science as Observation, Knowledge, and Truth derived from the Lived Experiences of the human race.

The best way to find and know the truth is to live it and experience it for yourself, or to choose to trust someone who has. You will never find the truth by trying to explain it away as the Materialists and Skeptics tend to do. You will never find the truth by rejecting observational evidence and experiential evidence as the Materialists, Naturalists, and Atheists do. The ONLY way to find and know the truth is to live it and experience it for yourself, or to choose to trust someone who has.

When it comes to Out-of-Body Experiences (OBEs), I have chosen to trust William Buhlman, because he has lived it and experienced it and observed it for himself.

Remember, Science 2.0 is based upon observations and lived experiences, not scientific inferences and personal speculation. Remember, Materialism and Naturalism and

Atheism have NO observational evidence supporting them, and they never will because they can't. Materialism, Naturalism, and their derivatives such as Darwinism, Nihilism, and Atheism are based upon scientific inferences rather than observational evidence, because there can never be any observation evidence supporting their primary assumptions or major premises which claim that the non-local or the non-physical does not exist.

Remember, it's philosophically and logically impossible to support the non-existence of something with observational evidence. Under Science 2.0, Materialism and Naturalism are unscientific, because their major premises lack observational support. Instead, ALL of the Lived Experiences of the human race tell us clearly and conclusively that Materialism and Naturalism are FALSE. Materialism, Naturalism, and their Derivatives FAIL to meet their burden of proof, through a preponderance of the evidence.

I quote from William Buhlman's book in other essays and books which I have written, so I won't do so here. You are going to have to look up the evidence for yourself and decide for yourself whether we meet our burden of proof through a preponderance of the evidence, or not. At least we have evidence supporting our chosen point of view, which is something that I didn't have when I was a Materialist, Nihilist, and Atheist.

I have changed. I'm no longer willing to exclude anything that might help me to get at and find the truth. I have upgraded my science to Science 2.0, which means that I now allow all of the evidence into evidence. I'm perfectly capable of interpreting that evidence on my own, thank you; and, so are you.

Honorable Mention in this Category

Rivas, T., Dirven, A., & Smit, R. H. (2016). *The Self Does Not Die: Verified Paranormal Phenomena from Near-Death Experiences*. Durham, NC: IANDS Publications.

This book turns Near-Death Experiences (NDEs) into a verified and validated Science, by providing external confirmation that the NDEs and Out-of-Body Experiences (OBEs) really took place. In complete harmony with Science 2.0, this book relies upon veridical evidence for NDEs and OBEs. Veridical means "truthful" and is defined as something that matches with reality or coincides with reality. In other words, in this book, their evidence which is obtained from NDEs and OBEs has been verified and confirmed as being real and true by observing and verifying that the objects, information, people, and events – that were seen while the patient was out-of-body traveling the astral plane – are in fact real and truly exist and truly did happen. In other words, these scientists and medical doctors got confirmational evidence or proof that what was seen or discovered through NDEs and OBEs really does exist and really is true. It's been verified, in proper scientific fashion.

The truthfulness and usefulness of Science 2.0 and the associated Ultimate Cause Model of Reality hinges upon the fact that the human psyche survives bodily death and brain death according to the empirical sciences of NDEs, (Shared-Death Experiences) SDEs, OBEs, and other types of Non-Local Experiences.

If there were no empirical evidence for psyche surviving bodily death and separation from the physical body, then Science 2.0 and the associated Ultimate Cause Model of Reality would be no better than Materialism and Naturalism, which also have no evidence to support their primary premises. But, since there is plenty of evidence to support Psyche as the Ultimate Cause and no evidence to support Materialism's claim that psyche or spirit does not exist, Science 2.0 and the Ultimate Cause Model of Reality ends up being infinitely superior and a whole lot more complete, compelling, believable, and useful than Naturalism or Materialism will ever be, as a model of reality.

Materialism and Naturalism don't match with reality, so they make for a very poor model of reality. Materialism and Naturalism are NOT veridical. They don't match with reality, nor the Lived Experiences of the human race as a whole. Something is said to be scientifically accurate if it matches with Reality. Since Materialism and Naturalism don't match with reality, they are by definition in principle unscientific – nothing but pure philosophy, sophistry, self-deception, and metaphysics.

## Quantum Field Theory Falsifies the Second Law of Thermodynamics

I'm not the first person on the planet to discover Quantum Field Theory, but as far as I know, I am the first person to identify the Perpetual Motion Cycle for what it truly is. Nobody thinks about these kinds of non-physical things, which is why nobody discovers them.

Quantum Field Theory and the Perpetual Motion Cycle are two of the greatest discoveries in Science and Physics that I have ever encountered. They falsify the second law of thermodynamics, and they are the perfect replacement for the falsified second law of thermodynamics.

The second law of thermodynamics erroneously claims that the amount of disorder (or entropy) in our universe is constantly increasing and that it can never decrease. That is NOT what we experience and observe. We don't observe proton decay, as the second law predicts. We don't observe an ever-encroaching gray goo coming in at us from all sides, as the second law predicts. Instead, we experience and observe constantly conserved order and organization.

Why?

Why do we ONLY experience constantly conserved order and organization? There has to be a reason.

Science gives us the reason, and it falsifies the second law of thermodynamics which states that this constantly conserved order and organization should not exist.

The reason we experience constantly conserved order and organization in our part of the multiverse is thanks to the Quantum Fields and the Perpetual Motion Cycle $E = mc^2$.

Disorder and disorganization ceased to exist in our part of the multiverse the very moment that the Gods and the Controlling Psyches designed, created, and made the Quantum Fields. According to Quantum Field Theory, the Gods and Controlling Psyches had to make the Quantum Fields BEFORE they could make and sustain physical matter. No Quantum Fields, then NO physical matter. It's that simple. The thing that we call the "Big Bang" took place when the Gods and the Controlling Psyches organized and made the Quantum Fields in our part of the multiverse. Only after making the Quantum Fields was it finally possible to make mass, mass's heat storage capacity, heat, and physical matter. Anything that is obviously made obviously has a maker who made it. The Quantum Fields, quantum waves, photons, and physical particles are obviously made. That means that they obviously have some type of Maker who made them. Our job as scientists is to try to determine the True Cause of everything that exists. The Psyches or Intelligences or Gods who made the Quantum Fields ARE the True Cause of everything that exists in our part of the multiverse.

Photons are massless, heatless, chargeless, entropyless, syntropic, non-physical, infinite acceleration quantum waves. The Quantum Fields are massless, heatless, entropyless, syntropic, non-physical perpetual motion machines. The Quantum Fields represent perfect order and organization. We will NEVER have true disorder in our part of the multiverse as long as the Quantum Fields exist. The second law of thermodynamics will NEVER be true so long as the Quantum Fields exist. That's just the way it is. The Quantum Fields are perpetual motion machines, and they will function perfectly, flawlessly, and eternally thanks to the Conservation of Energy and the Perpetual Motion Cycle $E = mc^2$. The Quantum Fields ARE the Perpetual Motion Cycle ($E = mc^2$) in action. They will go on forever, until the Gods or the Controlling Psyches decide to destroy or rearrange the Quantum Fields. You would have to destroy or disorganize the Quantum Fields in order to make the second law of thermodynamics true.

We will NEVER have true disorder and disorganization in our part of the multiverse as long as the Quantum Fields exist. The second law of thermodynamics will NEVER be true so long as the Quantum Fields exist. That's just the way it is. The Syntropic Entropyless Quantum Fields, Quantum Field Theory, Feynman Diagrams, Quantum Mechanics, and the Perpetual Motion Cycle $E = mc^2$ FALSIFY the second law of thermodynamics.

That explains why we ONLY experience and observe conserved order and organization in our observable universe. Order and organization is the ONLY thing that exists at the quantum level in our part of the multiverse, thanks to the perpetual motion machines that we call the Quantum Fields. The Quantum Fields function perfectly and eternally through the Conservation of Energy and the Perpetual Motion Cycle, which is $E = mc^2$. This Perpetual Motion Cycle is what we actually experience and observe.

Within our sun, Nature's Psyche is constantly transforming mass, heat, and entropy into massless, heatless, chargeless, entropyless, non-physical, infinite acceleration photons or quantum waves. It's happening right now; and, all the while the energy is conserved. Energy cannot be made, and it cannot be destroyed. Energy or Psyche or Light has always existed, and it will always exist. According to the Law of Psyche, every psyche has a certain amount of energy that's under its control, and that controlling psyche can form or transform the energy under its control into anything it wants that energy to be anytime and anywhere that it chooses to do so.

Therefore, every time a photon or quantum wave chooses to stop, it transforms its infinite acceleration into whatever it wants its energy to be. According to Quantum Field Theory, a photon or quantum wave can transform itself into some type of mass, if it wants to. Mass IS heat storage capacity and resistance to acceleration. Every time a photon or quantum wave chooses to stop, it transforms itself into mass or heat. Therefore, what do we observe? When a photon chooses to stop on your skin, do you observe an ever-increasing amount of mass forming onto your skin, or do you experience heat? Well, we experience heat. A photon or quantum wave doesn't have to transform itself into heat. It chooses to transform itself into heat when it lands on your skin because it KNOWS that it has landed on your skin. A photon can just as easily transform itself into mass, such as electrons, quarks, and positrons. That's what it seems to do when it stops on your solar array on your roof. Eventually it finds a way to transform itself into mass, mass's heat storage capacity, or electrons. Electrons have mass, or heat storage capacity, or resistance to acceleration. A photon or quantum wave can transform itself into anything that it wants to be, including heat, mass, quarks, electrons, and physical matter. A photon is intelligent, conscious, psychic, sentient, and alive. It chooses what it wants to be.

This is the Perpetual Motion Cycle in action. All the while, the energy is constantly transformed and constantly conserved. The Perpetual Motion Cycle $E = mc^2$ works perfectly, flawlessly, and endlessly both coming and going. Mass, heat, and entropy are being transformed into massless, chargeless, heatless, and entropyless photons and quantum waves all the time in our sun; and, massless, heatless, chargeless, entropyless, syntropic, exergic, non-physical, infinite acceleration quantum waves and photons are constantly transforming themselves into heat, mass, mass's heat storage capacity, and resistance to acceleration whenever they choose to stop. This is the Perpetual Motion Cycle, and it will go on forever for all eternity thanks to the Quantum Fields, the Conservation of Energy, and $E = mc^2$. You would have to destroy the Quantum Fields in order to stop the Perpetual Motion Cycle and make the second law of thermodynamics true instead.

A photon or quantum wave doesn't have to stop if it doesn't want to. A photon or quantum wave can pass through our earth, our sun, your physical body, and a black hole as if they weren't even there. WE KNOW that this is so because we observe photons or quantum waves passing through water and glass all the time. A photon doesn't have to stop if it doesn't want to stop. A photon or quantum wave can maintain its infinite acceleration and chosen speed-of-light velocity for as long as it desires to do so. A photon or quantum wave can also quantum tunnel if it chooses to do so. It has NO speed limit except for the speed limit that it chooses to impose upon itself.

The verified and proven existence of massless, entropyless, non-physical, syntropic, conserved energy, exergy, quantum waves, photons, and quantum fields FALSIFIES the second law of thermodynamics which states that they do not exist and cannot exist.

Once the scientists in our world figure out what's truly going on, then they will vote to replace the falsified second law of thermodynamics with the Perpetual Motion Cycle $E = mc^2$ and Quantum Field Theory instead. Quantum Field Theory and the Perpetual Motion Cycle have actually been experienced and observed; whereas, ever-increasing disorder, proton decay, and an ever-encroaching gray goo have NOT.

The one that has actually been experienced and observed ends up being the one that is actually real and true. That's the way that science is supposed to work. If we successfully identify and eliminate from science everything that is false and everything that has been falsified, then ONLY the truth will remain. The truth that remains will consist of the things that have actually been experienced and observed. The false is falsified by the truth, and the truth is constantly and repeatedly experienced and observed.

Mark My Words

## 3. Solving the Mind-Brain Problem:

Scientism is the philosophical belief that Science and the Scientific Method is the ONLY way to find and know the truth, or at least the best way to find and know the truth. As a practitioner of Scientism and a believer in Scientism, I have always considered myself a scientist. I'm not a Seer, Clairvoyant, nor Astral Traveler – I haven't had those kinds of experiences in great abundance. Such has not been my talent or gift. Typically, I have to get at the truth through other means.

After a lifetime of Scientific Research and study, I finally found an Interpretation of Quantum Mechanics that explains what Psyche or Non-Local Consciousness is and how it works, in a logical scientific fashion that actually made sense to me.

Schwartz, J. M., Stapp, H. P., & Beauregard, M. (2004). *Quantum Physics in Neuroscience and Psychology: A Neurophysical Model of Mind-Brain Interaction*. Published Online: Phil. Trans. R. Soc. B.

http://www-physics.lbl.gov/~stapp/PTRS.pdf

http://mypsyche.us/wp-content/uploads/2017/10/PTRS.pdf

https://www.researchgate.net/publication/7613549_Quantum_physics_in_neuroscience_and_psychology_A_neurophysical_model_of_mind-brain_interaction

http://escholarship.org/uc/item/4w8665vk

This paper summarizes everything I have been looking for during the past fifty-five years of my life. This paper solves the Mind-Brain Problem; and, it uses the Science of Quantum Mechanics to do so. For a scientist like me, it doesn't get better than this.

I found it impossible to choose my favorite book and article in this category or theme, because once I finally set out to SOLVE the Mind-Brain Problem, I found dozens of books and articles that did just that – a number of which have become my favorites.

**A Winner in the Mind-Brain Solution Category**

Beauregard, M. & O'Leary, D. (2007). *The Spiritual Brain: A Neuroscientist's Case for the Existence of the Soul*. New York: HarperCollins.

THIS!

Science 2.0 needs a believable, observable solution to the Mind-Brain Problem – an "unsolvable problem" which was created by Materialism and Naturalism to begin with.

This book resolves the Mind-Brain Problem.

Neuroscience has lots of different practical applications; however, from a theoretical and philosophical standpoint, the purpose of Neuroscience is to resolve the Mind-Brain Problem which was created by Materialism and Naturalism in the first place. From the philosophical perspective, if Neuroscience cannot resolve the Mind-Brain Problem, then it's not much use to us. I'm not the first scientist to come to this realization.

You have got to solve the Mind-Brain Problem, or you are never going to make any progress in science. That's just the true reality of the situation. Materialism and Naturalism and Atheism have held back science more than anything else on this earth, because the people behind these things reject any observational evidence that they don't like.

Remember, Materialism and Naturalism are based upon "scientific inference", meaning that they have NO observational evidence supporting them.

The solution to the Mind-Brain Problem is to eliminate Materialism and Naturalism. Materialism and Naturalism caused the Mind-Brain Problem in the first place; and, eliminating Materialism and Naturalism solves it. Simple! This book is brutal against Materialism and Naturalism.

Science 2.0 is based upon a resolution of the Mind-Brain Problem, through observational evidence.

There are two ways to resolve the Mind-Brain Problem.

**1. Eliminative Materialism: The mind-matter problem is resolved by denying that mental processes exist in their own right. "Consciousness" and "mind" (intentions, desires, beliefs, etc.) are prescientific concepts that belong to**

unsophisticated ideas of how the brain works, sometimes called "folk psychology." They can be reduced to whatever the neurons happen to be doing (neural events). "Consciousness" and "mind" as concepts will be eliminated by the progress of science, along with such ideas as "free will" and the "self." Current key exponents of this view include philosophers Paul and Patricia Churchland and Daniel Dennett.

http://vedicilluminations.com/downloads/Philosophy/Beauregard_Mario,_O'Leary_De nyse_-_The_Spiritual_Brain__A_Neuroscientist%92s_Case_for_the_Existence_of_the_Soul.pdf

**2. Eliminate the Exclusionary and Eliminative Philosophies:** Since Non-Local Consciousness and Eliminative Materialism are mutually exclusive, one of them has to go, if you want to solve the Mind-Brain Problem; consequently, the other way to solve the Mind-Matter Problem is to eliminate Materialism, Naturalism, Nihilism, and Atheism, which caused the Mind-Brain Problem in the first place. Simple. You have a choice to make, if you truly want to resolve the Psyche vs. Brain Dilemma. Make it a good one.

I will discuss my notes, from the margins of my copy of this book, *The Spiritual Brain*, in other essays and books, so I won't do so here. I simply submit the whole book into evidence with the claim that it solves the Mind-Brain Problem. You will have to decide for yourself, by examining the evidence, whether my claim is true or not.

Meanwhile, I simply let you know that the book exists, along with my claim that this book has proven to be one of the best Introductions and Introductory Solutions to the Mind-Brain Problem that I have found so far.

Next!

## A Winner in the Mind-Brain Solution Category

Jeffrey M. Schwartz, J. M., & Begley, S. (2002). *The Mind and the Brain: Neuroplasticity and the Power of Mental Force.* New York: HarperCollins.

A Sampling:

http://mypsyche.us/wp-content/uploads/2017/10/The-Mind-and-the-Brain-Neuroplasticity-and-the-Power-of-Mental-Force.pdf

Source:

https://www.amazon.com/Mind-Brain-Neuroplasticity-Power-Mental/dp/0060988479

A full and complete understanding and acceptance of Neuroplasticity SOLVES the Mind-Brain Problem. I quote from and use this book elsewhere, so I won't do so here. I simply announce it as a winner in this category, submit it into evidence, and then I leave it up to you to decide for yourself if I am right, or not.

## Winners in the Mind-Brain Solution Category

Stapp, H. P. (2007, 2011). *Mindful Universe: Quantum Mechanics and the Participating Observer* (2nd ed.). New York: Springer-Verlag.

Van Lommel, P. (2010). *Consciousness Beyond Life: The Science of the Near-Death Experience*. New York: HarperCollins.

I mentioned these two books in the Quantum Mechanics category; but, they also need to be mentioned here because they solve the Mind-Brain Problem. That's the reason they are winners in the Quantum Mechanics category.

The following free article is a winner in this Mind-Brain category or theme.

Schwartz, J. M., Stapp, H. P., & Beauregard, M. (2004). *Quantum Physics in Neuroscience and Psychology: A Neurophysical Model of Mind-Brain Interaction*. Published Online: Phil. Trans. R. Soc. B.

http://www-physics.lbl.gov/~stapp/PTRS.pdf

http://mypsyche.us/wp-content/uploads/2017/10/PTRS.pdf

https://www.researchgate.net/publication/7613549_Quantum_physics_in_neuroscience_and_psychology_A_neurophysical_model_of_mind-brain_interaction

http://escholarship.org/uc/item/4w8665vk

This paper summarizes everything I have been looking for during the past fifty-five years of my life. This article not only solves the Mind-Brain Problem, but it also solves the Free Will vs. Determinism Debate; and, it uses the Science of Quantum Mechanics to do so. This article successfully forms the foundation of Science 2.0, all by itself, if you are looking for such a thing.

The Materialists and Naturalists deliberately hide these books and papers from you, so if you want to save yourself years of research and tons of money, go straight to the end of the line and read these books and this article first. You'll be glad you did, if you are in fact looking for the truth.

Going to these books and article first will save you a great deal of time and effort and money, because the quickest way to find and know the truth is to find and study everything that the Materialists, Naturalists, Darwinists, Behaviorists, Nihilists, and Atheists are trying to hide from you. There are as many interpretations of Quantum Mechanics as there are physicists on this planet, so finding the best and most scientifically accurate interpretation first will be of great benefit to you. I'm not going to apologize for saving you some time and effort and money.

Remember, the correct and most-useful interpretations of Quantum Mechanics actually SOLVE the Mind-Brain Problem and the Free Will vs. Determinism Problem.

While reading these books and this article, I actually felt like I was in the presence of Scientific Truth. In contrast, while reading books that are written by Materialists, Naturalists, Nihilists, Atheists, Behaviorists, and Darwinists, I typically get the feeling that they are trying to trick me and deceive me. Obviously, your mileage may vary, because you are a completely different person or psyche – we each tend to find what we go searching for, and no more.

**How often have I said to you that when you have eliminated the impossible [and the false], whatever remains, *however improbable*, must be the truth? — Sherlock Holmes.**

During my research in recent years, I have come to the conclusion that the BEST way to find and know the truth is to live it and experience it for yourself, or to choose to trust someone who has.

The second-best way to find and know the truth is to eliminate everything that is false and everything that has been falsified, so that ONLY the truth remains. This process involves eliminating the Impossible, which means eliminating ALL of the Falsified Philosophies such as Creation Ex Nihilo, Spontaneous Generation, Abiogenesis, Macro-Evolution, Creation by Rocks, Design and Creation by Mutation/Selection, Materialism, Naturalism, Darwinism, the Theory of Evolution, Behaviorism, Determinism, Atomistic Reductionism, Scientism, Nihilism, and Atheism. If you successfully eliminate all of these, then whatever remains must be the truth. In fact, you will find The Truth staring you in the face.

Remember, the correct and most-useful interpretations of Quantum Mechanics actually SOLVE the Mind-Brain Problem.

How do they do this?

They SOLVE the Mind-Brain Problem by falsifying Materialism, Naturalism, and their derivatives, and then replacing these Falsified Philosophies with some kind of Non-Local Consciousness Interpretation or **Psyche Interpretation of Quantum Mechanics**.

Remember and be aware, you will NEVER learn about nor hear about these kinds of **Non-Local Interpretations of Quantum Mechanics** in our public schools, because the Materialists, Naturalists, and Atheists who run our public schools are trying to hide this kind of scientific knowledge from the public at-large, because the verified and proven existence of Quantum Non-Locality, Quantum Entanglement or Action at a Distance, and the Quantum Zeno Effect FALSIFIES Materialism and Naturalism leaving the Atheists without a leg to stand on. These people don't want you to know about this stuff. They have a vested interest in preventing you from finding out that these types of books and articles exist.

The Truth has been known for a couple of decades now, and the Truth has become particularly abundant and available during the past decade; but here it is November 2017, and people like me are just starting to find out about it, even though we have been scientists all of our lives.

## The Physical Brain Is a Transceiver

Memory engrams: Engrams are theorized to be means by which memories are stored as biophysical or biochemical changes in the brain (and other neural tissue) in response to external stimuli.

> **Karl Lashley, one of the world's foremost brain researchers, tried to locate the area in the brain where engrams or memory traces were stored. He sliced or removed sections of rat brains after teaching the rats to run mazes. None of the brain injuries abolished the "maze-running habit," although Lashley tried removing tissue in almost every area that allowed the rat to remain alive. Lashley concluded that memories had to be spread all over the brain, throughout the tissue.**
>
> http://www.intropsych.com/ch06_memory/lashleys_research.html

One possible interpretation of these science experiments is that there is NO such thing as memory engrams, nor will it ever be possible to make a recording of a person's brain state which includes a lifetime's worth of memories and experiences, as is done in the science fiction series "Extinct" on television. There's NO such thing as memory engrams. There's nowhere near enough capacity to store all of our memories and experiences within out physical brain. The vast majority of our memories are stored non-locally in the transdimensional realm, where there are NO physical limitations to limit the amount of memories and experiences that can be stored.

The impression given by Lashley and some of the other researchers who did similar experiments seemed to suggest that they had come to the conclusion that they could cut out the whole brain and yet the rat would still remember where the food was or how to run the maze.

As I repeatedly mention, when it comes to each science experiment and every piece of scientific evidence, there can theoretically be an infinite number of different possible interpretations that can be given to that specific piece of scientific evidence. So, when it comes to Lashley's experiments, you can interpret it any way that you want, as thousands of others have done. There are even materialistic and naturalistic interpretations given to Lashley's experiments which make the claim that ALL of our memories and experiences are stored within out physical brain, and that when our brain dies ALL of those memories are lost forever.

However, I personally choose to interpret Lashley's experiments in light of the observational and experiential evidence obtained from Near-Death Experiences (NDEs) – especially the panoramic Life Review which many of the NDErs receive. People who experience brain death typically have a Life Review wherein the memories from their whole mortal life passes before their eyes; and, they also witness how their chosen actions affected others and even at times feel and observe these events from the other person's perspective. This seems to be God's way of letting these people KNOW that they still exist and that all of them is still there even though their physical body and physical brain is dead now. These people are given their Life Review before being offered a choice to stay in heaven or a chance to return back to their physical body and back to mortality.

What we learn and KNOW from NDEs is that ALL of our memories and experiences are stored somewhere non-locally in the spirit realm, where they can later be retrieved and experienced again and again. The vast majority of our memories are NOT stored in our physical brain. They are stored non-locally in the non-local realm.

This Reality is a very good thing, because some scientists have estimated that our physical brain can only store about a second's worth of memories if our memories are stored as bits and bytes in our physical brain, and maybe up to seven minute's worth of memories if our memories are stored holographically in our physical brain. Our physical brain is nothing but a buffer capable of storing up to about seven minute's worth of memories before the older memories have to be purged in order to make room for new memories and experiences. While emptying the buffer, a physical brain that's functioning normally has been given the ability to upload its memories into non-local long-term storage, where these memories later can be accessed and retrieved later on.

What does this mean in practice?

It means that our physical brain is a very complex bio-engineered transceiver. There's no way on earth that evolution (genetic drift) and random mutations could have designed and created such a thing. We human beings couldn't design and create such a thing, so why on earth should anyone believe that evolution and random mutations could have designed and created such a thing? It makes no sense.

A transceiver can transmit radio signals to another radio receiver; and, a transceiver is also a radio receiver which can receive radio signals from another transmitter. Our physical brain, when functioning properly, is a transceiver – capable of transmitting memories to the non-local realm and capable of receiving stored memories from non-local consciousness.

Remember, our physical brain is a very complex bio-engineered transceiver.

How do we know?

First of all, we come to know the truth by getting rid of Materialism and Naturalism, because the observational evidence and experiential evidence and experimental evidence won't make any sense to us until we do. ALL of the observational evidence that we have on hand as a race falsifies Materialism and Naturalism. We have to get rid of these things if we want to find and know the truth.

Then we study the evidence and think about it logically and open-mindedly.

The human brain has something that has been called "working memory" or "short-term memory". It has been observed that working memory or short-term memory uses ALL of the physical brain – every system in the physical brain concurrently. Yes, there are parts of the brain that specialize at specific tasks; but, we use ALL of our brain throughout the day. The idea that we use only 10% of our brain is a fictional myth developed for the making of interesting science fiction stories, as is the idea that all of our memories and experiences can be downloaded from our physical brain and stored in a computer data bank.

The truth is that ALL, but about seven minute's worth, of our memories are stored non-locally in the transdimensional realm which has NO physical limitations and therefore no memory storage limitations. If our short-term memories are stored holographically, our working memory has about seven minute's worth of capacity before physical brain needs to start emptying the buffer as it goes along.

In contrast, the psyche or non-local consciousness is an infinite singularity capable of infinite memory storage capacity within something that takes up NO physical space whatsoever. Remember, space-time is a physical concept and a physical limitation, which really doesn't exist in the same form or manner in the non-local realm or transdimensional realm. Astral travelers and out-of-body travelers testify that there is linearity and sequentiality in the non-local realm, but that time and space feel different and ARE different in the transdimensional spirit realm. Space and time function differently in the Non-Physical Realm. The Non-Local Spirit Realm exists without ANY of the physical limitations that we are familiar with, here in mortality.

Let's restate this in different ways so that the lessons become clearer.

The estimates vary depending upon the task being performed and the hypothesized mode or method of storage (as bits and bytes or holographically); but, it has been estimated that the whole physical brain has enough memory storage capacity to store anywhere from one second to about seven minutes of memory before its capacity is exceeded and it needs to empty the buffer.

There's simply way too much information for ALL of our memories and experiences to be stored in our physical brain or even our DNA, because physical storage of information and data is a highly inefficient way of storing information, knowledge, memories, and full surround-sound Imax experiences. Objects that take up physical space, such as computer hard drives and physical brains and DNA, are subject to physical limitations which greatly limit the amount of information that can be stored in physical space as bits and bytes within atoms and molecules. Physical storage is highly inefficient storage.

Consequently, it has been estimated that our physical brain has enough memory storage capacity to store up to about 7 minute's worth of memories and experiences, before its buffer or working memory needs to be emptied. Our working memory or short-term memory tends to empty the buffer on a first-in first-out basis, with the older memories getting purged first.

While emptying the buffer, a fully functional human brain is capable of transmitting the information which your psyche wants to remember into long-term storage or non-local consciousness, where it can later be retrieved as needed. You see, when it comes to non-local consciousness or psyche, which is an infinite singularity, it's theoretically possible to store an infinite amount of information, data, and experiences in something that takes up NO physical space whatsoever. Now, that's highly dense or infinitely dense information storage capacity, which can NEVER be achieved through physical means.

Under the guidance of psyche or non-local consciousness, a normal fully functional brain is capable of transmitting its memories, knowledge, and experiences into long-term memory storage while emptying the buffer. During neurodegenerative disease or brain damage, however, working memory and working capacity slowly get destroyed until these people lose the ability to organize and transmit new memories into long-term storage in the non-local realm.

No matter how severe their brain damage or brain deterioration while these people experience neurocognitive diseases, they seem to be able to remember and retrieve everything that they learned or stored in long-term memory BEFORE the brain was damaged and lost the ability to form new memories and thereby lost its ability to transmit those new memories into long-term storage. No matter how damaged their brain gets to be, they can still access or retrieve everything they have stored in long-term memory because those memories are stored outside of the physical brain in the non-local realm. Brain deterioration does not affect nor change non-local memories. The only thing that gets destroyed during neurocognitive disorders is their short-term memory or working memory which is in fact a function of the physical brain. When the ability to form and store new memories gets destroyed by brain damage, then there's really nothing to transmit into the non-local realm, assuming of course that the brain's transmitter hasn't been damaged as well.

However, since these people typically retain the ability to retrieve or receive memories from non-local consciousness, this kind of suggests that the transceiver portion of the physical brain is still functional, but that the working memory or working capacity as a whole has been limited or destroyed. In other words, as the brain rots and deteriorates, their memory buffer gets destroyed and the net result is that there's no longer much if anything to transmit to the non-local realm by the time the buffer is full and needs to be emptied.

Study someone who has Alzheimer's Disease, and notice how they can remember everything from their childhood, and everything they learned in college, and everything they learned to do on their job, and everywhere they used to drive, and everything they knew before getting Alzheimer's; yet, they can't remember what they said or did seven or eight minutes ago. These people have full access to their non-local long-term memory storage, because brain damage doesn't touch nor affect any of that; but, they have no access to new memories because their physical brain or working memory has been destroyed and can't make new memories.

The only way to explain this evidence and Reality logically is through the existence of non-local transdimensional long-term storage, because if those long-term memories were actually stored within the physical brain, then they would have been the first memories lost or purged in order to make room for the new memories being generated by short-term working memory. Within these people there would be NO long-term memories whatsoever, if all of their memories were being stored in their physical brain. The fact that most people with neurocognitive degeneration and brain damage have full access to their long-term memories while being unable to generate and retrieve new memories suggests clearly and conclusively that their long-term memories are in fact being stored somewhere outside of their physical brain. There's no other logical explanation, because the physical brain is damaged so much that it can no longer form and transmit new memories into long-term non-local storage.

These realities provide evidential proof that the physical brain is a transceiver – transmitting memories and experiences into long-term non-local storage and then later retrieving or receiving memories and experiences from long-term memory storage as needed or desired. And, when the physical brain is damaged and loses its ability to form and store short-term memories in local capacity or local storage, then there's not much if anything to be transferred to long-term memory in the non-local realm.

Finally, we KNOW that this model of reality is real and true from all the different Life Reviews which people have experienced and observed during their Shared-Death Experiences, Out-of-Body Experiences, and Near-Death Experiences. It has been observed that all of our memories continue to exist and are retrievable AFTER our physical brain is dead and gone.

Eye-witness evidence or empirical evidence trumps scientific inferences and personal guesswork every time. Remember, Science 2.0 is based upon the Lived Experiences of the human race, including all of our non-local experiences.

## The Raspberry Award and the Rotten Tomato

Nowadays, I'm only interested in finding the truth, rather than being systematically indoctrinated and brainwashed into Darwinism, Materialism, and Naturalism which have been falsified by Eye-Witness Evidence, Observations, Experiences, Empirical Evidence, Corroborated Veridical Evidence, Science, and the Scientific Methods.

In this Mind-Over-Matter Category, the Raspberry and Rotten Tomato goes to:

Pinel, J. (2014). *Biopsychology* (9th ed.). New York: Pearson.

Note: There are thousands of college textbooks that could have been winners in this Rotten Tomato category, but *Biopsychology* has the dubious distinction of being the one that I was forced to buy and read as part of a college course in Neuroscience.

After reading the first couple of chapters of this book, *Biopsychology*, I was motivated to find some REAL Neuroscience that actually distinguished between the local and the non-local, or the physical and the non-physical.

I had a great deal of fun debunking the first couple of chapters of this book; and, I learned a lot about how the Materialists, Naturalists, and Darwinists go wrong, through their scientific inferences and logic fallacies. But, when it came time to find the truth, I had to turn elsewhere. This book, *Biopsychology*, motivated me to do so in a very big way.

The following are some more of the Books that I found as a result of my efforts to debunk and falsify the claims which were made in the opening chapters of *Biopsychology* by John Pinel.

## A Clear Winner in this Mind-Brain Category

Penfield, W. (1975). *The Mystery of the Mind*. Princeton, New Jersey: Princeton University Press.

https://muse.jhu.edu/book/39368

http://www.chabad.org/library/article_cdo/aid/113106/jewish/Appendix-5-Neurology-Medicine-and-the-Soul.htm

http://mypsyche.us/wp-content/uploads/2017/10/Neurology-Medicine-and-the-Soul.pdf

This book resolves the Mind-Brain Problem, using Scientific Experiments and Scientific Observations to do so. Wilder Penfield has the distinction of getting there first.

By using scientific experimentation, scientific methods, brain stimulation, and observation of the results, the Mind-Brain Problem was officially solved in 1975 by Wilder Penfield. Others have replicated and confirmed Wilder Penfield's findings; but, he got there first, because he was brave enough to report what he was finding.

The Materialists and Naturalists spend all of their time suppressing and ridiculing this type of observational evidence in an attempt to explain it away or convince us that it does not exist.

Remember, other neuroscientists like John Eccles, Mario Beauregard, and Eben Alexander have come to the very same conclusions, but Wilder Penfield has the distinction of getting there first. These men tend to quote Wilder Penfield and this book, *The Mystery of the Mind*, in many of their books and presentations.

The following is some of what they quote:

> **Is the mind merely a function of the brain? Or is it a separate but closely related element?**
>
> **Throughout my own scientific career, I, like other scientists, have struggled to prove that the brain accounts for the mind.**

(*The Mystery of the Mind*, Preface.)

This is the million-dollar question and the ultimate quest.

> **Is there any evidence of the existence of neuronal activity within the brain that would account for what the mind does?**
>
> **Before venturing to answer, it may be of interest to refer again to action that the mind seems to carry out independently, and then to reconsider briefly our experience with stimulation of the cortex of conscious patients and our experience of what effects are produced by epileptic discharge in various parts of the brain. This should give some clue if there is a mechanism that explains the mind.**
>
> **"(a) What the Mind Does**
>
> **It is what we have learned to call the mind that seems to focus attention. The mind is aware of what is going on. The mind reasons and**

makes new decisions. It understands. It acts as though endowed with an energy of its own. It can make decisions and put them into effect by calling upon various brain mechanisms. It does this by activating neuronal mechanisms. This, it seems, could only be brought about by expenditure of energy.

(b) What the Patient Thinks

When I have caused a conscious patient to move his hand by applying an electrode to the motor cortex of one hemisphere, I have often asked him about it. Invariably his response was: "I didn't do that. You did." When I caused him to vocalize, he said: "I didn't make that sound. You pulled it out of me." When I caused the record of the stream of consciousness to run again and so presented to him the record of his past experience, he marveled that he should be conscious of the past as well as of the present. He was astonished that it should come back to him so completely, with more detail than he could possibly recall voluntarily. He assumed at once that, somehow, the surgeon was responsible for the phenomenon, but he recognized the details as those of his own past experience. When one analyzes such a "flashback" it is evident, as I have said above, that only those things to which he paid attention were preserved in this permanently facilitated record.

(c) What the Electrode Can Do

I have been alert to the importance of studying the results of electrode stimulation of the brain of a conscious man, and have recorded the results as accurately and completely as I could. The electrode can present to the patient crude sensations. It can cause him to turn head and eyes, or to move the limbs, or to vocalize and swallow. It may recall vivid re-experience of the past, or present to him an illusion that present experience is familiar, or that the things he sees are growing large and coming near. But he remains aloof. He passes judgment on it all. He says "things seem familiar," not "I have been through this before." He says, "things are growing larger," but he does not move for fear of being run over. If the electrode moves his right hand, he does not say, "I wanted to move it." He may, however, reach over with the left hand and oppose his action.

There is no place in the cerebral cortex where electrical stimulation will cause a patient to believe or to decide...

I am forced to conclude that there is no valid evidence that either epileptic discharge or electrical stimulation can activate the mind...

The mind seems to act independently of the brain in the same sense that a programmer acts independently of his computer...

For my own part, after years of striving to explain the mind on the basis of brain-action alone, I have come to the conclusion that it is simpler (and far easier to be logical) if one adopts the (and far easier to be logical) if one adopts the hypothesis that our being does consist of two fundamental elements. If that is true, it could still be true that energy required comes to the mind during waking hours through the highest brain-mechanism.

> Because it seems to me certain that it will always be quite impossible to explain the mind on the basis of neuronal action within the brain, and because it seems to me that the mind develops and matures independently throughout an individual's life as though it were a continuing element, and because a computer (which the brain is) must be programmed and operated by an agency capable of independent understanding, I am forced to choose the proposition that our being is to be explained on the basis of two fundamental elements. This, to my mind, offers the greatest likelihood of leading us to the final understanding toward which so many stalwart scientists strive...
>
> The nature of the mind presents the fundamental problem, perhaps the most difficult and most important of all problems. For myself, after a professional lifetime spent in trying to discover how the brain accounts for the mind, it comes as a surprise now to discover, during this final examination of the evidence, that the dualist hypothesis seems the more reasonable of the two possible explanations.
>
> Since every man must adopt for himself, without the help of science, his way of life and his personal religion, I have long had my own private beliefs. What a thrill it is, then, to discover that the scientist, too, can legitimately believe in the existence of the spirit!
>
> (From Wilder Penfield, *The Mystery of the Mind*.)
>
> Source:
>
> http://www.chabad.org/library/article_cdo/aid/113106/jewish/Appendix-5-Neurology-Medicine-and-the-Soul.htm

Wilder Penfield set out to prove Materialism and Naturalism true; but, he ended up FALSIFYING Materialism and Naturalism instead.

This is the smoking gun – catching psyche in action completely separate from the physical brain. This effectively solves the Mind-Brain Problem.

However, if science continues to be defined as Materialism, Naturalism, Darwinism, Nihilism and Atheism, then science will NEVER be able to solve the Mind-Brain Problem, because Materialism and Naturalism cause the Mind-Brain Problem, not solve it.

But if instead, we choose to upgrade our science to Science 2.0 and choose to define science as Observation, Experience, Knowledge, Truth, Fact, and the Lived Experiences of the human race, then we quickly realize that the Mind-Brain Problem has already been solved by the collective Lived Experiences of the human race as a whole, including those presented to us by Wilder Penfield.

Good enough.

Since I have upgraded my science to Science 2.0, in all things I choose to go with the observational evidence rather than the philosophical guesswork of the Materialists and Naturalists. I simply find and present the observational evidence, and then leave it up to you my reader to decide for yourself what to make of it. Wilder Penfield made me a believer in the non-local or the non-physical, but your mileage may vary because you are a completely different person than I am.

## Another Winner in this Mind-Brain Category – Brain Wars

Beauregard, M. (2012). *Brain Wars: The Scientific Battle Over the Existence of the Mind and the Proof That Will Change the Way We Live Our Lives*. New York: HarperOne.

This book resolves the Mind-Brain Problem.

This book successfully debunks Materialistic Naturalism and provides Scientific Proof of Psyche or Scientific Proof of Non-Local Consciousness.

This book is an excellent introduction to Science 2.0. Beauregard states that we need to update our science and start including all of the evidence into our science, methods, interpretations, and scientific theories. Real Science is based upon evidence or direct observations by real living beings or psyches.

If we are going to upgrade our science to Science 2.0, such a change should be massive, significant, and noticeable. By deliberately including all of the evidence into evidence, this will indeed be a monumental change and something that science and scientists have never done before. Towards that goal, this book from Mario Beauregard is a good place to start.

Science 2.0 begins by eliminating Materialism, Naturalism, and any other Exclusionary Religions or Philosophies, which are based exclusively upon a complete denial and rejection of any evidence that contradicts the primary assumptions or major premises of Materialism and Naturalism.

Eliminating Materialism and Naturalism automatically solves the Mind-Brain Problem. Materialism and Naturalism create a whole host of different unsolvable problems; and therefore, getting rid of Materialism and Naturalism resolves them all. Materialism and Naturalism are the problem. This book successfully debunks Materialism and Naturalism, by providing scientific evidence or observational evidence that these philosophical worldviews are false.

## Another Winner – How the Self Controls Its Brain

Eccles, J. C. (1994). *How the Self Controls Its Brain*. New York: Springer-Verlag.

### The Dualist-Interactionist Theory

There are two wonderful concepts, 'the Self' and 'the Brain', that appear

as titles of three books: (1) *The Self and Its Brain* (Popper and Eccles 1977),

the title being created by Karl Popper; (2) *Evolution of the Brain, Creation*

*of the Self* (Eccles 1989); (3) *How the Self Controls Its Brain*, the present

**book.** (*How the Self Controls Its Brain*, p. x.)

https://www.scribd.com/document/334415538/eccles-john-how-the-self-controls-its-brain-doc

https://books.google.com/books?id=z8HqCAAAQBAJ&printsec=frontcover#v=onepage&q&f=false

The purpose of this book is to solve the Mind-Brain Problem, as is the purpose of all of John C. Eccles' books.

**Generation of Neural Events by Mental Events**

**Ingvar stated that a study of brain structures activated by pure ideation therefore appears to open up a new approach to understanding the human psyche.**

Source of this quote:

*How the SELF Controls Its BRAIN*

https://books.google.com/books?id=z8HqCAAAQBAJ&pg=PA159&lpg=PA159&dq#v=onepage&q&f=false

https://www.scribd.com/document/334415538/eccles-john-how-the-self-controls-its-brain-doc

*The Emergence of Complexity in Mathematics, Physics, Chemistry and Biology*

https://books.google.com/books?id=CpE-jUG214kC&pg=PA384&lpg=PA384&dq#v=onepage&q&f=false

*Rethinking Neural Networks: Quantum Fields and Biological Data*

https://books.google.com/books?id=urBQAwAAQBAJ&pg=PA11&lpg=PA11&dq#v=onepage&q&f=false

Anything within the physical brain that is activated by Psyche, or Pure Thought, or Pure Ideation will provide Signs of Psyche and catch Psyche in the act. It will also resolve the Mind-Brain Problem.

The Mind-Brain Problem, Quantum Non-Locality, and Non-Local Consciousness are the most interesting and fascinating aspects of science to study and learn about, but invariably you are forced to leave your Materialism and Naturalism at the door.

Remember, Materialistic Naturalism or Scientific Naturalism is the enemy to all science, because it is based upon the suppression of evidence and the denial of evidence. Science is Evidence; therefore, the suppression of evidence is unscientific.

**We regard promissory materialism as superstition without a rational foundation. The more we discover about the brain, the more clearly do we distinguish between the brain events and the mental phenomena, and the more wonderful do both the brain events and the mental phenomena become. Promissory materialism is simply a religious belief held by dogmatic materialists . . . who often confuse their religion with their science.**

― John C. Eccles, *The Wonder of Being Human: Our Brain and Our Mind*

Materialistic Naturalism or Scientific Naturalism is a dogmatic religion, which has to be taken on blind faith as being true, because there is NO observational evidence supporting it.

Any book from John Eccles will teach you this truth and reality; and, it will also solve the mind-brain problem.

Notice how each of these authors has their own set of talking points from their own science discipline, which successfully falsify Materialism and Naturalism while also solving the Mind-Brain Problem.

As I see it, getting rid of Psyche, Non-Local Consciousness, Mind, Self, Personality, Free Will, Intelligence, Awareness, and Soul perpetuates the Mind-Brain Problem; whereas, getting rid of Materialism and Naturalism and their derivatives solves it. Of course, the tenets of Science 2.0 teach that you are going to have to make up your own mind as to what you want to accept and believe; so, the ball is now in your court.

## Honorable Mention in this Mind-Brain Category

Barlow, D. H. & Durand, V. M. (2015). *Abnormal Psychology: An Integrative Approach* (7th ed.). Stamford, CT: Cengage Learning.

http://mypsyche.us/wp-content/uploads/2017/10/Abnormal-Psychology.pdf

A careful and open-minded reading of this book solves the Mind-Brain Problem with an Integrative Approach to Psychology, or a BioPsychoSocial Model of Psychology.

However, the authors of *Abnormal Psychology* did NOT set out to solve the Mind-Brain Problem and essentially don't discuss the topic at all. Instead, it is left up to you to use ALL of the Scientific Research within their book to solve the Mind-Brain Problem for yourself, if you want to do so – which I do in the next few sections of this book.

*Abnormal Psychology* is the first and only college textbook that I have come across so far where I never felt any need to buy additional books or do additional research in order to debunk and correct the flaws within the book, because this book was SOLID SCIENCE and unbiased science pretty much all the way through. They just present the evidence and scientific research, and then let you make up your own mind how to interpret it. They repeatedly point out that every mental illness has BioPsychoSocial causes; and, whenever they started discussing causes for a particular mental illness, they usually separated the causes of mental illnesses into Biological causes and PsychoSocial causes, always emphasizing that each mental illness is caused by a bit of both.

So far, this has been my most favorite college textbook. Reading college textbooks is a great deal of work; but, it can be extremely interesting and informative, especially if you are into science like I am. Of all the college textbooks that I was forced to buy, read, and memorize, *Abnormal Psychology* is to me the clear winner.

Again, Barlow and Durand did NOT set out to resolve the Mind-Brain Problem, and they never mention the Mind-Brain Problem in their book; but, their Integrative Approach and the BioPsychoSocial Model do in fact resolve the Mind-Brain Problem, so long as you don't allow the Materialists and Naturalists and Darwinists to intervene and mess everything up with their scientific speculation and denial of reality.

Measurements of heritability, derived through twin studies and adoption studies, provide evidence and proof of a genetic (Biological) component for any particular mental illness that you might select, which means that the other part of the equation is PsychoSocial in nature and origin. Although Barlow and Durand never mention anything about psyche or non-local consciousness, I quickly realized that when it comes to the

PsychoSocial causes of mental illnesses, chosen behaviors and chosen actions are always the result of Psyche; whereas, social influences, cultural influences, and environmental influences are the result of Other Psyches and the choices of others.

Developmental Disorders and Personality Disorders, which are BioPsychoSocial in nature and origin, are composed of all three different components or factors – biology, psychology, and sociology (environmental influences). The Mind-Brain Problem is automatically and naturally resolved simply by taking all of the factors into consideration whenever one is trying to determine the causes of any mental illness, psychopathology, or abnormal behavior.

It's the Materialists, Naturalists, Darwinists, Nihilists, and Atheists who caused and created the Mind-Brain Problem in the first place; and, belief in Materialism, Naturalism, Darwinism, Nihilism, and Atheism will prevent you from ever finding a workable realistic solution to the Mind-Brain Problem. Getting rid of these speculative philosophies or false religions and replacing them with observational evidence and research evidence is the only real solution to the Mind-Brain Problem.

Materialism, Naturalism, and their derivatives such as Darwinism, Nihilism, and Atheism have done more to stifle scientific progress than anything else on this earth. They are worthless. It's time for them to go; but, clearly they aren't going to leave without a fight. For many people, Darwinism, Naturalism, and Atheism is their personal religion. They aren't going to give it up without a fight. I KNOW, because I used to be a Materialist, Nihilist, and Atheist.

A careful reading of *Abnormal Psychology: An Integrative Approach* demolishes Materialism, Naturalism, and Determinism. A careful reading of this book also solves the Mind-Brain Problem, which was created by the Materialists and Naturalists in the first place.

As I do in this introduction, I kept returning to this book *Abnormal Psychology* over and over and over again – rethinking it and restating it in as many different ways and from as many different perspectives as possible. I consider this to be the pre-eminent Science Book in my study of Psychology – I took my bachelor's degree in Psychology, which is why I spent so much time with this book.

Barlow and Durand use the BioPsychoSocial Model, an integrative approach, to explain abnormal behavior and psychopathology.

Barlow and Durand still have residual Materialism and Naturalism conditioned into them, but they can't completely avoid Psyche no matter how much they might want to, assuming that they want to. Whether these authors are avoiding Psyche or not, I typically have to go out of my way to identify each and every time that they employ Psyche during their explanations of psychopathology and abnormal behavior. My original complaint against these authors and their book was that they NEVER mention psyche or mind explicitly; and, it took me awhile to realize that they were always using psyche or mind implicitly instead.

**Psychopathologists are exploring the causes of psychopathology, whether in the brain [Bio] or in the environment [Social], people [Psyches] are suffering and require the best treatments we have.** (*Abnormal Psychology*, p. 49.)

Initially, I actually believed that these authors left out Psyche; but, eventually I realized and learned that people are human beings or human psyches at their very core. Throughout this book, Barlow and Durand simply lump psyches and society into one united category which they call "PsychoSocial". It actually works, though, because the "Psycho" portion of the model is the Human Psyche and the "Social" portion of the model is Other Psyches.  Good enough!

What these authors do (along with the researchers that they quote) is use Identical Twin Studies, Family Studies, and Adoption Studies to measure and determine the heritable component (genetic component) of each type of mental illness, psychopathology, and abnormal behavior.  Once they have identified what proportion of the disease or mental illness is genetically caused or can be attributed to genetics (Bio), they then assign the remaining percentage or proportion to PsychoSocial causes.  They never make any attempt to resolve the Mind-Brain Problem in their book, and simply leave it up to their students to discover for themselves that the Psycho portion of the model is caused or controlled by the Human Psyche and the Social part of the BioPsychoSocial Model is caused or controlled by Society, the environment, or Other Psyches.

These authors also tend to lump all Developmental Effects into the PsychoSocial category where they technically belong.  Developmental defects and brain damage can often be the result of mommy's bad behavior (Alcohol Fetal Syndrome or Crack Babies), as well as being a result of a person's (psyche's) bad personal choices such as neurodegenerative diseases (brain damage) caused by alcoholism, drug abuse, risky daredevil behaviors, syphilis, eating toxins, and HIV, which can all cause brain damage or neurodegenerative diseases.

They repeated this process of separating each mental illness into heritable causes (Bio) and chosen causes (PsychoSocial causes) throughout the whole book.  This is by far my most favorite college textbook that I have encountered so far, because the authors are balance and even-handed and honest throughout the whole book.  This book changed the way that I look at Science.  These authors present the evidence or tell you that there is no evidence, then at times they tell you what that evidence means to them personally; but in the end, they leave it up to their students to decide for themselves what to make of all that evidence or research.

Remember, the MAIN PROBLEM of science and the scientific methods is that, for every scientific observation and for every piece of scientific evidence, there can be literally thousands of different possible interpretations given to that evidence.  There's NO way to tell which interpretation is correct.

The fact that these authors use scientific research and scientific measurements to separate the Biological component of a mental illness from the PsychoSocial component is a major step forward, because it is a major step away from Behaviorism and Materialism and Determinism which rely exclusively on the Biological component and completely ignore and reject everything else.

**Here, we address the issue of causation.  This chapter examines the specific components of a multidimensional integrative approach to psychopathology.  Biological dimensions include causal factors from the fields of genetics and neuroscience.  Psychological dimensions include causal factors from behavioral and cognitive processes, including learned helplessness, social learning, prepared learning, and even unconscious processes (in a different guise than in the days of Sigmund Freud).**

[Unconscious Processes are Psyche Processes within the person or personality or psyche; and, so are Learning Processes and Cognitive Processes and Thinking Processes and Emotional Processes and Memory Processes, because these types of processes survive the death of our physical body and physical brain according to the observational evidence obtained from Near-Death Experiences and Out-of-Body Experiences.]

Emotional influences contribute in a variety of ways to psychopathology, as do social and interpersonal influences. Finally, developmental influences figure in any discussion of causes of psychological disorders. You will become familiar with these areas as they relate to psychopathology and learn about some of the latest developments relevant to psychological disorders.

But keep in mind what we confirmed in the previous chapter: No influence operates in isolation. Each dimension – biological or psychological – is strongly influenced by the others and by development, and they weave together in various complex and intricate ways to create a psychological disorder. Here, we explain briefly why we have adopted a multidimensional integrative model of psychopathology. Then we preview various causal influences and interactions. After that, we look more deeply at specific causal influences in psychopathology, examining both the latest research and integrative ways of viewing what we know.

To say that psychopathology is caused by a physical abnormality or by conditioning is to accept a linear or one-dimensional model, which attempts to trace the origins of behavior to a single [biological] cause. In psychology and psychopathology, we still encounter this type of thinking occasionally, but most scientists and clinicians believe abnormal behavior results from multiple influences. A system, or feedback loop, may have independent inputs at many different points, but as each input becomes part of the whole it can no longer be considered independent.

This perspective on causality is *systemic* [and holistic], which derives from the word *system*; it implies that any particular influence contributing to psychopathology cannot be considered out of context. Context, in this case, is the biology and behavior of the individual [psyche], as well as the cognitive, emotional, social, and cultural environment, because any one component of the system inevitably affects the other components. This is a multidimensional model.

Source: (*Abnormal Psychology*, p. 29.)

https://books.google.com/books?id=00nAAgAAQBAJ&pg=PT63&lpg=PT63&dq#v=onepage&q&f=false

This is a BioPsychoSocial Model, a holistic model, an integrative multidimensional model; wherein the whole ends up being greater than the sum of the individual parts. By adopting a holistic model or an integrative model, we end up getting closer to the truth of the situation.

We have just reviewed three traditions or ways of thinking about causes of psychopathology: the supernatural, the biological, and the psychological (further subdivided into two major historical components: psychoanalytic and behavioral).

Supernatural explanations of psychopathology are still with us. Superstitions prevail, including beliefs in the effects of the moon and the stars on our behavior. This tradition has little influence on scientists and other professionals, however. Biological, psychoanalytic, and behavioral models, by contrast, continue to further our knowledge of psychopathology as you will see.

Each tradition has failed in important ways. Health professionals tend to look at psychological disorders narrowly, from their own point of view alone.

In the 1990s, two developments came together as never before to shed light on the nature of psychopathology:

(1) the increasing sophistication of scientific tools and methodology.

(2) the realization that no one influence – biological, behavioral, cognitive, emotional, or social – ever occurs in isolation.

Our thoughts, feelings, and actions inevitably influence the function and even the structure of the brain, sometimes permanently. [Neuroplasticity.]

In other words, our behavior, both normal and abnormal, is the product of a continual interaction of psychological, biological, and social influences.

The view that psychopathology is multiply determined had its early adherents. Adolf Meyer steadfastly emphasized the equal contributions of biological, psychological, and sociocultural determinism.

By 2000, a veritable explosion of knowledge about psychopathology was occurring. It was clear that a new model was needed that would consider biological, psychological, and social influences on behavior. This approach to psychopathology would combine findings from all areas.

In the remainder of this book, we explore the reciprocal influences among neuroscience, cognitive science, behavior science, and developmental science and demonstrate that the only currently valid model of psychopathology is multidimensional and integrative.

Source: (*Abnormal Psychology*, pp. 23-24.)

https://books.google.com/books?id=2sq5DQAAQBAJ&pg=PA27&lpg=PA27&dq#v=onepage&q&f=false

Barlow and Durand deliberately dismissed and excluded the Supernatural Model within their book, which means that they never explicitly discussed the Mind-Brain Problem or the Human Psyche. I took exception to this at first, as documented in another book that I wrote; but in the end, Barlow and Durand implicitly included the Human Psyche and the Supernatural Model within the PsychoSocial part of their integrative approach to psychopathology, possibly without even realizing that they were doing so.

Remember, when it comes to the BioPsychoSocial Model of reality and abnormal behavior, the "Psycho" portion represents the Human Psyche, and the "Social" portion represents Other Psyches including other people, the animals, and the environment around us. Consequently, Barlow and Durand ended up implicitly including Psyche, Non-Local Consciousness, and the Supernatural Model into the PsychoSocial portion of their BioPsychoSocial Model, because they simply couldn't avoid doing so and still have their models make any sense in the end. The Human Psyche and Other Psyches are always implicitly included in every PsychoSocial Model that is ever developed, whether the Materialists and Naturalists acknowledge that fact or not.

I learned to love their integrative approach to Abnormal Psychology, because it was so, well, scientific! They admitted ALL of the evidence into evidence, and then let the evidence lead us wherever it was trying to take us. That, to me, is the proper way to do science.

Over and over again, chapter after chapter, from one psychological disorder to the next, they first used twin studies and adoption studies to determine what portion of the specific mental illness under consideration was heritable or genetic in origin. Then in the next part of the same chapter, they discussed the possible PsychoSocial causes and influences for that very same specific mental illness. In the process, they trained me to do the same.

Now, let's dig into it!

**Signs of Psyche – BioPsychoSocial**

If I live long enough, I'm going to write a book entitled, "Signs of Psyche". For now, you'll just have to be satisfied with my book entitled, "BioPsychoSocial".

https://www.amazon.com/gp/product/B0713NDHVW

Barlow, D. H. & Durand, V. M. (2015). *Abnormal Psychology: An Integrative Approach* (7th ed.). Stamford, CT: Cengage Learning.

As a general rule, this book defines the **Bio** portion of the BioPsychoSocial Model as genetics and heritability.

All throughout this book, they confound or equate "Psyche" and "Environment" into one great whole which they call **PsychoSocial**. They had to do something like this, because this book is a college textbook after all. If they made "Psyche" explicit, then the Materialists, Naturalists, and Atheists would have taken up a campaign and had the book censored, blocked, and banned – that's what these people do.

Consequently, throughout this whole book, since they NEVER mention Psyche or Non-Local Consciousness explicitly, I found myself looking for and noting **Signs of Psyche** hidden within their book. This became a fun little task, which I found myself noting in the margins of my book.

> **Second, as noted earlier, it has become increasingly clear that genetic contributions cannot be studied in the absence of interactions with events in the environment [psyche] that trigger genetic vulnerability or "turn on" specific genes.**
>
> **The process of learning affects more than behavior. The very genetic structure of cells may change as a result of learning, if genes that were inactive or dormant interact with the environment [psyche] in such a way that they become active. In other words, the environment [psyche] may occasionally turn on genes. This type of mechanism may lead to changes in the number of receptors at the end of a neuron, which in turn, would affect biochemical functioning in the brain.**

> **The competing idea is that the brain and its functions are plastic, subject to continual change in response to the environment [psyche], even at the level of genetic structure. Now there is strong evidence supporting that view.** (*Abnormal Psychology*, p. 34.)
>
> https://books.google.com/books?id=yITxaqlS_yMC&pg=PT62&lpg=PT62&dq#v=onepage&q&f=false

Most of the time within this book, one can simply replace the word "environment" with the word "psyche" and the sentence will make just as much sense as it did before.

Why?

It's because these authors lump Psyche and Environment together into one united whole, which they call **PsychoSocial**. They are following the BioPsychoSocial Model of Reality, without mentioning Psyche and Non-Local Consciousness explicitly – in other words, they are catering to the Materialists and Naturalists while also at the same time trying to remain close to the truth of the matter. Psyche and "interaction with our environment" can turn on and turn off genes, and thus alter the way that physical cells function and develop. This is a scientifically observed and scientifically proven fact, which has become the sciences of Neuroplasticity and Epigenetics. Psyche can turn on and turn off genes.

They continue in this same vein throughout the book.

> **For years, scientists have assumed a specific method of interaction between genes and environment [psyche]. According to this diathesis-stress model, individuals inherit tendencies to express certain traits or behaviors, which may then be activated under conditions of stress. Each inherited tendency is a diathesis, which means literally a condition that makes someone susceptible to developing a disorder. This tendency is the diathesis or vulnerability. The diathesis is genetically based, and the stress is environmental [and/or psychological], but they must interact to produce a disorder.**
>
> **This study was very important in demonstrating clearly that neither genes nor life experiences (environmental events) can solely explain the onset of a disorder such as depression. It takes a complex interaction of the two factors.** (*Abnormal Psychology*, pp. 34-35.)
>
> https://books.google.com/books?id=yITxaqlS_yMC&pg=PT62&lpg=PT62&dq#v=onepage&q&f=false

These authors always deliberately leave out the "psyche" component and subsume it into the "environmental" component; but, it's there and real, just hidden away where the Materialists and Naturalists cannot see it nor find it.

If you study all of this closely, you will observe that it is Psyche who chooses to overcome a disorder that was originally produced by genetics and/or the environment.

**Life Experiences** ARE a function of Psyche, because a person continues to have Life Experiences in the Non-Local Realm or Spirit World long after his or her physical brain is dead and gone, according to the empirical evidence from Near-Death Experiences (NDEs), Shared-Death Experiences (SDEs), Out-of-Body Experiences (OBEs), and other types of Non-Local Experiences or Spiritual Experiences. There are THREE FACTORS at play – BioPsychoSocial – not two factors.

Notice how these authors carefully and deliberately suppress the "Psycho" or "Psyche" factor so as not to ignite the ire and the resistance (the boycott and the ban) of the Materialists and Naturalists – yet anyone who knows the facts simply KNOWS that Life Experiences are a function of Psyche, and simply facilitated by one's current environment. Technically, the environment [physical matter] can't force you to go to work or go to school and have a Life Experience; but, Other Psyches like your mother certainly can. You see, the environment also consists of Other Psyches, and some of them do in fact have a powerful impact on your Life Experiences. It's Psyche who does the ACTING, and the environment or physical matter simply REACTS.

### Gene-Environment Correlation Model

**These people [psyches], then, might have a genetically determined tendency to create the very environmental risk factors that trigger a genetic vulnerability.** (*Abnormal Psychology*, p. 36.)

https://books.google.com/books?id=reC5DQAAQBAJ&pg=PA40&lpg=PA40&dq#v=onepage&q&f=false

Notice once again the specific need for some very careful reading.

People are human psyches. Psyches design and create. Psyches ACT. Some of these human psyches might have inherited a genetic tendency to create the very environmental factors [physical factors] that trigger or turn-on one of their genetic vulnerabilities. ALL THREE FACTORS – BioPsychoSocial – interact and feed upon one another in one united integrated whole.

Psyche is there hidden all throughout their book – just off our radar – so that you actually have to KNOW about Psyche and believe in the existence of Non-Local Consciousness BEFORE you can completely and adequately interpret and understand what these authors are really writing and saying. That's how these authors were able to get their college textbook past the materialistic, naturalistic, and atheistic censors who determine which college textbooks end up being used in their public schools.

Remember and observe, our public schools are temples and shrines to Atheism, Materialism, Naturalism, Behaviorism, Nihilism, Scientism, and Darwinism. These people, these human psyches, hold all the keys now and they have the full backing of the US Supreme Court. If you don't walk a thin line, your books will be blocked, censored, and banned.

The next section of this book *Abnormal Psychology* was entitled, **Epigenetics and the Nongenomic "Inheritance" of Behavior**. As you can probably imagine, Barlow and Durand had to be careful with this section also, because the Materialists, Naturalists, and Darwinists teach that there is NO such thing as "non-gene inheritance" because these people believe that everything is caused by our genes and that we are automatons and puppets "dancing to our DNA".

**The environmental effects of early parenting seem to override any genetic contribution to be anxious, emotional, or reactive to stress.**

**Extremely chaotic early environments can override genetic factors and alter neuroendocrine function to increase the likelihood of later behavioral and emotional disorders.**

**It seems that genes are turned on or off by cellular material that is located just outside the genome ("epi," as in the word epigenetics, means on or around).**

> It seems that environmental manipulations, particularly early in life, may do much to override the genetically influenced tendency to develop undesirable behavioral and emotional reactions. Although current research suggests that environmental influences, such as peer groups and schools, affect this genetic expression, the strongest evidence exists for the effects of early parenting influence and other early experiences. (*Abnormal Psychology*, p. 37.)

This is the benefits of a calm and peaceful Psyche and Other calm and peaceful Psyches – they can override our genetics.

Again, these authors include Other Psyches simply by using the word "parenting", without actually mentioning the word "psyche", suggesting that parenting can override one's genetic inheritance. Our parents, neighbors, and nation – Other Psyches – are the ones who create these chaotic early environments and environmental manipulations that are mentioned. Physical environments, cities, nations, schools, roads, and homes don't just create themselves from scratch – Other Psyches put all of these things together for us, before we were born.

Most everything we learn comes from following the example of our family members, school teachers, peer group, and the animals around us – OTHER PSYCHES!

Manipulations imply hands, or Other Psyches. Peer groups and schools ARE Other Psyches who influence us. And obviously, parenting is the result of Other Psyches. Our whole environment is comprised of Other Psyches and what they do for us and to us. The physical matter is secondary and typically unimportant, which is something that the Materialists and Naturalists don't want you to realize or understand.

Unlike other college textbooks, which preach Materialism and Darwinism religiously and exclusively, the authors of this college textbook actually managed to trigger my thought processes and taught me how to think critically and deeply about what the research evidence is really trying to tell us and teach us – simply by demonstrating and teaching that each mental disorder or mental illness has BOTH a Biological component and a PsychoSocial component – neither is complete without the other.

The research evidence is always GOOD SCIENCE, as long as effective and proven methodologies are employed, and the rules of science are followed. It's always the materialistic and naturalistic interpretation, which is given to that evidence, that turns all of that GOOD SCIENCE into BAD SCIENCE.

That's why under Science 2.0, I deliberately encourage my readers to do their own interpretation of the Scientific Evidence, Eye-Witness Evidence, Empirical Evidence, Non-Local Evidence, and Research Evidence. Chances are good that their interpretation of the evidence will end up being infinitely more realistic and believable and robust than the Official Materialistic Interpretation which is typically given to research evidence. I KNOW that my own personal interpretations of the Research Evidence end up being infinitely better and more believable than the Official Naturalistic Interpretations which the scientists tend to give to Research Evidence; so, I figure that the same reality will be true for you as well, especially if you make the effort to develop better interpretations of the evidence as I have done.

Under Science 2.0 there is NO official interpretation of the evidence. Yes, it's perfectly permissible for me to present my own personal interpretation of the Scientific Evidence; but, I have NO right nor responsibility to make my personal interpretation the Official Interpretation. Under Science 2.0, you are encouraged to make your own interpretation of the evidence. You are likely to do so anyway, so why not encourage you to do so right from the start?

Barlow, D. H. & Durand, V. M. (2015). *Abnormal Psychology: An Integrative Approach* (7th ed.). Stamford, CT: Cengage Learning.

Through its positive example, this book changed the way I look at Science and do Science.

I went throughout their whole books making notes in the margins of my book, whenever I encountered Signs of Psyche. These authors NEVER mentioned psyche explicitly, but it was there to be found hidden in the sub-text all throughout their book, if one was willing and able to go looking for it.

Here's a sampling of some of what I wrote in the margins of this book, which I hope some people might find useful and interesting.

The unconscious is the Psyche.

Emotions have a strong Psyche component.

**Psychological science verifies the existence of the unconscious.** (*Abnormal Psychology*, p. xxi.)

We are composite beings, so our issues are composite as well.

**A psychological disorder or abnormal behavior is a psychological dysfunction within an individual** [psyche] **that is associated with distress or impairment of functioning** [bio] **and a response that is not typical or culturally expected** [social]. (*Abnormal Psychology*, p. 2.)

**Psychological dysfunction refers to a breakdown in cognitive** [psyche or thoughts], **emotional** [bio/psyche combination], and **behavioral functioning** [social interaction]. (*Abnormal Psychology*, p. 2.)

Psychological disorders, psychopathology, and abnormal behaviors are caused by the interplay between different BioPsychoSocial Factors.

Thoughts and dreams are a function of Psyche, because their contents cannot be recorded by our physical instruments.

Cognitions are thoughts; and, thoughts are a function of Psyche, because we continue to think long after our physical brain is dead and gone, according to the empirical evidence from Near-Death Experiences (NDEs) and other types of Non-Local Experiences.

Our chosen behaviors are a function of Psyche. Anytime we encounter choice, we have in fact encountered Psyche or Non-Local Consciousness, because Psyche can choose to act in any realm of existence or any reality, including the Non-Local or Transdimensional Realms.

Our reflexes tend to be a function of Biology, although Psyche can at times override our reflexes and our biology.

Public behaviors, and behavioral functioning and behavioral conditioning, tend to require some type of social interaction with Other Psyches.

When it comes to an organic human being, the whole thing is BioPsychoSocial in nature, origin, manifestation, and function. One of these factors can't truly exist without the others, and still end up with a human being. A human being is a composite of Biology, Psychology, and its Social Interaction with Other Psyches in the environment.

**Etiology, or the study of origins, has to do with why a disorder begins (what causes it) and includes biological, psychological, and social dimensions. Because the etiology of psychological disorders is so important to this field, we devote an entire chapter to it.** (*Abnormal Psychology*, p. 6.)

Once again, the whole discipline of Abnormal Psychology is BioPsychoSocial in nature, origin, and explanation; and, so is the rest of Reality as we know it.

In this book, *Abnormal Psychology*, they discount and dismiss and debunk the supernatural model or the Supernatural Tradition, in an attempt to appease the Materialists and Naturalists. But, Psyche or Non-Local Consciousness continues to be sub-text running all throughout the book, being subsumed into the PsychoSocial aspect of the BioPsychoSocial Model which they employ within their book.

**Psychosis – psychological disorders characterized in part by beliefs that are not based in reality (delusions), perceptions that are not based in reality (hallucinations), or both.** (*Abnormal Psychology*, p. 11.)

This is ironic, because Materialism, Naturalism, Darwinism, Behaviorism, Determinism, Atomism, Nihilism, and Atheism are based upon philosophical beliefs that are not based in reality nor supported by empirical evidence. These Falsified Philosophies are simply a figment of the imagination and have to be taken on blind-faith as being true, because there is NO observational evidence supporting them and never will be.

When it comes to science, scientific discovery, and scientific understanding, Materialism and Naturalism set us back every time.

I equate the "unconscious" and consciousness with the Psyche – not the physical brain. Why? It's because the Observational Evidence and Experiential Evidence from NDEs, OBEs, SDEs, and other Non-Local Experiences tells me that it must be so. I choose to follow the evidence, and not the philosophical speculation. People remain conscious long after their physical body and physical brain are dead and gone, according to the Empirical Evidence from NDEs and OBEs.

Distraction – learning something new – is my favorite coping mechanism or defense mechanism.

**Careful scientific studies of psychopathology have supported the observation of unconscious mental processes.** (*Abnormal Psychology*, p. 20.)

Unconscious Mental Process are Psyche Processes.

Observations and Research have supported the existence of Unconscious Mental Processes.

**Humanistic therapists believed that relationships, including the therapeutic relationship, were the single most positive influence in facilitating human growth.** (*Abnormal Psychology*, p. 21.)

Friendship Therapy works and ends up being an extremely effective form of Psyche-Therapy. Psyche can affect and influence psyche, in every realm of existence or every reality.

Phenomenology is the study of Life Experiences, Events, or Phenomena. Emmanuel Levinas developed a Phenomenological Philosophy and Personality Theory based upon the

Self's interaction with "The Other". The Self is the human psyche. "The Other" ends up being Other Psyches; and, we interact with Other Psyches through our Life Experiences, Life Events, and other Phenomena that are sometimes of a Non-Local nature and origin.

Barlow and Durand repeatedly equate Psyche with the Environment and Society, so as not to raise any flags in the minds of the Materialists, Naturalists, and Atheists.

They talk about all sorts of different types of psychotherapy, but none of them talk about Christ-Centered Therapy. I personally discovered that the Atonement of Christ is the BEST and the ULTIMATE form of Psyche-Therapy. It works every time and it lasts.

In order to cater to the Materialists, Naturalists, and Behaviorists, these authors discount and dismiss the supernatural, which means that they discount and eliminate Psyche, Non-Local Consciousness, and the Atonement of Christ. Then they sneak Psyche in the back door through inference and sub-text, and by equating psyche with the environment.

Imagination is a function of psyche. It all takes place inside your head and doesn't impinge on the physical world at all.

The Materialists, Naturalists, Determinists, Atomists, and Behaviorists are trying to hide any evidence for the Non-Physical or the Non-Local, so that we can't find it and know about it. They have become very good at what they do – suppressing evidence. They will tell you right to your face that no such evidence exists. They are wrong. But, there was a time when I fell for it – hook, line, and sinker. There was a time when I was a Materialist, Nihilist, and Atheist.

**Skinner coined the term *operant conditioning* because behavior operates on the environment and changes it in some way. For example, the boy's behavior affects his parents' behavior and probably the behavior of other customers. Therefore, he changes his environment. Most things that we do socially provide the context for other people to respond to us in one way or another, thereby providing consequences for our behavior. The same is true of our physical environment, although the consequences may be long term.** (*Abnormal Psychology*, p. 23.)

Notice that even something like Behaviorism has Psyche or Non-Local Consciousness hidden within its teachings and principles. Chosen behaviors are a hallmark of Psyche. Behaviorism attributes everything to society, or environmental causes, aka. environmental conditioning or operant conditioning. The thing that the Behaviorists don't recognize or realize is that the environment is created by and controlled by Other Psyches. The boy's chosen behaviors as well as some of his reflexes influence OTHER PSYCHES. Therefore, he (his human psyche) changes his environment or surroundings. Most things that we human psyches choose to do gives Other Psyches the context or reason to respond to us in one way or another. Our physical environment also has an influence on us and our Psyche, but not as powerfully as Other Psyches. There's nothing in the physical environment that can force you to get out of bed and go to work or go to school; but, your mother or your wife certainly can, because she is another Psyche. You don't want to mess with the woman, because she can make things miserable for you if you do.

Remember, "changing our environment" is a function of Psyche.

Psychologists keep returning to Behaviorism and Behavioral Conditioning because it works. Why does it work? It's because Other Psyches make it work. Psyche is hidden all throughout Behaviorism, even though the Cognitive Behaviorists deny that psyche exists and simply view us as mindless automatons.

Much like Watson, Skinner did not see the need to go beyond the observable and the quantifiable to establish a satisfactory science of behavior. He did not deny the influence of biology or the existence of subjective states of emotion or cognition: he simply explained these phenomena as relatively inconsequential side effects of a particular history of reinforcement.

The behavior model has contributed greatly to the understanding and treatment of psychopathology. Nevertheless, this model is incomplete and inadequate to account for what we now know about psychopathology. In the past, there was little or no room for biology in behaviorism, because disorders were considered, for the most part, environmentally determined reactions. This model also fails to account for development of psychopathology across the life span. Recent advances in our knowledge of how information is processed, both consciously and subconsciously [by Psyche], **have added a layer of complexity. Integrating all these dimensions requires a new model of psychopathology** [the Integrative Approach of the BioPsychoSocial Model]. (*Abnormal Psychology*, p. 23.)

The psychodynamic model (Freud), the behavioral model (Skinner), the humanistic model (Rogers and Maslow), and even the cognitive model (Bandura) are all being slowly replaced by and upgraded to the BioPsychoSocial Model, because the BioPsychoSocial is more explanatory and more all-inclusive than the other models were. In 2016, I had a visiting Clinical Psychologist from the University of Utah tell me that when it comes to abnormal behavior and psychopathology, it's ALL BioPsychoSocial now. When properly defined and properly used, the BioPsychoSocial Model actually takes the influence of Psyche or Non-Local Consciousness into account. If you start looking for signs of psyche, eventually it dawns on you that every Choice, every Intention, every Thought, every Action, and every Chosen Behavior is caused by Psyche or Non-Local Consciousness. Once you start keeping score, you quickly realize that Psyche actually does more to influence and change and control our behavior than nature and nurture combined; yet, the Materialists and Behaviorists insist that psyche does not exist.

Barlow and Durand mention, debunk, and dismiss the Supernatural Model. As far as I know, Barlow and Durand mention the word "psyche" once – in Chapter One – and then they dismiss it along with the Supernatural Model. They mention devils, evil spirits, torture chambers and ALL of the bad things typically associated with the Supernatural Model rather than focusing on the Atonement of Christ. The Supernatural Model should focus its attention on the Atonement of Christ – the good stuff – rather than mucking about in demonology and torture chambers as they typically do when discussing anything supernatural. It's a CHOICE. In the first chapter of their book, they chose to focus on the bad, rather than focusing on the good, while discussing the Supernatural Model. Their purpose was to dismiss the Supernatural Model as unscientific and leave it at that.

http://mypsyche.us/wp-content/uploads/2017/10/Abnormal-Psychology.pdf

In chapter 2, while developing and presenting their Integrative Approach to Psychopathology, Barlow and Durand deliberately suppress and refuse to mention anything about Psyche or the Supernatural. They subsume the whole thing into what they call the PsychoSocial aspect of abnormal behavior, or environmental influences. I assume they had to do this in order to get their college textbook past the materialistic, naturalistic, and atheistic censors who determine which textbooks are going to be used in our public schools. This means that if you want to find Signs of Psyche in Chapter 2 and their Integrative Approach, you are going to have to read between the lines and insert that information yourself, if you want it to be there.

That's what I do next. Here are some of my notes.

### Signs of Psyche – BioPsychoSocial – Chapter 2

Barlow, D. H. & Durand, V. M. (2015). *Abnormal Psychology: An Integrative Approach* (7th ed.). Stamford, CT: Cengage Learning.

The Materialists, Naturalists, and Atheists assign our thoughts, feelings, behaviors, choices, and emotions to our physical brain. The rest of us assign these things to Psyche or Non-Local Consciousness, because WE KNOW from the empirical evidence or lived experiences of the human race that our Psyche continues to think, feel, choose, and behave long after our physical brain is dead and gone.

The way to distinguish between Psycho and Social influences on behavior is to observe and record Chosen Behaviors. Any time we witness a Choice of some kind, we have in fact caught Psyche in action, whether we are talking about our own Psyche or the Other Psyches. Psyche chooses or acts. Physical matter reacts. That's how you tell the difference between the two.

**Judy's case is a good example of biology influencing behavior. But, behavior, thoughts, and feelings can also influence biology, sometimes dramatically. Emotions can affect physiological responses such as blood pressure, heart rate, and respiration.** (*Abnormal Psychology*, p. 30.)

This is mind-over-matter or psyche-over-matter – the documented and proven Placebo Effect. This is Psyche in action.

We can learn to CONTROL our thoughts, choices, behaviors, feelings, actions, reactions, emotions, and physiological responses. Who is "we"? "We" consists of Psyche or Non-Local Consciousness. We can learn to school our thoughts.

Anytime YOU witness or experience Mind-Over-Matter, you have in fact witnessed or experienced Psyche or Non-Local Consciousness in action, and therefore have effectively proven Psyche's existence. Remember, the BEST way to find and know the truth is to live it and experience it and witness it for yourself, or to choose to trust someone who has.

**It is the issue of etiology or causation that concerns us here. As you can see, finding the causes of abnormal behavior is a complex and fascinating process. Focusing on the biological or behavioral factors would not have given us the full picture of the causes of Judy's disorder; we had to consider a variety of other influences and how they might interact. A discussion in more depth follows, examining the research underlying the many biological, psychological, and social influences that must be considered as causes of any psychological disorder.** (*Abnormal Psychology*, p. 31.)

The causes of any psychological disorder are BioPsychoSocial in nature and origin. The one cannot exist without the other; and, every single one is incomplete without the others. They influence each other.

**Our genes seldom determine our physical development in any absolute way. They do provide some boundaries [limitations] to our development. Exactly where we go within these boundaries depends on environmental influences.** (*Abnormal Psychology*, p. 32.)

And the choices that the individual Psyche makes.

**Because there is plenty of room for the environment to influence our development with the constraints set by our genes, there are many reasons for the development of individual differences.**

**This question of nature (genes) versus nurture (upbringing and other environmental influences) is age-old in psychology, and the answers beginning to emerge are fascinating.**

**The first 22 pairs of chromosomes provide programs or directions for the development of the body and brain; and the last pair, called the sex chromosomes, determines an individual's sex.** (*Abnormal Psychology*, p. 32.)

This last bit isn't exactly correct. The chromosomes and genes provide programs or directions for the formation of proteins and enzymes, which go into the construction of the body and the brain. There's not enough information storage capacity in the genome for storing the bodily blueprint, brain blueprint, 3D location addressing of all 100 trillion cells in the physical body, the sequential development instructions of the physical body, cell assignments or differentiations or specialization information for each cell, molecular assignments and 3D addressing and appointments with destiny within the cells, and micromanagement instructions, housekeeping instructions, and repair instructions for each and every different type of physical cell. ALL of this information outstrips the genomic storage capacity trillions-to-one. That kind of information has to be stored someplace else besides the physical genome.

**Much of our development and, interestingly, most of our behavior, our personality, and even our intelligence quotient (IQ) score are probably polygenetic – that is, influenced by many genes, each contributing only a tiny effect, all of which, in turn, may be influenced by the environment.** (*Abnormal Psychology*, p. 32.)

Not to mention the Psyche's choices.

**Hundreds of genes can contribute to the heritability of a single trait.**

**Although all cells contain our entire genetic structure, only a small proportion of the genes in any one cell are "turned on" or expressed. In this way, the cells become specialized, with some influencing liver function and others affecting personality. What is interesting is that environmental factors, in the form of social and cultural influences, can determine whether genes are "turned on".** (*Abnormal Psychology*, p. 33.)

What determines this? Who or what throws the switch, turns a gene on, and differentiates a cell or forces a cell to specialize? NONE of that information and influence is found in the GENOME, if it in fact comes from social and cultural influences. Influences by definition in principle are a function of Psyche or Non-Local Consciousness.

Remember, there's only enough code in the genome for the maintenance of the of the body and the brain, through enzymes and proteins, and switches to turn genes on or off. All of the 3D blueprint, 3D addressing, 3D cell differentiation information, 3D environmental influences, and 3D cell functioning information has to be stored someplace else, because there is not enough storage capacity for all of that information within the genome itself. The genome is linear and sequential, which means that most of the 3D information associated with our cells, brain, and physical body has to be stored someplace else.

> **We present more examples later in the chapter when we discuss the interaction of genes and the environment. The study of gene expression and gene-environment interaction is the current frontier in the study of genetics.** (*Abnormal Psychology*, p. 33.)

Imagine what we will have to deal with when they finally start taking Psyche into consideration, because the human psyche can also turn on and turn off its genes – one cell at a time and one specific gene at a time.

Yikes!

All of the micro-management, location addressing, bodily blueprint, sequential assembly instructions, molecular assignments, gene addressing, and cellular differentiation – within each and every physical cell – is done by Psyche or Non-Local Consciousness according to Quantum Mechanics. There's no way in this universe that all of that information can be stored in our genes. That kind of cellular micromanagement information outstrips the storage capacity of our genome trillions-to-one. That kind of information has to be stored somewhere else non-locally, because something like an infinite singularity or psyche can really only exist non-locally because it has no size and takes up no space while at the same time by definition in principle having an infinite memory storage capacity. Physical matter greatly limits our informational storage capacity; but, there are NO physical limitations in the Non-Local Realm or Spirit World or Human Psyche.

In this book, they do a lot of research on twins, including monozygotic twins. These twins have the same genetics and often have the same environment; but, they have distinct psyches or personalities, and distinct courses and outcomes when it comes to abnormal behavior and psychopathology and psychological disorders. Personality is a function of Psyche, because our personality or intelligence continues onward after our physical brain and physical environment are dead or gone.

Naturalism and Materialism are too simplistic. They lack explanatory power.

Remember, Psyches are doing the interaction, and NOT the environment. The physical environment is just there as a foundational construct. It's the Psyches acting within that environment who choose to interact or choose not to interact with the physical environment and with Other Psyches. Psyche chooses and acts. Physical matter simply reacts. That's how we tell the difference between the two.

Barlow and Durand have a section on epigenetics.

> **Environmental events, in turn, seem to affect our very genetic structure by determining whether certain genes are activated or not.**
>
> **Strong environmental influences alone may be sufficient to override genetic diatheses. Thus, neither nature (genes) nor nurture (environmental events) alone, but rather a complex interaction of the two, influences the development of our behavior and personalities.**
>
> **The brain uses an average of 140 billion nerve cells, called neurons, to control every thought and action.** (*Abnormal Psychology*, p. 38.)

Again, they are deliberately ignoring Psyche, which means that this last sentence is incorrect as a result.

The neurons do NOT control thoughts. The thoughts control neurons, which then control actions of a physical nature. The physical brain is the way that the Psyche interacts with its assigned physical body and physical environment. The physical brain is a transceiver.

Thoughts or Psyche or Choice exists separate from the physical brain.

How do we know?

We KNOW from the Empirical Evidence associated with Near-Death Experiences (NDEs) and Out-of-Body Experiences (OBEs) that thoughts, psyche, choices, and control continue onward long after our physical brain and our neurons are dead or offline. The Empirical Evidence is the ONLY way to resolve this controversy. It can't be done with Materialism and Naturalism.

Materialism and Naturalism simply lack explanatory power. We cannot use Materialism and Naturalism to explain how Psyche or Non-Local Consciousness works.

Some epigenetic changes are heritable, meaning that they can be passed onto subsequent generations.

I read somewhere long ago that the human female has all of her eggs when she is born. If true, then epigenetic changes which take place in her DNA after her eggs are created cannot be passed on to her children. This would seem to indicate that most if not all of the epigenetic changes that a child inherits comes from the father's sperm – both X and Y. In contrast, mitochondrial changes are allegedly passed on through the mothers and their X chromosomes.

The Materialists, Naturalists, and Darwinists want you to believe that random mutations and natural selection designed, programmed, engineered, planned, developed, field-tested, fine-tuned, and constructed all of this from scratch out of thin air. We human beings can't do such a thing, so why on earth should anyone believe that evolution or genetic drift could have done the job? They believe it simply because they want to believe it. There's no other compelling reason to do so.

**Scientists have also discovered that we are not necessarily aware of a great deal of what goes on inside our heads. Technically, these cognitive processes are unconscious.** (*Abnormal Psychology*, p. 53.)

It's because these things are non-local and related to Psyche; and, they actually have to be retrieved into awareness from non-locality.

Our thoughts and cognitions are a function of Psyche or Non-Local Consciousness, not matter and not our physical brain.

How do we know?

It's because NONE of our thoughts can be recorded by our physical instruments, and because our thoughts have been observed to continue long after our physical brain is dead or offline, according to the empirical evidence from Near-Death Experiences (NDEs) and Out-of-Body Experiences (OBEs).

**Different people process information about events in the environment in different ways. These cognitive differences are an important component of psychopathology.** (*Abnormal Psychology*, p. 53.)

Cognitive processes are a function of Psyche. "Different people" are different human psyches. Our environment can be both local (physical) and non-local (spiritual). This means that our psychological disorders, personality disorders, and personality can continue onward AFTER our physical brain is dead and gone.

It has been estimated that there is only enough memory storage capacity in our physical brain to store one second's worth of life experiences if it is stored sequentially as bits and bytes, and maybe up to seven minutes of life experiences if that information and memories are stored holographically – this storing all the information that is currently available to our brain in working memory. All the rest of our memories and experiences have to be stored non-locally in a memory storage device that takes up no physical space and has no physical limitations whatsoever – our Psyche. Remember, these are estimates, because there is no way to know for sure. You will notice that each scientist will come up with a different estimate, which is why I didn't bother to mention a source. I have seen estimates ALL over the map from one extreme to the other; and, NO consensus.

There's another issue that comes into play when it comes to estimating the brain's storage capacity – *confirmation bias*. *Confirmation bias* is a logic fallacy, and it works. The estimates from the Materialists and Naturalists will be on the astronomical side, because these people have to make our physical brain capable of storing ALL of our memories. In contrast, other scientists tend to be more realistic and conservative. You can have any result you want simply by changing your assumptions.

So, who's right – the Materialists or the Spiritualists?

We simply KNOW from all the thousands of different life reviews which have taken place during NDEs, that ALL of our memories and experiences along with their ripple effects onto other psyches are stored Non-Locally in the Spirit Realm. Again, we have to resort to Observational Evidence and Experiential Evidence – Empirical Evidence – in order to resolve this controversy. The Spiritualists are demonstrably right, when all of the Observational Evidence is taken into account. In contrast, the Materialists and Naturalists tend to always be wrong, because they have NO observational evidence to support their guestimates. These people are simply guessing, and it shows.

**Advances in cognitive science have revolutionized our conceptions of the unconscious. We are not aware of much of what goes on inside our heads. We seem able to process and store information, and act on it, without having the slightest awareness of what the information is or why we are acting on it.**

**The existence of unconscious processes has been demonstrated, and we must take them into account as we study psychopathology.** (*Abnormal Psychology*, p. 55.)

The unconscious is non-local Psyche.

**Emotions play an enormous role in our day-to-day lives and can contribute in major ways to the development of psychopathology.**

**This kind of** [emotional] **reaction seems to be programmed in all animals, including humans, which suggests that it serves a useful purpose.** (*Abnormal Psychology*, p. 56.)

Who did the programming? ALL programming requires a Programmer! Random mutations and natural selection and genetic drift cannot function as a Programmer, so who did?

We don't observe the rocks getting emotional, so raw physical matter can't do emotions. So, where do these emotions come from?

**Emotion scientists now agree that emotion is composed of three related components – *behavior* (SOCIAL), *physiology* (BIO), and *cognition* (PSYCHO).** (*Abnormal Psychology*, p. 57.)

So, that's how it is done! Psyche and Other Psyches come into play!

I knew there had to be a logical explanation.

Emotion is a subjective feeling produced by Psyche, and then the physical matter or biology responds.

Cognition is a function of Psyche.

How do we know that emotion and cognition is a function of Psyche or Non-Local Consciousness?

Again, we have to resort to Empirical Evidence from NDEs and OBEs and simply observe that human beings (human psyches) continue to have feelings and emotions and cognitions long after their physical brain is dead or offline.

**On studying the cognitive aspects of emotions, Richard S. Lazarus proposed that changes in a person's environment are appraised in terms of their potential impact on that person. The type of appraisal you make determines the emotion you experience.**

**All components of emotion – behavior, physiology, and cognition – are important, and theorists are adopting more integrative approaches by studying their interaction.** (*Abnormal Psychology*, pp. 57-58.)

Subjective feelings are a function of Choice, which is a function of Psyche!

Persons are Psyches.

Appraisal is a function of Psyche. Psyche evaluates context and chooses the desired response. Desire is a function of Psyche. Appraisals are choices, which means that appraisals are a function of Psyche. Emotions are a function of Psyche.

Your chosen appraisals determine the emotion you will experience.

A person's environment (a psyche's environment) can be either local (physical) or non-local (spiritual).

Cognition or thought is a function of Psyche.

Chosen Behaviors or Chosen Actions are a function of Psyche. Automatic reflexes tend to be more the result of physiological programming along with interaction with the physical environment, although Psyches can learn to override their autonomic functioning.

Society or culture is a function of Other Psyches.

Physiology and physical development tend to be the result of a combination of genetics, environmental mishaps, and bad choices made by Psyche.

**Interestingly, it seems that adopting a forgiving attitude can neutralize the toxic effects of anger on cardiovascular activity.** (*Abnormal Psychology*, pp. 59.)

Adopting an attitude is a function of Psyche. Forgiveness is purely a function of Psyche. There's nothing in society or biology that can force you to forgive. That is indeed interesting, but only if Psyche is taken into account. ONLY you, your psyche, can choose to forgive and really mean it. That's something that your physiology and society cannot do for you.

Our chosen cognitions or thoughts or attitudes affect our moods and emotions.

**Emotions and mood also affect our cognitive processes: if your mood is positive, then your associations, interpretations, and impressions also tend to be positive.** (*Abnormal Psychology*, pp. 59.)

Emotions, moods, cognitive processes, associations, interpretation, impressions, and choices are a function of Psyche. Look and ye shall see, knock and it shall be opened unto you. If you can't see it, then you will never understand it. If you never go looking for it, then you will never find it. That doesn't mean it doesn't exist. It only means that you are not looking for it.

**Like a fever, a particular behavior or disorder may have a number of causes. The principle of Equifinality is used in developmental psychopathology to indicate that we must consider a number of paths to a given outcome.** (*Abnormal Psychology*, pp. 63.)

Every psychopathology is BioPsychoSocial in nature and origin.

CHOICE equals Psyche or the Psychological Factor.

ONLY Psyche can have experiences in every realm of existence, including the non-local realms, and actually remember having done so.

Science 2.0 allows ALL of the evidence into evidence. See what happens to our Science when ALL of the evidence is permitted into evidence and taken into consideration. The result is something much more expansive, interesting, and robust than can ever be accomplished by Materialism and Naturalism alone.

The Materialists, Naturalists, and Darwinists define science as Materialism, Naturalism, Darwinism, Nihilism, Behaviorism, Scientism, and Atheism. You can't get very far with these because they are based upon a refusal to look at evidence and the suppression of evidence. You can't get very far with these because these people won't let you. Remember, Materialism and Naturalism lack explanatory power, especially when it comes to the Non-Local or the Non-Physical or the Invisible.

If you understand what I have written here, you can probably see now why I abandoned My Materialism and My Nihilism and why I upgraded my science to Science 2.0 instead – Science 2.0 allows all of the evidence into evidence and allows me to take it into consideration while doing science.

That's just the first two chapters of the book, *Abnormal Psychology*. There 14 other chapters, pages of notes, and hundreds of pages of reading material and associated notes to explore; but, what I have written here will suffice for this particular book. I have to leave something to talk about for my other books, so I will.

**Schizophrenia**

For example, using twin studies, they determined that up to 48% of schizophrenia is heritable or genetic or Biological; and then, they just KNEW that the other 52% of schizophrenia was the result of PsychoSocial and Developmental Causes, which they then explored in the remainder of the chapter on schizophrenia.

The Genain quadruplets were one of the case studies that they used.

https://en.wikipedia.org/wiki/Genain_quadruplets

These four sisters were identical twins, having the same genetics; and, each one of them ended up developing schizophrenia. However, due to psychosocial factors and developmental factors, their schizophrenia ended up manifesting in different ways and had different outcomes in the end for each of the four sisters. This case study demonstrated perfectly that each mental illness has both a genetic component (Bio) and a societal and developmental and psychological component (PsychoSocial). This case study also demonstrated that the integrative holistic BioPsychoSocial Model is the best way to study and treat mental illnesses, psychopathology, personality disorders, and abnormal behavior.

## Alzheimer's Disease

The neurodegenerative disorder (brain damage) caused by Alzheimer's disease has a heritable component as large as 79% on average; however, this still leaves plenty of room for environmental triggers and variability in the progress and outcome of the disease.

The genetics of Alzheimer's disease:

https://www.hindawi.com/journals/scientifica/2012/246210/

There are probably around two dozen different genes that can contribute to Alzheimer's disease. Certain genes are 100% determinative for Alzheimer's disease, but these genes are relatively rare within the human population. In contrast, the most commonly inherited genes for Alzheimer's disease have to be employed in pairs or combinations and have to be triggered by the environment in order to manifest Alzheimer's disease in the individual, thereby leaving lots of room for PsychoSocial influences to come into play.

It's simply wrong and false to conclude that genetics or our reinforcement history causes ALL of our mental illnesses and abnormal behaviors, as the Materialists and Naturalists and Behaviorists do, because there are times when the Human Psyche can trump nature and nurture AT WILL. I KNOW this is true, because there was a time when I was addicted to half a dozen different prescription drugs, and I CHOSE to go cold turkey and suffer the consequences in order to get well in the end. That CHOICE was nothing that my genetics or my environment did for me or made me do.

In fact, my genetic inheritance was fueling my addictions because we have a tendency towards addiction on my mother's side; and, the medical doctors were feeding my addiction by pumping a dozen different drugs into me in the first place. These things were causing my addiction; therefore, it was left up to me, my psyche, to CHOOSE to end my addiction. It was something that I, my psyche, CHOSE to do for itself. It was nothing that they (my genetics and medical doctors) did for me. My Psyche had to trump my nature and my nurture in order to successfully overcome all my different addictions and survive the six-month long withdrawal process that was the result of my CHOICE.

Neuroscience and Genetics have demonstrated that "psyche" can actually turn genes on or off, and that "psyche" actually changes and shapes our physical brain – processes and disciplines known as Epigenetics and Neuroplasticity.

Any time you witness chosen behaviors, chosen decisions, chosen actions, chosen vices, chosen crimes, chosen sins, chosen malevolence, chosen love, chosen addictions, chosen neurodegenerative diseases, chosen illnesses, and a choice to go cold turkey, you are in FACT observing the Human Psyche in action. As human beings, we suffer (or enjoy) the consequences of the things which we have chosen for ourselves. We reap what we sow. Reaping and sowing are functions of Psyche or Choice. Psyche is synonymous with choice, because our Psyche is the thing that does all our choosing and remembers having done so after our physical brain is dead and gone.

Remember, whenever you witness inherited illnesses, you are witnessing genetics in action. Whenever you witness developmental illnesses, medically induced illnesses, and/or environmentally caused illnesses, you are witnessing societal, cultural, and environmental factors in action. Whenever you witness CHOICE or FREE WILL or INTENT, you are witnessing Psyche in action.

Over and over again, any time you witness CHOICE, or free will, or chosen behaviors, or intent, you are in fact observing some kind of PSYCHE in action; and, aside from God, the human psyche is the most powerful psyche on this planet. You will catch the human psyche CHOOSING THINGS INTO EXISTENCE all the time, if you take the time to stop, observe, and learn.

**Materialism and Naturalism Cause the Mind-Brain Problem**

Remember, materialism and naturalism cause the Mind-Brain Problem, which means that getting rid of materialism and naturalism solves the Mind-Brain Problem.

Materialism and Naturalism create a whole host of unsolvable problems within the science of Psychology – things like the Mind-Brain Problem, the Binding Problem, the Nature vs. Nurture Problem, the Consciousness and Sentience Problem, the Survival of Death Problem, and the Free Will vs. Determinism Problem. Consequently, getting rid of materialism, naturalism, radical reductionism, and determinism automatically solves these unsolvable problems.

The Quantum Consciousness Interpretation of Quantum Mechanics and the Orthodox Interpretation of Quantum Mechanics SOLVES the Mind-Brain Problem as well as all these other unsolvable problems which I have listed above. These particular interpretations of quantum mechanics rely upon observational evidence and experiential evidence in order to make their case. The same can be said of Science 2.0, which means that Science 2.0 is based upon the Quantum Consciousness and Orthodox Interpretations of Quantum Mechanics, which successfully falsify Materialism and Naturalism. Science 2.0 is based upon the Lived Experiences of the human race as a whole, and not the philosophical musings of the materialists and naturalists.

## 3. Mind-Over-Matter Resolves the Mind-Brain Problem

Anytime YOU witness or experience Mind-Over-Matter, you have in fact witnessed and experienced Psyche or Non-Local Consciousness in action, and therefore have effectively proven Psyche's existence. Remember, the BEST way to find and know the truth is to live it and experience it and witness it for yourself, or to choose to trust someone who has.

There are many different books out there to be found and read, which successfully document and demonstrate Mind-Over-Matter; and, anytime you witness or experience Psyche-Over-Matter, you have in fact solved the Mind-Brain Problem in the only manner that it technically can be solved.

In this section, I will list a few of the books I own, which successfully demonstrate the existence of Mind-Over-Matter.

As usual, Science 2.0 allows ALL of the evidence into evidence; but for the most part, I will leave it up to you to decide for yourself what to make of that evidence and what to do with that evidence.

The Materialists and Naturalists disparagingly call this stuff "Fringe Science"; but, it is science none-the-less, because it is a source of Observational Evidence, Experiential Evidence, Eye-Witness Evidence, and Empirical Evidence, which means that in the end it is Scientific Evidence.

### Foundational Tenets of Science 2.0

The BEST way to find and know the truth is to live it and experience it for yourself, or to choose to trust someone who has. If you are going to choose to trust someone, use your brains and choose to trust someone whom you deem to be absolutely trustworthy. You should do this with any type of evidence.

Although Science 2.0 allows all of the evidence into evidence, since the personal interpretation of the evidence is left up to YOU under Science 2.0, carefully examine that evidence and decide for yourself whether it is trustworthy or not. Under Science 2.0, YOU are called upon to decide for yourself whether evidence is reliable or not.

Science 2.0 is based upon the Burden of Proof Methodology and encourages you to pursue and follow the Preponderance of the Evidence.

Since there is NO officially sanctioned interpretation of the evidence under Science 2.0, you are left free to judge for yourself whether evidence is trustworthy and reliable, or not. This means that you are completely free to reject evidence if you deem it untrustworthy, unreliable, faked, or false – or if it fails to meet its Burden of Proof through a Preponderance of the Evidence.

The SECOND-BEST way to find and know the truth is to eliminate everything that is False and everything that has been Falsified, so that ONLY the Truth remains.

We start this process by eliminating Materialism, Naturalism, Darwinism, Nihilism, Behaviorism, Hard Determinism, Reductionistic Atomism, and Atheism because they are False and have been Falsified by Science and Observation. If you can successfully eliminate everything that is false and everything that has been falsified, then logically only the truth will remain standing.

Even though the Scientific Methods cannot be used to prove anything true, they can indeed be used to prove things false. When you falsify something, you do in fact prove that it is false. The Scientific Methods along with Direct Observations are extremely powerful tools for falsifying False Philosophies and False Theories such as Materialism, Naturalism, Darwinism, Nihilism, and Atheism.

It works.

How often have I said to you that when you have eliminated the impossible [and the false], whatever remains, *however improbable*, must be the truth? — Sherlock Holmes.

Eliminating the impossible and the false is what Science 2.0 is supposed to do, but only if you make it work for you. Just like in a court of law, Science 2.0 leaves it up to you to decide for yourself if the evidence meets its Burden of Proof through a Preponderance of the Evidence.

**The Placebo Effect**

Anytime YOU witness or experience Mind-Over-Matter, you have in fact witnessed or experienced Psyche or Non-Local Consciousness in action, and therefore have effectively proven Psyche's existence. Remember, the BEST way to find and know the truth is to live it and experience it and witness it for yourself, or to choose to trust someone who has.

The Placebo Effect is the Psyche's ability to treat, cure, and heal its assigned physical body. The Placebo Effect is literally defined as Mind-Over-Matter or Psyche-Over-Matter, because by definition in principle there is NO physical explanation for the Placebo Effect and the things which are observed and experienced through the Placebo Effect. The Placebo Effect is proven science, and the Placebo Effect is proof of Psyche or Non-Local Consciousness. There is NO physical explanation for the Placebo Effect, which means that the very existence of the Placebo Effect FALSIFIES Materialism, Naturalism, and Physicalism

Dr. Joe Dispenza has produced a book on this subject.

Dispenza, J. (2014). *You Are the Placebo: Making Your Mind Matter*. New York: Hay House, Inc.

I will leave it up to you to read this book and decide for yourself whether Dr. Joe meets his burden of proof through a preponderance of the evidence. You already know what I believe; otherwise, I wouldn't have included his book in this list.

Almost every book mentioned in these pages has a section on the Placebo Effect, because the Placebo Effect is possibly the strongest proof we have of Mind-Over-Matter or Psyche-Over-Matter.

Under Science 2.0, the evidentiary books rise to prominence over the philosophically speculative books. Science 2.0 defines science as Observational Evidence, Experiential Evidence, Eye-Witness Evidence, Experimental Evidence, and Empirical Evidence. Science 2.0 defines science as Knowledge of Evidence. Follow the evidence wherever it might lead you.

Science 2.0 teaches that Observational Evidence and Experiential Evidence dwarf Materialism and Naturalism, which have NO evidence to support their major premises or primary assumptions. Materialism, Naturalism, Darwinism, Nihilism, Behaviorism, and Atheism are pure philosophy – pure speculation – which is why they always FAIL to stand in the Light of Truth that's being produced by Empirical Evidence of some kind. In other words, you FALSIFY Materialism and Naturalism through Observational Evidence and Experiential Evidence, because Materialism and Naturalism are based exclusively upon blind-faith and a refusal to look at evidence.

Remember, the BEST way to find and know the truth is to live it and experience it for yourself, or to choose to trust someone who has.

## Neuroplasticity

Anytime YOU witness or experience Mind-Over-Matter, you have in fact witnessed or experienced Psyche or Non-Local Consciousness in action, and therefore have effectively proven Psyche's existence. Remember, the BEST way to find and know the truth is to live it and experience it and witness it for yourself, or to choose to trust someone who has.

Neuroplasticity is the Psyche's ability to change and re-wire brain circuitry. Obviously, the practitioners of Scientific Naturalism define and describe Neuroplasticity in materialistic terms and from a naturalistic perspective, which is fine because the brain is a physical object after all. This is simply a case where Materialism and Naturalism are incomplete and fail to give us the whole picture.

Doidge, N. (2007). *The Brain That Changes Itself: Stories of Personal Triumph from the Frontiers of Brain Science*. New York: Penguin Books.

This fascinating book is all about Neuroplasticity; and, it clearly and conclusively demonstrates that scientists around the world have used Neuroplasticity to treat and cure the incurable.

Doidge stays deliberately close to the materialistic and naturalistic explanation for Neuroplasticity, but occasionally he slips up and can't prevent Signs of Psyche from making an appearance. Doidge doesn't set out to solve the Mind-Brain problem, but he ends up finding evidence for Psyche nonetheless while studying and researching Neuroplasticity.

Anytime you witness or experience Neuroplasticity, Psyche isn't far away, because Neuroplasticity IS Mind-Over-Matter after all.

## Epigenetics

Anytime YOU witness or experience Mind-Over-Matter, you have in fact witnessed or experienced Psyche or Non-Local Consciousness in action, and therefore have effectively proven Psyche's existence. Remember, the BEST way to find and know the truth is to live it and experience it and witness it for yourself, or to choose to trust someone who has.

Epigenetics is the Psyche's ability to modify our genome and the Psyche's ability to turn on and turn off genes. Epigenetics is also defined as the nongenomic inheritance of behavior. Some different ways have been designed and programed into our genome which allows some of our epigenetic changes to be passed on to future generations.

Obviously, the practitioners of Scientific Naturalism define and describe Epigenetics in materialistic terms and from a naturalistic perspective, which is fine because our genes are physical objects after all. This is simply a case where Materialism and Naturalism are incomplete and fail to give us the whole picture.

Every day the scientists discover new and different types of Epigenetics, some of which are proving to be purely Mind-Over-Matter.

Carey, N. (2012). *The Epigenetics Revolution: How Modern Biology Is Rewriting Our Understanding of Genetics, Disease, and Inheritance*. New York: Columbia University Press.

The explosion of knowledge, information, and evidence during the past ten years has been simply amazing. NONE of this existed forty or fifty years ago when I first went looking for it. Back then it was easy to have doubts, because we didn't know better.

**Prayer**

Anytime YOU witness or experience Mind-Over-Matter, you have in fact witnessed or experienced Psyche or Non-Local Consciousness in action, and therefore have effectively proven Psyche's existence. Remember, the BEST way to find and know the truth is to live it and experience it and witness it for yourself, or to choose to trust someone who has.

Prayer is Action at a Distance, or Telepathy.

Larry Dossey has done a lot of scientific research into the efficacy of prayer.

Dossey, L. (1997). *Prayer Is Good Medicine: How to Reap the Healing Benefits of Prayer*. New York: HarperCollins.

Dossey, L. (2013). *One Mind: How Our Individual Mind Is Part of a Greater Consciousness and Why It Matters*. New York: Hay House Inc.

Dossey, L. (2011). *Healing Words: The Power of Prayer and the Practice of Medicine*. New York: HarperCollins.

> A few wise books on meditative practice had just begun to emerge, and I put their instructions to good use. With immense difficulty and struggle, I gradually adopted an eclectic philosophy that was more spiritually satisfying than anything I had grown up with.
>
> Even so, the experimental data on prayer that I turned up caught me off guard. I really wanted nothing to do with it. Meditation was acceptable, but the thought of "talking to God" in prayer was reminiscent of the fundamental Protestantism I felt I had laid to rest. Yet the results of the prayer experiments kept forcing themselves into my psyche.
>
> These studies showed clearly that prayer can take many forms. Results occurred not only when people prayed for explicit outcomes, but also when they prayed for nothing specific. Some studies, in fact, showed that a simple "Thy will be done" approach was quantitatively more powerful than when specific results were held in the mind. In many experiments, a simple attitude of prayerfulness — an all-pervading sense of holiness and a feeling of empathy, caring, and compassion for the entity in need — seemed to set the stage for healing.
>
> Experiments with people showed that prayer positively affected high blood pressure, wounds, heart attacks, headaches, and anxiety. The subjects in these studies also included water, enzymes, bacteria, fungi, yeast, red blood cells, cancer cells, pacemaker cells, seeds, plants, algae, moth larvae, mice, and chicks; and among the processes that had been influenced were the activity of enzymes, the growth rates of leukemic white blood cells, mutation rates of bacteria, germination and growth rates of various seeds, the firing rate of pacemaker cells, healing rates of wounds, the size of goiters and tumors, the time required to awaken from anesthesia, autonomic effects such as electrodermal activity of the skin, rates of hemolysis of red blood cells, and hemoglobin levels.
>
> Remarkably the effects of prayer did not depend on whether the praying person was in the presence of the organism being prayed for, or

whether he or she was far away; healing could take place either on site or at a distance. Nothing seemed capable of stopping or blocking prayer. Even when an "object" was placed in a lead-lined room or in a cage that shielded it from all known forms of electromagnetic energy, the effect still got through.

These experiments prompted me to continue saying to myself: "The evidence seems to show that prayer works. You claim to be a scientific doctor. Are you going to follow these scientific directions and actually use prayer?"

Over time I decided that not to employ prayer with my patients was the equivalent of deliberately withholding a potent drug or surgical procedure. I felt I should be true to the traditions of scientific medicine, which means going through scientific data and not around it, no matter how uncomfortable it might be to do so and no matter how it might shake up one's favored beliefs. I simply could not ignore the evidence for prayer's effectiveness without feeling like a traitor to the scientific tradition. And so, after weighing these factors for many months, I concluded that I would pray for my patients.

Dossey, Larry. *Healing Words: The Power of Prayer and the Practice of Medicine* (Kindle Locations 118-141). HarperCollins. Kindle Edition. HarperCollins permits quotation of their books in reviews and promotional literature.

A few years ago, I would have questioned all of this, but now I KNOW from personal experience that there's something to it.

Larry Dossey has a dozen other books for you to read and consider, which provide evidentiary proof of Mind-Over-Matter or Psyche-Over-Matter, including *What is Consciousness?: Three Sages Look Behind the Veil*.

See also:

*Biophysicist discovers new life after death*: Joyce Hawkes at TEDxBellevue.

https://www.youtube.com/watch?v=MyaBeHeRK6M&t=2s

Joyce Hawkes discusses the efficacy of prayer in her TED talk.

Both Ian McCormack and Howard Storm – along with many others – used prayer to get Jesus Christ to come and retrieve them from hell. If you ever find yourself trapped in hell, remember that Jesus Christ can get you out of there for the asking.

https://www.youtube.com/watch?v=UPj4wci_bcI&t=338s

https://www.youtube.com/watch?v=Vm647n1360A&t=1904s

https://www.youtube.com/watch?v=HbTAmN4m2lQ&t=3s

https://www.youtube.com/results?search_query=NDE+Jesus

*Taught by Jesus Christ*: Ralph Jensen Shares NDE (Near Death Experience)

https://www.youtube.com/watch?v=uWshfNnyEQA&t=39s

Notice how all these different things have come to light in the past ten or twenty years, in a big way. The Biblical God Jesus Christ is starting to make his arm bare and his plans known, in the sight of all the nations of this world. He's no longer hiding. The game's afoot.

"I Stand All Amazed: Love and Healing From Higher Realms" by Elane Durham.

Durham, E. (1998). *I Stand All Amazed: Love and Healing from Higher Realms*. Orem, UT: Granite Publishing and Distribution.

Prayer involves asking God questions, and then getting answers in return, as Elane Durham does in the following quote, during her Near-Death Experience (NDE).

### An Explosion of Questions

I wanted to know which was the correct Bible or the right church.

### The Perfect Church

I was informed that my term the "right" church was not a proper designation. Rather it was the "perfect" church, which was a part of heaven and had been created there. It was designated as perfect because it was composed of a perfect organizational structure and taught as saving doctrines and ordinances the perfect or complete truth as known by God the Father and His Beloved Son.

In its perfected state Christ's church had been brought to earth and given to mortals. However, it didn't take long for us to begin to change the Church and cause it to lose truth.

In that condition the "perfect" church became so weakened that it fragmented into numerous less perfect churches. This had not only happened once, I was told, but on many different occasions and in many different lands. I was told by the angel that all of these churches have a portion of the truth.

He was referring specifically to the various churches that had fragmented away from Christ's perfected church, but who nevertheless continue to seek diligently for greater oneness with God. These have a portion of the truth. But only the Church as it was created in heaven has all of it.

Both our speed [of spiritual progress] and our distance [covered] are determined by the amount of God's truth taught by the church we choose to join.

Only Christ's perfect church — represented by the rocket I had seen — could take us all the way to the highest heavenly realm in the shortest amount of time.

### I'm Not Informed Which Is Christ's Perfect Church

Though I was not informed by the angel which church it was, he did tell me that Christ's heavenly church is on the earth again today, organized just like it was organized in days of old. He also told me that if I truly desired to find it, I would.

Durham, E. (1998). *I Stand All Amazed: Love and Healing from Higher Realms*. Orem, UT: Granite Publishing and Distribution. (pp. 37, 65-68).

(The copyright permits quoting passages as part of articles and critical reviews. I picked one of her questions and followed it to its conclusion, leaving out the unnecessary bits as I went along, creating one seamless whole in the process.)

Remember, the BEST way to find and know the truth is to live it and experience it for yourself, or to choose to trust someone who has.

My best friend is a Seer, and he has experienced a wide variety of different Spiritual Experiences, Visions, Revelations, Miracles, Theophanies, Out-of-Body Experiences, Near-Death Experiences, Encounters with the Spirits of the Dead, as well as heaven and hell experiences that go beyond anything mentioned in these YouTube videos. My best friend is a member of the Church of Jesus Christ of Latter-day Saints – some of those people have had some amazing Spiritual Experiences. He's my best friend. I simply trust him. I KNOW that he's telling me the truth. I can tell.

The BEST way to find and know the truth is to live it and experience it for yourself, or to choose to trust someone who has.

The evidence from these different experiences eventually had an influence on me and got me to change my mind. I used to be a Materialist, Nihilist, and Atheist; but, I'm not anymore. There's just way too much evidence telling me that Materialism, Naturalism, Darwinism, Nihilism, and Atheism are FALSE. I can't deny the evidence and still remain true to myself. And why should I want to? The Non-Local and the Transcendental are infinitely more interesting than My Materialism and My Atheism ever were.

I, too, have experienced miraculous healing – both physically and spiritually – through the power of prayer, despite my initial lack of belief. As wonderful as the physical healing has been, the spiritual healing through the Atonement of Christ was infinitely more beneficial and life-changing than the physical healing was. But, you are going to have to live it and experience it for yourself, in order to KNOW that it is real and true. There is no other way. There is no better way.

Prayer is such a simple thing, but the results can be extremely powerful and life-changing.

Science 2.0 is a version of science that's based upon evidence, rather than the suppression of evidence as is done with Materialism, Naturalism, and Atheism.

Remember, Materialism, Naturalism, Darwinism, Nihilism, and Atheism are based upon the denial of evidence, the rejection of evidence, the suppression of evidence, and a refusal to look at evidence. These dogmatic orthodox religions define science axiomatically as Materialism, Naturalism, Darwinism, Nihilism, and Atheism.

## Hypnosis

Anytime YOU witness or experience Mind-Over-Matter, you have in fact witnessed or experienced Psyche or Non-Local Consciousness in action, and therefore have effectively proven Psyche's existence. Remember, the BEST way to find and know the truth is to live it and experience it and witness it for yourself, or to choose to trust someone who has.

Remember, Science 2.0 allows ALL of the evidence into evidence, including evidence from hypnosis. However, Science 2.0 leaves it up to you to decide for yourself whether that evidence is trustworthy, or not. Because there is NO officially sanctioned or officially produced interpretation of the evidence under Science 2.0, YOU are encouraged to use your brain and common sense to decide for yourself what to make of the evidence and how to use it in your own life.

Science 2.0 also permits me to share my own personal interpretation of the evidence, even though I have NO right to declare my own interpretation to be the Official Interpretation. I will do so now.

Many different mind-over-matter effects have been demonstrated and proven through hypnosis.

Hypnosis IS Mind-Over-Matter, or Psyche-Over-Matter, or Psyche-Without-Matter.

Caveat emptor!

There is a scientifically proven phenomenon called "False Memory Syndrome" which is generated and produced by hypnosis.

Through hypnosis, it is possible to manufacture false memories out of thin air; and, these fake memories will seem more real to you than your actual memories from your physical life. Consequently, though hypnosis or past-life regression, it is possible to manufacture and produce many past lives or scenes of sexual abuse that you never really lived nor experienced.

Instinctively, I have never trusted hypnosis and hypnotists; and, this was before I learned about False Memory Syndrome. Thank God there aren't too many of us, but I was a Natural Born Skeptic. I have had trust issues all of my life, until God and the Local Christians taught me how to trust. I have never trusted hypnosis.

http://www.fmsfonline.org/

Remember, the Human Psyche not only experiences alternative realities, but it also manufactures and creates alternative realities – study the difference between Consensus Realities and Non-Consensus Realities in order to see and understand how this works in practice. The creation of reality and the experience of reality is an innate inherent function of Psyche or Non-Local Consciousness.

Hypnosis is similar to hallucinations. It seems real, but it isn't.

Through hypnosis and past-life regression you can live and experience any kind of life that you desire. It never really happened, but it will seem absolutely real and true. In fact, the false memories generated under hypnosis seem more real and true than our actual memories from our actual physical mortal life.

Likewise, Astral Travelers and Near-Death Experiencers testify that the Non-Local Realm is more real, clear, concrete, sharp, focused, and true than their physical life and mortal memories have ever been. Physical matter puts a veil or a cloud over our Psyche, so nothing in the physical realm is anywhere near as clear and sharp and real as things are in the Non-Local Spirit Realm. That is what has been experienced by Out-of-Body Travelers.

My best friend tells me from his own experiences that the foggy, cloudy, soft-light portrayals of the Spirit World or Non-Local Realm given to us in the movies are the exact opposite of the true reality of the situation. He tells me that the Spirit World is infinitely more real, crisp, clear, solid, vibrant, and true than anything we experience here in this physical realm. This physical reality is The Illusion, not the spirit world.

My best friend also tells me that while exploring the Transdimensional Realm, or Non-Local Realm, or Spirit World, you absolutely want the Holy Ghost and/or Jesus Christ to be your guide and protector, rather than Satan or one of the evil spirits.

My best friend is also aware of the power of hypnosis, and its ability to generate false memories and fake past-lives. In fact, it's possible to channel another person's life through hypnosis, so that you come out of the experience really truly believing that you were Hitler, or Napoleon, Antony, or Cleopatra in one of your former lives.

I'm highly skeptical of any "past-life" or "reincarnation account" that was manufactured and produced through hypnosis. Through hypnosis, you can generate and produce anything you want, and it will seem absolutely real, even if it is pure fiction.

Nevertheless, hypnosis can indeed be used and has been used to demonstrate and prove Mind-Over-Matter and similar phenomena. Never underestimate the power of Psyche.

## Psychedelics

Anytime YOU witness or experience Mind-Over-Matter, you have in fact witnessed or experienced Psyche or Non-Local Consciousness in action, and therefore have effectively proven Psyche's existence. Remember, the BEST way to find and know the truth is to live it and experience it and witness it for yourself, or to choose to trust someone who has.

Stanislav Grof is the leader in this category. The dude is into shrooms and LSD. Grof is a prolific writer and researcher, and he has written a lot of books on the subject of psychedelics.

Although I own many of his books, as a general rule I find them ugly, dark, and depressing most of the time, because Grof was fascinated or even fixated on the Bad Trip, the Dark Trip, Blood, the Birth Canal, Voodoo, the Shadow, the Evil Mother, and the Evil Realm which psychedelics often introduce a person to.

Thankfully, Grof seemed to mellow with age and eventually turned to Psyche-Therapy instead – taking a more positive and constructive approach to the subject through Transpersonal Psychology which he helped to develop.

Grof, S. (2010). *Holotropic Breathwork: A New Approach to Self-Exploration and Therapy* (SUNY series in Transpersonal and Humanistic Psychology). New York: State University of New York Press.

A lot of evidence for Mind-Over-Matter and Psyche-Separate-From-Matter comes from Transpersonal Psychology.

Although I own a lot of books about Transpersonal Psychology, for me personally this is largely an untapped resource, which I hope to explore in some of my future books and research.

I'm not a big fan of drugs. Many people on my mother's side were alcoholics and drug addicts. I also experienced addiction to prescription drugs along with six-months of withdrawal when I finally decided to go cold turkey and get off the crap. It was hell – two years of hell. You don't have to die in order to go to hell. I went to hell here in the flesh. I pray to God every day that I never have to go back there ever again.

Many people claim to have had transcendent experiences with psychedelics; but, I don't want to try it and see. Their experiences and accounts are good enough for me.

There's a lot of weird and bizarre stuff associated with ayahuasca, mushrooms, peyote, and LSD – the hallucinogens – including a lot of BAD TRIPS. Spock did a lot of LDS in the sixties, after all. You can search for each one of these on YouTube if you want to experience Lucy in the Sky with Diamonds, vicariously.

Hallucinogens cause hallucinations, delusions, and psychosis. It seems real, but that doesn't mean it's real. The official term for what I experienced is Substance-Induced Psychotic Disorder – an official mental illness. Addiction is a mental illness. Ironically and unfortunately, in typical materialistic and naturalistic fashion, the psychiatrists often try to treat your mental illnesses by giving you another mental illness – a drug addiction. It doesn't work. The drugs can't do Psyche-Therapy, and only make things worse.

I experienced psychotic hallucinations and delusions, as well as suicidal ideation, for six months while going through withdrawal; and, it was hell. Mine was a BAD TRIP. I really don't want to go back there ever again. It was painful and confusing, magical and lonely, scary and strange. Nothing made sense.

**Mediums and Channeling**

Anytime YOU witness or experience Mind-Over-Matter, you have in fact witnessed or experienced Psyche or Non-Local Consciousness in action, and therefore have effectively proven Psyche's existence. Remember, the BEST way to find and know the truth is to live it and experience it and witness it for yourself, or to choose to trust someone who has.

Mediums communicate with the spirits of the dead.

Although I have looked, I haven't yet found a medium that I completely trust. There's a lot of showmanship, and a lot of room for trickery and deception where mediumship is concerned.

If forced to pick a winner in this category, I would go with Gordon Smith. I like his presentation.

> **Lots of people think that being a medium is just about talking to the dead, and dealing with all kinds of spooky things like ghosts and poltergeists, but I know after 18 years of working as one and giving sittings for thousands of people around the world, that my contribution is really about healing the living. I have worked as a healer with people who have physical health problems, but that's not what I mean. The ultimate healing a medium can offer is to those who have lost someone they loved, and who they felt meant more to them than life itself.**
>
> **I'm trying to lift the bereaved out of the fear, anger and sadness that crowd their every thought, and not only to help them overcome those in their daily lives, but also to show them that there is another life beyond death – a life after life, if you like.**

Smith, Gordon. *Through My Eyes* (p. 3). Hay House. Kindle Edition.

Here's another medium to consider. I think you can tell a lot depending upon what they choose to lead with.

Praagh, J. V. (1997). *Talking to Heaven: A Medium's Message of Life after Death*. New York: A Signet Book.

> **I am often asked if I was born a medium or if I was transformed into one by a terrible illness, or a freak accident that caused some sort of head**

trauma, or a near-death experience. As hair-raising as those possibilities may be, I cannot claim any one of them as the dramatic moment that introduced me to my life's work.

I am not unlike anyone else. We are all born with some level of psychic ability. The question is: Do we recognize our psychic abilities and act upon them? Like many others, I didn't know what it meant to have psychic ability. It was probably on some TV game show that I first heard the term "psychic." I was lucky enough to pronounce it, let alone understand its definition. It was a word that came closest to explaining why I knew things about people when they walked into a room. It was also the reason my first-grade Catholic school teacher kept me after school one day.

Van Praagh, James. *Talking to Heaven: A Medium's Message of Life After Death* (p. 3). Penguin Publishing Group. Kindle Edition.

Again, I encourage and recommend caution where mediumship and channeling are concerned, because there can be and often is a lot of deception going on here. If these people make contact with Satan or one of the evil spirits, then these people are going to end up being deceived in one form or another.

Channeling is when one of the spirits of the dead or one of the evil spirits takes temporary control of your body or possession of your body, and then speaks and acts through you. Channeling IS Mind-Over-Matter; wherein, somebody else's Psyche takes control of your physical body.

As far as I know, the most popular channeler is Jane Roberts, who channeled an entity or dead spirit who eventually called himself Seth, as documented in many of her Seth books.

Roberts, Jane. *The Seth Material*. New Awareness Network, Inc.

She contacted Seth through an Ouija board. I never found Seth completely trustworthy. In fact, he was trying to trick them, deceive them, and mess with them at first, until he developed a bond with Jane Roberts. It's impossible to tell when, or if, Seth started to tell them the truth. We hope he did.

**The first chapters of this book will deal with the emergence of Seth's personality and the impact he had on our lives as we tried to understand what was happening. Out of nowhere it seemed, I found myself having experiences that I considered nearly impossible. Never in our lives had we found ourselves so caught between curiosity and caution, so fascinated and baffled.**

**Excerpts from some of the early sessions will also be included in the first chapters, since Seth's ideas were then as new and strange to us as the sessions themselves. But the main emphasis will be on the story itself, from the first Ouija board experiment through the first instance when I startled Rob and myself by speaking for Seth; and the changes in our attitudes as further developments occurred. I'll also include examples of Seth's clairvoyant abilities.**

**The bulk of the book will deal with Seth's ideas on various subjects, such as life after death, reincarnation, health, the nature of physical reality, the God concept, dreams, time, identity, and perception. I'm sure that these excerpts from the material itself and some sample reincarnational readings**

**will give most readers greater insights into their own personalities and the situations in which they find themselves. I hope that Seth's theories on health will benefit all my readers, and that the material on personality will help each discover for himself the multidimensional reality that is his heritage.**

**The philosophical and psychological implications of mediumship and ESP phenomena and the possible origins of the Seth Material, along with several questions concerning Seth's independent reality, will be considered. I'll also give Seth's advice as to the development of psychic abilities.**

Roberts, Jane. *The Seth Material* (Kindle Locations 341-354). New Awareness Network, Inc. Kindle Edition.

Although I found the Seth material interesting, I didn't always find it trustworthy. Even Jane Roberts was skeptical at times. This thing was clearly trying to deceive them at first, and we have no idea when or if Seth ever started to tell them the truth. My personal opinion is that Seth was a purged spirit who had gone through Purgatory or Gehenna, atoned for its own sins, and had had most of its memories of its mortal lives purged or removed. That's the impression I got – your mileage may vary.

Although Science 2.0 allows all of the evidence into evidence, since the personal interpretation of the evidence is left up to YOU under Science 2.0, use your brains and common sense, carefully examine the evidence, and decide for yourself whether it is trustworthy or not. Does it fit with everything else that you know to be true? Under Science 2.0, YOU are called upon to decide for yourself whether evidence is reliable or not. That's something that none of the rest of us can do for you.

My observation and conclusion is that when it comes to Mediumship and Channeling, there can be a lot of deception, trickery, and lies going on – especially if the person is interacting with an evil spirit or Satan. Although I believe the phenomenon to be real – caveat emptor!

**Mysticism and Clairvoyance**

Anytime YOU witness or experience Mind-Over-Matter, you have in fact witnessed or experienced Psyche or Non-Local Consciousness in action, and therefore have effectively proven Psyche's existence. Remember, the BEST way to find and know the truth is to live it and experience it and witness it for yourself, or to choose to trust someone who has.

According to Uncle Google, mysticism is popularly known as becoming one with God or the Absolute, but may refer to any kind of ecstasy or altered state of consciousness which is given a religious or spiritual meaning.

Mysticism tends to be defined as Mind-Separated-From-Matter, or Psyche experiencing the Spiritual. If any of these experiences are real and true, they definitely falsify Materialism and Naturalism, solve the Mind-Brain Problem, and verify the existence of Psyche or Non-Local Consciousness.

A group of mystics, the Rosicrucians, are giving away all of their books for free on Kindle. I think there was over 70 books the last time I tried to count them.

Poole, C. A. (1980). *In Search of Reality*.

Mysticism is a type of mind-over-matter and/or mind-separate-from-matter.

Both Jeffery Schwartz and Joseph Selbie rely upon the Hindu Mystics and Buddhist Monks for the experiential evidence that they need to make their case or meet their burden of proof through a preponderance of the evidence.

Jeffrey M. Schwartz, J. M., & Begley, S. (2002). *The Mind and the Brain: Neuroplasticity and the Power of Mental Force*. New York: HarperCollins.

A Sampling:

http://mypsyche.us/wp-content/uploads/2017/10/The-Mind-and-the-Brain-Neuroplasticity-and-the-Power-of-Mental-Force.pdf

Source:

https://www.amazon.com/Mind-Brain-Neuroplasticity-Power-Mental/dp/0060988479

Joseph Selbie also relies upon his own transcendental experience that he experienced through psychedelics.

Selbie, J. (2018). *The Physics of God: Unifying Quantum Physics, Consciousness, M-Theory, Heaven, Neuroscience, and Transcendence*. Wayne, NJ: Career Press.

> **Then something happened that forever altered my course in life: I had a transcendent experience. Like many of my generation, I experimented with psychedelic drugs. In one life-transforming "trip" I had an utterly beguiling experience. I became — to my delight — highly intuitive, serenely calm, and warmly heart-centered. My awareness subtly expanded to take in everything from the glow of life in a plant to the hidden feelings of my companions. I felt myself to be more than a body. At the heart of my experience were expansive feelings of boundless peace, joy, and well-being, feelings that seemed completely natural, as if the person I had always been simply woke up. I had never felt more joyful, more alive, or more at peace in all my life. It was beyond wonderful.**

Joseph, Selbie. *The Physics of God* (pp. 15-16). Career Press. Kindle Edition.

My favorite New-Age Clairvoyant is Barbara Martin.

Barbara Martin put out a lot of different books, which I thought were interesting and informative. She even admits at times that she could in fact be interpreting her experiences incorrectly, but that she's doing the best she can with what she has been given to work with.

Martin, B. Y., & Moraitis, D. (2014). *Communing with the Divine: A Clairvoyant's Guide to Angels, Archangels, and the Spiritual Hierarchy*.

Martin, B. Y., & Moraitis, D. (2010). *Karma and Reincarnation: Unlocking Your 800 Lives to Enlightenment*.

Martin, B. Y., & Moraitis, D. (2006). *The Healing Power of Your Aura: How to Use Spiritual Energy for Physical Health and Well-Being*.

Martin, B. Y., & Moraitis, D. (2007). *Change Your Aura, Change Your Life: A Step-by-Step Guide to Unfolding Your Spiritual Power*.

My best friend is a Seer; but, he has been commanded by God not to share everything that he has seen and experienced. Still, what he has been permitted to share with me is a lot more powerful and instructive and convincing than what I have gotten from the Mystics and New-Age Clairvoyants.

**Quantum Zeno Effect**

The Quantum Zeno Effect is observational and experimental proof of Mind-Over-Matter or Psyche-Over-Matter.

The Quantum Zeno Effect is Telepathy, or Action at a Distance.

The Quantum Zeno Effect is the Mind/Brain Interface.

Each scientist and book describes the Quantum Zeno Effect in different ways, but it's always interesting.

> One quantum mechanics discovery that may help us understand how we decide is the quantum Zeno effect. Physicists have found that if they observe an unstable elementary particle continuously, it never decays — even though it would almost certainly decay if it were not observed. In quantum physics, it is not possible to separate the observer entirely from the thing observed. They are part of the same system. The physicists are, essentially, holding the unstable particle in a given state by the act of continuing to measure it.
>
> In the same way, experiments have shown that, because your brain is a quantum system, if you focus on a given idea, you hold its pattern of connecting neurons in place. The idea does not decay, as it would if it were ignored. But the action of holding an idea in place truly is a decision you make, in the same way that the physicists hold a particle in place by deciding to continue to observe it.
>
> Beauregard, Mario. The Spiritual Brain: A Neuroscientist's Case for the Existence of the Soul (Kindle Locations 824-831). HarperCollins. Kindle Edition.

Focusing on an idea holds the neurons in place, through the Quantum Zeno Effect. This is truly Mind-Over-Matter in action. It all comes down to choice – Psyche's choice.

> So, what happens in our brains when we make a decision? According to the model created by H. Stapp and J. M. Schwartz, which is based on the Von Neumann interpretation of quantum physics, conscious effort causes a pattern of neural activity that becomes a template for action. But the process is not mechanical or material. There are no little cogs and wheels in our brains.
>
> There is a series of possibilities; a decision causes a quantum collapse, in which one of them becomes a reality. The cause is the mental focus, in the same way that the cause of the quantum Zeno effect is the physicists' continued observation. It is a cause, but not a mechanical or material one. One truly profound change that quantum physics has made is to verify the existence of nonmechanical causes. One of these is the activity of the human mind, which, as we will see, is not identical to the functions of the brain.

Beauregard, Mario. The Spiritual Brain: A Neuroscientist's Case for the Existence of the Soul (Kindle Locations 840-847). HarperCollins. Kindle Edition.

Psyche's decision or choice causes a quantum collapse resulting in the manifestation of a physical reality. Mind or Psyche is a Causal Agent. Psyche is the Ultimate Cause – the Ultimate Causal Agent. Psyche is an immaterial, non-local, non-physical, transdimensional Causal Agent. Psyche is a cause, but not a mechanical or material one. Quantum physics has verified the existence of Psyche, immaterial non-local causal agents, and nonmechanical causes. The Quantum Zeno Effect is scientific proof of Telepathy or Action at a Distance.

**That is what Henry Stapp began to do: explore the physics by which mind can exert a causal influence on brain. To do so, he focused on an odd quantum phenomenon called the Quantum Zeno Effect. Named for the Greek philosopher Zeno of Elea, the Quantum Zeno Effect was introduced to science in 1977 by the physicist George Sudarshan of the University of Texas at Austin and colleagues. If you like nonlocality, you'll love Quantum Zeno, which puts the spookiness of nonlocality to shame: in Quantum Zeno, the questions one puts to nature have the power to influence the dynamic evolution of a system. In particular, repeated and closely spaced observations of a quantum property can freeze that property in place forever, or at least much longer than it would otherwise stay if unwatched.**

**Consider an atom that has absorbed a photon of energy. That energy has kicked one of the atom's electrons into what's called a higher orbital, kind of like a supermassive asteroid's kicking Mercury into Venus's orbit, and the atom is said to be "excited." But the electron wants to go back where it came from, to its original orbital, as it can do if the atom releases a photon. When the atom does so is one of those chance phenomena, such as when a radioactive atom will decay: the atom has some chance of releasing a photon (and allowing the electron to return home) within a given period. Thus, the excited atom exists as a superposition of itself and the unexcited state it will fall into after it has released a photon. Physicists can measure whether the atom is still in its initial state or not. If they carry out such measurements repeatedly and rapidly, they have found, they can keep the atom in its initial state. This is the Quantum Zeno Effect: such a rapid series of observations locks a system into that initial state. The more frequent the observations of a quantum system, the greater the suppression of transitions out of the initial quantum state. Taken to the extreme, observing continuously whether an atom is in a certain quantum state keeps it in that state forever. For this reason, the Quantum Zeno Effect is also known as the watched pot effect. The mere act of rapidly asking questions of a quantum system freezes it in a particular state, preventing it from evolving as it would if we weren't peeking. Simply observing a quantum system suppresses certain of its transitions to other states.**

**And so, by repeated observations at short intervals, one can prevent the nitrogen atom from ever leaving the top position. If you rapidly and repeatedly ask a system, Are you in this state or are you not? and make observations designed to ascertain whether or not the nitrogen atom is where it began, the system will not evolve in the normal way. It will become, in a sense, frozen. As Stapp puts it, "An answer of 'yes' to the posed question [in this case, Is the nitrogen atom on top?] will become fixed and unchanging. The state will be forced to stay longer within the realm that provides a yes answer."**

Quantum Zeno has been verified experimentally many times. One of the neatest confirmations came in a 1990 study at the National Institute of Standards and Technology. There, researchers measured the probability that beryllium ions would decay from a high-energy to a low-energy state. As the number of measurements per unit time increased, the probability of that energy transition fell off; the beryllium atoms stayed in their initial, high-energy state because scientists kept asking them, "So, have you decayed yet?" The watched pot never boiled. As Sudarshan and Rothman conclude, "One really can stop an atomic transition by repeatedly looking at it."

The Quantum Zeno Effect "fit beautifully with what Jeff was trying to do," recalls Henry Stapp. It was clear to Stapp, at least in principle, that Quantum Zeno might allow repeated acts of attention—which are, after all, observations by the mind of one strand of thought among the many competing for prominence in the brain—to affect quantum aspects of the brain. "I saw that if the mind puts to nature, in rapid succession, the same repeated question, 'shall I attend to this idea?' then the brain would tend to keep attention focused on that idea," Stapp says. "This is precisely the Quantum Zeno Effect. The mere mental act of rapidly attending would influence the brain's activity in the way Jeff was suggesting."

The power of the mind's questioning ("Shall I pay attention to this idea?") to strengthen one idea rather than another so decisively that the privileged idea silences all the others and emerges as the one we focus on — well, this seemed to be an attractive mechanism that would not only account for my results with OCD patients but also fit with everyone's experience that focusing attention helps prevent the mind from wandering.

A quantum theory of mind, incorporating the discoveries of nonlocality and the Quantum Zeno Effect, offers the hope of mending the breach between science and moral philosophy. It states definitively that real, active, causally efficacious mind operates in the material world. The shift in understanding inspired by neuroplasticity and the power of mind to shape brain undermines the claim of materialist determinism.

Schwartz, Jeffrey M. *The Mind and the Brain: Neuroplasticity and the Power of Mental Force* (pp. 350-352, 374). HarperCollins. Kindle Edition.

The Quantum Zeno Effect explains how Conscious Choices become one's physical reality – it's Mind-Over-Matter, Telepathy, and Telekinesis. It's Action at a Distance.

### The Quantum Zeno Effect

Within the Orthodox Quantum Mechanical description of nature, this physical power of your conscious thoughts can arise from a well-known rigorous property of quantum mechanics known as the Quantum Zeno Effect, and sometimes as the Anti-Quantum Zeno Effect.

Suppose a physical/material system is being probed by an observer whose mental aspect, his ego, is free to choose a sequence of probing Yes-No questions that will elicit responses, Yes or No from nature.

Thus, by a suitably rapid choice of probing questions the observer can, by its choices of these questions, effectively control both the perceived responses and the associated material reality that is being perceived.

Because the observer's choices are stemming from the mental realm, there is no known limit on exactly how rapid these free choices can be. But it is reasonable to suppose that survival consideration makes this effect far easier to use if the action is directly an action on the observer/ actor's sensitive brain than on a perceived brute external system.

Thus, the behavior of the brain, according to Orthodox Quantum Mechanics, is not completely determined by prior physically described properties of the universe alone, but can be significantly influenced by "free choices" made by human observers pertaining to which probing action to instigate, and when to do so. Here, again, the "free" in "free choice" means, specifically, that this choice is not determined by prior physically-described aspects of the universe alone. Our conscious free choices and mental efforts enter naturally, according to the quantum mechanical dynamical laws, into the evolution of the psycho-physical universe.

I shall describe next how, within Orthodox Quantum Mechanics, the simple holding-in-place action produced by the Quantum Zeno Effect can tend to make a person's physical actions conform to that person's mental intent.

Stapp, Henry P. *Quantum Theory and Free Will: How Mental Intentions Translate into Bodily Actions* (Kindle Locations 1006-1024). Springer International Publishing. Kindle Edition.

The Quantum Zeno Effect is how our Non-Local Intentions become physical reality. The Quantum Zeno Effect truly is Mind-Over-Matter.

**In this model all significant effects of consciousness upon brain activity arise exclusively from a well-known and well-verified strictly quantum effect known as the 'quantum Zeno effect' (QZE).**

**If one considers only passive events, then it is very difficult to identify any empirical effect of process 1, apart from the occurrence of awareness. In the first place, the empirical averaging over the 'Yes' and 'No' possibilities in strict accordance with the quantum laws tends to wash out all effects that depart from what would arise from a classic statistical analysis that incorporates the uncertainty principle as simply lack of knowledge. Moreover, the passivity of the mental process means that we have no empirically controllable variable.**

**However, the study of effortfully controlled intentional action brings in two empirically accessible variables, the intention and the amount of effort. It also brings in the important physical QZE. This effect is named for the Greek philosopher Zeno of Elea and was brought into prominence in 1977 by the physicists Misra & Sudarshan (1977). It gives a name to the fact that repeated and closely spaced observational acts can effectively hold the 'Yes' feedback in place for an extended time-interval that depends upon the rapidity at which the process 1 actions are happening. According to our model, this rapidity is controlled by the amount of effort being applied. In our notation, the effect is to keep the 'Yes' condition associated with states of the form PSP in place longer than would be the case if no effort were being made. This 'holding' effect can override very strong mechanical forces arising from process 2.**

The 'Yes' states PSP are assumed to be conditioned by training and learning to contain the template for action which if held in place for an extended period will tend to produce the intended experiential feedback. Thus, the model allows intentional mental efforts to tend to bring intended experiences into being. Systems that have the capacity to exploit this feature of natural law, as it is represented in quantum theory, would apparently enjoy a tremendous survival advantage over systems that do not or cannot exploit it.

See: J. M. Schwartz, H. Stapp, and M. Beauregard, "Quantum Theory in Neuroscience and Psychology: A Neurophysical Model of Mind/Brain Interaction," *Philosophical Transactions of the Royal Society B: Biological Sciences 360* (2005): 1309–27.

http://www-physics.lbl.gov/~stapp/PTRS.pdf

http://mypsyche.us/wp-content/uploads/2017/10/PTRS.pdf

https://www.researchgate.net/publication/7613549_Quantum_physics_in_neuroscience_and_psychology_A_neurophysical_model_of_mind-brain_interaction

http://escholarship.org/uc/item/4w8665vk

The Quantum Zeno Effect is telepathy and telekinesis. Your Non-Local Consciousness or Psyche literally freezes the neurons and atoms into place.

**Quantum Zeno Effect verified – atoms won't move while you watch them.**

https://phys.org/news/2015-10-zeno-effect-verifiedatoms-wont.html

http://mypsyche.us/wp-content/uploads/2017/10/Zeno-effect-verified—atoms-wont-move-while-you-watch.pdf

One verified experimental observation from the Quantum Zeno Effect is how atoms freeze into position whenever you, your psyche or non-local consciousness, observes them or looks at them or uses them. The Quantum Zeno Effect only makes sense from a non-local perspective or a spiritual perspective, because there is telepathic "action at a distance" taking place between your psyche and the psyche within nature or physical matter according to the proven Quantum Zeno Effect.

Verification of the Quantum Zeno Effect provides PROOF of **action at a distance** or telepathy between you and the atoms, because there is NO physical contact between you and the atoms being observed, yet the atoms freeze or stop moving when you look at them.

Furthermore, experimental verification of the Quantum Zeno Effect PROVES that telepathy is a normal, typical, ubiquitous function in the Non-Local Realm or the Transdimensional Spirit Realm. According to the Quantum Zeno Effect, your psyche and the psyche within a physical atom are literally communicating with each other telepathically. This IS Action at a Distance or Telepathy, which is being witnessed and proven by the Quantum Zeno Effect. There is NO physical contact between your physical body and that distant atom which you are observing while they are measuring the Quantum Zeno Effect. The whole effect is the result of psi or telepathy. It's truly mind-over-matter!

**Conclusions Regarding Mind-Over-Matter**

The explosion of knowledge, information, and evidence during the past ten years has been simply amazing. NONE of this existed forty or fifty years ago when I first went looking for it. Back then it was easy to have doubts, because we didn't know better.

Anytime YOU witness or experience Mind-Over-Matter, you have in fact witnessed or experienced Psyche or Non-Local Consciousness in action, and therefore have effectively proven Psyche's existence. Remember, the BEST way to find and know the truth is to live it and experience it and witness it for yourself, or to choose to trust someone who has.

I realize that I have just begun to scratch the surface where Psyche-Over-Matter is concerned. There are many other books and videos out there, yet to be found and explored.

## 4. Scientific Proof that Evolution (Genetic Drift), Natural Selection, Random Mutations, and Entropy Cannot Design and Create and Build Anything at All:

Sanford, J. (2014). *Genetic Entropy* (4th ed.). Cornell University: FMS Foundation.

The science in this book convinced me scientifically and conclusively that Random Mutations and Natural Selection (Evolution) can't design and create anything at all, even if given an infinite amount of time to do so. After reading this book, that was the complete END of my belief in Materialism, Naturalism, Darwinism, and the Theory of Evolution. There was no going back for me. The truth won't allow me to return to the deceptions and the lies of Darwinism.

I had been told for fifty years of my life that there's no way to falsify Materialism, Naturalism, Darwinism, Nihilism, and Atheism. I was told wrong. ALL of the observational evidence and experiential evidence that we have on hand as a race tells us clearly and conclusively that Materialism, Naturalism, Darwinism, Nihilism, Scientism, and Atheism are FALSE.

There are two main ways to FALSIFY Materialism, Naturalism, and Darwinism, Creation by Rocks, Creation Ex Nihilo, and the Theory of Evolution.

One way is to use science, observation, and common sense to demonstrate beyond any reasonable doubt that Macro-evolution, Abiogenesis, and Spontaneous Generation do not exist and could never have created the genomes and the life forms in the first place.

The other way is to use science, observation, and common sense to demonstrate beyond any reasonable doubt that Evolution (genetic drift), Random Mutations, and Natural Selection couldn't have done the job that needed to be done in order to produce the first genomes and first life forms on this planet. This book, *Genetic Entropy*, takes this second approach to FALSIFYING Darwinism, Naturalism, Materialism, and the Theory of Evolution.

Remember, Evolution (genetic change), Natural Selection, and Random Mutations did not exist until AFTER God designed and created the genomes and the life forms in the first place; therefore, there's no way on earth that Evolution, Natural Selection, and Random Mutations could have designed and created the first genomes and life forms on this planet.

For me personally, this reality and truth ends up being Scientific Proof of God's Necessity, which is essentially Scientific Proof of God's Existence.

## Evolution Is Entropy

Now pay attention!

Despite the FACT that I have and will have debunked and falsified the Theory of Evolution thousands of times in hundreds of different ways, don't ever once let me convince you that there is NO such thing as evolution or random mutations.

Evolution is REAL, very REAL. Evolution or random mutation is entropy or a version of the second law of thermodynamics. It's REAL. Entropy is very REAL at the physical level; and, it works as advertised at the physical level.

Genetic entropy is a real phenomenon.

While writing the book on Syntropy, I one day realized that Evolution is Entropy, particularly the Creation by Disorder definition for entropy or the Creation by Chance definition for entropy. There are dozens of different mutually exclusive definitions for entropy, which is why nobody seems to know what entropy is. But, genetic entropy or the deterioration of the human genome is a real phenomenon. Our genome was designed with backup genes in place, but as our human genome continues to deteriorate, the human race will become sterile and eventually go extinct. The experts who have been studying genetic entropy and gauging the rate of deterioration have concluded that the human genome has a lifespan or a shelf life of about 9,000 years total, before the collective human genome is so damaged that it can no longer produce viable offspring.

That kind of suggests that something unseen and supernatural has been propping up and sustaining the human genome, if the current rate of measurable deterioration is any indication.

There are literally dozens of different definitions for entropy. The Materialists, Naturalists, Darwinists, Nihilists, and Atheists believe that ONLY physical matter, entropy, death, and chance exist. Consequently, the mechanism behind evolution has to be one of these. Natural selection is creation by death. Therefore, evolution in general ends up being creation by chance or creation by entropy. This essay is based upon the Genetic Entropy definition for entropy, or the constant deterioration of our genomes.

Whenever the Materialists, Naturalists, Darwinists, Nihilists, Behaviorists, and Atheists start talking about evolution, they get most everything wrong, because evolution or entropy cannot design and create. However, these people do indeed get one thing perfectly right. Evolution, or random mutation, or entropy is indeed the CAUSE of ALL of our heritable diseases, developmental diseases, and heritable mental illnesses.

Remember, the Theory of Evolution is FALSE because random mutations or entropy cannot design and create genes, proteins, and life forms. However, evolution or random mutation or entropy is very REAL; and, it can indeed destroy genes, proteins, and life forms. Do you see how that works? It's important to understand.

The Theory of Evolution is a fictional story that they made up out of thin air. There is NO empirical evidence supporting any version of it. In fact, ALL of the empirical evidence and experimental evidence that we have on hand as a race FALSIFIES the different versions of the Theory of Evolution and VERIFIES Quantum Mechanics instead. Quantum Mechanics is Supernatural. Quantum Mechanics is the Priesthood Power of God. Quantum Mechanics and Psyche are Pure Syntropy. There is NO entropy in the spirit realm, the non-local realm, the quantum realm, or the transdimensional realm.

The very existence of something like the Orthodox Interpretation of Quantum Mechanics from Henry P. Stapp FALSIFIES Materialism, Naturalism, and the various versions of the Theory of Evolution. Fictional stories like the Theory of Evolution cannot stand in the light of truth.

The Theory of Evolution is the very pinnacle of fictional ad hoc just-so story telling; and, *ad hoc just-so stories* are logic fallacies. The Theory of Evolution is fictional, because it never happened – none of it happened! The chemical evolution of proteins and genes from atoms is physically impossible. Abiogenesis, spontaneous generation, and the various different forms of macro-evolution are physically impossible. The fictional nature of the story becomes most egregious whenever they try to guesstimate how many millions of years it took for evolution to do something for us. They are making it up as they go along. It's a fictional story, and nothing more.

Since the whole Theory of Evolution is nothing but a fictional story, you can successfully and rightfully make up fictional stories of your own to debunk it. That's the way fiction works!

Comparative Psychology and Evolutionary Psychology are based upon Darwinism and the Theory of Evolution, which means that they too are nothing more than *fictional ad hoc just-so stories* that these people have made up out of thin air.

In fact, in his book *Biopsychology*, John Pinel tells us as much when he tells us that his "evolutionary perspective" is based upon *scientific inferences*. *Scientific inferences* are fictional ad hoc just-so stories. *Scientific inferences* are logic fallacies. The whole of their evolutionary perspective, evolutionary psychology, and comparative psychology is based upon *scientific inferences* or stories that they have manufactured out of thin air. Making up stories is what makes being a scientist fun, according to John Pinel.

From *Biopsychology* page 13, John Pinel writes:

> **Scientific inference is the fundamental method of biopsychology and of most other sciences – it is what makes being a scientist fun. This section provides further insight into the nature of biopsychology by defining, illustrating, and discussing scientific inference.**
>
> **The scientific method is a system for finding things out by careful observation, but many of the processes studied by scientists cannot be observed. For example, scientists use empirical (observational) methods to study ice ages, gravity, evaporation, electricity, and nuclear fission – none of which can be directly observed; their effects can be observed, but the processes themselves cannot. Biopsychology is no different from the other sciences in this respect. One of its main goals is to characterize, through empirical methods, the unobservable processes by which the nervous system controls behavior.**
>
> **The empirical method that biopsychologists and other scientists use to study the unobservable is called <u>scientific inference</u>. Scientists carefully measure key events they can observe and then use these measures as a basis for logically inferring the nature of events that they cannot observe.**
>
> **Like a detective carefully gathering clues from which to recreate an unwitnessed crime, a biopsychologist carefully gathers relevant measures of behavior and neural activity from which to infer the nature of the neural processes that regulate behavior.**

**The fact that the neural mechanisms of behavior cannot be directly observed and must be studied through scientific inference is what makes biopsychological research such a challenge – and as I said before, so much fun.** (p. 13.)

*Scientific inference* is a logic fallacy; yet, he erroneously calls it an empirical method. The whole Theory of Evolution is based upon *scientific inferences* or *fictional ad hoc stories* that these people have manufactured out of thin air. There's nothing empirical about it! There is NO empirical evidence demonstrating the creative powers of entropy, evolution, or random mutations. Chemical evolution of proteins and genes from atoms is physically impossible thanks to entropy or the second law of thermodynamics. It can't happen, which means that it never happened. Evolution is entropy, or the second law of thermodynamics. It can't design and create. It's physically impossible.

Furthermore, evolution (genetic change), random mutations, and natural selection didn't even exist until AFTER God designed and created the proteins, genes, genomes, brains, eyes, and life forms in the first place. It's physically impossible for something that doesn't even exist yet to design, program, engineer, manufacture, and create proteins and genes out of thin air. Spontaneous generation or abiogenesis is physically impossible. Entropy or the second law of thermodynamics prevents it from happening. Evolution is entropy, which means that it can't design and create anything. Evolution or entropy can only deteriorate and destroy things. It can't design and create. It's physically impossible for evolution or entropy to design and create proteins, genes, genomes, brains, eyes, and life forms. That's just the way it is, because evolution of any kind is entropy.

It took me years, even decades, to discover that evolution or random mutation is entropy, or the second law of thermodynamics. That discovery also came with a powerful gift. I finally realized that Quantum Mechanics is the exact opposite of Materialism, Naturalism, Darwinism, Classical Physics, and Entropy. Quantum Mechanics and Psyche are Pure Syntropy. Quantum Mechanics is the Power of God, or the Priesthood Power of God. Psyche or Non-Local Consciousness is the only thing that can control Quantum Mechanics at the quantum level or the psyche level.

There is NO aging or entropy in the Quantum Realm, Psyche Realm, Spirit Realm, or Transdimensional Realm. Transdimensional means non-physical and non-local – not located in our physical 3D space-time realm. Everything in the Quantum Realm is Pure Syntropy. It is endless, timeless, eternal, and everlasting because there is NO entropy meaning that nothing ages, gets old, or dies in the Non-Local Realm.

The Gods create physical matter by infusing a particle of spirit matter with space-time and the ability to acquire entropy. The Gods create a particle of physical matter by taking a particle of spirit matter, filling it full of space, slowing it down to sub-light speeds, and making it subject to entropy or the passage of time. According to the theory of relativity, the particles of spirit matter existing at velocities faster than the speed of light experience NO passage of time, meaning that they do not age and are not subject to entropy. Entropy is a function of time or an aging process. Spirit matter and physical matter are the same thing – they are quantum objects. However, spirit matter is pure syntropy; whereas, physical matter has been slowed down by being infused with space-time and made subject to entropy or the passage of time.

I talk about all of this in great detail in my book, *Quantum Neuroscience: The Answer to Life, the Universe, and Everything*. If a comparison between Evolution and Quantum Mechanics interests you, I recommend you take a look at that book:

https://www.amazon.com/dp/B079Z6QQQB

The book, *Quantum Neuroscience: The Answer to Life, the Universe, and Everything*, makes a detailed comparison between Neuroscience and Quantum Neuroscience, which means that it makes a detailed comparison between Classical Physics and Quantum Mechanics.

When properly understood, Quantum Mechanics FALSIFIES Classical Physics, Materialism, Naturalism, Darwinism, Nihilism, Scientism, Behaviorism, Determinism, and even Atheism. Quantum Mechanics is Supernatural. Quantum Mechanics or Syntropy is the exact opposite of Classical Physics, Entropy, Random Mutations, Materialism, Naturalism, Darwinism, and the Theory of Evolution.

Quantum Mechanics is a proven and verified science. Quantum Mechanics is the best-proven and most-used science that we have. In contrast, the Theory of Evolution has NO empirical evidence supporting it. In fact, ALL of the empirical evidence that we have on hand as a race, including Quantum Mechanics, FALSIFIES Materialism, Naturalism, Darwinism, and the Theory of Evolution. Do you see how that works? It's important to understand.

Quantum Mechanics, Spirit Matter, and Quantum Non-Local Consciousness (Psyche or Intelligence) are PURE SYNTROPY. The syntropy has to exist somewhere someplace somehow, because according to the Law of Entropy and the Second Law of Thermodynamics, the physical Multi-verse should have burned out and suffered heat death an eternity or two ago; and, there should be NO more physical universes anywhere, but here we are nonetheless. The very existence of this physical universe – it's beginning full of syntropy and its ongoing existence billions of years later – is positive proof that Someone Psyche knows how to do syntropy or Someone Psyche is syntropy. Quantum Mechanics is syntropy or the Priesthood Power of God. God's Psyche knows how to do syntropy or Quantum Mechanics; otherwise, this physical universe would not exist.

This is what Quantum Mechanics or Transdimensional Physics is trying to teach us. Quantum Mechanics, Spirit Matter, and Psyche are pure syntropy. Evolution, Random Mutations, Physical Matter, and Classical Physics are entropy.

**Genetic Entropy**

Web Page: https://evolution-is-entropy.com/genetic-entropy/

Genetic Entropy is one of my most favorite scientific discoveries. This scientific discovery doesn't belong to me, though. It originates with John Sanford. After reading his book, *Genetic Entropy*, I simply KNEW that the theory of evolution is false because I now KNOW why it is false.

The theory of evolution is typically defined as Creation by Mutation/Selection – particularly, 'the origin of species by means of natural selection'. The first part of the title of Darwin's book is, "On the Origin of Species by Means of Natural Selection".

Creation by Natural Selection IS science fiction. Natural selection doesn't touch our genes! It can't. It's physically impossible for natural selection to get at our genes and change them. Natural selection cannot design and create anything, let alone a genome.

In truth, natural selection is NOT the mechanism of change behind the theory of evolution. It's the random mutations that produce genetic change, NOT natural selection! Natural selection or survival of the fittest doesn't do anything. It just waits for you to die. Natural selection is entropy or death. They built a whole "science" on a fictional, immaterial, invisible process that doesn't even touch our genes – natural selection! And, they literally give natural selection ALL the credit for designing, programming, creating, and producing our genomes and our physical bodies. For these people, natural selection is their god. They worship it with a passion. Natural selection is a man-made god, an idol.

> **Natural Selection: The evolutionary process by which heritable traits that best enable organisms to survive and reproduce in particular environments are passed to ensuing generations.**
>
> **Everyone who has taken introductory psychology has learned that nature and nurture together form who we are. As the area of a rectangle is determined by both its length and its width, so do biology and experience together create us.**
>
> **As *evolutionary psychologists* remind us, our inherited human nature predisposes us to behave in ways that helped our ancestors survive and reproduce. We carry the genes of those whose traits enabled them and their children to survive and reproduce. Thus, evolutionary psychologists ask how natural selection might predispose our actions and reactions when dating and mating, hating and hurting, caring and sharing. Nature also endows us with an enormous capacity to learn and to adapt to varied environments. We are sensitive and responsive to our social context.**
>
> **To explain the traits of our species, and all species, the British naturalist Charles Darwin (1859) proposed an evolutionary process. Follow the genes, he advised. Darwin's idea, to which philosopher Daniel Dennett (2005) would give "the gold medal for the best idea anybody ever had," was that natural selection enables evolution.**
>
> **Natural selection implies that certain genes — those that predisposed traits that increased the odds of surviving long enough to reproduce and nurture descendants — became more abundant.**
>
> **Natural selection, long an organizing principle of biology, has recently become an important principle for psychology as well. *Evolutionary***

*psychology* studies how natural selection predisposes not just physical traits suited to particular contexts — polar bears' coats, bats' sonar, humans' color vision — but also psychological traits and social behaviors that enhance the preservation and spread of one's genes. We humans are the way we are, say evolutionary psychologists, because nature selected those who had our traits — those who, for example, preferred the sweet taste of nutritious, energy-providing foods and who disliked the bitter or sour flavors of foods that are toxic. Those lacking such preferences were less likely to survive to contribute their genes to posterity.

As mobile gene machines, we carry not only the physical legacy but also the psychological legacy of our ancestors' adaptive preferences. We long for whatever helped them survive, reproduce, and nurture their offspring to survive and reproduce.

"The purpose of the heart is to pump blood," notes evolutionary psychologist David Barash. "The brain's purpose," he adds, is to direct our organs and our behavior "in a way that maximizes our evolutionary success. That's it." (*Social Psychology*, p. 8, 159.)

Everything they wrote here is false or incomplete.

It's NOT a rectangle, it's a triangle! It's not just nature and nurture that form us. There's an essential third component!

Do you know what it is?

**NATURE vs. NURTURE vs. NIRVANA: An Introduction to Reality**

https://www.amazon.com/dp/B01JWRCSVA

https://www.amazon.com/dp/1521132615

The third component is deliberately eliminated from science by the Materialists, Naturalists, Darwinists, Nihilists, and Atheists. These people state that it does not exist. The BioPsychoSocial Model tells us that it does exist. Somebody is right, and somebody is wrong. They both can't be right.

These people have been teaching for over 150 years that Natural Selection made you, that evolution made you; but, that's physically impossible. Natural selection can't make anything. Natural selection and the theory of evolution are based exclusively on entropy. Natural selection results in entropy or death. Evolution is entropy. Entropy is death. Entropy cannot make anything at all. Entropy or death can only destroy. The different types of evolution can only destroy. The different types of evolution or entropy can only produce death and extinction.

Natural selection doesn't predispose anything! Natural selection doesn't organize anything! It can't. Natural selection doesn't touch our genes! Natural selection doesn't endow us with anything! Natural selection has NO ability to learn anything. Natural selection doesn't enable anything. Natural selection doesn't do anything. Natural selection is supposed to be dumb and blind without a soul or a mind, according to the Darwinists. Creation by Natural Selection wins the rotten tomato for the most stupid, illogical, irrational, and ineffective idea ever created.

The theory of evolution is correlational, NOT observational. NO type of evolution has ever been caught in the act of design and creation. It's physically impossible for natural selection and random mutations to design and create something. Entropy prevents them from doing so. Chemical evolution or macro-evolution is prevented from happening by random diffusion or entropy. Macro-evolution is also prevented from happening by genetics. The genes are there to prevent macro-evolution from happening. Evolution of any type is entropy and death. Death cannot create life!

In fact, evolution (genetic change), random mutations, and natural selection didn't even exist until AFTER God designed, programmed, engineered, field-tested, fine-tuned, manufactured, created, and produced the proteins and their matching genes in the first place.

The theory of evolution is a fictional story that they made up out of thin air after-the-fact to fit the facts. NO part of it can actually design and create. It's a fictional story, not science.

**We are sensitive and responsive to our social context, NOT our genes!**

Who is this **WE** that they keep talking about in our Social Psychology textbooks? **WE** can't be our society, environment, or social context that **WE** are sensitive to and responsive to! And, it's definitely NOT our genes. Our genes aren't sensitive and responsive to anything according to the Evolutionists! The genes, natural selection, and random mutations are supposed to be dumb and blind without a soul or a mind. Our genes can't be sensitive nor responsive to anything.

Personal pronouns imply a person or a psyche – NOT our genes (nature) and NOT our environment (nurture).

Natural selection and your genes DO NOT and CANNOT pass your Psyche Legacy (psychological legacy) from one generation to the next! Natural selection doesn't touch your genes, and it definitely doesn't touch nor change your Psyche either. It's science fiction to imply that it does. The theory of evolution is science fiction. Design and creation by natural selection is science fiction. The idea that your genes carry your "longings" or "desires" from one generation to the next is science fiction. There's no such thing as genetic memory, at least not at the physical level. All of our thoughts and memories are carried as quantum waves from one generation to the next through our Psyche or Quantum Non-Local Consciousness.

Your genes don't care whether you live or die. Only YOU care whether you live or die. Your Psyche cares whether you live or die; but, your genes do not. In order for your genes to care, they would have to have some sort of Psyche, or Intelligence, or Consciousness, or Awareness. But, if your genes have a Psyche, then the very existence of that Psyche falsifies Materialism, Naturalism, and Darwinism which claim that Psyche does not exist. Physicalism, Naturalism, and the Theory of Evolution are self-defeating. They don't work as advertised because they can't work as advertised.

Natural selection results in entropy and death, NOT X-Men and new unique life forms. Natural selection cannot design and create and program genomes. Natural selection doesn't touch our genes.

Random mutations are also entropy; but, at least random mutations by definition in principle actually change our genes. However, random mutations cannot design and create anything either.

Entropy is death. Death cannot design and create new unique genomes and life forms. That's physically impossible! Mutation and Selection can only produce entropy or death. They are based exclusively on entropy or death. Death cannot design and create life. Death can only end life.

Remember, the theory of evolution is Creation by Entropy or Creation by Death. That's NEVER going to work because it's physically impossible!

Isn't it refreshing to finally have access to the truth, rather than all the science fiction that the Evolutionists have been feeding us throughout our lives?

Well, I think it is.

I used to be a Materialist, Naturalist, Nihilist, and Atheist until I finally started to study the evidence. The evidence and the truth set me free! The Science and the Scientific Evidence convinced me that God must exist in order to have done all the Science and Fine-Tuning which natural selection and evolution could NEVER have done.

Mark my Words

—

## Can Natural Selection Create?

The Physicalists, Naturalists, Darwinists, Nihilists, and Atheists teach that natural selection can create anything that it sets its mind to. These people teach that natural selection made you.

Are they right?

They are not!

They are deceiving themselves and trying to trick us and deceive us as well.

Creation by Natural Selection is demonstrably false, which means that it has been falsified by Scientific Evidence. Natural selection doesn't do anything. Natural selection doesn't touch our genes, nor does it pre-determine our future. Natural selection doesn't have a mind. There is NO intelligence or psyche within natural selection. There may (or may not) be some type of intelligence or psyche within our genes; but, there is NOTHING there when it comes to Natural Selection or Evolution. Natural selection is a fictional concept that they made up out of thin air. It doesn't really exist as a person or an entity. The same can be said of evolution. I'm not the only scientist to have figured this out by now. Natural selection is worthless as a creative agent and can't function as a creative agent.

[Editorial Note: I have written permission from John C. Sanford to use all of the quotes from John C. Sanford which I use in my books, so long as I cite the sources which I have done.]

START OF THE QUOTE FROM "GENETIC ENTROPY" BY JOHN SANFORD — USED BY PERMISSION FROM THE AUTHOR JOHN SANFORD.

Chapter 9: Can Natural Selection Create?

Newsflash — Mutation/Selection cannot even create a single gene.

We have been examining the problem of genomic degeneration and have found that deleterious mutations occur at a very high rate. Natural selection can only eliminate the worst of these, while all the rest accumulate — like rust on a car. Might beneficial mutations at other sites in the genome compensate for this continuous and systematic erosion of genetic information? The answer is that beneficial mutations are much too rare and are much too subtle to keep up with such relentless and systematic erosion of information. This is carefully documented by Sanford et al. (2013), and Montañez et al. (2013). It is very easy to systematically destroy information, but apart from the operation of intelligence it is very hard (arguably impossible) to create information.

This problem overrides all hope for the forward evolution of the whole genome. However, some limited traits might still be improved via Mutation/Selection. Just how limited is such progressive ("creative") Mutation/Selection? By now it should be clear that random spelling errors in an instruction manual could never give rise to an airplane component (say a molded aluminum part), which then resulted in a significantly improved overall performance of a jet plane. Not even with an unlimited number of flight trials/crashes and an unlimited budget. So, it is certainly reasonable to ask the parallel biological question, "Could Mutation/Selection create a single functional gene from scratch?"

A gene is like a book, book chapter, or an executable program — and minimally consists of a text string with 1,000 characters. Mutation/Selection could not create a single gene because of the enormous preponderance of deleterious mutations, even within the context of a single gene. The net information must always still be declining, even within a single gene or linkage block. Even if a gene was 50% established, deleterious mutations would degrade the completed half of the gene much faster than beneficials could create the missing half of the gene. However, to better understand the limits of forward selection, let us for the moment discount all deleterious mutations and only consider beneficial mutations. Could Mutation/Selection then create a new and functional gene?

1. Defining our first desirable mutation. The first problem we encounter in trying to create a new gene via Mutation/Selection is defining our first beneficial mutation. By itself, no particular nucleotide (A, T, C or G) has more value than any other, just as no letter in the alphabet has any particular meaning outside of the context of other letters. So, selection for any single nucleotide can never occur except in the context of the surrounding nucleotides (and in fact, within the context of the whole genome). A change of a single letter within a word or chapter can only be evaluated in the context of the surrounding block of text. This brings us to an excellent example of the principle of "irreducible complexity" within the genetic realm. In fact, it is irreducible complexity at its most fundamental level. We immediately find we have a paradox. To create a new function, we will need to select for our first beneficial mutation, but we can only define that new nucleotide's value in relation to its neighbors — and we are going to have to be changing most of those neighbors also. We create a circular path for ourselves. We will keep destroying the "context" we are trying to build upon. This problem of the fundamental inter-relationship of nucleotides is called epistasis. True epistasis is almost infinitely complex, and virtually impossible to analyze, which is why geneticists have always conveniently ignored it. Such bewildering complexity is exactly why language and information (including genetic language and genetic information) can never be the product of chance, but always requires intelligent design. The genome is literally a book, written literally in a language, and short sequences are literally sentences. Having random letters fall into place to make a single meaningful sentence, by accident, would require more tries (more time), than earth history can provide (i.e., "methinks it is like a weasel" would take $27 \wedge 28$ tries — that is 10 followed by 40 zeros). The same is true for any functional string of nucleotides. If there are more than a dozen nucleotides in a functional string, we know that realistically they will never just "fall into place". This has been mathematically demonstrated repeatedly.

But as we will soon see, neither can such a sequence arise by selecting one nucleotide at a time. A pre-existing "concept" is required as a framework upon which a sentence or a functional sequence must be built. Such a concept can only pre-exist within the mind of the author. Starting from the very first mutation, we have a fundamental problem even in trying to define what our first desired beneficial mutation should be.

2. Waiting for the first mutation. Let's assume we can know the first desired mutation. How long do we have to wait for it to happen? Human evolution is generally assumed to have occurred in a small population of about 10,000 individuals. The mutation rate for any given nucleotide, per person per generation is exceedingly small (very roughly about one mutation per 30 million individuals, for a given nucleotide site). Within a population of 10,000, one would have to wait 3,000 generations (at least 60,000 years) to expect a specific nucleotide to mutate. But two out of three times, it will mutate into the "wrong" nucleotide. So, to get a specific desired mutation at a specific site just in one individual will take three times as long, or at least 180,000 years. Once the mutation arises in one individual, it has to become "fixed" (such that each individual in the population will eventually have a double dose of that mutation). Because a newly arisen mutation arrives in a population as just a single copy, it arrives on the brink of extinction. The vast majority of new mutations soon drift back out of the population, even the ones that are beneficial. So, any specific desired mutation must arise many times before it "catches hold" in the population. Only if the mutation is dominant and has a very distinct benefit does selection have any reasonable chance to rescue it from random elimination via drift. According to population geneticists, apart from effective selection, in a population of 10,000, our given new mutant has only one chance in 20,000 (the total number of non-mutant nucleotides present in the population) of NOT being lost via drift. Even with some modest level of selection operating, there is a very high probability of random loss, especially if the mutant is recessive or is weakly expressed (we actually know that most mutations will be both recessive and nearly neutral). Therefore, even a beneficial mutation will be randomly lost due to genetic drift most of the time. Our numerical simulations suggest a weakly beneficial mutant will be lost about 99 out of 100 times. So, a typical mildly-beneficial mutation must happen about 100 times before it is likely to "catch hold" within the population. So, on average, in a population of 10,000 we would have to wait 180,000 × 100 = 18 million years to stabilize our first desired beneficial mutation, to begin building our hypothetical new gene. So, in the time since we supposedly evolved from chimp-like creatures (6 million years), there would not be enough time to realistically expect our first desired mutation to go to fixation in the genomic location where our required gene is hopefully going to arise. A vast amount of mutations would arise during 18 million years, but only once would that specific nucleotide mutate to that specific new nucleotide — such that it's not lost due to genetic drift and is fixed.

3. Waiting for the other mutations. After our first desired mutation has been found and fixed, we need to repeat this process for all the other nucleotides encoding our hoped-for gene. A gene is minimally 1,000 nucleotides long. More realistically, a human gene is on average about 50,000 nucleotides long, when regulatory elements and introns are included. To be extremely generous we will only consider a gene of 1,000 nucleotides (and we assume each nucleotide is by itself selectable). If this process was a straight, linear, and sequential process, it would require about 18 million years × 1,000 = 18 billion years to create the smallest possible gene. This is more than the time since the reputed Big Bang! So, it is a gross understatement to say that the rarity of desired mutations limits the rate of evolution. Furthermore,

single nucleotides do not carry any information by themselves, and cannot be selectively favored. Specified information requires many characters (minimally, a sentence or similar text string is needed). Like any message, a genetic message which specifies some life function requires many nucleotides to reach its "functional threshold". Functional threshold is the minimal number of characters (or nucleotides) needed to convey a meaningful message. Below the functional threshold, individual letters or nucleotides have no benefit and cannot be favored by selection. This means that realistically, waiting time will be much, much longer — because no selection can happen until the minimum string of nucleotides falls into place by chance. If the functional threshold for selection is 12 (no selection until all 12 letters are in place), the waiting time in our hypothetical human population becomes trillions of years.

Sanford, John (2015-02-23). Genetic Entropy (Kindle Locations 1684-1755). FMS Publications. Kindle Edition. USED BY PERMISSION.

END QUOTE.

—

Trillions of years!

Well, that's the END of the Theory of Evolution, isn't it?

The Darwinists NEVER use their God-given brains to stop and think about these kinds of things. At the best possible average pace, with God making sure that there are NO deleterious mutations and NO devolution taking place, it would take on average 18 million years to fixate and stabilize a SINGLE beneficial mutation through "Natural Means" into a population of 10,000 apes which God has already designed and created in the first place and kept alive and functional during those 18 million years, just so that population of 10,000 God-created apes can achieve their first beneficial mutation through "Natural" Hands-off Mutation and Selection. 18 million years on average per beneficial mutation! Think about it!

If those apes need 1,000 such beneficial mutations in order to become men, then you are looking at 18 billion years on average to produce those targeted 1,000 beneficial mutations; and, that's with a population of 10,000 apes that God has already designed and created in the first place and that God is making sure receive ONLY beneficial mutations and NO devolution or deleterious mutations. And, that's also with God keeping that population of 10,000 apes alive during those 18 billion years so that they can indeed "evolve" their necessary 1,000 beneficial mutations and become men all on their own through "Natural Means".

Furthermore, it has been estimated that it would in fact take at least 20 million such beneficial mutations to convert chimpanzees into humans through "Natural Means". With that targeted goal in mind and assuming NO deleterious mutations or extinctions along the way, how long would it take on average to convert 10,000 chimpanzees into 10,000 humans using Natural Selection and Random Mutations to do the job? So, what do you get if you multiply 20 million beneficial mutations with 18 million years per beneficial mutation? At the BEST possible pace, with God keeping those 10,000 chimpanzees alive all along the way, and with God making sure that there is NO devolution, NO extinction, and NO deleterious mutations taking place, the quickest on average that Mutation and Selection could convert a chimpanzee into a human through "Natural Means" is 360 trillion years.

John Sanford isn't exaggerating whenever he says that it could take trillions of years for Mutation/ Selection to design and create something useful "naturally". And, it really isn't Natural Evolution if God has to design and create the 10,000 chimpanzees in the first place, and then keep them alive for 360 trillion years by blocking ALL deleterious mutations and preventing ALL extinctions that might take place during that period of time, just so He can convert 10,000 chimpanzees into 10,000 humans "naturally" or through "Natural Means".

Think about it! At the BEST possible average pace, it would take at least 360 trillion years for Mutation and Selection to convert a population of 10,000 chimpanzees into 10,000 humans through "Natural Means", with God keeping those 10,000 mutants alive and preventing deleterious mutations during the whole time. And, that's with 10,000 chimpanzees that God has already designed and created in the first place! How old did they say our universe is? How long would it take to convert a bacterium into a human through "Natural Means" when billions of beneficial mutations are needed? Wouldn't it be easier and faster to just let God design and create those 10,000 humans in the first place?

YES, it would be!

The Darwinists and Materialists NEVER stop and use their God-given brains to think about and calculate these kinds of things. You will NEVER get these kinds of calculations and truths from the Darwinists because they don't DO this kind of science. It's too difficult and painful for them. I'm willing to wrap my mind around these kinds of things. Your typical Darwinist isn't. Your typical Darwinist is afraid of it because they don't want to be proven wrong. For the Materialists and Darwinists, ignorance is bliss! But, ignorance is the reason why the Darwinists and Materialists truly believe that Mutation/Selection can design and create anything that it sets its mind to. The rest of us KNOW BETTER!

If you think about it, this is radically advanced science — the best that humans are able to come up with! Can you see and understand now why the 9th Chapter of "Genetic Entropy" put an END to the Theory of Evolution for me? It's because I understood what John Sanford was talking about and chose to believe that it is true. Now the onus is on you.

What do the Darwinists typically do when presented with these kinds of Statistical Models of the Mutation/Selection Process?

Assuming that they don't go head-in-the-sand and actually study them instead, the Darwinists try to shave the figures in half or by one-tenth, which is exactly what they do when designing Mutation/Selection Models of their own. They cheat. They choose parameters that are scientifically inaccurate and don't match with Reality in order to shave those estimates down to something that they might be willing to accept. They keep shaving and cheating until they get the numbers that they want, and then they call the results "Science".

If they really take John Sanford's Models seriously, the Darwinists will demand that God artificially accelerate the Mutation/Selection Process so as to make it possible for the Theory of Evolution to be true. But, even if God were to speed up the process a thousand-fold, it's still going to take 360 billion years for chimpanzees to evolve into humans through "Natural Means"; and, the Darwinists are still going to complain, even though we are starting with Chimpanzees that God has designed and created in the first place.

It can be fascinating and entertaining to watch a Darwinist try to shave trillions of years off a Statistical Estimate that he doesn't like, all in an attempt to increase the possibility that the Theory of Evolution might be true.

And, that's just the beginning of the problems for the Theory of Evolution. It only gets worse from there on forward, because John Sanford actually has TEN points in chapter

9 of "Genetic Entropy", each of which decreases the likelihood of Mutation/Selection creating anything at all, even if it has an infinite number of years to do so. Evolution by random mutations and evolution by natural selection CANNOT design, create, and deploy anything! It has been conclusively and finally demonstrated that it is so. It has been empirically and logically observed to be so. The Theory of Evolution doesn't work and can't create new unique genomes from scratch, so we have no choice but to declare the whole thing to be FALSE.

Many different scientists taught me that Random Mutations and Natural Selection (Evolution) cannot design and create genomes and life forms. They met their burden of proof and demonstrated to me that it must be so. (See the partial list of Reference Materials below for a selection of some of the best scientific evidence that falsifies the Theory of Evolution.)

I chose to believe the scientific evidence rather than the claims of the Materialists, Naturalists, and Darwinists who teach that Psyche, Intelligence, and Syntropy do not exist.

**Evolution Is Creation by Death**

I did make a scientific discovery in recent days (May 2018) that I think has merit and value to the Scientific Community.

I do seem to be the first person on the planet to realize that NO type of evolution can do selection. Selection of any kind involves choice; and, choice is exclusively the product of a Psyche or a Mind. Evolution of any type by definition, in principle, is dumb and blind without a soul or a mind. The very definition of Evolution eliminates its ability to do selection, a priori. Evolution can't do selection because evolution can't do Psyche or Choice. Evolution doesn't exist as a Psyche, Person, Intelligence, Mind, or Soul capable of doing choice or selection.

Furthermore, any attempt to imbue evolution with a soul, psyche, or mind automatically FALSIFIES Materialism, Naturalism, Darwinism, Nihilism, and the Theory of Evolution which claim that Psyche or Syntropy does not exist. The theory of evolution is a non-starter because evolution of any kind cannot do selection or choice.

Always remember, creation by natural selection is science fiction and wishful thinking because evolution of any kind cannot do selection or choice.

Materialism, Naturalism, Darwinism, Nihilism, Behaviorism, Determinism, Physical Reductionism, Atheism, Classical Physics, and the Theory of Evolution are based exclusively on entropy. Entropy is death. Death cannot design, create, and produce life. Such a thing has never been experienced nor observed. Darwinism or the theory of evolution is Creation by Death, or Creation by Entropy. Creation by Death or the theory of evolution is impossible. It can't happen, which means that it didn't happen.

Therefore, the theory of evolution cannot be used to explain the origin of life. The most that the theory of evolution can explain is death, extinction, devolution, and genetic entropy. Evolution of any kind is a function of death or entropy. Evolution is entropy. Evolution is death. Evolution or death, of any kind, can't be used to explain the origin of life. The truthfulness of this reality becomes obvious, once a person realizes and accepts the fact that the Theory of Evolution, Physicalism, Naturalism, Atheism, Nihilism, Classical Physics, and Darwinism are based exclusively on entropy. Entropy or death cannot produce life. The different types of evolution or entropy cannot produce life. It's impossible for entropy or death to produce life.

At times I've wondered why nobody else has been able to see and understand these obvious truths; but then, I used to be a Materialist, Naturalist, Nihilist, and Atheism, and there I find my answer. At the time, I wasn't able to see nor understand these obvious scientific truths because I didn't want to see them, understand them, nor accept them. No seeking, then no finding. I wasn't looking for any of this, so I never found it. I only found it after I started looking for it. I had convinced myself that this type of information doesn't exist. Self-deception works, and it works every time, especially when it comes to scientists like me.

The axiom stating that evolution is dumb and blind without a soul or a mind, if taken as being true, prevents evolution of any type from being able to do selection. In other words, the very definition of evolution as being dumb and blind FALSIFIES the Theory of Evolution by preventing evolution of any kind from being able to do selection or choice. Evolution is entropy; and, entropy is death. Death cannot do selection or choice. Death cannot do life. Entropy or death can only destroy. That is what has been experienced and observed.

Meanwhile, the Materialists, Naturalists, Darwinists, Nihilists, and Atheists DEMAND that you accept on blind faith that their claims – that death, entropy, or evolution can produce life – are true.

We scientists have FALSIFIED the Theory of Evolution trillions of times in thousands of different ways, but we choose to ignore the evidence because it isn't telling us what we want to hear. That's the way we do science in this world – by ignoring the evidence, discounting the evidence, banning the evidence, and destroying the evidence. We do our science this way so that we can prove to ourselves and to others that the Theory of Evolution is true.

There is another way to do science, though – a better way of doing science. I call it Science 2.0; and, it involves allowing ALL of the evidence into evidence. Once we choose to do so, then ALL of the evidence that we have on hand as a race FALSIFIES the claims of Materialism, Naturalism, Darwinism, Nihilism, Behaviorism, Determinism, Physical Reductionism, and Atheism which claim that this evidence does not exist.

Once we have eliminated all of the falsehoods such as Materialism, Naturalism, Atheism, Nihilism, and Darwinism, then we are left staring at THE TRUTH, which is that ONLY Psyche can design, program, engineer, field-test, fine-tune, manufacture, create, and do science.

By eliminating all of the falsehoods or pseudo-sciences, it becomes obvious that God's Psyche must of necessity exist in order to have DONE all of the Science that needed to be done, which evolution and the rocks could NEVER have done. Remember, evolution and the rocks can't do science because they can't do selection or choice.

The observation and realization, that evolution of any type cannot do selection, just might be one of my greatest scientific discoveries even if I don't end up being the first person on the planet to have made this discovery.

Obviously, everyone across the world is now starting to make these kinds of scientific discoveries right and left because we have finally started to take our blinders off. More and more of us are willing to see, which makes us able to see. Nowadays, it's obvious that the Theory of Evolution is false; whereas, we couldn't see it before because we didn't want to see it, and we didn't know where to look.

All you want is the truth. Everything else is worthless in the end.

Mark My Words

—

## Source

### Science 2.0: I Upgraded My Science

https://www.amazon.com/dp/B0771K6WTX

### The Scientific Method Proves That the Theory of Evolution Is False

https://www.amazon.com/dp/B01IAAIRT2

https://www.amazon.com/dp/1521133611

### NATURE vs. NURTURE vs. NIRVANA: An Introduction to Reality

https://www.amazon.com/dp/B01JWRCSVA

https://www.amazon.com/dp/1521132615

Myers, D. G. (2010). *Social Psychology* (10th ed.). New York: McGraw-Hill.

## Reference Materials

Wells, J. (2000). *Icons of Evolution: Science or Myth? Why Much of What We Teach About Evolution Is Wrong*. Washington, DC. Regnary.

Sanford, J. (2014). *Genetic Entropy* (4th ed.). Cornell University: FMS Foundation.

Sanford, J. C., Marks, R. J., Behe, M. J., Dembski, W. A., & Gordon, B. L. (Eds.). (2013). *Biological Information: New Perspectives*. Hackensack, NJ: World Scientific.

Meyer, S. C. (2010). *Signature in the Cell: DNA and the Evidence for Intelligent Design*. New York: HarperCollins.

Meyer, S. C. (2013). *Darwin's Doubt: The Explosive Origin of Animal Life and the Case for Intelligent Design*. New York: HarperCollins.

Mark My Words. (2016). *The Scientific Method: Proves That the Theory of Evolution Is False*. Kindle. Retrieve from: https://www.amazon.com/dp/B01IAAIRT2

Mark My Words. (2016). *The Theory of Evolution Proved to Me that God Exists: Why I Am No Longer an Atheist and Why I No Longer Believe in the Theory of Evolution*. Kindle. Retrieve from: https://www.amazon.com/dp/B01HZYBZ7K

# Kin Selection

Web Page: https://evolution-is-entropy.com/2018/05/18/kin-selection/

Natural Selection is science fiction. By definition, in principle, Natural Selection can't do CHOICE or selection. Natural selection can't do what they say it does. By definition, in principle, natural selection or survival of the fittest is supposed to be dumb and blind without a soul, psyche, or mind. Therefore, natural selection can't do choice or selection. Instead, natural selection results in entropy or death. Natural selection, or entropy and death, cannot design and create and produce anything. Design and creation by Natural Selection is prevented from happening by entropy, random diffusion, or the second law of thermodynamics. Random mutations are also entropy. Entropy is death. The theory of evolution is Creation by Entropy, or Creation by Death. That's not going to work because it's physically impossible. Death or entropy cannot design and create life. This reality is obviously true.

Do you want the truth, or do you want the science fiction?

Can you handle the truth? Most of our scientists can't.

Kin Selection is another smoking gun where Psyche is concerned.

> **Kin Selection: The idea that evolution has selected altruism toward one's close relatives to enhance the survival of mutually shared genes.**
>
> **Our genes dispose us to care for relatives. Thus, one form of self-sacrifice that *would* increase gene survival is devotion to one's children. Compared with neglectful parents, parents who put their children's welfare ahead of their own are more likely to pass their genes on. As evolutionary psychologist David Barash wrote, "Genes help themselves by being nice to themselves, even if they are enclosed in different bodies." Genetic egoism (at the biological level) fosters parental altruism (at the psychological level). Although evolution favors self-sacrifice for one's children, children have less at stake in the survival of their parents' genes. Thus, according to the theory, parents will generally be more devoted to their children than their children are to them.** (*Social Psychology*, p. 452.)

By definition, in principle, evolution is dumb and blind without a soul or a mind. This means that evolution of any type cannot DO choice or selection. It's a deceptive lie to say that evolution can do choice or selection. It's science fiction.

Without realizing it, these people have painted themselves into a corner with this one. The idea of genes helping each other even if they are enclosed in different bodies is physically impossible. There's NO physical mechanism in place whereby the genes can communicate with each other, especially the genes in different bodies! Such an idea is ludicrous. If the genes are communicating with each other and recognizing each other, then they are doing so at the quantum level or the psyche level because they can't do so at the physical level.

The ONLY way to make Kin Selection true is if we all axiomatically agree in advance that the genes (and evolution) are psychic and are therefore capable of determining telepathically which genes are related to them and which genes are not. In order for the genes to be nice to each other, they have to know each other, perceive each other, recognize each other, and show favoritism to each other. They have to be psychic and have some kind of psyche. The genes can't be dumb and blind if we want Kin Selection to work as advertised. The genes have to communicate with each other, know each other, and recognize each other in order for them to be able to help themselves and be nice to themselves. The genes also have to be subliminally communicating their desires to the Human Psyche in order to make the Human Psyche be nice to the genes too. The only way to make Kin Selection work as advertised is if we agree in advance that the genes have some sort of Psyche or Mind and are telepathically connected with each other.

However, if we agree in advance axiomatically that genes have a psyche, that genes perceive each other telepathically, that genes know each other and recognized each other, and that genes are therefore psychic, then we have in fact FALSIFIED Materialism, Naturalism, Darwinism, Nihilism, Atheism, even the Theory of Evolution in the process.

These people personify the genes – imbue them with Psyche and Intelligence – and in the process falsify the major premises of Materialism, Naturalism, Darwinism, Nihilism, and Atheism which state that Psyche or Syntropy does not exist. Thereby, these falsified philosophies or falsified religions end up being self-defeating. The truth cannot be built upon falsehoods

When it comes to Science, we observe that Psyche, Quantum Mechanics, or Syntropy always ends up being the best possible explanation that can be given to ALL of the evidence that we are observing and experiencing, including the physical evidence. Remember, entropy and physical matter would not exist without a massive initial infusion of Syntropy, Intelligence, Power, Quantum Mechanics, or Psychic Intervention somewhere sometime along the way.

Remember, perception is a function and a product of Psyche. At the physical level, the genes have no way of knowing or perceiving which genes are related to them and which genes are not; and at the physical level, the genes have NO way to pass that information on to the physical body or physical brain even if the genes were to know who is related to them and who is not. At the physical level, the genes have no way to perceive each other, thereby falsifying the claims of Kin Selection. By restricting and limiting everything to the physical level, the Materialists and Naturalists automatically falsify anything and everything that needs Psyche or Intelligence or Perception in order to become true – things such as Kin Selection and Creation by Mutation/Selection.

Remember, evolution can't do selection! By definition, in principle, evolution cannot do CHOICE! Selection requires some type of choice, and choice requires some type of Psyche or Mind! By definition, in principle, evolution is dumb and blind without a soul or a mind. The theory of evolution is self-defeating because evolution of any kind can't do selection or choice. Evolution doesn't exist as some type of Psyche or Person who is capable of making choices or doing selection. The false is falsified by the truth; and, the truth is repeatedly experienced and observed. The theory of evolution is obviously false because evolution of any type by definition in principle cannot do selection, psyche, or choice.

Evolution of any type cannot do science, but the Human Psyche and God's Psyche certainly can. They've been caught in the act of doing so.

With just a bit of scientific observation, it's easy to see that Kin Selection is a *fictional ad hoc just-so story* that they made up out of thin air after-the-fact to match with what we have experienced and observed from Intelligent Beings or the Human Psyche. They took Kin Selection, a philosophical idea or religious idea, and they personified it, humanized it, anthropomorphized it, and deified it. Kin Selection and the Theory of Evolution are man-made idols or man-made gods. The Theory of Evolution is our modern-day form of idolatry, which is a belief in false gods that are incapable of delivering the goods.

The theory of evolution is self-defeating because it can't do what they say it does. The genes can't be nice to each other without a psyche or a soul; and, if the genes have a psyche or a soul, then the very existence of Psyche or Syntropy or Soul falsifies Materialism, Naturalism, Darwinism, Nihilism, Behaviorism, Determinism, Physical Reductionism, Atheism, and the Theory of Evolution. The false is falsified by the truth; and, the truth is repeatedly experienced and observed.

Mark My Words

—

**Source**

*God Is in the Light: God is light, and in Him is no darkness at all.*
https://www.amazon.com/dp/B07168S37N

**Reference**

Myers, D. G. (2010). *Social Psychology* (10th ed.). New York: McGraw-Hill.

## 5. The Beginning of Anything Always Points Us to the Beginner:

Schroeder, G. L. (1997). *The Science of God: The Convergence of Scientific and Biblical Wisdom*. New York: The Free Press.

The Theory of Relativity or time dilation IS the solution to the discrepancy of between the seven days of creation found in Genesis Chapter One and the scientific observations and scientific measurements of the cosmos which establish this physical universe as being 13.7 billion years-old and 93 billion light years in diameter.

This book contains an ingenious reconciliation between the seven days of creation in Genesis and the scientific observations of the cosmos, using time dilation and the Theory of Relativity to do so.

There's no way to know for sure if his theory is true; but, it does have promise and I found it extremely fascinating to study and read about. This is the best explanation of the Big Bang Theory that I have encountered so far, and one that has been made completely compatible with the Biblical account of creation found in the book of Genesis.

After reading this book, I finally felt like I had a pretty good feel for the Theory of Relativity and how it really works.

This book, *The Science of God*, provides scientific evidence or observational evidence for the Kalām cosmological argument for the existence of God. Any observed beginning, such as the Big Bang, must of necessity have had a Beginner to design it, organize it, gather it, and trigger it. Intelligent design, intervention, and creation by God's Psyche is the BEST, most logical, and most parsimonious explanation for the Big Bang Singularity and for the triggering of the Big Bang in the first place, along with ALL of the subsequent fine-tuning and physical matter that came out of it.

Remember, when it comes to the Big Bang or the Beginning of this Physical Universe, physical matter couldn't have done the job, because physical matter in principle didn't exist yet, until after the Big Bang! Since physical matter couldn't have done the job, who did? If you accept the observational evidence found in this book, then you simply KNOW who did the job, because He is the only one we know of who could have done the job – the Biblical God Jesus Christ.

### Obviously Made

Anything that is obviously made obviously has a Maker who made it.

Energy, Light, or Psyche is the ONLY thing in our universe that has always existed and will always exist. Everything else was designed and made by some type of Intelligence or Psyche.

The Law of Psyche states that every psyche has a certain amount of energy that's under its control, and that controlling psyche can form or transform the energy under its control into anything that it wants that energy to be, anytime and anywhere that it chooses to do so. Energy is always conserved, so every psyche will always have a certain amount of energy that's under its control.

According to Quantum Field Theory, the Gods and the Controlling Psyches had to design and make the massless, entropyless, exergic, syntropic Quantum Fields BEFORE they could make and sustain physical matter. NO quantum fields, then NO physical matter. It's that simple.

The thing that we call the Big Bang took place in our part of the multiverse when the Gods and the Controlling Psyches designed and made the Quantum Fields. According to the Big Bang Theory and Quantum Field Theory, the Gods and Controlling Psyches couldn't make mass, resistance to acceleration, mass's heat storage capacity, nor heat until after they had designed and made the Quantum Fields. First things first!

The Quantum Fields were obviously designed and made so that physical matter and heat could be designed and made. The purpose of science is to find the True Cause of everything that exists. Find the True Cause or the True Makers of the Quantum Fields, and you find the True Cause and the True Maker of everything that currently exists in our universe.

For all you know, your psyche may be one of the controlling psyches who was instrumental in designing and creating the Quantum Fields. Something massless, entropyless, heatless, and non-physical had to design and make the Quantum Fields; and, it could as well have been you. That's what Psyches do. They design and make things out of energy.

## 6. Scientific Proof of the Invisible and the Immaterial:

McTaggart, L. (2002). *The Field: The Quest for the Secret Force of the Universe*. New York: HarperCollins.

Lynne McTaggart interviewed over 70 different scientists, including many Quantum Physicists, who were being persecuted, ridiculed, mocked, censored, banned, and blocked by the Materialists and Naturalists for doing Fringe Science which these Atheists didn't find acceptable.

The result was one of the most interesting science books ever written.

If you want to find and KNOW the truth, then study and read everything that the Materialists and Naturalists are trying to censor, ban, and block. These people have a talent or a knack for identifying and rejecting anything that is real and true.

I discuss this book in greater detail in other books that I have written, so I won't do so here; but, it is a winner!

Honorable Mention in this Category

Talbot, M. (1992, 2011). *The Holographic Universe: The Revolutionary Theory of Reality*. New York: HarperCollins.

Talbot, M. (1997). *Beyond the Quantum*. New York: Scribner.

Talbot had direct experience with the paranormal and the supernatural, so he tended to look at and describe physics and quantum physics differently than most authors. Talbot was ahead of his time.

Honorable Mention in this Category

Kelly, E. F., Kelly, E. W., Crabtree, A., Grosso, M., & Gauld, A. (2007). *Irreducible Mind: Toward a Psychology for the 21st Century*. Plymouth, United Kingdom: Rowman and Littlefield.

This book is heavy-duty science. This book can get extremely complex, complicated, confusing, and even boring at times. Therefore, this really isn't the right book to be used as a primer to the non-physical, immaterial, and non-local realm, although this book does indeed demonstrate the existence of these things.

When it comes to scientific research and "scientific proof" of the invisible, immaterial, non-local, spiritual, and non-physical, this is the book that most scientists and medical doctors turn to when they are seeking evidence and information regarding these non-local phenomena which have either been indirectly detected or inferred to exist from the scientific research and scientific evidence they have been dealing with in their work.

Remember, in order to find and know the truth about objects and realities which cannot be detected nor recorded by our physical instruments, one must live and experience and observe these non-local realities for himself or herself in order to KNOW that they are real and true. Or, one must choose to trust someone who has lived, experienced, and observed these non-local realities. The truth is KNOWN by living it and experiencing it, rather than philosophizing about it and speculating about it as the Materialists and Naturalists do.

This book treats Non-Local Experiences as observational evidence, experiential evidence, empirical evidence, and scientific evidence; and, so does Science 2.0.

**Falsifying the Philosophy of Mechanism**

Nichol, L. (2003). *The Essential David Bohm*. London: Routledge.

Back in the 1950's, David Bohm was trying to falsify the Philosophy of Mechanism and replace it with Quantum Mechanics instead. Bohm recognized the NEED, and he was trying to provide the world with a better Philosophy of Science. Bohm was seventy years ahead of his time.

Within my books, "NATURE vs. NURTURE vs. NIRVANA: An Introduction to Reality and Quantum Mechanics" and "God Is in the Light: God is light, and in Him is no darkness at all", I included a 200-page book review of David Bohm's book *The Essential David Bohm*, demonstrating how David Bohm has been vindicated by Science during the subsequent decades, after writing the essays in that book.

https://www.amazon.com/dp/B01JWRCSVA

https://www.amazon.com/dp/1521132615

https://www.amazon.com/dp/B07168S37N

In case you don't know, the Philosophy of Mechanism is what we moderns call Materialism, Naturalism, Darwinism, Nihilism, Atheism, Creation Ex Nihilo, Creation by Chance, Classical Realism, the Theory of Evolution, or the Second Law of Thermodynamics. It's all Creation by Chance or the Null Hypothesis, and it has ALL been FALSIFIED by Statistics, Quantum Mechanics, Quantum Field Theory, Quantum Waves, Photons, Quantum Fields, the Quantum Zeno Effect, Quantum Tunneling, Action at a Distance, Quantum Non-Locality, Dark Matter or Spirit Matter, Dark Energy, and every other immaterial or non-physical thing that has ever been experienced or observed such as Gravity, Magnetism, Radio Waves, X-Rays, Microwaves, Energy, and Light.

https://philosophy-of-science.com/The-Essential-David-Bohm

https://epdf.tips/the-essential-david-bohm.html

Until I came along, *The Essential David Bohm*, was the best Philosophy of Science book on this planet. This is one of the most significant, interesting, useful, and truthful books on physics and science that has ever been made.

During his lifetime, David Bohm's ideas and theories were rejected, suppressed, ridiculed, and hidden by the Materialists, Naturalists, Darwinists, and Atheists. David Bohm has been vindicated by Science, the Scientific Methods, and Science Experiments; whereas, Materialism, Naturalism, Darwinism, Nihilism, and Atheism have been soundly defeated and falsified.

Mark My Words

## 7. Scientific Proof that Macroevolution Could Never Have Done the Job of Design and Creation:

Rana, F. (2008). *The Cell's Design: How Chemistry Reveals the Creator's Artistry*. Grand Rapids, MI: Baker Books.

Macro-Evolution is defined as Spontaneous Generation, Abiogenesis, Creation by Physical Matter, Something from Nothing, Creation Ex Nihilo, Pseudo-Science, Science Fiction, and Creation by Rocks. Whenever the Evolutionists and Darwinists tell you that evolution is true or that evolution is a proven science, they are surreptitiously talking about Macro-Evolution; and, they are making a claim that is clearly and obviously FALSE – a claim that has in fact been FALSIFIED by Science itself.

In 1859, using scientific methods to do so, Louis Pasteur FALSIFIED spontaneous generation, macro-evolution, abiogenesis, Darwinism, materialism, and naturalism. Ironically, this is the very same year that Charles Darwin published "On the Origin of Species", thereby introducing the theory of evolution to the world. Spontaneous generation or macro-evolution has never been observed nor replicated, which makes it pseudo-science at best and a down-right hoax and fraud at worst.

The Materialists, Naturalists, Darwinists, Nihilists, and Atheists deliberately **conflate** Micro-Evolution (which is true observed science) with Macro-Evolution (which is false and falsified science), hoping that you won't notice the ruse and hoping that you will thereby be tricked into believing in the Theory of Evolution. It's all about trickery and deception; and, it works. Millions have fallen for it. Remember, these people want to trick you and deceive you, just as they have been tricked and deceived.

Ironically, Micro-Evolution (natural selection and random mutations) cannot design and create anything, either, even though Micro-Evolution is an observed true science. The Materialists, Naturalists, Darwinists, Nihilists, and Atheists deliberately **confound** Micro-Evolution with Design and Creation in the hope that they will be able to trick you into believing that Evolution is your designer and creator. These people have been very successful in doing so, with upwards of half the people believing in evolution in some areas of the world. Those who have been tricked and deceived truly believe that Evolution is their Designer, Creator, and God. Materialism, Naturalism, Darwinism, and the Theory of Evolution are the modern-day form of idolatry – the worship of false and falsified gods.

The whole of the Theory of Evolution is a deliberate and cunning hoax, which attempts to convince you that Evolution is your designer and creator, a deception which has been extremely popular and successful. The Theory of Evolution just might be the greatest hoax and fraud ever perpetrated against the human race. It's radically successful, because millions of people desperately want the Theory of Evolution to be true.

Interestingly enough, Evolution (genetic change), Natural Selection, and Random Mutations didn't even exist until AFTER God designed and created the genomes and the life forms in the first place; consequently, there's no way on earth that Evolution could have been your designer and creator. It's fraudulent to claim otherwise. Evolution of any type is actually a function of disease, degeneration, chaos, entropy, death, and extinction – not a source of design, order, and creation.

Many books have been written trying to expose this fraud and deception to the world. Some of these books are extremely powerful and interesting science. The following is one of them.

Rana, F. (2008). *The Cell's Design: How Chemistry Reveals the Creator's Artistry*. Grand Rapids, MI: Baker Books.

This book is a modern up-to-date tour de force in observational science and microbiology.

Rather than trying to explain how mutation/selection fails, this book goes into the observational evidence from micro-biology, making it clear and obvious that such micro-machines and nano-technology as found in any living cell could have only come from some kind of highly intelligent creator and designer, even smarter and more intelligent and experienced than human beings are. This book is one of the best in observational science.

These highly-complex, ingenious, molecular machines cannot assemble themselves from scratch, just like your car or computer couldn't have assembled and programmed itself from scratch either. Clearly, molecular machines and genomes far more complex and interdependent than cars and computers couldn't have designed and created themselves – just like your car and computer couldn't have designed and created themselves. Such things require a highly intelligent designer, programmer, engineer, and manufacturer in order for them to come into existence in the first place.

Remember, spontaneous generation, macro-evolution, materialism, naturalism, and abiogenesis were falsified by science and scientific methods in 1859 by Louis Pasteur – ironically, the very same year that Charles Darwin published "On the Origen of Species". The theory of evolution, as your designer and creator, was falsified by science the very same year that it was presented to the world. Remember, evolution (genetic change), natural selection, and random mutations did not exist until after God designed and created the genomes and the life forms in the first place.

Which came first, evolution (genetic change) or genomes? That question has only one logical answer. Clearly the genomes came first, because evolution, natural selection, and random mutations did not exist until after the genomes had been designed and created and brought into existence in the first place.

I'm not going to apologize for being very impressed by this book from Fazale Rana, because this book is complex observational science at its very best.

Honorable Mention in this Category

Denton, M. (1986). *Evolution: A Theory in Crisis*. Chevy Chase, MD: Adler & Adler.

As far as I know, this was the first book on the planet to successfully falsify Macro-Evolution, Abiogenesis, and Spontaneous Generation by using observational science or observational evidence from micro-biology in order to do so.

This was the first science book that let me know what's wrong with the Theory of Evolution or Macro-Evolution and why it fails. I didn't read this book until ten years after it was written; and even then, it was a lone witness and the only book of its type at the time; and, this book took the creative powers of natural selection and random mutations as a given proven axiomatic truth, so it wasn't completely effective at debunking the Theory of Evolution. This book was half-in and half-out. But, this science book did demonstrate that macro-evolution is problematic, even though this book did seem to continue to support the creative powers of micro-evolution or natural selection.

This book was the first book that I ever read that cast doubt on the Theory of Evolution, because the author of this book allowed himself to be skeptical of the Theory of Evolution and to question the Theory of Evolution. I still like to go back and read this book from time to time.

There are two main ways to FALSIFY Materialism, Naturalism, and Darwinism, Creation by Rocks, Creation Ex Nihilo, and the Theory of Evolution.

One way is to use science, observation, and common sense to demonstrate beyond any reasonable doubt that Evolution (genetic drift), Random Mutations, and Natural Selection couldn't have done the job that needed to be done in order to produce the first genomes and first life forms on this planet.

The other way is to use science, observation, and common sense to demonstrate beyond any reasonable doubt that Macro-evolution, Abiogenesis, and Spontaneous Generation do not exist and could never have created the genomes and the life forms in the first place. This book, *Evolution: A Theory in Crisis*, takes this second approach to FALSIFYING Darwinism, Naturalism, Materialism, and the Theory of Evolution. I believe it succeeds in doing so.

**Mathematical Modeling Proves that Abiogenesis of Proteins and Genomes Is Impossible**

Sanford, J. (2014). *Genetic Entropy* (4th ed.). Cornell University: FMS Foundation.

Some people have suggested that mathematics is the purest, most reliable, most convincing, and best form of science. In other words, mathematics is science. I have studied mathematics, statistics, and probability; so, this wouldn't be the first time that mathematics made me a believer.

I have observed that the Young-Universe Creationists (YUCs) and Young-Earth Creationists (YUCs) are deeply disturbed by the idea that our earth is millions of years old and our universe is billions of years old, because in their minds that gives evolution long enough to work its magic.

These people would be wrong to have such concerns and beliefs.

All Life on Earth Is Left-Handed Chirality

https://futurism.com/left-handed-life/

https://www.forbes.com/sites/brucedorminey/2013/06/29/lifes-left-handed-amino-acids-remain-astrobiological-head-scratcher/#53a458c53b6d

Mathematical Modeling has demonstrated that even with an ocean full of the twenty essential correct-chirality (left-handed) amino acids, it would still take trillions of trillions of years on average for that ocean full of amino acids to produce one single functional useful protein, naturally by chance; and, that's with an ocean full of the correct amino acids being created by God in the first place, as a given. The moral of the story is that self-assembly of functional proteins is impossible, which means that abiogenesis of proteins is impossible. Abiogenesis can't happen, and it never happened.

Origin: Probability of a Single Protein Forming by Chance

https://www.youtube.com/watch?v=W1_KEVaCyaA&t=261s

https://www.youtube.com/watch?v=X7VqomfivC0&t=393s

This video demonstrates that it would take $10^{164}$ attempts on average to produce one functional protein that's 150 amino acids long, by chance; and, that's with an ocean full of the right amino acids which was put together by God in the first place. Even with an ocean full of the right amino acids and God protecting that ocean for trillions of trillions of years, abiogenesis by natural means is still impossible. In mathematical terms, $10^{164}$ is synonymous with impossible. Abiogenesis of proteins is impossible, as is an ocean full of the right amino acids. It didn't happen, and it can't happen.

Furthermore, you have to have functional proteins and DNA molecules before you can design, program, engineer, and create functional genomes. Self-assembly of functional information-rich genomes is impossible. It didn't happen. It can never happen. Remember, abiogenesis is impossible. Life-forms cannot put themselves together from scratch. It would literally take an infinite number of years for a functional genome to self-assemble from scratch, which means that the process is impossible.

These scientific realities end up being Scientific Proof of God's Necessity, which effectively ends up being Scientific Proof of God's Existence. The only thing you really have to figure out is which God, or which Engineer, or which Scientist, or which Intelligent Being did the job. WE KNOW that macro-evolution, abiogenesis, and spontaneous generation could NEVER have done the job, so who did?

## 8. Fine-Tuning, Extra-Dimensionality, Transdimensionality, Non-Locality, and String Theory Point Us Indirectly to God

Ross, H. (1996). *Beyond the Cosmos: The Extra-Dimensionality of God: What Recent Discoveries in Astronomy and Physics Reveal about the Nature of God*. Colorado Springs, CO: NavPress.

This book attempts to use the extra-dimensionality within String Theory as scientific proof of God's existence, and makes the claim that God is an extra-dimensional being. I like this book a lot, even though it is pure speculation for the most part.

It's a winner.

If you choose to go with speculation and scientific inferences, then go with something that actually tries to build and construct and edify, rather than Materialism and Naturalism which tear everything down and try to convince you that strings and extra-dimensionality and non-locality do not exist.

Science 2.0 has a unique contribution to make to extra-dimensionality, string theory, and m-theory, because Science 2.0 allows ALL of the observational evidence into evidence. The truth is that millions of different people have had Near-Death Experiences (NDEs) and Out-of-Body Experiences (OBEs); and, each one of these people experienced some kind of extra-dimensional, trans-dimensional, non-physical, non-local reality or realm. NDEs and OBEs can serve as observational evidence and experiential evidence that the hypothetical extra dimensions or branes of string theory and m-theory do in fact exist and are in fact real.

Hugh Ross updated this book in 2017. The new title follows:

Ross, H. (2017). *Beyond the Cosmos: The Transdimensionality of God* (3rd ed.). Orlando, FL: Signalman.

Honorable Mention in this String Theory Category

Michio Kaku's books are an honorable mention in this category, although Kaku doesn't typically use string theory and m-theory as scientific proof of the non-local or scientific proof of God's existence.

However, I personally took Michio Kaku's books and articles, and expanded the logic and theoretical science within them so as to demonstrate God's Necessity, which ends up becoming Scientific Proof of God's Existence if you do in fact choose to accept string theory, m-theory, and extra-dimensionality as the true Reality of our existence.

If you think about everything logically and without personal bias, it all ends up pointing to God's Psyche as the only realistic and logical explanation or the BEST explanation, especially since physical matter didn't exist until AFTER the Big Bang – meaning that the Big Bang Singularity was in fact an extra-dimensional non-local transdimensional supernatural object to begin with, and the Big Bang was in fact a supernatural non-physical extra-dimensional non-local transdimensional event. There's no other logical explanation, because physical matter didn't exist yet, in principle, before the Big Bang.

Remember, string theory, m-theory, strings, branes, and the Big Bang Singularity – if they are in fact real and truly exist – are by definition in principle extra-dimensional supernatural objects and realities. If you are willing to accept strings and branes as real, then you must of necessity accept the supernatural and the non-local as real, because strings and brains and cosmic music are extra-dimensional, supernatural, and non-local in principle, meaning that they can't be observed nor detected by our physical instruments.

If you are willing to admit strings, branes, and big bang singularities into existence, then you must of necessity admit the supernatural, extra-dimensional, trans-dimensional, non-local, non-physical, and spiritual into existence as well because strings, branes, and big bang singularities are by definition in principle supernatural objects to begin with because physical matter didn't exist until after the Big Bang.

It all ends up pointing us to Non-Local Consciousness, God, and God's Psyche or the Mind of God in the end. If you accept the Big Bang Theory at face-value, you simply KNOW that there was NO physical matter before the Big Bang, which means that physical matter couldn't have designed and created our physical universe to begin with.

In contrast, the Biblical God Jesus Christ tells us that Non-Local Consciousness or God's Psyche has always existed and will always exist – without beginning or end. That's one of the prominent messages within the Bible; and, it's unavoidably necessary and true, because physical matter didn't exist until after the Big Bang or the Beginning of this Physical Universe, which means that the Big Bang Singularity was in fact a non-local supernatural trans-dimensional object like God's Psyche.

Like can manipulate and control like; and, the smaller can reside within and control the larger. That's just the reality of our existence.

**Fine-Tuning**

Ross, H. (1991). *The Fingerprint of God: Recent Scientific Discoveries Reveal the Unmistakable Identity of the Creator* (2nd ed.). Orange, CA: Promise Publishing Co.

Ross, H. (2008). *Why the Universe Is the Way It Is*. Grand Rapids, MI: Baker Books.

Ross, H. (2008). *Improbable Planet: How Earth Became Humanity's Home*. Grand Rapids, MI: Baker Books.

In many of his books, Hugh Ross discusses cosmic fine-tuning, including the fine-tuning of the cosmological constants.

The Fine-Tuning exists, is observable, and is undeniable.

When it comes to Fine-Tuning, Hugh Ross is probably the world-wide expert on the subject, having done the most research on the topic of anyone that I know.

http://www.reasons.org/explore/publications/tnrtb/read/tnrtb/2010/11/16/rtb-design-compendium-2009

http://www.reasons.org/search-results?searchQuery=fine-tuning&mode=1

Look at the references at the bottom of some of these to see what I'm talking about.

Part 1. Fine-Tuning for Life in the Universe

http://d4bge0zxg5qba.cloudfront.net/files/compendium/compendium_part1.pdf

Part 2. Fine-Tuning for Intelligent Physical Life

http://d4bge0zxg5qba.cloudfront.net/files/compendium/compendium_part2.pdf

Part 3. Probability Estimates for the Features Required by Various Life Forms

http://d4bge0zxg5qba.cloudfront.net/files/compendium/compendium_Part3_ver2.pdf

Part 4. Probability Estimates on Different Size Scales for the Features Required by Advanced Life

http://d4bge0zxg5qba.cloudfront.net/files/compendium/compendium_Part4_ver2.pdf

This is brutal. You can't make this stuff up out of thin air as the Materialists and Naturalists do whenever they claim that these things do not exist. There are thousands of these acts of fine-tuning, which demonstrate clearly and conclusively the need for a Fine Tuner.

Remember, Fine-Tuning requires a Fine-Tuner who knows what he is doing. There's no other way to get Fine-Tuning. A Fine-Tuner must exist; otherwise, there would be NO fine-tuning for us to discover and observe. The Fine-Tuning Argument is one of the most convincing Scientific Proofs of God's Existence that we have as a race.

According to common sense logic, anything that has a beginning has a Beginner or something that brought it into existence. Each particle of physical matter and each physical universe had a beginning, which means that they needed some kind of Beginner. It also means that, since they had a beginning, they can theoretically come to an end.

Likewise, any time we successfully demonstrate the presence of fine-tuning, we have in fact demonstrated the existence of some kind of Fine-Tuner. Physical constants, physical matter, and physical universes cannot fine-tune themselves. WE KNOW that this is so due to Chaos Theory, the Law of Entropy, and even the scientific falsification of such things as Spontaneous Generation, Abiogenesis, and Macro-Evolution. There's NO such thing as Creation Ex Nihilo, or Creation by Chance, or Abiogenesis, or Spontaneous Generation where physical matter is concerned. If we encounter fine-tuning here in this physical realm, we have indeed encountered evidence of some kind of Fine-Tuner. There's no other logical explanation for it, because things don't just put themselves together. Here in his physical realm, everything tends towards Chaos, Entropy, Decay, Extinction, and Disintegration.

This means that our earth and our sun and our galaxy had a Beginner or an Organizer, because they had a beginning.

A genome also represents a highly complex and detailed amount of fine-tuning and programming, because a genome is a "four-dimensional computer operating system." This means that when it comes to each genome, there had to have been a Programmer or a Fine-Tuner involved in the system, because Spontaneous Generation, Abiogenesis, and Macro-Evolution have been FALSIFIED by science, and because there is NO such thing as Creation Ex Nihilo or Creation by Chance. Therefore, each and every physical genome has to have had a Beginner, or a Fine-Tuner, or Programmer. There's no other logical explanation for their existence, because these things don't just put themselves together out of thin air. They have never been observed doing so.

Likewise, our physical body is a complex piece of hardware similar to a computer; and, WE KNOW that hardware and computers needed some kind of Designer, Creator, Engineer, and Manufacturer. Hardware and computers don't just put themselves together Ex Nihilo. It's a logic fallacy to assume that they do.

Furthermore, there was NO such thing as evolution (genetic change), natural selection, or random mutations until AFTER God designed, programmed, engineered, and created the genomes and their associated life forms in the first place. Evolution couldn't have done the job, because evolution didn't exist until after the first genome was designed and created. This is just common-sense logic. The first genome had to have had a Beginner, a Programmer, or a Creator because anything that has a beginning has to have had a Beginner.

Physical matter couldn't have designed and created and organized our physical universe as the Materialists and Naturalists claim, because physical matter didn't exist until AFTER our physical universe was designed and created in the first place. By definition in principle, the Big Bang or the Beginning of Our Physical Universe was a supernatural event, because there was NO physical matter until AFTER the Big Bang – 380,000 years after according to some estimates.

https://en.wikipedia.org/wiki/Chronology_of_the_universe

The Big Bang, if accepted as being real and true, is actually Scientific Proof of the Supernatural. Likewise, since the Big Bang had a Trigger or a Beginning, the person who triggered the Big Bang had to exist before the Big Bang and outside the Big Bang Singularity in order to trigger it or begin it. The trigger of the Big Bang or the Beginner of Our Physical Universe has to have transcended and preceded our Physical Universe. These physical things don't just create themselves from nothing as the Materialists, Naturalists, and Atheists claim. Atheism is Creation Ex Nihilo; and, there is no such thing as Creation Ex Nihilo.

It is common sense logic or a given that something like a physical universe, which had a beginning, must have had some kind of Trigger or Organizer or Beginner who existed OUTSIDE of the system, transcended the system, and existed BEFORE the system was brought into being.

The Biblical God Jesus Christ has confessed to thousands of different people that He was that Trigger, Organizer, Fine-Tuner, Beginner, and Creator where our physical universe and our physical earth are concerned. We actually have someone and know of someone who confessed to doing the job. Confessions are admissible as evidence in a court of law, unless some tricky lawyer or materialist finds a way to get them thrown out.

I don't believe in Creation Ex Nihilo, or Creation by Chance, or Creation by Nothing, or Atheism, because it has never been experienced and never been observed. I do believe that every beginning needs a Beginner, because such a truth has been observed, experienced, and verified as being necessary and real. I choose to follow the evidence on this one, and not the philosophical speculations and scientific inferences of the Materialists and Naturalists, who are simply making up things out of thin air and then presenting them as scientific fact.

Remember, *scientific inference* is a logic fallacy – an *ad hoc just-so story*, and nothing more. Scientific inferences beg the question. Remember that.

## Unifying Quantum Physics, Consciousness, M-Theory, Heaven, Neuroscience, and Transcendence

Selbie, J. (2018). *The Physics of God: Unifying Quantum Physics, Consciousness, M-Theory, Heaven, Neuroscience, and Transcendence*. Wayne, NJ: Career Press.

This book from Joseph Selbie is a last-minute winner in this Transcendental String Theory category or theme.

Unifying Quantum Physics, Consciousness, M-Theory, Heaven, Neuroscience, and Transcendence is the whole purpose of Science 2.0. I wouldn't have needed to upgrade my science to Science 2.0, if Materialism and Naturalism had been doing the job of unifying all these things into one logical whole that made sense and matched with the Empirical Evidence.

If you are one who defines science as Materialism, Naturalism, Darwinism, Behaviorism, Nihilism, and Atheism, then your understanding of Quantum Physics, Consciousness, M-Theory, Heaven, Neuroscience, Transcendence, and God will be limited to and defined by *scientific inferences*, which are logic fallacies. Materialism, Naturalism, Darwinism, Behaviorism, Nihilism, and Atheism are also *scientific inferences*, because by definition in principle there is NO observational evidence supporting them, and can never be any observational evidence supporting their major premises or hidden assumptions.

By definition in principle, there is NO physical observation and can be NO physical observation of these things – whether we are talking about M-Theory or Materialism – because these things are Transdimensional, Non-Physical, and Immaterial. Yes, Materialism and Naturalism are immaterial – they are nothing but pure philosophy or pure speculation – thoughts and ideas and scientific inferences with NO observational evidence to support them.

In order to KNOW that Quantum Non-Locality, Non-Local Consciousness, Psyche, String Theory, M-Theory, Transdimensionality, Extra-Dimensionality, the Non-Local Realm, the Spirit World, the Afterlife, Heaven, Hell, the Psyche Basis of Neuroscience, Teleportation, Transcendence, and God are real and true, one has to live and experience these things, or choose to trust someone who has. There is NO other way, because these things cannot be detected nor recorded by our physical instruments.

Scientific inferences, personal interpretation, and philosophical speculation are absolutely worthless when it comes to these types of things, especially since we have the Materialists, Naturalists, and Atheists telling us that they know for a fact that NONE of these things exist. How do they know? They can't know. They make a blind leap of faith – a scientific inference – and simply assume that these things do not exist, while at the same time making another blind leap of faith assuming that Materialism, Naturalism, Nihilism, and Atheism are somehow magically real and true. It's all wishful thinking, and nothing more. Remember, *scientific inferences* and *wishful thinking* are logic fallacies; and, these are the things upon which Materialism and Naturalism are based.

This book, *The Physics of God*, is brutal against Scientific Materialism, as is any book that finds and explores THE TRUTH.

The only weakness of this book is that he sometimes relies upon *scientific inferences* rather than observational evidence; but then considering the subject matter, so does everyone else. His scientific inferences were a whole lot more realistic, plausible, and believable than the scientific inferences that are provided to us by the Materialists and Naturalists; so in the complete absence of observations, when it comes to *scientific inferences* we are forced to fall back on the Best Explanation Rule or the Parsimony Rule of Science. I have observed that everyone has a better explanation than the one provided by the Materialists and Naturalists.

Both Scientific Materialism and String Theory rely upon *scientific inferences*, which are technically logic fallacies; so, why should we give precedence and priority and value to String Theory over Materialism and Naturalism? It's because Materialism and Naturalism are trying to tell us that nothing non-physical exists; whereas, String Theory is trying to explain things that are already KNOWN to exist. There's a big difference there, even if some people can't see it or don't want to see it.

This book, *The Physics of God*, is based upon David Bohm's interpretations of Quantum Mechanics along with Bohm's Holographic Principle. I quote:

> **Holograms**
>
> **As his work progressed, Bohm also discovered that the mathematics governing how a hologram works provided an exceptionally useful mathematical model for how the implicate order, enfolded in nonlocal two-dimensional pre-space, allows the explicate order to unfold in local three-dimensional space. Holograms are two-dimensional. Holographic projections are three-dimensional. Holographic images are stored in flat, two-dimensional, media. When light interacts with the flat, two-dimensional media, however, a three-dimensional holographic projection appears. A**

holographic projection does not simply appear to be three-dimensional, it is: If one walks around a holographic projection one sees different sides of a three-dimensional object — all generated from two-dimensional media.

Bohm's work with the math behind holography has become known in physics as the *holographic principle*. It is not a fringe theory. Bohm's equations have become widely used in many branches of physics. After Bohm's passing, the holographic principle became a major feature of string theory. Leonard Susskind, the Felix Bloch professor of theoretical physics at Stanford University and one of the fathers of string theory, fleshed out Bohm's holographic principle in the context of string theory. Gerard 't Hooft, a Dutch theoretical physicist at Utrecht University and co-winner of the Nobel Prize in physics in 1999, worked with Susskind to expand further on the holographic principle. In 1997, Juan Maldecena, professor of physics at Princeton's Institute of Advanced Studies, published a paper in which the holographic principle took center stage. By 2010, Maldecena's paper had been cited in papers by other physicists more than 7,000 times, and thus it became the most frequently cited paper in the field of high energy physics.

The holographic principle, as it is used in string theory, states that the information determining the behavior of the three-dimensional volume of space we call the universe is "pasted on" the "boundary" between our three-dimensional universe and a two-dimensional brane. Put another way, the way the universe works — from the Big Bang to the present — is the result of information existing outside the universe itself. Put yet another way, light energy, interacting with a two-dimensional hologram in a two-dimensional brane, results in the colossal three-dimensional holographic projection we call the universe.

One's first encounter with the idea that the universe is a holographic projection can engender bemused disbelief or outright skepticism. The holographic projections with which we are familiar tend to be fuzzy and insubstantial, whereas the world we know is both finely detailed and solid. The fact that the world appears solid and finely detailed, however, does not negate the possibility that it is a holographic projection.

According to holographic string theory, if we could use an imaginary magnifying glass with unlimited magnification, we would be able to see that even space is made of dots. The physical universe, in the language of quantum physics, is discontinuous. It is not a seamless, continuous whole. We can fancifully imagine that what would "show through" between the dots, if we use our magnifying glass, is nonlocal, two-dimensional pre-space.

Einstein had the conviction that God does not play dice, that reality is ordered and determinate. He hoped to find the laws for such deterministic behavior within the local physical universe — but quantum mechanics ruled out the possibility. Bohm's deeper explorations of the weird side of quantum physics, and string theory's adoption of his holographic principle, strongly suggest that the order Einstein sought not only exists but is nonlocal.

[Nonlocal means Non-Physical, Transdimensional, Spiritual, and not-located in our physical 3D space-time realm. According to the holographic principle, our whole physical universe is like the holodeck in Star Trek, but much bigger of course. Our physical 3D universe is a holographic projection of a two-dimensional brane – a membrane from M-Theory and String Theory. These membranes or branes are not located in our 3D physical space-time but are nonlocal instead being located someplace else besides here in this physical universe.]

**Perhaps the most significant implication of the holographic principle is that the universe is being continuously created. Most conceptions of the creation of the universe — whether, for example, science's Big Bang or Christianity's seven days of creation — suggest that, after an initial creative event, physical creation remains as a permanent and independent reality. The holographic principle suggests otherwise. It suggests that if the energy that is interacting with the two-dimensional hologram in pre-space were withdrawn, the three-dimensional holographic projection of the universe would cease to exist — instantly. Furthermore, it suggests that the physical universe has no independent and enduring reality, that it is wholly dependent, moment by moment, on the information and energy originating from the nonlocal, two-dimensional energy-verse. You may be surprised to learn that saints and scientists alike concur on this point.**

Joseph, Selbie. *The Physics of God* (pp. 97-99, 101). Career Press. Kindle Edition.

As usual, I just present the evidence, and then let you decide for yourself what to make of it and how to interpret it.

I just want to say that God is Light, and that God is in the Light.

The Materialists and Naturalists will tell you that God, creation, strings, branes, nonlocality, non-physicality, extra-dimensions, holographic universes, non-local consciousness, psyche, non-local information, and saints DO NOT EXIST.

But, WE KNOW from observational evidence, experiential evidence, empirical evidence, science experiments, and eye-witness evidence that the Materialists and Naturalists are wrong. ALL of the empirical evidence tells us that these people are wrong. WE KNOW that light exists, that other dimensions exist, that quantum non-locality exists, that psyche exists, that the spirit realm exists, that non-physical things like dark energy, dark matter, magnetism, gluons, bosons, gravitons, and photons exist, and that God exists. These things have either been observed and/or detected experimentally; so, WE KNOW they exist, and now we are trying to explain them scientifically as best as we can.

I have observed that if we want to find and KNOW the truth, all we really have to do is find and study the things which the Materialists and Naturalists are trying to hide from us and assure us do not exist. Alas, the Materialists and Naturalists are always wrong, or so it seems.

The best way to find and KNOW the truth is to live it and experience it for yourself, or to choose to trust someone who has.

The second-best way to find and KNOW the truth is to eliminate everything that is false and has been falsified – such as Materialism and Naturalism and their derivatives – so that ONLY the truth remains when you are done.

# 9. Using the Scientific Method to Falsify Materialism, Naturalism, Darwinism, and Atheism

Rychlak, J. F. (1981). *A Philosophy of Science for Personality Theory* (2nd ed.). Malabar, FL: Robert E. Krieger Publishing Company.

My bachelor's degree is in Psychology. Psychology, as it was originally conceived by the father of American Psychology William James, was the study of the human psyche. A decade or two later, the Behaviorists appropriated or absconded with Psychology and changed it into the science of observed behavior; and, Psychology euphemistically lost its mind during the process of becoming less offensive to the Behaviorists. Fifty years later, as a reaction to the absurdities of Behavioristic Determinism, the Humanists like Joseph Rychlak tried to restore the study of personality or the study of psyche back into the Science of Psychology. That restoration is still ongoing a hundred years after Behaviorism was first introduced and became dominant.

This book (and some of Joseph Rychlak's other books) gave me my FIRST thorough understanding of the Scientific Method and the Philosophy of Science behind the Scientific Method, letting me know in detail the strengths and the weaknesses of the Scientific Method – what method can and cannot do; and, that new understanding of science and method gave me the inspiration and the ability to use the Scientific Method to FALSIFY Materialism, Naturalism, and their derivatives such as Darwinism and Atheism. My Materialism, My Nihilism, and My Atheism were based upon ignorance and a refusal to look at evidence. But, with the KNOWLEDGE of Science, the Scientific Method, and the Philosophy of Science provided by this book, there was no going back to my ignorance; and, I was finally free and able to develop some convincing falsifications of Materialism, Naturalism, Darwinism, and Atheism by using the Scientific Method itself to do so.

One way to find the truth is to use the Scientific Methods to eliminate all the falsehoods such as Materialism, Naturalism, Darwinism, and Atheism. By now (September 2017), I have done so millions of times, in dozens of different ways. Application of the philosophical and scientific principles in this book completely changed my life and the way that I look at science and reality, for the better. I have finally found the truth. I KNOW what the Scientific Methods are good for, and where they completely fail.

In many of my books, I have successfully used the Scientific Method to falsify Materialism, Naturalism, Darwinism, and Atheism. It's easy to do if you understand the Philosophy of Science behind the Scientific Method.

When it comes time to falsify Nihilism, we turn to observational evidence, eye-witness evidence, and the Lived Experiences of the human race in order to do so. When it comes to science, you use the best tool or the best method for the job at hand. That's the way it should be.

Honorable Mention in this Philosophy of Science Category

For a good and honest introduction to the Philosophy of Science, especially in relationship to science and human behavior and psychology, see also:

Slife, B. D. & Williams, R. N. (1995). Science and Human Behavior. In *What's Behind the Research? Discovering Hidden Assumptions in the Behavioral Sciences*, (pp. 167–204). Thousand Oaks, CA: SAGE Publications.

http://mypsyche.us/wp-content/uploads/2017/04/Science.pdf

They were handing this one out in class, so they want people to find it and read it. It was worth my time to do so.

Remember, the ultimate goal in all of this is to find and know the truth. If it's a lie, then ultimately it has no value to us.

Observational evidence, experiential evidence, eye-witness evidence, and the Lived Experiences of the human race should be given precedence over experimental evidence.

Why?

It's because experimental evidence can be interpreted incorrectly, and typically is. For every piece of evidence that we might ever encounter, there are theoretically an infinite number of possible different interpretations which can be given to that single piece of evidence. God knows which one of them is true!

The best way to find and know the truth is to live it and experience it for yourself, or to choose to trust someone who has. I have learned that when it comes to experimental evidence, scientific evidence, observational evidence, experiential evidence, and empirical evidence, I INTERPRET IT FOR MYSELF, rather than relying upon the "official interpretation" which is provided by the Materialists, Naturalists, Darwinists, Nihilists, and Atheists because these people have a knack for always providing a false interpretation to every piece of evidence they encounter. Personal interpretation of the evidence IS the primary flaw of the scientific methods, because personal interpretation of the evidence is based upon a bunch of different logic fallacies such as *affirming the consequent, begging the question, jumping to conclusions, scientific inferences, circumstantial evidence, pre-conceived conclusions, evidence manufactured to fit one's pre-conceived conclusions, ad hoc just-so story-telling,* and *confirmation bias*. A study of the Philosophy of Science can help us learn to recognize and identify these different logic fallacies whenever we come across them.

<u>Honorable Mention in this Philosophy of Science Category</u>

A Paradigm Shift represents a fundamental foundational change in one's Philosophy of Science and becomes a new and different (and hopefully) better way of doing science. For me personally, Science 2.0 ended up being a new and better way of doing science – a paradigm shift in my own way of thinking and doing science.

In his ground-breaking book, *The Structure of Scientific Revolutions*, Thomas Kuhn quoted Max Planck:

> **A new scientific truth does not triumph by convincing its opponents and making them see the light, but rather because its opponents eventually die, and a new generation grows up that is familiar with it.**

Materialism, Naturalism, Darwinism, Nihilism, and Atheism are at an end; but, they have been "at an end" for a few decades now because we are still waiting for these people to die. I just about died, which successfully ended My Materialism, My Nihilism, and My Atheism. These false philosophies have been falsified by Observational Evidence and Experiential Evidence, but the Materialists and Naturalists haven't gotten the memo yet; and, they probably never will, because they aren't looking for it and don't want it.

As far back as 1985, Stanislav Grof announced a paradigm shift in many of his books. It never took. Why? It's because the Materialists, Naturalists, Darwinists, Nihilists, and Atheists didn't want it to take.

Through the 1970's, 1980's, and 1990's, Joseph Rychlak similarly pointed us to the need for a new and better Paradigm and Philosophy of Science, within his many different books. Obviously, it never took.

Charles Tart made a similar announcement in 2009 with his book, *The End of Materialism*, a book that was over fifty years in the making. It never stuck, because the Materialists and Naturalists don't want it to stick.

Here I am in 2017 announcing a new paradigm, which I call Science 2.0. It will never take, at least not during my lifetime. Why? It's because the Materialists and Naturalists won't let it take. They won't want it to be the new paradigm, because it demonstrates and proves that they are wrong. As Thomas Kuhn so adequately observes, Materialists and Naturalists will resist and fight something like Science 2.0 to their very death; whereupon, they will discover to their surprise that they were wrong after all, as so many of us have.

It is what it is.

**Personality Theory – The Philosophy of Psychology**

Your personality is your Psyche or Non-Local Consciousness.

How do we know?

WE KNOW from the Empirical Evidence provided by Near-Death Experiences (NDEs), Shared-Death Experiences (SDEs), Out-of-Body Experiences (OBEs), Seeing Dead People, and other types of Non-Local Experiences that your personality, psyche, individuality, or identity survives the death of your physical body and physical brain.

Personality Theory IS the philosophy behind Psychology.

My college course in Personality Theory was the most fascinating and informative class in Psychology that I have taken so far. It all ends up pointing to Psyche or Non-Local Consciousness.

I pursue my notes and ideas from this course in some of my other books, so I won't do so here. However, I found the concepts fruitful and interesting.

Books that I have written explaining in part what I learned from my course in Personality Theory:

7. *NATURE vs. NURTURE vs. NIRVANA: An Introduction to Reality*

https://www.amazon.com/dp/B01JWRCSVA

https://www.amazon.com/dp/1521132615

9. *The Ultimate Model of Reality: Psyche Is the Ultimate Cause*

https://www.amazon.com/dp/B071NC9JK6

10. *Putting Psyche Back into Psychology: Restoring Science to Consciousness*

https://www.amazon.com/dp/B071NC987S

11. *BioPsychoSocial: Including Psyche or Light into our Theoretical Models*

https://www.amazon.com/dp/B0713NDHVW

Other books about Personality Theory:

Engler, B. (2009). *Personality Theories: An Introduction* (8th ed.). Boston, MA: Houghton Mifflin Harcourt Publishing Company.

Boeree, C. G. (2006). *Personality Theories*. Psychology Department: Shippensburg University. (Retrieved from http://webspace.ship.edu/cgboer/perschapterspdf.html).

## Honorable Mention in this Philosophy of Science and Personality Theory Category

Studying Personality Theory and Natural Theology opens the door to Psyche or Non-Local Consciousness. Personality Theory is the study of the human psyche or the human personality. Natural Theology is an attempt to use Nature in order to develop Scientific Proof of God's Existence.

The Materialists, Darwinists, and Naturalists told me that there can never be any scientific proof of God's existence, because science proves that God does not exist. They were wrong; but, I didn't know that at the time. There was a time when I used to believe them. I don't anymore. The Empirical Evidence proved to me that they are wrong. Nature as a whole and Science as a whole ARE powerful proof of God's existence. Your very existence here, right now, reading this sentence is powerful proof of God's existence. Your genome is God's signature. God has signed each and every one of your cells.

https://en.wikipedia.org/wiki/Natural_theology

https://en.wikipedia.org/wiki/Natural_Theology_or_Evidences_of_the_Existence_and_Attributes_of_the_Deity

The following books deal with the Philosophy of Science, Personality Theory, and Natural Theology:

Engler, B. (2009). *Personality Theories: An Introduction* (8th ed.). Boston, MA: Houghton Mifflin Harcourt Publishing Company.

Boeree, C. G. (2006). *Personality Theories*. Psychology Department: Shippensburg University. (Retrieved from http://webspace.ship.edu/cgboer/perschapterspdf.html).

Beauregard, M., & O'Leary, D. (2007). *The Spiritual Brain: A Neuroscientist's Case for the Existence of the Soul*. New York: HarperCollins.

Beauregard, M. (2013). *Brain Wars: The Scientific Battle over the Existence of the Mind and the Proof that Will Change the Way We Live Our Lives*. New York: HarperOne.

Buhlman, W. L. (1996). *Adventures Beyond the Body: How to Experience Out-of-Body Travel*. New York: HarperCollins Publishing.

Callister, T. R. (2006). *The Inevitable Apostasy and the Promised Restoration*. Salt Lake City, UT: Deseret Book Company.

Carter, C. (2012). *Science and Psychic Phenomena: The Fall of the House of Skeptics*. Rochester, VT: Inner Traditions.

Carter, C. (2010). *Science and the Near-Death Experience: How Consciousness Survives Death*. Rochester, VT: Inner Traditions.

Carter, C. (2012). *Science and the Afterlife Experience: Evidence for the Immortality of Consciousness*. Rochester, VT: Inner Traditions.

Craig, W. L. (2014). *Does God Exist?* Pine Mountain, GA: Impact 360 Institute.

Feser, E. (2007). *Locke*. London, England: Oneworld Publications.

Feser, E. (2008). *The Last Superstition: A Refutation of the New Atheism*. South Bend, IN: St. Augustine's Press.

Feser, E. (2009). *Aquinas (A Beginner's Guide)*. Oxford, England: Oneworld Publications.

Feser, E. (2014). *Scholastic Metaphysics: A Contemporary Introduction*. Piscataway, NJ: Editions Scholasticae.

Geisler, N., & Turek, F. (2004). *I Don't Have Enough Faith to Be an Atheist*. Wheaton, IL: Crossway Books.

James, W. (1912). *Essays in Radical Empiricism*. London: Longmans, Green, and Co.

James, W. (1902). *Varieties of Religious Experience: A Study in Human Nature*. Harvard University.

Lennox, J. C. (2009). *God's Undertaker: Has Science Buried God?* Oxford, England: A Lion Book.

Lennox, J. C. (2011). *God and Stephen Hawking: Whose Design Is It Anyway?* Oxford, England: A Lion Book.

Lennox, J. C. (2011). *Gunning for God: Why the New Atheists Are Missing the Target*. Oxford, England: A Lion Book.

Richards, P. S., & Bergin, A. E. (2004). *Casebook for a Spiritual Strategy in Counseling and Psychotherapy*. Washington DC: American Psychological Association.

Richards, P. S., & Bergin, A. E. (2005). *A Spiritual Strategy for Counseling and Psychotherapy* (2nd ed.). Washington DC: American Psychological Association.

Rychlak, J. F. (1979). *Discovering Free Will and Personal Responsibility*. New York: Oxford University Press.

Rychlak, J. F. (1981a). *A Philosophy of Science for Personality Theory* (2nd ed.). Malabar, FL: Robert E. Krieger Publishing Company.

Rychlak, J. F. (1981b). *Introduction to Personality and Psychotherapy: A Theory-Construction Approach* (2nd ed.). Boston, MA: Houghton Mifflin Company.

Rychlak, J. F. (1982). *Personality and Life-Style of Young Male Managers: A Logical Learning Theory Analysis*. New York: Academic Press.

Rychlak, J. F. (1988). *The Psychology of Rigorous Humanism* (2nd ed.). New York: New York University Press.

Rychlak, J. F. (1991). *Artificial Intelligence and Human Reason: A Teleological Critique*. New York: Colombia University Press.

Rychlak, J. F. (1994). *Logical Learning Theory: A Human Teleology and Its Empirical Support*. Lincoln, NE: Nebraska University Press.

Rychlak, J. F. (1997). *In Defense of Human Consciousness*. Washington DC: American Psychological Association.

Rychlak, J. F. (2003). *The Human Image in Postmodern America*. Washington DC: American Psychological Association.

Slife, B. D. (1993). *Time and Psychological Explanation*. Albany, New York: SUNY Press.

Slife, B. D. (2012). *Taking Sides: Clashing Views on Psychological Issues* (17th ed.). New York: McGraw-Hill.

Slife, B. D., & Hopkins, R. H. (2005). Alternative Assumptions for Neuroscience: Formulating a True Monism. In B.D. Slife, J. Reber, & F. Richardson (Eds.), *Critical Thinking about Psychology: Hidden Assumptions and Plausible Alternatives*, (pp. 121-147). Washington, D.C.: American Psychological Association Press.

Slife, B. D., & Williams, R. N. (1995). *What's Behind the Research? Discovering Hidden Assumptions in the Behavioral Sciences*. Thousand Oaks, CA. Sage Publications.

Slife, B. D. & Williams, R. N. (1995). Science and Human Behavior. In *What's Behind the Research? Discovering Hidden Assumptions in the Behavioral Sciences*, (pp. 167-204). Thousand Oaks, CA: SAGE Publications.

Slife, B. D., Reber, J., & Richardson, F. (Eds.). (2005). *Critical Thinking about Psychology: Hidden Assumptions and Plausible Alternatives*. Washington, D.C.: American Psychological Association Press.

Slife, B. D., Williams, R. N., & Barlow, S. H. (Eds.). (2001). *Critical Issues in Psychotherapy: Translating New Ideas into Practice*. Thousand Oaks, CA: Sage Publications.

Tart, C. T. (2009). *The End of Materialism: How Evidence of the Paranormal Is Bringing Science and Spirit Together*. Oakland, CA: New Harbinger Publications.

Turek, F. (2014). *Stealing from God: Why Atheists Need God to Make Their Case*. Colorado Springs, CO: NavPress.

This is huge!

You have to understand the Philosophy of Science as well as what's wrong with the Scientific Method, before you can attempt to fix it and repair it.

I discuss the Philosophy of Science, Personality Theory, and Natural Theology in many of my other books, so I won't do so here. Judging by the size of this list, you might understand why I chose to take up this subject in many other books which I have written.

## 10. The Unexplainable and Unexplained Aspects of the Big Bang Theory Point Us Directly to God, Because He Is the Only One Who Could Have Done All the Science that Needed to Be Done at that Point in Time:

Hartnett, J. & Williams, A. (2005). *Dismantling the Big Bang: God's Universe Rediscovered*. Green Forest, AR: Master Books.

Every scientific theory and philosophical idea needs a dedicated opposition to give it some teeth and staying power – Science 2.0 is no exception.

Since Materialism, Naturalism, Darwinism, Nihilism, and Atheism have been FALSIFIED trillions of times in thousands of different ways, they really cannot serve as the opposition which Science 2.0 needs, because they have NO observational evidence supporting them. Science 2.0, which is based exclusively on the observational evidence and lived experiences of the human race, can adequately serve as a replacement for Materialism, Naturalism, Darwinism, Nihilism, and Atheism because Science 2.0 automatically subsumes ALL physical observation or physical evidence that has ever been produced; but, Science 2.0 can't actually derive "inspiration" or "opposition" from the falsified philosophies of Materialism and Naturalism because they have been FALSIFIED already.

Instead, Science 2.0 repeatedly turns to Young-Universe Creationism (YUC) and Young-Earth Creationism (YEC) for the dedicated and cunning opposition that it needs to get established. The YUCs and the YECs are extremely intelligent; and, they actually teach science that can be debunked or falsified with observational evidence, which is exactly the purpose of Science 2.0 – getting rid of falsehoods through observational evidence and the lived experiences of the human race. Over and over again, Science 2.0 turns to the YUCs and the YECs in order to get the opposition that it needs.

Consequently, my tenth book and my tenth slot has been given to my dedicated and highly intelligent opposition, the YUCs and the YECs. *Dismantling the Big Bang* just might be the best book that the YUCs and the YECs have produced so far; or at least, certain chapters of it are.

Remember, evolution (genetic change), natural selection, random mutations, and physical matter couldn't have designed and created and produced the Big Bang Singularity, because these things didn't exist until long after the Big Bang. The Big Bang Singularity was a non-physical supernatural object; and therefore, the Big Bang was in principle a non-physical supernatural event.

When it comes to science, it's hard to pick a top-ten list; but, I have my reasons for listing this book as one of them. This book, *Dismantling the Big Bang*, made me stop and think. It's been a long time since any book has done that to me. This book forced me to think about the Beginning of our Physical Universe, or the Prime Event.

I started to lose my faith in Materialism and the creative powers of Natural Selection back in 2014. I'm a decade or two behind everyone else, I know; but, we each have to start where we are and go on from there.

However, my faith in, belief in, and promotion of the Big Bang Theory continued until the summer of 2017, when I finally got to the point in my science, study, and knowledge where I could actually see what's wrong with the Big Bang Theory and what's weak about it. This book helped a lot in that regard. This ended up being another life-changing book; and at first, I didn't think it would be.

This book is produced by Young-Universe Creationists (YUCs) and Young-Earth Creationists (YECs). Technically, I am not a creationist; but, that doesn't mean that I can't learn something new from the creationists. I'm a scientist; and, I truly believe that the Gods are scientists. I love to study the YUCs and YECs because they always dig up science that I have never heard of nor thought about before. These people are some of the very best scientists on the planet, because they are extremely skeptical and critical of everything else. I simply love studying them, learning from them, and then trying to debunk them. It's what I do.

No science book is completely perfect, especially when it comes to the Origin Sciences or the Historical Sciences. The main flaw of this book is their repeated and stubborn insistence that our earth is only 6,000 years old (a needless and worthless contention in my humble opinion); but, periodically throughout this book they do concede that our universe **appears** to be billions of years old, as it truly is.

Their debunking of the Big Bang Theory is some of the best and most interesting science that I have ever seen or encountered in my life. I was highly impressed with a few of their chapters where they dismantle the Big Bang Theory and examine it skeptically and critically. I love thinking critically about science!

Alas, the dozen or so alternative models that they present within this book as replacements for the Big Bang Theory are all extremely weak and unimpressive, because the creators of those models don't seem to understand the theory of relativity in the least; consequently, even the most plausible alternative models presented in this book based upon time dilation don't make any sense whatsoever if you actually understand how the theory of relativity works in principle.

Moving clocks are measured to tick more slowly than an observer's "stationary" clock. Remember, as physical objects approach the speed of light, time slows down for those physical objects; and, at the speed of light time stops and there is NO passage of time and thus nothing to experience and remember. Remember, clocks also run slower in deeper gravitational wells such as black holes. Within a black hole the passage of time literally stops, meaning that the contents of a black hole do not age. If time stops for a physical object, then no aging takes place because no passage of time takes place. This is what the theory of relativity is trying to teach us.

https://en.wikipedia.org/wiki/Theory_of_relativity

Within this book, *Dismantling The Big Bang*, the alternative models based upon time dilation actually turn the theory of relativity on its head – the ONLY way that they could possibly work as explained is if the earth was traveling at nearly the speed of light or in orbit of a black hole during the expansion of our universe, so that only six 24-hour solar days pass by on our earth while 13.7 billion years pass by for the rest of the expanding universe.

I have an extremely hard time visualizing these alternative earth-centric models, because they don't make any logical sense to me, especially where the speed of light is involved. These people don't seem to realize that their models are flawed, because they never once suggest that the earth had to have been traveling near the speed of light while our universe was expanding, so that our present earth ends up being only 6,000 years old as a result while our universe ends up being at least 13.7 billion years old as it **appears** to be. With the YUCs and the YECs, it's all about appearances.

Remember, the YUCs and YECs produce earth-centric or earth-centered models, because they believe that our physical universe and physical earth started or were brought into existence at the same time and that our earth is the only inhabited planet in the universe at the very center of our universe. I just can't visualize how our earth could be traveling at the speed of light while at the same time our physical universe expands much more slowly in every direction around our earth while remaining centered upon our earth. It's a paradox that I can't wrap my mind around, because it doesn't make any sense to me.

It would make a lot more sense to believe that our earth was in orbit of a black hole while our universe expanded around it; but, that isn't the way that they present their models. Where is that black hole now, and how did it become our sun?

These people do have some non-typical models based upon a white hole near our earth during the creation of this physical universe, which didn't quite make sense to me at first because we haven't studied white holes all that much, and thus I couldn't initially sense whether such a solution would be plausible or not.

Black holes have been detected and "observed" in great abundance, whereas white holes have not. There's only one white hole candidate that I'm aware of.

https://en.wikipedia.org/wiki/GRB_060614

With further study, I eventually concluded that a white hole would indeed be a plausible definition for the original Primal Singularity or Big Bang Singularity, even though the authors of these white hole models oppose the Big Bang Theory in principle.

**In general relativity, a white hole is a hypothetical region of spacetime which cannot be entered from the outside, although matter and light can escape from it. In this sense, it is the reverse of a black hole, which can only be entered from the outside and from which matter and light cannot escape.**

https://en.wikipedia.org/wiki/White_hole

Some of the alternative models in this book, *Dismantling the Big Bang*, are even crazy enough to suggest that our universe is only 6,000 years old, even though all the observational evidence and measurable evidence tells us that it is not. I, too, am an astronomer and cosmologist; and, I have spent hours looking through telescopes and truly believe that the speed of light is telling us the truth – namely that our universe is in fact at least 13.7 billion years old. God can design and create a 13.7 billion-year-old universe just as easily as a 6,000 year-old universe; but, a 13.7 billion-year-old universe would have given the Gods the opportunity to teach us, their spirit children, how everything works through the process along the way, so that someday we too can do the same thing for ourselves if we prove worthy.

I personally believe that the Gods deliberately slow everything down, so that we can participate in the process, live the experience, learn from the experience, and remember having done so. Without the passage of time, there is NO experience and therefore no memories to be had. A photon traveling at the speed of light experiences NO passage of time and never ages; consequently, from its perspective it simply teleports to its destination instantaneously even though from our much slower perspective it took 13.7 billion years for that same photon to get to us. The Gods deliberately slowed things down, including the creation of this physical universe, which is still ongoing, so that we their spirit children can learn from the process, experience it, live it, and then remember having done so.

If the Gods had simply snapped their fingers and brought everything into existence instantaneously 6,000 years ago as the Young-Earth Creationists (YECs) and Young-Universe Creationists (YUCs) repeatedly suggest, then we wouldn't have been able to participate in the process nor would we have learned anything from it because our spirits would have had NO memories of it.

Remember, as physical objects approach the speed of light, time slows down for those physical objects; and, at the speed of light time stops and there is NO passage of time and thus nothing to experience and remember.

Remember, if the Gods had done everything at the speed of light or faster, then nobody would have experienced it and there would have been nothing to remember because there would have been NO passage of time.

Repeatedly, the Biblical God suggests that He created this physical universe and physical earth to be a school for us. The Gods would have failed in their purpose and design if they had done everything at the speed of light, because there would have been NO passage of time and therefore nothing to experience, remember, or learn from. It would have truly been creation ex nihilo, which is in fact Atheism – creation from nothing by nothing – because nothing would have experienced it or remembered it. Remember, creation ex nihilo is another Satanic deception designed to mess us up as much as materialism and naturalism were designed to mess us up.

When it comes to the vast majority of us, the spirits within us instinctively KNOW and REMEMBER that this physical universe is much older than 6,000 years. Many of us can sense by the Spirit that this earth is much older than 6,000 years. Follow the Spirit, because it knows the truth. Furthermore, the ice cores we have dug up so far suggest that our earth is at least conservatively a million years old.

> **An ice core is a core sample that is typically removed from an ice sheet or a high mountain glacier. Since the ice forms from the incremental buildup of annual layers of snow, lower layers are older than upper, and an ice core contains ice formed over a range of years. Cores are drilled with hand augers (for shallow holes) or powered drills; they can reach depths of over two miles and contain ice up to 800,000 years old. The Dome C core had very low accumulation rates, which mean that the climate record extended a long way; by the end of the project the usable data extended to 800,000 years ago. There are plans to retrieve ice cores that reach back over 1.2 million years, in order to obtain multiple iterations of ice core record for the 40,000-year long climate cycles known to have operated at that time. Current cores reach back over 800,000 years and show 100,000-year cycles. The oldest core was found to include ice from 2.7 million years ago — by far the oldest ice yet dated from a core.**
>
> https://en.wikipedia.org/wiki/Ice_core

I always give precedence to observational evidence, and I choose to take that observational evidence or experiential evidence at face-value. I do the same exact thing in relationship to the observational evidence obtained from Near-Death Experiences (NDEs) and Out-of-Body Experiences (OBEs). I treat ALL of the observational evidence as evidence, and I make no attempt to try to explain it away.

I have upgraded my science to Science 2.0; and, in Science 2.0 ALL of the evidence is admitted into evidence.

Since science was first created, it has been defined as Materialism, Naturalism, Darwinism, Nihilism, and Atheism. Materialism and Naturalism and Atheism are based upon a one-sided skepticism. These people doubt everything, except for Materialism and Naturalism. They are closed-minded when it comes to Materialism, Naturalism, Darwinism, Nihilism, and Atheism – applying no critical thinking whatsoever to these particular philosophical ideas.

Under Science 2.0, science has been upgraded and is now defined as observation and experience, including all of our non-local experiences or out-of-body experiences as well as our physical experiences. Science IS Observation. Science consists of the lived experiences of the human race, including our non-local experiences or spiritual experiences. Science 2.0 gives preference to observation and experience rather than the philosophical speculation of "scientific inferences" which the materialists and naturalists rely upon. With scientific inferences, you can have any conclusion or result that you want, because there is NO observation evidence to contradict your chosen conclusions and assumptions. I call this the *scientific inference logic fallacy* – having no observational evidence to confirm nor falsify one's chosen beliefs.

Remember, there is NO observational evidence supporting the primary assumptions or major premises of Materialism, Naturalism, Nihilism, and Atheism; and, there never will be. The whole thing is based upon scientific inferences, and it ends up being a direct application of the *scientific inference logic fallacy* because there's NO observational evidence to support it.

Remember, Young-Universe Creationism (YUC) and Young-Earth Creationism (YEC) are Satanic deceptions, which Satan successfully uses to drive people away from Christianity, because practically everyone including the Materialists, Naturalists, and Atheists simply KNOW by the Spirit and common sense within them that this universe is much older than 6,000 years and so is this earth. Remember, there is NO such thing as creation ex nihilo, which is another Satanic concept which the YUCs and YECs and Catholics try to promote. Remember, Creation Ex Nihilo is the same thing as Atheism – Creation by Nothing from Nothing. Creation Ex Nihilo is another Satanic deception, so don't fall for it, because it doesn't make any logical sense and it's atheistic.

When it comes to the creation of this physical universe and this physical earth, the Gods deliberately slowed everything down so that we can experience it, live it, learn from it, and then remember having done so. Our universe is at least 13.7 billion years old, and our earth is at least a million years old, just as the observational evidence suggests. In other words, our universe and our earth are just as old as they appear to be, because the Gods are trying to teach us, not trying to deceive us.

From these logical truths and observational realities, you simply KNOW that any YUC and YEC interpretation of the Bible is in fact a faulty interpretation of the Bible, based upon blind dogmatic faith and desire rather than any actual observational evidence or workable science. The Biblical Model presented in this book, *Dismantling the Big Bang*, is based exclusively on YUC and YEC which makes it fundamentally flawed and no better than the Big Bang Theory, because it violates the observational evidence.

I always give precedence to observation and experience, because Science is Observation and Experience. Real Science is based upon eye-witness accounts.

Remember, circumstantial evidence is evidence that relies on an inference to connect it to a conclusion of fact — like a fingerprint at the scene of a crime, or a YUC and YEC Model for Origins. Every scientific inference is based upon circumstantial evidence – these people are simply guessing; and, this reality applies just as much to the YUCs and the YECs as it does to the Materialists, Naturalists, Darwinists, Nihilists, and Atheists. By contrast, direct evidence supports the truth of an assertion directly, without need for any additional evidence or inference. Remember, scientific inference of any kind is a logic fallacy and circumstantial evidence, because it lacks observational support. Truth is knowledge, because it's based upon direct observational evidence or an eye-witness account. The best source of truth and knowledge is direct observational evidence.

https://en.wikipedia.org/wiki/Circumstantial_evidence

http://www.ldsendowment.org/creation.html

Who are the Gods, the Elohim?

Psalm 82: 6:

**6 I have said, Ye are gods; and all of you are children of the most High.**

Who are the Gods?

John 10: 33-36:

**32 Jesus answered them, Many good works have I shewed you from my Father; for which of those works do ye stone me?**

**33 The Jews answered him, saying, For a good work we stone thee not; but for blasphemy; and because that thou, being a man, makest thyself God.**

**34 Jesus answered them, Is it not written in your law, I said, Ye are gods?**

**35 If he called them gods, unto whom the word of God came, and the scripture cannot be broken;**

**36 Say ye of him, whom the Father hath sanctified, and sent into the world, Thou blasphemest; because I said, I am the Son of God?**

**37 If I do not the works of my Father, believe me not.**

**38 But if I do, though ye believe not me, believe the works: that ye may know, and believe, that the Father is in me, and I in him.**

**39 Therefore they sought again to take him: but he escaped out of their hand.**

Who are the Gods?

Luke 3: 38:

**38 Which was the son of Enos, which was the son of Seth, which was the son of Adam, which was the son of God.**

Both Adam's spirit body and Adam's physical body were literally the son of God, descended from the Gods. Adam and Eve's surname was God.

Who are the Gods? Who are the Elohim?

WE ARE!

If you can read this, then you are a child of God; and, both your spirit body and your physical body are descended from the Gods. Your surname is God. That means that YOU participated in the creation of the heavens and the earth, to one extent or another, because you are part of the Elohim. This I KNOW, because the Bible tells me so. I like to think that I helped to create the turtles, frogs, and dogs.

This whole process of creation was slowed down so that we could participate in it, live it, experience it, process it, learn from it, and eventually remember it when our memories of our pre-mortal life are finally restored to us after the resurrection from the dead; and furthermore, the Creation is still ongoing. It's not a done deal as the YUCs and the YECs repeatedly suggest.

The Gods are scientists, not magicians! The Gods labor. The Gods work their creation, not just simply blink it into existence from nothing. We work the creation so that we can learn from it, experience it, and eventually remember it – that's the very definition of life. If there's no passage of time, then there's really no life, because there are no experiences to remember. That's one of the main reasons why the Biblical God introduced time into His physical creation. God slowed everything down so that we could live it, and experience it, participate in it, and remember it – including the creation of this physical universe and the creation of this physical earth.

As usual, you will have to decide for yourself whether I know what I'm talking about or not. That's something that only you can do for yourself. It's not anything that I can do for you.

I included this book, *Dismantling the Big Bang*, in my top-ten list because their critique of the Big Bang Theory is some of the best and most interesting science that I have ever encountered in my life. I included this book in my top-ten list because it made me stop and think, as this essay attests.

However, I put this book last on my list because the alternative models for the origin of this universe presented in this book are fundamentally flawed – with maybe the exception of the White Hole Theory being equivalent to the Big Bang Singularity, so long as we allow for the possibility that the matter that went into the creation of our earth could be as old as 13.7 billion years now and allow for the possibility that our earth is now 13.7 billion light years away from that white hole or primal singularity, which the YUCs and YECs don't allow. In my humble opinion, these people should have been just as skeptical and critical of their alternative models as they were of the Big Bang Theory, but they were not.

As human beings, we are always blind to the faults within our own personal theories and own personal lives. We can easily see what's wrong with someone else and their ideas; but, we are usually blind to what's wrong with us, unless we have the Spirit within us telling us what's wrong with us.

I have no idea what else I am missing because I'm naturally blind to it; but, I do get the impression that I have thought about these things and studied these things more than some of these others have.

Only the YUCs and the YECs are going to introduce something as bizarre as a white hole into their Model of Creation and Model of Origins; yet ironically, I have come to see a white hole as the perfect candidate and the perfect definition for the Big Bang Singularity. If the Big Bang Singularity did in fact exist, it was most likely a white hole of some kind. How interesting is that? God would have had to have loaded the thing up on the other side, and then triggered it causing our Physical Universe to come into existence as a result.

This book spurred in into the study of lots of related topics on origins, creation, purpose, and design. I'm not done with this book, *Dismantling The Big Bang*, because their critique of the Big Bang Theory pointed me directly to God's Necessity, which ended up being Scientific Proof of God's Existence based upon the Big Bang Theory itself – a topic which I intend to take up in greater detail in another essay.

### Honorable Mention in this Category

Only God could have done the design and creation of this physical universe, physical matter, physical genomes, and physical life forms. Evolution, random mutations, and natural selection couldn't have done the job, because they didn't exist yet. That's the "bootstrap problem" or the "chicken and egg problem" because something that doesn't exist yet – such as evolution or natural selection or random mutations – can't possibly be the designer and the creator of physical genomes and physical life forms. Evolution, natural selection, and random mutations did NOT exist until AFTER God brought the genomes and the life forms into existence in the first place.

ICR. Institute for Creation Research. (2013). *Creation Basics & Beyond: An In-Depth Look at Science, Origins, and Evolution*. Dallas, TX: Institute for Creation Research.

This is another book from the Young-Earth Creationists (YECs) and the Young-Universe Creationists (YUCs). I love to study these guys because they always introduce me to science that I have never heard of before nor ever thought about before. This book is no exception.

This is the best, most comprehensive, and in-depth introduction to YUC and YEC as a whole that I have encountered so far. Science 2.0 admits ALL of the evidence into evidence. I study the YUCs, YECs, Materialists, and Naturalists in order to know what they believe and why they believe it, so that I can figure out what's wrong with it, if anything.

I like these people. At least they try to use Science to support and explain their Biblical interpretations. This book is weaker than some of the other books on this list due to their stubborn insistence that this earth is only 6,000 years old, a claim which is scientifically and observationally unsustainable; but, this book is chock full of science, and this book represents the best that the Young-Earth Creationists (YECs) have been able to produce so far, in my humble opinion. It's fascinating to observe what they get right and what they get wrong, while they try to use science to support their personally chosen interpretation of Biblical passages such as the first chapter of Genesis.

The main advantage of the Young-Earth Creationists (YECs) and the Young-Universe Creationists (YUCs) is that they successfully and quickly point us to the unresolved controversies in science, which are some of the most interesting aspects of science to study and learn about.

Thousands of different scientists have falsified the Theory of Evolution in millions of different ways; and, this group is no exception. The Theory of Evolution is indeed falsifiable, and it has indeed been proven false in more ways than any one of us can possibly discover in our lifetime.

Materialism, Naturalism, Darwinism, and Atheism have been FALSIFIED by Science, Scientific Observations, Lived Experiences, Non-Local Experiences, and the Scientific Methods trillions of times in thousands of different ways; but, there are still millions of Materialists and Atheists who haven't gotten the memo yet. That's old news by now (2017). Like I said, I'm decades behind the others when it comes to this discovery and this scientific reality; but, I finally got there in the end. Now I finally have some interesting things to contribute.

However, when it comes to the Origin Sciences, or Forensic Sciences, or the Historical Sciences, there are still unresolved controversies that grab our attention and can be extremely fascinating to study, and read, and learn about. This book points us to some of those things.

Only God could have designed and created this physical universe and physical matter, because physical matter didn't exist until after God brought it into existence in the first place. Physical matter (materialism) couldn't have created physical matter, because physical matter didn't exist before God designed and created the first particle of physical matter.

Only God could have designed and created the physical genomes and physical life forms on this planet, because evolution, natural selection, and random mutations did not exist until after God brought the genomes and life forms into existence in the first place. The authors of this book caught onto this reality in a very big way. The PhD scientists who helped to write this book caught onto this truth as well as anyone that I have encountered so far. Evolution of any kind cannot explain the origin and the existence of life on this planet – that's what our scientific observations are trying to tell us.

Science 2.0 allows all of the evidence into evidence, which means that we allow all of the evidence from Young-Earth Creationists (YECs), Young-Universe Creationists (YUCs), Materialists, Naturalists, Darwinists, Out-of-Body Astral Travelers, Research Scientists, and the Prophets of God into evidence. The goal of Science 2.0 is to present ALL of the evidence, and then let you the reader decide for yourself what to make of it. In other words, I want to leave it up to my reader to decide for himself or herself how he or she wants to interpret the evidence rather than always providing my personal interpretation of the evidence as the ONLY possible and the ONLY authorized interpretation of the evidence, as the Materialists and Naturalists do.

There are times, though, when I will present my own personal interpretation of the evidence and tell you what it means to me personally, a process which is permitted and encouraged under Science 2.0; but, all the responsibility still falls upon you, my reader, to decide for yourself whether my personal interpretation of the observational evidence and experiential evidence makes any sense to you or not. If not, you are perfectly free to provide your own interpretation of the evidence, which you will do anyway, whether I try to prevent you from doing so or not.

### Problems with the Big Bang

All throughout my studies of science in all my different books, I often stray into the evidence that is used to falsify the Big Bang Theory.

Why?

It's because it points me, as a scientist, to the quantum or the non-physical. It points me to intelligence or psyche.

In "5 Major Problems with the Big Bang Theory", Joe explains what's wrong with the Big Bang Theory as well as providing links to a couple of alternative explanations that seem plausible.

https://www.youtube.com/watch?v=dbm3M9Bz4RE

As Joe repeatedly hints in his videos, when the Gods (Joe's all-knowing aliens) finally sit down with us and tell us how they did it and how everything works, it's not going to end up being anything like what we thought it was supposed to be.

Links to alternative possibilities.

http://nautil.us/issue/15/turbulence/do-we-have-the-big-bang-theory-all-wrong

http://discovermagazine.com/2008/apr/25-3-theories-that-might-blow-up-the-big-bang

https://www.space.com/24781-big-bang-theory-alternatives-infographic.html

http://www.learning-mind.com/5-most-interesting-alternative-theories-about-the-universe/

https://science.howstuffworks.com/dictionary/astronomy-terms/big-bang-theory7.htm

http://wwwphy.princeton.edu/~steinh/npr/

https://en.wikipedia.org/wiki/Ekpyrotic_universe

## It's Time for Science to Up Its Game

Modern-day scientists have proven beyond a shadow of a doubt that over 95% of our physical universe is in fact non-physical or immaterial. In other words, modern-day science has falsified Materialism, Naturalism, and Physical Reductionism.

Dark matter, a mysterious form of matter that has not yet been identified, accounts for 26.8% of the cosmic contents. Dark energy, which is the energy of empty space and is causing the expansion of the Universe to accelerate, accounts for the remaining 68.3% of the contents. That leaves 4.9% of our physical universe's contents as ordinary physical matter.

https://en.wikipedia.org/wiki/Universe

That means that 95.1% of our physical universe is in fact non-physical or spiritual in nature and origin. Get used to it because it's true and it's a proven fact. The contents of our own physical universe falsify Materialism, Naturalism, Nihilism, and even Atheism.

Whether they realize it or not, the Materialists, Atheists, Naturalists, Nihilists, and Behaviorists teach and believe that Quantum Mechanics or Spiritual Mechanisms do not exist. They are demonstrably wrong. The vast majority of our universe is non-physical or immaterial. The vast majority of our universe is quantum or spiritual.

It's time for these people to face the facts, admit that they were wrong, and get with the program. Materialism and Naturalism have been falsified by Science itself, especially by Quantum Mechanics or Spiritual Mechanics.

The Big Bang Theory is taught as a physical theory; but since over 95% of our physical universe is in fact non-physical, the Big Bang Theory is at best incomplete and at worst completely false. It's time for the scientists to up their game. We've got to start finding ways to grasp and understand the non-physical, the intangible, or the immaterial. We've got to embrace Quantum Mechanics or Spiritual Mechanisms and go with it to its logical conclusion.

It's time for us to launch out into the other dimensions rather than just limiting ourselves to the physical dimension. Materialism, Naturalism, Nihilism, Darwinism, and Atheism were designed to dumb us down and keep us stupid. It's time to jettison them and move on to something better.

These falsified philosophies need to be eliminated from Science. They are holding us back. In their place, we scientists need to learn how to develop and promote a better, more all-inclusive, and more realist philosophical foundation for Science as a whole. It's time to up our game. Materialism and Naturalism are dead. It's time for a paradigm shift.

## A Theory of Everything Will Have to Include the Non-Physical

The main reason the scientists can't develop a Theory of Everything is because they are determined to restrict themselves exclusively to the physical realm; and clearly, everything isn't physical.

In this video, Joe touches upon the Theory of Everything or Quantum Gravity.

https://www.youtube.com/watch?v=Rqu_uV-gIcU

Technically, quantum mechanics, magnetism, light, energy, radio waves, microwaves, dark matter or spirit matter, dark energy, space, time, quantum fields, thought, consciousness, intelligence, psyche, and even gravity are non-physical. Everything is non-physical except for physical matter. Get used to it. We scientists need to adjust philosophy and up our game. Materialism and Scientific Naturalism have been falsified by Science itself. It's time for a new and better way of doing science. It's time to include the non-physical or the immaterial into Science itself. It's time for Science 2.0.

https://www.youtube.com/watch?v=vJi3_znm7ZE

https://www.youtube.com/watch?v=w0ztlIAYTCU

https://www.youtube.com/watch?v=CbPWYjnQIO8

https://www.youtube.com/watch?v=y-Gk_Ddhr0M

https://www.youtube.com/watch?v=wfALJzn1hE8

https://www.scientificamerican.com/article/garrett-lisi-e8-theory/

### Quantum Fields Are Non-Physical in Nature and Origin

In this video Joe goes after Quantum Field Theory.

https://www.youtube.com/watch?v=at-mgqQg9Ds

Quantum Fields are non-physical and pre-physical. God needed to make the Quantum Fields before physical matter became possible to make. First things first. The non-physical or the spiritual had to be organized first before the physical even became possible. The energy or the light had to be organized first into Quantum Fields in order to make physical matter possible. Physical matter can't exist without Quantum Fields; and, Quantum Fields are non-physical to begin with.

Once again, we see how Science itself falsifies Materialism, Naturalism, and Physical Reductionism. Physical matter is NOT the fundamental unit of reality. Quantum Fields are not the fundamental unit of reality either. Quantum Fields are made from light or energy; and, every bit of light or energy is psychic, intelligent, conscious, sentient, omniscient, self-aware, and alive. So, either light is the fundamental unit of reality, or the consciousness (psyche) within the light is the fundamental unit of reality. They are probably two sides of the same coin.

https://www.youtube.com/watch?v=zNVQfWC_evg

https://www.youtube.com/watch?v=hYkaahzFWfo

https://www.youtube.com/watch?v=FBeALt3rxEA

Light, psyche, intelligence, consciousness, energy, and quantum fields are spiritual or non-physical in nature and origin. Here, with these, we finally start to get to the heart of reality itself; and, it's clearly and obviously non-physical. In the end, I chose to go with the Science itself rather than the philosophical speculation and wishful thinking of the Atheists, Naturalists, Materialists, Nihilists, Darwinists, Behaviorists, and Physical Reductionists. I used to be a materialist, naturalist, nihilist, and atheist; but, one day I saw the light, and I finally saw and accepted the fact that I was wrong.

Can you see how we modern-day scientists are finally coming online and starting to see and understand how things really work? At the fundamental core of everything around us is the non-physical or the spiritual or the quantum. It pervades everything. Reality is a mesh of at least 17 different non-physical quantum fields. Quantum fields are non-physical or spiritual in nature and origin; and, scientists like me are starting to get used to the fact, even though that's not what we are being taught in college by our atheistic, naturalist, and materialistic teachers. There's a paradigm shift coming that will signal the end of Materialism and Naturalism, as Science itself slowly falsifies Materialism and Naturalism.

Scientists are starting to catch the wave, the quantum wave. Quantum waves are Wi-Fi at the quantum level or the psyche level. Quantum waves are thoughts. Quantum waves can also be formed or organized into physical matter by psyche or thought.

### Applying Evidence Instead of Passion

We modern-day scientists are finally starting to see and understand what the Metaphysicists have known all along – namely, that the fundamental nature of reality is non-physical. In fact, the recently created Sciences of Near-Death Experiences and Out-of-Body Experiences are admitting into evidence actual experiences of the non-physical or the spiritual that the Materialists, Atheists, and Naturalists have successfully banned and hidden from society at-large for most of human history.

One of the primary maxims of Science 2.0 is to allow ALL of the evidence into evidence, and then to pursue a preponderance of that evidence. The following information from William Buhlman was a turning point for me personally. At the end of 2015, when I read his book *Adventures Beyond the Body: How to Experience Out-of-Body Travel*, that's when I officially abandoned My Materialism and My Naturalism and went in pursuit of something better instead.

https://psyche-ontology.com/buhlman/

After five decades of being lied to and deceived by the Materialists, Naturalists, Nihilists, Darwinists, and Atheists, all I wanted was the truth; and, I finally found it in the quantum or the non-physical.

## Conclusion

I review and critique these books in my other books, so I won't do so here, because I intend for this book to be a relatively short introduction to Science 2.0 and what it's all about.

Remember, Science 2.0 allows all of the evidence into evidence.

In contrast, I have observed that Materialism, Naturalism, Darwinism, Nihilism, and Atheism are based upon the suppression of all evidence which they personally don't like. That process of censorship and suppression is dogma and religious fanaticism, not science! Furthermore, I have observed that the Young-Earth Creationists (YECs) and the Young-Universe Creationists (YUCs) also deliberately reject observational evidence or try to explain it away, and they favor instead and give priority to their personal interpretation of the Bible.

Since Science 2.0 is based upon Observational Evidence, Experiential Evidence, Eye-Witness Evidence, Scientific Research, and the Lived Experiences of the human race,

Science 2.0 naturally finds itself in opposition to any philosophy that gives precedence to "personal interpretations" or "scientific inferences" over observational evidence – philosophies such as Materialism, Naturalism, Darwinism, Scientism, Nihilism, Atheism, Young-Earth Creationism (YEC), and Young-Universe Creationism (YUC).

Science 2.0 allows all of the evidence into evidence, and then leaves it up to you to decide for yourself what to make of it all.  If you study my writings long enough, you will probably notice or observe that I spend more time studying and critiquing the evidence provided by my opposition than I do studying the evidence that I actually believe in.  I allow all of the evidence into evidence, including the evidence from my opponents.  By studying my opponents, they help me to decide for myself what I truly believe and what I don't.

You have to KNOW what the Young-Earth Creationists (YECs), Young-Universe Creationists (YUCs), Materialists, Naturalists, Darwinists, Nihilists, and Atheists believe and why they believe it BEFORE you can figure out for yourself what's wrong with it, if anything.  This reality demands that I allow all of the evidence into evidence; and then, Science 2.0 becomes a weighing of the evidence.

The purpose of Science 2.0 is to eliminate all of the falsehoods, trickery, sophistry, deceptions, and lies that derive from the *scientific inference* logic fallacy, *affirming the consequent* logic fallacy, the *begging the question* logic fallacy, and the *jumping to conclusions* logic fallacy which are easily and automatically produced by the fourth and final step in the Scientific Method – the "interpretation phase" of the scientific data or scientific evidence.  *Personal interpretation* or *scientific inference* IS a logic fallacy – the logic fallacy upon which Materialism and Naturalism are built.

Remember, *scientific inference* or "someone else's personal interpretation of the scientific evidence" IS a logic fallacy, which introduces a whole host of falsehoods, deceptions, and lies into science and the scientific methods.  There are NO authorized credentialed authorities when it comes to one's personal interpretation of the observational evidence or scientific evidence.  You will decide all on your own what the observational evidence means to you personally, anyway, so why not let you do so from the beginning?  Your interpretation of the evidence is just as good as anyone else's interpretation.

Consequently, the primary purpose of Science 2.0 is simply to produce and present the eye-witness evidence or observational evidence, and then let you the reader decide for yourself what to make of it and how to interpret it.  At times, I will tell you what the observational evidence means to me personally; but, you remain completely free to disagree with me and to come up with your own personal interpretation of the observational evidence.  You are going to do so anyway, so why not let you?

It is October 2017 as I write this sentence.  Science 2.0 has been at least five years in the making.  I have been observing myself during the past five years, as I study every scientific discipline that I can get my hands on.  I have been a seeker in pursuit of the truth.

During the past five years of my life, I have observed that when it comes to any piece of scientific evidence, scientific data, correlation data, or research experiment, I always study the data and the results of the experiment; but then, I provide my own personal interpretation of the data rather than relying blindly upon the "official interpretation" provided by the Materialists, Naturalists, Darwinists, Nihilists, and Atheists.  In other words, I make up my own mind and decide for myself what that scientific evidence and the observational evidence means to me personally.  I provide my own interpretation of the data and the evidence, because I have observed that I tend to provide a better, more believable, more parsimonious, more logical, and more convincing interpretation than the "official interpretation" which is given to us by the Materialists and Naturalists.

Thereby, Science 2.0 was born. I accept ALL of the observational evidence, experimental evidence, and experiential evidence into evidence, study it all, and then decide for myself what it means to me. Rather than accepting someone else's interpretation of the evidence, I always provide my own; and, I encourage my readers to do the same. As a result, I effectively eliminate or skip past the major flaw hidden within the Scientific Method, which is the last and final step called "interpretation of the experiment", or more accurately "scientific inference". By providing my own interpretation of the data and the evidence, I eliminate all of their "scientific inferences," "philosophical conclusions", and "personal speculation or guesswork". I believe that I get closer to the truth as a result. My personal interpretations make more sense to me than theirs do, anyway, so I just naturally go with my own interpretation of the evidence and see where it takes me. I decided to follow the evidence wherever it might lead me. In my opinion, that's the true spirit of science, so why not pursue it?

Kalat, J. W. (2008). *Introduction to Psychology* (9th ed.). Belmont, CA: Wadsworth, Cengage Learning. **"The brain is the product of Evolution." (p. 12).**

Pinel, J. (2014). *Biopsychology* (9th ed.). New York: Pearson. **"Genetic endowment is a product of its evolution." (p. 24).**

In his book, James Kalat repeatedly told his students to "examine the evidence". I did; and by doing so, I decided that his personal interpretation of the evidence is often wrong. I have written over a dozen books simply by examining the evidence, which James Kalat presented to me in his book, in great detail; and then, while writing my books, I tell my readers what all of that evidence means to me personally.

In his book, John Pinel repeatedly tells his students to "think critically about the evidence". I did; and by doing so, I observed that his personal interpretation of the evidence is often wrong.

It was actually John Pinel who taught me that these people actually use "scientific inferences" as scientific evidence and scientific proof for their chosen conclusions. Although he wasn't intending to do so, it was John Pinel who first introduced me to the **Scientific Inference Logic Fallacy**.

From *Biopsychology* page 13, John Pinel writes:

> **Scientific inference is the fundamental method of biopsychology and of most other sciences – it is what makes being a scientist fun. This section provides further insight into the nature of biopsychology by defining, illustrating, and discussing scientific inference.**
>
> **The scientific method is a system for finding things out by careful observation, but many of the processes studied by scientists cannot be observed. For example, scientists use empirical (observational) methods to study ice ages, gravity, evaporation, electricity, and nuclear fission – none of which can be directly observed; their effects can be observed, but the processes themselves cannot. Biopsychology is no different from the other sciences in this respect. One of its main goals is to characterize, through empirical methods, the unobservable processes by which the nervous system controls behavior.**
>
> **The empirical method that biopsychologists and other scientists use to study the unobservable is called <u>scientific inference</u>. Scientists carefully measure key events they can observe and then use these measures as a basis for logically inferring the nature of events that they cannot observe.**

> **Like a detective carefully gathering clues from which to recreate an unwitnessed crime, a biopsychologist carefully gathers relevant measures of behavior and neural activity from which to infer the nature of the neural processes that regulate behavior.**
>
> **The fact that the neural mechanisms of behavior cannot be directly observed and must be studied through scientific inference is what makes biopsychological research such a challenge – and as I said before, so much fun.** (*Biopsychology*, p. 13.)

I couldn't believe what I was reading. I still find it amazing.

**Creating fictional stories is what makes science so much fun!**

Shoot me already!

I'm sorry, but I hate *scientific inferences*, especially when they are being constantly falsified by real observational evidence and real experiential evidence. I got tired of the medical doctors and the psychiatrists practicing on me at my expense. I was ready for a real cure rather than scientific inferences and guesswork. Scientific inferences are worthless in the end. They are instantly falsified by real experiences, real events, and real observations. Observation, experience, and eye-witness knowledge trumps scientific inferences every time.

In fact, *scientific inference* is a logic fallacy – a form of *begging the question* and *ad hoc just-so story-telling*.

With scientific inference you can provide proof for any conclusion that you want simply by making up the evidence out of thin air as you go along, and then using that evidence and submitting that evidence into your future arguments as proof that your chosen conclusions are true – this process is called *begging the question*, *jumping to conclusions*, and *affirming the consequent*. *Scientific inference* is just another version of these logic fallacies; and, it's through scientific inference that the Materialists, Naturalists, and Darwinists prove that the theory of evolution is true. And, this is precisely what John Pinel does – uses *scientific inferences* – while promoting Evolutionary Psychology and the Theory of Evolution throughout his book, in order to prove that the Theory of Evolution is true.

These people as a whole taught me to examine the evidence critically and extensively; and by doing so, I have successfully observed and demonstrated that many of their chosen conclusions – their "official interpretations of the evidence" – are flawed, biased, and demonstrably false since their chosen conclusions are based upon logic fallacies to begin with and have no observational evidence to support them. I proceeded to use science, the scientific methods, and observational evidence to FALSIFY their hidden assumptions and chosen conclusions. I was successful in doing so; and, it's what I have been doing for the past couple years of my life.

If you keep studying, learning, and thinking, I believe that you will probably come to the same exact scientific conclusions that I have come to in the end, but most likely through a completely different set of books, because truth is mutually supportive across the board. When pursuing the truth, there are a multitude of different books that can lead us to the same conclusions; but, these books in this list were the books that did the trick for me. Your mileage may vary, though.

These were the books that helped me to upgrade my science to Science 2.0; and, these were the books that helped me to establish the foundation for Science 2.0.

Good enough!

And sometimes, good enough really is good enough.

# PART VI — SCIENTIFIC PROOF OF GOD'S EXISTENCE

Science 2.0 is based upon the Observational Evidence and the Lived Experiences of the human race. The best way to find and know the truth is to live it and experience it for yourself, or to choose to trust someone who has. Science 2.0 is based exclusively upon the Law of Witnesses.

**2 Corinthians 13: 1: "In the mouth of two or three witnesses shall every truth be established."**

Based upon this particular definition for science, Science 2.0 can actually provide proof of God's existence through a preponderance of the evidence by admitting ALL of the evidence into evidence and by relying upon the Observational Evidence and the Lived Experiences of the human race to provide that eye-witness evidence.

Unlike Materialism, Naturalism, Darwinism, Nihilism, and Atheism which are trying to hide from the evidence, Science 2.0 pursues ALL of the evidence and then leaves it up to you to decide for yourself what to make of that evidence. It's what you are doing already, so why not take advantage of it?

When I first realized that Natural Selection, Random Mutations, and Evolution DID NOT EXIST until after God had designed and created the first genomes and life forms, that reality and epiphany became my first convincing Scientific Proof of God's Existence. In a very real sense, the Theory of Evolution proved to me that God exists and that God must exist in order to have done all of the design, programming, engineering, creation, field-testing, manufacturing, and deployment of genomes and life forms which Natural Selection, Random Mutations, and Evolution could NEVER have done. God must of necessity exist in order to have done ALL of the Science which Natural Selection, Random Mutations, Evolution, and the Rocks could NEVER have done. After that realization, I simply KNEW that God exists, because I KNEW why He must exist.

## Scientific Proof of God's Existence

Scientific proof of God's existence?

They ALL told me that it can't be done and that it will never be done.

I believed them.

For over fifty years of my life, I believed them.

THEY WERE WRONG!

What good are Science and the scientific methods if they can't be used to find and know the truth? What good are Science and the scientific methods if they can't be used to prove anything?

I have spent the last two years studying Science, the scientific methods, theology, theism, psychology, the philosophy of science, and natural theology – developing Scientific Proofs of God's Existence as I go along.

As far as I know, I'm the ONLY person on the planet who is doing this – using various different scientific methods to prove that God exists and to prove that God must of necessity exist. Why am I the ONLY person doing this? It's because everyone else believes that it can't be done, so they don't even try. That's the way I used to be.

I used to be a Materialist, Nihilist, Atheist, and Skeptic; but, I'm not anymore, because Science and the scientific methods proved to me that God exists and why He must exist.

I wasn't going to believe in God until after I had scientific proof of God's existence. I didn't think it was possible, which meant that I was destined to remain Agnostic and Skeptical for the rest of my life. BUT over time, after I finally started looking for it and asking God for it, He gave me scientific proof of His existence. Now, I have no choice but to believe in God's existence, because He gave me exactly what I was looking for.

So, how was this possible, if Science and the Scientific Methods can't be used to prove anything true? Well, let me start at the beginning and explain the process that I went through, in detail.

It all started by getting a truthful and accurate definition for Materialism and Naturalism. When I started this process two years ago, I didn't know what Materialism and Naturalism were. I was a scientist to my very core and a believer in Scientism, and I didn't even know what philosophical foundations Science is built upon. But, I'm not alone. Most scientists don't have a clue when it comes to the Philosophy of Science. They don't teach you this stuff in school. In fact, most of your college professors either don't know anything about the Philosophy of Science or they are trying to hide that information from you if they do know. I had a steep learning curve ahead of me!

I didn't know where to start my search, so I started by studying Atheism which quickly led me to Philosophical Proofs of God's existence – basically the opposite of Atheism. I KNEW what Atheism was, because I had spent time as an atheist. I didn't believe that there could be a Proof of God's existence, so the whole Natural Theology realm and Philosophical Proofs of God were completely new to me. I didn't believe such a thing was possible and that such a thing exists. But, I was wrong! We Materialists and Atheists are always wrong!

I started the whole learning process with Antony Flew, allegedly the world's most notorious atheist, and his book, *There Is a God: How the World's Most Notorious Atheist Changed His Mind*. I found out that he gave up his atheism because of Intelligent Design Theory. His reason for giving up his atheism was different than mine; and, I started to take Intelligent Design Theory seriously.

My search for "proof of God" on Amazon led me to John Lennox and a few of his books.

Lennox, J. C. (2009). *God's Undertaker: Has Science Buried God?* Oxford, England: A Lion Book.

Lennox, J. C. (2011). *God and Stephen Hawking: Whose Design Is It Anyway?* Oxford, England: A Lion Book.

Lennox, J. C. (2011). *Gunning for God: Why the New Atheists Are Missing the Target*. Oxford, England: A Lion Book.

This stuff is Natural Theology, or Philosophical Proof of God's Existence, wherein nature and science are used as premises while constructing Proofs of God.

This whole thing slowly led me to believe that it might in fact be possible to use the scientific methods to prove that God exists. This was before I knew that the scientific methods can't be used to prove anything to be true; and, this was long before I discovered that the scientific methods can in fact be used to prove things false. At the time, I simply thought that Science and the scientific methods ARE the TRUTH. I was a practitioner and adherent of Scientism, and I had a great deal of faith in Science and the scientific methods. I simply believed that if I could use Science and the scientific methods to PROVE God's existence, then I would have had found the Holy Grail for real.

I had a lot to learn!

I started to study Science and the Scientific Methods, in earnest, from a completely new perspective with a completely new goal in mind. I was looking for Scientific Proof of God's Existence.

Seek, and ye shall find. Knock, and it shall be opened unto you.

## Using the Scientific Methods to Prove God's Existence

Okay, so how do you use the scientific methods to prove God's existence, if the scientific methods can't be used to prove anything to be true?

When you take up a serious study of the Philosophy of Science, the first thing you discover is that the scientific methods can't be used to prove anything true, due to the *affirming the consequent*, the *begging the question*, and the *jumping to conclusions* logic fallacies which are built directly in to the Scientific Method.

For any piece of scientific data or scientific information or scientific evidence, there are theoretically an infinite number of possible interpretations or explanations which can be given to that scientific data. By *affirming the consequent*, you can pick any conclusion or consequent that you want and use the Scientific Method to affirm it or to "prove" it to be true, which is what the Materialists, Atheists, and Darwinists do when it comes to the Theory of Evolution. They *affirm the consequent*, or unilaterally declare their personal interpretation of the scientific evidence to be axiomatically true. It's a logic fallacy which they are employing in order to "prove" that the Theory of Evolution and Darwinism are true. It's a FALSE PROOF, or a FALSE POSITIVE!

For example, the same scientific evidence which the Darwinists use to "prove" that the Theory of Evolution is true CAN BE USED by *affirming the consequent* to prove that the flying spaghetti monster designed and created ALL of the genomes and life forms on this planet.

The same scientific evidence which the Darwinists use to "prove" that the Theory of Evolution is true CAN BE USED and HAS BEEN USED to prove that Intelligent Design Theory IS TRUE! But, it's ALL based upon the *affirming the consequent* or the *jumping to conclusions* logic fallacy! So, it really isn't a proof of these things! It only appears to be.

So again, how do you use the scientific methods to prove God's existence, if the scientific methods can't be used to prove anything to be true? This IS the great mystery, and God slowly showed me how to solve it.

In order to accomplish this task, one needs a solid understanding of the Philosophy of Science and the Scientific Method. You need to KNOW what the scientific method really is, how it works, what it does, what it is good for, and what it can't do! You also need a solid understanding of what Materialism, Naturalism, Darwinism, Nihilism, and Atheism REALLY ARE. You have to have a true and accurate definition for these philosophical concepts, or you will never be able to develop a credible argument for God's existence. I didn't even know what I was looking for at first; and, I spent nearly two years studying Science, the Scientific Methods, and the Philosophy of Science before it all came together for me.

Rychlak, J. F. (1981). *A Philosophy of Science for Personality Theory* (2nd ed.). Malabar, FL: Robert E. Krieger Publishing Company.

I used this book from Joseph Rychlak as my introduction to the Philosophy of Science. This was the first book to introduce me to the *affirming the consequent* logic fallacy and the primary weakness of the scientific methods – namely, that the scientific methods can't be used to prove anything to be true. You have to have a solid understanding of the Scientific Method and the Philosophy of Science if you are going to develop a sound and logical philosophical proof and scientific proof of God's existence.

Why is this important? Why is it important to understand the Philosophy of Science and how the Scientific Methods really work?

Well, I run into genius PhD scientists all the time who use the scientific methods every day in their work, and they have NO clue whatsoever about the limitations of the scientific methods. These people literally believe that the Scientific Method is God and that it can do anything and accomplish anything it sets its mind to. These people believe that the Scientific Method is Superman and that they are Supermen, and that they are going to save the world. These people have no real understanding of the Philosophy of Science, what the Scientific Method really is, what it's good for, what it can do, and what it can't do. These people have no idea what the limitations of the Scientific Method are and what logic fallacies are built into the scientific methods. These people are practitioners of the Religion of Scientism, which means that these people treat Science and the Scientific Method as if these things were God. I KNOW, because I used to be one of them.

I was a bit rusty when I first started developing Scientific Proofs of God's Existence. I went about it the hard way and the long way at first, but we each have to start where we are and go forward from there.

My first real breakthrough came when God revealed to me what Materialism, Naturalism, Darwinism, and Atheism really ARE. I needed a definition for these philosophical concepts that I could plug into the Scientific Method as a hypothesis; and, eureka! It just popped into my mind one day after I had started looking for it and asking for it.

**Materialism IS design and creation by physical matter.** Materialists really truly believe that the rocks or raw physical matter designed, programmed, engineered, created, manufactured, and deployed the first genomes and life forms on this planet. This IS what these people truly believe and what these people teach our children in our public classrooms. Immediately, it became obvious to me that Naturalists and Darwinists are also Materialists. The Materialists, Naturalists, and Darwinists really truly believe that the rocks designed and created ALL of the genomes and life forms on this earth.

Nihilism is the philosophical belief that there is no afterlife, no reason or purpose for living, and no such thing as a non-local realm or a spirit realm. Nihilism is the belief that we cease to exist when we die. Nihilism is another form of Materialism and Atheism.

Atheism is design and creation by nothing, or design and creation by chance. The Atheists really truly believe that nothing designed and created everything, including all of the genomes and life forms on this planet.

Due to the *affirming the consequent* logic fallacy, which is built into the scientific methods, you CANNOT USE the scientific methods to prove anything true directly, which means that the scientific methods cannot be used for a direct proof of God's existence. This reality has got most people believing that we can't use the scientific methods to prove anything at all. I have had many Scientists, Materialists, and Atheists tell me this. They have told me that Science and the Scientific Methods can't be used to prove anything. THEY ARE WRONG! It's annoying, because many of the Atheists, Naturalists, and Materialists use this false belief as a crutch or a shield to hide behind, as a reason for believing in the impossible and the patently absurd, and as a reason for doing sloppy science. These people don't understand the Scientific Method nor the Philosophy of Science, if they truly believe that the scientific methods can't be used to prove anything.

Nevertheless, there are millions of scientists and Darwinists out there in the world right now who are using their ignorance of the scientific method and their belief that the scientific methods can't prove anything as an excuse or a reason for continuing to hope that someday – some way – someday – Science will finally find a way to prove the theory of evolution true. It's never going to happen, because the scientific methods can't be used directly to prove anything to be true – certainly not something that is demonstrably false.

So, how do you use the scientific methods as a proof of God's existence, if the scientific methods can't be used to prove anything to be true? That is the million-dollar question, isn't it; and, I get the impression that I'm the ONLY person on the planet who has solved this one! It took a lot of prayer and study, but eventually God showed me how to do it. He showed me how things REALLY WORK!

When developing a Scientific Proof of God's Existence, the process starts by using the scientific methods to **falsify** Materialism, Naturalism, Darwinism, and Atheism. And, you also have to choose to use the Scientific Observations, or the Direct Observations, or the Lived Experiences of the human race to **falsify** Nihilism, Naturalism, and Atheism. That's the KEY, and that that's the thing which EVERYBODY refuses to do. I don't think that it's that they don't know how to do it or can't know how to do it. I think it's because they don't want to do it and refuse to do it. In fact, it probably NEVER enters their mind to even try to do it. As far as I know, I'm the ONLY person on the planet who has ever done it. I have used various different scientific methods and scientific observations to **falsify** Materialism, Naturalism, Darwinism, Scientism, Nihilism, and Atheism. It's really easy to do, once you learn how to do it. Everything **falsifies** these things!

But, let's run through the whole process in a bit more detail, rather than just making blanket statements.

First, you have to realize and accept the fact that the scientific methods can indeed be used to PROVE theories false! It's called falsifying a theory or *negating the consequent*; and, it's philosophically and logically sound. When you successfully *negate the consequent*, you have in fact PROVEN a theory false, or falsified a theory. This IS Philosophy of Science 101, and the core essential understanding of the Scientific Method one must have if one is going to use the scientific methods to PROVE anything. You have to KNOW that the scientific methods can be used to PROVE things false. You have to KNOW that the scientific methods can be used to falsify theories, hypotheses, and philosophical concepts that are in fact false.

Finally, you have to be willing to use the scientific methods and scientific observations to **falsify** Materialism, Naturalism, Darwinism, Nihilism, Scientism, and Atheism. If you are unwilling to do this, then you will NEVER be able to develop a Scientific Proof of God's Existence, and you will NEVER be able to find the truth. This is crucial. You have to be open-minded enough and smart enough to use the scientific methods for what they are good for and for what they were designed for, if you are going to get Science and the scientific methods to PROVE anything to you.

**Let's use the Scientific Method** for what it was designed for. Here's how falsifying a theory, or *negating the consequent*, works in principle using the Scientific Method:

**Scientific Hypothesis: If Theory X is true, then we will observe Y.**

**Scientific Observations: We don't observe Y.**

**Scientific Conclusion: Therefore, Theory X is false and has been falsified by the Scientific Method.**

Here's how it works in practice:

**Scientific Hypothesis: If the Theory of Evolution, Materialism, Naturalism, and Darwinism are true, then we will observe the rocks and**

physical reactions designing, creating, and manufacturing genomes and life forms from scratch.

**Scientific Observations: We have NEVER observed the rocks and physical reactions designing and creating genomes and life forms; and, we NEVER will. They can't.**

**The Scientific Conclusion: Therefore, the Theory of Evolution, Materialism, Naturalism, and Darwinism are false and have been falsified by the Scientific Method or Scientific Observation.**

I just successfully falsified Materialism, Naturalism, Darwinism, and the Theory of Evolution; and, I used the Scientific Method to do so! Best of all, my argument is philosophically and logically sound. I have in fact **falsified** Materialism, Darwinism, Naturalism, Creation by Rocks, and the Theory of Evolution for REAL! I have PROVEN them false! It's that simple! I successfully used the Scientific Methods for what they are good for – falsifying theories which are false. I wish I would have known how to do that forty years ago. It would have saved me a lot of confusion, frustration, time, and grief.

Now, let's run Atheism through the Scientific Method:

**Scientific Hypothesis: If Atheism is true, we will observe "nothing" designing, creating, and manufacturing everything, including genomes and life forms. It would be chaos, but that's exactly what we should be observing if Atheism were true.**

**Scientific Observations: Obviously, we have NEVER observed "nothing" designing and creating anything; and, we NEVER will. It's patently absurd.**

**The Scientific Conclusion: Therefore, Patent Absurdity or Atheism is false and has been falsified by the Scientific Method and Scientific Observations.**

I just falsified Atheism for REAL; and, it's philosophically and logically sound. I have in fact falsified Atheism, using the Scientific Method and Scientific Observations to do so! But, the process of designing and creating a Scientific Proof of God's existence isn't complete yet. Technically, falsifying Atheism is not the same thing as proving Theism. So, how do we proceed from here, and continue to use the Scientific Methods as we go along?

Well, let's see if you can follow this logic or reasoning.

Since we KNOW by the Scientific Method that Materialism, Naturalism, Atheism, Darwinism, Creation by Rocks, and the Theory of Evolution ARE FALSE, we also KNOW through the process of elimination that some form of their opposite MUST BE TRUE. When it comes to the scientific methods, we arrive at the truth through a process of elimination. By eliminating ALL of the falsehoods, we are left with the truth. This is Logic 101!

Here we are operating on the AXIOM or the GIVEN or the LAW that Truth, Reality, and Knowledge do in fact exist, if we know how and where to find them. Science, Knowledge, and Reality rely upon this basic essential assumption; otherwise, there's no sense doing science, if truth and reality and knowledge are in fact impossible to find and impossible to use once we have found them. We have to start with the primary assumption that THE TRUTH can indeed be found and does indeed exist!

Let's state this again, because it is essential and important! Since we KNOW by the Scientific Method that Materialism, Naturalism, Atheism, Darwinism, Creation by Rocks, and the Theory of Evolution ARE FALSE, we also KNOW through the process of elimination that some form of their opposite MUST BE TRUE. Using the scientific methods, we narrow in on the TRUTH through a process of elimination, by using the scientific methods to eliminate everything that is false. Once you have eliminated EVERYTHING that is false, then you are left staring at THE TRUTH. It's elementary my dear friend.

It's so obvious, that I sometimes wonder how the human race has been able to overlook it for thousands of years!

Once you have eliminated EVERYTHING that is false, then you are left with THE TRUTH. That's what Sherlock Holmes does.

**How often have I said to you that when you have eliminated the impossible [and the false], whatever remains, *however improbable*, must be the truth? — Sherlock Holmes.**

Sherlock Holmes slowly arrives at the truth by falsifying his theories and his imagined evidence. That's how I finally arrived at THE TRUTH, by falsifying Materialism, Naturalism, Darwinism, Nihilism, and Atheism. THE TRUTH is the opposite of falsehood and falsified theories!

Now, here comes the kicker and the final knock-down blow! THE TRUTH is repeatedly verified by scientific observations and scientific methods. By eliminating the impossible and ALL of the falsehoods, whatever remains must be THE TRUTH! That's how we USE the scientific methods or scientific observations to find THE TRUTH, and to KNOW the truth, and to PROVE the truth! The truth is going to be the opposite of falsified theories, or the opposite of the things which have been proven false by the scientific methods and the observations of the human race. Checkmate!

THIS IS THE KEY!

So again, how do you use the scientific methods to prove God's existence, if the scientific methods can't be used to prove anything to be true?

You USE the scientific methods to eliminate EVERYTHING that is false, so that ONLY the TRUTH remains! If you successfully falsify and eliminate ALL of the falsehoods such as Materialism, Darwinism, Naturalism, Scientism, Nihilism, and Atheism, then THE TRUTH is the only thing left standing. It's logical, and it works! Best of all, it works! THE TRUTH is constantly VERIFIED by the scientific methods and never falsified. In fact, there's NO way to use the scientific methods to falsify THE TRUTH. It can't be done. The scientific methods or scientific observations continuously VERIFY THE TRUTH. Cool, huh?

It's elementary my dear friend! And, it works!

Let's provide some examples of how this works in practice.

**What's the opposite of Materialism?** The non-local, the spiritual, the transdimensional, and the immaterial are the opposite of Materialism. Psyche or Intelligence or Non-Local Consciousness is the opposite of Materialism. Light and thoughts and dreams are the opposite of Materialism. (Spirit matter really isn't the opposite of Materialism, because spirit matter is matter, just like physical matter is matter. According to Quantum Mechanics, it takes conscious observation or the Word of Command to convert spirit matter into physical matter. It's ALL matter, just different phases of being or different states of existence.)

Observation, empirical evidence, direct experience, lived experiences, the scientific methods, direct observations, scientific observations, personal experiences, revelations from God, physical experiences, and spiritual experiences ARE the same exact thing. They are

ALL **the source** of Scientific Evidence or should be! Lived Experience includes our non-local experiences or our spiritual experiences. Lived Experience or Scientific Observation IS Scientific Evidence! Lived Experience or Scientific Observation is how scientific theories and scientific hypotheses are verified and falsified!

**THIS IS KEY!**

We KNOW from **observation**, direct experience, or the scientific methods that Intelligence (Psyche), Thought (Psyche), Consciousness (Psyche), Quantum Nonlocality, Light, Forces, Fields, Magnetism, Gravity, Dark Matter (Spirit Matter), Dark Energy (the Zero-Point Field of Light or the Light of Christ) do in fact exist because they have been observed, experienced, and **verified** trillions of trillions of times, just as we KNOW from the scientific methods or scientific observations that Materialism is false.

Darwinism and Naturalism are a form of Materialism. The very same scientific methods and scientific observations, which PROVE Materialism false, also simultaneously PROVE Darwinism and Naturalism false. Atheism is a form of Materialism and Naturalism. The very same scientific methods which PROVE Materialism and Naturalism false also PROVE Atheism false. They are all fruit from the same poisoned tree.

There is NO evidence and will NEVER be any evidence to support the major premises, primary assumptions, and main claims of Materialism, Darwinism, Naturalism, Nihilism, and Atheism. These philosophies or religions have to be taken on blind-faith as being true; and, blind-faith isn't scientific evidence, even though millions of Materialists and Atheists do in fact try to use their ignorance, flying spaghetti monsters, and blind-faith as scientific evidence.

What do we learn from this? We learn that **observations** or the **scientific methods** repeatedly falsify falsehoods such as Materialism, Naturalism, Darwinism, Nihilism, and Atheism – while at the very same time, the same exact observations or scientific methods repeatedly VERIFY the theories which are in fact TRUE.

**What's the opposite of Darwinism?** Intelligent Design Theory, or Design and Creation by Intelligent Beings, is the opposite of Darwinism and the Theory of Evolution. Darwinism IS Creation by Rocks. The opposite of that is Creation by Intelligent Psyches. The rocks or raw physical matter have NEVER been caught in the act of design and creation. In fact, in 1859, Louis Pasteur falsified Spontaneous Generation or Creation by Rocks. Louis Pasteur used a scientific method to falsify Materialism, Darwinism, Naturalism, Creation by Rocks, and the Theory of Evolution the very same year that Charles Darwin published "On the Origin of Species". How ironic is that?

Do you see how important it is to get a correct and useful definition for Darwinism, Materialism, Atheism, and Naturalism which can then be run through the Scientific Method and have Scientific Observations or Lived Experiences applied directly to those hypotheses or definitions? It makes ALL the difference in the world! As far as I know, I'm the ONLY person on the planet to have done such a thing; yet, God has been doing this same exact thing for trillions of years. God is the Ultimate Scientist! Science IS observation or lived experience; and, God observes!

**What's the opposite of Naturalism?** The supernatural, the psychic, the spiritual, the non-local, Theism, and God ARE the opposite of Naturalism. We KNOW from observations or the Lived Experiences of the human race that the Biblical God Jesus Christ exists and rose from the dead. Since we KNOW from the scientific methods that Naturalism IS FALSE, we should be looking for some type of Naturalism's opposite to be TRUE.

This is where the vast SUPERIORITY of Lived Experience or Direct Observation comes in handy and really shines. Science IS Observation, or it should be! God IS a Scientist, the Ultimate Scientist! God Observes! Through Direct Observation or Lived Experiences, we can go directly to KNOWING the TRUTH without having to rely upon the scientific methods or philosophical interpretation of scientific data. We KNOW from the Lived Experiences of the human race that the Biblical God does in fact exist and really did rise from the dead.

Furthermore, there are over 13 million Near-Death Experiences on record. These people KNOW that Naturalism and Materialism ARE FALSE, because these people have left their physical body and have gone to the spirit world and seen for themselves that it is real and truly exists. These people have FALSIFIED Nihilism through their Lived Experiences or Direct Observations. Nihilism is the philosophical belief that there is no such thing as a spirit world and an afterlife.

Thousands of different people KNOW that Jesus Christ lives and rose from the dead, because they have seen Him and touched Him both while out-of-body and while in the flesh. These people have also FALSIFIED Nihilism through their Lived Experiences or Direct Observations.

Lived Experiences or Direct Observations ARE Scientific Evidence! That's the best and most effective definition for scientific evidence that we can have. With Lived Experiences or Direct Observations, we human beings (human psyches) go directly to KNOWING the TRUTH! Lived Experiences or Scientific Observations include our Spiritual Experiences and our Out-of-Body Experiences. This is REAL SCIENCE! And, it's all based upon Lived Experience or Direct Observation of THE TRUTH.

Knowing by the scientific methods that Naturalism is false and why it is false is just icing on the cake, or confirmation of what we already KNOW from the Lived Experiences of the human race as recorded in the Bible, the Book of Mormon: Another Testament of Jesus Christ, the Doctrine and Covenants, the Pearl of Great Price, Near-Death Experiences, Out-of-Body Experiences, and other types of Spiritual Experiences or Non-Local Experiences.

Lived Experiences or Direct Scientific Observation of both the spirit realm and the physical realm are vastly SUPERIOR to the traditional scientific methods or the materialistic methods which restrict us exclusively to studying physical matter with physical methods. With Lived Experience or Direct Scientific Observation, we human beings or human psyches can go directly to KNOWLEDGE of the TRUTH without getting mired down in Materialism, Scientism, the Physical Sciences, and the limitations of the traditional materialistic methods which are limited to and limit us to studying physical matter and physical processes.

Finally, if Materialism and Naturalism were actually true, scientists and psychologists would be putting rocks into Skinner boxes in order to observe their behavior and shape their behavior. Why aren't the Behaviorists putting rocks into Skinner boxes for some behavioral technology or behavioral conditioning? It's because Materialism and Naturalism are FALSE. It's so obvious, that I sometimes wonder why nobody has ever thought of it before.

**What's the opposite of Atheism?** Theism and God are the opposite of Atheism. We KNOW through the scientific methods or scientific observations that the major premises or the primary assumptions of Atheism are patently absurd. So, we should be looking to Theism, Intelligent Beings, Intelligent Psyches, and some kind of God or Ultimate Scientist as the designer, programmer, creator and manufacturer of ALL the genomes and life forms on this planet.

We should also be looking to find some kind of God, or Intelligent Psyche, or Powerful Non-Local Consciousness, or Intelligent Being to be the Conscious Observer who is necessary for converting spirit matter into physical matter according to the Science of

Quantum Mechanics. God must of necessity exist, or there would be NO physical matter! God must of necessity exist, or there would be no genomes and no physical life forms. Right there IS Scientific Proof of God's Existence! It's not that great of a leap, once we have repeatedly used the scientific methods or scientific observations to falsify Materialism, Naturalism, Nihilism, Darwinism, and Atheism.

Once I KNEW through the scientific methods that Materialism, Darwinism, Naturalism, and Atheism ARE FALSE, then I simply KNEW that God or the Ultimate Scientist must of necessity exist in order to have done ALL of the Science which the rocks or raw physical matter could NEVER have done. I used Science and the Scientific Methods for what they are good for – falsifying theories – and through a process of elimination I arrived at THE TRUTH, namely that God or the Ultimate Scientist must of necessity exist, or there would be no physical matter, there would be no genomes, there would be no physical life forms, and there would be no science. If God or the Ultimate Scientist didn't exist, you and I would not exist; and, you wouldn't be reading this right now. It's elementary my dear friend!

We KNOW from the Lived Experiences and Near-Death Experiences of the human race that the Biblical God Jesus Christ does in fact exist. Knowing through the various scientific methods that Atheism is false and why it is false lends confirmation to the fact that the Lived Experiences or Direct Observations of the human race are in fact true.

After I had successfully used the scientific methods to falsify Materialism, Naturalism, Darwinism, and Atheism a dozen different ways, I simply KNEW that God or the Ultimate Scientist of necessity must exist in order to have done all of the different types of Science, which the rocks or physical matter could never have done. It was logical, and it was obvious. Suddenly, I simply KNEW that God exists; and, I had used Science and the Scientific Methods to gain that KNOWLEDGE.

I had effectively used the Scientific Method to prove to myself that God exists, by using the scientific methods to **falsify** everything that was the opposite of God, Non-Locality, Intelligence, Non-Local Consciousness, and Psyche. This process became SCIENTIFIC PROOF OF GOD'S EXISTENCE, through a process of falsifying everything that is false.

The beauty of this whole process is that I finally realized that the scientific methods and scientific observations have falsified Materialism, Naturalism, Darwinism, Nihilism, and Atheism trillions of trillions of times in thousands of different ways – while at the same time, scientific methods, observations, and Lived Experience continue to VERIFY the truth over and over again. Thousands of people have seen and touched our Resurrected Lord Jesus Christ. That REALITY keeps getting verified by scientific observations or direct observations or Lived Experiences over and over and over again. THE TRUTH is always VERIFIED by the scientific methods or direct observations or Lived Experience.

Lived Experience, including our non-local experiences or spiritual experiences, IS Scientific Evidence and becomes Scientific Proof of God's Existence. This IS Science at its very best! Through Lived Experience or Direct Scientific Observations, we can go directly to TRUTH, KNOWLEDGE, and PROOF of God's Existence. Powerful, huh?

**Quantum Non-Locality**, when properly understood scientifically, FALSIFIES Materialism, Darwinism, Naturalism, Nihilism, and even Atheism. Quantum Non-Locality IS the opposite of Materialism, Darwinism, Naturalism, Nihilism, and Atheism.

Non-Local means Non-Physical, Transdimensional, or Spiritual. Materialism, Darwinism, Naturalism, and Atheism have at their very core the major premise and the primary philosophical assumption which claims that the non-local, the non-physical, the immaterial, and the spiritual do not exist. These false philosophies and false religions are FALSIFIED each and every time we see a ray of light or feel the effects of gravity, which are in fact non-local and non-physical spiritual phenomena. Every time you have a thought or a dream, or feel the effects of time, you have in fact FALSIFIED Materialism, Naturalism, Darwinism, Nihilism, and even Atheism by experiencing something non-physical. These false materialistic philosophical concepts are FALSIFIED every time that Quantum Non-Locality is verified scientifically or experienced directly.

According to the Bible, the human spirit body is a God, sired and parented by God and the Goddesses. According to the Bible, your physical body is descended from the Gods, because Adam and his physical body were literally the son of God. Your surname is God! Every time we humans have a Near-Death Experience, Out-of-Body Experience, Shared-Death Experience, or a Spiritual Experience we have in fact FALSIFIED Materialism, Naturalism, Darwinism, Nihilism, and Atheism which do in fact make the claim as their primary assumption that the Non-Local, the Non-Physical, and the Spiritual do not exist. The Lived Experiences, or the Direct Observations, or the Scientific Observations of the human race as a whole FALSIFY Materialism, Naturalism, Nihilism, and Atheism. Lived Experiences including Spiritual Experiences ARE Scientific Observations and Scientific Evidence, or at least they should be if we are truly Scientists as we claim to be.

Near-Death Experiences and Out-of-Body Experiences are direct verification of Quantum Non-Locality and the Spirit Realm or the Non-Local Realm. These types of Lived Experience or Scientific Evidence also falsify Nihilism.

Quantum Non-Locality, Quantum Entanglement, and "Spooky Action at a Distance" have been repeatedly VERIFIED through scientific experimentation. Quantum Non-Locality IS the spirit realm, where people go whenever they separate from their physical body. This Non-Locality stuff has been OBSERVED and VERIFIED trillions of times. Every time you have a thought or a dream, you have in fact had a spiritual experience, because there is NO way for our scientists to record your thoughts and dreams using physical instruments to do so. The contents of your thoughts and dreams cannot be detected nor recorded by physical instruments, because they are non-local or spiritual in nature and origin.

There is NO evidence and will be NO evidence to support Materialism, Naturalism, Darwinism, Nihilism, and Atheism because they ARE FALSE. Instead, ALL of the evidence and ALL of the scientific observations and ALL of our Lived Experiences keep pointing us to non-locality, psyche, non-local consciousness, the non-local realm, the quantum realm, the spirit realm, the transdimensional realm, spiritual beings, intelligent beings, and psyche beings. The evidence keeps pointing us to some Grand Intelligence, Super Psyche, God, or Ultimate Scientist as our designer and creator and as the person who consciously commanded physical matter into existence in the first place.

It's literally impossible to falsify THE TRUTH using the Scientific Methods. The scientists will never be able to falsify Intelligent Design Theory, or design and creation by Intelligent Beings, or design and creation by Intelligent Psyches, because it's true. Design and Creation by Intelligence or Psyche can be verified and replicated on demand! The truth never fails and is never falsified!

ONLY Psyche – ONLY Intelligence – ONLY Intelligent Beings can design and create things, and do science. Psyche or Intelligence is the ONLY thing we have ever observed doing so. You can verify this right now for yourself, by choosing to design and create something for someone, because the rocks can't do it for you. It's elementary my dear friend.

As far as I know, I'm the ONLY person on the planet who is using Science and various different Scientific Methods directly and purposefully to FALSIFY Materialism, Darwinism, Naturalism, Scientism, Nihilism, and Atheism. Nobody else has ever thought to do so! I've never seen it done by anyone else, because everyone else has chosen to believe that it can't be done and that it will never be done. I have had dozens of Atheists and Materialists tell me that it can't be done, and for a while there, I used to believe them. THEY WERE WRONG!

Lived Experiences, including Spiritual Experiences or Out-of-Body Experiences, ARE Scientific Observations and Scientific Evidence. It has to be this way, or we will never be able to use Scientific Observations or Scientific Methods in order to get at THE TRUTH of our reality and our existence. Remember, Direct Observation is an integral part of the Scientific Methods – the most important part of the Scientific Methods! Science is ALL about Lived Experience or Direct Observation of any kind, or at least it should be, if we are in fact Scientists and not simply sophists and philosophers as the Materialists, Naturalists, and Atheists are. Lived Experience, including our Spiritual Experiences or Non-Local Experiences, IS Scientific Evidence which is based upon Scientific Observations!

See: Van Lommel, P. (2010). *Consciousness Beyond Life: The Science of the Near-Death Experience*. New York: HarperCollins.

Quantum Mechanics IS the Science of Near-Death Experiences. Quantum Mechanics is Spiritual Mechanics, or the way that spirit matter really works. Quantum Mechanics falsifies Materialism, Naturalism, Darwinism, Nihilism, and even Atheism.

Lived Experience, including our non-local experiences or spiritual experiences, IS Scientific Evidence and becomes Scientific Proof of Quantum Non-Locality. This IS Science at its very best! Through Lived Experience or Direct Scientific Observations, we can go directly to TRUTH, KNOWLEDGE, and PROOF of the Non-Local Realm or the Spirit Realm. Powerful, huh?

**Scientific Proof of God's Existence**

Through the direct process of eliminating ALL the falsehoods that we know about and falsifying ALL of the false theories that we know about such as Materialism and Naturalism and Atheism, this whole process becomes Scientific Proof of God's Existence. THE TRUTH is what remains after ALL of the false theories have been FALSFIED and eliminated from consideration. THE TRUTH is the thing which the scientific methods or direct observations repeatedly verify and never falsify. TRUTH and KNOWLEDGE are based upon Lived Experience, which is in fact Scientific Evidence. THE TRUTH is the thing which the scientific methods can't falsify! Over and over again, the scientific methods and scientific observations VERIFY THE TRUTH! It becomes obvious and clear, once a person chooses to look and see and trust and believe. Lived Experience IS Scientific Observation and Scientific Evidence. The Scientific Methods always verify the truth, if they are done right.

How often have I said to you that when you have eliminated the impossible [and the false], whatever remains, *however improbable*, must be the truth? — **Sherlock Holmes.**

So, what is this GRAND TRUTH, which the scientific methods can't falsify and which the scientific methods and scientific observations continuously verify?

This GRAND TRUTH is that ONLY intelligent beings or psyche beings can design, program, engineer, create, manufacture, and deploy genomes, physical life forms, computers, software, hardware, glassware, plastics, cars, houses, airplanes, rockets, space ships, internets, radio, bridges, skyscrapers, theories, ideas, plans, reasons, and blueprints. ONLY Psyche or Intelligence can do teleology, final causality, cosmic fine-tuning, thinking, consciousness, dreaming, imagination, intentionality, awareness, reasoning, knowledge, truth, and life. ONLY Psyche or Intelligence can ACT as an independent Agent. ONLY Psyche can do moral agency. ONLY Psyche can do Out-of-Body Experiences and remember having done so. ONLY Psyche can have Near-Death Experiences and see and embrace the Biblical God Jesus Christ in the spirit world and remember having done so. ONLY Psyche or Non-Local Consciousness can convert spirit matter into physical matter. This IS a Psyche Ontology, or the Ultimate Model of Reality, wherein Psyche is the fundamental unit of reality!

ONLY Psyche or Non-Local Consciousness can do scientific observations or Lived Experiences! ONLY Psyche can go to the spirit realm or the non-local realm and make scientific observations or have lived experiences, which are in fact Scientific Evidence and could be used as evidence in a court of law. ONLY the human psyche can record its lived experiences or its scientific observations which it has in the spirit realm, and then share that scientific evidence or those lived experiences with other human beings or other human psyches. When taken as a whole, these observations, knowledge, truth, lived experiences, spiritual experiences, direct observations, direct experiences of the non-local, and scientific evidence ultimately become Scientific Proof of God's Existence.

Why?

Only Psyche can do lived experience, observation, and science. Only God's Psyche or the Ultimate Scientist could have done ALL of the science that needed to be done, which the rocks or raw physical matter could NEVER have done. ONLY God's Psyche or the Ultimate Scientist could have done the conscious observation necessary to convert spirit matter into physical matter. If God didn't exist, there would be NO physical matter, and there would be NO genomes and NO physical life forms. This IS Scientific Proof of God's Existence!

Science is supposed to be about Observation, Lived Experience, and Knowledge of the Truth. God's Psyche or the Ultimate Scientist IS the only thing left standing when everything else has been falsified and eliminated. The Biblical God Jesus Christ IS the Being of Light whom we encounter during our Near-Death Experiences and Spiritual Experiences – during our Direct Scientific Observation of the Non-Local Realm! We KNOW this to be TRUE from the Lived Experiences or the Scientific Observations of the human race! Lived Experiences, including Spiritual Experiences or Non-Local Experiences, ARE Scientific Observations and Scientific Evidence.

If we are going to do Science for REAL, then we need a new definition for Science. Science is NOT Materialism; and, Science is NOT Naturalism. Materialism, Naturalism, Darwinism, Nihilism, and Atheism are metaphysics and philosophy, NOT Science. Science IS Lived Experience, including our Spiritual Experiences or Non-Local Experiences. Science IS Lived Experience or Direct Observation of THE TRUTH. Science IS KNOWLEDGE, which is based upon our Lived Experiences or our Scientific Observations, including our Spiritual Experiences and Non-Local Observations. That's what Science really is and really should be – the sum total of our Lived Experiences or Scientific Observations, including our Spiritual Experiences and our Out-of-Body Experiences. Lived Experience or Psyche Experience IS Scientific Evidence. I'm talking about a Paradigm Shift here – switching over to a Psyche Ontology and a Psyche Epistemology as the Ultimate Model of Reality.

Within this Ultimate Model of Reality or Ultimate Paradigm, psyche or non-local consciousness is the fundamental unit of reality. This is a Psyche Ontology.

Within this Ultimate Model of Reality or Ultimate Paradigm, lived experiences or psyche experiences or direct observations including spiritual experiences are the best way of finding and knowing the truth. This is a Psyche Epistemology.

I have had many different Atheists, Naturalists, Darwinists, Nihilists, and Materialists tell me that I don't understand Science and that I don't understand the Scientific Method. You will have to decide for yourself whether you agree with them or not. Now you know what I think, and that's all that really matters to me.

In summary, God or the Ultimate Scientist must of necessity exist, or your genome would not exist, the physical matter in your physical body would not exist, and you would not exist as a physical living being. This is scientifically accurate and scientific truth. This IS Scientific Proof of God's Existence! If God did not exist, you would not exist, and you wouldn't be reading this right now. That's what Science and the scientific methods are trying to tell us; but, the Materialists, Naturalists, and Atheists refuse to look, listen, and learn. It's their loss. I KNOW, because I used to be a Materialist, Nihilist, and Atheist. The grass really IS greener on the Theistic side of the fence, because I have been there now and KNOW that to be true as well.

Materialism, Naturalism, and Atheism are based upon a refusal to look at evidence and a denial of Reality. These people are in denial! There is NO evidence and will be NO evidence to support Materialism, Naturalism, Nihilism, and Atheism because these things contradict Lived Experiences, Empirical Reality, and Experiential Evidence; but, there IS tons of Lived Experience, or Scientific Observation, or Direct Observation, or Scientific Evidence supporting the existence of and the resurrection of our Lord Jesus Christ – if we choose to equate Lived Experience with Scientific Evidence as we should do if we are truly Scientists.

Remember, if it matches with Reality then it is declared to be scientifically accurate. If it has been falsified and contradicts Reality, then it is unscientific and false. If it contradicts the Lived Experiences of the human race, then it is FALSE. Materialism, Naturalism, Nihilism, Darwinism, Scientism, and Atheism contradict Reality and contradict the Lived Experiences or the Scientific Observations of the human race; therefore, we KNOW that these things are FALSE.

Throughout the rest of this book, I will go back to the beginning and explain in greater detail how I came to these conclusions. It didn't come to me all at once. It was at least two years in the making. It has taken me a while to refine my presentation. All I can say is that this whole thing is a lot more fun and infinitely more interesting than My Materialism and My Atheism ever were!

Go with the best and get rid of all the rest. That's what I finally decided to do.

Best wishes,

Mark My Words

May 2017

## The Materialists Don't Want You Using Scientific Theories to Prove the Truth and to Find the Truth

I have had Materialists and Atheists tell me that theories cannot be used to prove anything, because they are theories. That was kind of confusing and frustrating. There's some really strange ideas and information floating around concerning Science and the Scientific Method. I think what these people were trying to tell me is that Science and the scientific methods cannot be used to prove anything. These people ARE WRONG! What good is Science and the scientific methods if they can't be used to find and know the truth? What good is Science and the scientific methods if they can't be used to PROVE things?

I have had multiple Scientists, Atheists, Naturalists, and Materialists tell me that I don't understand Science and the Scientific Method. I have observed that the Materialists, Naturalists, Nihilists, and Atheists are always wrong. It's these people who really don't understand Science and the Scientific Methods. These people don't have a clue. I KNOW, because I used to be one of them. There used to be a time when I believed them, because I used to be a Materialist, Nihilist, and Atheist. These people had successfully convinced me that I would never be able to use Science nor Theories nor Methods to prove anything. I believed them, because I used to be a practitioner and adherent of Scientism. I saw no other way and knew of no other way.

These people had convinced me that Science and the Scientific Methods could never be used to prove Materialism, Naturalism, Darwinism, and Atheism false. THEY WERE WRONG! Falsifying Theories is precisely what the Scientific Method is good for and precisely what the Scientific Method was designed for!

After I repeatedly caught these people trying to trick us and deceive us, I stopped trusting them. Since then, Science and Logic and the Scientific Methods have proven to me that the Materialists and the Atheists are ALWAYS wrong. Once I got rid of My Materialism and My Atheism, Science and Scientific Methods have been proving things to me right and left. Materialism and Naturalism were holding me back and stunting my intellectual growth; and, I certainly am not the first victim of these false and unproductive and unscientific ideologies.

Materialism is the chosen philosophical belief that the Non-Local, or the Non-Physical, or the Spiritual does NOT exist. Ironically, there is NO Scientific Evidence and can be NO Scientific Evidence to support such a belief. Belief in Materialism requires a monumentally huge blind-leap of faith, a blind faith which is in fact Non-Physical or Spiritual in nature. Belief in Materialism requires a Spiritual and Non-Physical and Philosophical blind leap of faith, the very fact of which proves that the Spiritual exists and proves that Materialism is fundamentally FALSE.

I eventually realized that Materialism, Darwinism, Naturalism, and Atheism CANNOT BE used to prove anything, because they are FALSE. We cannot use anything that is false to prove something to be true; so, in the end, the Materialists and Atheists were right in that THEIR THEORIES cannot be used to prove anything, because their theories are false. However, once we know that ALL of their theories are false, we can indeed use their atheistic and materialistic falsehoods, deceptions, and lies to point us towards THE TRUTH. I do it all the time now.

Nowadays, I actively USE the Materialists, Darwinists, Naturalists, and Atheists to point me to the things that they don't want me seeing, reading, and understanding because I KNOW that the Science and the Theories which the Materialists and Atheists have formally rejected are in fact THE TRUTH and will always end up being THE TRUTH. I learned to use their atheistic theories and materialistic theories to point me to THE TRUTH. Genius, huh?

Consequently, I use their false theories to point me to the truth and to prove the truth to myself and to my readers and my friends.

If your Science isn't proving stuff to you, then you ain't doing it right or your theories are false. Remember, a false theory can be used to point us to the truth, because the truth is often the opposite of any theory which has been demonstrated to be false. Therefore, false theories can be used to point us to the truth and thereby prove the truth. Remember, Science is supposed to be the pursuit of truth, meaning that Science is supposed to eliminate the False Theories so that ONLY the True Theories remain.

This book is about using SCIENCE and THE SCIENTIFIC METHOD to eliminate the falsehoods and the impossible, in an attempt to demonstrate and prove THE TRUTH! What good is Science if Science can't be used to prove stuff?

In this book, I apply THE SCIENTIFIC METHOD, common sense logic, abductive reasoning, and deductive reasoning to various Scientific Theories in an attempt to do the real Science of eliminating False Hypotheses while at the same time keeping the True Hypotheses.

The goal is to find THE TRUTH among all of the materialistic and atheistic falsehoods that are typically presented to us for our consideration.

This was a fun book to research and write, because it made logical sense to me from beginning to end.

THE TRUTH, whenever I find it and wherever I find it, IS parsimonious and makes logical sense. THE TRUTH tastes good and feels good. There is NO confusion or doubt associated with THE TRUTH. You just know that it is TRUE.

The Materialists and Atheists are right. There is NO way to use Materialism and Atheism to prove that God exists. But, imagine what you could do if you got rid of Materialism and Atheism, as I have done!

The Spiritual Sciences or the Non-Physical Sciences become logical and rational possibilities once we have successfully eliminated Materialism and Atheism and Naturalism from our worldview or personal philosophy of life. It's like stepping through a door into the light. And, once a person has successfully made that transition into light and truth, he or she typically doesn't want to go back into the darkness and the lies. I don't want to go back to My Atheism and My Materialism, now that I KNOW how limiting, irrational, and boring they really were. That's just the true reality of the situation!

Non-Local means not located in our Physical 3D Space-Time. Non-Local means Non-Physical or Trans-Dimensional or Spiritual. Materialism, Naturalism, Darwinism, and Atheism cannot be used to study Nonlocality or Trans-Dimensionality, because the Materialists formally reject the Non-Local Sciences! There are Atheists and Materialists online right now mocking and ridiculing String Theory because it is Non-Local, Trans-Dimensional, and Spiritual in nature.

The Materialists and Atheistic Scientists are trying to severely limit how Science is used and how Science can be used; and, these people use force, intimidation, mocking, and ridicule in order to do so. This is one of the main reasons why I have lost all respect for Materialism, Darwinism, Naturalism, and Atheism. I don't like the tactics used by the people who promote these falsehoods and lies.

Furthermore, their materialistic rejection of the Spiritual Sciences or the Non-Physical Sciences (such as Quantum Nonlocality, Light, Time, String Theory, Trans-Dimensionality or Non-Local Reality, Quantum Mechanics or Spiritual Mechanics, Origins, Gravity, Infinite Velocity, Mathematics, Dark Matter or Spirit Matter, Dark Energy or the Zero-Point Field or the Light of Christ or the Quantum Sea of Light, Friendship and Love and Justice and Mercy, Philosophy of Mind, Psychology or the Study of the Human Spirit, and Non-Local Consciousness) LED ME to reject My Materialism and My Atheism so that I would then be free to study and learn about the Spiritual Sciences or the Non-Physical Sciences which exist in great abundance. Materialism is completely worthless when it comes to the Non-Physical Sciences!

YOU will NEVER be able to use Materialism or Atheism to prove that God exists or to prove that the Spiritual or the Non-Physical exists. Materialism is completely worthless when it comes to the Non-Local or the Non-Physical. Materialism can't be used to prove things, so what good is it? Materialism is Bad Science! Once I realized how much Materialism was holding me back and keeping me stupid and blind, then I was glad to be rid of it. The Scientist in me demanded no less! We each have a God-given right to reject the falsehoods and lies and to embrace THE TRUTH instead. The Materialists and Atheists are trying to take that away from us.

I have proven to myself that Materialism if FALSE and why it is false, which means that I have proven to myself that its opposite is TRUE and really exists. Your mileage may vary, especially if you haven't seen the evidence that I have seen.

The Materialists refuse to believe in the Non-Physical or the Non-Local, which means that the Materialists believe in Creation by Physical Matter or Creation by ROCKS. The Materialists really truly believe that the ROCKS designed and created it all, including everything that existed before the first physical matter was designed and created. The reason that we go to the public schools is so that we can be brainwashed by the Materialists and Atheists into believing in the IMPOSSIBLE; and, these people are very successful at what they do because they will fail you or fire you if you refuse to go along with their brainwashing procedures. Being awarded a PhD in Materialism of some kind is the ultimate form of brainwashing that our public schools offer us.

I have formally rejected Materialism and have chosen to go in pursuit of the Spiritual Sciences or the Non-Physical Sciences instead.

In this book, I discuss the various Scientific Theories and Scientific Evidence which proved to me that God exists. I found the Scientific Proofs of God's Existence infinitely more believable and convincing than the Philosophical Proofs of God's Existence. The Philosophical Proofs of God's Existence are only convincing if they have tons of Scientific Evidence supporting the truthfulness of the Premises and the Conclusion.

I no longer find any proofs based upon Materialism to be the least bit convincing, because I have trained myself to see and understand the falsehoods, deceptions, and lies upon which their Premises are based. When it comes to Materialism, Darwinism, Naturalism, and Atheism, since most of their Premises can be demonstrated to be false, that means that ALL of their Conclusions are false as well. Instead, I USE the falsehoods and lies of Materialism and Naturalism to point me to THE TRUTH. Science is supposed to be the pursuit of truth after all! Since I KNOW that Materialism, Atheism, Darwinism, and Naturalism are FALSE, I KNOW that their opposites are likely to be true; and, that ends up being a good place to start my pursuit of THE TRUTH.

# PART VII — THE BOOKS I HAVE WRITTEN IN AN ATTEMPT TO SUPPORT THESE THEMES

1. "Summary Of: The Theory of Evolution Proves that God Exists"

   https://www.amazon.com/dp/B01GQCWED6

   https://www.amazon.com/dp/1521130485

An introductory summary to my much larger book, "The Theory of Evolution Proved to Me that God Exists: Why I Am No Longer an Atheist and Why I No Longer Believe in the Theory of Evolution".

The second edition of this book contains an introduction to my much larger book, "The Ultimate Model of Reality: Psyche Is the Ultimate Cause", which is my core message to the world.

07JUN2016

—

2. "The Theory of Evolution Proved to Me that God Exists: Why I Am No Longer an Atheist and Why I No Longer Believe in the Theory of Evolution"

   https://www.amazon.com/dp/B01HZYBZ7K

   https://www.amazon.com/dp/1521131228

Over the period of a few months, I had an epiphany or a major breakthrough; and in the process, the Theory of Evolution proved to me clearly and conclusively that God Exists. I have had no doubt about God's existence, after the Theory of Evolution proved to me that God Exists. By that point in time, I was officially done with My Materialism and My Atheism and firmly in a different camp. I'm now a part of God's army. This huge book, "The Theory of Evolution Proved to Me that God Exists", explains in great detail why I am no longer an Atheist and why I no longer believe in the Theory of Evolution.

This book documents my journey of discovery and the defense of my thesis.

04JUL2016

—

3. "The Scientific Method Proves That the Theory of Evolution Is False"

   https://www.amazon.com/dp/B01IAAIRT2

   https://www.amazon.com/dp/1521133611

In this book, I document what I know about the Scientific Method, and I demonstrate how I use The Scientific Method in my pursuit of The Truth.

I have learned to trust The Scientific Method, The Rules of Science, Abductive Reasoning, and Deductive Reasoning to give me a clear and accurate assessment of the evidence at hand. In this book, I use Science, The Scientific Method, the Rules of Science and Scientific Research, Abductive Reasoning, and Deductive Reasoning to prove that the Theory of

Evolution is false and why it is false. You will have to judge for yourself if I meet my Burden of Proof through a preponderance of the evidence. I don't hold anything back!

Remember, we can indeed use the Scientific Methods to falsify theories or to prove theories false. Most people don't realize that, but it's true. That's what I did to the Theory of Evolution – I used various different scientific methods to falsify Darwinism, Naturalism, Materialism, Creation by Rocks, and the Theory of Evolution. Once you have successfully eliminated ALL of the falsehoods such as Materialism, Atheism, Naturalism, and Darwinism, you are left staring at THE TRUTH. It's elementary my dear friend.

The second edition of this book contains an introduction to my much larger book, "The Ultimate Model of Reality: Psyche Is the Ultimate Cause", which is my core message to the world.

11JUL2016

—

4. "The Second Comforter: Supping with Our Resurrected Lord Jesus Christ"

https://www.amazon.com/dp/B01IAKHTY6

https://www.amazon.com/dp/152113281X

1) – 3) The first three parts of this book, are my standard defense package against Materialism. I found that the best defense is a good offense. I first had to get rid of my residual Atheism, Materialism, and Darwinism, BEFORE I was free to pursue knowledge and information about Spirit, Spirituality, Quantum Mechanics or Spiritual Mechanisms, Revelations from God, Revelations of God, the Nature of God, the Scriptures of God, the Holy Ghost, and THE SECOND COMFORTER.
4) Exploring THE SECOND COMFORTER Experience and what it really means to sup with the Lord. The Focal Point of this book.
5) What is an "Anointed Savior", a "Jesus Christ"? Who can be called to fill that Position or Role?

The second edition of this book contains an introduction to my much larger book, "The Ultimate Model of Reality: Psyche Is the Ultimate Cause", which is my core message to the world.

I also included my essays on Lived Experience and My Scientific Discoveries, at the end of this book.

11JUL2016

—

5. "Quantum Mechanics from a Non-Physical Spiritual Perspective"

https://www.amazon.com/dp/B01J023TGU

https://www.amazon.com/dp/1521132380

I realized that Quantum Mechanics makes perfect logical sense from a Non-Physical Spiritual Perspective. Quantum Mechanics makes no sense whatsoever from a Materialistic Perspective. Materialism is the worst way and the most-limited way to interpret anything, especially Quantum Mechanics. The purpose of this treatise is to demonstrate clearly, logically, scientifically, and conclusively that the Trans-Dimensional Realm, or Non-Local Realm, or Non-Physical Realm, or Spirit Realm, or Quantum Realm truly exists, and that Materialism is FALSE. You will have to decide for yourself if I meet my Burden of Proof through a preponderance of the evidence. What I wrote in this book ended up becoming core foundational material for many of my other books.

The second edition of this book contains an introduction to my much larger book, "The Ultimate Model of Reality: Psyche Is the Ultimate Cause", which is my core message to the world.

24JUL2016

—

6. "Using the Scientific Method to Eliminate the Usual Suspects and to Prove the Truth"

https://www.amazon.com/dp/B01J6STHP0

https://www.amazon.com/dp/1521133581

Using Science and The Scientific Method to eliminate the Usual Suspects, in an attempt to demonstrate and prove THE TRUTH! In this book, I line up the Usual Suspects, gather and define the Evidence; and then, I apply The Scientific Method, common sense logic, abductive reasoning, and deductive reasoning to them, in order to see if any one of them could have done the job of designing and creating this Physical Universe, this Earth, and ALL of the Genomes and Life Forms on this Earth. The goal is to find the Correct Suspect among all of the Usual Suspects, who are typically presented to us for our consideration. This was a fun book to write, because it made logical sense to me from beginning to end.

Remember, we can indeed use the Scientific Methods to falsify theories or to prove theories false. Most people don't realize that, but it's true. That's what I did to the Theory of Evolution – I used various different scientific methods to falsify Darwinism, Naturalism, Materialism, Creation by Rocks, and the Theory of Evolution. Once you have successfully eliminated ALL of the falsehoods such as Materialism, Atheism, Naturalism, and Darwinism, you are left staring at THE TRUTH. It's elementary my dear friend.

The second edition of this book contains an introduction to my much larger book, "The Ultimate Model of Reality: Psyche Is the Ultimate Cause", which is my core message to the world.

27JUL2016

—

7. "NATURE vs. NURTURE vs. NIRVANA: An Introduction to Reality"

https://www.amazon.com/dp/B01JWRCSVA

https://www.amazon.com/dp/1521132615

In this book, I take the greatest and most controversial philosophical issues in human history, and I solve them all quickly and easily through a simple change in one's worldview or point-of-view. I introduce a bit of REALITY into the equation, which solves all of these different problems quite nicely.

This book is based upon one of my epiphanies; and thus, the contents of this book form part of my core foundational message to the world.

The second edition of this book contains an introduction to my much larger book, "The Ultimate Model of Reality: Psyche Is the Ultimate Cause", which is my core message to the world.

07AUG2016

—

8. "I Am Not a Creationist: So What Am I?"

https://www.amazon.com/dp/B071XTM8XY

Creationism is always presented by the Materialists, Naturalists, Behaviorists, and Atheists as some type of magic or voodoo.

My scientific observation is that Materialism, Naturalism, and Atheism are in fact THE MAGIC in all of this, because Materialism, Naturalism, and Darwinism reduce to "design and creation by rocks", which is synonymous with MAGIC, because the rocks have NEVER been caught in the act of design and creation and never will be.

ONLY intelligent beings or intelligent psyches have the capability to design, create, program, organize, engineer, field-test, manufacture, deploy, and do science.

Quantum Mechanics tells us that Psyche or Non-Local Consciousness and its associated Conscious Observation or Word of Command is required to convert spirit matter into physical matter. Furthermore, scientific observation tells us that ONLY intelligent beings or intelligent psyches have the ability to design, program, engineer, manufacture, field-test, organize, and create new and unique physical objects, including genomes and life forms.

Materialists, Naturalists, Darwinists, and Atheists ridicule and mock Theistic Creation while at the same time choosing to believe that the rocks or raw physical matter designed and created ALL of the genomes and life forms on this planet. As a result, these Materialists and Atheists present creation and creationism as some sort of magical Creation Ex Nihilo. I'm NOT that type of creationist. I do NOT believe in Creation Ex Nihilo or magic. And, I'm not the type of creationist who believes that the rocks or physical matter can design and create genomes and life forms. In other words, I'm NOT a Materialistic, Naturalistic, and Darwinistic Creationist either!

Instead, I choose to believe in the Truths which are presented to us by the Lived Experiences of the human race. We KNOW from the Lived Experiences of the human race that Psyche, Consciousness, Intelligence, or Life exists. We KNOW from the Lived Experiences of the human race that spirit matter is something completely different than Psyche. We KNOW from Quantum Mechanics that matter or Quantum Objects can be in either a spirit matter state of existence or a physical matter state of existence but NOT in both states simultaneously. We KNOW from Quantum Mechanics that Psyche or Conscious Observation is required to convert spirit matter into physical matter, which makes it kind of clear and obvious that Psyche or Non-Local Consciousness is immaterial and doesn't consist of spirit matter nor physical matter.

According to these Truths and KNOWLEDGE which we have gleaned from the Lived Experiences of the human race, we KNOW that Psyche or Non-Local Consciousness has always existed and will always exist. Psyche is immortal, immaterial, infinite, transcendent, and co-eternal with God. We also KNOW that spirit matter has always existed and will always exist; and, we KNOW that psyche is something completely different than spirit matter. Psyche acts. Matter, including spirit matter and physical matter, are acted upon by Psyche.

Again, we KNOW from Quantum Mechanics that ONLY Psyche can convert spirit matter into physical matter. Finally, we KNOW from the observations and lived experiences of the human race that ONLY Psyche or Intelligent Beings can organize physical matter into new, unique, and useful forms such a genomes, physical life forms, computers, cars, skyscrapers, bacteria, dinosaurs, planets, stars, galaxies, universes, and anything else that you can imagine. This scientific observation has been verified trillions of times with an infinite number of more times to go. ONLY intelligent beings or intelligent psyches can design and create and work with physical matter. Scientific methods or scientific observations will ALWAYS verify the truth. In contrast, Materialism and Naturalism and Atheism have been falsified by the scientific methods trillions of different times in thousands of different ways, because the rocks can't design and create.

So, technically, I am not a Creationist, because there really is no such thing as Creation Ex Nihilo; and, there's definitely NO such thing as Creation by Rocks or Creation by Physical Matter, because the rocks can't design and create.

So, what am I?

I'm more of an Organizationist and Scientist.

We KNOW from the Lived Experiences of the human race that Psyche and Spirit Matter have always existed and will always exist. We KNOW from Science, Logic, and Quantum Mechanics that ALL physical matter has a beginning when Psyche or Non-Local Consciousness commands spirit matter to convert into physical matter. But, converting spirit matter into physical matter is NOT creation ex nihilo – it is instead transmutation or transformation from one state of existence to a different state of existence.

Finally, when God's Psyche organizes physical matter into genomes and life forms, that process is NOT creation either, although most people would tend to classify it as a creative process. However, technically, God is organizing that physical matter into new and useful types and forms. God is NOT a Creationist either, because God doesn't do magic or creation ex nihilo. God is a Scientist, Engineer, and Manufacturer.

This book explains in part how I finally came to this Ultimate Conclusion. I am NOT a Creationist. I don't believe in Creation by Rocks or creation by physical matter, as the Darwinists, Materialists, Naturalists, and Atheists do. And, I don't believe in Creation ex Nihilo or Creation by Magic as many of the Theists and Christians do.

I'm an Organizationist or Scientist; and, I KNOW from observation and the Lived Experiences of the human race that the Biblical God Jesus Christ and His Father are the Ultimate Scientists and the Ultimate Organizationists. I'm NOT a Creationist, and God and Christ are NOT Creationists either. They are Scientists and Organizationists. They are in the Construction Business.

God and Christ took what already existed and lifted it to a greater and higher form or state of existence and organization. That's technically NOT creation. That's applied Science and Organization and Construction. God and Christ commanded physical matter and our physical space-time into existence; and, then they proceeded to organize some of that physical matter into planets, stars, galaxies, genomes, and physical life forms. That's the work of Scientists, not Creationists.

It's elementary my dear friend!

This book documents a lot of what I went through during 2015 and 2016 while I was trying to discover this information and gain this knowledge.

18APR2017

---

9. "The Ultimate Model of Reality: Psyche Is the Ultimate Cause"

https://www.amazon.com/dp/B071NC9JK6

This book introduces my Psyche Ontology or my Ultimate Cause Model of Reality, wherein Psyche is the fundamental unit of reality.

In this book, I present a new and unique Philosophy of Science which demonstrates the existence of Psyche and then proceeds to employ Psyche in ALL aspects of Philosophy, Science, Application, Spirituality, Lived Experience, and Knowledge.

This becomes a Psyche Paradigm or the Ultimate Model of Reality.

22APR2017

---

10. "Putting Psyche Back into Psychology: Restoring Science to Consciousness"

https://www.amazon.com/dp/B071NC987S

This book further develops my Psyche Ontology or Ultimate Cause Model of Reality, wherein Psyche is the fundamental unit of reality.

In this book, I document how Psyche was originally an integral part of Psychology, how Behaviorism caused Psychology to lose its mind, and the different types of Personality Theory or Psyche Theory which have attempted to put Psyche back into Psychology.

It is my intention and proposition to include Near-Death Experiences (NDEs), Shared-Death Experiences (SDEs), Out-of-Body Experiences (OBEs), and other types of Lived Experience into Psychology or the "Study of Psyche" as an integral and essential part of Psychology.

NDEs, OBEs, SDEs, Spiritual Experiences, and other types of Lived Experience ARE scientific proof of Psyche's existence. These things should be a part of the Science of Psychology or the Science of Psyche.

22APR2017

---

11. "BioPsychoSocial: Including Psyche or Light into our Theoretical Models"

https://www.amazon.com/dp/B0713NDHVW

This book further develops my Psyche Ontology or Ultimate Cause Model of Reality, wherein Psyche is the fundamental unit of reality.

In this book, my primary focus is on Mental Illness and various different forms of Psychotherapy. This book chronicles my search for the Ultimate Psychotherapy and will also chronicle in part my brush with mental illness and addiction. Addiction causes psychosis and mental illness. Everyone who goes through withdrawal will experience some kind of psychosis or mental illness.

I end this book by discussing what I believe to be the BEST forms of psyche-therapy.

22APR2017

—

12. "God Is in the Light: God is light, and in Him is no darkness at all."

https://www.amazon.com/dp/B07168S37N

In this book, I explore Light and God's Psyche. Psyche is Living Light, or light which is conscious, alive, intelligent, self-aware, and universally connected with its surroundings.

23APR2017

—

13. "Tripping the Light Fantastic: How Prescription Drugs Almost Killed Me"

https://www.amazon.com/dp/B071RJP9T8

I got ill.

I willingly let The Doctors get me addicted to half a dozen different prescription drugs; and at the peak of my misery, I was taking over a dozen different prescription drugs on any given day.

I went insane, and I tried to kill myself.

This book chronicles my addiction, and some of the things I went through trying to get sober. I discuss some of the strange things that I experienced during the withdrawal process while I was out of my mind. And, I discuss many of the interesting things that I learned along the way.

This is the book that my friends told me that I must write, because the world needs to hear my story.

23APR2017

—

14. "Scientific Proof of God's Existence: A Primer"

https://www.amazon.com/dp/B071713NNL

https://www.amazon.com/dp/1521325170

This book is a basic Primer explaining how I developed a Scientific Proof of God's Existence. The concepts and ideas were three or four years in the making. It wasn't obvious at first how this should be done or if it could be done.

18MAY2017
—

## The Grand Deception and the GRAND TRUTH

The Theory of Evolution: The philosophical belief that Random Mutations and Natural Selection designed and created all of the genomes and life forms on this planet from scratch. This philosophical belief has been repeatedly FALSIFIED by various different scientific methods and scientific disciplines. There's no compelling reason to believe in the Theory of Evolution anymore, because every aspect of it has been successfully FALSIFIED by Science and the scientific methods, as well as by logical common sense. Mutation/Selection have NEVER been caught in the act of design and creation, because it's impossible for these things to design and create anything.

Materialism: Design and creation by rocks or raw physical matter. Materialism is the philosophical belief that only physical matter (and its doppelganger, energy) exists. Materialism teaches that $E = mc^2$ is all that there is to life, the universe, and everything. Materialism is Abiogenesis, Spontaneous Generation, and Macro-Evolution – the idea that raw physical matter put itself together and designed and created everything from scratch out of thin air. Materialism, or design and creation by rocks, is impossible. It has NEVER been observed nor caught in the act.

Naturalism: Design and creation by Nature or Raw Physical Matter. Naturalism is the philosophical belief that the Supernatural, the Transdimensional, and the Non-Local do not exist. Naturalism is the philosophical belief that Psyche, or Non-Local Consciousness and Spirit Matter, do not exist. Naturalism is taken on blind faith as being true, because there can never be any Observational Evidence or Experiential Evidence to support this belief. Instead, ALL of the Observational Evidence, Experiential Evidence, Eye-Witness Evidence, Corroborated Veridical Evidence, and Empirical Evidence that we have on hand as a race FALSIFIES Materialism and Naturalism. The proven existence of Quantum Non-Locality and the Quantum Zeno Effect FALSIFIES Materialism and Naturalism. Furthermore, if you believe the Big Bang Theory to be true, as most scientists do, then you simply KNOW that the Big Bang Singularity was a non-local transdimensional supernatural object, and that the Big Bang itself was a Supernatural Event, because physical matter didn't come into existence until 380,000 years later according to the most recent estimates. There's no way in the universe that physical matter could be the ONLY thing that exists, because physical matter didn't exist until 380,000 years AFTER the Big Bang Supernatural Event. If the Big Bang truly happened, as the evidence suggests, then its very existence FALSIFIES Materialism and Naturalism and VERIFIES the existence of the Supernatural. It's impossible for physical matter or Nature to have designed and created our physical universe before physical matter and Nature even existed. Soon you come to realize that EVERYTHING FALSIFIES Materialism and Naturalism, except for Materialism and Naturalism.

Atheism: Design and creation by Nothing from Nothing. Atheism is the philosophical belief that God, or the Grand Designer and Creator, does not exist. Atheism is Creation Ex Nihilo. Science and scientific observations have proven this claim false. There's NO such thing as Abiogenesis, Spontaneous Generation, Macro-Evolution, Creation by Chance, or Creation Ex Nihilo. It has NEVER been observed and has NEVER caught in the act; and it never will be, because it's impossible. Atheism or Creation Ex Nihilo is scientifically impossible.

Nihilism: Design and creation by Chance. Nihilism is the philosophical belief that there is no such thing as Non-Locality, Transdimensionality, or an Afterlife. Chance cannot design and create, which means that someone or something Non-Local and Transdimensional had to exist BEFORE the design and creation of this physical universe in order to bring the first particle of physical matter into existence in the first place.

Everything that has a beginning has a Beginner, who brought it into existence. This reality and truth applies directly to physical matter and this physical universe. The proven existence of anything Non-Local or Transdimensional falsifies Nihilism. If you believe the Big Bang Theory to be true, as most scientists do, then you simply KNOW that the Big Bang Singularity was a non-local transdimensional supernatural object, and that the Big Bang itself was a Supernatural Event, because physical matter didn't come into existence until 380,000 years later according to the most recent estimates. There's no way in the universe that physical matter could be the ONLY thing that exists, because physical matter didn't exist until 380,000 years AFTER the Big Bang Supernatural Event. Scientific evidence supporting the Big Bang Theory successfully falsifies Materialism, Naturalism, and Nihilism because the Big Bang Singularity was a supernatural transdimensional non-local object and the Big Bang itself was a Supernatural Event. That's what all the scientific evidence is telling us. The proven existence of Quantum Non-Locality, Quantum Entanglement, and the Quantum Zeno Effect falsifies Materialism, Naturalism, and Nihilism. Furthermore, Nihilism has been successfully FALSIFIED by Observational Evidence and Corroborated Veridical Evidence from Near-Death Experiences (NDEs) and Out-of-Body Experiences (OBEs). The way that we successfully falsify falsified philosophies, such as Nihilism, is with Experiential Eye-Witness Evidence or Observational Evidence which has been corroborated and verified as being true. Nihilism has been repeatedly FALSIFIED with corroborated veridical evidence thousands of times and experiential evidence millions of times. You only need one; but, thousands and millions of NDEs and OBEs will do, especially when they have been verified and corroborated.

Behaviorism: Behaviorism is the philosophical belief that Psyche, Mind, Personality, Individuality, or Non-Local Consciousness does not exist. Behaviorism teaches that only physical behaviors exist. This philosophy has been FALSIFIED numerous different ways, through Observational Evidence and Experiential Evidence and Logical Common Sense. This false philosophy is particularly FALSIFIED by chosen behaviors which take place in the Non-Local Realm, or Transdimensional Realm, or Spirit Realm as experienced by Out-of-Body Travelers.

Scientism: Scientism is the philosophical belief that the Scientific Method is the ONLY way to find and know the truth. This philosophical idea was FALSIFIED by discovering other ways to find and know the truth. The Truth is KNOWN best by living it and experiencing it for yourself, or by choosing to trust someone who has. Science and the scientific methods pale in comparison.

Hard Determinism: Determinism is the philosophical belief that there is no such thing as Free Will, Freedom, or Choice. This philosophical idea is FALSIFIED every time your dog or your children choose to disobey you. This philosophical idea is FALSIFIED simply by choosing to eat pizza one morning and choosing to eat something different next, or choosing to eat something different the next morning. This philosophical idea is FALSFIED by choosing to let a coin toss make your choices for you for a while, and then arbitrarily out-of-the-blue choosing to take back control of your choices when you don't like what the coin is doing to you.

Reductive Materialism and Atomism: This is the philosophical belief that ONLY atoms or physical matter exist in the void. The proven existence of Light, Magnetism, Dark Energy, Forces, Fields, Dark Matter, Non-Locality, Quantum Entanglement, and the Quantum Zeno Effect FALSIFIES Atomism, Materialism, and Physical Reductionism. These falsified philosophies can't stand in the Light of the Scientific Evidence.

These Falsified Philosophies failed to meet their burden of proof, because the preponderance of the evidence stands firmly against them and tells us that they are false. All of these Falsified Philosophies are denialistic exclusionary philosophies which claim that something or someone does not exist. There's NO way to verify or observe the existence of something that by definition does not exist. There's no way to verify or observe the non-existence of something. It simply has to be taken on blind faith as being true. ALL of these Falsified Philosophies have to be taken on blind faith as being true, because ALL of the Observational Evidence, Experiential Evidence, and Corroborated Veridical Evidence tells us that they are false. The moral of the story is that philosophy FAILS in the Light of Evidence. Every false and falsified philosophy was taken down by Observational Evidence, Eye-Witness Evidence, Experiential Evidence, Empirical Evidence, Corroborated Veridical Evidence, Experimental Evidence, Scientific Evidence, and THE TRUTH.

These false and falsified philosophies are the Keystone of the Grand Deception. Everything in the Grand Deception focuses upon them, leans against them, and rests upon them.

**How often have I said to you that when you have eliminated the impossible [and the false], whatever remains, *however improbable*, must be the truth? — Sherlock Holmes.**

By eliminating everything that is false and everything that has been falsified, ONLY the Truth remains. This is logical common sense.

So, what is this GRAND TRUTH, which the scientific methods can't falsify and which the scientific methods and scientific observations continuously verify?

This GRAND TRUTH is that ONLY intelligent beings or psyche beings can design, program, engineer, create, manufacture, and deploy genomes, physical life forms, computers, software, hardware, glassware, plastics, cars, houses, airplanes, rockets, space ships, internets, radio, bridges, skyscrapers, theories, ideas, plans, reasons, and blueprints. ONLY Psyche or Intelligence can do teleology, final causality, cosmic fine-tuning, thinking, consciousness, dreaming, imagination, intentionality, awareness, reasoning, knowledge, truth, and life. ONLY Psyche or Intelligence can ACT as an independent Agent. ONLY Psyche can do moral agency. ONLY Psyche can do Out-of-Body Experiences and remember having done so. ONLY Psyche can have Near-Death Experiences and see and embrace the Biblical God Jesus Christ in the spirit world and remember having done so. ONLY Psyche or Non-Local Consciousness can convert spirit matter into physical matter. This IS a Psyche Ontology, or the Ultimate Model of Reality, wherein Psyche is the fundamental unit of reality!

ONLY Psyche or Non-Local Consciousness can do scientific observations or Lived Experiences! ONLY Psyche can go to the spirit realm or the non-local realm and make scientific observations or have lived experiences and do chosen behaviors, which are in fact Scientific Evidence and could be used as evidence in a court of law. ONLY the human psyche can record its lived experiences or its scientific observations which it has in the spirit realm, and then share that scientific evidence or those lived experiences with other human beings or other human psyches. When taken as a whole, these observations, knowledge, truth, lived experiences, spiritual experiences, direct observations, direct experiences of the non-local, and scientific evidence ultimately become Scientific Proof of God's Necessity and Scientific Proof of God's Existence.

The purpose of the Books that I have written has been to falsify and debunk the Grand Deception and to replace that thing with THE TRUTH. Eventually I decided that THE TRUTH is the only thing that has value to me in the end. Scientific inferences and philosophical speculation are worthless in comparison.

## You Have to Believe in the Theory of Evolution

Back in 1980, my Christian biology teacher in college told me that I have to believe in the Theory of Evolution because the scientists have no other explanation for the origin of life on this planet. His message to me was that I have to believe in the Theory of Evolution – I have no choice but to believe in it, because there is nothing better. The whole of Science tells us that the Theory of Evolution is true.

This was when the Theory of Evolution, Materialism, and Naturalism were at their height of dominance, the pinnacle of their prominence, and the Grand Deception was in full force. Back then, there was NO way to find and know the truth, because Science and the Scientists hadn't uncovered the truth yet.

Looking back at the History of Science, I can see that my Christian biology teacher was right. Science hadn't progressed enough nor learned enough by 1980 to be able to successfully debunk and falsify the Theory of Evolution. Back in 1980, the Grand Deception reigned supreme.

Things have changed massively since then.

Since 1980, starting with the 1986 book *Evolution: A Theory in Crisis* by Michael Denton, Science and the various different scientific methods and scientific disciplines – especially Quantum Mechanics – have been successfully used to FALSIFY Materialism, Naturalism, Darwinism, Behaviorism, Determinism, the Theory of Evolution, Physical Reductionism, Atomism, Creation by Mutation/Selection, Creation by Chance, Creation Ex Nihilo, and Abiogenesis or Macro-Evolution.

When those went down, Atheism went down with them; and, the Grand Deception had finally been broken. Unlike forty or fifty years ago, anyone who is looking for the truth can actually find it and know it now, if they want to. There's NO compelling reason to believe in the Theory of Evolution anymore, because Science and the scientific methods have FALSIFIED every aspect of it. Science tells us clearly and conclusively that Evolution (genetic change, random mutations, and natural selection) couldn't have done the job of design and creation, so who did?

Unlike forty or fifty years ago, there's now volumes of information that have been researched and written which provide Scientific Proof of God's Necessity and therefore Scientific Proof of God's Existence.

So, what is this Grand Truth which the Grand Deception so successfully hid from us for millennia?

The answer to that question comes in large part by studying what was needed to debunk and falsify Nihilism, or the belief that there is no such thing as an afterlife.

Starting sometime after 2005, observational evidence and experiential evidence corroborating and verifying Near-Death Experiences (NDEs) and Out-of-Body Experiences (OBEs) finally hit critical mass on YouTube. There are now thousands of first-hand NDEs and OBEs on YouTube with dozens more coming every day. Each and every NDE and OBE falsifies Nihilism, Behaviorism, Materialism, and Naturalism. Again, when those go down, Atheism, Darwinism, and the Theory of Evolution go down with them, as fruit from the poisoned tree.

The Grand Truth, which the Grand Deception so successfully hid from us for millennia, is that Psyche or Non-Local Consciousness is the fundamental unit of Reality and that the Transdimensional Realm, Non-Local Realm, or Spirit World is the fundamental construct of Reality.

When Materialism and Naturalism and Nihilism were finally falsified with Observational Evidence, Eye-Witness Evidence, Experiential Evidence, Scientific Evidence, Empirical Evidence, Corroborated Veridical Evidence, and Compelling Evidence, then suddenly WE KNEW THE TRUTH. When the Grand Deception was broken, then ONLY the Truth remained.

## Signs of Psyche – BioPsychoSocial

Barlow, D. H. & Durand, V. M. (2015). *Abnormal Psychology: An Integrative Approach* (7th ed.). Stamford, CT: Cengage Learning.

http://mypsyche.us/wp-content/uploads/2017/10/Abnormal-Psychology.pdf

I was impressed with this college textbook, overall. They never once mentioned Psyche or Mind directly; but, they went out of their way to allow ALL of the research evidence into evidence; and, they refrained from defining Psyche or Consciousness as an epiphenom of the physical brain. They allowed all of the evidence for Neuroplasticity into evidence. Neuroplasticity is observational evidence and experimental evidence which demonstrates and proves that the adult physical brain can be modified and changed by PsychoSocial factors.

Their Integrative Approach is the BioPsychoSocial Model in action.

The BIO portion of the BioPsychoSocial Model examines the effects of biological and genetic influences upon Psychology and Abnormal Behavior.

The PSYCHO portion of the BioPsychoSocial Model examines the effects of Psyche, Choice, Will Power, Mind, and Non-Local Consciousness on Psychology and Abnormal Behavior.

This book taught me a lot about how Science should be done. So far, this book has ended up being my most favorite "college textbook that I was forced to buy and read".

**Because we are never 100% confident that our experiments are internally valid – that no other explanations are possible – we must be cautious about interpreting our results.** (*Abnormal Psychology*, p. 105.)

That's the primary flaw of the Scientific Method, the fact that each piece of scientific evidence can theoretically have an infinite number of different plausible explanations or interpretations.

**An Integrated Model**

**Putting the factors together in an integrated way, we have described a theory of the development of anxiety called the *triple vulnerability theory*.**

**The first vulnerability (or diathesis) is a *generalized biological vulnerability*.** [BIO – genetic inheritance.] **We can see that a tendency to be uptight or high-strung might be inherited. But a generalized biological vulnerability to develop anxiety is not sufficient to produce anxiety by itself.**

**The second vulnerability is a *generalized psychological vulnerability*.** [PSYCHO – chosen beliefs and chosen behaviors.] **That is, you might also grow up believing the world is dangerous and out of control and you might not be able to cope when things go wrong base on your early experiences. If this perception is strong, you have generalized psychological vulnerability to anxiety.**

**The third vulnerability is a *specific psychological vulnerability* which you learn from early experience, such as being taught by your parents, that some situations or objects are fraught with danger (even if they really**

aren't). [SOCIAL – things that we are taught and learn from other psyches.] For example, if one of your parents is afraid of dogs, or expresses anxiety about be evaluated negatively by others, you may well develop a fear of dogs or social evaluation. These triple vulnerabilities are presented in Figure 5.3 and revisited when we describe each anxiety and related disorder.** (*Abnormal Psychology*, p. 127.)

Our Biology or physical body is constructed from our genome and from some kind of Non-Local Blueprint that is stored or remembered somewhere in Non-Locality, possibly within our spirit body itself. The Gods designed and created everything spiritually, before they organized it or created it physically. The Gods are scientists.

Psyche or Non-Local Consciousness is identified by chosen beliefs and chosen behaviors, both here in the Physical Realm and there in the Non-Local Supernatural Transdimensional Spirit Realm. Learning to control your environment is a function of Psyche. Environment and genes cannot control what you choose to believe and choose to do. Environment and genes can influence your chosen behaviors but cannot cause your chosen behaviors. Belief is a choice; and, choice is a function of Psyche. Chosen behaviors are also a function of Psyche, because only Psyche can choose what to make of its environment and genetic inheritance. Remember, Psyche can choose to trump nature and nurture at will.

Social, Environmental, and Cultural Influences are identified by the things that we are taught and learn from our Society, our Human Culture, and our Environment, including other psyches or the animals that we interact with.

Developmental Disorders are BioPsychoSocial in nature and origin, having come from all three aspects or factors. There can be genetic disorders during our development (BIO), personality disorders during our development (PSYCHO), and social or environmental disorders and mishaps during our development (SOCIAL). When it comes to Psychology (the Study of the Human Psyche), the BioPsychoSocial Model is the best and most realistic model for Psychology and the study of Abnormal Behavior or Psychopathology.

**Current thinking – based upon growing evidence from highly sophisticated research techniques – points to at least three specific neurochemical abnormalities simultaneously at play in the brains of people with schizophrenia.**

**Evidence for neurological damage in people with schizophrenia comes from a number of observations.** (*Abnormal Psychology*, pp. 495-496.)

Think about it!

Just three neurotransmitters or neurochemicals just slightly off or out-of-balance, and these people are insane for most of their lives. When it comes to schizophrenia, there's more than one problem going on here. I keep asking myself, "How did random mutations and natural selection develop and fine-tune this thing?" It's all about fine-tuning. If these neurotransmitters are out-of-balance, you are going to go nuts. The evidence makes it obvious and clear that there had to be some kind of Designer and Fine-Tuner behind our neurotransmitter system. It's too complex and required way too much fine-tuning for any other explanation to be possible, logical, or rational.

**How Many Types of Neurotransmitters Are There in a Human Brain?**

**It depends on how you count, but maybe 30 - 100 different molecule types, with 10 of them doing 99% of the work.**

In the big scheme of things, there are three main categories of neurotransmitters:

"Small molecule" neurotransmitters (glutamate, GABA, dopamine, serotonin, noradrenaline, acetylcholine, and histamine)

Neuropeptides (endorphins, oxytocin, and possibly 100 more)

Other (small molecules like nitric oxide, adenosine, ATP, glycine; or larger molecules like endocannabinoids)

In terms of total number of neurotransmitters, maybe 10 small molecules (less than 30 atoms each) and 100+ neuropeptides (50 - 200 atoms each), with 20 or so neuropeptides getting most of the research attention.

https://www.quora.com/How-many-types-of-neurotransmitters-are-there-in-a-human-brain

As I studied neurotransmitters and the human brain, one thing became abundantly clear and obvious to me. There's no way in the universe that this irreducibly complex neurotransmitter system could have developed by chance – step-by-step from one generation to the next through random genetic mutations. How on earth did natural selection and random mutations program for all the different neurotransmitters and neuropeptides and their complex interactions in the neurons, axons, dendrites, and synapses? If one of these is off, out-of-balance, or missing, then all sorts of nasty things can and do happen. The whole system had to be functional and balanced from the beginning or all of our ancestors would have killed themselves off due to all the insanity, assuming of course that they weren't born dead to begin with. Yet, the Materialists, Naturalists, and Darwinists want you to believe that random mutations and natural selection designed and created and fine-tuned the whole shebang, including your genome, your physical body, your neurotransmitter and neuropeptide systems, and your physical brain. But, it's impossible. Mutation and selection do not design and create; instead, they lead us to chaos, entropy, disorder, death, destruction, and extinction.

Observations, experiences, eye-witness accounts, and scientific research TRUMP scientific inferences, wishful thinking, and philosophical speculation every time, in my humble opinion.

**An Integrative Theory**

**How do we put all this together?**

**Basically, depression and anxiety may often share a common, genetically determined biological vulnerability that can be described as an overactive neurobiological response to stressful life events.**

**Again, this vulnerability is simply a general tendency to develop depression (or anxiety) rather than a specific vulnerability for depression or anxiety itself.**

**To understand the causes of depression, we must look at psychological vulnerabilities as well as life experiences that interact with genetic vulnerabilities.**

**People who develop mood disorders also possess a psychological vulnerability experienced as feelings of inadequacy for coping with**

difficulties confronting them as well as depressive cognitive styles. When these vulnerabilities are triggered, the pessimistic "giving up" process seems crucial to the development of depression.

Researchers have learned a great deal about the neurobiology of mood disorders during the past several years. Findings on the complex interplay of neurochemicals are beginning to shed light on the nature of mood disorders.

As we have noted, the principal effect of medications is to alter levels of these neurotransmitters and other related neurochemicals. Other biological treatments, such as electroconvulsive therapy, dramatically affect brain chemistry. A more interesting development, however, alluded to throughout this book, is that powerful psychological treatments [psyche treatments] also alter brain chemistry. (Abnormal Psychology, pp. 245-246.)

**To understand the causes of depression, we must look at psychological vulnerabilities as well as life experiences that interact with genetic vulnerabilities.** Here again, they are talking about the BioPsychoSocial Model of Reality and Abnormal Behavior. **To understand the causes of depression** [or any other mental illness], **we must look at psychological vulnerabilities** [Psyche or the PSYCHO part of the BioPsychoSocial Model] **as well as life experiences** [Social Interaction or the SOCIAL part of the BioPsychoSocial Model] **that interact with genetic vulnerabilities** [Biology and Genetics or the BIO part of the BioPsychoSocial Model].

I like this kind of science and scientific evidence; but then again, your mileage may vary, because you are a completely different person or psyche.

**I can suggest that just because certain circuits of memory or swiftness of synapses may fail, thought and awareness and consciousness do not.** (Abnormal Psychology, pp. 245-246.)

Thought, dreams, awareness, cognitions, and consciousness are functions of Psyche and Non-Local Consciousness, which means that they continue long after our physical body and physical brain are dead and gone, according to the empirical evidence derived from Near-Death Experiences (NDEs), Shared-Death Experiences (SDEs), Out-of-Body Experiences (OBEs), and other types of Non-Local Experiences or Spiritual Experiences.

Psychology was originally defined as "the study of the human psyche".

Thoughts, mind, and unconscious processes are a function of Psyche or Non-Local Consciousness. Remember, the medications or drugs can't do Psyche-Therapy, because they can't directly affect the human psyche. Only social interaction and psyche-to-psyche interaction can do Psyche-Therapy successfully. In fact, the medications and the drugs CAUSE many of our mental illnesses, including Substance-Induced Psychosis.

Remember, under Science 2.0, observational experiences TRUMP scientific inferences and philosophical speculations every time. The one is EXPERIENCE or EVIDENCE; whereas, the other is simply a *guess* or *wishful thinking*. Science 2.0 goes with the experience and the evidence every time.

Trust, but verify. The only caveat when it comes to Psyche Evidence or Non-Local Evidence is that you must learn to verify the evidence. Corroborated Veridical Evidence is the best type of Psyche Evidence or Non-Local Observations.

When it comes to observational evidence, experiential evidence, empirical evidence, non-local evidence, psyche evidence, and scientific evidence of any kind, it MUST meet its burden of proof through a preponderance of the evidence.

One Near-Death Experience (NDE) is an anomaly; but, the thousands of NDEs and OBEs reported on YouTube combined with the millions of NDEs documented or reported world-wide become EMPIRICAL EVIDENCE and SCIENTIFIC EVIDENCE through a preponderance of the evidence.

Barlow, D. H. & Durand, V. M. (2015). *Abnormal Psychology: An Integrative Approach* (7th ed.). Stamford, CT: Cengage Learning.

http://mypsyche.us/wp-content/uploads/2017/10/Abnormal-Psychology.pdf

I was impressed with this college textbook, overall. They never once mentioned Psyche or Mind directly; but, they went out of their way to allow ALL of the research evidence into evidence; and, they refrained from defining Psyche or Consciousness as an epiphenom of the physical brain. They allowed all of the evidence for Neuroplasticity into evidence. Neuroplasticity is observational evidence and experimental evidence which demonstrates and proves that the adult physical brain can be modified and changed by PsychoSocial factors.

As part of their Integrative Approach, these scientists go out of their way to include the PsychoSocial factors into their explanations and interpretations of the Scientific Research. This is Science, not the philosophical exclusions and intricate denials of the Materialists and Naturalists.

## Science 2.0 Formally and Functionally Opposes the Denialistic Philosophies and the Exclusionary Philosophies

As an ALL-INCLUSIVE way of doing science, Science 2.0 formally and functionally opposes the Falsified Philosophies, the Denialistic Philosophies, and the Exclusionary Philosophies – such as Materialism, Naturalism, Darwinism, Scientism, Behaviorism, Determinism, Nihilism, and Atheism. Science 2.0 does so by allowing ALL of the evidence into evidence. Materialism, Naturalism, and their derivatives are based upon a denial of evidence, a rejection of evidence, and the exclusion of evidence. In contrast, Science 2.0 allows all of the evidence into evidence. Naturally, Science 2.0 is a bit more robust than Materialism and Naturalism, which deny the evidence that Science 2.0 accepts.

The best way to find and know the truth is to live it and experience it for yourself, or to choose to trust someone who has. Truth is found and KNOWN through observation, experience, and eye-witness accounts.

The second-best way to find and know the truth is to falsify and eliminate everything that is false such as Materialism, Naturalism, Darwinism, Scientism, Behaviorism, Determinism, Physical Reductionism, Nihilism, and Atheism. If we successfully eliminate everything that is false and everything that has been falsified, then ONLY the truth will remain. Technically, the truth cannot be falsified without lying in the process. The only way to falsify the truth is through prevarication or deception of some kind.

Remember, there is no observational evidence and can be no observational evidence for the things that do not exist. Likewise, there can be no observational evidence or experiential evidence for the non-existence of something. You cannot experience the non-existence of something nor observe that something does not exist. Materialism, Naturalism, Darwinism, Scientism, Behaviorism, Hard Determinism, Physical Reductionism, Nihilism, and Atheism are based exclusively on the non-existence of someone or the non-existence of something, which means that there can never be any observational evidence to support these things. Instead, these things are in fact FALSIFIED by observational evidence, experiential evidence, eye-witness evidence, experimental evidence, scientific evidence, and empirical evidence.

Over and over again, Materialism and Naturalism are falsified by observational evidence and experiential evidence, which means that Materialism and Naturalism are impossible to support, sustain, and verify because they have no observational evidence supporting them and because all of the observational evidence and experiential evidence that we have on hand as a race falsifies Materialism and Naturalism and their derivatives.

**How often have I said to you that when you have eliminated the impossible [and the false], whatever remains, *however improbable*, must be the truth? — Sherlock Holmes**

Eliminate the FALSIFIED PHILOSPHIES, the DENIALISTIC PHILSOPHIES, and the EXCLUSIONARY PHILOSOPHIES such as Materialism, Naturalism, Darwinism, Scientism, Nihilism, Atheism, **Behaviorism, Determinism, Physical Reductionism,** Creation Ex Nihilo, Creation by Chance, Spontaneous Generation, Abiogenesis, Creation by Physical Matter, Creation by Mutation/Selection, Creation by Macro-Evolution, and Creation by Rocks; and, we are left staring at the truth. It's obvious, and it's logical. It's elementary my dear reader!

Once you have eliminated everything that is FALSE, whatever remains, however improbable, must be the truth!

Science 2.0 goes for the TRUTH rather than philosophical speculation, sophistry, denial, exclusion, scientific inferences, credentialism, and the personal ad hoc interpretations of the data from the Materialists and the Naturalists.

I'm not going to apologize for being impressed by Observational Evidence, Experiential Evidence, Empirical Evidence, Experimental Evidence, and Eye-Witness Accounts because these are the best way to do science and to find the truth. It's certainly a lot better and more convincing than claiming that something or someone does not exist, as the Materialists and Naturalists do.

Observational Evidence and Experiential Evidence TRUMPS scientific inferences, personal speculation, wishful thinking, and personal interpretations of the data every time; or at least, they should.

**Applied Science**

Jeffrey M. Schwartz, J. M., & Begley, S. (2002). *The Mind and the Brain: Neuroplasticity and the Power of Mental Force*. New York: HarperCollins.

If you love applied science as I do, you will find the Introduction to this book, *The Mind and the Brain: Neuroplasticity and the Power of Mental Force*, particularly fascinating.

> **The will, I was starting to believe, generates a force. If that force could be harnessed to improve the lives of people with OCD, it might also teach them how to control the very brain chemistry underlying their disease.**
>
> **It is not merely that the will is not free, in the modern scientific view; not merely that it is constrained, a captive of material forces. It is, more radically, that the will, a manifestation of mind, does not even exist, because a mind independent of brain does not exist. My deep doubts that human actions can be explained away through materialist determinism simmered just below the surface throughout my years of medical school.**
>
> **The most noteworthy result of mindfulness, which requires directed willful effort, is the ability it affords those practicing it to observe their sensations and thoughts with the calm clarity of an external witness: through mindful awareness, you can stand outside your own mind as if you are watching what is happening to another rather than experiencing it yourself. In Buddhist philosophy, the ability to sustain Bare Attention over time is the heart of meditation. The meditator views his thoughts, feelings, and expectations much as a scientist views experimental data — that is, as natural phenomena to be noted, investigated, reflected on, and learned from. Viewing one's own inner experience as data allows the meditator to become, in essence, his own experimental subject.**
>
> **In what came to be called the Four Steps regimen of cognitive-behavioral therapy for OCD, patients gain insight into the true nature and origin of the bothersome OCD thoughts and urges. They *Relabel* their obsessions and compulsions as false signals, symptoms of a disease. They *Reattribute* those thoughts and urges to pathological brain circuitry ("This thought reflects a malfunction of my brain, not a real need to wash my hands yet again"). They *Refocus*, turning their attention away from the**

pathological thoughts and urges onto a constructive behavior. And, finally, they *Revalue* the OCD obsessions and compulsions, realizing that they have no intrinsic value, and no inherent power. If patients could systematically learn to reassess the significance of their OCD feelings and respond differently to them through sustained mindful awareness, I reasoned, they might, over time, substantially change the activity of the brain regions that underlie OCD. Their mind, that is, might change their brain.

[Using their Psyche, Mind, Non-Local Consciousness, or Will, these patients successfully *Re-Wired* their brains.]

There, I propose that the time has come for science to confront the serious implications of the fact that directed, willed mental activity can clearly and systematically alter brain function; that the exertion of willful effort generates a physical force that has the power to change how the brain works and even its physical structure. The result is directed neuroplasticity. The cause is what I call directed mental force.

Through the mental act of focusing attention, mental effort becomes directed mental force.

For it is now clear that the attentional state of the brain produces physical change in its structure and future functioning. The seemingly simple act of "paying attention" produces real and powerful physical changes in the brain. In fact, Stapp's work suggests that there is no fully defined brain state until attention is focused. That physical activity within the brain follows the focus of attention offers the clearest explanation to date of how my hypothesized mental force can alter brain activity. The choice made by a patient — or, indeed, anyone — causes one physical brain state to be activated rather than another. A century after the birth of quantum mechanics, it may at last be time to take seriously its most unsettling idea: that the observer and the way he directs his attention are intrinsic and unavoidable parts of reality.

Schwartz, Jeffrey M. *The Mind and the Brain: Neuroplasticity and the Power of Mental Force* (pp. 7, 8, 11, 14, 17, 18, 19). HarperCollins. Kindle Edition.

http://mypsyche.us/wp-content/uploads/2017/10/The-Mind-and-the-Brain-Neuroplasticity-and-the-Power-of-Mental-Force.pdf

Source:

https://www.amazon.com/Mind-Brain-Neuroplasticity-Power-Mental/dp/0060988479

HarperCollins permits quotes for review and promotional purposes. I'm telling people to buy and read this book.

I personally found Cognitive Therapy and Mindfulness useful and effective in my own life while dealing with my own issues.

The book, *Feeling Good*, by David Burns saved my life, because it saved me from the debilitating effects of perfectionism. This book is still the best introduction to Cognitive Therapy and Applied Psychology and Self-Help Psychology, in my humble opinion.

My therapist and trainer called Mindfulness, "Being Present in the Present". You learn to let go of the Past and stop worrying about the Future. Be Present. And, that's the end of depression, anxiety, and fear.

## My Autobiography

I have a bit of a story to tell, which explains in part why I no longer trust and believe in Materialism, Naturalism, Nihilism, Darwinism, Scientism, Nihilism, and Atheism.

It's simply wrong and false to conclude that genetics or our reinforcement history causes ALL of our mental illnesses and abnormal behaviors, as the Materialists and Naturalists and Behaviorists do, because there are times when the Human Psyche can trump nature and nurture AT WILL.

I KNOW this is true, because there was a time when I was addicted to half a dozen different prescription drugs, and I CHOSE to go cold turkey and suffer the consequences in order to get well in the end. That CHOICE was nothing that my genetics or my environment did for me or made me do.

In fact, my genetic inheritance was fueling my addictions because we have a tendency towards addiction on my mother's side; and, the medical doctors were feeding my addiction by pumping a dozen different drugs into me in the first place. These things were causing my addiction; therefore, it was left up to me, my psyche, to CHOOSE to end my addiction. It was something that I, my psyche, CHOSE to do for itself. It was nothing that they (my genetics and medical doctors) did for me. My Psyche had to trump my nature and my nurture in order to successfully overcome all my different addictions and survive the six-month long withdrawal process that was the result of my CHOICE.

Neuroscience and Genetics have demonstrated that "psyche" can actually turn genes on or off, and that "psyche" actually changes and shapes our physical brain – processes and disciplines known as Epigenetics and Neuroplasticity.

Any time you witness chosen behaviors, chosen decisions, chosen actions, chosen vices, chosen crimes, chosen sins, chosen malevolence, chosen love, chosen addictions, chosen neurodegenerative diseases, chosen illnesses, and a choice to go cold turkey, you are in FACT observing the Human Psyche in action. As human beings, we suffer (or enjoy) the consequences of the things which we have chosen for ourselves. We reap what we sow. Reaping and sowing are functions of Psyche or Choice. Psyche is synonymous with choice, because our Psyche is the thing that does all our choosing and remembers having done so after our physical brain is dead and gone.

Remember, whenever you witness inherited illnesses, you are witnessing genetics in action. Whenever you witness developmental illnesses, medically induced illnesses, and/or environmentally caused illnesses, you are witnessing societal, cultural, and environmental factors in action. Whenever you witness CHOICE or FREE WILL or INTENT, you are witnessing Psyche in action.

Over and over again, any time you witness CHOICE, or free will, or chosen behaviors, or intent, you are in fact observing some kind of PSYCHE in action; and, aside from God, the human psyche is the most powerful psyche on this planet. You will catch the human psyche CHOOSING THINGS INTO EXISTENCE all the time, if you take the time to stop, observe, and learn.

The following book explains how I changed my mind and greatly improved my physical health as a result.

**Book 13. "Tripping the Light Fantastic: How Prescription Drugs Almost Killed Me"**

https://www.amazon.com/dp/B071RJP9T8

I got ill.

I willingly let The Doctors get me addicted to half a dozen different prescription drugs; and at the peak of my misery, I was taking over a dozen different prescription drugs on any given day.

I went insane, and I tried to kill myself.

This book chronicles my addiction, and some of the things I went through trying to get sober. I discuss some of the strange things that I experienced during the withdrawal process while I was out of my mind. And, I discuss many of the interesting things that I learned along the way.

This is the book that my friends told me that I must write, because the world needs to hear my story.

23APR2017

## Science 2.0 Is Based Upon the Non-Local Interpretations of Consciousness and Quantum Mechanics

I want to figure out how things really work rather than denying their existence, like I used to do when I was a Materialist, Naturalist, and Atheist. Thank God my days of hiding from evidence and suppressing the evidence are over.

Under Science 2.0, any evidence supporting Quantum Non-Locality, Non-Physicality, Spirituality, and Non-Local Consciousness is allowed into evidence. Science 2.0 allows all of the evidence into evidence. Nowadays, I examine all the evidence that comes my way; and then, I decide for myself what that evidence means to me personally rather than blindly accepting the materialistic and naturalistic interpretation which is typically provided with the evidence. I choose to doubt the interpretation, not the evidence. The Materialists and Naturalists reverse the process and doubt all the evidence, suppress and hide and eliminate the evidence they don't like, and choose only one allowable interpretation for any evidence that might come their way.

Remember, Science 2.0 allows all of the evidence into evidence.

Science 2.0 would also allow observational evidence supporting Materialism, Naturalism, Nihilism, and Atheism into evidence if there were any empirical evidence that actually supported these philosophical concepts. Materialism, Naturalism, Nihilism, and Atheism are easily falsified, because there's no observational evidence that supports them and their major premises, primary assumptions, and hidden assumptions.

Remember, Science 2.0 allows all of the evidence into evidence. Pseudo-sciences such as Materialism and Naturalism which reject observational evidence, experiential evidence, eye-witness evidence, and empirical evidence are of no value to us in the end. We want evidence – lots and lots of evidence. Science 2.0 is based upon a preponderance of the evidence rather than the rejection and denial of evidence.

Under Science 2.0, Materialism, Naturalism, Nihilism, and Atheism are considered unscientific, because these people deliberately reject observational evidence and experimental evidence that falsifies Materialism, Naturalism, Darwinism, Scientism, Nihilism, and Atheism. These people prefer instead to believe in philosophical conclusions that have never been experienced, never been observed, and can never be experienced nor observed.

These people deliberately suppress, censor, ridicule, and ban any scientific evidence or observational evidence that falsifies Materialism and Naturalism, including the various different **Non-Local Consciousness Interpretations of Quantum Mechanics** or **Psyche Interpretations of Quantum Mechanics** which they don't like and refuse to allow into evidence. Instead, they label these observational sciences and experimental sciences as unscientific pseudo-sciences, because these people literally define science as Materialism, Naturalism, Darwinism, Scientism, Nihilism, and Atheism. Anything that doesn't fit with Materialism and Naturalism, they label as pseudo-science and automatically dismiss it without giving it any consideration whatsoever.

I eventually chose to follow the evidence wherever it might lead me, because I wanted to figure out for myself how things really truly work. I first had to get rid of My Materialism, My Nihilism, and My Atheism before I was able to find what I was looking for; but, these philosophies and religions are really easy to falsify once a person actually starts looking at evidence and thinking about it rationally.

Science 2.0 allows all of the evidence from Quantum Mechanics into evidence, including any Non-Local or Spiritual Interpretations of Quantum Mechanics with their associated observational evidence and experimental evidence.

The following book, *Quantum Mechanics from a Non-Physical Spiritual Perspective*, documents my first foray into this Science Discipline; and, my efforts were based upon what I had learned while reading *Consciousness Beyond Life: The Science of the Near-Death Experience* by Pim van Lommel.

Quantum Mechanics and Quantum Non-Locality are the Science behind Near-Death Experiences (NDEs), Out-of-Body Experiences (OBEs), and Non-Local Consciousness or Psyche.

I have learned to trust and believe in the Non-Local Interpretations of Quantum Mechanics, because they match with ALL the observational evidence, science experiments, and the Lived Experiences of the human race.

The Non-Local Consciousness Interpretation of Quantum Mechanics, the Idealism Interpretation of Quantum Mechanics, the Quantum Consciousness Interpretation of Quantum Mechanics, the Cosmic Consciousness Interpretation of Quantum Mechanics, and the Orthodox Interpretation of Quantum Mechanics FALSIFY Materialism, Naturalism, Nihilism, and eventually Darwinism and Atheism as well.

This is why the Materialists, Naturalists, and Atheists deny, reject, ridicule, censor, ban, mock, and suppress these various Non-Local Consciousness interpretations of Quantum Mechanics, because they successfully falsify Materialism, Naturalism, and their derivatives such as Darwinism and Atheism.

**Book 5. "Quantum Mechanics from a Non-Physical Spiritual Perspective"**

https://www.amazon.com/dp/B01J023TGU

https://www.amazon.com/dp/1521132380

It's all about perspective and interpretation.

I eventually learned why I couldn't understand Quantum Mechanics for most of my life – the first fifty-five years of my life. Quantum Mechanics doesn't make any sense whatsoever from a physical perspective. From a materialistic and naturalistic perspective, Quantum Mechanics is ludicrous and nonsensical.

After reading *Consciousness Beyond Life*, I realized that Quantum Mechanics makes perfect logical sense from a Non-Physical, Non-Local, Spiritual Perspective; whereas, Quantum Mechanics makes no sense whatsoever from a Materialistic Perspective. Materialism or Naturalism is the worst way and the most-limited way to interpret anything, especially Quantum Mechanics. I simply wanted it to make sense. It's no good if it doesn't make any sense.

The purpose of this treatise, *Quantum Mechanics from a Non-Physical Spiritual Perspective*, is to demonstrate clearly, logically, scientifically, and conclusively that the Trans-Dimensional Realm, or Non-Local Realm, or Non-Physical Realm, or Spirit Realm, or Quantum Realm truly exists, and that Materialism is FALSE. You will have to decide for yourself if I meet my Burden of Proof through a preponderance of the evidence. What I wrote in this book ended up becoming core foundational material for many of my other books.

The second edition of this book contains an introduction to my much larger book, "The Ultimate Model of Reality: Psyche Is the Ultimate Cause", which is my core message to the world.

24JUL2016

## Science 2.0 Is Based Upon Eye-Witness Evidence of All Kinds

As a new and upgraded science, Science 2.0 allows all of the evidence into evidence. This is the way that Science should have always been done but wasn't.

Materialism, Naturalism, Darwinism, Scientism, Nihilism, and Atheism are exclusionary philosophies and exclusionary religions. They are based upon the dogmatic exclusion of evidence and a refusal to look at evidence.

The Materialists, Naturalists, and Atheists define science as Materialism, Naturalism, Darwinism, Scientism, Nihilism, and Atheism; and then, these people exclude everything else. Those are the actions of religious fanatics, dogmatic zealots, and blind faith; and not the actions of true scientists. True scientists remain open-minded throughout their lives.

In contrast, Science 2.0 defines science as Knowledge, Observational Evidence, Experiential Evidence, Empirical Evidence, Eye-Witness Evidence, Physical Evidence, Experimental Evidence, Veridical Evidence, and Non-Local Evidence of any kind including Visions of God, Revelations from God, Near-Death Experiences (NDEs), Out-of-Body Experiences (OBEs), Shared-Death Experiences (SDEs), Verified and Confirmed Contact with the Spirits of the Dead, as well as Confirmed and Verified Reincarnation Accounts.

Science 2.0 is an Inclusionary Science – all of the evidence is allowed into evidence – and then that evidence must meet its burden of proof through a preponderance of the evidence. The preponderance of the evidence must convince me that that evidence is real and true.

Even though Science 2.0 allows all of the evidence into evidence, I no longer trust anyone or anything on blind faith alone like I used to do when I was a Materialist, Nihilist, and Atheist. My days of Materialism, Blind Faith, and Rejection of Evidence are over; but, that doesn't mean that I simply check my brain at the door and admit everything into evidence in my own personal life.

**Trust, but verify.**

I want veridical evidence, proven evidence, verified evidence, and confirmed evidence. I want the people who claim Reincarnation or Spiritual Experiences to meet their burden of proof through a preponderance of the evidence; or, I'm going to be skeptical. The same thing would apply to Near-Death Experiences and Out-of-Body Experiences, especially if they were being manufactured under hypnosis or drugs or some such. Verify and confirm whenever possible. Pursue the eye-witness evidence and a preponderance of the evidence.

**2 Corinthians 13: 1: "In the mouth of two or three witnesses shall every truth be established."**

Pursue the eye-witness accounts and the verified accounts whenever possible, no matter whether we are talking about Near-Death Experiences (NDEs) or Reincarnation Accounts. There are a few of them out there that are quite convincing, because they have been verified and proven to be true.

Furthermore, some of these accounts claim to have been given by God as visions or revelation of a previously lived life. If you trust God, then you know that they are true, because God doesn't lie and because God has ways of verifying that it's true.

1. *Reincarnation: Manika, the Girl Who Remembered Her Past Life*:

https://www.youtube.com/watch?v=-L4hXXSq2O8&t=3s

2. Stevenson, I. (2001). *Children Who Remember Previous Lives: A Question of Reincarnation*. Jefferson, NC: McFarland.

3. Rivas, T., Dirven, A., & Smit, R. H. (2016). *The Self Does Not Die: Verified Paranormal Phenomena from Near-Death Experiences*. Durham, NC: IANDS Publications.

4. Anonymous. (2013). *Teachings of the Doctrine of Eternal Lives*. Salt Lake City, UT: Digital Legend Press.

Here are a few others that were a bit less convincing but interesting, nonetheless.

"Beyond the Ashes: Cases of Reincarnation from the Holocaust" by Yonassan Gershom.

"From Ashes to Healing: Mystical Encounters with the Holocaust" by Yonassan Gershom.

"Coming Back: A Psychiatrist Explores Past-Life Journeys" by Paul Perry and Dr. Raymond Moody.

"Children's Past Lives: How Past Life Memories Affect Your Child" by Carol Bowman.

All you need is one verified and proven account of Reincarnation or NDE in order to know that the phenomenon is real and truly happens from time to time; but, here we have a few to think about and consider. Some of them are quite convincing, because they have been verified and proven to be true.

I'm no longer hiding from evidence and avoiding the evidence; and, you shouldn't either, in my humble opinion.

I wrote the following books in an attempt to allow all of the evidence into evidence. I was looking for a preponderance of the evidence.

**Book 4. "The Second Comforter:  Supping with Our Resurrected Lord Jesus Christ"**

https://www.amazon.com/dp/B01IAKHTY6

https://www.amazon.com/dp/152113281X

This book discusses some of the Observational Evidence associated with the Resurrection of Jesus Christ; and, we try to discuss the Second Comforter and what might be needed in order to sup with our Resurrected Lord in the flesh.

1) – 3) The first three parts of this book, are my standard defense package against Materialism. I found that the best defense is a good offense. I first had to get rid of my residual Atheism, Materialism, and Darwinism, BEFORE I was free to pursue knowledge and information about Spirit, Spirituality, Quantum Mechanics or Spiritual Mechanisms, Revelations from God, Revelations of God, the Nature of God, the Scriptures of God, the Holy Ghost, and THE SECOND COMFORTER.

4) Exploring THE SECOND COMFORTER Experience and what it really means to sup with the Lord.  The Focal Point of this book.

5) What is an "Anointed Savior", a "Jesus Christ"? Who can be called to fill that Position or Role?

The second edition of this book contains an introduction to my much larger book, "The Ultimate Model of Reality: Psyche Is the Ultimate Cause", which is my core message to the world.

I also included my essays on Lived Experience and My Scientific Discoveries, at the end of this book.

11JUL2016

**Book 8. "I Am Not a Creationist: So What Am I?"**

https://www.amazon.com/dp/B071XTM8XY

In this book, I discuss the Seth Books from Jane Roberts, which are often used as proof of life after death, proof of the spirit world, and as proof of origins or some kind of pre-mortal life as a spirit.

From what I can tell, Seth is a purged spirit.

We are the spirit children of God the Father and Heavenly Mother. If you can read this, you are a child of God and descended from the Gods, both spiritually and physically.

Satan and his followers are our brothers and sisters who openly rebelled against God and formally rejected His plan of salvation. These spirits didn't want to have anything to do with God or the Atonement of Christ, so they became the demons and evil spirits, and God uses them to try us and test us. They never did receive a physical body nor a mortal life, because they refused to participate in God's plan from the very moment they were given a choice.

Seth doesn't seem to be one of these evil spirits or demons.

Instead, Seth seemed to be a purged spirit, who had gone through an unsuccessful mortal life, had gone through Purgatory or Gehenna, had atoned for his own sins, and then had his memories of his former life purged or eliminated so that he would now be in a position of innocence so that he could be reborn or reincarnated, and try the whole process all over again to see if he could do better next time.

Even though Seth didn't seem to have many memories of his own mortal life, his personality or inclinations seemed to remain intact. He was a liar and a trickster when Jane first contacted him through the Ouija board. We assume that he eventually came clean and started telling them the truth; but, we can't know for sure.

To me personally, Seth seemed to be a bit of a parasite – developing or redeveloping his own personality while interacting with and through Jane Roberts. He was feeding off of her. This observation led me to conclude that Seth is a purged spirit, with most of his memories having been purged during the atonement process in Purgatory or Gehenna. He was still a liar and trickster after the purging; but, he was a bit of a blank slate, having few if any memories of his former lives.

In this book, *I Am Not a Creationist*, I discuss origins and creation, including the idea that we each had some kind of pre-mortal spirit life, as well as possible previous mortal lives.

I am not a Creationist, because I don't believe in Creation by Evolution and I don't believe in Creation Ex Nihilo.

In this book, *I Am Not a Creationist*, I explore my journey of discovery trying to figure out what I am, what I believe, and why I believe it.

I'm not a Creationist in the typical evolutionary sense nor in the traditional Creation Ex Nihilo sense. Creation from Nothing by Nothing is Atheism; and, I'm no longer an Atheist, Materialist, or Nihilist.

Instead, I choose to follow the evidence wherever it might lead me – all of the evidence. I'm a Scientist in the truest sense of the word; and, I believe that the Gods are Scientists too.

18APR2017

## Book 10. "Putting Psyche Back into Psychology: Restoring Science to Consciousness"

https://www.amazon.com/dp/B071NC987S

This book is a work-in-progress.

I have collected dozens of books about Near-Death Experiences. I have listed most of them in this book; and as time allows while finishing this book, I hope to provide a brief review for each book within this book, *Putting Psyche Back into Psychology*.

The purpose of this book is to put Psyche back into Science and Psychology, and to use empirical evidence or experiential evidence in order to do so.

This book further develops my Psyche Ontology or Ultimate Cause Model of Reality, wherein Psyche is the fundamental unit of reality.

In this book, I document how Psyche was originally an integral part of Psychology, how Behaviorism caused Psychology to lose its mind, and the different types of Personality Theory or Psyche Theory which have attempted to put Psyche back into Psychology.

It is my intention and proposition to include Near-Death Experiences (NDEs), Shared-Death Experiences (SDEs), Out-of-Body Experiences (OBEs), and other types of Lived Experience into Psychology or the "Study of Psyche" as an integral and essential part of Psychology.

NDEs, OBEs, SDEs, Spiritual Experiences, and other types of Lived Experience ARE scientific proof of Psyche's existence. These things should be a part of the Science of Psychology or the Science of Psyche.

22APR2017

## Science 2.0 Is Based Upon All of the Evidence that Falsifies Materialism, Naturalism, Darwinism, Nihilism, and Atheism

The best way to find the truth is to identify and eliminate all the falsehoods and lies.

Materialism, Naturalism, Darwinism, Scientism, Nihilism, and Atheism are exclusionary philosophies and exclusionary religions. The Materialists, Naturalists, and Atheists define science as Materialism, Naturalism, Darwinism, Scientism, Nihilism, and Atheism; and then, these people exclude everything else from consideration.

Subsequently, these people suppress, vilify, censor, reject, exclude, ridicule, and mock anything and everything that successfully falsifies Materialism, Nihilism, Naturalism, and Atheism. Materialism, Naturalism, Nihilism, and Atheism of any kind are based upon a refusal to look at evidence and the deliberate exclusion of any evidence that they personally don't like.

Unlike Materialism and Naturalism, Science 2.0 allows all of the evidence into evidence, including all of the evidence that successfully falsifies Materialism, Naturalism, Darwinism, Scientism, Nihilism, and Atheism.

Science 2.0 is an Inclusionary Science – it allows all of the evidence into evidence. Under Science 2.0, we are no longer hiding from evidence, like I used to do when I was a Materialist, Nihilist, and Atheist.

Remember, Science 2.0 allows all of the evidence into evidence.

Now, just because I allow all of the evidence into evidence, that doesn't mean that I leave my brain and my common sense behind at the door. I demand that these people meet their burden of proof through a preponderance of the evidence.

One NDE, or OBE, or SDE, or Reincarnation Account could be a fluke or a fake; but, thousands or millions of them, not so much.

With thousands of NDEs on record and millions of NDEs across this planet in every nation on earth, NDEs as a whole meet their burden of proof through a preponderance of the evidence; and as true scientists willing to follow the evidence wherever it might lead us, we have no choice but to conclude that there must be something to it after all. Through a preponderance of the evidence, we simply KNOW that some of these phenomena are real and truly happened.

Nowadays, I study the evidence, all of the evidence, and then decide for myself how to interpret it and what it means to me personally. I encourage you to do the same.

Science 2.0 is a Paradigm Shift – a new and better way of doing Science – the way that Science should have been done all the way along but wasn't.

## Science 2.0 Is Based Upon All the Observational Evidence that Solves the Mind-Brain Problem

Unlike Materialism and Naturalism, Science 2.0 allows any evidence supporting Psyche or Non-Local Consciousness into evidence. The Materialists and Naturalists will simply tell you that such evidence does not exist and cannot exist; and, there were times when I used to believe them. They were wrong.

Psychology was originally defined as the Study of the Human Psyche.

I have written a number of different articles and books in an attempt to put Psyche back into Psychology.

You will have to read them for yourself and judge whether I have succeeded or not.

Materialism and Naturalism cause the Mind-Brain Problem in the first place; and therefore, the easiest and quickest way to solve the Mind-Brain Problem is to eliminate Materialism, Naturalism, Darwinism, Nihilism, and Atheism.

In the following books, I produced, developed, and refined a Psyche Ontology, which I eventually called **The Ultimate Model of Reality**. In a Psyche Ontology, Psyche or Non-Local Consciousness is the fundamental unit of reality and existence. I concluded that a Psyche Ontology is the ultimate model of reality, because a Psyche Ontology is the ONLY thing I have come across so far that is supported by ALL the Observational Evidence, Experiential Evidence, Scientific Evidence, Experimental Evidence, Empirical Evidence, and Corroborated Veridical Evidence.

I was led to a Psyche Ontology or the Ultimate Model of Reality, when ALL of the evidence started point me to Psyche as the Ultimate Causal Agent, or Psyche as the Ultimate Cause. I arrived at that conclusion simply by following all the evidence to where it wanted to lead me. That's what eventually happens to you when you start allowing all of the evidence into evidence.

I also developed a Psyche Epistemology when I finally realized that Psyche or Non-Local Consciousness is the ONLY way to find and know the truth in any realm of existence or dimension that we might encounter.

Science 2.0 is based upon The Ultimate Model of Reality, because Science 2.0 is based upon ALL the evidence that we have on hand as a race. I go for completion rather than speculation.

Let's finish this thing off!

### Book 7. "NATURE vs. NURTURE vs. NIRVANA: An Introduction to Reality"

https://www.amazon.com/dp/B01JWRCSVA

https://www.amazon.com/dp/1521132615

In this book, I take the greatest and most controversial philosophical issues in human history, and I solve them all quickly and easily through a simple change in one's worldview or point-of-view. I introduce a bit of REALITY into the equation, which solves all of these different problems quite nicely.

This book is based upon one of my epiphanies; and thus, the contents of this book form part of my core foundational message to the world.

The second edition of this book contains an introduction to my much larger book, "The Ultimate Model of Reality: Psyche Is the Ultimate Cause", which is my core message to the world.

07AUG2016

### Book 9. "The Ultimate Model of Reality: Psyche Is the Ultimate Cause"

https://www.amazon.com/dp/B071NC9JK6

This book introduces my Psyche Ontology or my Ultimate Cause Model of Reality, wherein Psyche is the fundamental unit of reality.

In this book, I present a new and unique Philosophy of Science which demonstrates the existence of Psyche and then proceeds to employ Psyche in ALL aspects of Philosophy, Science, Application, Spirituality, Lived Experience, and Knowledge.

This becomes a Psyche Paradigm or the Ultimate Model of Reality.

22APR2017

### Book 10. "Putting Psyche Back into Psychology: Restoring Science to Consciousness"

https://www.amazon.com/dp/B071NC987S

This book further develops my Psyche Ontology or Ultimate Cause Model of Reality, wherein Psyche is the fundamental unit of reality.

In this book, I document how Psyche was originally an integral part of Psychology, how Behaviorism caused Psychology to lose its mind, and the different types of Personality Theory or Psyche Theory which have attempted to put Psyche back into Psychology.

It is my intention and proposition to include Near-Death Experiences (NDEs), Shared-Death Experiences (SDEs), Out-of-Body Experiences (OBEs), and other types of Lived Experience into Psychology or the "Study of Psyche" as an integral and essential part of Psychology.

NDEs, OBEs, SDEs, Spiritual Experiences, and other types of Lived Experience ARE scientific proof of Psyche's existence. These things should be a part of the Science of Psychology or the Science of Psyche.

22APR2017

### Book 11. "BioPsychoSocial: Including Psyche or Light into our Theoretical Models"

https://www.amazon.com/dp/B0713NDHVW

This book further develops my Psyche Ontology or Ultimate Cause Model of Reality, wherein Psyche is the fundamental unit of reality.

In this book, my primary focus is on Mental Illness and various different forms of Psychotherapy. This book chronicles my search for the Ultimate Psychotherapy, and it will also chronicle in part my brush with mental illness and addiction. Addiction causes psychosis and mental illness. Everyone who goes through withdrawal will experience some kind of psychosis or mental illness.

I end this book by discussing what I believe to be the BEST forms of psyche-therapy.

22APR2017

## Science 2.0 Is Based Upon All the Scientific Evidence that Falsifies Darwinism and the Theory of Evolution

The BEST way to find and know the truth is to live it and experience it for yourself, or to choose to trust someone who has.

The Second-Best way to find the truth is to identify and eliminate all the falsehoods and lies.

Science 2.0 allows all of the evidence into evidence, including all the evidence that falsifies Materialism, Naturalism, Darwinism, Scientism, Nihilism, and Atheism. The way you falsify these falsified philosophies is with observational evidence and common-sense logic.

Scientism is the philosophical religion that claims that ONLY science and the scientific method in particular can be used to discover, find, and know the truth. These people literally worship science and the scientific methods as if these things were infallible Gods.

Nihilism is the philosophical belief that there is no afterlife and that the psyche or soul or consciousness ceases to exist when the physical body and physical brain die.

There are a bunch of different ways to FALSIFY Darwinism and the Theory of Evolution. It's easy to do because these things are false.

First, you can falsify Micro-Evolution (Natural Selection and Random Mutations) as our Designer and Creator by demonstrating that these physical processes can't design and create anything and by reminding people that Evolution (genetic change), Natural Selection, and Random Mutations didn't even exist until after God designed and created the first genomes and the first life forms to begin with. Which came first, evolution or the genomes? The only correct answer is the genomes, because evolution of any kind didn't exist and couldn't exist without them.

Second, you can remind people that Louis Pasteur falsified Macro-Evolution, Spontaneous Generation, Abiogenesis, Materialism, Naturalism, and Creation Ex Nihilo in 1859 – the very same year that Charles Darwin published "On the Origin of Species" and introduced the Theory of Evolution to the world. Macro-Evolution was falsified by Science the very same year that it was presented to the world. Isn't that amazing?

Third, you can use scientific evidence, observational evidence, experiential evidence, and experimental evidence to falsify Materialism and Naturalism. When those go down, Darwinism and the Theory of Evolution go down with them, as fruit from the poisoned tree.

Fourth, you can use Observational Evidence and Experiential Evidence and Eye-Witness Evidence of the Biblical God Jesus Christ to falsify Materialism, Naturalism, Nihilism, and Atheism. When those go down, the Theory of Evolution and Darwinism go down with them, as fruit from the poisoned tree. Furthermore, the Biblical God Jesus Christ repeatedly confesses to being the creator of this physical universe and the organizer of this physical earth, leaving nothing for evolution to have done.

Fifth, you can use Near-Death Experiences (NDEs), Shared-Death Experiences (SDEs), Out-of-Body Experiences (OBEs), and other types of Non-Local Spiritual Experiences to falsify Materialism, Naturalism, and Nihilism. When those go down, Darwinism and the Theory of Evolution go down with them, as fruit from the poisoned tree. Whenever you use observational evidence and scientific evidence to falsify Materialism and Naturalism, you effectively neuter their derivatives such as Scientism, Nihilism, Darwinism, and Atheism. These things can't produce fruits without the tree upon which they were built.

Some people have said that I'm like a dog with a bone – I keep chewing on it, processing it, and digesting it until it is gone. They didn't mean this in a good way, either. I'm a bit OCD; and, they can sense that. Once I get ahold of something, I can't seem to let go of it. When I finally discovered the truth, I wanted to see how many different ways that I could present it to the world and still remain consistent with the truth. I got pretty good at it, where the Theory of Evolution is concerned. I know what I know because the observational evidence and scientific evidence says that it's so.

Of course, when it comes to dogs and their bones, I'm an amateur compared with someone like Henry P. Stapp, who has spent his life developing and promoting his **Orthodox Interpretation of Quantum Mechanics**, which successfully falsifies Materialism and Naturalism. Notice all the different ways in which he presents basically the same message. He's consistent and stays on theme, which can also be said of me. We are scientists, and we love science.

http://mypsyche.us/wp-content/uploads/2017/10/Henry-P-Stapp-Articles.pdf

Remember, you have to eliminate all the falsehoods before you can successfully find and know the truth.

I spent the first year of my writing career debunking and falsifying Materialism, Naturalism, Darwinism, and the Theory of Evolution – as well as fighting against and debating against the people who didn't like what I was doing. The following books were some of the fruits of that process. You will have to decide for yourself if I meet my burden of proof through a preponderance of the evidence.

**Book 1. "Summary Of: The Theory of Evolution Proves that God Exists"**

https://www.amazon.com/dp/B01GQCWED6

https://www.amazon.com/dp/1521130485

An introductory summary to my much larger book, "The Theory of Evolution Proved to Me that God Exists: Why I Am No Longer an Atheist and Why I No Longer Believe in the Theory of Evolution".

The second edition of this book contains an introduction to my much larger book, "The Ultimate Model of Reality: Psyche Is the Ultimate Cause", which is my core message to the world.

07JUN2016

**Book 2. "The Theory of Evolution Proved to Me that God Exists: Why I Am No Longer an Atheist and Why I No Longer Believe in the Theory of Evolution"**

https://www.amazon.com/dp/B01HZYBZ7K

https://www.amazon.com/dp/1521131228

Over the period of a few months, I had an epiphany or a major breakthrough; and in the process, the Theory of Evolution proved to me clearly and conclusively that God Exists. I have had no doubt about God's existence, after the Theory of Evolution proved to me that God Exists. By that point in time, I was officially done with My Materialism and My Atheism and firmly in a different camp. I'm now a part of God's army. This huge book, "The Theory of Evolution Proved to Me that God Exists", explains in great detail why I am no longer an Atheist and why I no longer believe in the Theory of Evolution.

This book documents my journey of discovery and the defense of my thesis.

04JUL2016

### Book 3. "The Scientific Method Proves That the Theory of Evolution Is False"

https://www.amazon.com/dp/B01IAAIRT2

https://www.amazon.com/dp/1521133611

In this book, I document what I know about the Scientific Method, and I demonstrate how I use The Scientific Method in my pursuit of The Truth.

I have learned to trust The Scientific Method, The Rules of Science, Abductive Reasoning, and Deductive Reasoning to give me a clear and accurate assessment of the evidence at hand. In this book, I use Science, The Scientific Method, the Rules of Science and Scientific Research, Abductive Reasoning, and Deductive Reasoning to prove that the Theory of Evolution is false and why it is false. You will have to judge for yourself if I meet my Burden of Proof through a preponderance of the evidence. I don't hold anything back!

Remember, we can indeed use the Scientific Methods to falsify theories or to prove theories false. Most people don't realize that, but it's true. That's what I did to the Theory of Evolution – I used various different scientific methods to falsify Darwinism, Naturalism, Materialism, Creation by Rocks, and the Theory of Evolution. Once you have successfully eliminated ALL of the falsehoods such as Materialism, Atheism, Naturalism, and Darwinism, you are left staring at THE TRUTH. It's elementary my dear friend.

The second edition of this book contains an introduction to my much larger book, "The Ultimate Model of Reality: Psyche Is the Ultimate Cause", which is my core message to the world.

11JUL2016

### Book 6. "Using the Scientific Method to Eliminate the Usual Suspects and to Prove the Truth"

https://www.amazon.com/dp/B01J6STHP0

https://www.amazon.com/dp/1521133581

Using Science and The Scientific Method to eliminate the Usual Suspects, in an attempt to demonstrate and prove THE TRUTH! In this book, I line up the Usual Suspects, gather and define the Evidence; and then, I apply The Scientific Method, common sense logic, abductive reasoning, and deductive reasoning to them, in order to see if any one of them could have done the job of designing and creating this Physical Universe, this Earth, and ALL of the Genomes and Life Forms on this Earth. The goal is to find the Correct Suspect among all of the Usual Suspects, who are typically presented to us for our consideration. This was a fun book to write, because it made logical sense to me from beginning to end.

Remember, we can indeed use the Scientific Methods to falsify theories or to prove theories false. Most people don't realize that, but it's true. That's what I did to the Theory of Evolution – I used various different scientific methods to falsify Darwinism, Naturalism, Materialism, Creation by Rocks, and the Theory of Evolution. Once you have successfully eliminated ALL of the falsehoods such as Materialism, Atheism, Naturalism, and Darwinism, you are left staring at THE TRUTH. It's elementary my dear friend.

The second edition of this book contains an introduction to my much larger book, "The Ultimate Model of Reality: Psyche Is the Ultimate Cause", which is my core message to the world.

27JUL2016

## Science 2.0 Allows Evidence for the Non-Local, the Immaterial, the Non-Physical, and the Invisible into Evidence

Evidence is of no value to us, unless it is allowed into evidence.

Light, Gravity, Forces, Fields, Psyche, Gluons, Bosons, Dark Energy, and even Dark Matter or Spirit Matter are non-physical, immaterial, and invisible; but, we KNOW that they exist by the effects that they have on physical matter and our physical instruments. The very existence of these things FALSIFIES Materialism and Naturalism.

I explore these concepts in the following book.

### Book 12. "God Is in the Light: God is light, and in Him is no darkness at all"

https://www.amazon.com/dp/B07168S37N

In this book, I explore Light and God's Psyche. Psyche is Living Light, or light which is conscious, alive, intelligent, self-aware, and universally connected with its surroundings.

23APR2017

## Science 2.0 Is Based Upon All the Evidence that Falsifies Macroevolution, Abiogenesis, Spontaneous Generation, and Creation by Chance

The best way to find the truth is to identify and eliminate all the falsehoods and lies.

There are a bunch of different ways to FALSIFY Darwinism and the Theory of Evolution. It's easy to do because these things are false.

First, you can falsify Micro-Evolution (Natural Selection and Random Mutations) as our Designer and Creator by demonstrating that these physical processes can't design and create anything and by reminding people that Evolution (genetic change), Natural Selection, and Random Mutations didn't even exist until after God designed and created the first genomes and the first life forms to begin with. Which came first, evolution or the genomes? The only correct answer is the genomes, because evolution of any kind didn't exist and couldn't exist without them.

Second, you can remind people that Louis Pasteur falsified Macro-Evolution, Spontaneous Generation, Abiogenesis, Materialism, Naturalism, and Creation Ex Nihilo in 1859 – the very same year that Charles Darwin published "On the Origin of Species" and introduced the Theory of Evolution to the world. Macro-Evolution was falsified by Science the very same year that it was presented to the world. Isn't that amazing?

**Book 1. "Summary Of: The Theory of Evolution Proves that God Exists"**

https://www.amazon.com/dp/B01GQCWED6

https://www.amazon.com/dp/1521130485

An introductory summary to my much larger book, "The Theory of Evolution Proved to Me that God Exists: Why I Am No Longer an Atheist and Why I No Longer Believe in the Theory of Evolution".

The second edition of this book contains an introduction to my much larger book, "The Ultimate Model of Reality: Psyche Is the Ultimate Cause", which is my core message to the world.

07JUN2016

**Book 2. "The Theory of Evolution Proved to Me that God Exists: Why I Am No Longer an Atheist and Why I No Longer Believe in the Theory of Evolution"**

https://www.amazon.com/dp/B01HZYBZ7K

https://www.amazon.com/dp/1521131228

Over the period of a few months, I had an epiphany or a major breakthrough; and in the process, the Theory of Evolution proved to me clearly and conclusively that God Exists. I have had no doubt about God's existence, after the Theory of Evolution proved to me that God Exists. By that point in time, I was officially done with My Materialism and My Atheism and firmly in a different camp. I'm now a part of God's army. This huge book, "The Theory of Evolution Proved to Me that God Exists", explains in great detail why I am no longer an Atheist and why I no longer believe in the Theory of Evolution.

This book documents my journey of discovery and the defense of my thesis.

04JUL2016

### Book 3. "The Scientific Method Proves That the Theory of Evolution Is False"

https://www.amazon.com/dp/B01IAAIRT2

https://www.amazon.com/dp/1521133611

In this book, I document what I know about the Scientific Method, and I demonstrate how I use The Scientific Method in my pursuit of The Truth.

I have learned to trust The Scientific Method, The Rules of Science, Abductive Reasoning, and Deductive Reasoning to give me a clear and accurate assessment of the evidence at hand. In this book, I use Science, The Scientific Method, the Rules of Science and Scientific Research, Abductive Reasoning, and Deductive Reasoning to prove that the Theory of Evolution is false and why it is false. You will have to judge for yourself if I meet my Burden of Proof through a preponderance of the evidence. I don't hold anything back!

Remember, we can indeed use the Scientific Methods to falsify theories or to prove theories false. Most people don't realize that, but it's true. That's what I did to the Theory of Evolution – I used various different scientific methods to falsify Darwinism, Naturalism, Materialism, Creation by Rocks, and the Theory of Evolution. Once you have successfully eliminated ALL of the falsehoods such as Materialism, Atheism, Naturalism, and Darwinism, you are left staring at THE TRUTH. It's elementary my dear friend.

The second edition of this book contains an introduction to my much larger book, "The Ultimate Model of Reality: Psyche Is the Ultimate Cause", which is my core message to the world.

11JUL2016

### Book 6. "Using the Scientific Method to Eliminate the Usual Suspects and to Prove the Truth"

https://www.amazon.com/dp/B01J6STHP0

https://www.amazon.com/dp/1521133581

Using Science and The Scientific Method to eliminate the Usual Suspects, in an attempt to demonstrate and prove THE TRUTH! In this book, I line up the Usual Suspects, gather and define the Evidence; and then, I apply The Scientific Method, common sense logic, abductive reasoning, and deductive reasoning to them, in order to see if any one of them could have done the job of designing and creating this Physical Universe, this Earth, and ALL of the Genomes and Life Forms on this Earth. The goal is to find the Correct Suspect among all of the Usual Suspects, who are typically presented to us for our consideration. This was a fun book to write, because it made logical sense to me from beginning to end.

Remember, we can indeed use the Scientific Methods to falsify theories or to prove theories false. Most people don't realize that, but it's true. That's what I did to the Theory of

Evolution – I used various different scientific methods to falsify Darwinism, Naturalism, Materialism, Creation by Rocks, and the Theory of Evolution. Once you have successfully eliminated ALL of the falsehoods such as Materialism, Atheism, Naturalism, and Darwinism, you are left staring at THE TRUTH. It's elementary my dear friend.

The second edition of this book contains an introduction to my much larger book, "The Ultimate Model of Reality: Psyche Is the Ultimate Cause", which is my core message to the world.

27JUL2016

## Science 2.0 Allows All of the Evidence for Fine-Tuning, Transdimensionality, Non-Locality, and Light into Evidence

Fine-Tuning is a whole science all by itself based upon a lot of physical evidence. I have been fascinated by the Cosmological Constants from the very moment that I was exposed to them. Someone has been mucking about in the innards of this thing, if you ask me.

In contrast, things like Transdimensionality, Extra-Dimensionality, Non-Locality, and String Theory by definition and in principle have NO physical evidence supporting them, which means that we have to turn to eye-witness evidence, experiential evidence, empirical evidence, observational evidence, and even at times inferential evidence in order to support these types of science.

I have learned to hate scientific inferences, because they are logic fallacies, which is why I prefer to interpret the evidence myself and encourage my readers to do the same. Faulty interpretations are the source for most of the errors in Science. The weakest and most fatal part of the scientific methods is the part that calls for and relies upon the personal interpretations of the scientists. Personal interpretations and scientific inferences end up being logic fallacies most of the time.

When it comes to the Non-Local Sciences or the Spiritual Sciences, I prefer to present the observational and experiential evidence, and then let my reader decide what to make of it. There's tons of observational evidence supporting the Non-Local or the Non-Physical Sciences. Once you start looking, there's literally no end to it.

After allowing all the Observational Evidence into evidence, the only thing that's left that's based upon a lot of Inferential Evidence is String Theory and M-Theory, which don't seem to have any observational support of any kind, and end up being pure theory. Theory and inference will have to do if we have nothing else to fall back on; but, I always give precedence and superior respect to Observational Evidence and Empirical Evidence.

Observational Evidence and Experiential Evidence FALSIFY Materialism, Naturalism, Nihilism, Atheism, and even Darwinism. In contrast, from what I have been able to tell so far, Observational Evidence neither confirms nor contradicts String Theory and M-Theory, except for the fact that there's tons of observational evidence supporting the Non-Local Dimensions.

Check it out and decide for yourself what you want to believe:

https://www.youtube.com/results?search_query=NDE

So far, my primary focus in this area has been on Near-Death Experiences (NDEs) and exploring all the different ramifications associated with Light.

Remember, Light is immaterial and non-physical. The very existence of Light FALSIFIES Materialism and Naturalism. Psyche and Spirit Matter are some type of light, and the same reality holds true for physical matter as well.

David Bohm said that physical matter is condensed light or frozen light. Physical matter is light that has been slowed down to sub-light speeds. In contrast, Psyche and Spirit Matter are some type of light that exists natively at speeds or velocities faster than the speed of light.

I'm interested in the whole of Science, and not just the physical matter.

I explore some of these realities in the following book.

**Book 12. "God Is in the Light: God is light, and in Him is no darkness at all"**

https://www.amazon.com/dp/B07168S37N

In this book, I explore Light and God's Psyche. Psyche is Living Light, or light which is conscious, alive, intelligent, self-aware, and universally connected with its surroundings.

23APR2017

## Science 2.0 Uses the Scientific Methods to Falsify Materialism, Darwinism, and Naturalism

The best way to find the truth is to identify and eliminate all the falsehoods and lies.

I explore this reality in ALL of my books.

https://www.amazon.com/Mark-My-Words/e/B01IAEF2Y6/

After I falsified and abandoned My Materialism, My Nihilism, and My Atheism, then my ultimate goal was to find and know the Truth.

I have been busy doing research ever since.

## Science 2.0 Chooses to Believe All of the Evidence that Indicates that Our Physical Universe Had a Beginning and Could Someday Have an End

Anything that has a beginning can have an end. Furthermore, anything that has a beginning has someone or something that brought it into existence in the first place. The fact that we exist means either that we had a beginning or that we have always existed.

Our Psyche or Non-Local Consciousness has always existed and will always exist. Spirit matter has always existed and will always exist. There is no entropy or aging when it comes to the Non-Local Realm or Spirit Realm. Our spirit bodies and physical bodies had a beginning, which means that they had some kind of intelligent Organizer, Creator, or Beginner.

According to the Big Bang Theory and according to logical common sense, each physical particle, each physical universe, each physical galaxy, each physical genome, and each physical life form had a beginning, which means that each one of these had a Beginner, or a Creator, or an Organizer, or a Manufacturer.

Physical things don't just put themselves together spontaneously ex nihilo. They have NEVER been observed doing so. Order doesn't spontaneously burst forth from disorder and chaos. There's no such thing as spontaneous generation. There is no such thing as Creation Ex Nihilo. Creation Ex Nihilo is Atheism – Creation from Nothing by Nothing. It makes no logical sense. Every beginning has a Beginner.

This is Logic 101.

Obviously, you can choose to think illogically and irrationally if you want to, as the Materialists, Naturalists, and Atheists do where these topics are concerned. Or, you can choose to follow the logic and the evidence to wherever it might lead you. The choice is yours. It's a choice that nobody else can make for you.

--

**Book 16. "Scientific Proof of God's Existence: Finding God Where the Atheists Refuse to Look for Him"**

https://www.amazon.com/dp/B07B26CRHX

I have been working on this book for half a decade now. Anytime I have come across a Scientific Proof of God's Existence, I have tried to get a copy of it into this book.

The Materialists, Naturalists, Darwinists, and Atheists told me that there is NO scientific proof of God's existence and that there will NEVER be any scientific proof of God's existence. I believed them. I believed them for over fifty years of my life, until I discovered that they are wrong. These poor people are almost always wrong. Once I found my first convincing Scientific Proof of God's Existence, dozens more were forthcoming over the years, as I kept my eyes open and my mind open to the possibilities. I like to think that there are always possibilities.

25FEB2018

**Book 17. "Quantum Neuroscience: The Answer to Life, the Universe, and Everything"**

https://www.amazon.com/dp/B079Z6QQQB

https://www.amazon.com/gp/product/1521132380/

This book is my Magnum Opus. All of my research and all of the evidence led me to write this book. Everything in my life has led me to write this book. I made some unbelievable discoveries while researching and writing this book. I'm still having a hard time believing some of the things that I discovered about Quantum Mechanics and the Physical Brain. This book is guaranteed to shock you and surprise you. You are going to resist this one because you have been trained, conditioned, and brainwashed to do so. I KNOW, because I resisted it as well, until the preponderance of the evidence got through to me and convinced me that I was wrong. Enjoy the ultimate Mind Game!

    25FEB2018

## Science 2.0 States That There Is No Such Thing as Creation Ex Nihilo, Because It Has Never Been Experienced nor Observed

Non-Local means not located in our local 3D physical space-time realm. Quantum Non-Locality is typically associated with Quantum Entanglement, Action at a Distance, Telepathy, and Psychokinesis.

If it has never been experienced nor observed by anyone, that reality greatly increases the chance or probability that it doesn't exist. The only way to know for sure that Non-Local people and events are real and true is to experience them for yourself, or to choose to trust someone who has.

Creation Ex Nihilo, Creation by Nothing, Creation by Chance, Creation by Naturalism, Creation by Materialism, Creation by Rocks, Creation by Mutation/Selection, Spontaneous Generation, Abiogenesis, and Macro-Evolution have never been experienced nor observed, which greatly increases the chance or the probability that these things do not exist and are simply figments of the imagination.

One of the ways to find and know the truth is to find and eliminate ALL of the fictional philosophies and falsified philosophies, so that ONLY the truth remains. Falsified philosophies like Materialism, Naturalism, Darwinism, Behaviorism, Atomism, Determinism, Scientism, Nihilism, and Atheism have to go, if we ever want to find and know the TRUTH.

I believe that the Gods are scientists which, if true, is a claim that bodes well for Science 2.0 as the Ultimate Science. I wanted to find and develop the Ultimate Science, so that what I ended up with would require no more tweaking and adjusting. It should be good enough to stand until the end of time.

I discuss some of this in the following book:

**Book 8. "I Am Not a Creationist: So What Am I?"**

https://www.amazon.com/dp/B071XTM8XY

In this book, I try to explain why I don't believe in Creation Ex Nihilo or Creation by Physical Matter, otherwise known as Materialism, Naturalism, Darwinism, and Atheism. I'm not that type of Creationist. So, what am I?

Starting with this book, in ALL of my books I falsify Creation by Chance. ALL of the different false and falsified philosophies of science are Creation by Chance or Creation Ex Nihilo. I don't believe in that type of creationism anymore. Creation by Entropy (the second law of thermodynamics), Creation by Death (natural selection), Creation by Disorder and Chaos (creation ex nihilo), and Creation by Chance (modern-day science) are obviously false and have been clearly and conclusively FALSIFIED by Science, the Scientific Methods, the Null Hypothesis in statistics, and the Observations and Experiences of the human race as a whole.

## Science 2.0 Allows Any Evidence for God's Existence into Evidence

Unlike Materialism and Naturalism, Science 2.0 allows any evidence supporting Psyche, Non-Local Consciousness, Non-Locality, and God into evidence. The Materialists and Naturalists will simply tell you that such evidence does not exist and cannot exist; and, there were times when I used to believe them. But, they were wrong. The evidence for such things is copious and abundant.

I used to be a Materialist, Nihilist, and Atheist; but, when I finally started looking at the evidence, I eventually concluded that the Biblical God Jesus Christ has in fact met His burden of proof through a preponderance of the evidence.

Science 2.0 allows ALL of the evidence into evidence; but then, I typically like to leave it up to my reader to decide for himself or herself what to make of that evidence and how to use that evidence in their own lives.

Materialism, Naturalism, Nihilism, and Atheism are based upon a refusal to look at evidence. The Materialists and Naturalists refuse to allow any evidence for God's Existence into evidence; so that they can then conclude before examining any evidence that God does not exist.

This process is called *begging the question*, a logic fallacy. The Materialists and Naturalists conclude that God does not exist; and then, these people insert that conclusion into all their arguments as their major hidden premise or their primary assumption, and then use that conclusion as evidence that their conclusion is true. This is *begging the question* and *jumping to conclusions*, which are logic fallacies. You can't submit your pre-conceived conclusions as evidence that your chosen conclusions are true; but, that's exactly what the Materialists and Naturalists do.

Science 2.0 allows all of the evidence for God's Existence into evidence, and then leaves it up to the reader to decide for himself or herself whether it's real and true, or not. Allowing all of the evidence into evidence is simply a different way and better way of doing Science.

In contrast, Materialists and Naturalists define science as Materialism, Naturalism, Darwinism, Scientism, Nihilism, and Atheism; and then, these people proceed to exclude and reject and revile and mock any evidence which contradicts their pre-chosen conclusions. These are the actions of dogmatic religious fundamentalists and fanatics, and not scientists.

Hundreds of people have observed, during their Near-Death Experiences (NDEs), that Jesus Christ is the Being of Light and the Being of Love whom the righteous people and the redeemed typically encounter during their NDEs. Jesus Christ has even come and rescued some of these people from hell.

I submit all of these into evidence, and I leave it up to you to decide for yourself what to make of them.

NDE Encounters with Jesus:

https://www.youtube.com/results?search_query=NDE+Jesus

I submit the whole load into evidence. Remember, Science 2.0 allows all of the evidence into evidence, but leaves it up to you to decide for yourself what to make of it and how to interpret it. This is the way that science should be done, in my humble opinion.

Howard Storm:

https://www.youtube.com/watch?v=Vm647n1360A

https://www.youtube.com/watch?v=UPj4wci_bcI

Dr. Mary Neal:

https://www.youtube.com/watch?v=DX473dF7ChY

Ian McCormack:

https://www.youtube.com/watch?v=HbTAmN4m2lQ

Joe Hadwin:

https://www.youtube.com/watch?v=IOhOynR9Jxg

Another:

https://www.youtube.com/watch?v=N4ut09jDdV0

Remember, unlike Materialism and Naturalism, Science 2.0 allows all of the evidence into evidence. Furthermore, unlike Materialism and Naturalism, Science 2.0 leaves it up to you to interpret the evidence for yourself rather than imposing on you and enforcing upon you an "official interpretation" of the evidence that has been blessed and sanctified and purified by the Materialists and Naturalists who control the review process.

Since I'm no longer a Materialist, Nihilist, and Atheist, I'm no longer hiding from the evidence. I now allow all of the evidence into evidence, and then I decide for myself how to interpret it and what to make of it. I encourage you to do the same.

The following books also provide evidence for God's Existence; and since Science 2.0 allows all of the evidence into evidence, I must submit them into evidence as well.

The Holy Bible

https://www.lds.org/scriptures/bible?lang=eng

The Book of Mormon: Another Testament of Jesus Christ

https://www.lds.org/scriptures/bofm?lang=eng

The Doctrine and Covenants

https://www.lds.org/scriptures/dc-testament?lang=eng

The Pearl of Great Price

https://www.lds.org/scriptures/pgp?lang=eng

The Biblical God Jesus Christ had a hand in writing and producing these books. You can decide for yourself what to do with them and what to make of them.

Remember, Science 2.0 allows ALL of the evidence into evidence, and then leaves it up to you to decide for yourself what to make of it.

When it comes to Scientific Proof of God's Existence, I have been in the process of writing a few books about this subject. One of them will end up being over a thousand pages long, and I will end up splitting it into two volumes when I'm done.

So far, I have only released the Primer, for this topic. The magnum opus remains a work-in-progress, as of October 2017 as I write this.

**Book 14. "Scientific Proof of God's Existence: A Primer"**

https://www.amazon.com/dp/B071713NNL

https://www.amazon.com/dp/1521325170

This book is a basic Primer explaining how I developed a Scientific Proof of God's Existence. The concepts and ideas were three or four years in the making. It wasn't obvious at first how this should be done or if it could be done.

You will have to decide whether I succeeded or not. I think I did a pretty good job; but then, I'm biased, and your mileage may vary.

This book is guaranteed to make some people squirm.

18MAY2017

# Science 2.0 Massively Upgrades Science and the Philosophy of Science

After I released the original version of Science 2.0 on 31OCT2017, I then USED Science 2.0 to upgrade the Philosophy of Science and the whole of Science all across the board.

I introduced NEW Science to the world that it has never been thought about before.

https://science-2-0.com/

I went through Neuroscience correcting its falsehoods and errors, and Quantum Neuroscience was the result. I single-handedly brought Neuroscience into the quantum era. I demonstrated how Quantum Mechanics produces better and more believable scientific explanations of Neuroscience than Materialism and Naturalism and Darwinism do.

I'm not all that bright. I live in a haze most of the days. I was simply willing to follow the evidence wherever it was willing to lead me. My hope is that in the coming century after I am dead, if the world lasts that long, that people who are much more talented and skillful and intelligent than I am will take what I have started in Quantum Neuroscience, enhance it, correct it, and improve upon it. The book could have been twice as long as it was. The work is unfinished. In that particular book, I ran out of space and it was time to quit.

I'm the first doctorate in Quantum Neuroscience in a world that doesn't give doctorates for falsifying Materialism, Naturalism, Darwinism, Nihilism, and Atheism. Maybe in some future world I will be granted a doctorate in Quantum Neuroscience posthumously. I can only hope.

--

**Book 17. "Quantum Neuroscience: The Answer to Life, the Universe, and Everything"**

https://www.amazon.com/dp/B079Z6QQQB

https://www.amazon.com/gp/product/1521132380/

https://quantum-neuroscience.com/

This book is my Magnum Opus. All of my research and all of the evidence led me to write this book. Everything in my life has led me to write this book. I made some unbelievable discoveries while researching and writing this book. I'm still having a hard time believing some of the things that I discovered about Quantum Mechanics and the Physical Brain. This book is guaranteed to shock you and surprise you. You are going to resist this one because you have been trained, conditioned, and brainwashed to do so. I KNOW, because I resisted it as well, until the preponderance of the evidence got through to me and convinced me that I was wrong. Enjoy the ultimate Mind Game!

25FEB2018

--

The majority of scientists on this planet claim to believe in the Big Bang Theory, and they really truly believe that the Big Bang happened.

The Big Bang Theory has baggage. If you believe in the Big Bang Theory, then you have to believe in the God who designed the Big Bang Singularity, triggered it and made it go bang, controlled the faster-than-light inflation so that space could be produced and so that the universe could eventually produce physical matter "spontaneously", and then stopped the inflation so that slow sub-light physical matter could actually form or be formed.

Over and over again, ALL of the scientists keep saying that the Big Bang represented a massive reversal of entropy in our part of the multiverse. If the Big Bang Theory is true, then Someone Psyche or Someone Intelligent KNOWS how to reverse entropy at will. There has to be some type of Syntropy; otherwise, all of that subsequent physical matter and entropy wouldn't exist in the first place, and we wouldn't exist either. The Big Bang represented a massive infusion of Syntropy into our part of the multiverse. If the Big Bang Theory is true, and most scientists seem to believe that it is, then Syntropy has to be true and Syntropy has to exist. So, what is Syntropy? Where is it? It should be all around us and all through us, if the Big Bang Theory is true.

Since I had developed Science 2.0 and was willing to allow all of the evidence into evidence, I realized that WE NEED to determine what Syntropy is so that we can then use it to explain how the Big Bang Singularity was produced and how the Big Bang happened. Syntropy has to exist, or we wouldn't exist.

Consequently, I started to research Entropy and its opposite, Syntropy. I tried to figure out what these things are and how they really work. In 2017 and 2018, I wrote the book on Syntropy and introduced this NEW science concept to the world. I figured out what Syntropy is, and I figured out how everything works and how everything was done. I encountered Quantum Field Theory in the process.

--

**Book 18. "Syntropy in Defense of Quantum Mechanics: The Answer to Life, the Universe, and Everything"**

https://www.amazon.com/dp/B07BPT3W8R/

25MAR2018

Syntropy is the topic that nobody knows anything about. They have never heard of it before. They have no idea what it is. They don't know that it is the opposite of entropy and that it falsifies the second law of thermodynamics. The science community doesn't even know what entropy, so there's no way in the universe that they could possibly know what Syntropy is. There are literally two dozen different, contradictory, self-defeating, mutually exclusive definitions for entropy on the books. Nobody knows what entropy is. I had to figure out what Syntropy is BEFORE I could start to figure out what entropy truly is.

This is the book that I will be uploading for free to all of my different websites. It will take some time. This is the one that's going to be the hard sell, because nobody is looking for it and nobody wants it. It's completely foreign to anything that we currently have in Science, because our science community isn't ready for it yet.

I continue to work on it and will eventually finish it, although I introduced the original version of this book on 25MAR2018. Here are some of the websites that it created based upon the information that I initially produced and published in this book on Syntropy:

https://syntropy.site/

https://syntropy.website/

https://origin-science.org/

https://quantum-law-of-thermodynamics.com/

https://ultimate-law-of-thermodynamics.com/

https://philosophy-of-science.com/

https://evolution-is-entropy.com/

https://god-is-light.com/

https://scientific-proof-of-god.com/

https://psyche-ontology.com/

https://ultimate-model-of-reality.com/

Then, during the subsequent year and a half, I slowly started to figure out what entropy is and how it works. I eventually realized that most of modern-day science is wrong because we have a false and falsified definition for entropy that we have been using for a 150 years now; and, we have based most of this false and falsified "science" on the second law of thermodynamics, which is demonstrably false and doesn't match with what has actually been experienced and observed. In 2018 and 2019, I slowly learned how to falsify the second law of thermodynamics, and then I started looking for suitable replacements, including a couple that I had already found while writing the book on Syntropy.

https://quantum-law-of-thermodynamics.com/

https://ultimate-law-of-thermodynamics.com/

--

In 2018 and 2019, I was lucky and blessed to find David Bohm's book, *The Essential David Bohm* – a Philosophy of Science based upon Quantum Mechanics and Consciousness.

https://philosophy-of-science.com/The-Essential-David-Bohm

https://epdf.tips/the-essential-david-bohm.html

In the 1950's, Bohm was using Quantum Mechanics to falsify the Philosophy of Mechanism (Materialism and Naturalism). Bohm tried to introduce a NEW and BETTER Philosophy of Science to the world based upon Quantum Mechanics and Quantum Field Theory. Bohm was at least seventy years ahead of his time, so everything that he proposed and wrote was suppressed and hidden by our science community at large. Our science community NEVER talks about nor mentions the essays that were written in *The Essential David Bohm*. All of his work was completely ignored, and only his falsified Pilot-Wave Theory has made it into the scientific community at-large.

I took each chapter in this book, demonstrated how David Bohm's ideas had been vindicated by Science in 2019, and used the results to finish my book entitled, "NATURE vs. NURTURE vs. NIRVANA".

Book 7. "NATURE vs. NURTURE vs. NIRVANA: An Introduction to Reality and Quantum Mechanics"

https://www.amazon.com/dp/B01JWRCSVA

https://www.amazon.com/dp/1521132615

I finished and provided the enhanced edition of this book on 07JUN2019.

--

In 2018, through PBS Spacetime, I was introduced to Quantum Field Theory and Feynman Diagrams for the first time in my life, and I immediately realized that it was what I had been looking for all of my life.

https://www.youtube.com/playlist?list=PLsPUh22kYmNBpDZPejCHGzxyfgitj26w9

https://www.youtube.com/playlist?list=PLsPUh22kYmNA6WUmOsEEi32zi_RdSUF4i

https://www.youtube.com/playlist?list=PLsPUh22kYmNCLrXgf8e6nC_xEzxdx4nmY

https://www.youtube.com/playlist?list=PLsPUh22kYmNAHB1W2_Ka2F83sQbdczwKr

I eventually realized that $E = mc^2$ is the perpetual motion cycle and that the Quantum Fields are massless, entropyless, heatless, non-physical, syntropic perpetual motion machines. That was a major discovery and completely changed the way that I look at science and reality. I finally found out where ALL the Syntropy is. It's currently being stored in the Quantum Fields.

$E = mc^2$, the Perpetual Motion Cycle, Quantum Field Theory, and Feynman Diagrams FALSIFY the second law of thermodynamics.

--

In 2019, I USED the full weight of what I had learned about Quantum Mechanics, Photons, Quantum Fields, the Conservation of Energy, $E = mc^2$, and Quantum Field Theory to falsify the second law of thermodynamics. I eventually realized that everything falsifies the second law of thermodynamics, which claims that the total amount of entropy or disorder in our universe is constantly increasing and can never decrease.

We don't observe anything that is constantly increasing in amount, because the Conservation of Energy prevents that from happening. We don't observe an ever-increasing amount of disorder and chaos in our universe as the second law of thermodynamics predicts should be happening. We don't observe an ever-encroaching grey goo coming in at us from all sides as the second law predicts as our whole universe dissolves into chaos and disorder. Instead, we observe constantly conserved order and organization. Everything falsifies the second law of thermodynamics.

I took these observations from Quantum Field Theory, Feynman Diagrams, Quantum Mechanics, the Perpetual Motion Cycle $E = mc^2$, and the massless entropyless heatless non-physical Photons and Quantum Fields and USED THEM to falsify the second law of thermodynamics.

The results were published in the enhanced and final edition of my book, "Quantum Mechanics from a Non-Physical Spiritual Perspective", on 04JUL2019.

**Book 5. "Quantum Mechanics from a Non-Physical Spiritual Perspective"**

https://www.amazon.com/dp/B01J023TGU

https://www.amazon.com/dp/1521132380

That was the end of the second law of thermodynamics.

--

As if that weren't enough, then, throughout the summer of 2019, I worked on preparing the final version of my book Science 2.0, wherein I used everything that I had learned in my Statistics class in college to falsify the second law of thermodynamics. The whole of Science, Statistics, the Scientific Method, and the Null Hypothesis in science experiments FALSIFY the second law of thermodynamics. Everything falsifies the second law of thermodynamics. The fact that you are here and reading this right now FALSIFIES the second law of thermodynamics. There's nothing true about the second law of thermodynamics, and it's time for it to be replaced by something better. In my books, I suggest that we replace the second law of thermodynamics with Quantum Field Theory and the Perpetual Motion Cycle $E = mc^2$. It will happen when our science community finally figures out what's really going on in our universe and how things really work.

I finally figured out what entropy truly is, and I published the finished results in my book "Science 2.0".

**Book 15. "Science 2.0: I Upgraded My Science"**

https://www.amazon.com/dp/B0771K6WTX

This book, entitled "Science 2.0", provides a solid foundation for a NEW, BETTER, and UPGRADED Philosophy of Science. It is my gift to the world, where I hope that someday it will do some good.

# PART VIII — SCIENCE 2.0 NEEDS AN EYE-WITNESS ACCOUNT OF ORIGINS OR CREATION FROM SOMEONE WHO WAS THERE

### The Law of Witnesses:

2 Corinthians 13: 1: "In the mouth of two or three witnesses shall every truth be established."

Since Science 2.0 is based exclusively upon the Law of Witnesses, Science 2.0 needs Observational Accounts of origins and creation for both this earth and this universe.

Whether I'm writing about Creation Accounts or Near-Death Experiences, I like to treat them as if they were Ted Talks – simply present the evidence, and then leave it up to my audience to decide what to make of it. Whenever I pick a particular topic to discuss, I like to pick the three best samples or witnesses that I have found so far – according to the Law of Witnesses. I have tried to integrate the Law of Witnesses into everything that I do.

Three Favorite Ted Talks about NDEs:

Joyce Hawkes:

https://www.youtube.com/watch?v=MyaBeHeRK6M

Anita Moorjani:

https://www.youtube.com/watch?v=rhcJNJbRJ6U

Lewis Brown Griggs:

https://www.youtube.com/watch?v=bi_QsbnrTXo

### Origin Science:

In this section of this book, I will now present Origin Accounts or Creation Accounts that were produced or given to us by the Biblical God Jesus Christ himself.

Under Science 2.0, it's permissible for me to tell you what the evidence means to me personally; but, in this particular section, I will simply present the evidence, and then leave it up to you to decide for yourself what to make of it.

I will simply list and quote my favorite Origin Accounts or Creation Accounts. These are my favorites, because I found them to be the most informative, from the standpoint of Observation or Science. Remember, under Science 2.0, Science is defined as Observation and Experience – not Materialism and Naturalism.

What I tend to do with all of these different versions of Origins from the Biblical God Jesus Christ is to combine them together into a sort of amalgamated whole. You can do with them whatever you want to do with them, because under Science 2.0 my only obligation is to present the observational evidence, and then leave it up to my reader to decide for himself or herself what to make of that evidence.

But, notice an interesting fact about the three different Creation Accounts that I chose to present – they each claim to be eye-witness accounts of the events in question. Remember, in the mouth of two or three witnesses shall every truth be established.

Combined together, these three Creation Accounts conform to the Law of Witnesses, upon which Science 2.0 is based.

# Introducing Ultimate Cause and a Psyche Ontology

Psyche is the Ultimate Cause.

Psyche is the Ultimate Causal Agent.

Psyche is the fundamental unit of reality and existence.

This is a Psyche Ontology.

I spent a year upgrading the Philosophy of Science to include the Ultimate Cause and a Psyche Ontology. As far as I know, I'm the first person to introduce a Psyche Ontology to the world. I'm also the first person to introduce the Ultimate Cause to the world, which is an addition to Aristotle's four causes. The Ultimate Cause adds Quantum Mechanics, Quantum Fields, and Quantum Consciousness to Aristotle's four causes. Aristotle's four causes try to limit themselves exclusively to physical matter, and it was finally time to include Quantum Mechanics and Quantum Consciousness (Psyche or Intelligence) into the mix as one of the causal forces in this universe because it obviously exists. I'm not afraid to fix science and philosophy whenever it is obvious that they are broken. Psyche is the Ultimate Cause. Psyche is the Ultimate Causal Agent.

The Gods and Controlling Psyches had to design and make the massless, heatless, entropyless, non-physical, perpetual motion machines that we call the Quantum Fields BEFORE they could make and sustain heat and physical matter. NO Quantum Fields, then NO mass, NO heat storage capacity (entropy), NO resistance to acceleration, NO heat, NO thermodynamics, and NO physical matter. Without the Quantum Fields, it would be nothing but random chaos, and the inappropriately named second law of thermodynamics would actually be true.

I created a website where I try to explain some of these things.

https://psyche-ontology.com/

The quantum physicists, quantum theorists, and mathematicians who understand Quantum Mechanics and Quantum Field Theory keep saying that Quantum Mechanics can't possibly be true unless the LAW of Quantum Information Conservation is true. They keep saying that the Conservation of Quantum Information has to be true, or Quantum Mechanics can't be true.

That means that there has to be some type of conserved mechanism or conserved machine at the quantum level capable of transmitting, receiving, producing, using, processing, and storing Quantum Information. Psyche or Quantum Consciousness is that machine. Psyche, Intelligence, or Quantum Consciousness has to exist, or Quantum Mechanics can't possibly be true. Since Quantum Mechanics and Quantum Field Theory continue to be verified and proven true, that makes it obvious that the Quantum Law of Information Conservation is true and that some type of machine or mechanism exists at the quantum level that is capable of producing, using, and storing quantum information.

Psyche has to exist, or Quantum Information can't exist. Psyche is an infinite singularity. Psyche is the ultimate point particle. Psyche is a real quantum computer. Psyche has to exist, or Quantum Mechanics and Quantum Field theory can't possibly be true. First things first!

Likewise, Psyche or Intelligence or Quantum Consciousness has to exist, or the Quantum Fields wouldn't exist. Something immaterial and non-physical at the quantum level had to design and make the Quantum Fields BEFORE the first particle of physical matter could be made and sustained. According to Quantum Field Theory, the Gods and the Controlling Psyche had to design and make the Quantum Fields BEFORE they could make and sustain physical matter. No quantum fields, then no physical matter. It's that simple.

Psyche has been experienced and observed, so we KNOW that it exists. Even at the quantum level and the psyche level, a Psyche looks like a photon, a pinprick of light, or a spark. If a true point particle exists, Psyche is it. Psyche has actually been experienced and observed.

https://psyche-ontology.com/psyche-experienced-and-observed/

https://psyche-ontology.com/buhlman/

I like inputting the observed evidence as a given, and then I try to take that observed evidence to its logical conclusion if possible. It makes for better and more realistic science in the end. Science is observation and experience after all, or it should be.

Quantum Information can't exist without Psyche, Intelligence, or Quantum Consciousness. Quantum Information Conservation has to be true in order for Quantum Mechanics to be true. Therefore, there has to be some non-physical, immaterial, conserved machine or mechanism at the quantum level that is capable of making, using, and storing Quantum Information; otherwise, Quantum Information couldn't exist in the first place. Psyche is that machine. Quantum waves are thoughts. Psyches are quantum computers that make, transmit, receive, process, and store quantum waves or thoughts. Psyches have to exist, or physical matter wouldn't exist. Psyches have to exist, or we wouldn't exist.

The massless, entropyless, non-physical Quantum Fields had to be designed and made by the Controlling Psyches or Controlling Intelligences in our universe; otherwise, they wouldn't exist. Nature's Psyche makes the quantum waves in the first place, and then later, it is Nature's Psyche who collapses the wave function. This is the Perpetual Motion Cycle $E = mc^2$ in action. This is Quantum Field Theory. Physical matter couldn't exist and wouldn't exist without the Quantum Fields, and the Quantum Fields couldn't exist and wouldn't exist without the Psyches or Intelligences who designed them and made them.

This is a psyche ontology wherein Psyche, or Intelligence, or Life Force, or Quantum Consciousness is the fundamental unit of reality and existence. If there is such a thing as a point particle, Psyche is it. If there is such a thing as an infinite singularity, Psyche is it. By bringing Psyche and Quantum Mechanics in to play, we can literally explain everything that comes our way.

Anything that is obviously made obviously has a Maker who made it.

Quantum waves are obviously made. Quantum waves don't just spring into existence from thin air out of nothing as the Materialists, Naturalists, Darwinists, Determinists, and Atheists claim. Quantum waves are deliberately and intentionally made. That means that there has to be some type of entity or machine at the quantum level that is capable of making, transmitting, receiving, processing, and storing quantum waves. Psyche is that quantum machine or quantum mechanism. We NEED something that has actually been experienced and observed to serve as the transmitter and the receiver and the processor of quantum waves. Psyche is it.

All of your memories that survive the death of your physical brain – and show up in your after-death life review – are obviously stored within some type of Psyche. Thoughts and memories are quantum waves. They continue to happen and continue to exist long after our physical brain is dead and gone. This is what has actually been experienced and observed. Psyche is a quantum wave, quantum memory, or quantum thought Storage Device at the quantum level.

Psyche obviously exists because quantum waves or photons obviously exist. Psyche obviously exists because Quantum Information obviously exists.

According to the LAW of Psyche, every psyche has a certain amount of energy that's under its control, and that controlling psyche can form or transform the energy under its control into anything that it wants that energy to be, anytime and anywhere that it chooses to do so.

Psyche is the Ultimate Cause, which means that Psyche is the Ultimate Causal Agent and the Ultimate Actor and Chooser. Nature's Psyche makes the quantum waves in the first place, and then later, it is Nature's Psyche who collapses the wave function. It's NOT being done by physical matter nor by chance as the Materialists, Naturalists, Darwinists, Determinists, and Atheists claim. The Materialists, Darwinists, and Naturalists are always wrong, and they don't even know it, because they have no comprehension as to what the quantum or the non-physical is and how it works.

Nature's Psyche or the Intelligences in Nature made and MAPPED your physical brain. When you want to move your finger, Nature's Psyche reads your mind and then FIRES the specific neuron that moves your finger. Nature's Psyche made that neuron and assigned that neuron to that specific function. Nature's Psyche collapses the necessary wave functions needed to FIRE or TRIGGER that specific neuron. The Human Psyche isn't consciously aware of any of these things. It's all being handled by Nature's Psyche. The physical brain is MAPPED at the quantum level by Nature's Psyche in order to give us specific functionality at the physical level. Nature's Psyche KNOWS which neuron it needs to fire or trigger in order to make your finger move, because Nature's Psyche is the one who MAPPED that specific function onto that specific neuron in the first place. This is Quantum Neuroscience.

The brain is plastic. The brain can be REMAPPED by Nature's Psyche. When the brain is damaged, it can take a while for Nature's Psyche to REMAP that specific lost functionality to other nearby neurons. Sometimes it cannot be done if there are no other neurons nearby or if the nerves in the spine, arms, or legs have been severed.

Psyche, or Intelligence, or Quantum Consciousness, or Nature's Psyche is running and controlling everything at the quantum level, including the quantum fields, the quantum waves, and your physical brain. It's the only thing we know of that can.

You will NEVER get anything like this from the Materialists, Darwinists, Naturalists, and Atheists. Instead, these people keep telling us that quantum waves, quantum fields, photons, and psyches do not exist. They are obviously wrong. Psyches obviously exist because quantum waves, photons, and quantum fields obviously exist. Quantum Order, Organization, and Law obviously exist. These non-physical quantum objects or quantum machines couldn't exist without some type of non-physical Quantum Psyche or Quantum Intelligence to make them exist in the first place. Psyches obviously exist because the Perpetual Motion Cycle $E = mc^2$ and the Quantum Fields obviously exist. These non-physical cycles and machines cannot exist without some type of Psyche to run them or operate them. Random chaos is worthless if you want to get anything done. Psyches obviously exist because physical matter (collapsed wave functions) and physical brains (quantumly fired neurons) obviously exist. I mean, it's obvious.

Mark My Words

# A Philosophy of Science for Personality Theory or Psyche Theory

## INDIVIDUAL QUANTUM OBJECTS

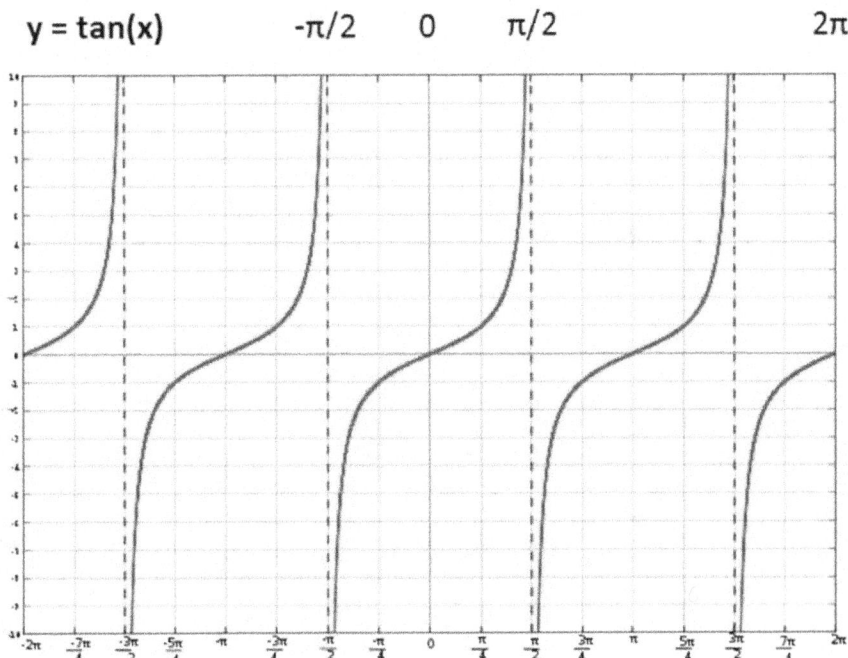

A Psyche Ontology is based upon the idea that Psyche, or Quantum Non-Local Consciousness, is the fundamental unit of reality. If there really is such a thing as a point particle or an infinite singularity, Psyche is it. Psyche or consciousness is the "elementary particle" upon which everything else is based. Psyche determines or chooses everything else.

I use **y = tan(x)** to mathematically model my Psyche Ontology or my Ultimate Cause Model of Reality, quantum objects, spirit matter, physical matter, spirit bodies, physical bodies, universes, infinite singularities, psyche, and non-local consciousness. Each wave on the graph is a different Quantum Object, or a different particle of matter, or a different universe.

This model extends all the way from the infinitely small to the infinitely large; and, I was extremely satisfied with its explanatory power when it was revealed to me.

A single Universe, or Quantum Object, or Particle of Matter, and/or its lifeline exists in the domain between the -π/2 and π/2 asymptotes. Matter or a Quantum Object has some limitations. Matter in any format, whether spiritual or physical, reacts. There's a time lag associated with Matter or a Quantum Object, because it waits for and then reacts to Psyche's Word of Command.

When a psyche or consciousness enters a non-consensus reality, that spiritual reality conforms itself to that psyche's demands and commands. In contrast, a physical reality is the ultimate consensus reality. It doesn't reshape itself automatically to the demands of our psyche. Instead, we have to work it in order to reshape it.

Again, a single Universe, or Quantum Object, or Particle of Matter (and its size or the amount of space it takes up) exists in the domain between the $-\pi/2$ and $\pi/2$ asymptotes. That Quantum Object, or matter, or particle, or wave, or material is spiritual and takes up little space when it is below (0, 0) on the graph; and, that same Quantum Object or matter is physical and can theoretically take up infinite space and become infinite in size when it is above (0, 0) on the graph above – although it's prevented from reaching infinite size and infinite velocity by the restraints of physical law, on the physical side of the equation.

Psyche resides ON the $-\pi/2$ and $\pi/2$ asymptotes. Think about it. That means that Psyche or Non-Local Consciousness has NO size and takes up NO space, yet is simultaneously infinite in size, range, scope, and presence. It's a paradox, but it's Reality. It's one eternal round! Psyche goes down the rabbit hole and climbs the stairway to heaven simultaneously, because Psyche exists at the infinite and IS infinite.

Psyche or Intelligence is something completely different from a spirit body; but, they can be combined together into a functional whole. The spirit body and psyche combination are typically called a Spirit, Ghost, or Soul.

At the $-\pi/2$ asymptote, Psyche or Quantum Non-Local Consciousness is an infinite singularity or point particle which has NO physical size and takes up NO physical space whatsoever; yet, simultaneously at the $\pi/2$ asymptote, Psyche is capable of being omnipresent and omniscient. Psyche resides in the realm of the infinite. Psyche is infinite and instantaneous. In its native format, there are no speed limits placed upon Psyche. Psyche is capable of being instantaneously and simultaneously everywhere. Psyche is infinitely small; but, it is also simultaneously capable of infinite omnipresence and infinite omniscience.

Psyche is viewed or seen as a spark or a pinpoint of light. Psyche is experienced as an immaterial viewpoint in space. Psyche or Intelligence is a whole other animal besides matter or dust; yet, both are comprised of different types and/or different frequencies of quantum waves.

Quantum waves are also thoughts and memories. They survive the death of our physical brain according to the observational evidence obtained from Near-Death Experiences (NDEs), Out-of-Body Experiences (OBEs), Shared-Death Experiences (SDEs), and our after-death Life Reviews. A Psyche is capable of transmitting, receiving, and storing quantum waves, which are thoughts and memories.

In contrast, Quantum Objects, particles, or matter have two states of existence according to the Quantum Law of Complementarity – a non-local spiritual state and a localized physical state. According to the Quantum Law of Complementarity, a Quantum Object can be either spirit matter or physical matter, but it can't be in both states simultaneously. A choice has to be made by Someone Psyche to determine which of the two states it will be in. Psyche ACTS. Quantum Objects REACT to psyche's demands and intervention.

According to the Quantum Law of Superposition, a Quantum Object or Particle of Matter can be phase-shifted, dimension-shifted, or frequency-shifted along the domain between the $-\pi/2$ and $\pi/2$ asymptotes. Matter is finite and reacts to psyche's demands, particularly to God's Psyche who provided matter with Laws, Order, Structure, Restrictions, and Purpose.

Thanks to the Quantum Law of Superposition, a psyche, a spirit body, and a physical body can occupy the same space at the same time because they are out of phase with each other, exist at different frequencies, and therefore exist in different dimensions, but in the very same space at the exact same time.

A Quantum Object or particle of matter can be transitioned, or phase-shifted and moved, all along the x-axis between the $-\pi/2$ and $\pi/2$ asymptotes. Out of body travelers have observed that on the spiritual side, their spirit body can phase-shift or go into different dimensions at will. Being in a different phase or at a different frequency, their spirit body can walk through physical walls and doors; and, they can levitate and rise into air through physical ceilings and roofs.

However, on the physical side, the physical laws that God put into place prevents us from phase-shifting our physical bodies at will and then walking through walls and doors. God retains the capability of phase-shifting physical matter for himself and apparently doesn't share that capability with mortals. The same thing applies to the teleportation or the quantum tunneling of physical matter. God put the physical laws into place in order to restrain it and contain it so that your physical body doesn't quantum tunnel away on you one atom at a time. God put the sub-light speed limitations and entropy into place on physical matter in order to create the Ultimate Consensus Reality for us to live and experience.

In a physical reality, I can depend on my house, my car, this paper, and my dog being there when I go looking for them tomorrow. A physical reality is the ultimate consensus reality because it is dependable, reliable, and predictable. It provides an excellent school-ground upon which to learn how to interact with other psyches in a social manner.

Physical matter is still capable of phase-shifting and teleportation – changing frequencies – but the whole thing is under God's control. God teleported me and my car to safety one time. It wasn't anything I did. It was something that God did for me to save me. My best friend has experienced phase-shifting. An elk passed through his truck un-phased. Again, it wasn't anything that my friend did. It was something that God did for him in order to save him, or the elk.

Quantum Objects – spirit matter and physical matter – reside in the realm of the finite. Psyche, Intelligence, Consciousness, or Life is the Master of the Infinite!

This IS the Ultimate Model of Reality and the Grand Unified Theory of Everything; and, it's really quite simple to visualize and understand once a person chooses to do so.

Mark My Words

# The Creation Drama

Science 2.0 is observation and experience. Science 2.0 needs an account of origins from the people who were actually there and saw how it was done.

I found the following Creation Story online, and I decided to take advantage of it to see what we can learn from it. Science 2.0 is about keeping an open mind. Science 2.0 allows all of the evidence into evidence, and then tries to pursue a preponderance of that evidence.

We need knowledge of the creation from someone who was actually there and saw how it was done.

The following is a play or drama that was revealed to us by the Biblical God Jesus Christ. I submit this one first as observational evidence supporting Science 2.0, because I found it the most unique and informative. Science 2.0 encourages you to make your own personal interpretation of any observational evidence. I will make comments at the end of each section; but in the final analysis, it will be left up to you, my reader, to interpret it and decide for yourself what to make of it.

Online source: http://www.ldsendowment.org/creation.html

## THE FIRST DAY

[Elohim, Jehovah, and Michael are heard speaking outside the Creation Room.]

ELOHIM: Jehovah, Michael, see – yonder is matter unorganized. Go ye down and organize it into a world like unto the worlds that we have heretofore formed. Call your labors the first day and bring me word.

JEHOVAH: It shall be done, Elohim. Come, Michael, let us go down.

MICHAEL: We will go down, Jehovah.

JEHOVAH: Michael, see – here is matter unorganized. We will organize it into a world like unto the worlds that we have heretofore formed. We will call our labors the first day and return and report.

MICHAEL: We will return and report our labors of the first day, Jehovah.

JEHOVAH: Elohim, we have been down as thou hast commanded, and have organized a world like unto the worlds that we have heretofore formed; and we have called our labors the first day.

ELOHIM: It is well.

(Notice that the Gods created and populated many other worlds within this fallen mortal physical universe, before the Gods got around to organizing and populating our earth. There was a large gap of time between the creation of our physical universe and the formation of our particular earth. This Observational Account of origins tells us that the Gap Theory is correct. The Gap Theory states that there was a huge gap of time between the creation of our physical universe and the formation of our earth, just as the observational evidence and the Big Bang Theory seem to suggest. The Gods formed and populated many other worlds before they got around to forming and creating our earth. The Gods were there, and they remember how it was done, because they made it happen in the first place.)

(Notice that nothing can substitute for nor surpass Eye-Witness Evidence. This observational evidence suggests that our physical universe really is billions of years old, just as our astronomical observational evidence and our measurements of the speed-of-light and light-years suggest.)

(The Gods are scientists, not magicians. The Gods can't do Creation Ex Nihilo. They are NOT magicians. Like us, the Gods take pre-existing energy and matter, and form the stuff into something new and useful. The Gods work the matter and the energy until it forms into what they want it to be. It takes hours and days and years for the Gods to do their work. There's no such thing as Creation Ex Nihilo or Magic. Even the Gods have to do work in order to get the results that they desire. Matter and energy are the media in which the Gods do their work. The same can be said for us, just on a much smaller scale. We work the matter and the energy until we get the results that we want.)

(Under Science 2.0, Observational Evidence, Experiential Evidence, and Eye-Witness Evidence takes precedence and priority over philosophical speculation, guesswork, wishful thinking, and scientific inferences. Observations and Eye-Witness Evidence resolve a lot of controversy and contention. Through observation, we simply KNOW how things work and how things were done. The purpose of this little exercise is to demonstrate Science 2.0 in action, because under Science 2.0 science is defined as Observational Evidence, Experiential Evidence, Empirical Evidence, Eye-Witness Evidence, and Knowledge.)

(This version of creation suggests that the Gods organized pre-existing matter in order to produce our earth. $E = mc^2$ and Quantum Field Theory tell us that energy and matter are the same thing. $E = mc^2$ is the perpetual motion cycle, and it works perfectly fine both coming and going. The Gods made the quantum fields, the Gods made the matter, and then the Gods organized that matter into the things that they wanted that matter to be. It was done this way because chance or random chaos prevents matter from organizing spontaneously into planets, stars, and galaxies. Random chance destroys things. It doesn't create them. It requires intelligence in order to organize and create.)

(According to the Perpetual Motion Cycle $E = mc^2$, energy and matter are the same thing. The Gods organize energy or matter into useful things. According to Quantum Mechanics and Quantum Field Theory, particles or matter are quanta or packets of energy. Psyche is the ultimate point particle. From a Quantum Perspective or a Spiritual Perspective or a Non-Physical Perspective, there really is NO difference between physical matter and non-physical particles. They are ALL made from quanta or organized packets of energy.)

(Not only is psyche, or intelligence, or quantum consciousness made from energy, but psyche is seen or viewed at the quantum level and the psyche level as being a pinpoint of light, a point particle, a spark, a particle of matter, or a photon. Psyche is not only omnipresent quantum waves, but psyche is also seen or viewed as being a point particle of matter, a quantum, or a packet of energy at the quantum level and the psyche level of existence. According to the Obvious Law of Physics, the smaller dwells within and controls the larger. Psyche dwells within and controls everything else. It's obvious. Psyche, or Intelligence, or Quantum Consciousness, or Life Force makes, transmits, receives, processes, and stores quantum waves, thoughts, or memories. This is what we actually experience and observe, both at the physical level and the quantum spiritual level! Our Psyche or Intelligence continues to think and live long after our physical brain is dead and gone. Our Psyche or Intelligence is eternal and everlasting, because like energy it is always conserved, because it is energy. Psyche or Intelligence cannot be made, and it cannot be destroyed. It has always existed, and it will always exist.)

(The thing that we call a ghost, or a spirit, is in fact a spirit body and a Psyche or Intelligence combined together into one entity or person. Spirit bodies are born or made. The Psyche or Intelligence is eternal and everlasting. The Gods did on our world precisely what they had done on previous inhabited worlds. The Gods organize matter, or packets of energy, or quanta into worlds. This matches perfectly with Quantum Mechanics, Quantum Field Theory, the Perpetual Motion Cycle $E = mc^2$, and what we have actually experienced and observed.)

### The Birth of Our Earth
https://ldssoul.com/the-birth-of-our-earth/

The whole purpose of Science 2.0 is to demonstrate that there are BETTER interpretations of the scientific evidence than the ones we are getting from the Materialists, Naturalists, Darwinists, Nihilists, and Atheists. With the exception of physical matter, everything that exists falsifies Materialism, Naturalism, Darwinism, Nihilism, and Atheism.

According to the Philosophy of Science, there are theoretically an infinite number of different, conflicting, and mutually exclusive personal interpretations of each piece of scientific evidence and scientific data that we produce or encounter. Personal Interpretation of the Evidence IS the fatal flaw and the fatal step in the Scientific Method. The purpose of a science experiment is to find the True Cause of an event; and, CHANCE cannot be the True Cause of anything significant.

### The Probability of Protein Synthesis by Chance Alone
https://origin-science.org/the-probability-of-protein-synthesis-by-chance-alone/

Creation by Chance cannot be the True Cause of the things that we observe all around us, according to the Rules of Science, Statistics, Common-Sense, and the Philosophy of Science. We must find a better and more believable causal source for origins than chance causality or creation by chance.

The Gods actually admit that they are the ones who organized or created our heavens and our earth. There's NO chance that they could be wrong, because chance can't design and create anything of significance. Chance can only corrupt, deteriorate, and destroy.

## THE SECOND DAY

ELOHIM: Jehovah, Michael, go down again. Gather the waters together and cause the dry land to appear. The great waters call ye Seas, and the dry land call ye Earth. Form mountains and hills, great rivers and small streams, to beautify and give variety to the face of the earth. When ye have done this, call your labors the second day, and bring me word.

JEHOVAH: It shall be done, Elohim. Come, Michael, let us go down.

MICHAEL: We will go down, Jehovah.

JEHOVAH: Michael, we will gather the waters together and cause the dry land to appear. The great waters we will call Seas, and the dry land we will call Earth. We will form mountains and hills, great rivers and small streams to beautify and give variety to the face of the earth. We will call our labors the second day and return and report.

MICHAEL: We will return and report our labors of the second day, Jehovah.

JEHOVAH: Elohim, we have been down as thou hast commanded, and have gathered the waters together, and have caused the dry land to appear. The great waters we have called Seas, and the dry land we have called Earth. We have formed mountains and hills, great rivers and small streams, to beautify and give variety to the face of the earth; and we have called our labors the second day.

ELOHIM: It is well.

(It seems to me that the Gods can do mind-over-matter. They can command the matter, and it chooses to obey. This ability is often called Priesthood Power or the Word of Command. In fiction, fantasy, and Marvel Comics it is called "superpowers" or "abilities". The Gods have superpowers or abilities. Obviously, we fallen mortal physical beings don't have this kind of ability, yet.)

(Their labors took however long they took. Each "day" could have lasted for thousands or millions of years, for all we know. The Gods simply called their labors "the second day", and they left it at that. This Observational Account of origins suggests that the Day-Age Theory is correct. The Day-Age Theory holds that the six days referred to in the Genesis account of creation are not ordinary 24-hour days, but are much longer periods, from thousands to billions of years. The observational evidence from the Antarctic ice cores and the fossil record seems to suggest that our earth is millions if not billions of years old. Astronomical observations and the speed-of-light definitely prove that our physical universe is billions of years old. It all wasn't done in a day. The Gods are scientists, not magicians.)

(The Gods seem to be able to manipulate matter telepathically with their minds, which is a psychic phenomenon that we call mind-over-matter. This phenomenon has been called the Word of Command, or simply The Word. In other words, the Gods can do Quantum Mechanics, whereas, we cannot. The Gods call it Priesthood Power, but it's what we call Quantum Mechanics.)

(The biggest problem that the Materialists, Naturalists, and Atheists have with Quantum Mechanics – and that necessary Conscious Observer – is that they can't make money from them, because they can't control Quantum Mechanics and other Conscious Observers. In other words, the Quantum cannot be controlled nor manipulated from the physical level. Thankfully, mortal fallen physical beings can't do Quantum Mechanics like the Gods can, because we would only use these supernatural abilities to destroy each other, as our superhero movies attest. Look at any movie or series from Marvel Comics and DC Comics to see what we fallen mortal beings do with quantum abilities or superpowers.)

(The Atheists are deathly afraid of Psyche, or Quantum Consciousness, or Intelligent Creators, or Gods. Are they not?)

**THE THIRD DAY**

ELOHIM: Jehovah, Michael, return again to the earth that you have organized. Divide the light from the darkness. Call the light Day, and the darkness Night. Cause the lights in the firmament to appear – the greater light to rule the day, and the lesser light to rule the night. Cause the stars also to appear and give light to the earth, the same as with other worlds heretofore created. Call your labors the third day and bring me word.

JEHOVAH: It shall be done, Elohim. Come, Michael, let us return again to the earth that we have organized.

MICHAEL: We will return again, Jehovah.

JEHOVAH: Michael, we will divide the light from the darkness, and we will call the light Day, and the darkness Night. We will cause the lights in the firmament to appear – the greater light to rule the day, and the lesser light to rule the night. We will cause the stars also to appear and give light to the earth, the same as with worlds heretofore created. We will call our labors the third day and return and report.

MICHAEL: We will return and report our labors of the third day, Jehovah.

JEHOVAH: Elohim, we have been down as thou hast commanded, and have divided the light from the darkness, and have called the light Day and the darkness Night. We have caused the lights in the firmament to appear – the greater light to rule the day, and the lesser light to rule the night. We have caused the stars also to appear and give light to the earth, the same as with worlds heretofore created; and we have called our labors the third day.

ELOHIM: It is well.

(Notice that the sun, moon, stars, galaxies, and physical universe were already there. The Gods simply caused these celestial objects to become visible on the surface of our earth on the Third Day. They did NOT say that they created the sun, moon, and stars on the third day. They said that they caused them to appear on the surface of the earth on the third day. This suggests that the Canopy Model is correct and that our earth's atmosphere was opaque until the Third Day. On the Third Day, the Gods made our earth's atmosphere translucent causing the already pre-existing stars in the heavens to appear. The Gods also emphasize once again that they are simply doing here what they have already done on other physical worlds that were formed and passed away and died, long before our particular earth was organized and made. Our earth isn't their first time at bat. The Gods have been making habitable worlds for a very long time.)

(What isn't said, but hinted at, is also interesting. There are other ways to interpret this besides the obvious way. It's possible that our earth was made somewhere else, and then quantum tunneled or teleported here into this particular solar system into a very specific spot in our solar system. The Gods can teleport or quantum tunnel physical bodies, physical matter, planets, stars, and galaxies, whenever they choose to do so. The Gods can also raise physical bodies from the dead.)

(There are indications from the different calendars that have been discovered that each planet in our solar system was teleported or quantum tunneled into its specific orbit at a specific time to serve as prophecy or a prediction for specific events that were yet to come. There's more than meets the eye, within these verses of scripture or revelation. There are many hidden wonders that have been waiting to be found.)

https://www.johnpratt.com/items/docs/lds/dates.html

(The Gods don't have the physical limitations that we do. The Apostle Paul in 1 Corinthians 15 says that the Gods are Spiritual Beings or have Spiritual Resurrected Bodies – the Gods have resurrected physical bodies that function like spirit bodies. Their physical bodies or resurrected bodies can walk through walls, teleport or quantum tunnel, levitate, yet at the same time eat physical food. Their physical bodies function as if they were spirit bodies. Their resurrected physical bodies are immortal like spirit bodies. Yet, at the same time, their resurrected physical bodies can do everything that our physical bodies can do, and more.)

(It's possible that the reason that the sun, moon, and stars appeared on the third day is because that was the day that the Gods teleported, or quantum tunneled, our earth from its construction yard into orbit of our current sun. There are lots of interesting possibilities, but only when we are willing to allow the quantum or the spiritual into the account. The quantum tunneling or the teleportation of mass or physical matter or physical particles has been experienced and observed. It's a real, proven, and verified phenomenon. We try to avoid it in our computer chips by building barriers wide enough to prevent it from happening, but where the Gods are concerned, there are NO barriers wide enough to prevent them from quantum tunneling mass or matter to wherever they want it to go, whenever they want to do so. We're not able to stop them. The Gods come down to our world and interact with us in physical form at will at any time they choose to do so.)

**THE FOURTH DAY**

ELOHIM: Jehovah, Michael, return again. Place seeds of all kinds in the earth that they may spring forth as grass, flowers, shrubbery, trees, and all manner of vegetation, each bearing seed in itself after its own kind, as on the worlds we have heretofore created. Call your labors the fourth day and bring me word.

JEHOVAH: It shall be done, Elohim. Come, Michael, let us go down.

MICHAEL: We will go down, Jehovah.

JEHOVAH: Michael, we will place seeds of all kinds in the earth that they may spring forth as grass, flowers, shrubbery, trees, and all manner of vegetation, each bearing seed in itself after its own kind, as on the worlds we have heretofore created. We will call our labors the fourth day and return and report.

MICHAEL: We will return and report our labors of the fourth day, Jehovah.

JEHOVAH: Elohim, we have been down as thou hast commanded, and have placed seeds of all kinds in the earth that they may spring forth as grass, flowers, shrubbery, trees, and all manner of vegetation, each bearing seed in itself, after its own kind, as on the worlds we have heretofore created; and we have called our labors the fourth day.

ELOHIM: It is well.

(Notice that the Gods planted seeds and waited for those seeds to become adult trees hundreds of our years old, on the Fourth Day. There are trees hundreds of years old in our fossil record; and, we assume that some of those plants and trees made it into the fossil record on the Fourth Day. Once again, this suggests that the Day-Age Theory is correct. The plants and trees were multiplying from seed and dying as they usually do, all on the Fourth Day. This also suggests that life and death, as well as reproduction, were taking place on this earth and throughout our physical universe long before our particular Adam and Eve were placed into the Garden of Eden. There's plenty of time here for the 24 mass extinctions in our fossil record to take place before our Adam and Eve were placed into the Garden of Eden.)

https://en.wikipedia.org/wiki/List_of_extinction_events

(Every time a mass extinction took place, the Gods would simply transplant replacements from other worlds to take their place. Chance cannot make proteins and genes, so new life forms on this earth had to come from other worlds, and they had to be teleported or quantum tunneled here by the Gods.)

(It's fascinating to notice that the Gods did NOT make genomes and life forms on the fourth day. The Gods simply transplanted seeds from other worlds to this world. Genes, genomes, proteins, and life forms were made long ago on other worlds, and then transplanted to this world as seeds. For all we know, genes, genomes, proteins, and our physical bodies were made trillions of trillions of years ago in other physical universes, and then subsequently quantum tunneled into this universe. The Gods are masters of Quantum Mechanics. The Gods can quantum tunnel or teleport physical matter just as easily as they can quantum tunnel or teleport photons or quantum waves. The Gods would have no problem quantum tunneling or teleporting seeds from other worlds to our earth. The Gods could be doing so right now, and we would never know it.)

(The Gods planted seeds from other worlds onto our world.)

## THE FIFTH DAY

ELOHIM: Jehovah, Michael, now that the earth is formed, divided, and beautified, and vegetation is growing thereon, return and place beasts upon the land: the elephant, the lion, the tiger, the bear, the horse, and all other kinds of animals – fowls in the air in all their varieties, fishes of all kinds in the waters, and insects and all manner of animal life upon the earth.

Command the beasts, the fowls, the fishes, the insects, all creeping things, and other forms of animal life to multiply in their respective elements, each after its kind, and every kind of vegetation to multiply in its sphere, that every form of life may fill the measure of its creation and have joy therein. Call your labors the fifth day and bring me word.

JEHOVAH: It shall be done, Elohim. Come, Michael, let us go down.

MICHAEL: We will go down, Jehovah.

JEHOVAH: Michael, now that the earth is formed, divided, and beautified, and vegetation is growing thereon, we will place beasts upon the land: the elephant, the lion, the tiger, the bear, the horse, and all other kinds of animals – fowls in the air in all their varieties, fishes of all kinds in the waters, and insects and all manner of animal life upon the earth.

We will command the beasts, the fowls, the fishes, the insects, all creeping things, and other forms of animal life to multiply in their respective elements, each after its kind, and every kind of vegetation to multiply in its sphere, that every form of life may fill the measure of its creation and have joy therein. We will call our labors the fifth day and return and report.

MICHAEL: It is well, Jehovah. Now that the earth is formed, divided, and beautified, with vegetation growing thereon, and provided with animal life, it is glorious and beautiful.

JEHOVAH: It is, Michael.

MICHAEL: Let us return and report our labors of the fifth day, Jehovah.

JEHOVAH: Elohim, we have been down as thou hast commanded. We have placed beasts upon the land: the elephant, the lion, the tiger, the bear, the horse, and all other kinds of animals – fowls in the air in all their varieties, fishes of all kinds in the waters, and insects and all manner of animal life upon the earth.

We have commanded the beasts, the fowls, the fishes, the insects, all creeping things, and other forms of animal life to multiply in their respective elements, each after its kind, and every kind of vegetation to multiply in its sphere, that every form of life may fill the measure of its creation and have joy therein. We have called our labors the fifth day.

ELOHIM: It is well.

(You and I are the children of the Gods. You and I helped to form this earth. We are Gods. Our surname is God.)

(The Gods placed animals onto our earth in a systematic fashion. The Gods are scientists, not magicians. The Gods terraformed, seeded, and then placed animals onto our earth in an orderly and systematic fashion. The Gods were teaching us and showing us how to do it. The animals were living, multiplying, and dying like they always do on a fallen mortal earth such as ours, before the Garden of Eden was planted and before our Adam and Eve were placed into the Garden of Eden.)

(The Gods commanded the animals to multiply after their own kind. The animals were born, and they died, and they ended up in our fossil record millions of years before our Adam and Eve were placed into the Garden of Eden.)

https://en.wikipedia.org/wiki/List_of_extinction_events

(Every time a mass extinction took place, the God would simply transplant replacements from other worlds to take their place. Chance cannot make proteins and genes, so new life forms on this earth had to come from other worlds, and they had to be teleported or quantum tunneled here by the Gods.)

(Our fossil record tells us that fourth day and the fifth day lasted for nearly a billion years, before the Gods planted the Garden of Eden and placed an immortal Adam and Eve into it. Our fossil record along with ice core samples from the polar ice sheets tell us that Young-Earth Creationism is false! Astronomy and the speed-of-light tell us clearly and conclusively that Young-Universe Creationism is false. Finally, ALL of these different eyewitness accounts of origins and creation tell us that Creation Ex Nihilo or Magic or Atheism is false.)

(We don't observe macro-evolution, chemical evolution, abiogenesis, or spontaneous generation. We don't observe chimpanzees giving birth to humans, which is macro-evolution. Instead, we observe animals multiplying, each after its own kind. We observe the effects of genetics and genomes! Sharks today are the same as what they were millions of years ago. The same reality applies to every species that hasn't gone extinct.)

(The false is falsified by the truth, and the truth is repeatedly and constantly observed. We don't observe chemical evolution nor macro-evolution. We don't observe spontaneous generation nor abiogenesis. We observe plants and animals reproducing after their own kind for millions of years at a time. We observe precisely what is talked about in these eyewitness accounts of creation and origins. So far everything in this creation account matches perfectly with what has actually been experienced and observed. Even the quantum tunneling of mass has been experienced and observed.)

(The fact that this particular creation account matches perfectly with what we have actually experienced and observed tells me that this is a TRUE account of origins or a TRUE revelation from Jesus Christ and God the Father, because this account was obviously produced by the people who were actually there and saw how it was really done. Simple though it is, this particular creation account comes directly into line with what we scientists or observers have actually experienced and observed. This account of origins doesn't contain any of that fictional and falsified creation ex nihilo, creation by chance, chemical evolution, spontaneous generation, or macro-evolution garbage that we get from the

Materialists, Naturalists, Darwinists, Nihilists, Atheists, and various Christian Denominations. To me, this seems like a true account of the origin of our earth and the life forms on this earth. It's sufficient for my needs as a scientist, because it matches with what has actually been experienced and observed.)

## THE SIXTH DAY

ELOHIM: Jehovah, Michael, is man found upon the earth?

JEHOVAH: Man is not found on the earth, Elohim.

ELOHIM: Jehovah, Michael, then let us go down and form man in our own likeness and in our own image, male and female, and put into him his spirit, and let us give him dominion over the beasts, the fishes, and the birds, and make him lord over the earth, and over all things on the face of the earth.

We will plant for him a garden, eastward in Eden, and place him in it to tend and cultivate it, that he may be happy and have joy therein. We will command him to multiply and replenish the earth, that he may have joy and rejoicing in his posterity.

We will place before him the tree of knowledge of good and evil, and we will allow Lucifer, our common enemy, whom we have thrust out, to tempt him and to try him, that he may know by his own experience the good from the evil.

If he yields to temptation, we will give unto him the law of sacrifice, and we will provide a Savior for him, as we counseled in the beginning, that man may be brought forth by the power of the redemption and the resurrection, and come again into our presence, and with us partake of eternal life and exaltation.

We will call this the sixth day, and we will rest from our labors for a season. Come, let us go down.

JEHOVAH: We will go down, Elohim.

(Finally, the Gods plant the Garden of Eden onto our fallen mortal earth; and, the Gods place Adam and Eve into it. Adam's physical body was organized in the image of Elohim's and Jehovah's physical body. The Gods can lay down their physical body, and take it up again, at will – such is the nature of the resurrection. Eve's physical body was organized in the image of Heavenly Mother's physical body. The Gods have wives; and, the Gods have resurrected physical bodies. There is no God without a Goddess.)

(The purpose of this mortal life is to be a trial or a test. We are given the chance to learn from our own experience the difference between the good and the evil that exist. The whole plan of happiness or plan of salvation is laid out in these verses of revelation or scripture. The Gods have a plan for us. The Gods have prepared a redemption, or a resurrection, or a restoration, or an Atonement for us. We come here to learn and to be tried, and a way has been prepared for us to return back to God's presence once again after our mission has been accomplished.)

(The Gods planted the Garden of Eden AFTER our fossil record had been established, after the plants and animals had been multiplying after their own kind for millions if not billions of years. This account of origins gets the timing of everything correct! That tells me that this is a true account of origins.)

(Notice that it is the Tree of the Knowledge of Good and Evil. We emphasize the evil, but the whole purpose for this mortal life on this fallen telestial world is to learn from our own experience to tell the difference between what is good and what is evil. If we don't learn that lesson, then our purpose for being here has been a failure. Another purpose for being here is to learn how to have joy. Joy is a part of the good that we are here to learn from our own experience.)

(Our whole fossil record and 24 mass extinctions and replantings took place before our Adam and Eve were placed into the Garden of Eden. Lots of things were happening on this earth before Adam and Eve were placed onto this earth.)

https://origin-science.org/Before-Adam-Hugh-Nibley

http://www.bhporter.com/Porter%20PDF%20Files/Before%20Adam.pdf

I relied on Hugh Nibley's account of origins and creation in his "Before Adam" article, before I started to embrace some of the others that I came across.

## THE CREATION OF ADAM

[Elohim, Jehovah, and Michael enter the Creation Room. Michael sits in a chair with his head on his chest, his eyes closed, as if he were lifeless.]

ELOHIM: Jehovah, see the earth that we have formed. There is no man to till and take care of it. We are here to form man in our own likeness and in our own image.

JEHOVAH: We will do so, Elohim.

[Elohim and Jehovah pass their hands in the air over Michael's body, from head to foot.]

ELOHIM: Jehovah, man is now organized, and we will put into him his spirit, the breath of life, that he may become a living soul.

[Elohim and Jehovah lift Michael's head into an upright position. He opens his eyes.]

(This account seems to imply that the Gods made a new physical body or a new type of physical body for Michael at this point in time. It doesn't tell us what they did with his other physical body. Michael was a God and a member of the Godhead, who condescended to come to our world and become the father of our human race. The Gods can lay aside their physical bodies or resurrected bodies at will. The Gods can choose to become mortal again. The Gods can also choose to be born again through a woman's womb, as Jesus Christ did. There are lots of different ways to be resurrected or reborn.)

## THE CREATION OF EVE

ELOHIM: Jehovah, is it good for man to be alone?

JEHOVAH: It is not good for man to be alone, Elohim.

ELOHIM: We will cause a deep sleep to come upon this man whom we have formed, and we will take from his side a rib, from which we will form a woman to be a companion and helpmeet for him.

[Elohim and Jehovah lower Michael's head to his chest again. He closes his eyes as if asleep.]

JEHOVAH: Brethren and sisters, this is Michael, who helped form the earth. When he awakens from the sleep which we have caused to come upon him, he will be known as Adam and, having forgotten all, will have become as a little child. Brethren, close your eyes as if you were asleep.

[While the brethren's eyes are closed, Elohim and Jehovah lead Eve into the Creation Room.]

ELOHIM: Adam, awake and arise.

JEHOVAH: All the brethren will please arise.

ELOHIM: Adam, here is a woman whom we have formed and whom we give unto you to be a companion and helpmeet for you. What will you call her?

ADAM: Eve.

ELOHIM: Why will you call her Eve?

ADAM: Because she is the mother of all living.

ELOHIM: That is right, Adam; because she is the mother of all living.

(Notice that Adam used to be one of the Gods, before he volunteered to come to this earth and become the father of our human race. Eve is the mother of all the physical human beings on this earth. Adam used to be the God Michael. The Gods condescend to our level in the hope of helping to raise us to their level. The resurrected Gods can also lay aside their physical body, meaning that they can die at will and then be reborn or reconstituted in physical form like us. Jesus Christ was able to die at will, and then resurrect, because He was the God Jehovah before being born of Mary on this physical fallen earth. Because Jesus Christ was an immortal God here in mortality, He had to willingly give up His life and die while on the cross, or He would still be hanging there today. The Gods condescend. The Gods put their own hides on the line in an attempt to teach us, save us, and exalt us to their level. They do this for us, because we are their children. Our spirit bodies and physical bodies were organized in their image. We are descended from the Gods both physically and spiritually. As human beings, our surname is God. We are gods.)

(This account of origins seems to imply that Eve is a female clone of Adam. She would have gotten one of her X chromosomes from Adam, which means that she got her other X chromosome from her heavenly mother. Again, the Gods are scientists, and NOT magicians. Eve was formed. She wasn't just snapped into existence by magic. It took however long it took to get the job done.)

(All of this forming and cloning could have taken place on some other world, but this account of origins seems to suggest that it took place here on this earth. The Gods did it however they chose to do it.)

## ADAM AND EVE ARE INTRODUCED INTO THE GARDEN

ELOHIM: Adam, we have organized for you this earth and have planted a garden, eastward in Eden. We will place you in the garden and will there command you and Eve to multiply and replenish the earth, that you may have joy and rejoicing in your posterity.

Jehovah, introduce Adam into the garden which we have prepared for him.

JEHOVAH: It shall be done, Elohim.

[Jehovah instructs the brethren to follow Adam, and the sisters to follow Eve, into the Garden Room.]

(Notice that the Gods planted the Garden of Eden at the end of the creation period, because the Garden of Eden wasn't already here on our fallen, mortal, death-filled world. Our whole universe was already physical and mortal, before the Gods got around to organizing our fallen mortal earth. The Garden of Eden is simply a semi-immortal oasis in the middle of all that life and death, which is taking place just outside of Eden.)

## THE GARDEN OF EDEN

(The drama or the play continues in the Garden of Eden.)

http://www.ldsendowment.org/garden.html

(This account makes it clear that the Garden of Eden co-existed with the fallen mortal world. The Garden of Eden was simply an oasis planted within our death-filled mortal world. The only known immortals on our earth at the time were Adam, Eve, the Tree of Life, and possibly the Tree of Knowledge of Good and Evil. Adam was no longer a God, but he was an immortal being, suggesting some type of intermediate terrestrial or translated or transfigured or paradisiacal state of physical being that is somewhere between Godhood and our fallen physical mortality that we are currently experiencing. Some people call these intermediate beings "translated beings" or "seraphim". They are NOT Gods, and they are not fallen mortal beings like us, either.)

(Obviously, I took this account of origins seriously, and tried to see what I could learn from it and derive from it. I chose to treat it as if it had actually happened, and then tried to see what I could learn from the evidence that it presented to us. Its whole meaning and value changes when we realize that it happened or choose to believe that it happened. Instantly, there are lots of other interesting things that we can surmise from this account simply by choosing to believe that it happened as described.)

## THE GODS WITHDRAW

ELOHIM: Adam, we have created for you this earth, and have placed upon it all kinds of vegetation and animal life. We have commanded all these to multiply in their own sphere and element. We give you dominion over all these things and make you, Adam, lord over the whole earth and all things on the face thereof. We now command you to multiply and replenish the earth, that you may have joy and rejoicing in your posterity.

We have also planted for you this garden, wherein we have placed all manner of fruits, flowers, and vegetation. Of every tree of the garden thou mayest freely eat, but of the tree of knowledge of good and evil thou shalt not eat; nevertheless, thou mayest choose for thyself, for it is given unto thee. But remember that I forbid it, for in the day thou eatest thereof thou shalt surely die.

Adam, remember this commandment which we have given unto you. Now go to – dress this garden, take good care of it, be happy and have joy therein.

We shall go away, but we shall visit you again and give you further instructions.

[Elohim and Jehovah withdraw from the Garden Room. Eve also withdraws from view, as if passing into a different portion of the garden. Adam is left alone.]

(Notice that the Gods withdraw, leaving us human beings to make our own choices and decisions. We have no idea how long Adam and Eve were in the Garden of Eden. We are never told; but, life and death were going on outside of the Garden of Eden throughout our physical universe, and even within the Garden of Eden, while the immortal Adam and Eve were residing in the Garden of Eden. Adam and Eve tended the garden, grew things, and ate from the garden, which means that some type of death was already taking place within the Garden of Eden right from the beginning. Only Adam, Eve, the Tree of Life, and the Tree of Knowledge seemed to be immortal.)

(Now, it's time for the dedicated opposition to enter the scene.)

## LUCIFER APPROACHES ADAM

[Lucifer enters.]

LUCIFER: Well, Adam, you have a new world here.

ADAM: A new world?

LUCIFER: Yes, a new world, patterned after the old one where we used to live.

ADAM: I know nothing about any other world.

LUCIFER: Oh, I see – your eyes are not yet opened. You have forgotten everything. You must eat some of the fruit of this tree.

[Lucifer pantomimes picking two pieces of fruit from the tree of knowledge of good and evil. He offers the fruit to Adam.]

LUCIFER: Adam, here is some of the fruit of that tree. It will make you wise.

ADAM: I will not partake of that fruit. Father told me that in the day I should partake of it, I should surely die.

LUCIFER: You shall not surely die, but shall be as the Gods, knowing good and evil.

ADAM: I will not partake of it.

LUCIFER: Oh, you will not? Well, we shall see.

[Adam withdraws from view.]

(Even though Lucifer is a disembodied non-physical spirit, he can interact directly with physical matter telekinetically. This reality matches with people's experiences with evil spirits and Satan. Lucifer doesn't have a physical body, but it seems as if he does. The concept of hard-light comes into play here, for anyone who has studied holography. Light can take on many different forms, including spirit matter and physical matter as well as some kind of telekinetic hard-light which simulates physical matter but isn't physical in nature. We can learn a lot about the Gods and the fallen evil spirits by studying and thinking about these different Observational Accounts, from the people who were there and remember how it really was.)

(Notice that Adam, Eve, and Lucifer used to live on a previous physical world similar to ours before being transplanted or teleported here to our world. Our earth is patterned after that previous world where we used to live. Of course, Adam was one of the Gods and a member of a Godhead before condescending to become Adam and the father of our race. This account of origins implies that Lucifer used to have a physical body.)

## EVE PARTAKES OF THE FRUIT

[Eve returns.]

LUCIFER: Eve, here is some of the fruit of that tree. It will make you wise. It is delicious to the taste and very desirable.

EVE: Who are you?

LUCIFER: I am your brother.

EVE: You, my brother, and come here to persuade me to disobey Father?

LUCIFER: I have said nothing about Father. I want you to eat of the fruit of the tree of knowledge of good and evil that your eyes may be opened, for that is the way Father gained his knowledge. You must eat of this fruit so as to comprehend that everything has its opposite: good and evil, virtue and vice, light and darkness, health and sickness, pleasure and pain. Thus, your eyes will be opened, and you will have knowledge.

EVE: Is there no other way?

LUCIFER: There is no other way.

EVE: Then I will partake.

[Eve pantomimes taking one of the pieces of fruit from Lucifer's hand and eating it.]

LUCIFER: There. Now go and get Adam to partake.

[Lucifer pantomimes placing the second piece of fruit in her hand. He withdraws from view.]

(Notice that Adam, Eve, the Gods, and Lucifer existed in one form or another long before our particular physical earth was organized, seeded, and populated. Psyche or Non-Local Consciousness is immortal, with no beginning and no end. You and I have always existed. Spirit matter has always existed. It's only physical matter and physical organization that has a beginning, and possibly an end if the Gods stop sustaining it and supporting it.)

## ADAM PARTAKES OF THE FRUIT

[Adam returns.]

EVE: Adam, here is some of the fruit of that tree. It is delicious to the taste and very desirable.

ADAM: Eve, do you know what fruit that is?

EVE: Yes. It is the fruit of the tree of knowledge of good and evil.

ADAM: I cannot partake of it. Do you not know that Father commanded us not to partake of the fruit of that tree?

EVE: Do you intend to obey all of Father's commandments?

ADAM: Yes, all of them.

EVE: Do you not recollect that Father commanded us to multiply and replenish the earth? I have partaken of this fruit and by so doing shall be cast out, and you will be left a lone man in the garden of Eden.

ADAM: Eve, I see that this must be so. I will partake that man may be.

[Adam pantomimes eating the fruit.]

(Notice, only the Tree of Knowledge is capable of making an immortal Adam and Eve mortal. The formerly immortal Adam and Eve were able to freely eat from anything else on this fallen mortal earth without it making them mortal. This is science in action. Something within the Tree of Knowledge had the ability to make Adam and Eve mortal.)

## THEIR NAKEDNESS

[Lucifer returns as Adam is eating the fruit.]

LUCIFER: That is right.

EVE: [To Adam.] It is better for us to pass through sorrow that we may know the good from the evil. [To Lucifer.] I know thee now. Thou art Lucifer, he who was cast out of Father's presence for rebellion.

LUCIFER: Yes. You are beginning to see already.

ADAM: What is that apron you have on?

LUCIFER: It is an emblem of my power and priesthoods.

ADAM: Priesthoods?

LUCIFER: Yes, priesthoods.

ADAM: I am looking for Father to come down to give us further instructions.

LUCIFER: Oh, you are looking for Father to come down, are you?

[Elohim and Jehovah are heard speaking outside the Garden Room.]

ELOHIM: Jehovah, we promised Adam that we would visit him and give him further instructions. Come, let us go down.

JEHOVAH: We will go down, Elohim.

ADAM: I hear their voices; they are coming.

LUCIFER: See – you are naked. Take some fig leaves and make you aprons. Father will see your nakedness. Quick! Hide!

[Lucifer withdraws from view.]

ADAM: Brethren and sisters, put on your aprons.

[He waits until they have done so.]

[To Eve.] Come, let us hide.

[Adam and Eve withdraw from view.]

**THE GODS RETURN**

[Elohim and Jehovah return to the Garden Room.]

ELOHIM: Adam!

Adam!

Adam, where art thou?

[Adam returns.]

ADAM: I heard thy voice and hid myself, because I was naked.

ELOHIM: Who told thee that thou wast naked? Hast thou partaken of the fruit of the tree of knowledge of good and evil, of which we commanded thee not to partake?

ADAM: The woman thou gavest me and commanded that she should remain with me – she gave me of the fruit of the tree, and I did eat.

ELOHIM: Eve!

[Eve returns.]

What is this that thou hast done?

EVE: The serpent beguiled me, and I did eat.

(They take turns blaming each other for what happened, but the truth is that they were given the Tree of Knowledge so that they could fall, become mortal, and have children. They were set up; and, there was no other way. The needed to become mortal and gain knowledge before they could become parents and have children. They needed to know how things work which they didn't know when they were blissfully innocent children in the Garden of Eden.)

(Notice that it is the Tree of Knowledge of Good and Evil. We always emphasize the evil aspect, but the tree also gave them the ability to KNOW good and understand the good, which they didn't have before they partook of the Tree of Knowledge. Like Adam and Eve, we each had to CHOOSE to come here and to be mortal. We each have to experience the bitter, so that we can KNOW the sweet. Like Adam and Eve, we are here on this fallen, mortal, physical world because we chose to be here. There was NO other way to gain knowledge, except to experience it firsthand for ourselves. That's why Science 2.0 is knowledge, or firsthand experience and observation. We can't know anything until after we have experienced it and observed it for ourselves. That's the way Science works. The Gods are scientists, after all – the Ultimate Scientists. We learn best and fastest by experience and observation.)

(Like with anything else, you will have to decide for yourself if any of this is useful to you or not. All we can do is to present the evidence, and then we have to leave it up to you to decide for yourself what you want to make of it.)

## LUCIFER IS EXPELLED FROM THE GARDEN

ELOHIM: Lucifer!

[Lucifer returns.]

ELOHIM: What hast thou been doing here?

LUCIFER: I have been doing that which has been done on other worlds.

ELOHIM: What is that?

LUCIFER: I have been giving some of the fruit of the tree of knowledge of good and evil to them.

ELOHIM: Lucifer, because thou hast done this, thou shalt be cursed above all the beasts of the field. Upon thy belly thou shalt go, and dust shalt thou eat all the days of thy life.

LUCIFER: If thou cursest me for doing the same thing which has been done on other worlds, I will take the spirits that follow me, and they shall possess the bodies thou createst for Adam and Eve!

ELOHIM: I will place enmity between thee and the seed of the woman. Thou mayest have power to bruise his heal, but he shall have power to crush thy head.

LUCIFER: Then with that enmity I will take the treasures of the earth, and with gold and silver I will buy up armies and navies, popes and priests, and reign with blood and horror on the earth!

ELOHIM: Depart!

[Lucifer passes from the Garden Room into the World Room.]

(This account of origins suggests that Lucifer and his followers are now disembodied spirits. There are also indications that Lucifer, and some of the spirits that followed him, had their physical bodies taken away from them when they rebelled against God and Christ. Even though this is one of the best accounts of origins on record, there's still a lot that remains unanswered. However, this account makes it clear that Satan or Lucifer is real and truly exists. In this modern age, the first thing Satan does in his attempts to deceive us is to convince us that he and God do not exist.)

(There's no poofing, or magic, or teleportation going on here. Lucifer and his followers have already been cast out of heaven to our fallen mortal earth. The portrayals of this drama simply show Lucifer walking out of the Garden of Eden into our fallen mortal world, which is already there just outside of Eden. The Garden of Eden is simply a paradisiacal oasis right in the middle of our fallen, mortal, death-filled, barren, desert-type world and universe, which this drama calls "the lone and dreary world". Just outside the Garden of Eden is of a different kind or category than within the Garden of Eden. There was life and death and reproduction taking place on our earth and throughout our physical universe, while our Adam and Eve were immortal inside the Garden of Eden.)

(What makes this important is that this drama claims to be an Observational Account of origins by the people who were there and actually saw and experienced how it was all done and/or actually made it happen in the first place. It's also consistent with the other Observational Accounts that I have come across from Christian sources. Under Science 2.0, Observations take precedence or priority over philosophical speculations, scientific inferences, and guesswork. As a former accountant, I was taught to always seek the primary source or the primary document and to give it precedence over everything else – go to the source, and not the speculation, for your answers. The purpose of this exercise is to demonstrate Science 2.0 in action. Science is observation and experience, or it is supposed to be.)

## CONDITIONS OF THE FALL

ELOHIM: Jehovah, let cherubim and a flaming sword be placed to guard the way of the tree of life, lest Adam put forth his hand, and partake of the fruit thereof, and live forever in his sins.

JEHOVAH: It shall be done, Elohim.

[Jehovah stretches his hand towards the tree of life.]

Let cherubim and a flaming sword be placed to guard the way of the tree of life, lest Adam put forth his hand, and partake of the fruit thereof, and live forever in his sins.

It is done, Elohim.

(The Tree of Life is immortal, and the Tree of Life is capable of making Adam and Eve immortal again. Now that Adam and Eve have fallen and become mortal like everything else on this earth, Adam and Eve are blocked from the Tree of Life so that they can't become immortal again. This whole thing comes across to me as science, not magic.)

(The Gods know that if Adam and Eve, and the rest of us, gain access to the Tree of Life, we will become immortal. The worst thing that can happen to us is to live forever in our sins. Look at something like "Marvel Comics" or "Altered Carbon" to see what happens to human beings whenever they figure out how to become immortal and learn how to live forever in their sins. They become monsters. There's no progress. There's no refinement. The immortals in "Altered Carbon" realized towards the end of the series that they had lived too long and that it was time for them to die. That's what happens when we mortal, physical, fallen human beings learn how to live forever in our sins. We degrade into monsters and become despicable human beings. There's no redemption or refinement taking place. Immortality is NOT a blessing, unless Christ and His Atonement are there to help us overcome and abandon our sins.)

## ADAM AND EVE ARE EXPELLED FROM THE GARDEN

ELOHIM: Jehovah, see that Adam is driven out of this beautiful garden into the lone and dreary world, where he may learn from his own experience to distinguish good from evil.

JEHOVAH: It shall be done, Elohim.

[Jehovah instructs the brethren to follow Adam, and the sisters to follow Eve, into the World Room. Elohim and Jehovah do not accompany initiates into the World Room.]

(Like Satan or Lucifer before them, Adam and Eve simply walk out of the Garden of Eden into the "lone and dreary world" where death, disease, and the weeds reign supreme. It was Adam and Eve who fell, and not anything else in this universe. Everything else on this earth was already Fallen or Telestial, and subject to Satan's influence. In fact, when Adam and Eve go where Satan has gone, into this fallen world where we now live, Lucifer declares himself to be the god of this world in which we now live, which explains in large part why our current physical fallen world is so messed-up. The Gods have left these people to do their own thing and to make their own choices, and to suffer the consequences of their choices.)

## THE TELESTIAL WORLD

http://www.ldsendowment.org/telestial.html

This (world room) represents the telestial kingdom, or the world in which we now live. Adam, on finding himself in the lone and dreary world, built an altar and offered prayer.

[Lucifer comes into view.]

LUCIFER: I hear you. What is it you want?

ADAM: Who are you?

LUCIFER: I am the god of this world.

ADAM: You, the god of this world?

LUCIFER: Yes. What do you want?

ADAM: I am looking for messengers.

LUCIFER: Oh, you want someone to preach to you. You want religion, do you? I will have preachers here presently.

(And, the drama or play continues. Both Satan and the GODS send messengers, teachers, and preachers to Adam and Eve in an attempt to convert them to their side – Adam and Eve end up choosing between the two competing factions, and so do the rest of us in the end. I found this a fascinating and informative Observational Account of origins. I hope you did too. The next ones are consistent with this one, but also completely different. That's as it should be. There's something new and interesting to be learned from each of them. Science 2.0, as an observational science, needs an account of origins that has actually been experienced and observed. The Gods were there and KNOW how it was done.)

(We are the children of the Gods. We were also there and saw how it was done. Like Adam or Michael, many of us helped the Gods to make and seed this earth.)

Online source: http://www.ldsendowment.org/

It's there to be found, so why not take advantage of it?

# The Book of Abraham

This is a Creation Account that was revealed to the prophet Abraham. I will leave it up to you, my reader, to interpret it and decide for yourself what to make of it – kind of like a Ted Talk – just present the evidence and leave it up to the audience to decide what to do with it.

This is one of my top three most favorite Origin Accounts on record. It portrays the Gods as scientists.

Remember, the best way to find and know the truth is to live it and experience it for yourself, or to choose to trust someone who has.

Remember, Science 2.0 allows all of the evidence into evidence; and, I often like to let my reader interpret the evidence for himself or herself, and decide for himself what to make of it. My only obligation as a Science 2.0 scientist is to present the evidence – the personal interpretation of the evidence, I leave up to you.

Abraham Chapter 3

1 And I, Abraham, had the Urim and Thummim, which the Lord my God had given unto me, in Ur of the Chaldees;

2 And I saw the stars, that they were very great, and that one of them was nearest unto the throne of God; and there were many great ones which were near unto it;

3 And the Lord said unto me: These are the governing ones; and the name of the great one is Kolob, because it is near unto me, for I am the Lord thy God: I have set this one to govern all those which belong to the same order as that upon which thou standest.

4 And the Lord said unto me, by the Urim and Thummim, that Kolob was after the manner of the Lord, according to its times and seasons in the revolutions thereof; that one revolution was a day unto the Lord, after his manner of reckoning, it being one thousand years according to the time appointed unto that whereon thou standest. This is the reckoning of the Lord's time, according to the reckoning of Kolob.

5 And the Lord said unto me: The planet which is the lesser light, lesser than that which is to rule the day, even the night, is above or greater than that upon which thou standest in point of reckoning, for it moveth in order more slow; this is in order because it standeth above the earth upon which thou standest, therefore the reckoning of its time is not so many as to its number of days, and of months, and of years.

6 And the Lord said unto me: Now, Abraham, these two facts exist, behold thine eyes see it; it is given unto thee to know the times of reckoning, and the set time, yea, the set time of the earth upon which thou standest, and the set time of the greater light which is set to rule the day, and the set time of the lesser light which is set to rule the night.

7 Now the set time of the lesser light is a longer time as to its reckoning than the reckoning of the time of the earth upon which thou standest.

8 And where these two facts exist, there shall be another fact above them, that is, there shall be another planet whose reckoning of time shall be longer still;

9 And thus there shall be the reckoning of the time of one planet above another, until thou come nigh unto Kolob, which Kolob is after the reckoning of the Lord's time; which Kolob is set nigh unto the throne of God, to govern all those planets which belong to the same order as that upon which thou standest.

10 And it is given unto thee to know the set time of all the stars that are set to give light, until thou come near unto the throne of God.

11 Thus I, Abraham, talked with the Lord, face to face, as one man talketh with another; and he told me of the works which his hands had made;

12 And he said unto me: My son, my son (and his hand was stretched out), behold I will show you all these. And he put his hand upon mine eyes, and I saw those things which his hands had made, which were many; and they multiplied before mine eyes, and I could not see the end thereof.

13 And he said unto me: This is Shinehah, which is the sun. And he said unto me: Kokob, which is star. And he said unto me: Olea, which is the moon. And he said unto me: Kokaubeam, which signifies stars, or all the great lights, which were in the firmament of heaven.

14 And it was in the night time when the Lord spake these words unto me: I will multiply thee, and thy seed after thee, like unto these; and if thou canst count the number of sands, so shall be the number of thy seeds.

15 And the Lord said unto me: Abraham, I show these things unto thee before ye go into Egypt, that ye may declare all these words.

16 If two things exist, and there be one above the other, there shall be greater things above them; therefore, Kolob is the greatest of all the Kokaubeam that thou hast seen, because it is nearest unto me.

17 Now, if there be two things, one above the other, and the moon be above the earth, then it may be that a planet or a star may exist above it; and there is nothing that the Lord thy God shall take in his heart to do but what he will do it.

18 Howbeit that he made the greater star; as, also, if there be two spirits, and one shall be more intelligent than the other, yet these two spirits, notwithstanding one is more intelligent than the other, have no beginning; they existed before, they shall have no end, they shall exist after, for they are gnolaum, or eternal.

19 And the Lord said unto me: These two facts do exist, that there are two spirits, one being more intelligent than the other; there shall be another more intelligent than they; I am the Lord thy God, I am more intelligent than they all.

20 The Lord thy God sent his angel to deliver thee from the hands of the priest of Elkenah.

21 I dwell in the midst of them all; I now, therefore, have come down unto thee to declare unto thee the works which my hands have made, wherein my wisdom excelleth them all, for I rule in the heavens above, and in the earth beneath, in all wisdom and prudence, over all the intelligences thine eyes have seen from the beginning; I came down in the beginning in the midst of all the intelligences thou hast seen.

22 Now the Lord had shown unto me, Abraham, the intelligences [psyches] that were organized [into spirit bodies] before the world was; and among all these there were many of the noble and great ones;

23 And God saw these souls that they were good, and he stood in the midst of them, and he said: These I will make my rulers; for he stood among those that were spirits, and he saw that they were good; and he said unto me: Abraham, thou art one of them; thou wast chosen before thou wast born.

24 And there stood one among them that was like unto God, and he said unto those who were with him: We will go down, for there is space there, and we will take of these materials, and we will make an earth whereon these may dwell;

25 And we will prove them herewith, to see if they will do all things whatsoever the Lord their God shall command them;

26 And they who keep their first estate shall be added upon; and they who keep not their first estate shall not have glory in the same kingdom with those who keep their first estate; and they who keep their second estate shall have glory added upon their heads for ever and ever.

27 And the Lord said: Whom shall I send? And one answered like unto the Son of Man: Here am I, send me. And another answered and said: Here am I, send me. And the Lord said: I will send the first.

28 And the second [Satan] was angry, and kept not his first estate; and, at that day, many followed after him.

Abraham Chapter 4

1 And then the Lord said: Let us go down. And they went down at the beginning, and they, that is the Gods, organized and formed the heavens and the earth.

2 And the earth, after it was formed, was empty and desolate, because they had not formed anything but the earth; and darkness reigned upon the face of the deep, and the Spirit of the Gods was brooding upon the face of the waters.

3 And they (the Gods) said: Let there be light; and there was light.

4 And they (the Gods) comprehended the light, for it was bright; and they divided the light, or caused it to be divided, from the darkness.

5 And the Gods called the light Day, and the darkness they called Night. And it came to pass that from the evening until morning they called night; and from the morning until the evening they called day; and this was the first, or the beginning, of that which they called day and night.

6 And the Gods also said: Let there be an expanse in the midst of the waters, and it shall divide the waters from the waters.

7 And the Gods ordered the expanse, so that it divided the waters which were under the expanse from the waters which were above the expanse; and it was so, even as they ordered.

8 And the Gods called the expanse, Heaven. And it came to pass that it was from evening until morning that they called night; and it came to pass that it was from morning until evening that they called day; and this was the second time that they called night and day.

9 And the Gods ordered, saying: Let the waters under the heaven be gathered together unto one place, and let the earth come up dry; and it was so as they ordered;

10 And the Gods pronounced the dry land, Earth; and the gathering together of the waters, pronounced they, Great Waters; and the Gods saw that they were obeyed.

11 And the Gods said: Let us prepare the earth to bring forth grass; the herb yielding seed; the fruit tree yielding fruit, after his kind, whose seed in itself yieldeth its own likeness upon the earth; and it was so, even as they ordered.

12 And the Gods organized the earth to bring forth grass from its own seed, and the herb to bring forth herb from its own seed, yielding seed after his kind; and the earth to bring forth the tree from its own seed, yielding fruit, whose seed could only bring forth the same in itself, after his kind; and the Gods saw that they were obeyed.

13 And it came to pass that they numbered the days; from the evening until the morning they called night; and it came to pass, from the morning until the evening they called day; and it was the third time.

14 And the Gods organized the lights in the expanse of the heaven, and caused them to divide the day from the night; and organized them to be for signs and for seasons, and for days and for years;

15 And organized them to be for lights in the expanse of the heaven to give light upon the earth; and it was so.

16 And the Gods organized the two great lights, the greater light to rule the day, and the lesser light to rule the night; with the lesser light they set the stars also;

17 And the Gods set them in the expanse of the heavens, to give light upon the earth, and to rule over the day and over the night, and to cause to divide the light from the darkness.

18 And the Gods watched those things which they had ordered until they obeyed.

19 And it came to pass that it was from evening until morning that it was night; and it came to pass that it was from morning until evening that it was day; and it was the fourth time.

20 And the Gods said: Let us prepare the waters to bring forth abundantly the moving creatures that have life; and the fowl, that they may fly above the earth in the open expanse of heaven.

21 And the Gods prepared the waters that they might bring forth great whales, and every living creature that moveth, which the waters were to bring forth abundantly after their kind; and every winged fowl after their kind. And the Gods saw that they would be obeyed, and that their plan was good.

22 And the Gods said: We will bless them, and cause them to be fruitful and multiply, and fill the waters in the seas or great waters; and cause the fowl to multiply in the earth.

23 And it came to pass that it was from evening until morning that they called night; and it came to pass that it was from morning until evening that they called day; and it was the fifth time.

24 And the Gods prepared the earth to bring forth the living creature after his kind, cattle and creeping things, and beasts of the earth after their kind; and it was so, as they had said.

25 And the Gods organized the earth to bring forth the beasts after their kind, and cattle after their kind, and every thing that creepeth upon the earth after its kind; and the Gods saw they would obey.

26 And the Gods took counsel among themselves and said: Let us go down and form man in our image, after our likeness; and we will give them dominion over the fish of the sea, and over the fowl of the air, and over the cattle, and over all the earth, and over every creeping thing that creepeth upon the earth.

27 So the Gods went down to organize man in their own image, in the image of the Gods to form they him, male and female to form they them.

28 And the Gods said: We will bless them. And the Gods said: We will cause them to be fruitful and multiply, and replenish the earth, and subdue it, and to have dominion over the fish of the sea, and over the fowl of the air, and over every living thing that moveth upon the earth.

29 And the Gods said: Behold, we will give them every herb bearing seed that shall come upon the face of all the earth, and every tree which shall have fruit upon it; yea, the fruit of the tree yielding seed to them we will give it; it shall be for their meat.

30 And to every beast of the earth, and to every fowl of the air, and to every thing that creepeth upon the earth, behold, we will give them life, and also we will give to them every green herb for meat, and all these things shall be thus organized.

31 And the Gods said: We will do everything that we have said, and organize them; and behold, they shall be very obedient. And it came to pass that it was from evening until morning they called night; and it came to pass that it was from morning until evening that they called day; and they numbered the sixth time.

Abraham Chapter 5

1 And thus we will finish the heavens and the earth, and all the hosts of them.

2 And the Gods said among themselves: On the seventh time we will end our work, which we have counseled; and we will rest on the seventh time from all our work which we have counseled.

3 And the Gods concluded upon the seventh time, because that on the seventh time they would rest from all their works which they (the Gods) counseled among themselves to form; and sanctified it. And thus, were their decisions at the time that they counseled among themselves to form the heavens and the earth.

4 And the Gods came down and formed these the generations of the heavens and of the earth, when they were formed in the day that the Gods formed the earth and the heavens,

5 According to all that which they had said concerning every plant of the field before it was in the earth, and every herb of the field before it grew; for the Gods had not caused it

to rain upon the earth when they counseled to do them, and had not formed a man to till the ground.

6 But there went up a mist from the earth and watered the whole face of the ground.

7 And the Gods formed man from the dust of the ground, and took his spirit (that is, the man's spirit), and put it into him; and breathed into his nostrils the breath of life, and man became a living soul.

8 And the Gods planted a garden, eastward in Eden, and there they put the man, whose spirit they had put into the body which they had formed.

9 And out of the ground made the Gods to grow every tree that is pleasant to the sight and good for food; the tree of life, also, in the midst of the garden, and the tree of knowledge of good and evil.

10 There was a river running out of Eden, to water the garden, and from thence it was parted and became into four heads.

11 And the Gods took the man and put him in the Garden of Eden, to dress it and to keep it.

12 And the Gods commanded the man, saying: Of every tree of the garden thou mayest freely eat,

13 But of the tree of knowledge of good and evil, thou shalt not eat of it; for in the time that thou eatest thereof, thou shalt surely die. Now I, Abraham, saw that it was after the Lord's time, which was after the time of Kolob; for as yet the Gods had not appointed unto Adam his reckoning.

14 And the Gods said: Let us make an help meet for the man, for it is not good that the man should be alone, therefore we will form an help meet for him.

15 And the Gods caused a deep sleep to fall upon Adam; and he slept, and they took one of his ribs, and closed up the flesh in the stead thereof;

16 And of the rib which the Gods had taken from man, formed they a woman, and brought her unto the man.

17 And Adam said: This was bone of my bones, and flesh of my flesh; now she shall be called Woman, because she was taken out of man;

18 Therefore shall a man leave his father and his mother, and shall cleave unto his wife, and they shall be one flesh.

19 And they were both naked, the man and his wife, and were not ashamed.

20 And out of the ground the Gods formed every beast of the field, and every fowl of the air, and brought them unto Adam to see what he would call them; and whatsoever Adam called every living creature, that should be the name thereof.

21 And Adam gave names to all cattle, to the fowl of the air, to every beast of the field; and for Adam, there was found an help meet for him.

# The Book of Moses

### The Law of Witnesses:

**2 Corinthians 13: 1: "In the mouth of two or three witnesses shall every truth be established."**

These three Creation Accounts or Origin Accounts are harmonious, consistent, and sustain one another. I only need three in order to conform to the Law of Witnesses.

Remember, Science 2.0 is based upon the Law of Witnesses. If you have got three separate witnesses saying basically the same exact thing, that's good enough. You find yourself looking at the truth.

The following is a Creation Account that was revealed to Moses. I assume that the Book of Genesis in the Bible was derived from this particular revelation to Moses. I will leave it up to you, my reader, to interpret it and decide for yourself what to make of it.

In other books and essays that I have written, I make commentary on parts of these three separate Creation Accounts, so I won't do so here.

Notice how the following Origin Account fits more closely with the Book of Genesis in the Bible, than the preceding ones did, because Moses is the source of them. I find the similarities and differences informative.

This particular account claims to be a revealed account of origins, from the original Eye-Witnesses themselves, the Gods.

Moses Chapter 1

1 The words of God, which he spake unto Moses at a time when Moses was caught up into an exceedingly high mountain,

2 And he saw God face to face, and he talked with him, and the glory of God was upon Moses; therefore, Moses could endure his presence.

3 And God spake unto Moses, saying: Behold, I am the Lord God Almighty, and Endless is my name; for I am without beginning of days or end of years; and is not this endless?

4 And, behold, thou art my son; wherefore look, and I will show thee the workmanship of mine hands; but not all, for my works are without end, and also my words, for they never cease.

5 Wherefore, no man can behold all my works, except he behold all my glory; and no man can behold all my glory, and afterwards remain in the flesh on the earth.

6 And I have a work for thee, Moses, my son; and thou art in the similitude of mine Only Begotten; and mine Only Begotten is and shall be the Savior, for he is full of grace and truth; but there is no God beside me, and all things are present with me, for I know them all.

7 And now, behold, this one thing I show unto thee, Moses, my son, for thou art in the world, and now I show it unto thee.

8 And it came to pass that Moses looked, and beheld the world upon which he was created; and Moses beheld the world and the ends thereof, and all the children of men which are, and which were created; of the same he greatly marveled and wondered.

9 And the presence of God withdrew from Moses, that his glory was not upon Moses; and Moses was left unto himself. And as he was left unto himself, he fell unto the earth.

10 And it came to pass that it was for the space of many hours before Moses did again receive his natural strength like unto man; and he said unto himself: Now, for this cause I know that man is nothing, which thing I never had supposed.

11 But now mine own eyes have beheld God; but not my natural, but my spiritual eyes, for my natural eyes could not have beheld; for I should have withered and died in his presence; but his glory was upon me; and I beheld his face, for I was transfigured before him.

12 And it came to pass that when Moses had said these words, behold, Satan came tempting him, saying: Moses, son of man, worship me.

13 And it came to pass that Moses looked upon Satan and said: Who art thou? For behold, I am a son of God, in the similitude of his Only Begotten; and where is thy glory, that I should worship thee?

14 For behold, I could not look upon God, except his glory should come upon me, and I were transfigured before him. But I can look upon thee in the natural man. Is it not so, surely?

15 Blessed be the name of my God, for his Spirit hath not altogether withdrawn from me, or else where is thy glory, for it is darkness unto me? And I can judge between thee and God; for God said unto me: Worship God, for him only shalt thou serve.

16 Get thee hence, Satan; deceive me not; for God said unto me: Thou art after the similitude of mine Only Begotten.

17 And he also gave me commandments when he called unto me out of the burning bush, saying: Call upon God in the name of mine Only Begotten, and worship me.

18 And again Moses said: I will not cease to call upon God, I have other things to inquire of him: for his glory has been upon me, wherefore I can judge between him and thee. Depart hence, Satan.

19 And now, when Moses had said these words, Satan cried with a loud voice, and ranted upon the earth, and commanded, saying: I am the Only Begotten, worship me.

20 And it came to pass that Moses began to fear exceedingly; and as he began to fear, he saw the bitterness of hell. Nevertheless, calling upon God, he received strength, and he commanded, saying: Depart from me, Satan, for this one God only will I worship, which is the God of glory.

21 And now Satan began to tremble, and the earth shook; and Moses received strength, and called upon God, saying: In the name of the Only Begotten, depart hence, Satan.

22 And it came to pass that Satan cried with a loud voice, with weeping, and wailing, and gnashing of teeth; and he departed hence, even from the presence of Moses, that he beheld him not.

23 And now of this thing Moses bore record; but because of wickedness it is not had among the children of men.

24 And it came to pass that when Satan had departed from the presence of Moses, that Moses lifted up his eyes unto heaven, being filled with the Holy Ghost, which beareth record of the Father and the Son;

25 And calling upon the name of God, he beheld his glory again, for it was upon him; and he heard a voice, saying: Blessed art thou, Moses, for I, the Almighty, have chosen thee, and thou shalt be made stronger than many waters; for they shall obey thy command as if thou wert God.

26 And lo, I am with thee, even unto the end of thy days; for thou shalt deliver my people from bondage, even Israel my chosen.

27 And it came to pass, as the voice was still speaking, Moses cast his eyes and beheld the earth, yea, even all of it; and there was not a particle of it which he did not behold, discerning it by the Spirit of God.

28 And he beheld also the inhabitants thereof, and there was not a soul which he beheld not; and he discerned them by the Spirit of God; and their numbers were great, even numberless as the sand upon the sea shore.

29 And he beheld many lands; and each land was called earth, and there were inhabitants on the face thereof.

30 And it came to pass that Moses called upon God, saying: Tell me, I pray thee, why these things are so, and by what thou madest them?

31 And behold, the glory of the Lord was upon Moses, so that Moses stood in the presence of God, and talked with him face to face. And the Lord God said unto Moses: For mine own purpose have I made these things. Here is wisdom and it remaineth in me.

32 And by the word of my power, have I created them, which is mine Only Begotten Son, who is full of grace and truth.

33 And worlds without number have I created; and I also created them for mine own purpose; and by the Son I created them, which is mine Only Begotten.

34 And the first man of all men have I called Adam, which is many.

35 But only an account of this earth, and the inhabitants thereof, give I unto you. For behold, there are many worlds that have passed away by the word of my power. And there are many that now stand, and innumerable are they unto man; but all things are numbered unto me, for they are mine and I know them.

36 And it came to pass that Moses spake unto the Lord, saying: Be merciful unto thy servant, O God, and tell me concerning this earth, and the inhabitants thereof, and also the heavens, and then thy servant will be content.

37 And the Lord God spake unto Moses, saying: The heavens, they are many, and they cannot be numbered unto man; but they are numbered unto me, for they are mine.

38 And as one earth shall pass away, and the heavens thereof even so shall another come; and there is no end to my works, neither to my words.

39 For behold, this is my work and my glory — to bring to pass the immortality and eternal life of man.

40 And now, Moses, my son, I will speak unto thee concerning this earth upon which thou standest; and thou shalt write the things which I shall speak.

41 And in a day when the children of men shall esteem my words as naught and take many of them from the book which thou shalt write, behold, I will raise up another like unto thee; and they shall be had again among the children of men — among as many as shall believe.

42 (These words were spoken unto Moses in the mount, the name of which shall not be known among the children of men. And now they are spoken unto you. Show them not unto any except them that believe. Even so. Amen.)

Moses Chapter 2

1 And it came to pass that the Lord spake unto Moses, saying: Behold, I reveal unto you concerning this heaven, and this earth; write the words which I speak. I am the Beginning and the End, the Almighty God; by mine Only Begotten I created these things; yea, in the beginning I created the heaven, and the earth upon which thou standest.

2 And the earth was without form, and void; and I caused darkness to come up upon the face of the deep; and my Spirit moved upon the face of the water; for I am God.

3 And I, God, said: Let there be light; and there was light.

4 And I, God, saw the light; and that light was good. And I, God, divided the light from the darkness.

5 And I, God, called the light Day; and the darkness, I called Night; and this I did by the word of my power, and it was done as I spake; and the evening and the morning were the first day.

6 And again, I, God, said: Let there be a firmament in the midst of the water, and it was so, even as I spake; and I said: Let it divide the waters from the waters; and it was done;

7 And I, God, made the firmament and divided the waters, yea, the great waters under the firmament from the waters which were above the firmament, and it was so even as I spake.

8 And I, God, called the firmament Heaven; and the evening and the morning were the second day.

9 And I, God, said: Let the waters under the heaven be gathered together unto one place, and it was so; and I, God, said: Let there be dry land; and it was so.

10 And I, God, called the dry land Earth; and the gathering together of the waters, called I the Sea; and I, God, saw that all things which I had made were good.

11 And I, God, said: Let the earth bring forth grass, the herb yielding seed, the fruit tree yielding fruit, after his kind, and the tree yielding fruit, whose seed should be in itself upon the earth, and it was so even as I spake.

12 And the earth brought forth grass, every herb yielding seed after his kind, and the tree yielding fruit, whose seed should be in itself, after his kind; and I, God, saw that all things which I had made were good;

13 And the evening and the morning were the third day.

14 And I, God, said: Let there be lights in the firmament of the heaven, to divide the day from the night, and let them be for signs, and for seasons, and for days, and for years;

15 And let them be for lights in the firmament of the heaven to give light upon the earth; and it was so.

16 And I, God, made two great lights; the greater light to rule the day, and the lesser light to rule the night, and the greater light was the sun, and the lesser light was the moon; and the stars also were made even according to my word.

17 And I, God, set them in the firmament of the heaven to give light upon the earth,

18 And the sun to rule over the day, and the moon to rule over the night, and to divide the light from the darkness; and I, God, saw that all things which I had made were good;

19 And the evening and the morning were the fourth day.

20 And I, God, said: Let the waters bring forth abundantly the moving creature that hath life, and fowl which may fly above the earth in the open firmament of heaven.

21 And I, God, created great whales, and every living creature that moveth, which the waters brought forth abundantly, after their kind, and every winged fowl after his kind; and I, God, saw that all things which I had created were good.

22 And I, God, blessed them, saying: Be fruitful, and multiply, and fill the waters in the sea; and let fowl multiply in the earth;

23 And the evening and the morning were the fifth day.

24 And I, God, said: Let the earth bring forth the living creature after his kind, cattle, and creeping things, and beasts of the earth after their kind, and it was so;

25 And I, God, made the beasts of the earth after their kind, and cattle after their kind, and everything which creepeth upon the earth after his kind; and I, God, saw that all these things were good.

26 And I, God, said unto mine Only Begotten, which was with me from the beginning: Let us make man in our image, after our likeness; and it was so. And I, God, said: Let them have dominion over the fishes of the sea, and over the fowl of the air, and over the cattle, and over all the earth, and over every creeping thing that creepeth upon the earth.

27 And I, God, created man in mine own image, in the image of mine Only Begotten created I him; male and female created I them.

28 And I, God, blessed them, and said unto them: Be fruitful, and multiply, and replenish the earth, and subdue it, and have dominion over the fish of the sea, and over the fowl of the air, and over every living thing that moveth upon the earth.

29 And I, God, said unto man: Behold, I have given you every herb bearing seed, which is upon the face of all the earth, and every tree in the which shall be the fruit of a tree yielding seed; to you it shall be for meat.

30 And to every beast of the earth, and to every fowl of the air, and to everything that creepeth upon the earth, wherein I grant life, there shall be given every clean herb for meat; and it was so, even as I spake.

31 And I, God, saw everything that I had made, and, behold, all things which I had made were very good; and the evening and the morning were the sixth day.

Moses Chapter 3

1 Thus the heaven and the earth were finished, and all the host of them.

2 And on the seventh day I, God, ended my work, and all things which I had made; and I rested on the seventh day from all my work, and all things which I had made were finished, and I, God, saw that they were good;

3 And I, God, blessed the seventh day, and sanctified it; because that in it I had rested from all my work which I, God, had created and made.

4 And now, behold, I say unto you, that these are the generations of the heaven and of the earth, when they were created, in the day that I, the Lord God, made the heaven and the earth,

5 And every plant of the field before it was in the earth, and every herb of the field before it grew. For I, the Lord God, created all things, of which I have spoken, spiritually, before they were naturally upon the face of the earth. For I, the Lord God, had not caused it to rain upon the face of the earth. And I, the Lord God, had created all the children of men; and not yet a man to till the ground; for in heaven created I them; and there was not yet flesh upon the earth, neither in the water, neither in the air;

6 But I, the Lord God, spake, and there went up a mist from the earth, and watered the whole face of the ground.

7 And I, the Lord God, formed man from the dust of the ground, and breathed into his nostrils the breath of life; and man became a living soul, the first flesh upon the earth, the first man also; nevertheless, all things were before created; but spiritually were they created and made according to my word.

8 And I, the Lord God, planted a garden eastward in Eden, and there I put the man whom I had formed.

9 And out of the ground made I, the Lord God, to grow every tree, naturally, that is pleasant to the sight of man; and man could behold it. And it became also a living soul. For it was spiritual in the day that I created it; for it remaineth in the sphere in which I, God, created it, yea, even all things which I prepared for the use of man; and man saw that it was good for food. And I, the Lord God, planted the tree of life also in the midst of the garden, and also the tree of knowledge of good and evil.

10 And I, the Lord God, caused a river to go out of Eden to water the garden; and from thence it was parted, and became into four heads.

11 And I, the Lord God, called the name of the first Pison, and it compasseth the whole land of Havilah, where I, the Lord God, created much gold;

12 And the gold of that land was good, and there was bdellium and the onyx stone.

13 And the name of the second river was called Gihon; the same that compasseth the whole land of Ethiopia.

14 And the name of the third river was Hiddekel; that which goeth toward the east of Assyria. And the fourth river was the Euphrates.

15 And I, the Lord God, took the man, and put him into the Garden of Eden, to dress it, and to keep it.

16 And I, the Lord God, commanded the man, saying: Of every tree of the garden thou mayest freely eat,

17 But of the tree of the knowledge of good and evil, thou shalt not eat of it, nevertheless, thou mayest choose for thyself, for it is given unto thee; but, remember that I forbid it, for in the day thou eatest thereof thou shalt surely die.

18 And I, the Lord God, said unto mine Only Begotten, that it was not good that the man should be alone; wherefore, I will make an help meet for him.

19 And out of the ground I, the Lord God, formed every beast of the field, and every fowl of the air; and commanded that they should come unto Adam, to see what he would call them; and they were also living souls; for I, God, breathed into them the breath of life, and commanded that whatsoever Adam called every living creature, that should be the name thereof.

20 And Adam gave names to all cattle, and to the fowl of the air, and to every beast of the field; but as for Adam, there was not found an help meet for him.

21 And I, the Lord God, caused a deep sleep to fall upon Adam; and he slept, and I took one of his ribs and closed up the flesh in the stead thereof;

22 And the rib which I, the Lord God, had taken from man, made I a woman, and brought her unto the man.

23 And Adam said: This I know now is bone of my bones, and flesh of my flesh; she shall be called Woman, because she was taken out of man.

24 Therefore shall a man leave his father and his mother and shall cleave unto his wife; and they shall be one flesh.

25 And they were both naked, the man and his wife, and were not ashamed.

# Margaret Barker

Methodist Minister and Biblical Scholar, Margaret Barker, has written many books and done many different presentations which deal directly with Origins, our Pre-Mortal Existence, and our Heavenly Mother. Margaret Barker and associates are in the process of turning the edited and adjusted "canonized scripture" back into Scripture.

I submit her work and her scholarship into evidence.

As usual, when it comes to Science 2.0, there is NO official interpretation of the evidence to be had or to be given. I simply present the evidence, and then I leave it up to you, my reader, to decide for yourself what to make of that evidence – if anything.

Here are a few of my favorites.


1) "The Mother in Heaven and Her Children" - Margaret Barker

The mentioned handout: http://mypsyche.us/wp-content/uploads/2017/10/Margaret-Barker-HANDOUT-2015.pdf

https://www.fairmormon.org/conference/august-2015/the-mother-in-heaven-and-her-children

http://mypsyche.us/wp-content/uploads/2017/10/The-Mother-in-Heaven-and-Her-Children.pdf

https://www.youtube.com/watch?v=ilF9NXEl6Xs


2) "The Woman Clothed with the Sun in the Book of Revelation" – Margaret Barker

https://www.youtube.com/watch?v=f6OTD7plobA

http://www.templestudies.org/wp-content/uploads/2013/10/Utah-The-Woman-clothed-with-the-Sun.doc

http://mypsyche.us/wp-content/uploads/2017/10/Utah-The-Woman-clothed-with-the-Sun.pdf

The whole conference:

http://www.templestudies.org/2013-the-lady-of-the-temple-conference/conference-videos/


3) "Restoring Solomon's Temple" – Margaret Barker

https://www.youtube.com/watch?v=xalAoRGsU7c

The whole conference:

http://www.templestudies.org/2012-conference/conference-videos/

http://www.templestudies.org/wp-content/uploads/2013/02/MormonismAndTheTemple.pdf

4) "Theosis & Divinization" – Margaret Barker

https://www.youtube.com/watch?v=nOnHDQgIoCU

http://www.templestudies.org/conferences/sacred-space-sacred-thread-a-global-conference-at-usc/

5) Margaret Barker's books on Amazon:

https://www.amazon.com/Margaret-Barker/e/B001IQWG34/

# John Pratt

John Pratt is PhD astronomer, mathematician, physicist, and expert on ancient calendars.

While studying calendars, he has uncovered some interesting coincidences and convergences. I found this science simply fascinating.

As usual, when it comes to Science 2.0, the purpose is to admit all of the evidence into evidence. Then I leave it up to you, my reader, to decide for yourself what to make of that evidence – if anything.

http://www.johnpratt.com/

http://www.johnpratt.com/items/docs/lds/dates.html

It's almost as if the Biblical God Jesus Christ has had his hand in it all, throughout the whole process. He seems to know what He is doing.

# A Revelation about Origins from the Biblical God Jesus Christ

Doctrine and Covenants 93: 1-53:

1 Verily, thus saith the Lord: It shall come to pass that every soul who forsaketh his sins and cometh unto me, and calleth on my name, and obeyeth my voice, and keepeth my commandments, shall see my face and know that I am;

2 And that I am the true light that lighteth every man that cometh into the world;

3 And that I am in the Father, and the Father in me, and the Father and I are one —

4 The Father because he gave me of his fulness, and the Son because I was in the world and made flesh my tabernacle, and dwelt among the sons of men.

5 I was in the world and received of my Father, and the works of him were plainly manifest.

6 And John saw and bore record of the fulness of my glory, and the fulness of John's record is hereafter to be revealed.

7 And he bore record, saying: I saw his glory, that he was in the beginning, before the world was;

8 Therefore, in the beginning the Word was, for he was the Word, even the messenger of salvation —

9 The light and the Redeemer of the world; the Spirit of truth, who came into the world, because the world was made by him, and in him was the life of men and the light of men.

10 The worlds were made by him; men were made by him; all things were made by him, and through him, and of him.

11 And I, John, bear record that I beheld his glory, as the glory of the Only Begotten of the Father, full of grace and truth, even the Spirit of truth, which came and dwelt in the flesh, and dwelt among us.

12 And I, John, saw that he received not of the fulness at the first, but received grace for grace;

13 And he received not of the fulness at first, but continued from grace to grace, until he received a fulness;

14 And thus he was called the Son of God, because he received not of the fulness at the first.

15 And I, John, bear record, and lo, the heavens were opened, and the Holy Ghost descended upon him in the form of a dove, and sat upon him, and there came a voice out of heaven saying: This is my beloved Son.

16 And I, John, bear record that he received a fulness of the glory of the Father;

17 And he received all power, both in heaven and on earth, and the glory of the Father was with him, for he dwelt in him.

18 And it shall come to pass, that if you are faithful you shall receive the fulness of the record of John.

19 I give unto you these sayings that you may understand and know how to worship, and know what you worship, that you may come unto the Father in my name, and in due time receive of his fulness.

20 For if you keep my commandments you shall receive of his fulness and be glorified in me as I am in the Father; therefore, I say unto you, you shall receive grace for grace.

21 And now, verily I say unto you, I was in the beginning with the Father, and am the Firstborn;

22 And all those who are begotten through me are partakers of the glory of the same, and are the church of the Firstborn.

23 Ye were also in the beginning with the Father; that which is Spirit, even the Spirit of truth;

24 And truth is knowledge of things as they are, and as they were, and as they are to come;

25 And whatsoever is more or less than this is the spirit of that wicked one who was a liar from the beginning.

26 The Spirit of truth is of God. I am the Spirit of truth, and John bore record of me, saying: He received a fulness of truth, yea, even of all truth;

27 And no man receiveth a fulness unless he keepeth his commandments.

28 He that keepeth his commandments receiveth truth and light, until he is glorified in truth and knoweth all things.

29 Man was also in the beginning with God. Intelligence, or the light of truth, was not created or made, neither indeed can be.

30 All truth is independent in that sphere in which God has placed it, to act for itself, as all intelligence also; otherwise there is no existence.

31 Behold, here is the agency of man, and here is the condemnation of man; because that which was from the beginning is plainly manifest unto them, and they receive not the light.

32 And every man whose spirit receiveth not the light is under condemnation.

33 For man is spirit. The elements are eternal, and spirit and element, inseparably connected, receive a fulness of joy;

34 And when separated, man cannot receive a fulness of joy.

35 The elements are the tabernacle of God; yea, man is the tabernacle of God, even temples; and whatsoever temple is defiled, God shall destroy that temple.

36 The glory of God is intelligence, or, in other words, light and truth.

37 Light and truth forsake that evil one.

38 Every spirit of man was innocent in the beginning; and God having redeemed man from the fall, men became again, in their infant state, innocent before God.

39 And that wicked one cometh and taketh away light and truth, through disobedience, from the children of men, and because of the tradition of their fathers.

40 But I have commanded you to bring up your children in light and truth.

41 But verily I say unto you, my servant Frederick G. Williams, you have continued under this condemnation;

42 You have not taught your children light and truth, according to the commandments; and that wicked one hath power, as yet, over you, and this is the cause of your affliction.

43 And now a commandment I give unto you — if you will be delivered you shall set in order your own house, for there are many things that are not right in your house.

44 Verily, I say unto my servant Sidney Rigdon, that in some things he hath not kept the commandments concerning his children; therefore, first set in order thy house.

45 Verily, I say unto my servant Joseph Smith, Jun., or in other words, I will call you friends, for you are my friends, and ye shall have an inheritance with me —

46 I called you servants for the world's sake, and ye are their servants for my sake —

47 And now, verily I say unto Joseph Smith, Jun. — You have not kept the commandments, and must needs stand rebuked before the Lord;

48 Your family must needs repent and forsake some things, and give more earnest heed unto your sayings, or be removed out of their place.

49 What I say unto one I say unto all; pray always lest that wicked one have power in you, and remove you out of your place.

50 My servant Newel K. Whitney also, a bishop of my church, hath need to be chastened, and set in order his family, and see that they are more diligent and concerned at home, and pray always, or they shall be removed out of their place.

51 Now, I say unto you, my friends, let my servant Sidney Rigdon go on his journey, and make haste, and also proclaim the acceptable year of the Lord, and the gospel of salvation, as I shall give him utterance; and by your prayer of faith with one consent I will uphold him.

52 And let my servants Joseph Smith, Jun., and Frederick G. Williams make haste also, and it shall be given them even according to the prayer of faith; and inasmuch as you keep my sayings you shall not be confounded in this world, nor in the world to come.

53 And, verily I say unto you, that it is my will that you should hasten to translate my scriptures, and to obtain a knowledge of history, and of countries, and of kingdoms, of laws of God and man, and all this for the salvation of Zion. Amen.

## Other Scriptures about Origins

Here are some other scriptures from the Biblical God Jesus Christ about Origins, which I found useful, consistent, and interesting.

According to the Bible, Adam (his physical body) was the son of God.

Luke 3: 38: **"Which was the son of Enos, which was the son of Seth, which was the son of Adam, which was the son of God."**

I take this literally, and you should too. The scriptures come alive if you take them literally, rather than interpreting them materialistically and atheistically as most people seem to do. It ALL points back to God and to God's Psyche. Psyche IS the fundamental unit of reality. Without Psyche or Intelligence, there would be NO physical matter, NO genomes, NO physical life forms, and there would be NO existence whatsoever. Without God's Psyche, the whole universe would still be spirit matter and nothing but unorganized Chaos.

The whole thing centers on Psyche or Intelligence, because ONLY Psyche can have Lived Experiences in ALL realms of existence and remember having done so. ONLY Psyche can do the Word of Command, which means that ONLY Psyche can ACT. ONLY Psyche can do science!

Doctrine and Covenants 76: 22-24:

**22 And now, after the many testimonies which have been given of him, this is the testimony, last of all, which we give of him: That he lives!**

**23 For we saw him, even on the right hand of God; and we heard the voice bearing record that he is the Only Begotten of the Father [in the flesh] —**

**24 That by him, and through him, and of him, the worlds are and were created, and the inhabitants thereof are begotten sons and daughters unto God [the Father].**

If you can read this, you are a son or daughter of God. You are descended from the Gods – both your spirit body and your physical body.

The Biblical God repeatedly tells us that our Spirits ARE the children of the Most High God. If you can read this, then you are ONE of the children of God. Our surname is God. This means that you did NOT descend from the apes or pond scum.

Furthermore, whenever the Biblical God talks about the flesh or "men", God is talking about our physical bodies.

God of the spirits of all flesh, Num. 16: 22.

Ye are the children of the Lord your God, Deut. 14: 1.

there is a spirit in man, Job 32: 8. (There is a spirit body within man's physical body.)

breath of the Almighty hath given me life, Job 33:4.

**Ye are gods ... children of the most High**, Ps. 82: 6.

the spirit shall return unto God who gave it, Eccl. 12: 7. (It was God who organized our spirit bodies and then place our Psyches within those spirit bodies.)

he that giveth breath ... and spirit to them that walk, Isa. 42: 5.

**Ye are the sons of the living God**, Hosea 1: 10.

Have we not all one father, Mal. 2: 10.

Be ye therefore perfect, even as your Father, Matt. 5: 48 (3 Ne. 12: 48).

Our Father which art in heaven, Matt. 6: 9 (3 Ne. 13: 9).

**we are the offspring of God**, Acts 17: 29.

Spirit itself beareth witness ... **we are the children of God**, Rom. 8: 16.

One God and Father of all, Eph. 4: 6.

be in subjection unto **the Father of spirits**, Heb. 12: 9.

he hath created his children, 1 Ne. 17: 36.

Spirits [psyche and spirit body combination] ... taken home to that God who gave them life, Alma 40: 11.

from God, for the benefit of **the children of God**, D&C 46: 26.

**inhabitants thereof are begotten sons and daughters unto God**, D&C 76: 24.

spirit of man in the likeness of His [God's] person, D&C 77: 2. (We are created in God's image, because we are the children of the Gods – Heavenly Father and Heavenly Mother).

your Father, who is in heaven, knoweth, D&C 84: 83.

the spirit and the [physical] body are the soul of man, D&C 88: 15.

I may testify unto your Father, and your God, D&C 88: 75.

man is spirit, D&C 93: 33. (Physical man has a Spirit residing within him. Remember, a Spirit or a Ghost is comprised of a spirit body and a psyche united together as one. The Most High God – God the Father and Heavenly Mother – organized our spirit bodies and placed our psyche or intelligence within those spirit bodies.)

God, had created all the children of men, Moses 3: 5. (Remember, God created or organized or brought into existence the physical matter within our physical bodies and the genomes within our physical cells. God's signature or God's programming is within each of our physical cells. God is written all over us.)

**I made the world, and men before they were in the flesh**, Moses 6: 51. (When it comes to this physical universe, God created everything spiritually, BEFORE God brought into existence the first particle of physical matter. God made this world and all men, BEFORE they were placed into physical bodies.)

intelligences that were organized before the world, Abr. 3: 22. (Remember, Intelligences or Psyches are organized by placing them into spirit bodies.)

Notice how the Biblical God goes out of His way to make sure that we get the FACT that we are descended from the Gods, both physically and spiritually. He's consistent. You've got to say that for Him.

We ARE gods. Our surname is God. If I'm going to accept the Bible into evidence and use it as Observational Evidence or Scientific Evidence, then I'm going to accept ALL of God's Word into evidence.

Remember God reveals Himself to us through Scripture and through Science.

John 10: 34-36:

34 Jesus answered them, Is it not written in your law, I said, **Ye are gods**?

35 If he called them gods, unto whom the word of God came, and the scripture cannot be broken;

36 Say ye of him, whom the Father hath sanctified, and sent into the world, Thou blasphemest; because I said, I am the Son of God?

Do you believe it? Do you embrace your heritage? God isn't stupid. He KNOWS how these things really work!

1 Corinthians 8: 5-6:

5 For though there be that are called gods, whether in heaven or in earth, (as there be gods many, and lords many,)

6 But to us there is but one God, the Father, of whom are all things, and we in him; and one Lord Jesus Christ, by whom are all things, and we by him.

In the eternal scheme of things, there are Gods many, Lords many, Saviors many, and Worlds many; but, from our local perspective here on this particular earth, we are accountable to ONLY one God the Father and only ONE Savior and Lord Jesus Christ.

What is God's ultimate plan for us, His children? He reveals that to us as well.

Revelation 3: 20-22:

**20 Behold, I stand at the door, and knock: if any man hear my voice, and open the door, I will come in to him, and will sup with him, and he with me.**

**21 To him that overcometh will I grant to sit with me in my throne, even as I also overcame, and am set down with my Father in his throne.**

**22 He that hath an ear, let him hear what the Spirit saith unto the churches.**

If we prove worthy or "overcome" all things, someday we will sup with God the Father and Jesus Christ, and we will sit down with Jesus Christ on His Father's throne. We will become one of the Elohim, or a member of the Council of the Gods.

Even the apostate Christians mock and ridicule this to no end; but, the Most High God and Jesus Christ ARE in fact God-Makers. They mean what they say. Jesus Christ invites each one of us to come and sit down with Him on His Father's throne!

Christ's message to each one of us is to come and follow Him – do what He has done so that we can go where He has gone.

<u>Moses 6: 7-9, 22-23</u>:

**7 Now this same Priesthood, which was in the beginning, shall be in the end of the world also.**

**8 Now this prophecy Adam spake, as he was moved upon by the Holy Ghost, and a genealogy was kept of the children of God. And this was the book of the generations of Adam, saying: In the day that God created man, in the likeness of God made he him;**

**9 In the image of his own body, male and female, created he them, and blessed them, and called their name Adam, in the day when they were created and became living souls in the land upon the footstool of God.**

**22 And this is the genealogy of the sons of Adam, who was the son of God, with whom God, himself, conversed.**

**23 And they were preachers of righteousness, and spake and prophesied, and called upon all men, everywhere, to repent; and faith was taught unto the children of men.**

Adam was the son of God, both physically and spiritually. This means that we are descended from the Gods, not the apes. It's physically impossible for us to be descended from the apes anyway, so why not go with the revealed truth instead?

We were created in the image of the Gods, both physically and spiritually, which means that there are Female Goddesses in the heavens. In other words, we have a Heavenly Father and a Heavenly Mother, and we are descended from their physical bodies as well as their spirit bodies.

<u>Doctrine and Covenants 88: 6-21</u>:

6 He that ascended up on high, as also he descended below all things, in that he comprehended all things, that he might be in all and through all things, the light of truth;

7 Which truth shineth. **This is the light of Christ.** As also he is in the sun, and the light of the sun, and the power thereof by which it was made.

8 As also he is in the moon, and is the light of the moon, and the power thereof by which it was made;

9 As also the light of the stars, and the power thereof by which they were made;

10 And the earth also, and the power thereof, even the earth upon which you stand.

11 And the light which shineth, which giveth you light, is through him who enlighteneth your eyes, which is the same light that quickeneth your understandings;

12 Which light proceedeth forth from the presence of God to fill the immensity of space —

13 The light which is in all things, which giveth life to all things, which is the law by which all things are governed, even the power of God who sitteth upon his throne, who is in the bosom of eternity, who is in the midst of all things.

14 Now, verily I say unto you, that through the redemption which is made for you is brought to pass the resurrection from the dead.

15 And the spirit and the body are the soul of man.

16 And the resurrection from the dead is the redemption of the soul.

17 And the redemption of the soul is through him that quickeneth all things, in whose bosom it is decreed that the poor and the meek of the earth shall inherit it.

18 Therefore, it must needs be sanctified from all unrighteousness, that it may be prepared for the celestial glory;

19 For after it hath filled the measure of its creation, it shall be crowned with glory, even with the presence of God the Father;

20 That bodies who are of the celestial kingdom may possess it forever and ever; for, for this intent was it made and created, and for this intent are they sanctified.

21 And they who are not sanctified through the law which I have given unto you, even the law of Christ, must inherit another kingdom, even that of a terrestrial kingdom, or that of a telestial kingdom.

This Light of Christ sounds like Dark Energy, or the Zero-Point Field of Light, or the Quantum Sea of Light to me. It permeates everything and fills the immensity of space. The Light of Christ might be the thing holding our sun together, regulating the nuclear reaction in our sun, and preventing our sun from going off like a hydrogen bomb. God says that the Light of Christ is within the sun – the law by which it is governed or controlled.

Remember, Science 2.0 allows all of the evidence into evidence. The Scriptures of God are evidence of His existence, and also evidence of YOUR origins. I place all of this into evidence, and then I leave it up to you to decide for yourself what to make of it.

Most people never find any of this stuff because they don't know that it exists. The same thing happened to me when I was a Materialist, Nihilist, and Atheist. I never thought to look for evidence of Near-Death Experiences on YouTube because I didn't know that they existed. My best friend had to lead me to them, because I wasn't looking for them.

Likewise, I didn't think that there could ever be any convincing proof of God's Existence, because I didn't believe that it existed or would ever exist, so I simply wasn't looking for it. I had bought into the Materialist Credo which states that there is no proof of God's Existence and that there will never be any convincing proof of God's Existence. They were wrong, but I didn't know that at that point in time. I simply believed them and assumed that they knew what they were talking about. I was wrong.

I was quite naïve at the time. I didn't even know what a Materialist or Naturalist was; but, I could see that some of these people were getting some things wrong, because I had a pretty good handle on science and the world religions so I KNEW that they were getting some things wrong; and, they didn't like it whenever I spoke out and voiced my opinion.

It took concentrated flaming, mocking, ridicule, and banning from some of my atheistic friends online to spur me into action and make me start looking for scientific proof of God's existence. They sicked their Flying Spaghetti Monster on me. I had never heard of such a thing. I simply had to find out who this Flying Spaghetti Monster is; and, why he is so important to these people, and why they seem to worship him.  I have been going at it ever since.

## My Personal Interpretation of this Evidence

My interpretation of this Observational Evidence or Eye-Witness Evidence is not and cannot be the Official Interpretation of this evidence. Under Science 2.0, there is NO Official Interpretation of the evidence – there are no PhDs authorized to interpret the evidence for you. Instead, you are left on your own to decide for yourself what to make of it and how to use it. Furthermore, the ONLY interpretation of this Observational Evidence that really matters is God's interpretation of the evidence; and, I leave it up to you to discover for yourself what that interpretation might be.

Okay, I can hear a few people complaining, "Why didn't you bother to take the time and provide more of your own interpretation for these creation and origins accounts? I want to see what you got out of it all."

Science 2.0 permits me to present my own personal interpretation of the evidence for your consideration; but, under Science 2.0 there is NO official interpretation of the evidence, because it's highly likely that your interpretation of the evidence might be better and more believable to you than mine.

I do interpret some of these creation accounts in my other books and essays, and did some interpretation within these selections; but in anticipation of any possible complaint, I will mention a few of the other highlights here for the sake of completeness. You will have to decide for yourself if any of this is useful to you, or not. That's something I can't do for you, anyway.

1. Whenever the Biblical God Jesus Christ starts talking about Intelligences, He is in fact talking about individual Psyches, or Non-Local Consciousness. For all eternity, we have been Psyches or Intelligences. This reality applies to God as well. We all have a Psyche or Non-Local Consciousness, which means that we all have Intelligence. God is simply more intelligent than all the rest of us combined. Intelligence is defined as light and truth. The glory of God is intelligence, which means that God has more light, truth, and glory than all the rest of us combined. That's why He is God, and we are not. He got there first. Remember, Psyche and Spirit Matter don't have any physical limitations, because they are non-physical or transdimensional in nature and origin.

2. Psyche and Spirit Matter have always existed, and they will always exist. They are eternal. There is NO entropy in the Non-Local Realm. Our primal elements, Psyche and Spirit Matter, do not age. Entropy is introduced into physical matter during the creation and organization of physical matter. Entropy is a physical limitation and has nothing to do with the Spirit Realm or the Non-Local Realm. Entropy is a physical law, not a non-local spiritual law. No aging takes place in the Non-Local Realm.

3. We are the spirit children of the Gods. We each have a Heavenly Father and a Heavenly Mother. If you can read this, your surname is God. We are gods, the children of the Gods. This means that your Heavenly Father and Heavenly Mother are literally the parents of your spirit body. It also means that your physical body is descended from the Gods, because Adam was a son of God. Our physical bodies are literally in the image of their physical bodies, because our physical bodies are descended from the Gods. Our spirit body is also in the image of the Gods' physical bodies and spirit bodies.

Methodist Minister, Margaret Barker, teaches and believes that Mary the Mother of Jesus was literally the physical incarnation of Heavenly Mother, who was also the mother of Jesus's spirit body. Jesus or Jehovah was the birthright son of Heavenly Mother and Heavenly Father – both spiritually and physically. That would also mean that Jesus Christ was the direct son of God the Father's physical body. Jesus was literally the Son of God, the Son of the Gods. Margaret Barker realizes that these kinds of ideas make some people squirm; but, this is what the Bible is trying to teach us, nonetheless. Jesus Christ was literally the Son of the Gods – both spiritually and physically. We are all descended from the Gods; even our physical bodies are descended from the Gods. If you can read this, your surname is God.

The scriptures seem to indicate that all of us on this particular earth have the same Heavenly Father; but, I think that it's logical and possible that each nation or race of people on this earth has a different Heavenly Mother. In other words, there are reasons why God commanded Abraham, Jacob, Moses, David, and Solomon to engage in polygamy. There was precedent. The Gods see this whole thing differently than we do.

4. When God "organized" the Intelligences, He clothed them in spirit bodies or gave them an inheritance. The Gods aren't magicians. They don't do Creation Ex Nihilo. There is NO such thing as Creation from Nothing, or Creation by Nothing. The Gods always organize and uplift and edify what already exists. The Gods find Intelligences or Psyches, and then clothe them in spirit bodies made from already pre-existing spirit matter. In other words, the Gods organize these things.

Spirit matter and psyches have always existed and will always exist; but, their organization into Spirits has a beginning point in time. Anything that has a beginning has a Beginner. Spirits, Ghosts, and Souls are in fact a holistic combination of spirit matter and a psyche; and, the organization of a Spirit takes place when the Gods combine a psyche or intelligence with a spirit body of some kind, which the Gods have also organized. The Gods are scientists, not magicians. Technically, the Gods organize things – already pre-existing things – rather than creating them out of thin air like magic. The Gods don't do Creation Ex Nihilo or magic, because there is no such thing as Creation from Nothing. The Gods are Organizationists, and not Creationists.

5. It is clear from these Creation Accounts that the Gods first created this physical universe space-time, and then brought into existence the physical matter in this universe along with the galaxies and the stars and gas clouds some 380,000 years later, according to the Big Bang Theory.

Hydrogen makes up approximately 75% of our universe. Many of the Materialists and Naturalists claim that Hydrogen was produced in the Big Bang. They are wrong. Hydrogen was brought into being 378,000 years after the Big Bang, according to the current Big Bang Theories.

https://en.wikipedia.org/wiki/Recombination_(cosmology)

https://en.wikipedia.org/wiki/Chronology_of_the_universe

Of course, the scientists are constantly adjusting their guestimates; but for the sake of argument, I'm going to take them at their word where the beginning of physical matter is concerned, because I have no better evidence to draw upon.

The Gods organized galaxies and stars in place by converting the necessary spirit matter or dark matter into physical matter in situ. The Gods had to intervene directly and purposefully in order to create stars and galaxies. How do we know this? We know this because it is physically impossible for hydrogen atoms to coalesce into stars, because raw hydrogen atoms naturally repel each other.

If you start pumping raw hydrogen (gas or liquid) into a tank, the pressure quickly builds, and you don't get much in there because all of the hydrogen atoms are repelling each other. Consequently, in order to get more hydrogen into a fuel cell, the hydrogen atoms are incorporated into a solid lattice framework within the tank or the fuel cell; and thereby, you can get a lot more of those hydrogen atoms into a tank or fuel cell by having them bind to a lattice, than by pumping them into an empty tank. It seems counter-intuitive, but the scientists assure us that this is true, which is why they make hydrogen fuel cells out of some kind of lattice framework instead of just using an empty tank.

Hydrogen atoms repel each other. They fight against each other, which is one of the things that makes them so buoyant and makes hydrogen balloons float. There's NO way in this universe that hydrogen gas clouds could ever coalesce into a sun or a star, naturally, without some kind of external force making them do so. That reality applies here on earth, and it also applies out there in our universe as well.

The Gods would have had to have converted all of that spirit matter or dark matter into hydrogen matter – in place – in order to produce each and every sun, because it's impossible for hydrogen atoms to coalesce naturally into stars despite all the fictional accounts being produced by the Materialists and Naturalists and Atheists. Yes, hydrogen atoms can be forced together, but they would require some kind of external force (God's Psyche or God's Spirit or the Light of Christ) in order to get all of them to combine together into a sun.

Supposedly with enough of them combined together, gravity will hold them together and start fusing them together within a sun; but, even this aspect of stellar fusion is being questioned by some of the scientists because gravity is so weak. It's like there still has to be some kind of external force currently holding our sun together due to all of the nuclear fusion allegedly going on inside, which should be blowing our sun apart into space like a nova. Think of a hydrogen bomb. It requires a massive external compressive directed explosion to fuse the hydrogen together in the first place; and then, when that hydrogen fuses and explodes with energy there is nowhere near enough gravity on this earth to keep that explosion contained in the core. That explosion blows apart as far as it can go.

Logically, our sun, too, should blow apart; but, it doesn't. Something else is going on keeping it contained and regulated. Many scientists don't believe that it's just gravity that's holding our sun together right now, because nuclear explosions from nuclear fusion are trillions of times more powerful than gravity. Gravity is the weakest force at about $6 \times 10^{-29}$ times that of the strong nuclear force. Our sun should simply explode like a hydrogen bomb; but, it doesn't. Somehow God is holding it together – controlling and containing the nuclear reaction as well. There's more going on there than meets the eye. Someone is twiddling with the innards keeping the nuclear reaction contained and regulated so that it doesn't blow that star apart into space; or at least, that's the way I see it. Someone is also holding all of that hydrogen in our sun in place so that it doesn't repel itself out into space, and someone is regulating and controlling and pacing the nuclear reaction, because it would be bad if all the hydrogen in our sun simply fused all at once.

Of course, the Materialists, Naturalists, and Atheists are still teaching and preaching on blind faith that stars and planets can simply coalesce all by themselves with no external help; but then, everything that the Materialists and Naturalists say is nothing but science fiction and wishful thinking. These people don't think logically and rationally about what they are talking about and preaching.

Hydrogen atoms repel each other. They don't coalesce. If hydrogen coalesced and suns spontaneously generated as the Materialists and Naturalists claim, then we would have hydrogen bombs going off all the time all around us spontaneously, and that would be bad. Thank God hydrogen atoms naturally repel each other. You don't want those things ever getting together and fusing spontaneously and naturally, while you are around. It would simply get too hot to handle. It would be like living at the center of a sun.

The Materialists and Naturalists and Atheists are driven by wishful thinking and blind faith, so they never stop to think about these things or what they are really saying and claiming to be true. I KNOW, because I used to be a Materialist, Nihilist, and Atheist.

If hydrogen gas clouds could spontaneously coalesce into stars and suns, as the Materialists and Naturalists and Atheists claim, then we would have the same effect taking place here on our earth, and we would have hydrogen bombs spontaneously going off all the time all around us as a result. Think about it the next time some Materialist or Naturalist or Atheist on television tries to convince you that the stars and the galaxies simply put themselves together from already pre-existing gas clouds of hydrogen. It's an ad hoc just-so story that these people are presenting to you – not science. These people don't know what they are talking about, even though they pretend that they do. These people have PhDs in Materialism and Naturalism – the philosophies of men – which simply blinds them to the truth.

Suns and galaxies and functional genomes don't just put themselves together. They can't. It's physically impossible. Hydrogen gas clouds need some kind of external force from God in order to make them come together and coalesce into stars and galaxies in the first place. That's just the way it is, because that's the way it was designed to be. You don't want hydrogen gas clouds coalescing around you spontaneously and igniting naturally into little suns. That would be bad. It wouldn't take too much of that stuff to cause a big bang.

Thank God hydrogen atoms repel each other rather than fusing together naturally. Thank God hydrogen atoms are attracted to oxygen atoms and become harmless $H_2O$ (water) as a result, where God can keep all of that energy contained and regulated in a harmless format. Without all of that oxygen and water, the hydrogen gas would simply float out into space and be gone, because hydrogen atoms naturally repel each other. Thank God that He has designed ways to keep those hydrogen atoms and hydrogen gas clouds from fusing together naturally into hydrogen bombs and suns, because you don't want a fusion bomb going off spontaneously in your backyard. That would be bad, don't you think? Thank God that He knows what He is doing, because the Materialists, Naturalists, and Atheists certainly don't.

Eventually, I noticed that everything testifies that God must exist – even the hydrogen atoms. God knows what He is doing, even if we don't.

Since Science 2.0 allows all of the evidence into evidence and permits you to make your own interpretation of the evidence, Science 2.0 is going to be a lot more extensive and expansive and robust than Materialism and Naturalism can ever be. That's just the way it is. We are talking about things here that the Materialists and Naturalists refuse to think about and consider. It's their loss, not ours.

When it comes to these types of comments and ideas, the Materialists and Naturalists will ridicule and mock, ban and block – it's what they do. Atheism of any kind is based upon a refusal to look at evidence. All of this is foolishness to these people. They don't want to know anything about it, because it FALSIFIES Materialism and Naturalism. Remember, FALSIFY means to prove something false. We can actually use scientific evidence, scientific methods, and observational evidence to prove things false, because *negating the consequent* is philosophically and logically sound. Materialism, Naturalism, Darwinism, Nihilism, Behaviorism, Scientism, and Atheism have been FALSIFIED trillions of times in thousands of different ways. If you successfully eliminate everything that is FALSE, you will end up with the TRUTH. And nowadays, the Truth is the only thing that is of interest to me, because I used to be a Materialist, Nihilist, and Atheist – been there, done that, and I don't want to go back. I know better now.

6. The Gods created, populated, and saved many worlds BEFORE they got around to organizing and populating our earth. This means that the putative Gap Theory is correct. There were billions of years between Genesis 1: 1 when the Gods created the physical matter for the heavens and the earth, and Genesis 1: 2 when the Gods finally got around to organizing our particular earth. Many earths had "passed away" – died and been resurrected – before the Gods got around to organizing our earth. Things were dying or passing away in this universe and on this earth long before our Adam and Eve were placed into the Garden of Eden. There were galaxies, stars, planets, and inhabited worlds in existence long before our earth was organized; and, some of those worlds had passed away or been saved and resurrected before our earth was brought into existence. The Gods eventually organized our earth.

There were waters above, which I take to mean that our earth's atmosphere was opaque; and, on the third day the God's made our earth's atmosphere transparent so that the pre-existing sun, moon, and stars appeared in the heavens of our earth during that transformation on the third day. No magic. The Gods simply state that the sun, moon, and stars **appeared** on the Third Day, which means that they existed but weren't apparent on the days before that. From these observational accounts, this tells me that there must be some kind of veracity to the Canopy Model and an Opaque Atmosphere, for explaining the events of the Third Day. The lights were already in the firmament or the universe BEFORE the Gods caused them to appear or be visible upon the face of our earth. The Gods are scientists, not magicians. Good enough.

7. Adam was a son of God. This means that both his spirit body and his physical body were descended from the Gods. We each have a Heavenly Father and a Heavenly Mother. These people are the parents of our spirit bodies; and, we are also descended from some of their physical bodies. The Gods have physical bodies, which they can allow to die or set aside, just like Jesus allowed His physical body to die while on the cross. Then the Gods can be born again into mortality like Jehovah (Jesus) and Michael (Adam) were born again. The Gods can condescend – come down to our level in an attempt to help raise us to their level. The Gods put their own hide and their own salvation on the line in order to save us and lift us. The Gods can also be resurrected from the dead, or take up their immortal physical body once again, if they prove faithful during their mission.

Some of the Gods have lived many different lives on many different physical worlds. This same reality applies to some of us. Some of us are old spirits, and some of us are not. For some of us, this might be our first go at mortality, and for others of us this might be our thousandth attempt to get it right. Most of us who are here have some issues that we need to work through and overcome; but, a few of us were already Gods, and we condescended to come here to help others to rise to a higher level, possibly in the hope of receiving a better resurrection for ourselves when we are done as a reward.

A mortal life was meant to be educational with lots of chances for growth and advancement along the way.

8. I encountered a woman online who told me that she is proud to teach her children that they are descended from apes and monkeys. That's physically impossible! NO pair of apes have ever copulated and produced a human as a result – that's physically impossible. It can't be done – let alone having two apes copulate and produce a matching genetically compatible male and female pair of humans. It can't be done. It's physically impossible. Apes never give birth to humans; so, there's no way on earth that any of us

could have descended from the apes. This woman had denied her inheritance. The truth is that we humans have descended from the Gods. Both our spirit body and our physical body are descended from the Gods. Our surname is God. We are Gods. Therefore, it doesn't matter how proud she is to teach her children that they descended from apes and moneys. She is wrong and doesn't know it. She didn't know what she was talking about. She didn't want to know it. She had denied her inheritance.

9. Michael was a God on the Council of the Gods. He condescended to become Adam, the father of our human race. Jehovah was a God on the Council of the Gods. He condescended to become Jesus Christ and atone for our sins. The Gods come down to our level, condescend, are born again in the flesh, and risk it all in an attempt to help lift some of us up to their former level. Some of us were Gods in the pre-mortal life; and, we condescended to be born again here in mortality in the hope of a better resurrection (a better station) in the next life and the hope of helping others to rise to our former level of Godhood and glory.

10. The Gods seeded our earth with plants and trees. The Gods used seeds, not adult trees. The Gods reveal themselves to us through Scripture and Science. The plants and trees were commanded to multiply after their kind. There are adult trees in the fossil record. That means that the Gods actually let those trees grow naturally from seed into adulthood and even let some of them reproduce and die, BEFORE the Gods called their labors of the fourth day done and completed. This tells me that the Day-Age Theory is correct. Each day in the Genesis Account is at least 100 years long, and most likely 1,000 years at the very least. Each day could have been longer.

I have noticed that ALL of the observational evidence provided to us by the Gods FALSIFY the claims of the Young-Universe Creationists (YUCs), the Young-Earth Creationists (YECs), the Materialists, the Naturalists, the Darwinists, and the Atheists. The Gods were scientists teaching us their children to be scientists – not magicians. There's no such thing as Creation Ex Nihilo. Creation Ex Nihilo is magic; and, Creation Ex Nihilo is Atheism. The YUCs, YECs, Materialists, Naturalists, and Atheists preach Creation Ex Nihilo and Magic to a great extent; and, they are wrong to do so. The Gods are scientists, not magicians.

11. The Gods placed animals on this earth on the Fifth Day. The Gods teleported these animals to this earth in a systematic fashion, when they were needed and when the earth was ready for them. The Gods reveal themselves to us through Scripture and Science. There are at least 25 mass extinctions, with subsequent reseeding and replacement of life forms, in the fossil record. The Bible contains the record of only one mass extinction event, not 25, which I take to mean that the other 24 mass extinctions took place BEFORE our Adam and Eve were placed into the Garden of Eden.

Again, this tells us that the Day-Age Theory is correct. As I see it, with at least a year between each mass extinction so that it can get into the fossil record, the fifth day has to be at least 50 of our years long; but more likely, the fifth day was as least a thousand of our years long.

The fossil record and all of these recorded mass extinctions tell us that reproduction and death were taking place on this planet BEFORE the Garden of Eden was planted by the Gods and BEFORE our Adam and Eve were placed into the Garden of Eden.

The observational evidence from the Gods make these realities and truths obvious and clear, for anyone who is willing to look, see, and understand.

Ice cores tell us that our earth is at least 800,000 years old. That means that each day of creation in Genesis could have been at least 100,000 years and more likely was at least 200,000 years. Adam and Eve could have been in the Garden of Eden for a 100,000 years for all we know, or a million years, while plants and animals were living and dying outside of the Garden of Eden. We don't know, because the Gods don't give us any details about what was happening on this earth on the Seventh Day while Adam and Eve were in the Garden of Eden and the Gods were resting from their labors.

12. When Adam and Eve fell, they were cast out of Eden into a lone and dreary world that already existed outside of Eden. They simply walked out of Eden into this lone and dreary world that already existed. It was Adam and Eve who fell – not the Tree of Life and not the plants and animals. Adam and Eve were cast out of Eden into something that was already fallen and was already subject to death – that's why it was called the fall of Adam and Eve – the fall of man. The plants and animals are not subject to sin, and they have no need for the atonement of Christ. They were already fallen, mortal, and subject to death to begin with. Only Adam and Eve and the Tree of Life were immortal, which is why Adam and Eve had to be blocked from partaking of the Tree of Life and becoming immortal again – a real possibility that the Gods took seriously at the time. Everything else on this planet was subject to death before Adam and Eve fell. They ate the fruit and it technically died. They tended the Garden keeping it going. If the Tree of Life were still around, you and I could partake of the fruit from that tree and become immortal. Again, the Gods are scientists, not magicians.

13. It's also interesting to note that the Gods created everything spiritually or non-locally BEFORE they created it physically.

**In a word, absolutely everything in nature, from the smallest to the greatest, is a correspondence. The reason correspondences occur is that the natural world, including everything in it, arises and is sustained from the spiritual world.** — Emanuel Swedenborg, Christian mystic

**Everything was created of spirit matter before it was created physically — solar systems, suns, moons, stars, planets, life upon the planets, mountains, rivers, seas, etc. This Earth is only a shadow of the beauty and glory of its spirit creation.** — Betty J. Eadie, author of Embraced by the Light

**The blueprints of everything in the physical universe have been astrally conceived — all the forms and forces in nature, including the complex human body, have been first produced in that realm where God's causal ideations are made visible in forms of heavenly light and vibratory energy.** — Paramhansa Yogananda, yoga master

This reality and truth also explains where the blueprint for the construction of our physical body is stored – non-locally.

Some of the scientists have observed that there is nowhere near enough information in our genome to construct a physical body from scratch. ONLY the information for manufacturing proteins is there, along with information for turning genes on and off.

According to google, "The 2.9 billion base pairs of the haploid human genome correspond to a maximum of about 725 megabytes of data, since every base pair can be coded by 2 bits."

It has been estimated that there is anywhere from 37.2 trillion to 100 trillion cells in the human body.

The information telling a single bone cell to divide into a bone cell and an adjacent muscle cell, along with the associated locational addressing, is not and cannot be stored in the DNA – it has to be stored in our non-local blueprint instead.

There's NO WAY that the 300 trillion 3D location addresses for each cell in the human body, along with the cell type or differentiations for each 100 trillion cells, along with the sequential ordering of their assembly, along with the addressing of genes on the chromosomes, along with repair instructions, along with assignments and tasks for each molecule in a cell, along with instructions as to where each amino acid and protein needs to go, along with how all the different neurotransmitters are supposed to function, can possibly be stored in our 725 megabyte genome. That information has to be stored someplace else, because our miniscule genomes couldn't even begin to contain it all. Physical matter is a highly inefficient way to store data and information. There has to be a better way.

Due to the fact that there are NO physical limitations in the non-local realm or spirit realm, it's theoretically possible to store an infinite amount of information in something that takes up NO space whatsoever, if that information is stored in the non-local realm or psyche. Those who have experienced their psyche or non-local consciousness tend to describe it as an infinite singularity or a spark with no size, no shape, and no form taking up no space whatsoever – just a viewpoint in space, yet containing all of their memories and personality and information and life force.

If you study carefully the different 3D models of the human cell, you quickly observe that each amino acid, mRNA molecule, gene transcriber, molecule, and protein within the cell knows where it is going and proceeds to go there and do its job and take its place in the cell. All this information telling all these critters what to do and where to go cannot by definition in principle be stored within our DNA or genome. Yet, the amino acids and molecules KNOW they have an appointment someplace, they know where they are going, and they know how to get there, they know what to do when they get there, and they seem to line up all by themselves in the right order when they get there BEFORE they are assembled by the ribosomes into proteins. All of this knowledge and information cannot be stored in our genomes or DNA. There's something spiritual or non-local going on here, because it is physically impossible for all of this to be controlled by DNA alone.

ONLY an infinite singularity would have the informational storage capacity necessary to store the trillions of trillions of bits necessary to contain something like the cell differentiations, maintenance plan, gene addressing, assignment codes and addressing for amino acids and proteins, neurotransmitter functioning, 3D blueprint, and all the locational addressing for something like our 100 trillion cell physical body. Yet, the Materialists, Naturalists, Darwinists, and Atheists want you to believe that Random Mutations and Natural Selection did all of this all by itself and continues to do so flawlessly every time. These people have way too much blind faith in the Theory of Evolution, because they never stop and think about these things.

14. The Biblical God Jesus Christ foresaw the upcoming war between the Young-Earth Creationists (YECs) and the Old-Earth Creationists; and, He took steps to prevent it by giving us at least three new Observational Accounts of the Origin of our Earth.

It's possible to get the correct interpretation of the Bible and Genesis as Hugh Ross tends to do; but, the Young-Earth Creationists (YECs), the Young-Universe Creationists (YUCs), as well as the Materialists, Naturalists, and Darwinists consistently and reliably choose, create, and promote the wrong interpretation of Genesis every time.

If one reads these newly revealed or restored Observational Accounts carefully, he or she quickly realizes that the last thing the Gods did was to plant the Garden of Eden onto our fallen mortal world in our fallen mortal universe, because the Garden of Eden didn't exist yet on this planet. The Garden of Eden was the exception, not the rule!

It also becomes clear that Adam, Eve, the Tree of Life, and possibly the Tree of Knowledge were the only things immortal in Eden.

When Adam and Eve fell, they were the only things that fell; and, then they simply walked out of the Garden of Eden into the "lone and dreary world" – the fallen and mortal world – which already existed and was there waiting for them.

When our Adam and Eve fell, our physical universe didn't fall with them, because our physical universe was already physical, fallen, and mortal with stars and worlds passing away all the time even while our particular Adam and Eve were in the Garden of Eden.

Our physical universe was created fallen and mortal from the very beginning; and, it remains so today. Adam, Eve, and the Tree of Life were the exception, NOT the rule. Someday in the future, during the Millennium after the Second Coming of Christ, our earth will finally receive its paradisiacal glory and will be transmuted, transfigured, translated, or made immortal at that point in time and not before. Our earth – NOT the whole universe. Our earth won't die, get resurrected from the dead, and then get transferred to the New Heavens where the Gods reside, until AFTER the Millennium is over.

A Millennium and a Gospel Dispensation is typically a Thousand Years. The Biblical God repeatedly tells us that ONE of His Days is at least a thousand of our years, which means that one of their weeks is 7,000 of our years. As promised, Adam really did die on the very same day that he partook of the Tree of Knowledge. However, there are many indications in the fossil record and the ice cores that the Gods took their time and took many of their weeks to organize, terraform, fine-tune, seed, populate, and produce our earth. The Gods took their time organizing this earth, because WE participated in the process, and they were teaching us how to do it.

There are at least 25 mass extinctions in the fossil record, followed by new environments and the reseeding of our planet. Since only one mass extinction is mentioned in the Bible, I assume that the other 24 mass extinctions and the subsequent repopulating of our earth took place BEFORE the Garden of Eden was planted and BEFORE our Adam and Eve were placed into that garden as immortal beings. The ice cores tell us that our earth is at least 800,000 years old, and more likely at least 1,400,000 years old. Many 7,000-year God weeks or dispensations went by BEFORE our particular Adam and Eve were placed into the Garden of Eden. The fossil record and sedimentary record tell us that there was death and mass extinctions taking place on this planet BEFORE the Garden of Eden was planted and Adam and Eve were placed into it.

There are Earths many, Gods many, Lords many, Saviors many, Adams many, Eves many, and Lucifers many.

The Gods, and Lucifer, simply did on our earth what they had already done on many other worlds before our earth was organized.

Satan and his followers – the demons or evil spirits – will never receive a physical body, because they formally rejected God's plan of salvation and the Atonement of Christ. They were damned – meaning that they had stopped their own progression – because they didn't want to have anything to do with God's plan nor the Atonement of Christ. They chose to forego the experience of mortality and chose to remain as spirits instead. God simply gave them what they wanted.

Our physical universe had a beginning, which means that it had a Beginner. The Big Bang Singularity was a supernatural object; and, the Big Bang was a supernatural event, because physical matter didn't exist yet.

Our physical universe really is at least 13.8 billion years old, as the Standard Light-Year seems to suggest. Worlds were organized and passed away, long before the Gods got around to organizing and populating our particular earth. The YUCs and the YECs get all of this wrong. The Gap Theory and the Day-Age Theory really are true.

Science 2.0 allows ALL of the evidence into evidence; and while I'm interpreting it, I tend to take it all at face-value. I really do believe that the Speed of Light is telling us that our universe is at least 13.8 billion years old; and, I really do believe that the ice cores are telling us that our earth is at least 800,000 years old. I try to follow the observational evidence and adjust my personal interpretations accordingly.

The Observational Evidence from those on the scene also tells us clearly and conclusively that the Gods really did create our heavens, our earth, and all of the life forms on our earth, even though it is clear that the seeds and the animals existed on other worlds before being teleported here. Our particular earth was a new planting.

15. This is some of what I got out of these Observational Accounts from the Gods who were there and did the work. Chances are good that you got something different out of the same information. That's the way it was meant to be. We each go only as far as we are prepared to go, and no more. Some of you are more advanced than I am; and consequently, you will see and understand a lot more than I did.

16. Again, under Science 2.0 it is permissible for me to present my own personal interpretation of the Observational Evidence; but, there is NO official interpretation of the evidence under Science 2.0. You are encouraged to make your own interpretation of the evidence and decide for yourself what to make of it and what to do with it. All I'm going to do is to present the Observational Evidence and Experiential Evidence, and then sometimes tell you what it means to me personally. I'm going to leave it up to you to decide for yourself what you want to believe and what you want to do with your beliefs. You are going to have to decide for yourself whether I know what I'm talking about, or not. That's something I can't do for you. You are going to have to do it for yourself.

The purpose of this exercise was to demonstrate how I use Science 2.0 or Observational Evidence to resolve the major controversies of our day. Rather than denying the evidence and hiding from the evidence – like I used to do when I was a Materialist, Nihilist, and Atheist – I now USE Observational Evidence, Experiential Evidence, Eye-Witness Evidence, Scientific Evidence, Experimental Evidence, and Empirical Evidence to debunk and falsify Materialism and Naturalism and their derivatives.

# Part VII — Conscious While Outside the Physical Body

The following are observational accounts and lived experiences from William Buhlman and others supporting Science 2.0. Although I might comment and tell you what some of this means to me, ultimately I will leave it up to you, my reader, to interpret it and decide for yourself what to make of it.

## The Astral Traveler

It can be fascinating to see how the Materialists and Naturalists try to explain this one away.

Twenty years ago, I firmly believed that the physical world we see and experience was the only reality. I believed what my eyes told me — life possessed no hidden mysteries, only countless forms of matter living and dying. The facts were clear; there was no evidence or proof of nonphysical worlds or our continued existence after death. I questioned the intelligence of anyone softheaded enough to accept the illogical concepts of heaven, God, and immortality. In my mind these were fairy tales created to comfort the weak and manipulate the masses. For me, life was simple to understand: the world consisted of solid matter and form, and the concepts of life after death and heaven were feeble human attempts to create hope where none existed.

I possessed the arrogant knowledge of a man who judges the world with his physical senses alone. I supported my conclusions with the overwhelming observations provided by science and technology. After all, if something mysterious was there, science would certainly be aware of it.

My firm convictions of reality and life continued until June of 1972. During a conversation with a neighbor, our discussion turned to the possibilities of life after death and the existence of heaven. I proceeded to present my agnostic viewpoints with vigor. To my surprise my neighbor didn't contest my conclusions; instead, he related an experience that he had had several weeks before. One evening just after drifting to sleep, he was shocked to discover himself floating above his body. Completely awake and aware, he became frightened and instantly fell back into his physical body. Excited, he told me it wasn't a dream or his imagination, but a fully conscious experience.

Intrigued by his experience, I decided to investigate this strange phenomenon for myself. After several days of research, I discovered numerous references to out-of-body experiences throughout history. With some searching I found a book on the topic that actually described how out-of-body experiences are induced. The entire subject seemed extremely weird, and I considered the book the result of an overly active imagination.

Out of curiosity, I decided to try one of the out-of-body techniques before sleep. After repeated daily attempts, I began to feel a little ridiculous. In three weeks, the only thing I experienced out of the norm was an increase in my dream recall. I became more and more convinced that this entire subject was nothing more than an intense or vivid dream stimulated by the so-called out-of-body techniques.

Then, one night about eleven o'clock I drifted to sleep during my out-of-body technique and began dreaming that I was sitting at a round table with several people. They all seemed to be asking me questions related to my self-development and state of consciousness. At that moment in the dream I began to feel extremely dizzy, and a strange numbness, like from Novocain, began to spread throughout my body. Unable to keep my head up, I passed out, hitting my head on the table. Instantly I was awake, fully conscious, lying in bed facing the wall. I could hear an unusual buzzing sound and felt somehow different. Extending my arm, I reached for the wall in front of me. I stared in amazement as my hand actually entered the wall; I could feel the vibrational energy of it as if I was touching its very molecular structure. Only then did the overwhelming reality hit me, *My God, I'm not in my body.*

Excited, my only thought was, *It's real. My God, it's real!* Lying in bed, I stared at my hand in disbelief. When I tried clenching my fist, I could feel the pressure of my grip; my hand felt completely solid, but the physical wall in front of me looked and felt like a dense, vaporous material with form.

Determined to stand, I began to move effortlessly to the foot of my bed, my mind racing with the reality of it all. Standing, I quickly touched my arms and legs, checking to see if I was solid, and to my surprise I was completely solid, completely real. But around me, the familiar physical objects in my room no longer appeared completely real or solid; instead, they now looked like three-dimensional mirages. Glancing down, I noticed a large lump in my bed. Amazed, I could see that it was the sleeping form of my physical body silently facing the wall.

As I focused my vision on the opposite side of the room, the wall seemed to fade slowly from view. In front of me I could see a wide, green field extending far beyond my room. Looking around, I noticed a figure silently watching me from about ten yards away. It was a tall man with dark hair, a beard, and a purple robe. Startled by his presence, I became frightened and instantly "snapped back" into my physical body. With a jolt I was in my body, and a strange feeling of numbness and tingling faded as I opened my eyes. Excited, I sat up, my mind exploding with the realization of what had just occurred. I knew it was absolutely real, not a dream or my imagination. My entire ego awareness had been present.

Suddenly, everything I had ever learned about my existence and the world around me had to be reappraised. I had always seriously doubted that anything beyond the physical world existed. Now my entire viewpoint changed. Now I absolutely knew that other worlds do exist and that people like myself must live there. Most important, I now knew that my physical body was just a temporary vehicle for the real me inside, and that with practice I could separate from it at will.

Excited about my discovery, I grabbed a pen and paper and wrote down exactly what had occurred. A flood of questions filled my mind. Why is the vast majority of the human race unaware of this phenomenon? Why aren't the various sciences and religions investigating it? Is it possible that this unseen world is the "heaven" referred to in religious texts? Why isn't our government exploring this apparent parallel energy world? Is it possible that our overwhelming dependence on physical perceptions has led us to overlook an incredible avenue of exploration and discovery?

As the initial shock of my first experience sank in, I realized that my life could never be the same again. The more I pondered the significance of my experience, the more profound I realized it to be. All my agnostic beliefs had been swept away in a single night. I knew that I had to reappraise everything that I had learned since childhood, everything that I had assumed to be true. My comfortable conclusions about science, psychology, religion, and my existence had obviously been based on incomplete information. I felt excited, but also uneasy — my familiar concepts of reality no longer seemed relevant. Increasingly, I felt in a void. On several occasions when I talked to friends about my experience, they found it too bizarre to take seriously. In 1972 the term *out-of-body experience* had not even been coined; back then, the most common description was astral projection. No one that I knew at the time had even heard of astral projection, and if you told people you had left your body, they immediately thought that you were on drugs or losing your mind. I quickly discovered that I had to keep my experiences to myself or face some degree of disbelief and even ridicule.

After my first out-of-body experience, my mind was overflowing with endless possibilities and questions. Desperate for information and guidance, I spent several weeks in libraries and bookstores searching for whatever knowledge was available on the topic. I quickly found that little was available; only a handful of books had been written on the subject, and some of these were decades old and out of print. By the end of July 1972, I realized that I was on my own.

I decided to focus on the one technique that had worked for me before. This technique involved visualizing a physical location that I knew well as I drifted off to sleep. As before, I pictured my mother's living room with as much detail as possible. At first it seemed difficult, but after a few weeks I could picture the room's details with increasing clarity; the furniture, patterns in fabrics, textures, even small imperfections in wood and paint began to be clear in my mind. I realized that the more I pictured myself within the room interacting with the physical objects, the more detailed my visualizations would become. With practice I learned to physically walk around the room and memorize specific items that it contained. I also learned the importance of "feeling" the environment with my mind: the feel of carpet on my feet; the sensation of sitting in a chair, walking, turning on a lamp, or even opening the door. The more detailed and involved I was within my visualization, the more effective were my results. Although it was challenging at first, after a while it became fun to make my visualizations come alive in my mind. At this point I decided to keep a journal to record my out-of-body experiences.

Buhlman, William L., "Adventures Beyond the Body: How to Experience Out-of-Body Travel" (pp. 3-8). HarperCollins. Kindle Edition.

## Experiencing the Light

After a couple of years of research, I finally realized that Lived Experience is a much better way of finding and KNOWING the TRUTH than the Scientific Method.

The following Lived Experience changed the way that I look at Reality. I'm no longer a Materialist, Nihilist, and Atheist. In the following Out-of-Body Experience, William Buhlman exited his physical body and went to the non-local realm or spirit realm. There in the spirit realm or transdimensional realm, he exited his spirit body, looked at his spirit body, and described his spirit body.

This Lived Experience is Scientific Evidence or Observational Evidence that our physical body, our spirit body, and our Psyche or Intelligence or Non-Local Consciousness ARE three completely different entities; and, this Lived Experience provides Scientific Evidence and Observational Proof that it's the immaterial Psyche who does all of the observing, thinking, and remembering. Through Lived Experiences, we human beings or human psyches can go directly to KNOWING the TRUTH and directly to PROOF of the TRUTH. Our Lived Experiences ARE the best and most efficient type of Scientific Evidence! Just because you and I haven't had some of these types of experiences doesn't mean that nobody else has.

This IS the Out-of-Body account that first made me realize that Psyche or Non-Local Consciousness is something completely different than spirit matter and physical matter – Substance Dualism. Spirit matter and physical matter are Quantum Objects or Particles of Matter. Matter of any kind reacts. Psyche or Non-Local Consciousness chooses and acts. Psyche appears to be an infinite singularity – a spark – an immaterial viewpoint in space having no size and taking up no space whatsoever. This Out-of-Body Experience changed the way that I look at Reality.

Journal Entry, October 2, 1982

I hear the buzzing, engine-like sounds and will myself out-of-body. I step to the bedroom door and automatically request "Clarity now!" My vision improves and I step through the door, into the living room. Still feeling a little out of sync, I verbally repeat my request with more emphasis, "Clarity now!" I feel my awareness and vision snap into place.

My thoughts are clear, and I make a verbal demand, "I need to see the form I'm in now!" Instantly I feel an intense sensation of being drawn within myself. I'm suddenly different, weightless as though I'm floating in space. As I look forward I see a sparkling, bluish white form. For some reason, I seem to know that I'm looking at my nonphysical body from a different perspective. I stare in amazement at this form before me that shines and flows with energy and light. It looks like an energy mold created from a million tiny points of light; it radiates a bluish glow but appears to have a defined outer structure. The body of light before me is naked and is identical to my physical form. Even though my body looks firm, there is a noticeable energy motion and radiation present. I can see what appears to be an ocean of blue stars throughout my body. It's difficult to describe because the stars are stable, yet moving at the same time; the light and energy of my body appear to change and flow almost like the waves of an ocean.

As I stare at the body of light, it hits me that I must be in another body. Yet I can't perceive any form or substance; I'm like a viewpoint in space without shape or form of any kind. As I reflect upon my new state of being, I feel a sensation of rapid motion and I'm instantly back within my physical body.

Lying still and reviewing my experience, I'm struck by an inescapable conclusion: I must possess multiple energy-bodies. The form I just experienced was noticeably lighter (less dense) than even my second nonphysical body. I realize that the traditional view of our possessing two bodies — a physical body and a spiritual body — is far too simplistic; we are much more complex than this. Just as there are multiple nonphysical energy dimensions within the universe, each of us must consist of multiple energy-bodies or vehicles of expression.

Now I seriously wonder just how many nonphysical bodies or forms this involves. I suspect that there must be one within each dimension of the universe and that all of these are interrelated and connected, just as the physical body is connected to its first nonphysical (spiritual) body.

Buhlman, William L., "Adventures Beyond the Body: How to Experience Out-of-Body Travel" (pp. 34-35). HarperCollins. Kindle Edition.

You won't get anything like this from the Materialists, Atheists, Naturalists, and Darwinists because they don't have anything like this to give you.  These people have chosen to believe that the Scientific Method is the only way to find and know the truth; but, they are wrong.  Materialism and Atheism of any kind are based upon a refusal to look at evidence.  Ironically Lived Experience, including our spiritual experiences, is an infinitely better source of Scientific Evidence than the Scientific Methods will ever be.  The Scientific Methods are based upon logic fallacies and are fundamentally flawed.  Not so with Lived Experience!  Through Lived Experiences, we human beings or human psyches can go straight to KNOWING the TRUTH.  Powerful, huh?

The real William Buhlman, his immaterial Psyche, is "like a viewpoint in space without shape or form of any kind".  His Psyche doesn't have any "form or substance".  It's completely immaterial.  His Psyche is a force or a field, and NOT MATTER.  His Psyche is PURE LIGHT or PURE CONSCIOUSNESS.  His Psyche is the Ultimate Force.  His Psyche is the Ultimate Causal Agent.

His spirit body is "energy and light", "identical to his physical form", and "like an ocean of stars".  His physical body is back home sleeping in his bed.  For me, this is Scientific Evidence and Scientific Proof that Psyche is something completely different than our spirit body, and that our spirit body is something completely different than our physical body.  They are ALL made up of light, but different types of Light with different assignments or tasks or aspects!

After reading this Lived Experience, I got the feeling that I had finally found the TRUTH about our existence and our reality as human beings and human psyches.  Psyche or Intelligence or Non-Local Consciousness is something completely different than spirit matter and spirit bodies.  Furthermore, our spirit body is something completely different than our physical body.  They are ALL made of light, but light that is in different states of being or different states and phases of existence.

Nowadays, I treat Lived Experience as Scientific Evidence.  Lived Experiences do have some of the same weaknesses and limitations that the Scientific Methods have, though.  Lived Experiences, whether we are talking about physical experiences or spiritual experiences, aren't always replicable on demand.  Furthermore, Lived Experiences or Direct Scientific Observations are subject to biases and faulty interpretations just as much as the Scientific Methods are – some people come up with some very strange interpretations and explanations for their spiritual experiences.  But, in every other respect, Lived Experiences or Psyche Experiences are vastly superior to the Scientific Methods.

I really liked William Buhlman's book, because he turned the whole Out-of-Body or Astral Projection thing into a Science and experimented with it while he was out of body.

I got really excited about this information and this KNOWLEDGE, and I started looking for independent confirmation of this Ultimate Reality or Psyche Ontology.  Seek, and ye shall find.  Knock, and it shall be opened unto you.

## Evidence of Psyche and Signs of Life

While researching the Lived Experiences of the human race, I now keep my eyes open for Evidence of Psyche. I now treat Lived Experience as Scientific Evidence, as any Real Scientist should. Lived Experience is the BEST type of Scientific Evidence, because with Lived Experience or Direct Observation we can go directly to KNOWING the TRUTH by experiencing and living the truth for ourselves. The TRUTH is KNOWN by living it and experiencing it for yourself, or by choosing to trust someone who has.

Lived Experience or Direct Observation IS the BEST and most efficient way for doing Science! I'm talking about a Paradigm Shift here, a Psyche Ontology and Psyche Epistemology, wherein Lived Experience or Psyche Experience becomes our primary source of Scientific Evidence and our primary way of doing Science. Lived Experience subsumes or includes Science Experiments and Scientific Observations of Science Experiments, because these are Lived Experiences as well. The major advantage is that Lived Experience also includes our non-local experiences or our spiritual experiences as Scientific Evidence. This all-inclusive Psyche Ontology, based upon Lived Experience, is the Ultimate Paradigm. This Ultimate Paradigm becomes the Ultimate Model of Reality.

Close your eyes. Where are you? The REAL YOU. You can sense where you are, where your living spark or your Psyche resides. It's not in the back of your head. It's not in your ears. It's not in your nose. It's not in your hands or your feet. It's not in your stomach. Your Psyche or Consciousness resides behind the bone in your forehead, in the Third Eye position. Yes, you have a spirit body. Yes, you have a physical body. But the REAL YOU, your personality, consciousness, intelligence, awareness, and life is a tiny immaterial Spark residing and emanating from the Third Eye Position in your forehead.

Every now and then, Out-of-Body Travelers will exit their spirit body and will look at their spirit body (or look at their physical body) from an external third-person perspective or vantage-point. This immaterial viewpoint in space, or spark on the ceiling, or pinpoint of light is Psyche or Non-Local Consciousness. It's your person and your personality, complete with all of your intelligence and memories. It's an infinite singularity. Psyche is infinite memory storage capacity taking up no space whatsoever. Whenever Psyche goes looking for itself, there's nothing there to be seen. Psyche is completely immaterial. Psyche IS the Ultimate Force. Psyche is consciousness and life. Psyche is Intelligence. Psyche IS Light.

Astral Travelers have observed that their spirit body and their physical body don't have experiences and don't remember anything while their Psyche is away from their spirit body and physical body. It's the Psyche who does the experiencing and the remembering, not the spirit body and certainly not the physical body. Psyche is the light of life or the spark of life. Your Psyche is your Ultimate Reality. This is the Ultimate Model of Reality or a Psyche Ontology, wherein Psyche is the fundamental unit of reality. As far as I know, I'm the only person on the planet to have developed and presented a Psyche Ontology. I sometimes wonder how everyone else could have missed it. But, they miss it because they don't want to see it, don't want to understand it, and don't want to accept it as being true. I KNOW, because I used to be one of them. I used to be a Materialist, Nihilist, and Atheist until I learned better.

Obviously, I use William Buhlman's out-of-body experiences as Scientific Evidence or Observational Evidence that Psyche is something completely different than one's spirit body.

On YouTube, you can find thousands of Near-Death Experiences (NDEs) and Out-of-Body Experiences (OBEs). Every now and then, you will find someone who has left their spirit body and has experienced their independent immaterial Psyche directly. We KNOW from the Lived Experiences or Scientific Observations of the human race that Psyche exists, and that Psyche is something completely different than our spirit body.

One woman described leaving her body and floating up near the ceiling. She said that she went to look for herself and couldn't see anything. She said that she was just a spark on the ceiling. Most people who go looking for themselves while out-of-body explain seeing parts of their spirit body. A few, however, see nothing at all. They are pure psyche, or pure light, or pure consciousness. They have no material form. They are just a spark on the ceiling, or a viewpoint in space, or a pinpoint of light. The accounts of those who experience their Psyche or Non-Local Consciousness separate from their spirit body are rare, but they do exist.

During my research, I have discovered that Lived Experience is the BEST type of Scientific Evidence, because when it comes to Lived Experience or Direct Observation, we KNOW the TRUTH by living the truth and experiencing the truth directly for ourselves. In fact, in most instances, Lived Experiences or Out-of-Body Experiences are the only way to gather Scientific Evidence from the Non-Local Realm or the Spirit Realm.

When Dr. Anthony Cicoria was struck by lightning, he died. He, his spirit body, was outside his physical body. As he walked up the stairs, his spirit body dissolved beneath him and transformed into an orb of light. He, his Psyche, was looking at his spirit body and later that orb of light from an immaterial third-person perspective or that Third Eye Perspective. This NDE is Scientific Evidence or Empirical Proof that Psyche or Intelligence is a different entity than our spirit body and our physical body.

https://www.youtube.com/watch?v=WUXzj0Tczz4

When people like Dr. Anthony Cicoria describe looking at their spirit body from within their spirit body, they describe the process from the Third-Eye Perspective. The interesting thing to note is that the Psyche can not only see, but that the Psyche can also see events from different perspectives, both from within the spirit body and from outside the spirit body. Psyche can project, which means that Psyche can exit its assigned spirit body. It's always the Psyche who does the experiencing and the remembering, and not the spirit body and not the physical body.

Often the Psyche is called the Spark. It's the Spark of Life. This indicates that the Psyche is made of energy or light. The following videos describe the Spark of Life that takes place at Conception:

https://www.youtube.com/watch?v=UUZfWB95I1A

https://www.youtube.com/watch?v=1fAfQlpPKK0

https://www.youtube.com/watch?v=DfH4dP26WsE

ONLY Psyche can do telepathy. Psyche, or mind-over-matter, is involved in physical healing. Psyche is a supernatural phenomenon; and, there are often many different supernatural healings and supernatural events which take place during Near-Death Experiences, that lend evidence to Psyche and the supernatural. These are Empirical Experiences which take place under the direction of God's Psyche; but, are impossible for the physical body and physical matter to perform or to do on their own. Psyche IS the Ultimate Causal Agent. Psyche is the Prime Mover or the Ultimate Cause.

The following Near-Death Experience (NDE) talks about the Bubble of Protection and a Physical Teleportation to safety. This is NOT something that we consciously do. It's something that God does for us.

https://www.youtube.com/watch?v=DmBnCTuQUOc

I have experienced this Bubble of Protection and Physical Teleportation myself, so I KNOW that this is real and can truly happen to us, under God's direction and protection.

I was going too fast and couldn't miss the car that had pulled out in front of me. Time stopped, and I seemed to have 360-degree awareness of my surroundings, and I saw or felt where I wanted to be. It felt as if the whole of eternity had gathered into me; and then, snap! I and my car were someplace else in the middle of a lawn twenty or thirty feet away with the car completely stopped and the engine shut off. I have NO memory of traveling the distance, stopping the car, and shutting the car off. I was on that other car's bumper one instant and someplace else entirely the next instant. Amazing! Teleportation is not anything that we do for ourselves or can do for ourselves as mortal fallen beings. Teleportation to safety is something that God has to do for us.

In the next NDE, Ana Christina was murdered, saw Christ, and was returned completely healed. God can heal our physical body miraculously, if we have been killed and He wants us to return to life.

https://www.youtube.com/watch?v=GDmtj6KHJNk

A favorite Ted talk. Anita Moorjani died of cancer. After we die, God can miraculously heal our physical body if He wants us to return to life, or God can teach our Psyche how to do so. Psyche is the part of us that learns and remembers, not our spirit body. Our spirit body is just another shell. Our spirit body and our physical body don't have experiences and don't make memories while our Psyche is away from our spirit body and our physical body.

https://www.youtube.com/watch?v=rhcJNJbRJ6U

Kim Rives died of cancer. Jesus absorbed all of her pain. ONLY Psyche can do telepathy.

https://www.youtube.com/watch?v=EnRZGLkp2M4

The following from Ian Wiltshire is the funniest NDE that I have come across so far:

https://www.youtube.com/watch?v=yjiSbpOJhZo

In this NDE, Ian describes himself [psyche] watching himself [spirit body] watch himself [physical body]. It's kind of confusing, unless you understand the fact that Psyche is something completely different than our spirit body and that our spirit body is something completely different than our physical body.

In the next NDE, Barbara Wilcox explains not having a body, or spirit, or a soul but being Total Intelligence.

https://www.youtube.com/watch?v=JQzap_jFXT8

In the following NDE, Jessica describes being a Point of Light, a Small Spec, and a Pinpoint of Light during her near-death experience. She, too, is describing Psyche or Non-Local Consciousness. Jessica also described the fact that we don't see everything and don't understand everything which is there to be seen and understood while out-of-body. Instead, there is ever-growing awareness and understanding, with still more to go that is never reached. Jessica could sense the presence of others but didn't engage with them during her NDE as much as others seem to do. Hers is the most "self-centered" NDE that I have encountered so far. Jessica's Psyche is very much the center of her universe.

https://www.youtube.com/watch?v=Ve6RG9K3qrA

https://www.youtube.com/watch?time_continue=211&v=VD0Pd7twFBY

http://ndestories.org/jessica-haynes/

God gives each person a piece of the puzzle, and we have to find a way to put all of those pieces together if we want to see the Whole Picture, or the True Reality of our existence.

There's tons of evidence for the existence of the spirit body. But, you have to pay close attention and do a bit of research to find evidence of Psyche existing and functioning separately from the spirit body. It's there to be found, but it's not as common.

I have faith that there are many more people who have experienced their Psyche, Intelligence, or Spark separate from their spirit body. That's one of my talents or gifts – the ability to notice patterns and trends. For me personally, this is one of my greatest and most useful Scientific Discoveries or Scientific Observations, namely that Psyche is something completely different than our spirit body and spirit matter, and something completely different than the typical forces, fields, and LAWS. Psyche is a god. This KNOWLEDGE and TRUTH is one of the greatest things that God has revealed to me through the Lived Experiences or the Scientific Evidence of the human race as a whole.

Yes, it's ALL made of light – our psyche, our spirit body, and our physical body – and every type of light has a bit of consciousness or awareness and can respond to stimuli. However, spirit matter and physical matter are the types of light that REACT, while the Psyche is the type of light that ACTS, CHOOSES, THINKS, EXISTS, REMEMBERS, and is CONSCIOUS, AWARE, SENTIENT, and fully ALIVE. Psyche or Non-Local Consciousness is our true essence.

$E = mc^2$. Matter contains tons of energy or light – physical matter more so than spirit matter. Remove the mass or the energy from physical matter, and it becomes spirit matter again. Whenever God chooses to pump some of His Light or Life or Consciousness or Energy into spirit matter, it takes on mass and transforms into physical matter. All Matter is made of light. Forces and Fields are also a type of energy or light, and complete this picture.

This whole thing is a Psyche Ontology wherein Psyche is the fundamental unit of Reality. Psyche is immaterial light, or Pure Light. Psyche is an infinite singularity. We could also call this a Light Ontology, because Psyche IS Light. Either way, this is the Ultimate Model of Reality and the Ultimate Paradigm.

God gives each person a piece of the puzzle. When we combine all of our pieces together, we find ourselves looking at the Whole Truth or the True Reality of Our Existence. That's why I call it the Ultimate Model of Reality, because it is based upon and built upon the Lived Experiences or the Direct Observations of the human race as a whole, including our Non-Local Experiences or our Scientific Observations while in the spirit realm. Within this Ultimate Model of Reality or Psyche Ontology, I treat Lived Experiences or Psyche Experiences as Scientific Evidence. This is Real Science, and not the brain-dead junk which the Materialists and Atheists try to force upon us.

Go with the best and get rid of all the rest. That's what I try to do.

## NDE Encounters with Jesus

This is a category of NDEs that I personally find fascinating, because they serve as evidential proof or observational proof that God really does exist.

http://ndestories.org/category/jesus/

https://www.youtube.com/results?search_query=NDE+Jesus

http://ndestories.org/category/god/

Remember, Science 2.0 allows all of the evidence into evidence, including the Lived Experiences and Out-of-Body Experiences of the human race.

I'll let you watch these for yourself and decide for yourself what to make of them. That's what I did. But, I warn you that I personally wasn't able to maintain my belief in Materialism, Naturalism, Nihilism, and Atheism after watching some of these videos, because I found the observational evidence and experiential evidence a bit too convincing. I'm a natural-born skeptic, but even I can be persuaded by evidence.

My best friend introduced me to these videos a couple of years ago. I wasn't looking for them, because I didn't know they existed and didn't believe that they existed. However, he was looking for them and shared them with me because he is a seer, has visions, has connected with other humans living throughout this universe on other planets, has been shown Satan, has seen outer darkness, and has had some NDEs and out-of-body experiences of his own including an encounter or two with Jesus our Savior. God won't permit my friend to share with me some of the things that he has seen and experienced while in the spirit; but, God encouraged my friend to share all of these different NDEs with me.

Good enough.

I no longer have my head in the sand and I'm no longer hiding from observational evidence like I was when I was a Materialist, Nihilist, and Atheist. As a result, I have upgraded my science to Science 2.0, and now I allow all of the evidence into evidence. You can too, if you want to.

Here are some others with the same theme:

Tamara Laroux:

https://www.youtube.com/watch?v=HGQDkCi-OIY

There are now hundreds of different books about Near-Death Experiences, if not thousands. It's impossible to list them all.

In my book, *Putting Psyche Back into Psychology: Restoring Science to Consciousness*, I tried to list the ones I own; and as time allows, I hope to comment on each one of them; so, I won't do so here in this book.

https://www.amazon.com/Putting-Psyche-Back-into-Psychology-ebook/dp/B071NC987S

Each Near-Death Experiences provides the Empirical Evidence needed to make Science 2.0 a real upgrade and improvement over Scientific Materialism or Scientific Naturalism.

# Part VIII — Conscious After Dying

Empirical Evidence: Empirical evidence, also known as sensory experience, is the knowledge received by means of the senses, particularly by observation, experimentation, and experience. The term "empirical" comes from the Greek word for experience.

Transdimensional: Relating to a dimension other than those of the normal three-dimensional world. Transdimensional means Non-Physical Dimension – a dimension other than our normal physical 3D space-time dimension. Transdimensional means Supernatural.

Non-Local: Non-Local means "not located in our physical 3D space-time dimension". Non-Local means Transdimensional, which means Non-Physical. The Non-Local Realm is the Spirit Realm, or the Transdimensional Realm.

By definition, in principle, the Big Bang Singularity was a transdimensional object, and the Big Bang itself was a supernatural event. Both of these things or events theoretically happened BEFORE physical matter first came into existence some 380,000 years later. Remember, the Primal Singularity was a transdimensional object, and the Beginning of our Physical Universe was a supernatural event. Physical matter couldn't have done the job as the Materialists and Naturalists claim, because physical matter didn't exist yet.

https://en.wikipedia.org/wiki/Chronology_of_the_universe

If the Big Bang truly happened, it was in fact a supernatural event and is in fact scientific proof of the supernatural or transdimensional, because the Big Bang or the Beginning of our Physical Universe happened before the first particle of physical matter was brought into existence, where our physical universe is concerned. It's only logical.

Science 2.0 is based solidly on the existence of Quantum Non-Locality or Transdimensionality. Science 2.0 is the Non-Local Consciousness Interpretation of Quantum Mechanics, or the Quantum Consciousness Interpretation of Quantum Mechanics, or the Orthodox Interpretation of Quantum Mechanics. Therefore, there absolutely must be some kind of eye-witness evidence, or observational evidence, or experiential evidence of Quantum Non-Locality and Non-Local Consciousness; or else, Science 2.0 would be no better than Materialism and Naturalism, which already claim that these things do not exist.

Remember, there must be some kind of sensory experience or empirical evidence of the Non-Local Realm; or, Science 2.0 has no reason to exist, and has no observational evidence to support it – just like Materialism, Naturalism, and Atheism.

According to this YouTube Ted Talk video, it has been estimated that there are now over 13 million different Near-Death Experiences (NDEs) on record.

https://www.youtube.com/watch?v=MyaBeHeRK6M

The following article states that over 13 million people in the US have had NDEs, according to a 1992 Gallup Poll.

https://www.theepochtimes.com/how-common-are-near-death-experiences-ndes-by-the-numbers_757401.html

http://mypsyche.us/wp-content/uploads/2017/10/How-Common-Are-NDEs.pdf

Here's another one:

https://www.nderf.org/NDERF/Research/number_nde_usa.htm

I would say that, where Near-Death Experiences (NDEs) and Out-of-Body Experiences (OBEs) are concerned, the observational evidence and experiential evidence have hit critical mass. It seems as if dozens of new NDEs are reported on YouTube every day now, with no end in sight. Jesus Christ has truly made His arm bare in the sight of the nations, as He said He would.

You could literally watch NDEs on YouTube around the clock for days on end and never run out of new ones to watch.

I started collecting all the different books that have been written about NDEs and OBEs; and eventually, I realized that it's an impossible task because there's probably thousands of them by now and more being printed every day.

The preponderance of the evidence stands squarely against the Materialists, Naturalists, Darwinists, Nihilists, and Atheists who tell us and assure us that NO such evidence exists. Who are you going to believe? Who's telling you the truth and who is lying to you? Who is trying to trick you and deceive you? I have answered these questions for myself to my satisfaction now; but, you are a completely different person, and you are going to have to make up your own mind what you want to accept and believe.

There used to be a time when I believed the Materialists, Naturalists, Nihilists, and Atheists whenever they told me that there would never be any proof of God's existence and could never be any proof of God's existence. I made a major conceptual breakthrough – a paradigm shift in my own way of thinking – the day that I finally realized and observed that they are wrong. I have never looked back ever since. I was done with My Materialism, My Naturalism, My Scientism, and My Atheism.

Now I want evidence, lots and lots of evidence!

NDE Research:

https://www.youtube.com/results?search_query=NDE+research

Yolaine Stout:

https://www.youtube.com/watch?v=BMfjM6TrkV8&t=2692s

Wayne Morrison:

https://www.youtube.com/watch?v=vw6W_gopaHM

There are now hundreds of different books about Near-Death Experiences, if not thousands. It's impossible to list them all.

In my book, *Putting Psyche Back into Psychology: Restoring Science to Consciousness*, I tried to list the ones I own; and as time allows, I hope to comment on each one of them; so, I won't do so here in this book.

https://www.amazon.com/Putting-Psyche-Back-into-Psychology-ebook/dp/B071NC987S

Each Near-Death Experiences provides the Empirical Evidence needed to make Science 2.0 a real upgrade and improvement over Scientific Materialism or Scientific Naturalism.

## Shared-Death Experiences

For me personally, Shared-Death Experiences (SDEs) are the most convincing proof of continuing consciousness after the death of one's physical body.

https://www.google.com/search?q=shared+death+experience+stories

https://www.youtube.com/results?search_query=shared+death+experiences

We have Raymond Moody to thank for revealing and popularizing Shared-Death Experiences.

## Reincarnation and Previous Lives

Science 2.0 allows all of the evidence into evidence.

Consequently, any independently verified veridical Reincarnation account that we come across is PROOF that one's psyche, soul, non-local consciousness, intelligence, or personality continues past physical death forward into other lives.

Even though Science 2.0 allows all of the evidence into evidence, I'm highly skeptical and doubtful of any Reincarnation accounts that are manufactured or produced under hypnosis.

Why?

My bachelor's degree is in Psychology, and I know for a fact that false memories and fake past lives can be easily manufactured while under hypnosis. This documented and proven reality is called False Memory Syndrome (FMS).

http://www.fmsfonline.org/

If the hypnotist tells you to remember the times when your parents physically and sexually abused you, your psyche or mind will manufacture the events and your memories of those events will be more real and more powerful than the true memories of your life. Likewise, if the hypnotist tells you to remember your past lives, your psyche or mind can generate a whole host of them, and they will seem more real to you than your own life. That's the power of hypnosis.

False Memory Syndrome is a real, documented, and proven phenomenon; and, it has caused some people a lot of pain and grief.

Therefore, I tend to dismiss every Reincarnation account or Past Lives account that is produced or manufactured by hypnosis. These accounts of Reincarnation have to come through some other means in order to merit my effort and time.

Young children, who spontaneously remember their previous life, are the best and most reliable Proof of Reincarnation, in my humble opinion – especially when their accounts are verified and proven to be real and true. I'm not the only one who is of this opinion.

Even though Science 2.0 allows all of the evidence into evidence, I no longer trust anyone or anything on blind faith alone like I used to do when I was a Materialist, Nihilist, and Atheist. My days of Materialism, Blind Faith, and Rejection of Evidence are over; but, that doesn't mean that I simply check my brain at the door and admit everything into evidence in my own personal life.

**Trust, but verify.**

I want veridical evidence, proven evidence, verified evidence, and confirmed evidence. I want the people who claim Reincarnation to meet their burden of proof through a preponderance of the evidence; or, I'm going to be skeptical. The same thing would apply to Near-Death Experiences and Out-of-Body Experiences, especially if they were being manufactured under hypnosis or drugs.

**2 Corinthians 13: 1: "In the mouth of two or three witnesses shall every truth be established."**

Pursue the eye-witness accounts and the verified accounts whenever possible, no matter whether we are talking about Near-Death Experiences (NDEs) or Reincarnation

Accounts. There are a few of them out there that are quite convincing, because they have been verified and proven to be true.

Furthermore, some of these accounts claim to have been given by God as visions or revelation of a previously lived life. If you trust God, then you know that it is true, because God doesn't lie and because God has ways of verifying that it's true.

1. *Reincarnation: Manika, the Girl Who Remembered Her Past Life*:

https://www.youtube.com/watch?v=-L4hXXSq2O8&t=3s

2. Stevenson, I. (2001). *Children Who Remember Previous Lives: A Question of Reincarnation*. Jefferson, NC: McFarland.

3. Rivas, T., Dirven, A., & Smit, R. H. (2016). *The Self Does Not Die: Verified Paranormal Phenomena from Near-Death Experiences*. Durham, NC: IANDS Publications.

4. Anonymous. (2013). *Teachings of the Doctrine of Eternal Lives*. Salt Lake City, UT: Digital Legend Press.

Here are a few others that were a bit less convincing but interesting nonetheless.

"Beyond the Ashes: Cases of Reincarnation from the Holocaust" by Yonassan Gershom.

"From Ashes to Healing: Mystical Encounters with the Holocaust" by Yonassan Gershom.

"Coming Back: A Psychiatrist Explores Past-Life Journeys" by Paul Perry and Dr. Raymond Moody.

"Children's Past Lives: How Past Life Memories Affect Your Child" by Carol Bowman.

All you need is one verified and proven account of Reincarnation in order to know that the phenomenon is real and truly happens from time to time; but, here we have a few to think about and consider. Some of them are quite convincing, because they have been verified and proven to be true.

# PART IX — QUANTUM CONSCIOUSNESS AND UNIVERSAL CONSCIOUSNESS

Quantum Consciousness is small enough to dwell within an atom, with plenty of room to spare.

One theory is that Non-Local Consciousness or Psyche is an infinite singularity having no size and taking up no space whatsoever. Imagine it! Something with infinite informational storage capacity within something that has no size and takes up no space. If true – if Psyche really is an infinite singularity – then we automatically KNOW from observation and experience that the smaller can always reside within or dwell within the larger and control the larger, which means that it's Psyche or Non-Local Consciousness who decides the size and shape of strings and the various different notes which the strings will play, when it comes to String Theory. We also KNOW that it's Psyche or Non-Local Consciousness who is residing within, controlling, shaping, and holding together each physical atom and each physical particle.

In contrast, Universal Consciousness or Cosmic Consciousness is something like the Cosmic Internet, the Universal Broadcast, Dark Energy, The Music of the Spheres, Brahman, the Holy Ghost, the Spirit of God, or the Light of Christ. Universal Consciousness is in all things and through all things. It's a universal psyche or consciousness.

Dozens of different books have been written about these two concepts in recent decades.

Whenever one of these things is lived, experienced, or observed, we end up with empirical evidence that proves that Quantum Non-Locality and Non-Local Consciousness are real and true.

Whenever one of these things survives or transcends the death of our physical body and physical brain, then we have empirical evidence that Non-Local Consciousness or Psyche exists separate from our physical body and physical brain – meaning that Psyche transcends our physical body and physical brain.

Science 2.0 is all about the observational evidence or eye-witness evidence associated with physical experiences and non-local experiences, because Science 2.0 is based upon empirical evidence rather than scientific inferences.

The Materialists and Naturalists define science as Materialism, Naturalism, Darwinism, Scientism, Nihilism, and Atheism; and then, these people deliberately reject and exclude anything that doesn't fit with their pre-conceived conclusions. I KNOW because I used to be a Materialist, Nihilist, and Atheist.

In contrast, Science 2.0 defines science as Observation, Experience, Knowledge, Truth, and Fact. Science 2.0 allows ALL of the evidence into evidence. Nothing is excluded.

Give me evidence – lots and lots of evidence!

## Books Supporting Quantum Consciousness

I own 48 Kindle books that have the word "Consciousness" in their title or as part of their subject matter.

There are many more on Amazon that I don't own.

https://www.amazon.com/s/ref=nb_sb_noss_2?url=search-alias%3Daps&field-keywords=Consciousness

Technically, Quantum Consciousness is the type of Psyche or Non-Local Consciousness that resides within each physical atom, each quark, and each vibrating string.

If you search for Quantum Consciousness on Amazon, dozens of different books come up. There's too much there to even begin to process all of it in a timely fashion.

https://www.amazon.com/s/ref=nb_sb_noss_2?url=search-alias%3Daps&field-keywords=Quantum+Consciousness

If time allows, I hope to read and review some of these books within other books that I am writing; but, the world doesn't need another Quantum Consciousness book because it has hundreds of them already.

My primary purpose here is to let you know that these books exist – to introduce ALL the evidence into evidence.

Empirical Evidence for Non-Local Consciousness or Quantum Consciousness provides evidence and justification for Science 2.0.

Psyche seems to be an infinite singularity with no size taking up no space, yet its range, velocity, sphere of influence, and scope seem to be infinite and instantaneous. Psyche is NOT a Quantum Object or Particle of Matter, and Psyche doesn't have any physical limitations. Psyche seems capable of an infinite amount of information storage capacity. Psyche seems to be an immaterial immortal recording device. Psyche is life and Psyche is alive. Since Psyche has no mass or size, it has no speed limit and is capable of being instantaneously and simultaneously everywhere.

Psyche is the Actor and the Chooser. In contrast, a Quantum Object or Matter of any kind reacts to Psyche's observation and command. Psyche is the thing that chooses the frequency at which Strings or Quarks vibrate. The smaller can always reside within and control the larger, which means that Psyche can reside within and control Strings and Quarks, Gluons and Photons, Electrons and Atoms. Psyche is infinite frequency light and living light. At an infinite frequency, Psyche transmutes into a Unity or an Infinite Singularity that has completely different properties than a Quantum Object, String, Quark, Electron, or Particle of Matter.

In some of my other books, I use **Y = TAN(X)** to model and explain Psyche and Quantum Objects, including the difference between the two. For example, *The Ultimate Model of Reality: Psyche Is the Ultimate Cause*.

https://www.amazon.com/Ultimate-Model-Reality-Psyche-Cause-ebook/dp/B071NC9JK6

Check it out if any of this is of interest to you.

Even though I didn't realize it at the time, everything I have written has been leading me to Science 2.0.

https://www.amazon.com/Mark-My-Words/e/B01IAEF2Y6

Quantum Consciousness is a popular topic on Amazon.

https://www.amazon.com/s/ref=nb_sb_ss_c_1_32?url=search-alias%3Daps&field-keywords=quantum+physics+of+consciousness&sprefix=quantum+physics+of+consciousness%2Caps%2C506&crid=5RFDU3WX8JHB

Here are a couple other books that I own, which mention Quantum Consciousness.

> It has been a dozen years since *Physics of the Soul* was first published and I am happy to say that the theory, data, and conclusions reported in this book remain ever more validated. In short, survival after death and reincarnation are valid scientific concepts. When you read *Physics of the Soul*, you will discover that the central theory of survival after death and reincarnation reported in the book crucially depends on a concept called quantum memory. The idea is that part of our memory (call it quantum memory), specifically that of our learning, is nonlocal, which means that this memory resides not locally in the brain but outside of space and time altogether.

Goswami, Amit. *Physics of the Soul: The Quantum Book of Living, Dying, Reincarnation, and Immortality* (Kindle Locations 58-63). Hampton Roads Publishing. Kindle Edition.

Consciousness after the death of our physical body:

**THE BIG QUESTION**

Does our consciousness — mind, soul, or spirit — end with the death of our body? Or does it continue in some way, perhaps in another realm or dimension of the universe? This is the "big question" thoughtful people have asked throughout the ages. Let us come down to the bottom line right away. Are we entirely mortal? Or is there an element or facet of our existence that survives the death of our body? This question is of the utmost importance for our life and our future.

We know that conscious experience can occur in the temporary absence of brain function: this is the case in so-called NDEs—near-death experiences. Could conscious experience occur also in the permanent absence of brain function—when the individual has died? It makes sense to ask this question as well, because it is important, meaningful, and not without observational evidence.

Mainstream science — the science taught in most schools and colleges — does not confront these questions: it denies the very possibility that consciousness could exist in the absence of the living organism.

Laszlo, Ervin. *The Immortal Mind: Science and the Continuity of Consciousness beyond the Brain* (p. 1-2). Inner Traditions/Bear & Company. Kindle Edition.

Henry P. Stapp makes another appearance:

The arguments given above rest heavily upon the contrast between the reason-based dynamics of the unfolding universe described by quantum mechanics and the physical-description-based dynamics of the block universe described by classical mechanics. In this final section I shall pinpoint the technical features of orthodox quantum mechanics that underlie this fundamental difference between these two theories.

Von Neumann created a formulation of quantum mechanics in which all the physical aspects of nature are represented by the evolving quantum mechanical state (density matrix) of the universe. Each subsystem of this physically described universe is represented by a quantum state obtained by performing a certain averaging procedure on the state of the whole universe. Each experience of a person is associated with an "actual event". This event reduces, in a mathematically specified way, the prior quantum state of this person's brain --- and consequently the quantum state of the entire universe --- to a new "reduced" state. The reduction is achieved by the removal of all components of the state of this person's brain that are incompatible with the increment of knowledge associated with the experience. The needed mapping between "experiential increments in a person's knowledge" and "reductions of the quantum mechanical state of that person's brain" can be understood as being naturally created by trial and error learning of the experienced correlations between intentional efforts and the experiential feedbacks that these efforts tend to produce.

A key feature of von Neumann's dynamics is that it has two distinct kinds of mind-brain interaction. Von Neumann calls the first of these two processes "process 1". It corresponds to a choice of a probing action by the person, regarded as an agent or experimenter. The second kind of mind-brain interaction was called by Dirac "a choice on the part of nature". It specifies nature's response to the probing action selected by a logically preceding process 1 action. Von Neumann uses the name "process 2" to denote the physical evolution that occurs between the mind-brain (collapse) interactions. I therefore use the name "process 3" to denote the reduction/collapse process associated with nature's response to the process 1 probing action.

The mathematical form of process 1 differs from that of process 3. This mathematical difference causes these two processes to have different properties. In particular, process-1 actions have only local effects, in the sense that the dependence of the predictions of quantum mechanics upon a process-1 action itself, without a specification of the response, is confined (in the relativistic version) to the forward light-cone of the region in which the process-1 physical action occurs: the empirically observable effects of a process-1 action never propagate faster than the speed of light. On the other hand, nature's response (process 3) to a localized process-1 action can have observable statistical effects in a faraway contemporaneous region. The no-faster-than-light property of the empirically observable effects of any process-1 action is what justifies the word "relativistic" in relativistic quantum theory, even though the underlying mathematical description involves abrupt process-1-dependent faster-than-light transfers of information in connection with nature's response to the process-1 action.

Process 2 is a generalization of the causal process in classical mechanics, and, like it, is deterministic: the state of the universe at any earlier time completely determines what it will evolve into at any later time, insofar as no process 1 or process 3 event intervenes. But if no process 1 or process 3 event intervenes then the process 2 evolution would take the initial "big bang" state of the universe into a gigantic smear in which, for example, the moon would be smeared out over the night sky, and the mountains, and the cities, and we ourselves, would all be continuously spread out in space.

It is the process 1 and process 3 actions that, in the orthodox ontology, keep the universe in line with human experiences. On the other hand, the von Neumann ontology certainly does not exclude the possibility that non-human-based analogs of the human-based process 1 and follow-up process 3 actions also exist. Rather, it explains why the existence of reduction processes associated with other macroscopic agents would be almost impossible to detect empirically. These features of the von Neumann ontology justify focusing our attention here on the human involvement with nature.

Process 1, unlike process 2, is not constrained by any known law. In actual practice our choices of our probing actions appear to us to be based on reasons. We open the drawer in order to find the knife, in order to cut the steak, in order to eat the steak, in order to satisfy our hunger. Whilst all of this chain of reasons would, within the deterministic framework of classical physics, need to be, in principle, explainable in mathematical ways based upon the physical description of the universe, there is no such requirement in orthodox quantum mechanics: the sufficient reasons could be "reasons"; reasons involving the experiential dimension of reality, rather than being fully determined within the physical dimension. And these reasons could be, at each individual moment of experience, sufficient to determine the associated process 1 choice, without those choices having been mathematically computable from the state of the universe at earlier times.

The process 3 selection on the part of nature, unlike the process 1 choice, is not completely unconstrained: it is constrained by a statistical condition. According to the principle of sufficient reason, the process 3 choice must also be, in principle, determined by a sufficient reason. But, as emphasized above, nature's choice is nonlocal in character. Thus, the reason for a process 3 choice need not be located at or near the place where the associated process 1 action occurs. Yet, as was clear already in classical statistical mechanics, there is an á priori statistical rule: equal volumes of phase space are equally likely. The (trace-based) statistical rule of quantum mechanics is essentially the quantum mechanical analog of this á priori statistical rule. The quantum statistical rule is therefore the natural statistical representation of the effect of a reason-based choice that is physically far removed from its empirical process 3 manifestation.

In closing, it is worth considering the argument of some physicalist philosophers that the replacement of classical mechanics --- upon which physicalism is based --- by quantum mechanics is not relevant to the resolution of the mind-matter problem for the following reason: that replacement has no bearing on the underlying problem of human freedom. The argument is that the essential difference between the two theories is (merely) that the determinism of classical mechanics is disrupted by the randomness of quantum mechanics, but that an introduction of randomness into the dynamics in no way rescues the notion of meaningful human freedom: a random choice is no better than a deterministic choice as an expression of meaningful human freedom.

This physicalist argument flounders on the fact that the element of quantum randomness enters quantum mechanics only via process 3, which delivers nature's choice. Man's choice enters via process 1, which is the logical predecessor to nature's process 3 "random" choice. In orthodox quantum mechanics, no elements of quantum randomness enter into man's choice. Nor is man's choice fixed by the deterministic aspect of quantum mechanics: that aspect enters only via process 2. Von Neumann's process 1 human choice is, in this very specific sense, "free": it is von Neumann's representation of Bohr's "free choice of experimental arrangement for which the quantum mechanical formalism offers the appropriate latitude" (Bohr 1958. p. 51). Human choices enter orthodox quantum mechanics in a way not determined by a combination of the deterministic and random elements represented in the theory.

Stapp, Henry. *Quantum Physics of Consciousness* (Kindle Locations 461-518). Cosmology Science Publishers. Kindle Edition.

Psyche or Non-Local Consciousness is something completely different than a Quantum Object or Particle of Matter. A Quantum Object has two known states of existence – a non-local spirit matter aspect and a localized physical matter aspect.

According to this Orthodox Interpretation of Quantum Mechanics, you (your human psyche) have consciousness, and nature (other psyches) have consciousness. They interact with each other through Quantum Objects or Particles of Matter.

You will have to decide for yourself if any of this is of use to you. Under Science 2.0, my only obligation is to present the evidence, or to allow someone else to do it for me. We leave it up to you to decide what to do with it and what to make of it.

I used to be a Materialist, Naturalist, and Atheist. In my ignorance, I didn't know that any of this evidence existed, so I wasn't looking for it, because I truly believed that it didn't exist.

After I abandoned My Atheism and My Materialism and My Nihilism, I have found the evidence for Non-Local Consciousness overwhelming and extremely difficult to assimilate in full.

I own hundreds of books that talk about, discuss, and provide evidence for Non-Local Consciousness and Life after Death. I could probably double the size of this book simply by quoting from each and every one of them.

It's huge; and, I never knew.

I've changed my mind now. Now I believe that the Biblical God Jesus Christ has indeed met His burden of proof through a preponderance of the evidence, because now I'm trying to look at and assimilate all of that evidence; whereas before, My Materialism, Nihilism, and Atheism were based upon a refusal to look at evidence, and things were much simpler back then because my ignorance made it so.

## Books Supporting Universal Consciousness or Cosmic Consciousness

I own 48 Kindle books that have the word "Consciousness" in their title or as part of their subject matter.

There are many more on Amazon that I don't own.

https://www.amazon.com/s/ref=nb_sb_noss_2?url=search-alias%3Daps&field-keywords=Consciousness

Technically, Universal Consciousness or Cosmic Consciousness is the type of Psyche or Non-Local Consciousness that fills the immensity of space and permeates the whole universe – something like Dark Energy, the Zero-Point Field of Light, the Quantum Sea of Light, the Holy Spirit, or the Light of Christ.

If you search for Cosmic Consciousness on Amazon, dozens of different books come up. There's too much there to even begin to process all of it in a timely fashion.

https://www.amazon.com/s/ref=nb_sb_noss_2?url=search-alias%3Daps&field-keywords=Cosmic+Consciousness

If you search for Universal Consciousness on Amazon, dozens of different books come up. There's too much there to even begin to process all of it in a timely fashion.

https://www.amazon.com/s/ref=nb_sb_noss_2?url=search-alias%3Daps&field-keywords=Universal+Consciousness

If time allows, I hope to read and review some of these books within other books that I am writing. I do mention a few of them in this book. My primary purpose here is to let you know that they exist – to introduce ALL the evidence into evidence.

Empirical Evidence for Non-Local Consciousness provides evidence and justification for Science 2.0.

I discuss the Light of Christ in some of my other books. For example, *God Is in the Light: God is light, and in Him is no darkness at all*.

https://www.amazon.com/God-Light-light-darkness-all-ebook/dp/B07168S37N

Even though I didn't realize it at the time, everything I have written has been leading me to Science 2.0.

https://www.amazon.com/Mark-My-Words/e/B01IAEF2Y6

Here are a few other books that I own, which mention Cosmic Consciousness and/or Universal Consciousness.

Goswami, A. (1993). *The Self-Aware Universe: How Consciousness Creates the Material World*. New York: Putnam Penguin.

Radin, D. (). *The Conscious Universe: The Scientific Truth of Psychic Phenomena*. New York: HarperCollins.

Here are a couple of others:

**WHAT is Cosmic Consciousness? The present volume is an attempt to answer this question; but notwithstanding it seems well to make a short prefatory statement in as plain language as possible so as to open the door,**

as it were, for the more elaborate exposition to be attempted in the body of the work. Cosmic Consciousness, then, is a higher form of consciousness than that possessed by the ordinary man. This last is called Self Consciousness and is that faculty upon which rests all of our life.

>Bucke, Richard Maurice. *Cosmic Consciousness* (Kindle Locations 81-84). Unknown. Kindle Edition.

I seem to have three different copies of this book from Bucke.

>**Issues related to consciousness in general and human mental processes in particular remain the most difficult problem in science. Progress has been made through the development of quantum theory, which, unlike classical physics, assigns a fundamental role to the act of observation. To arrive at the most critical aspects of consciousness, such as its characteristics and whether it plays an active role in the universe requires us to follow hopeful developments in the intersection of quantum theory, biology, neuroscience and the philosophy of mind. Developments in quantum theory aiming to unify all physical processes have opened the door to a profoundly new vision of the cosmos, where observer, observed, and the act of observation are interlocked. This hints at a science of wholeness, going beyond the purely physical emphasis of current science. Studying the universe as a mechanical conglomerate of parts will not solve the problem of consciousness, because in the quantum view, the parts cease to be measurable distinct entities. The interconnectedness of everything is particularly evident in the non-local interactions of the quantum universe. As such, the very large and the very small are also interconnected.**

>Clarke, Chris. *Cosmology of Consciousness: Quantum Physics & Neuroscience of Mind* (Kindle Locations 17-25). Cosmology Science Publishers. Kindle Edition.

Obviously, I'm not the first person to figure out this stuff; and, I won't be the last.

I own 48 Kindle books that have the word "Consciousness" in their title or as part of their subject matter. If I quoted from all of them, this section would never end.

Clearly, I have some reading and studying to do. What about you?

## Prejudice: The Good, the Bad, and the Ugly

The Militant Atheists are the force behind Communism. Darwinism was the force behind Nazism or Fascism. These people will kill you if you refuse to believe as they believe. This is the ugly and the evil side of prejudice.

There are certain aspects of prejudice that I want to remember and retain. Prejudice isn't always bad for us. There are times when prejudice can be good for us – it can actually save our lives.

**"Prejudice. A vagrant opinion without visible means of support."** — Ambrose Bierce, *The Devil's Dictionary*, 1911.

Can you sense the hidden message within this quote?

Our prejudices are NOT based upon physical matter! Our prejudices are a function of Syntropy – they are chosen into existence by the Human Psyche. Our genes and our cells aren't prejudiced against anyone nor are they biased toward anyone. Prejudice is NOT a function of our genetics; and, our society or those other psyches can't force us to be prejudiced either. Our prejudice is chosen into existence by our Psyche. Our prejudices reflect the personal preferences of our Psyche, Spirit, Mind, or Soul.

Fascinating, is it not?

Likewise, when it comes to Materialism, Naturalism, Darwinism, Nihilism, Behaviorism, and Atheism, there's NO visible means of support for their major premises or hidden assumptions which state that Syntropy or Psyche or Quantum Mechanisms do not exist. In other words, these materialistic and naturalistic philosophies are also a product of the Human Psyche – they have NO basis in physical reality.

This is where the tires really hit the pavement. It doesn't get any more real or essential than this!

There's NO physical reality and NO physical evidence backing up the naturalistic claims which state that Quantum Mechanics, Supernatural Mechanisms, Action at a Distance, Psyche, Non-Local Consciousness, Syntropy, Free Will, Forgiveness, Mercy, and Love DO NOT EXIST. These are philosophical beliefs or metaphysical beliefs that were chosen into existence by Someone Psyche.

Materialism, Naturalism, Darwinism, Nihilism, Behaviorism, Determinism, and Atheism are philosophical assumptions, and they only exist within the Human Psyche or the Human Mind. There's no physical evidence supporting them! In fact, ALL of the observational and experiential evidence falsifies them. If we choose to define Science as observation and experience rather than Materialism and Naturalism, then Science itself falsifies Materialism and Naturalism.

Fascinating, is it not?

Materialism, Naturalism, Darwinism, Nihilism, Behaviorism, Determinism, Physical Reductionism, and Atheism are prejudices! They are a vagrant opinion without any visible means of support. They are internally inconsistent or logically inconsistent. In other words, they are self-defeating.

This is one of the coolest and most useful scientific observations that I made while studying Social Psychology.

> **Prejudice is an attitude.** An attitude is a distinct combination of feelings, inclinations to act, and beliefs. It can be easily remembered as the ABCs of attitudes: *a*ffect (feelings), *b*ehavior tendency (inclination to act), and *c*ognition (beliefs). A prejudiced person may *dislike* those different from self and *behave* in a discriminatory manner, *believing* them ignorant and dangerous. Like many attitudes, prejudice is complex.
>
> A **stereotype** is a belief about the personal attributes of a group of people. Stereotypes are sometimes overgeneralized, inaccurate, and resistant to new information.
>
> *Prejudice* is a negative *attitude; discrimination* is negative *behavior.* Discriminatory behavior often has its source in prejudicial attitudes. (*Social Psychology*, p. 309.)

I don't buy into the idea that prejudice and discrimination are always negative. They can save your life. The idea that prejudice and discrimination are always negative is a leftist idea that the Materialists, Naturalists, Darwinists, Atheists, and Gay Activists have produced. They demand that everyone tolerate their actions and beliefs; but, they have no tolerance for anyone else's ideas and beliefs. They define prejudice narrowly as prejudice against Gays and Atheists; and, they define discrimination narrowly in the same manner.

However, my prejudice and discrimination against terrorists, drug dealers, psychiatric drugs, alcoholism, indiscriminate sex, Fascists, and Militant Atheists could actually save my life.

I have already decided that homosexuality is not for me; and, I'm extremely grateful that I'm not so inclined. My homosexual friends have suffered greatly throughout their lives; and, all of that lost potential starts to wear on them as the years go by. Some of them have no children that they can call their own. Others have children, but they all have abandoned the spouse who gave them their children. It's tragic! It's heart-rending. All those broken promises and ruined lives!

Free will is to be celebrated, for sure; but, same-sex attraction has nothing else to recommend it in my humble opinion. All that lost potential is sad and should be pitied. I feel sorry for them. There's no need to persecute them because their curse carries within it its own natural and automatic punishment. Same-sex attraction and homosexual behavior is punishment enough. It carries within it the seeds of its own destruction.

Prejudice is an attitude; and, attitudes are chosen into existence by the Human Psyche. Feelings, chosen behaviors, and beliefs are chosen into existence by the Psyche, Mind, Spirit, or Soul. It has nothing to do with our genes or our evolutionary history. It has little to do with our reinforcement history – the Human Psyche can trump or override its nature and nurture at will.

Our beliefs are chosen into existence by the Human Psyche. The decision to act or not to act is chosen into existence by the Human Psyche. Chosen behaviors are chosen into existence by the Human Psyche. Our genes don't make our choices for us. They can't! However, other psyches or society can make choices for us. Either way, though, choice is solely a function and a product of the Human Psyche as well as all those other psyches. Only Psyche can do choice.

We continue to have feelings and perceptions long after our physical brain is dead and gone according to the scientific evidence that has been obtained from Near-Death Experiences (NDEs), Shared-Death Experiences (SDEs), Out-of-Body Experiences (OBEs), and our after-death Life Reviews. Feelings are a product of the Human Psyche.

Materialism, Naturalism, Darwinism, Nihilism, Behaviorism, Determinism, Physical Reductionism, and Atheism are prejudices! They are a vagrant opinion without any visible means of support. They were chosen into existence by the Human Psyche. Prejudice is typically defined as a negative attitude, a false belief, or an unproductive attitude. Materialism, Naturalism, Darwinism, and their derivatives certainly qualify, don't they! Attitudes are produced by the Human Psyche, not our genes. Our genes don't have attitudes. Evolution doesn't have an attitude.

Materialism, Naturalism, Darwinism, Nihilism, Behaviorism, Determinism, Physical Reductionism, and Atheism were purposefully and deliberately designed to resist and reject new information – any information or evidence that falsifies Materialism, Naturalism, and their derivatives such as the Theory of Evolution. These things are based exclusively upon a refusal to look at evidence because ALL of the evidence that we have falsifies them! They are resistant to new information or the new scientific discoveries that falsify them. These people have convinced themselves that Psyche or Syntropy does not exist. Self-deception works, and it works every time.

Furthermore, the Materialists, Naturalists, Darwinists, and Atheists have chosen to discriminate against those who don't believe in their philosophy, dogma, or religion. They set up a complex peer review system, academic system, censorship systems, mind-guards, evangelists, and social pressure systems in order to enforce their beliefs onto others. Their ultimate goal is to make everyone believe as they do. The purpose of our public schools is to indoctrinate us into Materialism, Naturalism, Darwinism, Nihilism, Behaviorism, Determinism, Physical Reductionism, and Atheism. The laws of the United States were redesigned to protect them in these endeavors and to promote their cause. That's where all the money goes.

Fascinating, is it not?

Well, I think it is because I used to be a Materialist, Naturalist, Nihilist, and Atheist. The scientific evidence or the observational evidence convinced me that I was wrong. Quantum Mechanics, Action at a Distance, Non-Locality, Psyche, and Syntropy convinced me that I was wrong.

You can't be right when you are wrong.

The verified and proven existence of Action at a Distance and Quantum Mechanics falsifies Materialism, Naturalism, and their derivatives such as Darwinism, Nihilism, and Atheism. Quantum Mechanics is Syntropy; and, Syntropy is powerful science to know and understand. Physical matter is based upon entropy. Everything else is Syntropy. Psyche is Syntropy.

## My Observations and Conclusions

One of the most interesting scientific discoveries or scientific observations that I have made is that Materialism, Naturalism, Darwinism, Nihilism, Behaviorism, Determinism, Physical Reductionism, and Atheism are prejudices! They are a vagrant opinion without any visible means of support. In the complete absence of physical evidence demonstrating and proving their case, these people have pre-judged that they are right and that everyone else is wrong. Without doing any actual Science, these people have jumped to the conclusion that the quantum, or the non-local, or the supernatural, or the non-physical does not exist. That's prejudice! That's pre-judgment. That's blind faith!

As our textbook says, even social psychologists and Christians have and experience preferences or prejudices. It seems to be a natural innate part of our being. We are constantly doing social comparisons, both consciously and unconsciously. It's there, and it's real. The only thing you can do about it is to be aware of it and actively resist it within yourself when deemed appropriate to do so. You can't force others to stop being prejudiced, so there is no use trying. The best you can do is to show them another example or another way, and then let them make up their own minds what they want to do and what they want to believe.

I'm 57 now. I don't think about my prejudices all that much like I used to. Some of them are newly developed and I actually want to keep them. Some of our prejudices can actually be good for us.

It's obvious that I have become prejudiced against Materialism, Naturalism, Darwinism, Nihilism, Behaviorism, Determinism, Physical Reductionism, Atheism, and the Theory of Evolution because I have successfully used Science, Scientific Observations, and the Scientific Method to falsify them. That was a hard-won prejudice or discrimination, and I hope to never overcome it. All I ever really wanted is to know the truth, and now I do.

I now define Science as observation and experience, NOT Materialism and Naturalism. I've upgraded my science to Science 2.0, which allows ALL of the evidence into evidence and pursues a preponderance of the evidence. You can too, if you want to.

Mark My Words

—

**Reference Material**

Myers, D. G. (2010). *Social Psychology* (10th ed.). New York: McGraw-Hill.

***Science 2.0: I Upgraded My Science***

https://www.amazon.com/dp/B0771K6WTX

# PART X — THE SCIENTIFIC BREAKTHROUGHS FROM SCIENCE 2.0

Large parts of this book consist of the metaphysical breakthroughs and spiritual breakthroughs that are possible from choosing to allow all of the evidence into evidence.

However, if you are like me, a scientist, then you are mostly going to be interested in the scientific breakthroughs that have come from Science 2.0. If it doesn't produce any useful fruits, then it's worthless. I didn't want Science 2.0 to be hollow words or simply a gimmick. I want it to produce new and useful Science. I want it to produce demonstrable science, observed science, experiential science, and verified science. I want it to produce scientific breakthroughs.

Well, here are a few of the scientific breakthroughs that came from Science 2.0 – especially from my decision to allow the massless, or the non-physical, or the quantum into evidence.

## A Physical Atom Is Physically Impossible

Science 2.0 allows all of the evidence into evidence and then pursues a preponderance of that evidence.

By allowing ALL of the evidence into evidence, we quickly run into the physically impossible.

Quantum Tunneling is physically impossible. Quantum Tunneling is purely a psychic phenomenon, non-local phenomenon, or non-physical phenomenon. It defies the rules of Classical Physics or Newtonian Physics.

I eventually discovered that Nature or Nature's Psyche can do the physically impossible at will. The Big Bang is physically impossible; yet, most of the scientists in this world believe that it happened. Inflation was faster-than-light, which means that Inflation if it happened was by definition physically impossible.

By studying the massless, the non-physical, and the physically impossible, we are able to develop logical mechanisms or logical physics for the physically impossible – for such things as the Big Bang Singularity, the Big Bang, the Law of Psyche, the Quantum Law of Thermodynamics, the Ultimate Law of Thermodynamics, Ultimate Causality, Quantum Tunneling or Teleportation, Omnipresent Quantum Fields, Infinite Velocity Quantum Waves or Thoughts, Quantum Omnipresence or Quantum Superposition, Instantaneous Action at a Distance, Non-Locality or Transdimensionality, Quantum Entanglement, Spacetime, Gravity, Magnetism, Dark Energy or Quantum Fields, Dark Matter or Spirit Matter, the Behavior of Massless Photons, Syntropy or the Conservation of Energy, the Massless Definition for Entropy, and Psyche or Intelligence or Consciousness or Life Force.

I eventually discovered that the existence of physical atoms is in fact physically impossible. Physical atoms can't be explained in terms of Newtonian Physics, Classical Physics, or Classical Realism. In order to explain how a physical atom really works, we must delve into the massless, the non-physical, the transdimensional, the quantum, and the physically impossible from which it was made. We must understand Quantum Mechanics, Quantum Field Theory, Energy, and Nature's Psyche from a massless, non-physical, quantum perspective if we truly want to know how a physical atom works.

The same reality and truths apply to a photon. A massless photon, by definition, is physically impossible. We must be willing to study a photon from a massless, non-physical, quantum perspective if we want to know how the thing truly works and what it does. Most scientists aren't willing to do that. I'm one of the first.

Most of the scientists in this world are Materialists, Naturalists, Darwinists, Nihilists, and Atheists; and, these people are deathly afraid of the non-physical. By definition, these people are afraid of consciousness or psyche. They are afraid of God. The way that these people deal with their fears is to define them out of existence. So, according to these people, the non-physical, the massless, the quantum, the spiritual, dark energy or quantum fields, quantum waves or thoughts, dark matter or spirit matter, massless photons, and consciousness or intelligence or psyche DO NOT EXIST. They deal with their fears by defining their fears out of existence; and then, they officially never think about them or study them ever again, except that subconsciously they are constantly thinking about them. An atheist seems to do a lot more talking about God and why He doesn't exist than a theist ever does.

In contrast, a theist's greatest experiences with God usually remain unwritten and unspoken, usually by the command of God Himself or through the power of forgetfulness. I have observed that most out-of-body explorers and near-death experiencers are NOT permitted to remember everything that they learned and experienced while in the presence of God the Father, or Jesus Christ, or the Being of Light and Love. Some of them report that they knew everything and understood everything, but that it's now gone, now that they are back inside their physical body. Physical matter has a way of dumbing us down and making our Psyche forget the things it used to know. That's just the way it works.

https://dailygalaxy.com/2017/08/life-after-death-renowned-physicists-says-its-quantum-information-stored-at-a-sub-atomic-level-that/

Quantum Information or Intelligence is stored in all of that "empty space" between the nucleus of an atom and its orbiting electrons. There's theoretically enough room there to store ALL of the knowledge and information in the universe at the quantum level in a massless or non-physical holographic format. The Holographic Theory states that each physical atom contains the sum-total of ALL the knowledge and information in our whole universe. Each physical atom is a miniature universe unto itself – the very pinnacle of science, engineering, order, organization, and technology.

Anyway, back to the topic at hand – the Science of the Physically Impossible.

By allowing ALL of the evidence into evidence, Science 2.0 makes the "Science of the Physically Impossible" possible. It's a new and better way of doing science.

I found it extremely interesting to observe and study all the different aspects about the physical atom that are physically impossible. They completely falsify Newtonian Physics or Classical Physics because they are in fact physically impossible.

Let's start first by looking at electrons and their orbital shells. The whole system is physically impossible.

https://en.wikipedia.org/wiki/Atomic_orbital

Think about it logically.

In every physical atom, we have one or more electrons moving faster than the speed-of-light forming a solid shell or force field around the nucleus of an atom so that a physical atom can become "solid" or "physical". That's physically impossible!

According to Classical Physics, Relativity, and Newtonian Physics, nothing can travel faster than the speed-of-light; yet, it is obvious that a single lone electron in orbit of a hydrogen atom is moving so fast at nearly an infinite velocity in a circular or elliptical orbit that it actually forms a solid shell around the nucleus of that atom. Violating the laws of Classical Physics is precisely how physical matter is made in the first place.

The existence of physical matter is physically impossible. It requires faster-than-light Quantum Mechanics, Energy Mechanics, Psyche Mechanics, and Quantum Field Theory to make physical matter in the first place. Our scientists would KNOW that if they had spent any time studying the non-physical or the quantum aspects of a physical atom.

The current quantum theories state that the electron is moving so fast that it actually forms a solid shell or a forcefield around the nucleus of an atom. Our scientists have also observed that an electron is also half-in and half-out. Half of the time an electron is physical and has mass or resistance to acceleration, and the other half of the time it is quantum, massless, and capable of infinite acceleration and infinite velocity. By constantly changing the direction of its spin, an electron spends half of its time as a physical particle and half of its time as a non-physical particle. Intelligent, is it not? Interesting, is it not? Half of the time, an electron is actually a photon.

The Higgs Mechanism Explained

https://www.youtube.com/watch?v=kixAljyfdqU

Electrons are moving so fast that it's theoretically possible that there's only one of them in the whole universe. If there is in fact only one electron in our universe, the thing is clearly moving at an infinite velocity, in complete violation of Classical Physics or the Theory of Relativity.

The One Electron Universe

https://www.youtube.com/watch?v=9dqtW9MslFk

The electrons are moving so fast (faster the speed-of-light half the time) that they should tear themselves away from an atom and set off in a straight line through space; yet, some invisible and powerful force is holding them in orbit of an atom. Electrons are super smart. They actually change their behavior and customize their behavior depending upon whether they are in open space, orbiting an atom alone, or orbiting an atom with other electrons. The intelligence there seems to be beyond anything that we humans are capable of achieving. A photon is infinitely more intelligent than we human beings are, because a photon has NO physical limitations; and we do.

https://en.wikipedia.org/wiki/Gluon

The gluons are also invisible force fields that hold the quarks together and form them into neutrons and protons. The gluons are extremely short-range tractor beams. Yet in complete defiance of logic and physics, the farther away the quarks get from each other, the stronger the force of the gluon or the stronger the resistance to acceleration. The force of the gluon seems to drop to zero when the quarks are near to each other or in the right position. In other words, there's nothing in a gluon that will actually make quarks merge into a solid substance. However, if the quarks try to get away from each other, the power of the invisible intangible non-physical gluon tractor beam seems to approach infinity in order to prevent it from happening. Invisible non-physical force fields, tractor beams, and repulsor beams – ALL of this is physically impossible! Yet, all of this exists within a physical atom.

Within an atom we see extreme intelligence, law, enforced order, replication, duplication, and a whole bunch of different invisible intangible non-physical forces and fields. Within a physical atom, we observe a lot of invisible, intangible, and non-physical Action at a Distance or Quantum Mechanics or Quantum Field Theory.

A physical atom is physically impossible because a physical atom is made up from a bunch of different non-physical forces and fields. Classical Physics, Newtonian Physics, the Theory of Relativity, Materialism, Naturalism, Darwinism, and Classical Realism DO NOT ALLOW for the existence of a physical atom because a physical atom is comprised exclusively of the physically impossible or the non-physical.

Isn't that interesting?

The science tells us that the existence of physical matter is physically impossible. Yet physical matter exists. The best explanation for the existence of physical matter is the obvious fact that Nature or Nature's Psyche can do the physically impossible at will.

Science 2.0 is an upgrade to science because Science 2.0 allows ALL of the evidence into evidence, and then tries to pursue a preponderance of that evidence. It's a NEW and BETTER way to do science. It's the way that science should have always been done – but wasn't.

Mark My Words

## Falsifying Creation by Chance

Materialism, Naturalism, Darwinism, Nihilism, Atheism, Chemical Evolution, Macro-Evolution, Spontaneous Generation, Abiogenesis, Creation by Death or by Natural Selection, Behaviorism, Determinism, Creation by Disorder, Creation by Chaos, and the Second Law of Thermodynamics ARE CREATION BY CHANCE. They are chance causality or caused by chance alone. They are the Null Hypothesis in any science experiment.

The whole purpose of science, the scientific methods, statistics, and a science experiment is to falsify the Null Hypothesis, Chance Causality, or Creation by Chance. Statisticians, Logicians, and True Scientists KNOW that CHANCE is an invalid causal mechanism. In a science experiment, if the statistical analysis of the data is telling us that the results were caused by chance alone, then we end the experiment by retaining the Null Hypothesis and CONCLUDE that the independent variable had NO EFFECT on the dependent variable. In other words, we conclude that there is NO CAUSAL RELATIONSHIP between the independent variable and the dependent variable. We reject the alternative hypothesis. Chance cannot be the cause of anything that exists, according to the rules of Statistics and Science. The purpose of a science experiment is to find the True Cause of an event. Chance can never be the True Cause of anything.

Yet, the Materialists, Naturalists, Darwinists, Nihilists, and Atheists *start with the conclusion* that CHANCE caused everything that exists. Therefore, these people use CHANCE as the hidden variable or the independent variable in ALL of their science experiments and in ALL of their interpretations of the scientific evidence. Whenever people use CHANCE as their independent variable, as the Materialists, Naturalists, Darwinists, and Atheists do, then WE KNOW automatically that their independent variable has NO causal relationship with any dependent variable that they might choose to study. They have chosen an invalid causal mechanism by choosing CHANCE as their only causal mechanism. We have no choice but to conclude that Materialism, Naturalism, Darwinism, Nihilism, Atheism, the Theory of Evolution, and the Second Law of Thermodynamics have NO causal relationship with anything because they are in fact Creation by Chance or the Null Hypothesis to begin with. Creation by Chance is automatically false.

Where Materialism, Naturalism, Darwinism, and their derivatives are concerned, they are automatically FALSE because they are Creation by Chance. Because they ARE the Null Hypothesis or Creation by Chance, ANY alternative hypothesis derived from them is automatically FALSE. There is NO causal relationship between Materialism, Naturalism, Darwinism, Nihilism, Atheism, the Theory of Evolution, and the Second Law of Thermodynamics and ANYTHING ELSE because these naturalistic and atheistic philosophies of science are CREATION BY CHANCE or the Null Hypothesis to begin with. Every science experiment was designed to falsify Creation by Chance or the Null Hypothesis, which means that every science experiment was designed to falsify Materialism, Naturalism, Darwinism, Nihilism, Atheism, Creation Ex Nihilo, Chemical Evolution, the Theory of Evolution, the Second Law of Thermodynamics, and every other form of CREATION BY CHANCE that exists. Creation by Chance is automatically false. That's powerful Science, isn't it?

Remember, Materialism, Naturalism, Darwinism, and their derivatives are automatically FALSE because they are Creation by Chance or the Null Hypothesis.

Mark My Words

## Spirit Matter or Dark Matter Exists below the Ground State

I'm a Quantum Theorist, and I have been having a lot of fun trying to figure out what Dark Matter is and how it works. I'm not satisfied that we have enough evidence to know what it is, yet. There also seems to be different types of Dark Matter or Spirit Matter or Exotic Matter, with different parameters, capabilities, purposes, and functions.

Quantum fields form a lattice or a matrix that serves as the foundation for the construction, transmission, and sustenance of quantum waves, quanta, particles, spirit matter or dark matter, and physical matter. Dark energy is a type of quantum field that helps to form our cosmic density.

Spirit matter or the stuff that the astrophysicists call "dark matter" exists below the ground state. It may have lost its electrons altogether, and only the nucleus remains.

https://en.wikipedia.org/wiki/Ground_state

It's obvious that dark matter exists and that it has mass; but, does it have electrons? Can dark matter have mass or resistance to acceleration without its electrons or when its electrons are located below the ground state? Or, does a lone proton with some neutrons and no electrons, or very low-level electrons, become a purely quantum object with the ability to phase-shift and to quantum tunnel or teleport galactic distances at will?

I'm leaning towards the belief that the lone proton still has mass or resistance to acceleration and is capable of exerting an invisible and intangible influence on physical matter from the invisible, non-physical Quantum Realm or Spirit World. A lone proton with mass but no electrons is probably the stuff that we call dark matter or spirit matter. The thing I don't know is if it can have electrons and still be classified as spirit matter or dark matter.

There are still unanswered questions; but, it is becoming clear to me and many other scientists that dark matter or the stuff that's holding our galaxies together is in fact spirit matter or intangible, non-physical, quantum matter. Spirit matter or dark matter is like Shepherd Moons. It silently, invisibly, and successfully holds the galaxies together, and keeps the physical matter on its proper pre-designated course.

After some thought, I eventually realized that there are many different types or levels or dimensions of spirit matter, according to the out-of-body explorers.

Massless, omnipresent, quantum, wave-like photons and tachyons are clearly particles of spirit matter. They function and act just like spirit matter does. A tachyon is simply a photon that has chosen to travel faster than the speed-of-light. A photon has limited itself to traveling at the speed-of-light so that it can transform or transition into our physical realm whenever it decides to do so. The trajectory of photons is bent or changed by gravity; but, they don't produce gravity because they are massless or non-physical.

Therefore, we KNOW that dark matter is a completely different type of spirit matter, or non-physical intangible invisible particle, than massless photons. Dark matter provides additional gravity because it has mass and can resist acceleration. Dark matter acts like physical matter, but it's intangible, invisible, and spiritual at the same time. Clearly, dark matter is a type of spirit matter, but it's the type that has mass and that acts like physical matter. In a very real sense, according to Quantum Mechanics, dark matter is phase-shifted physical matter. It's phase-shifted which means that it acts like physical matter, but it exists in a different phase, frequency, level, or dimension. Dark matter is phase-shifted physical matter, which means that dark matter is spirit matter; but, it's spirit matter that produces a gravitational effect.  Dark matter and spirit matter seem to be protons that either have NO electrons, or protons that have electrons well below the ground state.

It's all matter, including the photons.  It's just that the vast majority of it is spirit matter, or it exists in a different "non-physical" dimension.  Any quantum is a particle of matter, whether that quantum chooses to remain as massless omnipresent spirit matter or eventually chooses to stop and manifest itself at the physical level in the physical realm as heat, mass, and entropy.  Dark matter seems to have stopped, and it seems to have chosen a location in our space-time realm, which makes me think that dark matter is physical matter that has been phase-shifted into the spirit world and has become a type of spirit matter as well.

Dark matter functions as if it were physical matter.  Dark matter produces extra mass and therefore extra gravity to help hold our galaxies together and keep them from flying apart.  Every molecule of physical matter and dark matter in our galaxy is tethered to the center of our galaxy by an invisible thread that we call gravity, and the result is angular momentum of the physical matter and dark matter rather than straight-lined vectored momentum.

https://en.wikipedia.org/wiki/Dark_matter

Instead of traveling in a straight line for all eternity according to Newton's First Law of Motion, physical matter and dark matter actually travels in a curve and rotates around the center of its galaxy thanks to the concerted efforts of the gravity produced by both dark matter and the physical matter.  That's NOT a violation of the First Law of Motion, because Newton's First Law allows invisible non-physical forces to alter the trajectory of physical matter; but now we actually have a name for the stuff and the forces that keep physical matter from traveling in a straight line for all eternity and is making the physical matter rotate around the center of its galaxy instead.

Gravity isn't sufficient enough to keep the physical matter, planets, and stars from flying-off out into space.  That's how the astrophysicists were able to discover the presence of and measure the density of the invisible, non-physical stuff that they call dark matter.  That's how they KNOW that the dark matter is collected in haloes around galaxies.  It needs to be there to keep the galaxies from flying apart.  Otherwise, Newton's First Law would prevail, and each physical atom would fly off in a straight course from the very moment of its birth, and the physical galaxies would literally fly apart in all directions.  Remember, hydrogen molecules and helium atoms naturally repel each other; so, something invisible and non-physical has to be overcoming their natural tendency to separate and go their own way in a straight line for all eternity.

Gravity is a start; but, dark matter or mass-based spirit matter is the thing that's actually holding our galaxy together.  The Biblical God Jesus Christ has said that the Gods created everything spiritually before organizing it or creating it physically.  Now we know why they had to do that.  They had to do that in order to keep the galaxies from flying apart.  They built the whole galaxy from mass-based spirit matter, and then they phase-shifted, transmuted, or transformed less than 5% of that spirit matter and quantum field energy (dark energy) into physical matter.

Moses 3: 5, 7:

**4 And now, behold, I say unto you, that these are the generations of the heaven and of the earth, when they were created, in the day that I, the Lord God, made the heaven and the earth,**

**5 And every plant of the field before it was in the earth, and every herb of the field before it grew. For I, the Lord God, created all things, of which I have spoken, spiritually, before they were naturally upon the face of the earth. For I, the Lord God, had not caused it to rain upon the face of the earth. And I, the Lord God, had created all the children of men; and not yet a man to till the ground; for in heaven created I them; and there was not yet flesh upon the earth, neither in the water, neither in the air.**

**7 And I, the Lord God, formed man from the dust of the ground, and breathed into his nostrils the breath of life; and man became a living soul, the first flesh upon the earth, the first man also; nevertheless, all things were before created; but spiritually were they created and made according to my word.**

We KNOW from the revelations from the Biblical God Jesus Christ that the Gods created everything spiritually BEFORE they created it physically, and now we KNOW why.

The galaxies were created spiritually and non-locally BEFORE five percent of their cosmic density was transformed or phased instantaneously into physical matter. We now have plenty of evidence suggesting that galaxies are born all at once, and then they start to wind up from there. The spiral galaxies and barred spiral galaxies that we observe were clearly born all at once instantaneously, and they have only had enough time to rotate maybe once or twice since they were born. We can see it with our own eyes. They didn't slowly coalesce out of gas clouds over eons of times. Their physical presence was made instantaneously, and they were birthed all at once whole and functional. That's how they got their shape. That's the only way they could have gotten their shape.

They couldn't be spiral shaped or barred shaped spirals by coalescing out of expanding clouds of hydrogen and helium gas. Furthermore, hydrogen and helium gas do not coalesce. We can't even get the stuff to coalesce or fuse in a tokamak, so there's no way in the universe that the stuff is going to coalesce naturally in the ever-expanding expanse of space. Hydrogen molecules and helium gas naturally repel each other.

Stars and galaxies were deliberately designed and made, or they wouldn't exist in the first place. They are made at the spiritual level, and then transformed or birthed instantaneously into the physical level.

The elliptical galaxies and globular clusters aren't so clear. It's as if those galaxies are older and have actually had time enough to homogenize and lose their spiral structure. But they still rotate, and they still hold their stars in place around their center thanks to the shepherding influence of dark matter or spirit matter. NO dark matter then NO galaxies. All of their guts and gas would be instantaneously expanding out in all directions into empty space without the shepherding influence of dark matter.

This is what we have actually experienced and observed. It's obviously true, or the galaxies wouldn't be there; and, our universe would be nothing but one huge homogeneous ball of gas if it even existed at all.

Planets, stars, and galaxies are made spiritually BEFORE they are birthed, or transformed, or phase-shifted into physical matter.

<u>The Strangest Star in the Universe</u>

https://www.youtube.com/watch?v=XyuXBYWZegY

Mira is a bat out of hell. Mira and its dwarf companion were birthed or made 30,000 years ago all at once and sent in a deliberate trajectory counter to the natural rotation of our galaxy at nearly the speed-of-light when they were made. These two stars obviously didn't coalesce out of gas clouds "naturally". Mira is one of the Signatures of God. Mira was made to be that way as if God were saying, "Look what I can do." Only a God could make such a thing and then send it at a velocity relatively close to the speed-of-light counter to the natural rotation of its galaxy.

Physical matter, planets, stars, and galaxies are made anytime that Nature's Psyche, the Controlling Psyches, or God's Psyche chooses to make them.

In our day and age, thanks to the internet, it's impossible to remain in ignorance, although many people like me were determined to do so. Eventually, though, the preponderance of the evidence got through to me as well. I used to be a materialist, naturalist, nihilist, and atheist; but now, it's obvious to me that I was wrong.

Enoch Had a Vision of the Birth of Our Earth

https://ldssoul.com/the-birth-of-our-earth/

Planets, stars, and galaxies are made spiritually, and then they are phase-shifted or transformed whole into physical matter and obtain a physical presence in our physical spacetime realm. It's the same exact process that happens every time that an omnipresent photon chooses to stop and transform itself into heat, mass, and entropy instead. It happens with photons, and it happens with planets, stars, and galaxies as well. It's the very same mechanism. It's called Quantum Phase-Shifting for those who are interested in knowing what it is.

Thanks to Quantum Phase-Shifting and Quantum Superposition, dozens of different physical earths like ours can actually exist in the same space at the same time without crashing into each other or being detectable by each other, because they are out-of-phase with each other. The spirit world and our spirit earth are right here right now in the very same space at the very same time – just out of phase with our physical dimension or physical spacetime. We have NO idea how many other physical earths exist right here right now in the same space at the same time because they are all out of phase with each other and exist in different dimensions.

All of this is physically impossible. But Nature or Nature's Psyche can do the physically impossible at will.

Producing Dark Matter?

Randall Mills claims to have developed a process that transforms hydrogen into a dark matter hydrino below the ground state, producing a lot of energy, heat, and light as a result.

https://brilliantlightpower.com/

**ELECTRICAL POWER FROM HYDROGEN TO DARK MATTER.**

I keep waiting for them to make this thing commercially viable. Year after year, they postpone their release date. There are many problems with the process that they have kept hidden.

Finally, after decades, in their latest video they wrote, "Pushing the limit of the reactor melting down within seconds".

https://www.youtube.com/watch?v=UwjyO8H0E6U

They've only started to admit it, but the thing generates too much energy and heat way too fast and starts to melt the container, so they have to turn it off before it even gets going. They are starting to submerge the thing in water and use the steam to generate energy as well, like they do with fission reactions in a nuclear reactor; but, the thing still melts down or seems to stop working.

I've known all along that there has to be some reason why they aren't able to sell the thing. I suppose their custom-designed container starts to melt and disintegrate under all the pressure and heat. It's only good for a couple of tries, and then they have to start over again, with a brand-new container. Or maybe ultracold neutrons are disintegrating and transmuting the container. There's some reason why they aren't able to mass-produce and sell the thing; but, they only recently started to give reasons why. It works too well and produces a meltdown in seconds.

A couple of years ago, they just keep saying that "they have to refine the engineering and that takes time"; but, now they are starting to publish some of the results. They've been refining the engineering and the process for decades, and they still have nothing that they can sell to the public. They need actual commercial designs to make the thing viable; and, they are not telling us when that will happen, but it hasn't happened yet.

https://brilliantlightpower.com/plasma-video/

Anyway, they are having fun; but, I don't know if they are getting anything done. It looks promising, and they keep refining and improving the process, but they don't have anything to sell to the public.

**The SunCell® was invented and engineered to harness the clean energy source from the reaction of the hydrogen atoms of water molecules to form a non-polluting product, lower-energy state hydrogen called "Hydrino" that is the dark matter of the universe wherein the energy release of H2O to Hydrino and oxygen is 100 times that of an equivalent amount of high-octane gasoline at an unprecedented high power density.**

https://brilliantlightpower.com/suncell/

It's only been recently (around 2017) that they started to claim that their process is transforming physical matter into dark matter (spirit matter). But, that only seems to be what it's doing. I personally don't believe that they are converting hydrogen into dark matter. The stuff would just disappear into thin air if they were in fact doing that. That's not what's happening.

The problem is that Randall Mills never believed in Quantum Mechanics, which means that he has no idea what dark matter is. Dark matter is a type of spirit matter. Dark matter is phase-shifted physical matter. There are different types of spirit matter and different types of phase-shifted physical matter, and dark matter is one of them. Randall Mills doesn't know this, because he has rejected Quantum Mechanics; so, he really doesn't know what dark matter is. Consequently, to say that his new invention is producing dark matter is off the mark, because his process is producing tangible physical results instead.

The "hydrino" that they are producing clearly is NOT dark matter. Just look at this video.

https://www.youtube.com/watch?v=Epenv-PPLJM

The stuff they call a hydrino is visible, tangible, and magnetic. It's like a superconductor. That's definitely NOT dark matter. That's a new, physical, solid form of hydrogen that they are producing – assuming that it is made of hydrogen.

If it were truly dark matter, the hydrino particles would disappear into thin air and pass through the walls of the container leaving nothing behind.

They don't understand the physics, because they don't understand what dark matter is. Dark matter or spirit matter is invisible and intangible at the physical level. It's out-of-phase with the physical level. They error because they don't have a definition for dark matter and spirit matter that actually matches with the observational and experiential evidence.

Nobody realizes that a simple, massless, omnipresent photon is in fact a type of spirit matter or dark matter. We can ONLY see it and touch it and interact with it when it stops and transforms itself into heat, mass, and energy. When a photon is in its omnipresent, dark matter, or spirit matter format, the stuff is intangible and undetectable by our physical instruments. It's ONLY when it stops and transforms itself into heat, mass, and entropy that it shows up on our physical detectors.

Out-of-body explorers have observed that the spirit world or quantum world has multiple different levels, layers, phases, frequencies, or dimensions. The take-home message from these observations is that there are many different types of spirit matter or dark matter – not just one type. We have a periodic table for all the different types of physical matter, so why should dark matter or spirit matter be any different? There are going to be many different types of spirit matter or dark matter, once we start to figure out what the stuff really is.

https://www.youtube.com/watch?v=Epenv-PPLJM

In the video above, their hydrino is visible and is visibly interacting with the magnet. It's NOT dark matter. It's simply a different type of physical matter.

The part of their experiment that is in fact being transformed into "dark matter" or "spirit matter" is in fact the photons that they are making. Photons are simply one type of dark matter or spirit matter. There are many different types of spirit matter. A type of intangible spirit matter is made every time a photon is made.

The photons in their experiment are hitting the walls of the container and are being transformed into energy, heat, and light instead of dark matter – just like what happens in a hydrogen bomb but in a much more controlled manner.

What they have here is not the transformation of hydrogen into the dark matter of the universe. They instead have a form of hot fusion that is in fact creating a new type of solid, visible, magnetic hydrogen as a by-product of the fusion reaction.

They have in fact created Mr. Fusion, from the "Back to the Future" movies. They are doing hot fusion with plasma in a relatively inexpensive container, instead of doing it in a 100-million-dollar tokamak. It is hot fusion, and not dark matter production.

I don't think any dark matter proper is being produced at all. The closest they get to dark matter or spirit matter in this process is through the massless photons that their process is generating in great abundance; and, photons brake or stop for physical matter and absorb themselves into the physical matter by transforming themselves into heat, mass, and entropy when they encounter the physical matter.

I think that's what's really happening in their SunCells, and it has nothing to do with dark matter, unless of course you want to include the wave-like, omnipresent, massless photons that their process is producing as a type of dark matter. Photons in their wavelike omnipresent format are a type of spirit matter, or a particle of spirit matter, whether we realize it or not. The photon only becomes physical or mass-based when it stops and transforms itself into heat, mass, and entropy.

This is how things really work according to Quantum Mechanics and Quantum Field Theory. Massless particles like photons are omnipresent in the non-local realm, until they stop and transform themselves into heat, mass, and entropy.

The dark matter that's being detected in haloes around galaxies seems to be a type of mass-based physical matter that is simply out-of-phase with our physical reality. It's not tangible or visible because it is out-of-phase or in a different dimension of the same space; but, it seems to have mass like physical matter does. Dark matter is just a different type of spirit matter. There are many different types of spirit matter – or many different dimensions and frequencies and phases of spirit matter.

Quantum Superposition and Quantum Phase-Shifting allow many different types of matter to exist in the same space at the same time – just out-of-phase with each other. There could be dozens of our earths occupying the same space at the same time, each invisible to the other because they are out-of-phase with each other and exist in different dimensions at different frequencies. The stuff we call dark matter simply exists in one of those other dimensions, slightly out-of-phase with our visible tangible physical dimension, but still able to interact with our physical dimension.

The dark matter and spirit matter are right here, right now, inside you and all around you. It's intangible and invisible because it's out-of-phase with our physical reality; but, in its dimension, it's just as real and just as tangible as physical matter is in our physical reality.

Mark My Words

## The Conservation Laws Falsify the Second Law of Thermodynamics

Every Conservation Law that exists falsifies the second law of thermodynamics.

Why?

The Conservation of Energy states that the amount of energy cannot increase and that it cannot decrease. The Conservation of Energy states that energy is a substance that cannot be made and cannot be destroyed. It has always existed, and it will always exist.

The Conservation of Charge states that the amount of charge remains constant. All of the positive charges match equally with all of the negative charges.

In complete violation of the Conservation Laws, the second law of thermodynamics states that entropy is a substance whose amount is always increasing and whose amount can never decrease. The second law of thermodynamics ends up being Creation Ex Nihilo, or the constant creation of entropy from nothing. The different substance-based definitions for entropy and the second law, that define entropy as all of the energy in the universe or all of the heat in the universe, end up being falsified by the fact that they violate the Conservation of Energy. It's impossible for the amount of entropy to be constantly increasing because it violates all the different Laws of Conservation, including the Conservation of Energy and Psyche.

The Quantum Law of Information Conservation is the only conservation law that I'm aware of that actually allows for the amount of information to increase over time. At the quantum level, information storage or data storage is based upon the rearrangement of energy. In other words, the energy is reorganized or rearranged in such a way that it stores that information permanently. The Quantum Law of Information Conservation is NOT based upon some substance being created out of thin air. Instead, it is a rearrangement or a reordering of the energy that already exists.

In contrast, the second law of thermodynamics has a new substance that they call entropy being made out of thin air from nothing so that its amount can constantly be deceasing and so that its amount can never decrease. The second law is Creation Ex Nihilo; and, creation from nothing violates and falsifies all the different Conservation Laws that exist, and vice versa. If the Conservation of Energy is true, then Creation Ex Nihilo, the Second Law of Thermodynamics, and Atheism are automatically false.

Finally, when it comes to the "disorder" definition for entropy, the second law of thermodynamics has the amount of disorder in the universe constantly increasing and never decreasing. This "disorder" definition for the second law is in fact the Conservation of Entropy or the Conservation of Disorder. This definition for entropy is also flawed. It has been observed that information, or intelligence, or knowledge, or psyche eliminates entropy and disorder at will.

Mark My Words

## Entropy Is the Heat Storage Capacity of Mass

**S = mc**

Entropy is the heat storage capacity of mass.

Entropy is mass or resistance to acceleration.

This is the mass-based definition for entropy.

This is possibly the most interesting and useful definition for entropy. It's based upon the "heat transfer" definition of entropy, and it concerns the definition for heat and the equations for entropy that are based upon heat. I've never seen this definition for entropy before, so I'm the one who discovered it or created it. It's there in the equation for heat, and therefore in the equations for entropy that are based upon heat.

The mass-based definition for entropy is derived from the heat-based definitions for entropy, starting with the following unique equation and definition for entropy.

**S = ΔQ**

I call this the thermodynamics definition for entropy. Entropy is an exchange of heat from a hot reservoir to a cold reservoir. No physical reservoirs, then NO entropy. This is a powerful, demonstrable, and true definition for entropy! It defines entropy by what it really is and by what entropy truly does. Centuries of superstition simply disappear, once we have a true definition for entropy.

Entropy (**S**) is an exchange of heat (**ΔQ**) from a hot reservoir to a cold reservoir. No physical reservoirs, then NO exchange of heat and therefore NO entropy.

However, for a complete proof of this mass-based definition for entropy (**S = mc**), we need to use the complete equations for heat (**Q**) and entropy (**S**).

**Q = mcΔT**

**S = Q/ΔT**

**S = mcΔT/ΔT**

**S = mc**

**c** = the specific heat capacity of the mass under consideration

**m** = mass

**M = mc**

Entropy (**S**) equals mass (**M**), particularly the heat storage capacity of mass (**mc**). Entropy is mass's heat storage capacity (**S = mc**). This is the mass-based definition for entropy. No mass, then NO entropy.

**S = Q = M = mc**

In the heat (**Q**) equation, (**c**) is the specificity of heat constant for a given material or mass. This specific heat constant is basically the amount of heat that a specific type of mass can hold. Regardless of what happens to the temperature (**ΔT**), entropy (**S**) equals heat (**Q**) which basically equals mass (**M**) or resistance to acceleration. More specifically, we are talking about the heat storage capacity of the mass (**mc**). This is a more conservative definition for entropy. In fact, it's a demonstrably true definition for entropy. It has actually been experienced and observed. Entropy is mass's heat storage capacity (**S = mc**). This is the mass-based definition for entropy. It is a true definition for entropy. It is a conservative definition for entropy because it does not define entropy as all of the energy in the universe, or all of the heat in the universe, or all of the disorder in the universe.

What makes this definition for entropy interesting is the fact that NO mass means NO entropy. No mass (**m = 0**), then no heat (**Q = 0**). Whenever heat is zero (**Q = 0**), then entropy is equal to zero (**S = 0**). Consequently, according to the equations for entropy and heat, whenever the mass is zero, then the entropy is also zero.

**M = 0 = Q = S = mc**

When mass or resistance to acceleration equals zero, heat equals zero, which means that entropy has gone to zero and ceased to exist.

In other words, massless particles like photons and quantum waves have NO entropy according to this definition for entropy. Likewise, the infinite, absolute-zero temperature, massless quantum fields have NO entropy because they have no mass or resistance to acceleration. No mass or no resistance to acceleration, then no friction or heat, which means NO entropy. Here we finally have a scientific explanation for how the Big Bang could have happened or how entropic physical matter was made by Nature or Nature's Psyche from massless, heatless, and entropyless Quantum Fields.

**Remove the mass from a system, and that system's entropy goes to zero! As mass goes to zero, heat goes to zero, which means that entropy goes to zero. The way we remove entropy from a system is to remove its mass or resistance to acceleration. That's something that Nature or Nature's Psyche can do at will, according to the photon and the quantum wave. Nature or Nature's Psyche is creating or instantiating massless particles all the time; and, they have no entropy because they have no mass or no resistance to acceleration.**

**A miniature Big Bang happens every time a photon chooses to stop and enter into our mass-based physical realm. That photon goes from zero mass and zero heat to a complete stop, which then produces heat, mass, and entropy as a result of its deceleration, or resistance to acceleration, or transition into our physical realm. This is what we actually experience and observe. The creation or transformation is ongoing to this very day. Massless, heatless, and entropyless photons are stopping and transforming into mass and heat all the time, every day.**

If mass equals zero (**m = 0**), then heat equals zero (**Q = 0**), which means that the amount of entropy has gone to zero (**S = 0**). This is scientific and mathematical proof that massless non-physical objects such as photons, quantum waves (thoughts), virtual particles, quantum fields, and psyches have NO entropy, which means that their energies are eternal, everlasting, and always conserved. A photon is scientific proof that Nature or Nature's Psyche can convert massless and heatless energy into mass and heat anytime that it chooses to do so. The truth has always been there, but we have been too stubborn to see it because we don't want it to be true.

**The process is reversible; and, that makes it Good Science. $E = mc^2$. Whenever a particle goes massless or non-physical, it loses ALL of its entropy, and that entropy ceases to exist. Its mass, heat, and entropy are transformed into infinite acceleration instead. This is Quantum Field Theory. This is what we actually experience and observe. Whenever the mass (or resistance to**

acceleration) ceases to exist, the entropy ceases to exist along with it, according to the equations for heat and entropy.

This is an extremely powerful definition for entropy because of all the different truths that it reveals to us. Massless particles have no entropy and can have no entropy because they have no mass or resistance to acceleration.

My discovery, that the massless or the non-physical FALSIFIES the second law of thermodynamics, may just be the greatest scientific discovery of all time; and there it is, plain as day, within the equations for heat and entropy. No mass, then no entropy. The reason that nobody has ever discovered it before now is because nobody has been willing to look at physics from a non-physical quantum perspective before I came along and started doing so.

### Entropy Is Mass or Resistance to Acceleration

This is by far the most useful, most powerful, most truthful, and most interesting definition for entropy that I have encountered so far, because it takes into consideration everything that has ever been experienced and observed.

The second law of thermodynamics is based upon an *exclusion of evidence*. Everything massless and non-physical is excluded from consideration where the second law of thermodynamics is concerned, which is why it is easily falsified by the massless or the non-physical. *Deliberate exclusion of evidence* is a logic fallacy. The second law of thermodynamics is based upon a logic fallacy because it excludes evidence.

Furthermore, the second law of thermodynamics is based upon the *affirming the consequent* or the *jumping to conclusions* logic fallacy. They pick the conclusion that they want in advance, and then they deliberately exclude any evidence that falsifies it. Anytime they encounter entropy decreasing to zero and ceasing to exist, they label it as invalid and toss it out. They ONLY affirm or acknowledge the instances and the evidence where entropy is increasing. Anytime entropy goes to zero and ceases to exist, as it does within any particle that becomes massless, then they call that evidence and situation invalid; and, they toss it out and pretend that it doesn't exist, so that the second law of thermodynamics can seem to remain true.

**If you want the truth, it is important to allow ALL of the evidence into evidence – and that includes ALL of the times when entropy and mass have gone to zero and ceased to exist!**

**When mass or resistance to acceleration goes to zero, the amount of entropy goes to zero.**

I'm very excited about this particular definition for entropy. I've been searching for it for fifty years. The world has been searching for it for millennia. It explains everything that has ever been experienced and observed. I finally KNOW what entropy is; and now, we have complete scientific, mathematical, and observational proof that this particular definition for entropy is valid, verified, and true.

### Deriving the Mass-Based Definition for Entropy

Let's examine the heat-based equation for entropy, and then use it to derive the mass-based definition for entropy.

**S = Q/ΔT**

Entropy (**S**) equals heat (**Q**) divided by a change in temperature (**ΔT**).

**Q = mc ΔT**

Heat (**Q**) equals mass (**m**) times that mass's heat storage capacity (**c**) times the change in temperature (**ΔT**).

**S = mcΔT/ΔT**

Solving for entropy in terms of the heat storage capacity of the mass.

**S = mc**

Obviously, by definition, a massless photon has zero mass (**m = 0**), which means that it has no heat (**Q = 0**) and no entropy (**S = 0**). Since a massless particle of any kind has zero heat (**Q = 0**), that means that its temperature in Kelvin is equal to zero (**T = 0**), and its change in temperature is also equal to zero (**ΔT = 0**). In other words, it's at thermal equilibrium. Therefore, anything at thermal equilibrium (**ΔT = 0**) has NO entropy, because it's unable to produce and store any heat or entropy. We get this additional amazing truth from the equation for heat (**Q = mcΔT**).

Entropy becomes a fleeting thing, because it ONLY exists in the presence of mass and heat. No mass, then no entropy. No heat, then no entropy. No change in temperature, then no entropy. In fact, we quickly observe that entropy is constantly decreasing, going to zero, and ceasing to exist – which is the exact opposite of what the second law of thermodynamics claims.

In this equation for heat (**Q = mcΔT**), (**c**) is the specific heat value constant. Solving the equation for **c** when everything else has gone to zero, I suspect that when mass equals zero, and temperature is equal to zero, and heat is equal to zero, then the specific heat value of a photon or massless particle of any kind is also equal to zero (**c = 0**). In other words, a photon has NO heat storage capacity because it has NO mass or resistance to acceleration.

So, once again, we make a fascinating scientific discovery simply by choosing to look at science and entropy from a massless, non-physical, quantum perspective. We discover that the specific heat value for massless particles and photons is equal to zero (**m = 0 = c**). Massless particles have NO capacity for storing heat! It requires mass to store heat. No heat, then no entropy. No mass then no entropy.

A massless photon has NO entropy; and, when it reaches its chosen destination and encounters some mass or physical matter, it stops, and it converts its energy into heat. This is precisely what we experience and observe! This is sunshine in action! This is proven and verified science. A massless photon has NO entropy, NO mass, NO resistance to acceleration, and NO heat until it encounters physical matter; whereupon, it stops and converts itself into heat, mass, and entropy. Here is the scientific explanation for sunlight! Sunlight has NO entropy. Photons have no entropy. Cold, absolute-zero temperature, massless Photons convert into mass and heat when they stop moving or reach their chosen destination. This is how everything really works. It's hidden in plain sight where nobody can see it because they aren't looking for it.

A traveling photon is massless syntropy or massless "no-entropy", and it literally converts itself into heat, mass, resistance to acceleration, and entropy when it chooses to stop moving. There is the explanation for how the Big Bang was done. If the Big Bang really happened, it was a deliberate quantum transformation of massless syntropy, or conserved energy, or photons, or absolute-zero temperature light INTO heat, mass, resistance to acceleration, and entropy. Nature or Nature's Psyche is doing that ALL the time, not just during the Big Bang. Nature's Psyche is constantly converting massless photons into mass, heat, and entropy all the time; and, Nature's Psyche can reverse the process as well, at will.

According to this definition for entropy, a massless photon has NO ENTROPY because it has no mass or no resistance to acceleration. Anytime the mass or resistance to acceleration is removed from a system, ALL of the entropy is removed from the system. NO mass or resistance to acceleration, then NO heat. No heat, then NO entropy. It's really simple to understand. The truth always is.

This is the Quantum Law of Thermodynamics which states "no mass then no entropy". If there is no mass, then there can be NO heat or resistance to acceleration or friction, which means that there can be NO entropy. No mass and no heat mean no thermodynamics, which means NO second law of thermodynamics and NO entropy in the Non-Local, Non-Physical, Massless Quantum Realm, or Spirit World.

I think the "entropy equals mass" definition for entropy is the most interesting definition for entropy because of what it means at the quantum level, or the massless level, or the non-physical level. This definition for entropy gives us the quantum perspective on entropy; and, it ends up being the Quantum Law of Thermodynamics which states that there is no mass, no heat, no heat transfer, and therefore NO entropy at the quantum level in the Massless Realm, the Non-Local Realm, the Transdimensional Realm, the Non-Physical Realm, the Absolute-Zero Realm, the Syntropic Realm, the Conserved Realm, or the Quantum Realm.

Massless ($m = 0$) means NO entropy ($m = 0 = Q = S$); and, this ends up being the quantum definition for entropy. There is no resistance to acceleration at the quantum level in the massless Quantum Realm, which means that there is NO entropy at the quantum level in the Syntropic Realm.

**Entropy is mass or resistance to acceleration. Entropy is the heat storage capacity of mass.**

This is the best definition for entropy because it allows ALL of the evidence into evidence; and consequently, it successfully explains everything that has ever been experienced and observed. This is the best definition for entropy, because it's really simple to understand. No mass, then no entropy. Simple. Logical. Obvious. True.

This definition for entropy has already been verified and proven true.

I like this definition for entropy because it reveals the truth to us that there is no entropy and can be no entropy in the Massless Realm or Non-Physical Realm among the quantum fields. This definition for entropy ends up producing the Quantum Law of Thermodynamics which states that there is no mass, no heat, no heat transfer, no thermodynamics, and therefore no entropy in the Massless Realm or the Syntropic Realm or the Conserved Realm or the Quantum Realm. This might in fact be my greatest scientific discovery of all time. I've been studying entropy for fifty years trying to figure out what it is, and this is the culmination of all my effort and research. I figured out what happens to entropy when it goes massless, or goes quantum and non-physical. It ceases to exist. It falsifies the second law of thermodynamics. It makes the Big Bang a realistic possibility. It makes the creation of stars and galaxies by Nature's Psyche – from absolute-zero temperature, non-local, non-physical energies and quantum fields – the best explanation for how the creation was actually done.

By studying Syntropy, the Conservation of Energy and Psyche, and the different equations for heat and entropy, I was finally able to MAP entropy or mass onto the quantum realm from whence it originally came. Heat, and mass, and entropy don't exist at the quantum level; and, they are made to exist at the physical level by Nature or Nature's Psyche, whenever Nature's Psyche decides to make them, according to the Big Bang Theory. Nature or Nature's Psyche can make mass, heat, and physical matter out of absolute-zero temperature quantum fields and energies, anytime Nature's Psyche chooses to do so; and, there's nothing we can do to stop it or to make it happen. It's completely out of our control, and totally in the control of Nature or Nature's Psyche.

The Law of Psyche states that every psyche or intelligence has a certain amount of energy that's under its control; and, that controlling psyche can form or transform the energy under its control into anything that it wants that energy to be anytime that it chooses to do so – including mass, heat, or physical matter.

This reality and truth explains everything that has ever been experienced and observed.

The Law of Psyche combined with the Quantum Law of Thermodynamics even provide a realistic and logical mechanism for the creation of stars and galaxies out in the absolute-zero temperatures of spacetime, among the expanding clouds of hydrogen and helium gas. It provides a mechanism for the Big Bang too, if that's the way that the Gods chose to do it.

https://ldssoul.com/the-birth-of-our-earth/

The account of our earth's construction, at the preceding link, provides a different version of the organization of our earth; and, it actually becomes a realistic possibility once we know about and accept the Law of Psyche and the Quantum Law of Thermodynamics. Nature, Nature's Psyche, and the Gods can make heat, mass, and physical matter from massless dark energy or massless quantum fields anytime they choose to do so; and, there's nothing we can do to stop them.

Entropy is mass or resistance to acceleration. This is the best definition for entropy that I have found so far. It's already been proven true by science, observations, logical common-sense, and math. It's been there all the way along waiting for someone to find it; and, the only reason nobody has found it until now is because they have chosen to believe in the second law of thermodynamics instead.

This definition for entropy falsifies the second law of thermodynamics and supports the Big Bang Theory instead. The Materialists, Naturalists, Darwinists, Nihilists, and Atheists aren't going to like that because this definition for entropy falsifies their beliefs that entropy, heat death, and death are conserved. In this definition for entropy, all of the associated entropy ceases to exist whenever a particle goes massless or non-physical. The energy is actually used or transformed to make the particle massless and non-physical.

This is not going to sit well with the Materialists and Naturalists and Darwinists, because these people preach, teach, and believe that the non-physical or the massless does not exist. This definition for entropy falsifies their beliefs and proves them wrong.

Mark My Words

## The Philosophy of Science Falsifies the Second Law of Thermodynamics

So, what do you think so far?

Does science need an upgrade to Science 2.0, or is science fine where it currently is – back in the dark ages with Newton, Galileo, Classical Physics, Materialism, Naturalism, and Darwinism?

In these introductory essays, I'm trying to demonstrate why science needs to be upgraded to Science 2.0, and why science needs to start studying the massless and the non-physical – or what we typically call the quantum, the spiritual, and the transdimensional.

So, why is Science 2.0 a NEW and better way to do science? Why is Science 2.0 a better Philosophy of Science?

Well, Science 2.0 allows ALL of the evidence into evidence, and then it pursues a preponderance of that evidence; and, that's precisely what's wrong with the second law of thermodynamics. The second law doesn't allow all of the evidence into evidence. Instead, the second law was deliberately and carefully designed to exclude evidence from consideration, particularly the evidence that falsifies it; and, the second law was deliberately designed so that it can't be falsified. They don't allow the second law to be falsified.

The second law of thermodynamics is based exclusively upon a few egregious logic fallacies, and they don't even know it, because they aren't looking for it. Instead, they ALL take it on blind faith that the second law of thermodynamics is axiomatically true. They turned the second law of thermodynamics into an axiom or a law so that it is exempt from falsification. Nobody is trying to falsify it. Nobody is trying to science it. That's why the second law is loaded with a bunch of hidden logic fallacies that nobody can see nor understand. They don't see it because they don't want to see it and aren't looking for it.

I'm a logician, mathematician, statistician, and a philosopher as well as being a scientist, physicist, and a quantum theorist. I look for these kinds of things. I see trends. I identify equations, and definitions, and theories that were designed to trick us and deceive us. I employ critical thinking because I have trained myself to do so. I've worked hard to master the Philosophy of Science and to understand what it is and how it's supposed to work. Science 2.0 is based upon the Philosophy of Science, and its proper application.

Careful examination reveals that the second law of thermodynamics is falsified by the different equations for entropy. I describe that reality in great detail within my book, "Quantum Mechanics from a Non-Physical Spiritual Perspective", so I won't do that here. All that suffices for my purposes today is to demonstrate why the second law of thermodynamics is false.

The second law of thermodynamics is an exclusionary science. That means that it deliberately excludes everything that falsifies it. That's the very definition of Bad Science or Biased Science; and, they don't even know it, because they aren't looking for it and refuse to think about it. They don't know it because they don't know anything about the Philosophy of Science. Self-deception works, and it works every time, especially when it comes to a genius level intelligence. *Exclusion of evidence* or *rejection of evidence* is a logic fallacy. These people make these kinds of logic errors because they have never studied nor thought about the Philosophy of Science.

By setting entropy (**S**) equal to ALL of the energy (**J**) in the universe (**S = J**), it makes it impossible to determine what entropy is not. That's a *category error* logic fallacy.

It's a *conflation error*. It makes it impossible to ascertain what entropy is and what entropy is not. Science desperately needs a better definition for entropy!

Nobody on the planet can see what's wrong with the second law of thermodynamics; but, I can see it because I have studied and worked hard attempting to master the Philosophy of Science. I've asked myself why the second law of thermodynamics is false. In other words, I have tried to falsify it like a good scientist should, and I have succeeded in falsifying the second law of thermodynamics in many different ways. The second law is demonstrably false. The second law of thermodynamics has been falsified, and nobody knows it, because nobody wants it.

Well, I'm no longer a nobody! One day I started finding different ways to falsify the second law of thermodynamics; and, I continue to do so to this very day. The second law is easy to falsify because the second law is incomplete and because the second law is demonstrably false. The second law is easy to falsify because it deliberately excludes evidence from being entered into evidence. Just find that evidence, and that evidence falsifies the second law of thermodynamics, clearly and conclusively.

So, what's wrong with the second law of thermodynamics?

The main problem with entropy and the second law of thermodynamics is that there is NO way to accurately measure the amount of entropy, due to the wide variety of conflicting and contradictory definitions for entropy. There are literally dozens of different definitions for entropy, and they actually falsify each other. They deliberately made the second law *too good to be true*, which ended up making it self-defeating; and, they don't even know it.

By setting entropy equal to all of the heat in the universe, or all of the energy in the universe, or all of the disorder in the universe, it becomes impossible to measure entropy and impossible to differentiate entropy from everything else. By defining entropy as "everything in the universe", it becomes impossible to determine what entropy is not. They did it that way on purpose in order to trick us and deceive us into believing that the second law of thermodynamics is true.

When it comes to most of the definitions for entropy, we can't measure entropy because the stuff isn't physically detectable, because the stuff is defined as being supernatural, because the stuff is defined axiomatically as being infinite and as being comprised of everything in the universe, and because the stuff has no definition since it has been defined as being "everything in the universe". The official definitions for entropy are worthless because they aren't measurable and because they aren't detectable, because they are infinite and supernatural to begin with.

Entropy has NO definition. Consequently, they make up a bunch of statistical and probability definitions for entropy that have nothing to do with reality and aren't really detectable or measurable anyway. Therefore, in our college physics books, they equate the amount of entropy with the number of heads and tails that show up in a set or a system, because entropy has already been defined axiomatically as "everything in the universe" including the number of heads and tails in the universe. In our college textbooks, entropy no longer has anything to do with thermodynamics. It's all about probabilities and statistics instead. In other words, entropy is about heads and tails instead of being about thermodynamics, when it comes to our college physics textbooks.

What science desperately needs is a definition for entropy that is actually detectable, measurable, definable, conservative, and has actually been experienced and observed. I discovered two; and, I introduce them in the following essays, so I won't do so here, except indirectly.

So, what's wrong with the second law of thermodynamics?

It's rigged.

It deliberately excludes all the evidence that falsifies it.

In ALL of their thought experiments into entropy, they start with a hot reservoir and a cold reservoir FOR FREE as a given. They put them into contact with each other so that there can be some entropy or a transfer of heat. NO hot reservoir, then NO entropy.

They always start with a hot reservoir as a given. They produce that hot reservoir for free as a given out of thin air ex nihilo. They never once realize that it required a massive reversal of entropy to make that hot reservoir in the first place.

They start their thought experiment into entropy with the Big Bang Singularity, and they start by defining that Big Bang Singularity as infinite heat and infinite temperature. They start with a hot reservoir as a given, and that hot reservoir is made out of thin air from nothing. They NEVER once realize that it took an infinite reversal of entropy or an infinite amount of syntropy to make that infinite heat and infinite temperature Big Bang Singularity in the first place.

Once you have the truth, though, you no longer need the Big Bang Singularity to start with infinite heat and infinite temperature. Once you have the truth, you no longer need the Big Bang Singularity.

So, what is this truth that they are missing and overlooking?

The Big Bang Singularity, if it ever existed, was cold. It was absolute zero temperature to begin with. It had no heat. In other words, their hypothetical Big Bang Singularity was MADE from heatless, massless, entropyless exergy or syntropic energy. If their Big Bang Singularity had infinite heat and infinite temperature, that infinite heat and infinite temperature was literally made from absolute zero temperature, massless, heatless, and entropyless ENERGY in the greatest reversal of entropy or the greatest source of syntropy in the whole of history.

They ALWAYS start with the hot reservoir for free as a given; and, they never once realize that it took an infinite reversal of entropy, or an infinite amount of negative-entropy, to make that hot reservoir in the first place. The hot reservoir is ALWAYS MADE by converting or transforming massless and heatless and entropyless absolute zero temperature exergy, syntropy, or energy INTO mass and heat and resistance to acceleration (entropy) instead.

Once we realize that the hypothetical infinite temperature and infinite heat Big Bang Singularity was made from absolute zero temperature exergy or syntropy through a HUGE and INFINTE reversal of entropy, then all sorts of new possibilities start to open up.

Some of us begin to realize that the Gods, Nature's Psyche, or the Controlling Psyches can literally TRANSFORM massless, entropyless, and heatless absolute zero temperature exergy or syntropy or quantum fields INTO mass, heat, and entropy AT WILL, anytime and anywhere.

There really wasn't a Big Bang. There really wasn't an infinite temperature Big Bang Singularity. There didn't need to be, because the "Big Bang" or the creation is ongoing to this very day. Anytime that the Gods need some mass, heat, or entropy (resistance to acceleration), they can simply TRANSFORM absolute zero temperature exergy, syntropy, energy, or quantum fields into mass, heat, and entropy AT WILL right here right now.

The "Big Bang" took place when the Gods, Nature's Psyche, or the Controlling Psyches organized the Quantum Fields; and, what do we KNOW about the Quantum Fields?

The Quantum Fields are massless, entropyless, and heatless absolute zero temperature exergy or syntropic energy. The "Big Bang" was actually cold – it was absolute zero temperature, and it's still that way today. The "Big Bang" was the organization of the Quantum Fields. It was ALL done at absolute zero temperature. Quantum Mechanics functions optimally at absolute zero Kelvin. Quantum Mechanics doesn't need heat to work. That's the KEY!

In their wave-like Quantum Format, photons are massless, heatless, and entropyless. They are pure syntropy. They are capable of infinite acceleration because they have NO mass or NO resistance to acceleration. ONLY when the photons choose to stop – ONLY when the photons choose to become part of our physical reality – do they in fact TRANSFORM themselves from infinite acceleration INTO mass, resistance to acceleration, friction, heat, and entropy instead. A miniature big bang takes place every time a photon chooses to stop and chooses to transform itself into mass and heat. You feel the new-made heat every time that a photon stops and lands on your skin. That's brand-new heat that didn't exist a second before. A second before, that photon was massless, heatless, and entropyless. A second before, that massless and heatless and entropyless photon was capable of passing through glass, water, and the earth as if they weren't even there. The mass and the heat were made the very moment that the photon chose to stop.

The hot reservoir, that they ALL start with as a given in ALL of their thought experiments into entropy, was made the very moment that Nature or Nature's Psyche decided to TRANSFORM absolute zero temperature photons, or quantum fields, or exergy, or syntropy INTO mass, heat, resistance to acceleration, and entropy. Some of that "Big Bang Heat" was made right now, right where you sit, as the photons chose to stop and chose to register on your skin and your eyes.

The heat WAS NOT MADE all at once at the beginning and stuffed into some kind of Big Bang Singularity. The making of the heat and the making of the hot reservoir is ongoing and happening right now as Nature's Psyche chooses to TRANSFORM small parts of the quantum fields from heatless, massless, and entropyless syntropy INTO mass, heat, and physical matter instead. The making of the heat and the making of the hot reservoir is ongoing and happening right now as photons or quantum waves CHOOSE TO STOP and land on your skin rather passing through your body as if you weren't even there.

The "Big Bang" is happening right now every time the Gods, Nature, Nature's Psyche, or the Controlling Psyches CHOOSE TO TRANSFORM massless, heatless, and entropyless absolute zero temperature syntropy or exergy INTO mass, heat, physical matter, and entropy instead. A miniature big bang happens every time a photon chooses to stop and chooses to TRANSFORM its infinite acceleration INTO mass, resistance to acceleration, and heat instead.

We don't need to start with an infinite heat and infinite temperature Big Bang Singularity for free as a given so that we can have some entropy. ALL we need to start with is the absolute zero temperature, massless, heatless, and entropyless Quantum Fields. Once the Quantum Fields are made and are in place, then Nature or Nature's Psyche can make mass, heat, and entropy (resistance to acceleration) from the Quantum Fields AT WILL. The Quantum Fields are necessary to sustain the existence of the mass, heat, and resistance to acceleration (entropy); but, heat and mass can be made AT WILL from absolute zero temperature energy or syntropy anytime and anywhere that Nature's Psyche chooses to do so.

This is what Quantum Mechanics and Quantum Field Theory are trying to tell us; but, nobody's listening.

We don't have to start with that infinite heat and infinite temperature hot reservoir for free as a given. ALL we need is energy and the quantum fields; and, once all of that is in place, then Nature's Psyche can make heat, mass, entropy (resistance to acceleration), planets, stars, and galaxies AT WILL from the massless, heatless, and entropyless Quantum Fields anywhere and anytime that Nature's Psyche chooses to do so.

We don't need the Big Bang nor the Big Bang Singularity, because a miniature big bang takes place every time that Nature or Nature's Psyche TRANSFORMS heatless, massless, and entropyless photons or quantum fields INTO mass, heat, physical matter, resistance to acceleration, and entropy instead. The Big Bang or the Creation is ongoing and happening all around you right now as Nature or Nature's Psyche continues to TRANSFORM massless and heatless photons and quantum fields into mass and heat whenever and wherever it desires to do so.

This ongoing and omnipresent TRANSFORMATION shows up in the cosmic microwave background radiation whenever and wherever Nature, Nature's Psyche, the Controlling Psyches, or the Gods CHOSE to transform some massless and heatless photons or quantum fields INTO mass, heat, and physical matter instead.

WE KNOW that this is true, because each one of us has experienced it and observed it whenever massless, heatless, and entropyless quantum waves or photons choose to stop and register on our skin as heat, and mass, and friction, and resistance to acceleration instead of passing through us as if we weren't even there. We each experience a miniature big bang every time that massless and heatless and entropyless quantum waves CHOOSE TO TRANSFORM themselves into mass, heat, and resistance to acceleration instead. We feel the heat on our skin whenever this takes place; and, we know that it is true.

Furthermore, through the process that we call entropy – through the transfer of heat from a hot reservoir to a cold reservoir – any heat that Nature's Psyche makes slowly makes its way back into the infinite, massless, entropyless, heatless, absolute zero temperature Quantum Reservoir or Quantum Fields from whence that heat originally came. This, too, we have experienced and observed.

The second law of thermodynamics can't explain any of this, because the second law of thermodynamics has been deliberately limited exclusively to the physical realm. Everything to do with the Quantum Realm, the massless and the non-physical, has been deliberately excluded from the second law of thermodynamics. The second law of thermodynamics has been deliberately crippled; and therefore, it lacks explanatory power where the massless, the entropyless, the heatless, the syntropic, or the quantum are concerned. The second law of thermodynamics is in desperate need of an upgrade to the massless, heatless, and entropyless Quantum Era in which we currently live.

The second law is classical physics or Newtonian Physics, and it needs to be updated so that it actually matches with Quantum Mechanics and Quantum Field Theory instead.

So, what's wrong with the second law of thermodynamics?

The Gibbs free energy equations allow for the existence of negative entropy; and, they actually allow negative entropy or syntropy to do work and to make things. The Gibbs free energy equations falsify the second law of thermodynamics.

https://www.khanacademy.org/science/chemistry/thermodynamics-chemistry/entropy-chemistry-sal/v/chem20-entropy

Although he sticks with the party line and even though he uses the standardized "disorder" definition for entropy, in this video, using chemistry, he falsifies the second law of thermodynamics, and he doesn't even know it. The Gibbs free energy equations falsify the second law of thermodynamics. The photons and the quantum fields and quantum waves falsify the second law of thermodynamics.

https://www.khanacademy.org/science/chemistry/thermodynamics-chemistry/gibbs-free-energy/v/gibbs-free-energy-and-spontaneity

Every time there is a reversal of entropy, the second law of thermodynamics is falsified. The Big Bang Theory falsifies the second law of thermodynamics. Every time there is negative entropy, the second law of thermodynamics is falsified. The Gibbs free energy equations allow for negative entropy; and, they actually allow that "negative entropy" to make things, to make order, and to do work. Of course, whenever the entropy is actually making things or doing work, they rename it heat or work or some such, so that it is no longer entropy in their minds. But, it's the same difference. Entropy can make things, and do work, and actually reduce heat whenever Nature or Nature's Psyche is reversing it or transforming it from entropy into something else instead. Entropy is just a different type of heat.

The second law of thermodynamics is a scam. It's a kludge. It deliberately excludes ALL the evidence that falsifies it. The second law of thermodynamics is based exclusively upon the *exclusion of evidence* logic fallacy. It excludes all the evidence that falsifies it; and, that's how it is able to survive.

### $dS \geq 0$

There it is, the great law – the second law of thermodynamics – the law that controls the destiny of our universe. Entropy (**S**) and change in entropy (**dS**) are always greater than or equal to zero. *Any process that reduces the entropy of the universe is forbidden.* In other words, the second law of thermodynamics deliberately excludes ALL of the evidence that falsifies it. It is forbidden to falsify the second law of thermodynamics.

Yet, the second law of thermodynamics is FALSIFIED by $E = mc^2$, every time that Nature, Nature's Psyche, God, or the Controlling Psyches CHOOSE to transform physical matter, mass, heat, or entropy back INTO massless, heatless, and entropyless exergy, syntropy, quantum waves, and quantum fields instead. Every time that Nature TRANSFORMS mass, heat, or entropy of any kind INTO massless, heatless, and entropyless photons, the second law of thermodynamics is FALSIFIED!

The second law of thermodynamics is easily falsified by the massless, the heatless, and the entropyless – which happen to be photons, quantum fields, quantum waves, and quantum mechanisms in general. Every time a massless and entropyless photon is made from mass, heat, or entropy, it falsifies the second law of thermodynamics. This is happening all the time; but, it is automatically excluded from consideration because it falsifies the second law of thermodynamics.

So, what's wrong with the second law of thermodynamics?

One of my many physics books defines entropy as "irreversibility".

This was the common definition for entropy back in the 1960's and 1970's, thanks to the film strips that we had in our public schools back then. The science teacher would run the film backwards, the kid would come flying up out of the pool, land on the diving board, and start running backwards. Then the science teacher would say that entropy prevents us from running the film backwards in real life.

That's simple enough to understand, but it's wrong.

It has absolutely nothing to do with thermodynamics or the transfer of heat from a hot system to a cold system. The reason we can't run the film backwards in real life is due to the arrow of time and has nothing to do with entropy or thermodynamics. This is yet another *category error* logic fallacy, where they equate entropy with the arrow of time as if they were the same thing. It just adds another layer of confusion to the thing that we call entropy or the second law of thermodynamics. The whole thing was designed to trick us and deceive us, and we don't even know it, because most of us want to be tricked and deceived.

Ironically, the whole thermodynamics process is completely reversible all the time according to $E = mc^2$, the Law of Conservation of Energy and Psyche, and the Quantum Law of Information Conservation. In other words, Nature or Nature's Psyche can TRANSFORM mass, heat, and entropy back INTO massless, heatless, and entropyless tachyons, photons, quantum waves, or quantum fields anytime and anywhere that it chooses to do so, according to $E = mc^2$. This is what Quantum Mechanics and Quantum Field Theory are all about!

$E = mc^2$ falsifies the second law of thermodynamics by making the whole thermodynamics process reversible by Nature or Nature's Psyche anytime and anywhere that Nature chooses to reverse the process. In other words, Nature can TRANSFORM our earth or our sun instantly into raw exergy or quantum fields, anytime and anywhere that it decides to do so. The only reason it doesn't happen is because Nature's Psyche or the Gods or the Controlling Psyches choose not to do so, for now. According to $E = mc^2$ and Quantum Field Theory, Nature or Nature's Psyche can TRANSFORM the mass and entropy of a black hole instantly INTO massless, heatless, and entropyless photons or quantum fields or quantum waves anytime it chooses to do so; and, there's nothing we can do to stop it, or to make it happen.

We have NO control here! God or Nature has all of the control. We have none! Humbling, is it not?

Nature or Nature's Psyche made the mass, heat, physical matter, planets, stars, and galaxies from the Quantum Fields in the first place; and, Nature or Nature's Psyche can TRANSFORM all of that mass, heat, and entropy back INTO massless, heatless, and entropyless Quantum Fields anytime and anywhere that it chooses to do so. Nature is doing this ALL the time.

The question we should be asking is why Nature or Nature's Psyche isn't doing these kinds of TRANSFORMATIONS at a larger scale with something like our earth, our sun, or our galaxy. God knows! Literally! According to different groups of Christians, Jews, Hindus, and Muslims, God is literally preventing Nature's Psyche from dissolving our physical bodies, our earth, our sun, our galaxy, and our physical universe back INTO the quantum fields or the chaos from which they originally came. Nature could do so anytime it chooses, so the theory goes that some type of God is encouraging Nature not to do so, for now. According to this theory, if God were to cease to be God, our physical universe would cease to exist instantly and be dissolved back into the Quantum Fields or even further back into the Chaos from whence it originally came.

The Big Bang Theory, if true, represents a HUGE reversal of entropy or a HUGE amount of exergy or syntropy. Anytime we encounter an $E = mc^2$ situation where mass, heat, and entropy are being converted or transformed back into massless, heatless, and entropyless Quantum Fields or Photons, we have in fact falsified the second law of thermodynamics, and its claim that entropy or thermodynamics cannot be reversed.

According to Quantum Field Theory, particles are born, and particles die. This is the primary axiom of Quantum Field Theory, and it falsifies the second law of thermodynamics. In other words, Nature or Nature's Psyche is constantly making and unmaking particles all the time, including particles of mass or particles of physical matter. Every time in Quantum Field Theory in the Feynman Diagrams where you see an electron TRANSFORMED into a photon or a virtual particle (quantum wave), you have in fact witnessed a falsification of the second law of thermodynamics. Falsification of the second law of thermodynamics is right there on our Feynman Diagrams all the time, and we don't even know it because we aren't looking for it and don't want to find it.

https://en.wikipedia.org/wiki/Feynman_diagram

Mass is resistance to acceleration or entropy. Mass is also heat storage capacity. NO mass, then no resistance to acceleration, no heat storage capacity, and NO entropy. It's that simple.

Every time that Nature or Nature's Psyche TRANSFORMS mass of any kind back into massless, heatless, and entropyless photons or quantum fields or quantum waves, we have in fact witnessed a falsification of the second law of thermodynamics.

You see, all of the entropy completely ceases to exist every time that mass is TRANSFORMED into a massless, heatless, and entropyless photon, quantum wave (virtual particle), or quantum field because the heat storage capacity of mass ceases to exist, which means that there's no place where entropy or heat can be stored within a massless photon or quantum wave. Massless, heatless, and entropyless photons, quantum waves, and quantum fields can have NO entropy, because they have no place where they can store that heat or entropy because they have NO mass or NO heat storage capacity. It's that simple.

Quantum Field Theory and the Feynman Diagrams are constantly falsifying the second law of thermodynamics, and we don't even know it because we have axiomatically defined the second law of thermodynamics as unfalsifiable. We don't allow the second law to be falsified, which is why nobody has ever figured out how to falsify it. But, it is really easy to use observed and experienced truths to falsify the second law of thermodynamics once a person finally chooses to do so. The problem is that nobody has ever chosen to do so until now.

## Quantum Field Theory Falsifies the Second Law of Thermodynamics

So, what's wrong with the second law of thermodynamics?

This Feynman Diagram falsifies the second law of thermodynamics, and we don't even know it because we aren't looking for it and don't want it to be true.

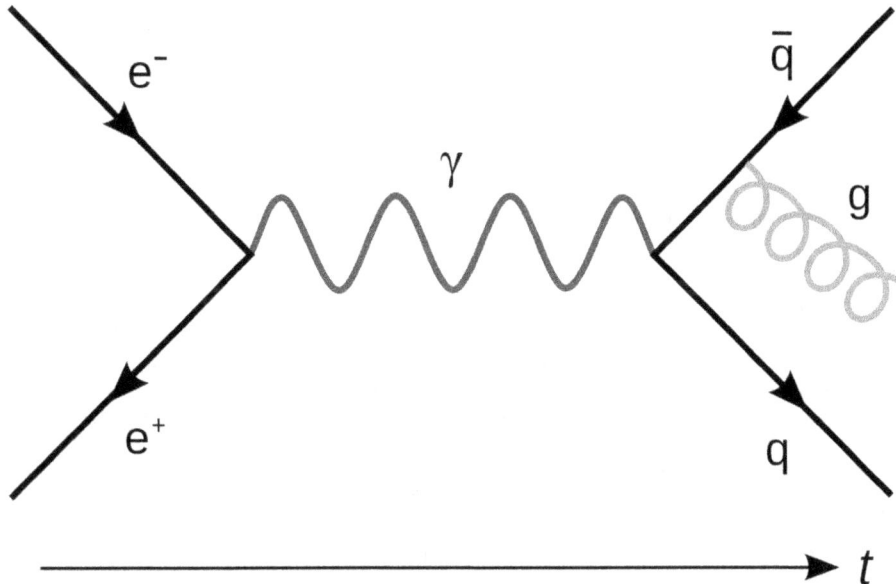

https://en.wikipedia.org/wiki/Feynman_diagram#/media/File:Feynmann_Diagram_Gluon_Radiation.svg

**In this Feynman diagram, an electron and a positron annihilate, producing a photon (represented by the sine wave) that becomes a quark–antiquark pair, after which the antiquark radiates a gluon (represented by the helix).**

In this Feynman Diagram, the mass, heat, and entropy are TRANSFORMED into massless, entropyless, and heatless infinite acceleration instead, when the mass or heat storage capacity is transformed into a massless, heatless, and entropyless photon. At the very moment that the electron and positron merge and the photon is made, the mass, heat, and entropy completely cease to exist; and, the second law of thermodynamics is falsified. It's that simple!

Isn't that amazing?

The truth has been before our very eyes the whole time, and nobody can see it because nobody is looking for it.

The very moment that the electron and the positron are TRANSFORMED into a massless, heatless, and entropyless photon, the associated mass, heat, and entropy completely cease to exist; and, they are transformed into infinite acceleration instead. The very moment that the electron and the positron are TRANSFORMED into a massless, heatless, and entropyless photon, the second law of thermodynamics is falsified. The truth has been hiding in plain sight the whole time, but nobody is looking for it, so nobody finds it.

In this diagram, it is the (**t**), the arrow of time, that prevents us from reversing this process, NOT the laws of thermodynamics! The whole thermodynamics process is reversible all the time according to $E = mc^2$, the Law of Quantum Information Conservation, Quantum Field Theory, and the Conservation of Energy and Psyche. The only reason we don't see it happening right now is because we aren't looking for it; but, it is happening all the time, and we choose to ignore it because it falsifies our pre-chosen beliefs.

Mass, heat, and entropy are being converted or transformed into massless, heatless, and entropyless photons, quantum waves (virtual particles), and quantum fields ALL THE TIME, and we don't even know it because we aren't looking for it and don't want it to be true. The entropy or heat completely ceases to exist, and the second law of thermodynamics is falsified, every time that Nature or Nature's Psyche chooses to TRANSFORM mass, heat, or entropy into massless, entropyless, and heatless quantum waves (virtual particles), photons, or quantum fields.

All of the mass, heat, and entropy are instantly transformed into infinite acceleration instead, whenever mass of any kind is transformed into a photon or a quantum wave.

Mass is literally heat storage capacity, resistance to acceleration, or entropy. Whenever the mass is transformed into infinite acceleration or photons, the heat storage capacity, the heat, and the entropy are transformed and cease to exist along with it. NO heat storage capacity – NO mass – then NO heat and NO entropy either. It's that simple. The only reason it hasn't been discovered before now is because nobody is looking for it and nobody wants it to be true. It's been true all the way along; and, it is really simple to understand. The truth always is.

In contrast, we have the second law of thermodynamics which states that the amount of entropy in the universe is always increasing and can never decrease. The second law turns entropy into the ultimate perpetual motion machine – the only thing in the universe that can constantly increase in amount without end, contrary to all the rules of science, logic, and common sense. The second law of thermodynamics breaks parsimony or common sense every time, which explains why nobody can figure out what entropy really is and why nobody can come up with a useful and true definition for entropy to begin with.

This also explains why most scientists believe that the second law of thermodynamics is axiomatically true. They have been deceived; and, they don't even know it, because they aren't looking for it and because they have been taught that the second law of thermodynamics can't be falsified. They believe in the second law of thermodynamics because they desperately want it to be true; but, wanting it to be true doesn't make it true.

Every Feynman Diagram where mass or heat storage capacity is being transformed by Nature into massless, heatless, and entropyless photons or quantum waves (virtual particles) falsifies the second law of thermodynamics clearly and conclusively. I chose to go with it, because it's obviously true.

So, what's wrong with the second law of thermodynamics?

The second law of thermodynamics is the Conservation of Entropy. By definition, according to the second law of thermodynamics, entropy consists of everything that is conserved. If it's conserved, then these people consider it to be entropy.

I was shocked and surprised when I first discovered how much they lie, cheat, deceive, and exaggerate in order to establish the truthfulness of the second law of thermodynamics. The second law of thermodynamics is the Conservation of Entropy, and entropy can literally be anything that you want it to be.

There are literally dozens of different definitions for entropy. Every day I seem to find a new definition for entropy. Entropy is often defined as ALL of the energy in the universe or ALL of the heat in the universe. Entropy is sometimes defined as ALL of the mass or physical matter in the universe. Entropy is also defined as ALL of the disorder in the universe, all of the chance or blind luck in the universe, or all of the randomness in the universe. Entropy has been defined as our Designer and our Creator. Quantum fluctuations or random fluctuations in entropy produced the big bang, the genomes, and all the physical matter, after all. Entropy can do anything that it sets its mind to. Entropy has also been defined as the number of heads and tails in a set of coin tosses. Entropy has been defined as the probability of getting snake eyes when tossing a pair of dice.

Entropy can be anything that you want it to be, so long as that thing is always increasing, always positive, and can never go below zero. Entropy is everything in the universe. Entropy can be anything you want it to be. Entropy is the most flexible substance in the universe. You want to start collecting entropy because the stuff is always conserved; and, no matter how much you collect, the universe will just make more, because the amount of entropy is constantly on the increase. According to these people, if we ever run out of entropy, the universe will just make more.

The SI (Standards International) defines entropy according to the following equation:

**$S = J/K$**

Entropy (**S**) is equal to ALL of the energy (**J**) in the universe divided by something as close to absolute zero (**K = 0**) as they can possibly get. By setting entropy equal to the ONLY truly conserved substance in the whole universe, these people are axiomatically declaring their belief that entropy is a conserved substance too. Thus, their equation for entropy and their definition for entropy – as a conserved substance that is equal to ALL of the energy in the universe – comes into line with the claim made by the second law of thermodynamics that entropy is always increasing, and that entropy can never decrease. The amount of energy can NEVER decrease, because energy is always conserved; so, what they did is that they made entropy equal to ALL the energy in the universe so that entropy is always conserved as well. They literally defined entropy as everything that is conserved so that the second law of thermodynamics could be the Conservation of Entropy.

The second law and the Standards International axiomatically define entropy as a conserved substance consisting of all the energy in the universe, before trying to increase the amount of entropy even more by dividing by something close to zero. The second law defines entropy axiomatically as a conserved substance that can never decrease. The second law *conflates* or equates entropy with energy, and makes them synonymous with each other, so that entropy is always conserved.

Surprised?

Well, I was. I know that these guys lie and cheat to prove the truthfulness of their ideas, so I shouldn't have been surprised; but, this has got to be the ultimate *category error* logic fallacy that has ever been produced. Setting entropy equal to "everything that is conserved" has got to be the ultimate quantum leap where the Materialists, Naturalists, Darwinists, and Atheists are concerned. It's *too good to be true*, which means that it can't possibly be true. It can't be true, according to the rules of science, because there is no way to falsify it. By setting entropy equal to everything in the universe, there's NO way to determine what entropy is NOT, because everything is entropy by definition to begin with.

The second law of thermodynamics is defined as the Conservation of Disorder or the Conservation of Entropy. Consequently, if you can falsify the Conservation of Disorder or the Conservation of Entropy by giving examples where entropy has gone to zero and ceased to exist – examples such as the hypothetical Big Bang Singularity, the initial Big Bang Syntropy, the Photons, the Gibbs Free Energy Equations, and the Quantum Fields – then you have in fact successfully falsified the second law of thermodynamics. Anytime you demonstrate the existence of the physically impossible, you have in fact falsified Materialism, Naturalism, Darwinism, and the Second Law of Thermodynamics. Photons are physically impossible, because they accelerate instantly with an infinite acceleration to the speed-of-light; and, they travel at the speed-of-light or even faster if they want to. Massless and entropyless photons falsify the second law of thermodynamics, and they don't even know it because they aren't looking for it and don't want to know it.

The second law of thermodynamics is FALSIFIED by massless and entropyless quantum waves, photons, tachyons, quantum phenomena, and quantum fields. The second law of thermodynamics is always FALSIFIED by the massless, the heatless, and the entropyless. Anything that loses all of its mass and entropy, by being converted into photons and raw energy instead, FALSIFIES the second law of thermodynamics which states that entropy is always increasing, and that entropy can NEVER decrease and cease to exist. Over 95% of our physical universe is defined as being non-physical, or quantum, or spiritual, which means that over 95% of our physical universe (the dark matter and the dark energy) FALISFIES the second law of thermodynamics in one way or another by having NO heat or NO entropy.

The quantum or the non-physical functions perfectly fine, in fact they function as eternal perpetual motion machines, within the absolute zero temperatures of the quantum fields where there is NO heat and NO thermodynamics whatsoever. No heat, then NO entropy!

Falsify the Conservation of Entropy and you falsify the second law of thermodynamics, because the second law is the Law of Conservation of Entropy.

So, what's wrong with the second law of thermodynamics?

The official equation for heat (**Q**) and the heat-based equation for entropy (**S**) FALSIFY the second law of thermodynamics, which states that the universal amount of entropy is always increasing and that the universal amount of entropy can never decrease to zero and cease to exist.

The official heat-based equation for entropy (**S**) is really simple and easy to understand.

**S = Q/T**

It doesn't take a genius to realize that when heat goes to zero ($Q \rightarrow 0$), then entropy goes to zero along with it ($S \rightarrow 0$). Remove all the heat, then you remove all the entropy! In other words, entropy completely ceases to exist and cannot exist at absolute-zero temperature (**T = 0**), among all the massless and entropyless photons and all the heatless and massless quantum fields. This is the Quantum Law of Thermodynamics. Entropy does not exist and cannot exist in the Quantum Realm or Spirit World where there is no heat and no thermodynamics. The truth is hiding in plain sight where nobody can see it, because they aren't looking for it and don't want to find it.

The official mass-based equation for heat (**Q**) is really simple and easy to understand.

**Q = mcΔT**

It doesn't take a genius to realize that when mass goes to zero (**m → 0**), then heat goes to zero along with it (**Q → 0**). Remove all the mass or resistance to acceleration (**m = 0**), then you remove mass's heat storage capacity (**mc → 0**), and you eliminate all the heat (**Q → 0**) and all of the entropy along with it (**S → 0**)!

No mass, then NO entropy! The amount of entropy goes to zero and entropy ceases to exist whenever physical particles are converted back into massless, heatless, and entropyless quantum fields and photons. This reality and truth FALSIFIES the second law of thermodynamics. Massless, heatless, and entropyless Photons and Quantum Fields FALSIFY the second law of thermodynamics. The second law can't be true if there are multiple different ways to falsify it.

**Remember, entropy completely ceases to exist and cannot exist within massless particles, massless quantum waves, massless photons, and massless quantum fields. Remove the mass, and you remove all the entropy, and then the entropy completely ceases to exist. The heat, mass, and entropy are converted or transformed into infinite acceleration instead, according to Quantum Mechanics and Quantum Field Theory.**

**Mass is by definition heat storage capacity. No mass, then NO heat storage capacity. No heat, then NO thermodynamics. No thermodynamics and no heat, then NO entropy. It's that simple. When mass goes to zero, entropy goes to zero and ceases to exist along with it. When mass and heat go to zero, then thermodynamics ceases to exist; and, there can be no entropy.**

**Massless photons and massless particles have NO heat storage capacity, which means that they have no entropy and can have no entropy. This is the Quantum Law of Thermodynamics. The Quantum Law of Thermodynamics falsifies the second law of thermodynamics.**

**Entropy does not exist and cannot exist in the Quantum Realm, the Massless Realm, the Syntropic Realm, the Conserved Realm, or the Spirit World. Mass and heat are what make the physical realm physical. Without them, there is NO physical realm and therefore there is NO entropy. This is really simple to understand, which is why everyone in the world has completely overlooked it and rejected it. They don't want it to be true because it falsifies Materialism, Naturalism, Darwinism (Creation by Entropy), Nihilism, and Atheism (Creation Ex Nihilo).**

https://quantum-law-of-thermodynamics.com/

The equations for heat and entropy falsify the second law of thermodynamics. That's a fatal flaw, and they don't even know it. They have NEVER discovered this simple truth, because they NEVER allow the amount of entropy to go to zero. They NEVER allow entropy to cease to exist. Instead, they stack the deck and load the dice so that it always comes up entropy. Through the second law of thermodynamics, they axiomatically define entropy as everything greater than zero, so that it can NEVER go to zero and cease to exist.

Through axiom and law, they cheat and make it so that the second law of thermodynamics can NEVER be falsified. That's Bad Faith and Bad Science.

Whenever Nature or Nature's Psyche converts mass or resistance to acceleration into photons and infinite acceleration, that conversion or transformation falsifies the second law of thermodynamics every time. In other words, **E = mc²** falsifies the second law of thermodynamics, and they don't even know it.

So, what's wrong with the second law of thermodynamics?

The Law of Psyche and the First Law of Thermodynamics falsify the second law of thermodynamics.

The Law of Psyche states that every psyche, spark, intelligence, consciousness, or photon has a certain amount of energy that's under its control. That controlling psyche can form or transform the energy under its control into anything that it wants that energy to be, anytime and anywhere that it chooses to do so. According to the First Law of Thermodynamics or the Conservation of Energy, psyches cannot make nor destroy energy. Psyches can ONLY form it and transform it. Nature's Psyche can transform Quantum Fields into physical matter, heat, mass, planets, stars, and galaxies anywhere and anytime that Nature's Psyche chooses to do so. Furthermore, Nature's Psyche can take any black hole, planet, star, or galaxy and transform it back into Quantum Fields and Pure Exergy anytime and anywhere that Nature's Psyche chooses to do so. Nature's Psyche can transform mass, heat, entropy, and physical matter into massless and entropyless photons and infinite acceleration anytime and anywhere that Nature's Psyche choose to do so.

The false is falsified by the truth; and, the truth is repeatedly experienced and observed. The truth falsifies the second law of thermodynamics; and, they don't even know it because they don't want to know it.

So, what's wrong with the second law of thermodynamics?

Through the statistical mechanics equation for entropy, **S = k$_B$ ln (Ω)**, entropy (**S**) is defined axiomatically as ALL of the disorder, chaos, randomness, and chance in the universe. By definition, the second law of thermodynamics is statistics, probability, luck, and chance.

Statistical mechanics defines entropy as the chance of getting heads or tails – as the chance of producing genomes, life-forms, big bangs, planets, stars, and galaxies. The chance is high because the entropy is always high. The chance is high because entropy or natural selection can design and create anything that it sets its mind to.

Defining entropy as the amount of disorder in a system is probably the most common definition for entropy. It's also a false and deceptive definition for entropy. It was designed to enshrine entropy or chance as the ultimate causal force in the universe – destined to control the fate of our universe. It's a scam.

Technically, it's physically impossible to measure the amount of disorder, randomness, chaos, or chance within a system. We would actually have to order, organize, structure, or freeze the system in order to do so. We would also have to observe and process ALL of the atoms in the system all at once in order to see how disorganized they really are. It can't be done – not at the physical level. Consequently, defining entropy as ALL of the disorder in the universe ends up being the fictional or hypothetical definition for entropy, because it's limited exclusively to statistics, probabilities, chance, thought experiments, and wishful thinking. It can't be measure, which means that it really can't be defined. Consequently, they define entropy as ALL of the disorder in the universe and leave it at that, because they can't measure it or do experiments with it.

Statistics is the art of learning how to lie with math and science.

In the hands of the Darwinists, Naturalists, and Atheists, statistics produces more errors and falsehoods than it ever solves; and, the honest scientists never catch the deceptions and the lies because they never learn how to use statistics and logic to eliminate the lies and because they trust the Darwinists, Naturalists, and Atheists to tell them the truth. I no longer trust these people because they have lied to me one too many times. I've caught these people producing thousands of falsehoods and lies. It's what they do. Their purpose in life is to trick you and deceive you, as they have been tricked and deceived.

According to the Materialists, Naturalists, Darwinists, Nihilists, and Atheists, chance or entropy can design and create anything that it sets its mind to, including mass, physical matter, genomes, life forms, heat, planets, stars, and galaxies. To these people, chance or entropy is their Creator and their God. These people really truly believe that chance or entropy designed and created their bodies, brains, eyes, and genomes. They truly believe that the design and creation of our physical universe was nothing more than random chance or a random fluctuation in entropy.

That's what they believe, is it not? The internet is saturated with their falsified beliefs, and so are our college classrooms.

The Philosophy of Science is supposed to study what these people have chosen to believe, and then try to find ways to falsify it. Science and the scientific method ARE falsification of the data and falsification of the ideas that are demonstrably false. If you are not falsifying the things that are false, then you are not doing science. You are doing religion and dogma instead. The second law of thermodynamics enshrines and sanctifies many different falsehoods, deceptions, and lies; and then, it goes one step further and calls them axioms, truths, and laws instead.

Contrary to what the Materialists and Naturalists claim, entropy, chance, or natural selection CANNOT BE the cornerstone of science. If it were, we wouldn't be here right now. Design and creation by entropy or chance is physically impossible. Entropy, chance, or natural selection can't be our Designer and our Creator, because it has NO physical mechanism for doing so. It doesn't even have a supernatural or quantum mechanism for doing so. Remember, the theory of evolution and the associated natural selection are in fact "design and creation by entropy and death". By definition, evolution and natural selection and entropy have NO psyche and NO intelligence, so there's no way in the universe that they can be our Designer and our Creator.

Likewise, if the second law of thermodynamics is true, then the Big Bang Theory can't possibly be true. They are mutually exclusive. If the Big Bang Theory is true, then the second law of thermodynamics can't possibly be true, because the Big Bang is by definition a complete and total reversal of entropy. The Big Bang, if it happened, was pure exergy and pure syntropy. By definition, the Big Bang was a complete reversal of entropy by Nature, Nature's Psyche, the Controlling Psyches, or the Gods. Somebody somewhere KNOWS how to reverse entropy and make it cease to exist. Entropy or chance cannot design and create at will as the Materialists, Naturalists, and Darwinists claim; but, psyche or intelligence certainly can. Psyche or intelligence has been observed doing so trillions of trillions of times over and over again. The false is falsified by the truth; and, the truth is repeatedly experienced and observed. This is the Philosophy of Science at its very best.

Creation by entropy and chance is a scam; and the geniuses in this world fall for it every time because they desperately want it to be true. The second law is a lie, and we don't even know it, because everyone is telling us that it is the truth, and we believe them without question.

So, what's wrong with the second law of thermodynamics?

The second law is based upon the exclusion of all the evidence that falsifies it. The second law is based upon the *exclusion of evidence* logic fallacy. The second law of thermodynamics deliberately excludes the massless entropyless photons, the massless entropyless quantum waves, the massless entropyless quantum fields, and the massless entropyless dark energy, as well as all the different massless and entropyless forces and fields that we have discovered. The second law of thermodynamics is exclusively based upon the *exclusion of all the evidence* that falsifies it.

Science 2.0 by definition allows ALL of the evidence into evidence, and then it tries to pursue a preponderance of that evidence. The second law of thermodynamics is a violation of Science 2.0 because the second law (through fiat or decree) axiomatically refuses to allow into evidence anything that falsifies it. The second law claims that entropy is always increasing, and that entropy can never decrease; therefore, anytime that entropy is caught decreasing or going to zero, they toss the evidence out and call it invalid because it violates and falsifies the second law of thermodynamics. That's *circular reasoning* and *affirming the consequent*, both of which are logic fallacies. They only affirm the evidence that verifies the second law, while simultaneously excluding and rejecting ALL of the evidence that falsifies the second law. That's the very definition of Bad Faith, Biased Science, and Bad Science.

The second law of thermodynamics is based upon a couple dozen different logic fallacies. Through the second law of thermodynamics, they load the dice and stack the deck so that it always comes up entropy. That's cheating! *Cheating* is a logic fallacy. If you observe carefully, you will soon notice that the Materialists, Naturalists, Darwinists, and Atheists are constantly cheating and lying in order to prove that their theories and ideas are true. They are constantly tossing out the evidence that falsifies their beliefs. That's what they do, is it not?

They deliberately designed the second law of thermodynamics so that it can't be falsified, which automatically falsifies it according to the rules of science. The second law of thermodynamics exists because it is a complete violation of the rules of science. It's granted a *special pleading*, or a *special exception*, or a *special exemption*. In other words, the second law is an exception to the rules of science. Its existence and "truthfulness" is based upon the *special pleading* logic fallacy because the second law of thermodynamics deliberately and axiomatically excludes all of the evidence that falsifies it. The second law is granted a *special pleading exemption* from the rules of science by being presented as a law, which means that it's presented as being axiomatically true and is not allowed to be falsified. That's cheating! The second law is a violation of the rules of science.

The same thing can be said about natural selection, which is a type of entropy. Evolution or natural selection is, by definition, entropy. Natural selection or the theory of evolution is creation by death and entropy. We know this is so because these people have chosen to believe that only entropy and physical matter exist. Some of these people have chosen to believe that rocks can design and create; and, others have chosen to believe that entropy or natural selection or death can design and create. They are willing to believe in anything but psyche, intelligence, or God.

The Materialists, Naturalists, Darwinists, Nihilists, and Atheists set it up so that their theories and ideas are axiomatic, exempt from scrutiny, automatically lawful and true, don't have to be verified in order to be true, exempt from the rules of science, exempt from falsification, and declared in advance to be immune to falsification and automatically true. They cheat. That's the way they do science. They cheat.

They use logic fallacies, mocking, ridicule, intimidation, and peer review to make their theories axiomatically true. They do their science by vote, by fiat, by decree, by forced consensus, by law, by intimidation, by peer review, and by committee. They decide what's true in advance, and then they adjust the evidence or destroy the evidence so as to force the evidence to match with what they have already decided is true. They turn science upside down and reform it on their own image. The second law of thermodynamics is a part of this system of forced consensus or religious dogma or peer review, which makes it automatically false because these people are deliberately violating the rules of science in order to prove that their theories and ideas are true.

This is part of what I learned from the Philosophy of Science. I learned how the Materialists, Naturalists, Darwinists, Nihilists, and Atheists manipulate the system and cheat in order to turn their lies and deceptions into axioms and laws instead. I learned how to identify it and see it. I learned how to falsify them or prove that they are false. I learned that with science and the scientific method you can PROVE that certain things are false. I learned how to falsify the second law of thermodynamics. I learned how to PROVE that it is false.

The proven and verified existence of massless and entropyless photons, quantum waves, forces and fields, dark energy, and quantum fields FALSIFIES the second law of thermodynamics which states that these things do not exist and cannot exist.

So, what's wrong with the second law of thermodynamics?

The second law of thermodynamics is Materialism, Naturalism, Darwinism, Nihilism, Behaviorism, Physical Reductionism, Atomism, Determinism, Classical Physics, Classical Realism, and Atheism all rolled into one. That automatically falsifies the second law of thermodynamics. The second law is fruit from the poisoned tree.

These people truly believe that they know everything that there is to know about physics and the universe because these people refuse to allow massless photons, entropyless quantum fields, massless quantum information, non-physical and massless psyche or intelligence, and heatless and massless dark energy into their Philosophy of Science, their personal religion, and their worldview. These people arrogantly and erroneously believe that they know everything, because they have excluded everything except entropic physical matter from their science, their purview, their religion, and their personal beliefs.

The second law of thermodynamics is Materialism, Naturalism, Darwinism, Nihilism, and Atheism – which is all that it needs to be in order to falsify it, because all of these falsified philosophies are demonstrably false bringing the second law of thermodynamics down with them.

Materialism, Naturalism, Darwinism, Nihilism, and Atheism ARE different philosophies of science, which means that the Philosophy of Science can be used to falsify them. By studying the Philosophy of Science, I learned how to use the Philosophy of Science and logic to falsify Materialism, Naturalism, Darwinism, Nihilism, Atheism, the Conservation of Entropy, and the Second Law of Thermodynamics. It's really easy to do once you know how.

These people erroneously claim that everything is made from atoms.

Contrary to their claims, everything is NOT made out of atoms! Atoms are NOT made from atoms! The existence of atoms, quarks, electrons, photons, quantum waves, and quantum fields is physically impossible according to classical physics and the second law of thermodynamics. Their very existence and the fact that they are NOT made from atoms is scientific proof that Materialism, Naturalism, Darwinism (Creation by Entropy), Nihilism, Atheism (Creation Ex Nihilo), and the Second Law of Thermodynamics are false.

Atoms are made from different forms of energy. We can safely claim that everything is made from energy, including entropic physical atoms. But, it is a scientific error to claim that everything is made from atoms. Quantum waves are NOT made from atoms. Photons are NOT made from atoms. Quantum Fields are NOT made from atoms. The different forces and fields are NOT made from atoms. The things that FALSIFY the second law of thermodynamics are in fact the non-physical entropyless things that are NOT made from atoms. They are made from Syntropy, Exergy, or Conserved Energy instead.

The proven and verified existence of massless and entropyless photons and quantum fields FALSIFIES Materialism, Naturalism, Darwinism, Nihilism, Atheism, and the Second Law of Thermodynamics along with them. The false is falsified by the truth; and, the massless and entropyless photons are constantly being experienced and observed.

Remember, every photon, and every quantum field, and every non-physical force and field FALSIFIES the second law of thermodynamics because these things have NO mass, NO resistance to acceleration, NO heat, NO thermodynamics, and therefore NO entropy. These things can't have entropy because they have NO mass, NO heat storage capacity, and NO resistance to acceleration. This is the Quantum Law of Thermodynamics.

Photons are capable of infinite acceleration, because they have NO entropy, mass, or resistance to acceleration. The photons go from zero to the speed-of-light instantly in no time at all, because for a split second they are accelerating at an infinite acceleration and traveling infinitely faster than the speed-of-light.

Photons can also travel faster than the speed-of-light if they choose to. If they stop their infinite acceleration at the speed-of-light rather than quantum tunneling directly to their destination, it's because they have chosen to do so and not because something is forcing them to do so. By definition, tachyons are photons that have chosen to travel faster than the speed-of-light. Any quantum wave or quantum particle who chooses to quantum tunnel instantly to its chosen destination is in fact a type of tachyon and does in fact travel faster than the speed-of-light. It does indeed travel the distance, and it travels that distance in a massless, entropyless, heatless, and non-physical form.

This is actually what we have experienced and observed.

The verified and proven existence of anything massless or non-physical FALSIFIES Materialism, Naturalism, Darwinism, Nihilism, Atheism, and the Second Law of Thermodynamics. The proven and verified existence of massless entropyless photons falsifies the second law of thermodynamics.

These people deliberately designed the second law of thermodynamics so that it can't be falsified, in order to stop you from falsifying it. Consequently, you have to be willing to pull out all the stops in order to falsify the second law of thermodynamics. That's eventually what I chose to do, once I started to find multiple different ways to falsify the second law of thermodynamics. I went after the truth instead. I went after what has actually been experienced and observed. Entropyless and massless photons have been experienced and observed. Photons have NO mass and NO entropy because they have NO heat storage capacity and NO resistance to acceleration. That's why they falsify the second law of thermodynamics, because the second law tells us that the photons can't exist and don't exist.

So, what's wrong with the second law of thermodynamics?

The second law of thermodynamics is Conservation of Entropy.

The purpose of the Philosophy of Science is to reveal and explore the different ways that they use science to trick us and deceive us. As usual, there are different definitions for conservation, some of which were designed to trick us and deceive us.

The second law of thermodynamics defines entropy and conservation as something that is always increasing and that can never decrease or go to zero. The second law of thermodynamics defines entropy as a substance that is quickly filling up a universal reservoir so that someday everything will be entropy. According to the second law of thermodynamics, everything in the universe is being converted into entropy. The amount of entropy is constantly increasing. That's what our physics teachers teach and believe, is it not?

According to this second law definition for entropy and conservation, entropy is ultra-conserved – the universe is constantly making more; and, there's no limit to the amount of entropy that the universe can make. Its amount is constantly increasing and can never decrease. Entropy is the ultimate perpetual motion machine according to the Materialists, Naturalists, Darwinists, and Atheists. The universe is constantly making more. Entropy is more than conserved – its amount is constantly increasing like magic – according to the second law of thermodynamics and the people who invented the concept.

Then there is the more conservative and truthful definition for conservation, like the one we get when conservation is applied to energy instead of entropy. When something like energy is conserved, its amount never increases, and its amount never decreases. When something like energy is conserved, it cannot be made, and it cannot be destroyed. Energy has always existed, and it will always exist, because energy is always conserved. Unlike entropy, it's impossible to make more energy, because the amount of energy is always conserved. This is the conservative and realistic definition for conservation.

In contrast, according to the second law of thermodynamics, entropy is a magical substance whose amount is constantly increasing; and, more entropy is being made all the time. Entropy is supernatural. Its amount is constantly increasing, and there's literally no limit to how high it can go. Entropy is conserved-plus or ultra-conserved. Entropy is better than energy, because entropy's amount is constantly increasing and will continue to increase forever. It will never stop. Entropy is the ultimate perpetual motion machine. The universe is constantly finding ways to make more entropy. This is the loose and liberal definition for conservation, and it becomes the Conservation of Entropy or the Second Law of Thermodynamics.

The universe is constantly finding ways to make more entropy. The amount of entropy is always increasing. That's what the Materialists, Naturalists, Darwinists, and Atheists teach, preach, and believe. They believe that the amount of entropy is constantly increasing.

Alas, the Conservation of Entropy or the second law of thermodynamics is the ultimate fiction and the premier lie. It's too good to be true. It's extravagant. It's physically impossible. It violates the Conservation of Energy. There is no such thing as the conservation of entropy. Entropy is not conserved. Its amount is constantly changing; and, whenever matter is converted into energy, the associated entropy completely ceases to exist, according to the official equations for heat and entropy.

$E = mc^2$ falsifies the second law of thermodynamics by allowing mass, heat, and entropy to be converted or transformed into massless, heatless, and entropyless quantum fields and photons, anytime and anywhere that Nature or Nature's Psyche chooses to do so. Instead of constantly increasing, the amount of entropy is constantly going to zero and entropy is constantly ceasing to exist, whenever mass or entropy is transformed into massless and entropyless quantum fields and photons.

According to Quantum Field Theory, particles are made by Nature or Nature's Psyche; and, Nature's Psyche can unmake the particles and reabsorb the particles back into the Quantum Fields anytime and anywhere that Nature's Psyche chooses to do so. This reality and truth applies to physical particles as well. Whenever Nature converts or transforms physical matter into photons, the associated heat, mass, and entropy are completely transformed into infinite acceleration instead. This is what we actually experience and observe. Is it not? Whenever anything mass-based or physical is converted into a photon, that thing completely loses its mass or resistance to acceleration; and, that photon accelerates instantly at an infinite acceleration directly to the speed-of-light.

Tachyons are photons that have decided to quantum tunnel or have decided to travel faster than the speed-of-light. The photons and tachyons have infinite acceleration capabilities and NO speed limits because they have transformed ALL of their mass, heat, and entropy into infinite acceleration instead. The photons and tachyons have infinite acceleration capabilities and NO speed limits because they have NO mass or resistance to acceleration, which means that they have NO entropy. This is obviously true, or massless photons wouldn't exist in the first place. Massless photons are physically impossible, which means that they are non-physical quantum phenomena to begin with and remain as such to this very day.

The Ultimate Law of Thermodynamics states that ONLY energy and psyche are truly conserved. The quantum information within Nature's Psyche is always conserved too, according to Quantum Mechanics; but, the amount of information or knowledge or experience is supposedly always increasing. The Quantum Physicists state that the Quantum Law of Information Conservation has to be true, or Quantum Mechanics and Quantum Field Theory can't be true. According to the strict or conservative definition for conservation, ONLY psyche and energy are conserved. Quantum Information is granted a *special pleading*, and its amount is allowed to be constantly on the increase because it is being stored within and used by Psyches, or Quantum Photons of Light.

In a strange sort of way, Quantum Information is the spiritualist's or quantum physicist's counterpart to the materialist's and naturalist's Entropy – both are defined as being constantly on the increase. However, Entropy represents the destruction of information, disorder, randomness, chance, and chaos; whereas, Quantum Information represents an increase in order, intelligence, knowledge, information, structure, syntropy, power, glory, love, and experience. Entropy is the Sith; whereas, Quantum Information is the Jedi.

In other words, Entropy and Quantum Information are mutually exclusive. The one represents Chaos, and the other represents Perfection. If Quantum Information is being conserved or preserved, then ever-increasing disorder becomes impossible to achieve, because the order or increase in knowledge is winning the war and will someday reign supreme, thanks to the conservation of energy and psyche. At the quantum level, order destroys or eliminates disorder, permanently. At the quantum level, intelligence or knowledge destroys or eliminates entropy, permanently.

Remember, the Conservation of Quantum Information eliminates entropy permanently at the quantum level and FALSIFIES the second law of thermodynamics at the quantum level.

The idea that the amount of disorder in our universe is constantly increasing is nothing but a myth, especially once we realize that the exact opposite is what's happening at the quantum level in the Syntropic Realm or the Quantum Realm.

The Chaos Realm or the Chaos Construct was the original primal construct. Then what happens is that one of the Gods steps into that and teaches the psyches or the intelligences within that how to organize themselves into quantum fields. Once the quantum fields are organized and fully functional, then mass, heat, physical matter, genomes, planets, stars, and galaxies can be made from the quantum fields anywhere and anytime that the Gods or Nature's Psyche choose to do so.

The Quantum Fields are perpetual motion machines. They are perfectly organized and remain so to this very day. Even if our whole universe were to attain heat death at the physical level, all the energy and all of the quantum fields will still be there; and, Nature's Psyche can reabsorb the mass and entropy and black holes back into the quantum fields instantly, and make brand-new planets, stars, and galaxies from all of that reabsorbed energy anytime and anywhere that Nature chooses to do so.

The Bible says that someday there will be a NEW heaven and a NEW earth. How do you think that's to be done? It will be done the same way that it's always been done. At God's command, the old will be absorbed back into the quantum fields, and the NEW will be made from there. Energy is perfectly recyclable; and, every time that energy is recycled, the associated entropy and disorder completely cease to exist. It's really easy to understand once a person chooses to do so.

It's obviously true. This science has never been discovered before now because nobody has ever thought about it before now. Yet, those crazy Christians, Hebrews, Jews, and Egyptians knew all about this thousands of years ago because God taught them about it.

Hugh Nibley, *The Message of the Joseph Smith Papyri: An Egyptian Endowment*, p.275:

**(134) The (creative) word which it (Wisdom) spoke is the outflowing of light which came (lit. drew) together at the Temple. (The personification of Wisdom, Sophia, is biblical as well as Egyptian [Maat]. Thus Proverbs 7:4; 8:1; Maat. 11:19; Luke 7:35; 11:49; specifically, as an agent of the creation: Jer. 51:15; Prov. 3:19.) (129) It was Michael and Gabriel who took charge, bringing the outpouring of light into disorganized matter (chaos) ... they were entrusted with the distribution (*aporrhoia*) of the Light, which I handed over to them to be applied (introduced into) to Chaos. (132) The outpouring of light entered into chaos and suffused itself over all the realm of blind chance (the *Authades*, that which is under no control but its own nature, the "self-willed"). And also, the Word of thy Power uttered through Solomon: It has drawn all things and brought them together at the Temple. ... And again: ... The emanations of Blind Chance (the *Authades*, who acknowledges no law) could not dominate over (*amahte*, contain, control) the outpouring of light within the confines (*sovt*) of the darkness of Chaos. (36) (After the Sun's light has been purified) left-over matter is brought down to this lower sphere and fashioned into men, and likewise into the souls of creeping things, and cattle and wild animals and birds ... all being sent down to this world of man, where they become souls ... (so also 35).**

**(128) I (Christ) called upon Gabriel from the midst of the worlds (aeons) along with Michael, pursuant to the command of my Father ... and I gave to them the task of outpouring of the light and caused them to go down into matter**

unorganized (Chaos) and assist Pistis Sophia. And as soon as the flow of light entered the Chaos, it became brighter and brighter, suffusing all Chaos, and spread out and occupied every part (topos) of it. Then Michael and Gabriel introduced the stream of light into the inert matter (hyle) of the body of Pistis Sophia, pouring back into her the light-powers which had been taken from her; and all her physical body became full of light ... the light having been given to them by me. And Michael and Gabriel who are charged with administering the light, they who introduce light into Chaos, will also give the ordinances of the light to them (the initiates?). (129f, 132-35) (It is Michael and Gabriel who pass between heaven and earth from time to time in their assignment of organizing the worlds out of Chaos.)

Once the light or the quantum fields have been organized, then physical matter and physical life forms can easily be made and sustained anytime and anywhere thereafter.

Once Chaos has been organized into quantum fields, then it sounds like additional levels or degrees of organization can take place thereafter. Also, whenever any part of it becomes disorganized or loses its potency, it can be reorganized or suffused with additional light anytime thereafter. In other words, the Gods and Nature's Psyche can reverse the entropy and eliminate the disorder at will, anytime and anywhere they choose to do so.

Information, or intelligence, or knowledge, or light is all that's needed to eliminate entropy for good. Organize Chaos and the associated blind luck, random chance, or entropy completely ceases to exist; and, the chaos or entropy ends up being replaced by the perfectly organized quantum fields instead. The Quantum Fields are always conserved, because they are pure energy or pure massless entropyless exergy or syntropy. It's entropic physical matter or mass that has to be periodically renewed, enhanced, or reorganized, which is easy to do thanks to the Conservation of Light or the Conservation of Energy.

The second law of thermodynamics is FALSIFIED by the Quantum Fields, by the Photons, by the Conservation of Quantum Information, by the Gods, by Nature's Psyche, and by the Conservation of Energy and Psyche. It's really simple to understand once a person chooses to do so.

So, what's wrong with the second law of thermodynamics?

The second law of thermodynamics is based upon the loose, unlimited, too good to be true, liberal definition for Conservation. The Ultimate Law of Thermodynamics and the Conservation of Energy are based upon the tight, the conservative, the realistic, and the demonstrable definition for Conservation.

The Ultimate Law of Thermodynamics states that mass, physical matter, heat, and entropy are NOT conserved. Their amount is constantly changing, which means that they are not conserved. They are always being transformed back into raw energy or exergy, which means that they are not conserved. Heat, mass, and the associated entropy are MADE by Nature's Psyche anytime and anywhere that Nature's Psyche decides to make them, which means that they are NOT conserved. Anything that is made is technically NOT conserved. According to Quantum Field Theory, mass, particles, physical matter, heat, and entropy are MADE by Nature, which means that Nature or Nature's Psyche can unmake them and absorb them back into the massless and entropyless and heatless Quantum Fields anytime and anywhere that Nature's Psyche chooses to do so.

https://ultimate-law-of-thermodynamics.com/

The second law of thermodynamics is constantly falsified by the truth.

The Gibbs Free Energy equations accurately and truthfully allow the entropy to decrease. They allow the mass, heat, and entropy to be transformed into massless, heatless, and entropyless photons instead. So does the Conservation of Energy, Quantum Mechanics, and Quantum Field Theory. In contrast, the second law of thermodynamics DOES NOT ALLOW physical particles to be converted back into massless, heatless, and entropyless photons. The second law violates the Conservation of Energy. The second law of thermodynamics violates $E = mc^2$, by not allowing mass to be converted back into the massless and the entropyless. To get around this violation, they axiomatically define entropy as ALL of the energy in the universe, because they know that energy is conserved.

Every hydrogen bomb falsifies the second law of thermodynamics by converting mass back into massless and entropyless photons instead, according to $E = mc^2$. The false is falsified by the truth; and, the truth is repeatedly experienced and observed.

The official equations for entropy and heat prove that when heat is zero, then entropy is zero; and, when mass is zero, then heat is zero, which means that entropy is zero. Entropy is NOT conserved. Entropy is constantly decreasing and ceasing to exist.

In other words, according to the equations for heat and entropy, photons and other massless particles have NO entropy. Mass is heat storage capacity, which means that massless particles and the massless quantum fields and the entropyless photons have NO heat storage capacity! That's why their temperature is absolute zero, because they have no heat storage capacity and have no entropy or resistance to acceleration. ALL of their heat, mass, and entropy has been transformed into infinite acceleration instead. This is what we have actually experienced and observed. We experience it every time that mass is converted or transformed into photons by Nature or Nature's Psyche. The Conservation of Energy and $E = mc^2$ falsify the second law of thermodynamics or the conservation of mass and entropy. Mass or physical matter can be transformed into entropyless and massless photons, quantum waves, and quantum fields anytime and anywhere that Nature chooses to do so.

The second law of thermodynamics or the Conservation of Entropy is falsified by the equations for entropy and heat. They are also falsified by the Gibbs Free Energy equations, by the first law of thermodynamics, and by the infinite expanse of massless entropyless Quantum Fields. Massless and entropyless Photons falsify the second law of thermodynamics. Everything falsifies the second law of thermodynamics – including mass or physical matter being converted back into entropyless and massless quantum waves, quantum fields, and photons.

So, what's wrong with the second law of thermodynamics?

Well, once I understood Quantum Field Theory, I quickly realized that the second law of thermodynamics deliberately excludes the infinite expanse of no-entropy, massless, absolute-zero temperature, thermal equilibrium, syntropic, exergic, and conserved Quantum Fields. At absolute zero among the Quantum Fields, there is perfect thermal equilibrium and no entropy, because there is NO mass, heat, or resistance to acceleration. It's perfection. It's perfectly organized; and, it's eternal because it's pure syntropy, pure energy, pure exergy, and always conserved. The physical realm is a drop in the ocean compared to the infinity of perfectly organized, no-entropy, pure exergy, syntropic Quantum Fields.

Remember, the Matrix of Quantum Fields is pure syntropy because quantum fields have NO entropy and are always conserved. In the Quantum Realm, ALL of the work is done for free, and ALL of the energy is available for use all of the time, which means that there is NO entropy in the Quantum Realm among the Quantum Fields. This is the

Quantum Law of Thermodynamics, which states that there is no heat, no mass, no resistance to acceleration, no thermodynamics, and no entropy in the Quantum Realm among the Quantum Fields.

The second law of thermodynamics is falsified by the infinitely ordered, perfectly organized, entropyless, perpetual motions machines that we call the Quantum Fields. The work of the Quantum Fields is done for free thanks to the Conservation of Energy. The verified and proven existence of entropyless or syntropic Quantum Fields falsifies the second law of thermodynamics which claims that such things do not exist and cannot exist.

We are easily tricked and deceived by the second law of thermodynamics in large part because we have infinitely ranged, unrealistic, and physically impossible definitions for entropy that can't possibly be true. We know that the "infinite" definitions for entropy, which are based upon the second law, are false once we start to take the Syntropy Realm or the Quantum Fields into consideration. The first law of thermodynamics falsifies the second law of thermodynamics or the Conservation of Entropy, once we realize that ONLY the energy is infinite and conserved, and that entropic physical matter is a drop in the ocean compared to all of that entropyless, massless, and heatless expanse of the Quantum Fields.

The primary axiom of Quantum Field Theory states that particles are born, and particles die. Particles, including physical particles, can and do die; and, when they die, their entropy or resistance to acceleration dies with them. Quantum Field Theory falsifies the second law of thermodynamics or the conservation of entropy. There is no entropy and can be no entropy in the photons, in the massless particles, in the quantum waves, and in the quantum fields. If you want to get rid of the entropy, then remove the mass or the resistance to acceleration. If you want to get rid of the entropy, then remove the heat. If you want to remove the entropy, then get rid of the heat storage capacity of mass or matter. If you want to get rid of the entropy, then let the system achieve thermal equilibrium. If you want to remove the entropy, then convert the physical particle into a photon instead.

### $E = mc^2$

If you want to get rid of the entropy, then convert the mass into energy. The truth of the matter is there in plain sight for anyone who is willing to look at it and see it. We KNOW precisely what happens when mass or physical matter is converted or transformed into massless non-physical photons; but, we are only able to understand it when we are willing to take the massless or the non-physical into consideration in the first place – which the Materialists, Naturalists, Darwinists, Nihilists, and Atheists are NOT willing to do. These people prefer to be deceived instead.

So, what's wrong with the second law of thermodynamics?

They can't see it because they aren't looking for it, but defining entropy as "all the heat in the universe", or as "all the energy in the universe", or "as all the disorder in the universe" are effectively mutually exclusive definitions for entropy, which means that they have a tendency to falsify each other. They can't all be true because they are each *too good to be true*. They produce infinities that are impossible to measure or observe. I mean, how are we going to measure the sum total of ALL the heat in the universe, or the sum total of ALL the energy in the universe, or the sum total of ALL the disorder in the universe? It can't be done.

**By setting entropy (S) equal to ALL of the energy (J) in the universe (S = J), it makes it impossible to determine what entropy is not. That's a *category error* logic fallacy. It's a *conflation error*. It makes it impossible to ascertain what**

entropy is and what entropy is not. Science desperately needs a better definition for entropy!

By defining entropy as infinity to begin with, there's NO way to determine what is NOT entropy, because everything is entropy by definition to begin with. That's *cheating*! It's *circular reasoning*. It's *affirming the consequent*. These are logic fallacies. The second law of thermodynamics is based upon a bunch of different logic fallacies, which is why the second law never makes any sense to a logician.

Energy is the ONLY thing that we can define as infinite, because energy is always conserved which means that energy was already infinite to begin with. Likewise, there is NO way to determine what is NOT energy, because everything was energy by definition to begin with. When it comes to physics, energy is the ONLY thing that we can safely define as infinite and still have the truth after we have made that definition – thanks to the Conservation of Energy.

Entropy obviously is NOT equal to all of the energy in the universe. Syntropy, by definition, is the thing that is actually equal to ALL of the energy in the universe! Syntropy is the Conservation of Energy; and therefore, syntropy is the exact opposite of entropy. Isn't it ironic and interesting that in order to make the second law of thermodynamics true, they set entropy equal to syntropy (or all the energy in the universe) before they do anything else? That's the ONLY way that they can make the second law of thermodynamics true, because the thing actually claims that entropy can NEVER decrease but always has to increase. The only way to do that is to set entropy equal to all of the syntropy or energy in the universe, because energy and therefore syntropy are always conserved and can never decrease. By setting entropy equal to all of the energy or syntropy in the universe, they commit a *category error* logic fallacy, in order to make the second law of thermodynamics axiomatically true.

Quantum Mechanics is Energy Mechanics or the way that energy really truly works at both the quantum level and the physical level. Quantum Mechanics is the scientific study of energy. Quantum Mechanics is the study of the infinite and the conserved. Quantum Mechanics and Quantum Field Theory falsify the second law of thermodynamics, and we don't even know it.

Remember, by defining entropy as all of the energy in the universe, we commit a *category error* logic fallacy, especially when we simultaneously define entropy as all of the disorder, randomization, and chaos in the universe. The different definitions for entropy end up being mutually exclusive, which means that they falsify each other. Defining entropy as infinite provides us with a meaningless definition for entropy, because then we have no way to determine what entropy is not.

The conservation of energy or the conservation of perfectly ordered quantum fields diametrically opposes the conservation of disorder or the conservation of entropy. They both can't be true, because they are mutually exclusive. If the one can be demonstrated to be true, then the other has been falsified. Disorder can't be conserved while the perfectly ordered and perfectly functional quantum fields continue to exist. Therefore, entropy cannot be conserved while the perfectly ordered entropyless quantum fields exist. The Quantum Fields FALSIFY the second law of thermodynamics, the conservation of disorder, and the conservation of entropy.

The different extravagant definitions for entropy end up being meaningless definitions for entropy, which explains why nobody seems to know what entropy really is, because all we have in science is the extravagant and the infinite definitions for entropy. Setting entropy equal to infinity before doing any science experiments makes no logical sense, which explains why nobody can define what entropy truly is. Entropy has dozens of different conflicting and contradictory definitions. It's Bad Science, as a result. I'm a logician; and, entropy never made any logical sense to me. No wonder. It is illogical. It's loaded. It's rigged. The way they set it up, entropy is philosophical metaphysics and supernatural mysticism masquerading as science and truth! It's magic!

So, what's wrong with the second law of thermodynamics?

*Overreaching* is a logic fallacy.

Through entropy and the second law of thermodynamics, they literally turn disorder and chance into a creative force and a living person who is capable of designing and creating anything at will. It's *too good to be true*.

For the Materialists, Naturalists, Darwinists, Nihilists, Behaviorists, and Atheists, entropy or natural selection is their Deus Ex Machina – their god in the machinery, or their god from the machine. Anytime they paint themselves into a corner and their science explanations don't make any sense, they turn to entropy or natural selection to solve their problem and to provide the answer to their conundrum. Anything they can't solve, they just turn to entropy or natural selection for the solution; and, it works every time.

Entropy or natural selection was designed to be the answer and the solution for everything that they don't understand or can't figure out. By definition, entropy can do anything and be anyone that they need it to be. They made entropy the cause of our physical universe, the controller of our physical universe, and the destiny of our physical universe. Anytime they need some magic, they call upon entropy or natural selection, and it provides them with exactly what they need. Entropy or natural selection or chance has become their God. It's the modern-day form of idolatry.

Unlike energy, which is simply conserved, entropy is supernatural in that its amount is always increasing and can never decrease. Energy has its limits; but, entropy is infinite and unrestricted and unlimited. Entropy can do anything and be anything. Entropy is their God in the machinery of the universe.

They turned entropy or natural selection or chance into our Designer, Creator, and God because they will have no other.

So, what's wrong with the second law of thermodynamics?

Statistics and the scientific method were designed to eliminate chance as an explanation for the data, the science, and the scientific evidence. In Real Science, chance is NEVER allowed to be the causal explanation for anything.

Yet, what do we observe where the official definitions for entropy and natural selection are concerned? We observe these people defining entropy axiomatically as design and creation by chance, or design and creation by randomization. We see Boltzmann Brains and Big Bang Singularities being made by random fluctuations in entropy. We see DNA, RNA, genes, and proteins being made by chance or by random fluctuations in entropy. We see natural selection, which is entropy and death, designing and creating genomes and life forms at will. In complete violation of statistics and the scientific method, we see these people using entropy, chance, randomization, or disorder as our Designer, Creator, and God. These people literally use chance as the causal agent behind the creation of everything that exists. Science, the scientific method, and statistics were designed to eliminate chance as a causal explanation; yet, these people turn around and use chance to cause and to explain everything that has ever been experienced and observed. It's Bad Faith and Bad Science. It's religion and dogma instead. It is philosophy, metaphysics, and magic. These people literally turn entropy or chance into a supernatural being who can design and create anything that it sets its mind to, in complete violation of statistics, the scientific method, and everything that science is supposed to stand for.

The disorder definition for entropy, or the second law of thermodynamics, or materialism, naturalism, and atheism IS the hidden lie that has been enshrined within physics. The whole thing is a lie, and we don't even know it because we have been brainwashed to believe it.

Likewise, creation by entropy or creation by natural selection is the hidden lie that has been enshrined and sanctified and axiomized within Biology, Materialism, Naturalism, and Darwinism; and, we are just beginning to realize that it is false, because we were brainwashed and intimidated into believing that it is true while we were in school.

Materialism, Naturalism, Darwinism, Nihilism, and Atheism were designed to trick us and deceive us. They turn everything into a lie, including biology and physics.

Entropy, and entropy defined as disorder, cannot design and create anything. Natural selection is by definition "design and creation by entropy" because these people really truly believe that ONLY entropy and physical matter exist. Natural selection is creation by entropy and death. They've been deceived, and they don't even know it, because they have chosen to believe that entropy, or heat death, or natural selection can design and create anything that it sets its mind to.

This becomes possible within their minds because they have in fact defined entropy as all of the energy in the universe, all of the heat in the universe, and all of the disorder in the universe. Once they have defined entropy as everything in our universe, then it becomes obviously clear to them that entropy can design and make anything in this universe that it decides to create.

This is a *category error* logic fallacy. They *conflate* entropy with energy and syntropy and psyche, so that entropy can then do anything that energy and syntropy and psyche can do. They cheat in order to make entropy, natural selection, and the second law of thermodynamics our Creator and our God.

Self-deception works, and it works every time, especially where genius level Einsteins are concerned. They want entropy and the second law of thermodynamics to be true, so they start by defining entropy as everything in the universe so that entropy is axiomatically true right from the very beginning before they start to try to find ways to make even more entropy from there.

However, these people break parsimony whenever they erroneously define entropy as something that it is not. These people break parsimony whenever they define entropy as randomness, disorder, or chance. These people break parsimony whenever they turn chance or entropy into our Designer and our Creator. By defining entropy as chance and by making entropy or chance our Creator, these people break science and parsimony, and they don't even know it. They make entropy or chance into something that it is not, and then they go one step further and have entropy or chance designing and creating physical matter, genes, proteins, genomes, eyes, brains, planets, stars, and galaxies at will. It violates parsimony, because we ALL instinctively KNOW that chance or entropy or disorder can't organize and create anything at all. These people constantly have entropy or chance doing what it can't do, so that entropy or chance can be their Designer, Creator, and God. But, it breaks parsimony every they choose to do so.

Entropy should be defined as what it really is – thermodynamics – and leave it at that. But nope! These people have to turn entropy into our Designer, Creator, and God. Remember, nevertheless, that it breaks logic, parsimony, physics, and the rules of science every time they turn entropy into something that it is not – just so that they can have entropy or chance be our Designer and our Creator. These people don't want the truth. They want the deceptions and the lies instead.

Entropy is thermodynamics. No thermodynamics, then no entropy. Simple. Parsimonious. Logical. True.

This is NEW and BETTER Science here. This is UPGRADED Science.

As you can see, I have worked it and reworked it over and over again in my mind trying to make sure that I didn't miss anything. In the process, I have had to break my conditioning or break my brainwashing. The Human Psyche can break its conditioning at will; but, it isn't always easy. I used to be a materialist, naturalist, nihilist, and atheist; and, these NEW scientific discoveries continue to cause me a lot of cognitive dissonance. They run counter to everything that I have been trained, and brainwashed, and conditioned to believe. These NEW scientific discoveries are exciting for me and scary for me all at the same time. The thing that drives me forward is that I sincerely want to find and know the truth, more than anything else.

Now, don't worry about the second law of thermodynamics. It's perfectly safe. It's not going anywhere. The science community made the second law of thermodynamics into an axiom and a law so that it can't be falsified. You can falsify it all you want, and it will still be true. The second law of thermodynamics is untouchable. There's nothing in the world that you can do to falsify it because it will always be true. It was designed to be that way, and it will stay that way.

Mark My Words

# The Correlation Coefficient of Zero Falsifies the Second Law of Thermodynamics

One of the primary rules of Science, Statistics, and the Scientific Methods states that correlation is not causation.

Thermodynamics is defined as a transfer of heat from a hot reservoir to a cold reservoir. Thermodynamics is the scientific study of heat – what it is and how it works.

The second law of thermodynamics defines entropy as all of the disorder in this universe.

There's a conceptual problem with that particular definition for entropy and thermodynamics. Disorder by definition has absolutely NO correlation whatsoever with heat, thermodynamics, or entropy. Disorder, by definition, is a correlation coefficient of zero ($r = 0.0$). Disorder, by definition, has NO correlation with anything.

Whenever the second law of thermodynamics is defined as "disorder", we are dealing with a purely *a priori* situation, which means that NO data has been collected. The second law was established without collecting any data because it is impossible to collect meaningful data on disorder! The second law defined as disorder is purely a thought experiment, and nothing more, because they can't collect any meaningful data on disorder! It's unscientific because there is no data that can be collected and no experiment that can be run in real life. They are determining probability without collecting any data, and in this case, the probability of "pure disorder" correlating with anything is ZERO according to the rules of correlation and statistics. It's impossible to collect data on disorder, because pure disorder by definition has a correlation coefficient of zero, which means that it doesn't correlate with anything. Whenever we collect data on "disorder", by definition and in principle or practice the data is completely meaningless if you are in fact dealing with pure disorder.

**Remember, the second law of thermodynamics or "entropy defined as disorder" was determined axiomatically without collecting any data. It was deduced by reason alone. It is nothing but a thought experiment. The second law of thermodynamics has NO correlation with reality.**

It's against the rules of Science and Statistics to correlated "disorder" or "random chance" with heat, thermodynamics, entropy defined as thermodynamics, life, genomes, planets, universe, or anything else that you can possibly imagine, because each of these things have an inherent amount of order built directly into them. Yet, that's precisely what the Materialists, Naturalists, Darwinists, and Atheists do. These people deliberately correlate or equate "disorder" or "chance" with entropy and natural selection, and then these people go one step further and make chance, entropy, natural selection, or disorder the CAUSE of everything that exists in our universe. As a result, their "science" and their interpretations of science end up being nothing more than unscientific philosophy or unscientific religious dogma.

Showing that a correlation exists between two variables is often the first step towards proving that they are causally related. In contrast, if a correlation does not exist between the two variables, a causal relationship can be ruled out. By definition there is NO correlation between disorder and anything else. Once we start detecting correlation, then we are in fact detecting some type of order or organization in the system. By the laws of statistics and correlation, disorder has NO causal relationship with anything because it has NO correlation with anything. According to statistics, the rules of causation, and the rules of correlation, "disorder" and "entropy defined as disorder" can't be the cause of anything because they don't correlate with anything.

Whenever two variables are correlated, there are four possible explanations for that correlation. First, the correlation or relationship between X and Y is spurious or caused by random chance. This is an invalid correlation or a false correlation. Second, X is the cause of Y. Third, Y is the cause of X. Fourth, a third variable is the cause of the correlation between X and Y. We cannot use correlation to establish causality. Correlation is not causation! This is one of the primary rules of Science, Statistics, and the Scientific Methods.

Remember, the whole purpose for Science, Scientific Experiments, Statistics, Independent Variables, Dependent Variables, and the Scientific Methods is to eliminate chance as the cause of any correlations we might observe. Chance is defined axiomatically in science and statistics as an invalid cause or a spurious cause. There can be NO causal relationship between chance and anything else. The rules of logic, science, and statistical analysis don't allow chance to do causality. There can be chance correlations, but there can NEVER be chance causations, because chance isn't reliable nor replicable. Chance cannot do causality. And, that's the fatal flaw in Materialism, Naturalism, Darwinism, Nihilism, Behaviorism, Determinism, Atheism, Classical Realism, and the Second Law of Thermodynamics – these people make chance the cause of everything that they do in science, in complete violation of the rules of statistics, science, and scientific methodology. These people literally make disorder, chance, or entropy the cause of everything that exists.

These people cheat in an attempt to convince us that their false and falsified ideas are true.

Correlation does NOT imply causation, except when you are dealing with entropy, or natural selection, or creation by chance, or creation by random disorder, or the second law of thermodynamics, because these things have been granted a *special exemption* and are immune to the rules of Science, Statistics, Logic, and the Scientific Method. By axiom and by decree, it is forbidden to falsify "entropy defined as disorder", natural selection, creation by chance, or the second law of thermodynamics. These are declared immune to falsification by axiom, by law, and by decree so that they no longer have to correlate with anything in order in order to be the cause of everything.

In real life, however, to establish causation, one of the variables must be independently manipulated and its effect on the other variable measured. This IS the Scientific Method. This is how science experiments are supposed to be done.

Notice carefully here that there is NO way to independently manipulate disorder so that we can then measure its effect on heat, or entropy, or thermodynamics, or life! Once we have manipulated disorder, then it is NO longer disorder but is instead some type of order because it has been manipulated. There's NO way to do science or the Scientific Method with disorder or "entropy defined as disorder". Once we start to manipulate it, then it is no longer disordered. Once the system has magically attained "pure disorder" or a correlation coefficient of zero, then it has nowhere to go but up into some type of order. Contrary to the claims of the Materialists, Naturalists, Darwinists, and Atheists, disorder is NOT self-sustaining! Once we start to manipulate it, then we introduce some type of order into it!

Likewise, once Nature or Nature's Psyche starts to manipulate the energies and starts to form that energy into Quantum Fields, Photons, Electrons, Dark Matter or Spirit Matter, and Physical Matter, then we are no longer looking at some type of chaos or disorder but are looking at some type of order instead. And, that QUANTUM ORDER remains in existence even at thermal equilibrium, absolute zero temperature, or the complete randomization of the atoms or molecules within a gas. Heat death or maximum disorder at the physical level doesn't touch nor change the QUANTUM ORDER that exists at the quantum level. They are not the same thing at all. They don't correlate with each other one bit. The QUANTUM ORDER continues to exist and function perfectly at the quantum level even when the system is in maximum disorder, thermal equilibrium, or heat death at the physical level.

**Remember, according to $E = mc^2$ and Quantum Field Theory and the Conservation Laws, Psyche or Intelligence or Quantum Consciousness can eliminate disorder and any type of entropy AT WILL, both at the quantum level and at the physical level.**

Disorder completely ceased to exist in our part of the multiverse when the Gods, Controlling Psyches, and Nature's Psyche made the Quantum Fields. The Quantum Fields are pure syntropy, pure order, and pure organization; and, the Quantum Fields continue to exist and continue to function perfectly at absolute zero, at the heat death of a physical system, within stars, within black holes, and whenever a physical system somehow manages to achieve maximum disorder or maximum randomization of its atoms. No matter what happens at the physical level, the perfect order of the Quantum Fields continues to exist at the quantum level.

**Remember, disorder was completely eliminated in our part of the multiverse when the Gods, the Controlling Psyches, and the Psyches within Nature created the Quantum Fields. The Quantum Fields are pure order or pure syntropy. The Quantum Fields are the ultimate perpetual motion machines. Now that they have been made or organized, they will never wear out.**

Physical matter is a drop in the ocean compared to the Quantum Fields; and, the Quantum Fields are pure, massless, heatless, and entropyless syntropy, order, exergy, and perfection. The Quantum Fields are a correlation coefficient of 1.00. They correlate with everything and sustain everything in our part of the multiverse. The Quantum Fields are the construct from which everything is made. They are pure order or pure organization.

The Quantum Fields are the exact opposite of "disorder" and the "second law of thermodynamics defined as disorder". Disorder, the second law of thermodynamics, and entropy defined as disorder have a correlation coefficient of ZERO with everything in the universe. In contrast, the Quantum Fields correlate perfectly with everything in our universe. The massless, heatless, entropyless, and non-physical Quantum Fields and Quantum Field Theory FALSIFY Materialism, Naturalism, Darwinism (Creation by Chance or Creation by Entropy), Nihilism, Atheism (Creation Ex Nihilo), and the Second Law of Thermodynamics. The two are mutually exclusive. If one of them can be demonstrated to be true, then the other has been falsified. This is Science. This is the Scientific Method in action!

Disorder can't be scienced because once we start to science it, then we introduce some type of order into it. In other words, once a physical system has hit maximum disorder, there is no place to go but up towards some type of order or organization.

Science is an attempt to establish causation, and that typically starts by establishing some type of correlation. There is NO way in the universe to establish correlation with disorder or a correlation coefficient of zero (**$r = 0.0$**). If there is NO correlation between variables, then we KNOW for a fact that there is no causal relationship between the variables; and then, we KNOW for a fact that it is impossible to do science with those two variables because NO causal relationship can ever be established between those two variables. Whenever one of those variables is "disorder", there is NO correlation and can be NO correlation with that disorder and anything else.

By definition, disorder IS a correlation coefficient of zero. It is disorder because it doesn't correlate with anything. It is a *category error* logic fallacy to define anything as disorder, because in order to define something scientifically or statistically it has to correlate with something in the first place. Disorder, by definition, correlates with nothing. Therefore, there's no way in the universe that disorder can correlate with entropy or thermodynamics, because there is NO causal relationship between disorder and thermodynamics. Once you have thermodynamics, you automatically have some level of order or some type of order within the system already.

Disorder does NOT cause heat, and heat does NOT cause disorder. There is NO correlation between the two! Heat of any kind represents some type of order or organization! Absolute zero temperature Quantum Fields also represent perfect order and organization – perfect perpetual motion machines. Pure disorder or pure chaos is impossible in our part of the multiverse so long as the Quantum Fields exist.

Remember, disorder is the wrong definition for entropy and thermodynamics, because you can't science it or do science with it, because disorder doesn't correlate with anything and therefore disorder doesn't cause anything. It's impossible to establish a causal relationship between disorder and anything else, which means that it is impossible to do science with disorder or anything that is defined as disorder. It is impossible to establish a causal relationship between heat and disorder, or thermodynamics and disorder. Remember, the purpose of Science and the Scientific Method is to establish a causal relationship between two variables; and, that is impossible to accomplish when one of those variables is disorder or a correlation coefficient of zero, where there is NO correlation to be found.

Here we use Statistics, the Philosophy of Science, and a Correlation Coefficient of Zero to demonstrate clearly and conclusively why the second law of thermodynamics and "entropy defined as disorder" is a scientifically invalid definition for entropy and thermodynamics. Entropy defined as disorder or the second law defined as disorder is self-defeating or self-falsifying; and, our genius scientists don't even know it because they don't understand Statistics and the Philosophy of Science. They have never done the math and figured out how things really work where Statistics and the Scientific Method are concerned.

According to the Scientific Method and the Rules of Statistics, any correlation obtained by chance is an invalid correlation and falsifies the alternative hypothesis and forces us to conclude by retaining the null hypothesis, which means that we are forced to conclude that the independent variable (chance or disorder) IS NOT the cause of the dependent variable (entropy or thermodynamics or genomes or life).

The Rules of Statistics and the Scientific Method automatically FALSIFY the second law of thermodynamics whenever entropy is defined as chance or disorder. The Rules of Statistics and the Scientific Method automatically FALSIFY Materialism, Naturalism, Darwinism, Creation by Natural Selection, Chemical Evolution, Abiogenesis, Spontaneous Generation, Macro-Evolution, or any other Alternative Hypothesis whenever these things use chance as their causal agent or causal explanation. All of these Alternative Hypotheses are supposed to be FALSIFIED because it has been claimed and allegedly demonstrated that they were caused by chance. That's the way that Statistics and the Scientific Method works or is supposed to work. But these people grant themselves a *special exemption* from the Rules of Statistics and the Rules of Science in order to prove to themselves and the world that Darwinism, Materialism, Naturalism, and Atheism are true. They cheat.

This is powerful science because this is the Scientific Method! The explanatory power goes through the roof because I'm finally dealing with the truth.

Here I use the Rules of Statistics and the Scientific Method to FALSIFY creation by chance and the second law of thermodynamics. I use the Scientific Method the way that it is supposed to be used – to falsify and eliminate from science anything that was caused by chance. In science and statistics, we are supposed to falsify and eliminate everything that is explained by chance or is caused by chance. We are supposed to falsify entropy defined as chance or disorder, and we are supposed to falsify natural selection and the theory of evolution whenever they are defined as Creation by Chance. I'm using Statistics and the Scientific Method the way that they are supposed to be used, by using them to identify and eliminate from science anything that is obviously false. This is Pure Science, Good Science, or True Science that I'm trying to do here.

This is also forbidden science. This is the science that our atheistic and naturalistic college professors forbid us from doing or thinking about; and, that's one of the ways that we KNOW that it is true. These people have a knack for identifying and rejecting the truth. It's what they do. From what I can tell, all of the errors and falsehoods in science come from Creation by Chance, Creation by Natural Selection, Creation by Entropy, or Creation by Disorder. All of the errors in science come from the second law of thermodynamics and the theory of evolution. All of the errors in science come from Materialism, Naturalism, Darwinism, Nihilism, Scientism, Behaviorism, Determinism, Physicalism, Atheism, Creation by Natural Selection, Creation by Chance, Creation by Death or Entropy, and the Second Law of Thermodynamics.

Science defined as disorder FALSIFIES science defined as order. The two are mutually exclusive. If one of them can be demonstrated to be true, then the other one has automatically been falsified or proven false. So, which is it? Do we see an ordered and organized universe, or do we see complete and total random chaos instead? The two are mutually exclusive. Only one of them can truly exist.

## Quantum Field Theory Represents the Truth in Science

ALL of the different LAWS of Conservation falsify the second law of thermodynamics, which states that the amount of disorder or entropy in our universe is constantly increasing and can never decrease.

Quantum Field Theory and $E = mc^2$ represent the truth in science. You see, $E = mc^2$ is a completely reversible process according to Quantum Field Theory. That makes it good science. It's reversible and replicable. It's demonstrable and observable. It is order and organization, rather than complete and random chaos. It has actually been experienced and observed.

**According to the Law of Psyche, $E = mc^2$, and Quantum Field Theory, every psyche has a certain amount of energy that's under its direct control. That controlling psyche can form or transform the energy under its control into anything that it wants that energy to be, anytime and anywhere that it chooses to do so. The controlling psyche chooses what it wants to do with the energy that's under its control. This is the LAW of Psyche, Intelligence, or Quantum Consciousness. Psyche, intelligence, or consciousness has actually been experienced and observed. It's Nature's Psyche who collapses the wave function! The Psyches within Nature were collapsing wave functions long before the first particle of physical matter was designed and made.**

According to $E = mc^2$ and Quantum Field Theory, the Controlling Psyches in Nature can form or transform energy, quantum waves (virtual particles or thoughts), photons, and quantum fields INTO mass and heat anytime and anywhere that they choose to do so. They are doing so all the time, at will, according to Quantum Field Theory. And, the process is fully reversible. The Controlling Psyches in Nature can form or transform mass, heat, physical matter, entropy, resistance to acceleration, and heat storage capacity INTO massless, heatless, and entropyless photons, quantum waves (virtual particles), and quantum fields anytime and anywhere that they choose to do so. They are doing so all the time, at will, according to Quantum Field Theory. $E = mc^2$ is fully reversible according to the Law of Psyche and Quantum Field Theory.

This is what we actually experience and observe!

Whenever a heatless, massless, and entropyless photon chooses to stop, it transforms itself into heat and mass and resistance to acceleration instead. Then in reverse, according to the observations in Quantum Field Theory, whenever an electron and positron merge to form a photon, all of their mass, heat, and entropy is TRANSFORMED into infinite acceleration instead. Anytime mass, heat, and entropy are converted into massless, heatless, and entropyless photons, quantum waves, or quantum fields, we observe a complete reversal of $E = mc^2$, where mass, heat, and entropy are converted or transformed back into massless, heatless, and entropyless quantum fields, quantum waves (virtual particles), and photons. This is happening all the time all around us, and most of us are completely unaware of this process because we have chosen to believe that ONLY physical matter and entropy exist. Yet, $E = mc^2$ and Quantum Field Theory are scientific proof that this transversal process or reversal process is real and true.

According to Quantum Field Theory and $E = mc^2$, Nature's Psyche (or the Controlling Psyches in Nature) can transform mass, heat, and entropy into massless, heatless, and entropyless photons, quantum fields, and infinite acceleration anytime and anywhere that it chooses to do so; and, there's nothing we can do to stop it. Likewise, it is obvious that every time a photon chooses to stop, it literally transforms its infinite acceleration and speed-of-light velocity INTO mass and heat and resistance to acceleration instead. This transformation back and forth is constantly being experienced and observed. It didn't happen just one time during the Big Bang. It's constantly ongoing to this very day, according to Quantum Field Theory and $E = mc^2$.

Nature's Psyche is constantly transforming mass, heat, and entropy back INTO massless, heatless, and entropyless quantum waves (virtual particles), quantum fields, and photons all the time whether we realize it or not. It's happening at the quantum level every second of the day, according to Quantum Field Theory and $E = mc^2$. Where do you think the light comes from in the first place? It doesn't just spring into existence from thin air as the Materialists and Naturalists claim. Light is made whenever Nature's Psyche transforms mass, heat, and entropy back into massless, heatless, and entropyless photons and infinite acceleration instead. Photons go to the speed-of-light instantly with an infinite acceleration; and, this is possible because they have transformed their mass, heat, entropy, and resistance to acceleration into massless, heatless, and entropyless infinite acceleration instead. This is obviously true, because it's constantly being experienced and observed every second of our lives. The only reason nobody has ever discovered it before now is because everyone has convinced themselves that only physical matter and entropy exist.

https://www.khanacademy.org/science/physics/electric-charge-electric-force-and-voltage/charge-electric-force/v/conservation-of-charge

ALL of the different LAWS of Conservation falsify the second law of thermodynamics, which states that the amount of disorder or entropy in our universe is constantly increasing and can never decrease.

Quantum Field Theory, $E = mc^2$, and the Conservation Laws made it clear to me why the second law of thermodynamics is false. According to the second law, entropy is axiomatically defined as some type of substance or entity whose amount is always increasing and whose universal amount can never decrease. Such a thing has never been experienced nor observed anywhere in nature or in our universe because it is impossible for such a thing to exist. It violates the conservation laws. The second law is Creation Ex Nihilo. It's magic. It's the creation of something from nothing by nothing. It violates the conservation laws. The first law of thermodynamics is mutually exclusive to the second law of thermodynamics. If one of them can be demonstrated to be true, then the other has been automatically falsified, because they can't both be true at the same time.

According to the Conservation of Energy and the Conservation of Charge, there can be NO substance or entity whose amount is constantly increasing. There can be NO creation ex nihilo or creation from nothing. Everything is made out of energy, and the amount of energy can never increase. Energy can only be transformed from one thing to another. If the second law of thermodynamics is true, then all of the conservation laws are automatically false. It's that simple.

So, which is it?

Do we have some invisible, immaterial, hypothetical, theoretical, fictional entity or substance whose amount is constantly increasing; or, do we have the Conservation of Energy which says that the amount of energy can never increase and can never decrease? Do we have true conservation, or do we have creation ex nihilo? Which one has actually been experienced and observed? Which one has actually been verified by science? Which one is replicable, demonstrable, and usable on demand? Which one does work?

Conservation means that its amount cannot increase, and its amount cannot decrease. The second law of thermodynamics and creation ex nihilo violate the Conservation Laws.

Quantum Field Theory, $E = mc^2$, and the Conservation Laws made it clear to me why the second law of thermodynamics is false. And, that's precisely what we need in science. We need the false and the fictional being falsified and eliminated by the things that have actually been experienced and observed. According to Science 2.0, science is eye-witness observation and first-hand experience. I'm going with the Conservation of Energy; and then, I'm using that to falsify the second law of thermodynamics. The two are mutually exclusive after all. Energy and the conservation of energy have actually been experienced and observed; whereas, creation ex nihilo and an ever-increasing amount of substance made out of nothing has never been experienced nor observed. Not even God can do creation ex nihilo, so there's no reason to believe that the second law of thermodynamics can.

**Quantum Field Theory, Conservation of Charge, Conservation of Energy and Psyche, the Law of Psyche, and $E = mc^2$ explain to us how mass, heat, resistance to acceleration, heat storage capacity, charge, and entropy get transformed INTO massless, heatless, entropyless, and chargeless photons, quantum waves, and quantum fields. It's happening all the time according to Quantum Field Theory. Just look at our sun! The Quantum Fields are perfectly conserved. The Quantum Fields are the ultimate perpetual motion machines. $E = mc^2$ works perfectly both**

**directions all the time thanks to the Quantum Fields; and, ALL of the energy is conserved.**

This has got to be my most favorite science discovery of all time.

By falsifying the second law of thermodynamics and replacing it with Quantum Field Theory, $E = mc^2$, the Law of Psyche, and the Conservation of Energy and Psyche, instantly we KNOW that the Big Bang was not a one-time event but is ongoing to this very day. We KNOW that a miniature big bang happens every time a photon chooses to stop and transforms its massless, chargeless, heatless, and entropyless infinite acceleration and speed-of-light velocity into mass, heat, charge, resistance to acceleration, and heat storage capacity instead. This process of creating new mass and new heat out of massless, entropyless, and heatless photons, quantum waves, and quantum fields is ongoing to this very moment in time. It didn't happen just once during some hypothetical Big Bang. It's happening all the time right now all around us and every time we step outside into the sun. As those massless, heatless, chargeless, and entropyless photons stop, they transform themselves into heat and mass (or resistance to acceleration), and we can feel it on our skin. This is not theory. It has actually been experienced and observed. It's already proven and verified science; and, it actually falsifies the second law of thermodynamics.

Instantly, we have a scientific explanation for how something like the Big Bang – or the creation of mass, heat, entropy, and physical matter from massless, entropyless, heatless, syntropic Quantum Fields – could actually be accomplished anytime and anywhere in this multiverse.

Furthermore, according to Quantum Field Theory, $E = mc^2$, the Law of Psyche, and the Conservation of Energy and Psyche, we KNOW for a fact that Nature or Nature's Psyche can transform mass, heat, charge, and entropy back into massless, heatless, chargeless, and entropyless photons, quantum waves (virtual particles or thoughts), and quantum fields anytime and anywhere that it chooses to do so. It is Nature's Psyche who collapses the wave function; so, this whole $E = mc^2$ process is reversible at will under the control of Nature's Psyche. Entropy loses its significance and its sting the very moment we realize that Nature's Psyche can transform mass, heat, charge, and entropy into massless, heatless, chargeless, and entropyless photons, quantum waves, and quantum fields AT WILL. Instantly we KNOW where the light comes from. Instantly, we KNOW how our sun converts mass, heat, and entropy into massless, heatless, and entropyless photons or light. Instantly, everything in science is explained.

**Entropy Defined as Disorder Is Self-Defeating**

One day, I finally chose to believe the science rather than believing the wishful thinking and philosophical speculation of the Materialists, Naturalists, Darwinists, and Atheists. One day, I finally realized that these people are always wrong. One day, I finally decided that it was time for me to abandon the falsehoods and lies, and to go with the truth instead. I used to be a materialist, naturalist, nihilist, and atheist until Science, Statistics, the Scientific Methods, the Conservation Laws, and Quantum Field Theory convinced me that I was wrong.

Nature or Nature's Psyche tends toward or strives toward order and organization, NOT disorder and disorganization. That's the nature of true intelligence. Intelligence or psyche or quantum consciousness is constantly seeking for higher orders of purpose, order, organization, and structure; and, so is the human psyche.

The second law, "entropy defined as disorder", or "thermodynamics defined as disorder" has absolutely nothing to do with science, because science or physics is the study of the order and organization within our universe. Science is an attempt to establish causation between two variables, and that starts by finding a correlation or a relationship between the two variables. It is impossible to establish a correlation between disorder (a correlation coefficient of zero) and anything else. It can't be done, which means that disorder can't be scienced! Science, entropy, and thermodynamics "defined as disorder" is self-defeating or self-falsifying, whether we realize it or not. Statistics and the lack of correlation falsifies it. That's what statistics is supposed to do. Statistics is supposed to eliminate from science anything that was caused by chance!

Here's some of the best science and statistics and theory that I have ever done in my life. The explanatory power goes through the roof when we finally have the truth.

According to the Philosophy of Science, the FLAW within the Scientific Method consists of all those different and contradictory personal interpretations or wishful thinking that takes place at the end of the Scientific Method when it comes time to interpret the meaning of the data that has been derived from the science experiment. Statistics was designed to mitigate, lower, or minimize the FALSE personal interpretations associated with the Scientific Methods, and to help point us towards a true interpretation of the data that IS NOT based upon chance or wishful thinking. Obviously, you can't do science with disorder because you can't correlate disorder with anything. You can't run science experiments on disorder for real, because you would first have to organize it or order it in some way before you could find some kind of correlation in the data. There can be NO causal relationship between "disorder or chance" and anything else, because disorder or chance can't be reliably correlated with anything. By definition, there's NO replication possible with disorder or chance. They can't run reliable science experiments on chance or disorder. Creation by disorder or creation by chance can't be scienced because they are automatically FALSIFIED or automatically REJECTED according to the Rules of Science, the Rules of Statistics, the Rules of Causation, and the Rules associated with the Scientific Method.

You can't correlate disorder with anything; so, what the Materialists, Naturalists, Darwinists, and Atheists do is that they correlate or equate disorder with entropy, and then they make that entropy or disorder or chance the cause of everything that exists. These people literally circumvent Statistics, Logic, the Scientific Method, and the Rules of Science by making chance, disorder, natural selection, or entropy the cause of everything that exists.

They lie and cheat in order to make us think that their personal interpretations are true. They resort to magic and trickery and deception in order to make us believe that Materialism and Naturalism and Darwinism are true. They also hide and destroy and ban any scientific evidence that falsifies their personal interpretations. They forbid us from falsifying creation by chance, creation by natural selection, or creation by entropy. They cheat. They label everything that falsifies their beliefs, and they call it pseudo-science. They cheat.

Disorder, by definition, has NO correlation with anything; so, what these people do is turn everything upside-down and make disorder or entropy the cause of everything. They cheat! That is what they do, is it not? They break all the rules of Logic, Statistics, Correlation, Causality, and Science in order to trick us into believing that Materialism, Naturalism, Darwinism, Nihilism, and Atheism are true and the cause of everything that exists. They literally correlate disorder with thermodynamics, thereby giving us the false impression that thermodynamics causes disorder or that disorder somehow causes thermodynamics. Then they go one step further and make disorder or entropy or chance the cause of everything that was ever made. It's a kludge. It's a scam. We fall for it because we desperately want to believe that it is true. We want to fit in and be "scientists" like them.

There is NO relationship whatsoever between "disorder" and heat, which means that there is no correlation between "disorder" and thermodynamics. Disorder doesn't cause heat; and, disorder doesn't eliminate heat either! The amount of disorder doesn't do anything to change the amount of heat! There is NO correlation between the two. Highly disordered gas or plasma can contain a lot of heat; yet, highly disordered gas can also exist at absolute zero. The amount of heat has absolutely nothing to do with amount of disorder in a system. Likewise, highly ordered and structured mass can exist in a hot dynamic system like our sun, or it can exist closer to absolute zero within a neutron star or black hole. Disorder has absolutely nothing to do with heat or thermodynamics. There's NO correlation between the two. Disorder doesn't cause heat or thermodynamics, and thermodynamics or heat doesn't cause disorder either. There's no correlation between the two.

The payoff is strong with this one because we are dealing with statistics, correlation, causation, and the Scientific Methods as well. We are dealing with Science! Entropy and the second law of thermodynamics defined as "disorder" FALSIFIES entropy and the second law defined as "thermodynamics" or the "transfer of heat". The two are mutually exclusive. They falsify each other. If one of them can be demonstrated to be true, then the other one has been falsified or proven false. This is intense, is it not? That's what happens when we finally find the truth. It gets really intense because it starts to falsify our favorite beliefs.

Anything in Science defined as chance or disorder is self-defeating or self-falsifying according to Statistics, the Rules of Causality, and the Rules of Science. If the results of a science experiment were caused by chance or disorder, we are supposed to retain the null hypothesis and conclude that the independent variable (chance or disorder) had NO EFFECT on the dependent variable (entropy or thermodynamics). That's the way Statistics and the Scientific Method work.

The Materialists, Naturalists, Darwinists, Nihilists, and Atheists circumvent Statistics, the Philosophy of Science, and the Scientific Method by axiomatically defining chance, entropy, or disorder as the cause of everything that exists in our universe. These people literally reject the null hypothesis and jump straight to the conclusion that the independent variable (chance or disorder) is the CAUSE of the dependent variable (everything that exists).

They are wrong! Statistics, the Rules of Causality, and the Scientific Method prove that they are wrong.

Disorder doesn't cause heat, and disorder doesn't eliminate heat. There's no correlation between them. There's NO correlation between disorder and thermodynamics. There's NO correlation between disorder and anything. Therefore, there is NO causal relationship between disorder and thermodynamics. They are not the same thing at all. Thermodynamics of any kind represents a certain amount of order and organization to begin with, because thermodynamics correlates with heat and temperature! There is order and organization and correlation there within mass, thermodynamics, and heat. Whereas, disorder doesn't correlate with anything. Once disorder starts to correlate with something, then it is NO longer disorder but is some type of order instead. Using disorder or chance as the causal explanation for the scientific data or the dependent variable is self-defeating and self-falsifying; or, it's supposed to be.

This means that defining entropy or the second law of thermodynamics as "disorder" is the flawed, unscientific, uncorrelated, and illogical definition for entropy and the second law. There's NO correlation between "disorder" and "entropy defined as heat" or "entropy defined as thermodynamics". The proven and observable fact that there is NO correlation between thermodynamics and disorder FALSIFIES the second law of thermodynamics whenever the second law axiomatically defines entropy or thermodynamics as "disorder".

There is no correlation between entropy and disorder whenever entropy is defined as thermodynamics or the "transfer of heat from a hot system to a cold system". That hot system and cold system represent a great amount of order and organization to begin with, and it stays that way even at thermal equilibrium or absolute zero. The physical matter, or mass, or heat storage capacity continues to exist, even at thermal equilibrium or absolute zero. Does it not? The order and organization in the system continues to exist, and the physical matter continues to exist, even at thermal equilibrium or absolute zero. Does it not? That very fact and reality falsifies the second law of thermodynamics and tells us clearly and conclusively that entropy and thermodynamics have absolutely nothing to do with disorder, and disorganization, and the randomization of a physical system.

Disorder doesn't correlate with anything. It can't. Once disorder starts to correlate with something, then it is NO longer disorder. Disorder can't correlate with thermodynamics, heat, or "entropy defined as thermodynamics". Once a correlation has been made, then we are no longer looking at disorder but some type of order or organization instead. Disorder can't correlate with entropy or thermodynamics; yet, what do we observe. We observe the Materialists, Naturalists, Darwinists, Nihilists, Behaviorists, and Atheists correlating or equating entropy and thermodynamics with "disorder" and calling it science. Then they go one step further and they turn this disorder or entropy or chance into the Ultimate Causal Agent of all time. Then before you know it, they have entropy or disorder or chance making Big Bang Singularities, Physical Matter, Planets, Stars, Galaxies, Genomes, Boltzmann Brains, Proteins, and Life Forms out of thin air from nothing.

This is what they do, is it not?

The Darwinists believe that ONLY entropy and physical matter exist. Consequently, the theory of evolution automatically devolves into Creation by Entropy or Creation by Chance. Evolution is entropy. That's the ONLY thing it can be according to the Darwinists. According to these people, evolution or entropy made everything that exists. Whenever we are selected against, we die (natural selection) or we are never born in the first place (sexual selection). Natural selection is literally Creation by Entropy, Creation by Chance, or Creation by Death. That's the ONLY thing it can be because entropy, death, and physical matter are the ONLY things that the Darwinists believe in.

The whole thing violates correlation, causality, science, statistics, common sense, logic, and the scientific method.

But that's what these people do, is it not?

What they do is they take something that has NO correlation with anything – namely disorder or chance – and then they magically turn it into the causation of everything, in complete violation of the scientific and statistical LAW that states that correlation is not causation. According to Science, the Scientific Methods, and Statistics, if there is no correlation, then there can be no causal relationship; yet, they take something that has no correlation with anything ("entropy defined as disorder" or "chance") and turn it into the cause of everything that exists, because they have chosen to believe that only physical matter and entropy exist.

Since they have chosen to believe that ONLY physical matter and entropy exist, they have automatically and erroneously chosen to believe that entropy or disorder or chance is the cause of the physical matter and the cause of everything that Nature or Nature's Psyche has made from that physical matter or mass. They have erroneously chosen to believe that entropy or chance is the cause of everything, in complete defiance of the rules of Science, Statistics, Logic, and Scientific Methodology.

If you are willing to observe and think about it carefully, it soon becomes obvious that the relationship between heat and disorder is non-existent. Disorder doesn't make heat; and, disorder doesn't eliminate heat either. There's zero correlation between the two. That means that we have proven scientifically and statistically that there is NO causal relationship between disorder and heat. Disorder can exist in both hot and cold systems. Supreme order and organization can exist in both hot and cold systems. There's absolutely NO correlation between disorder and thermodynamics. Yet, the Darwinists, Naturalists, and Atheists correlate or equate the two and make them one through the second law of thermodynamics, so that disorder, chance, or entropy can be our Designer, our Creator, and our God. Creation by Chance or Creation by Entropy becomes their causal mechanism for the construction of everything that exists in our universe, including Nature's Psyche, Quantum Mechanics, Energy, and the Quantum Fields. These people literally make chance the cause of everything.

If you know anything about correlation, the null hypothesis, science, statistics, and the scientific method then you KNOW that it is wrong for them to do so. The null hypothesis in a science experiment states that the results of the experiment were caused by chance or created by chance, which means that the independent variable had NO EFFECT on the dependent variable. That means that there is NO causal relationship between the variables, and that means that your alternative hypothesis is most likely false.

It is also obvious that Nature or Nature's Psyche tends towards order or makes order at will, especially at the quantum level where the Quantum Fields are concerned. The Quantum Fields represent perfect order, and they were made by Psyches within Nature who control the energy from which the Quantum Fields were made. The Quantum Fields are now eternal perpetual motion machines thanks to the Conservation of Energy and Psyche. ONLY the psyches within nature can actually destroy them or dissolve them back into the chaos from whence they came.

It is Nature's Psyche (or the Controlling Psyches within Nature) who collapses the wave function, which means that it is Nature's Psyche who transforms at will the energies within the Quantum Fields INTO quantum waves, particles, photons, quarks, electrons, dark matter or spirit matter, mass, and physical matter. It's the ONLY thing that can. Physical matter wouldn't exist without Nature's Psyche and the Quantum Fields. It is obvious that the physical matter didn't cause Nature's Psyche or the Quantum Fields, because physical matter can't exist without the Quantum Fields, and the Quantum Fields wouldn't exist without Nature's Psyche. First things first. Psyche and energy have always existed, and they will always exist. They cannot be made, and they cannot be destroyed, because they are constantly conserved. Conservation of Energy and Psyche is the REAL law of thermodynamics, quantum mechanics, quantum field theory, and everything else. Conservation of Energy and Psyche is where science begins and chance or chaos ends.

Where there is NO correlation, it is obviously wrong to conclude that there is a causal relationship between disorder and thermodynamics. Where there is NO correlation, it is obviously wrong to conclude that there is a causal relationship between entropy and disorder. Where there is NO correlation between disorder and anything else, the Materialists and Naturalists and Atheists establish a causal relationship between disorder (defined as entropy or chance) and everything that exists in our universe; and by axiom or decree through the second law of thermodynamics, they make that disorder or chance or entropy the cause of everything that exists in our universe. Correlation does not imply causation; yet, the Materialists and Naturalists take something that has zero correlation with anything (disorder or "entropy defined as disorder and chance"), and they literally turn it into the cause of everything.

They cheat in order to establish the truthfulness of their beliefs. It's what they do. It's their reason for being. They desire to trick you and deceive you just as they have been tricked and deceived. Self-deception works, and it works every time, especially when it comes to PhD scientists who believe that they know everything already.

Study statistics and correlation to find out how science is supposed to work.

By definition, maximum disorder means NO correlation or NO relationship between the variables. If you can find correlation between two variables, you have in fact detected some type of order or organization within the system being examined. Yet, what do we see from our materialistic, naturalistic, and atheistic scientists? They literally take disorder and correlate it with thermodynamics or entropy, and then they call it science. They do this so that chance, disorder, or entropy can magically become our Designer and our Creator. These people literally imbue disorder, chance, natural selection, or entropy with the ability to create order and organization out of thin air at will. These people turn entropy or chance into our Creator and our God; and then, they accuse everyone else of being irrational if they don't follow suit.

One day, after fifty years of thinking about it all, I finally realized that we will never have the truth in science as long as it remains forbidden to falsify the falsehoods that they have enshrined into science by turning them into unfalsifiable axioms or laws. We have to be willing to identify and falsify the falsehoods in science if we truly want to find and know the truth. That's the way that science is supposed to work – but it doesn't work that way in actuality thanks to the Materialists, Naturalists, Darwinists, Nihilists, Behaviorists, Physicalists, Classical Realists, and Atheists who have stolen science and axiomatically defined science as Materialism, Naturalism, Darwinism, Nihilism, Behaviorism, Determinism, Atheism, and Creation by Chance instead.

We don't observe or experience Creation by Chance, Creation by Entropy, or Creation by Death in nature and in our laboratories; yet by axiom or by decree, the Materialists, Naturalists, and Darwinists have made chance or entropy or death the cause of everything that exists.

Here I have used Science, Statistics, and the Scientific Method to FALSIFY creation by chance, creation by entropy, creation by disorder, creation by natural selection, and the second law of thermodynamics. That's what science, statistics, and the scientific methods were designed to do, and we should let them do it. We should follow the evidence wherever it might lead us. The false if falsified by the truth; and, the truth is repeatedly and constantly experienced and observed.

Mark My Words

## The Second Law of Thermodynamics Is Protected and Unfalsifiable

Falsifying the second law of thermodynamics is a fun little hobby of mine, but you don't need to worry about the second law. It's not going anywhere. There's nothing you can do to falsify it because it is defined as being axiomatically true.

The second law of thermodynamics is untouchable and unfalsifiable, so my hobby is simply a way for me to exercise my brain cells and my writing skills, and nothing more. So, have fun with it! Just remember that the second law of thermodynamics remains true and remains safe, no matter how many different ways that we might find to falsify it. That's the magic of the second law of thermodynamics. It remains true, axiomatically true, no matter how many different ways we might falsify it. It's unfalsifiable because it's axiomatically true. It's a law that's not allowed to be broken. Anything that falsifies the second law of thermodynamics or natural selection is declared to be false. That's how these people operate. They actively and deliberately turn falsehoods into axioms, truths, and unfalsifiable laws. That's the very definition of Bad Faith and Bad Science.

In the summer of 2019, I used Statistics and my statistics course in college to falsify the second law of thermodynamics multiple different times in multiple different ways. By the end of the summer, I had written a couple of books on the subject. It is easy to falsify the second law once a person knows how to do it. Once I discovered how to do it, I started doing it right and left, coming and going.

Here's one of the ways that I did it.

I realized that "disorder" is represented graphically as a correlation coefficient of zero ($r = 0.0$). Whenever they define entropy or the second law of thermodynamics as "disorder", they are effectively defining entropy as a correlation coefficient of zero. Then you effectively have maximum disorder or maximum entropy, anytime they or you produce a scatter plot that has a correlation coefficient of zero. Defining entropy as maximum disorder or as a correlation coefficient of zero has statistical significance, as I will now try to demonstrate. It actually falsifies the second law of thermodynamics by destroying the causal relationship that they have established between disorder and entropy.

Entropy (n.) 1868, from German *Entropie* "measure of the disorder of a system," coined 1865 (on analogy of *Energie*) by German physicist Rudolph Clausius.

Ironically, it's physically impossible to measure the disorder of a system, because a perfectly disordered system has nothing to measure. You would actually have to organize it before you could measure it. A perfectly disordered system has a correlation coefficient of zero ($r = 0.0$), which means that there's nothing to correlate with it or nothing within it to measure of any significance. Defining entropy as disorder removes its claws and teeth because it can no longer be tested, measured, verified, or falsified.

Entropy, the disorder of a system, and the second law of thermodynamics are recent inventions, and they are demonstrably false. Entropy, defined as the measure of disorder, is an oxymoron – a contradiction in terms. Disorder can't be measured because, by definition, it has a correlation coefficient of zero!

I'm a generalist. I'm good at everything and master of nothing.

I've worked hard studying philosophy, science, physics, math, statistics, logic, quantum field theory, action at a distance, the non-physical or the immaterial, and quantum mechanics so that I can see and understand things that others refuse to see and understand.

When I first saw a scatter plot with a correlation coefficient of zero, I immediately recognized that I was looking at a graphical representation of disorder or a graphical representation of "entropy defined as disorder". It looked just like all those canisters of gas particles that I had been looking at in the different entropy videos that I had been watching online.

The main or primary definition for entropy, and consequently for the second law of thermodynamics, is "disorder", or "randomness", or "chaos", or "chance". In biology, entropy is defined axiomatically as "natural selection". It's the same thing in the end.

Whether they call it natural selection, chemical evolution, or the big bang, it is Creation by Entropy or Creation by Chance no matter what you choose to call it.

If you have ever studied statistics, then you KNOW that a perfectly disordered system or a perfectly randomized system has a correlation coefficient equal to zero ($r = 0.0$). Disorder is portrayed graphically as a correlation coefficient of zero. Entropy is defined axiomatically as "disorder" through the second law of thermodynamics. Therefore, entropy is also represented graphically as a scatter plot that has a correlation coefficient of zero. Anytime they encounter two variables that have zero correlation with each other, they automatically call it entropy. Thereby, entropy becomes anything that you want it to be.

Remember, entropy and maximum disorder are represented graphically as a system of molecules or atoms that have a correlation coefficient of zero ($r = 0.0$). By axiom or by decree, entropy and maximum disorder are declared to be the same thing, according to the second law of thermodynamics and according to Rudolph Clausius who coined the term "entropy". Something like this:

### ▸ No Correlation (horizontal line)

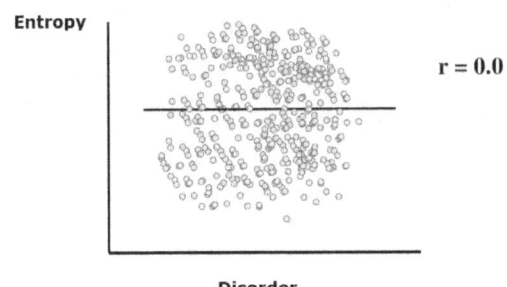

The scientists in general have axiomatically defined entropy as maximum disorder, maximum randomness, or maximum chaos. By axiom or law, they have equated entropy with disorder. These people have put these things on the same graph together and treat them as one. Through the second law of thermodynamics, they have literally defined entropy as a correlation coefficient of zero ($r = 0.0$), or as maximum disorder. As a result, entropy is meaningless and doesn't correlate with anything, because entropy can then be forced or made to correlate with everything.

After they have established this relationship between entropy and disorder and have defined entropy as all of the disorder in our universe, then they take the next quantum leap and define entropy as all of the energy in the universe or they define entropy as all of the heat in the universe.

By axiom or by decree, they make a causal relationship between disorder and entropy where there is none and can be none. Remember, there's nothing empirical about disorder. There can't be. If there were, then it would no longer be disorder. Consequently, it is inappropriate to call thermodynamics, heat, mass, heat storage capacity, or the "transfer of heat from a hot system to a cold system" by the name "disorder" as they do in the second law of thermodynamics. They are

NOT the same thing at all. If you can establish a correlation between two variables, you have in fact detected some type of order or organization within the system you are examining.

There is NO causal relationship between disorder and heat, but we scientists correlated and equated them anyway; and, we called the result entropy or the second law of thermodynamics. There is NO causal relationship between disorder and thermodynamics, but we scientists correlated and equated them anyway; and, we called the result entropy of the second law of thermodynamics.

The scientists cheated, and we allowed them to get away with it, because we don't understand what they are doing and take in on blind faith that it must be true.

By axiom, law, or decree, they establish a correlation between "disorder" and "thermodynamics or entropy" where there is none and can be none.

They do this implicitly, but here I do it explicitly in order to expose their deceptions and lies. That's the way we falsify anything in science – by exposing the deceptions, falsehoods, and lies. That's what scientists are supposed to do. But they don't do this in our science classes in college, because our science teachers in college aren't really scientists – they are preachers and teachers of Naturalism, Darwinism, Scientism, and Atheism instead.

They will never admit that this is what they are doing, but this is precisely what they are doing. They are taking variables that have zero correlation, and they are calling them entropy or correlating them with entropy. Since they have equated entropy with disorder, entropy can literally be anything that you want it to be. That's why there is a couple dozen different definitions for entropy on the books. For five decades, I could never figure out what entropy is because I could never get a consistent definition for entropy. Entropy ends up being anything they want it to be, because they make entropy correlate with everything.

Anything that has zero correlation is labeled as being entropy, or disorder, or chaos, or chance. Then they turn zero correlation into the cause of everything. It's a scam, and nobody realizes it, because nobody has figured out that this is what they are doing. These people have literally turned "no correlation" into causation. Correlation is one of the strongest and most useful concepts in statistics. NO correlation then NO causation. Yet, these people have taken something that has NO correlation with anything – disorder, randomness, or chance – and they have literally turned it into the cause of everything. They have done precisely what they are not supposed to do, and have turned entropy into causation, in order to make natural selection and the second law of thermodynamics axiomatically true. It's a con. It's a ruse. We fall for it, because we don't understand it.

Once people have seen this correlation between entropy and every type of disorder, with their own eyes, they can automatically see what's wrong with it – if they understand anything about correlation and statistics at all.

These people correlate entropy with disorder, and then they go one step further and turn that correlation into causation. They violate the whole of statistics when they do so; and, they call it science. Since maximum disorder, by definition, has a correlation coefficient of zero; its prediction value is no better than astrology. The prediction value of "entropy defined as disorder" or the "second law of thermodynamics" is no better than astrology or chance. It's a massive logic error to turn chance or disorder into causation as the Materialists, Naturalists, and Darwinists do. Creation by Entropy or Creation by Chance is astrology! There's NO real effect and NO correlation between disorder and thermodynamics, so what they did is to equate disorder with thermodynamics; and then, they called the result entropy or natural selection. We fall for the ruse because we give these people credit for being smarter than they really are. We fall for the deception because we never question the science, or the statistics, or the logic of their claims.

These people literally take "zero probability" and turn it into "infinite certainty", whenever they turn natural selection or entropy or chance into our Designer and Creator and God.

These people take everything that has zero correlation, and they call it entropy by fiat or decree. It's unbelievable, but this is precisely what they do, and most of us don't even know it. These "scientists" or religious fanatics correlate disorder with entropy, and then they go one step further and turn chance, disorder, or "random fluctuations in entropy" into our Designer and Creator, and the Ultimate Causal Agent of all time. They imbue disorder, death, natural selection, or entropy with the power and the ability to design and create at will. They literally correlate disorder and entropy with Design and Creation. Look up Boltzmann brains, or quantum fluctuations, or random fluctuations in entropy, or natural selection, or chemical evolution if you don't believe me. See how they have imbued these different types of entropy or different types of disorder with the ability to design and create at will. These people have literally turned "no correlation" into complete causation; and then, they called the result "science".

These people worship entropy or heat death as if it were their Designer, Creator, and God. The reverence with which they speak of it is amazing to behold.

In one of the greatest deceptions of all time, they made it technically impossible for entropy to be correlated with anything, by defining entropy as "disorder" – so that they could then turn around and correlate entropy with anything that they desire to correlate it with and get away with it. In other words, they deliberately correlated entropy with everything that has no correlation with anything. Anywhere they see disorder, randomness, chaos, or chance, they call it entropy. Entropy becomes the catch-all term for anything and any system that has zero correlation with anything else. Then entropy takes on a life of its own.

It can't be falsified because it doesn't correlate with anything. It can't be verified or predict anything because it doesn't correlate with anything. It can be anything you want it to be because it doesn't correlate with anything. It's sleight-of-hand. It's magic. It's the grand illusion. It's idolatry. Yet, we fall for it every time because we desperately want to believe that it is true.

Amazing, is it not?

By correlating or equating entropy with disorder, they falsify the second law of thermodynamics, and they don't even know it, because disorder doesn't correlate with anything, especially thermodynamics. To call thermodynamics disorder is an *oxymoron*, a contradiction in terms. It's a *category error* logic fallacy, which means that the second law of thermodynamics is based upon a bunch of different falsehoods or logic fallacies; and, we don't even know it because it is presented to us axiomatically as an absolute truth.

When dealing with maximum disorder or a correlation coefficient of zero, there's NO way to make any predictions. In other words, there's NO way to do science. There's nothing empirical about disorder. In fact, a correlation coefficient of zero tells us that there is NO relationship between the variables at all. There's no relationship between thermodynamics and disorder at all. There can't be! So, what our scientists do is to make them equal to each other by fiat, through the second law of thermodynamics.

It's impossible to run actual science experiments with disorder, which also makes it impossible to run science experiments with thermodynamics while thermodynamics is defined as "disorder". The science stops when the correlation coefficient is zero, which is why they have to define entropy as "disorder" through fiat, axiom, law, dogma, and decree. There's no way to do actual science with it, because there is no way for us to get our hands on entropy, natural selection, chemical evolution, or complete disorder. It's untouchable and unfalsifiable; and from their perspective, it's automatically true because it can't be falsified. They stack the deck and load the dice so that it always comes up entropy. Anytime there is zero correlation between anything, it is always defined as disorder or entropy. They can't lose, because the second law of thermodynamics has been rigged so that it always comes up true.

There's NO causal relationship between thermodynamics and disorder, so they manufacture one out of thin air and call it entropy. It's a kludge, and we don't even know it, because it's labeled as being science while anything that falsifies it is called a pseudo-science. The second law is protected by axiom, law, or decree so that anything that falsifies it automatically becomes pseudo-science. I'm doing pseudo-science here by falsifying the second law of thermodynamics, according to the makers and the inventors of the second law of thermodynamics. That's the way these people operate. They take lies and make those lies axiomatically true. The lies and falsehoods within the second law are protected by axiom and law. We can falsify the second law of thermodynamics thousands of different ways, yet it will still remain axiomatically true in our science community at-large. It's protected.

The second law of thermodynamics is the greatest scam in the whole of human history; and, it continues to this very day. It's protected and untouchable and unfalsifiable by decree. We can falsify it all we want, and it will still remain true.

I have been a scientist all of my life, for nearly sixty years. I love to study and think about science. I love discovering methods for falsifying the things that are false. In all of my years of study and research, I have observed that most of the lies, falsehoods, and deceptions in Science stem from the second law of thermodynamics and natural selection – which are Creation by Entropy, Creation by Disorder, or Creation by Chance. Nowadays, whenever I encounter natural selection or the second law of thermodynamics, a yellow flag pops us saying, "Caution! Beware! These people are trying to trick you and deceive you." Invariably, if I think about it long enough or study it sufficiently enough, I can see that it is true. These people are trying to trick us and deceive us.

So, what do I want you to see from this essay, or what do I want you to be warned about by this essay?

Don't blindly believe everything that the scientists tell you, because most of them have an agenda, and because many of them are actually trying to trick you and deceive you. It's how they make their living. Learn to think for yourself. Learn to identify falsehoods wherever you might encounter them.

I want you to know and remember that entropy, the second law of thermodynamics, and maximum disorder are represented graphically as a system of molecules or atoms that have a correlation coefficient of zero ($r = 0.0$). By axiom or by decree, entropy and maximum disorder are declared to be the same thing, according to the second law of thermodynamics and according to Rudolph Clausius who coined the term "entropy".

These people have literally taken something that has a correlation coefficient of zero and correlated it directly to the amount of heat in a system or the thermodynamics in a system, in complete violation of logic and the rules of statistics and science. It is a *category error* logic fallacy to classify entropy, heat, or thermodynamics as "disorder". They don't have anything to do with each other at all.

If we study entropy defined as "heat" or a "transfer of heat", we quickly observe that the amount of heat is constantly decreasing and going towards zero. In contrast, entropy defined as "disorder", or the second law of thermodynamics, tells us that the amount of entropy or disorder is constantly increasing and can never decrease. So, which is it? Is the amount of "heat or entropy" constantly decreasing, or is the amount of "disorder or entropy" constantly increasing? They are mutually exclusive, which means that they falsify each other. Which is the true definition for entropy, or are they both false?

Isn't that interesting?

The second law defined as heat or thermodynamics FALSIFIES the second law defined as disorder. They are mutually exclusive. They move in opposite directions. Yet, our atheistic and naturalistic scientists have equated them by axiom, law, or decree. It's one of the dumbest things that they have ever done in science. Heat, after it has been made by Nature's Psyche, is constantly decreasing towards absolute zero. Yet, the second law of thermodynamics by axiom or by decree states that the amount of entropy or disorder is constantly increasing and can never decrease. Entropy defined as heat or a transfer of heat FALSIFIES entropy defined as disorder, randomness, or chance. They move in opposite directions which means that they can't be the same thing no matter how much we might decree that it is so.

So, which is it? Is the amount of "heat or entropy" constantly decreasing, or is the amount of "disorder or entropy" constantly increasing? They are mutually exclusive, which means that they falsify each other. Which is the true definition for entropy, or are they both false?

From statistics, I KNOW which one is most likely to be false, because I know what disorder or a correlation coefficient of zero looks like graphically. It's obvious to me which one is the scam or the fraud. The Law of the Conservation of Energy, which states that the amount of energy remains the same, FALSIFIES the second law of thermodynamics which says that the amount of entropy is constantly increasing. The second law of thermodynamics violates the Conservation of Energy or the first law of thermodynamics, which tells me that there is something wrong with the second law of thermodynamics. The second law is a kludge. It was purposefully designed to falsify and surpass the first law of thermodynamics. The second law is *too good to be true*. The second law is magic.

Statistically, disorder can't be correlated with anything; yet, our science community at-large has correlated entropy with disorder; and then, they have gone one step further and correlated that entropy or disorder with all of the energy in the universe or all of the heat in the universe. Then they called that monster the second law of thermodynamics – the great law, the law that determines the destiny of our lives and our universe. They have personified, reified, and deified disorder or entropy as our Designer, Creator, and God by axiom and by decree and by law. That's quite a coup, and the average Joe falls for it hook, line, and sinker because he blindly trusts and believes that the Atheists, Naturalists, Darwinists, Nihilists, Behaviorists, and Atheists are telling him the truth. What reason would they have to lie to us? They have no motive to lie to us, now do they? They are scientists after all. Everyone knows that they are completely objective and would never tell a lie. They are not to be questioned.

Well, they should be questioned, and the false should be falsified by whatever methods can successfully falsify it. This was the original purpose for science after all – to falsify the false – and we should really start by falsifying the science that is demonstrably false, rather than turning the falsehoods into axioms, truths, laws, and science. For an open-minded and objective scientist like me, it can be disheartening to see all the different falsehoods, deceptions, and lies that they have introduced into science through natural selection and the second law of thermodynamics – which are Creation by Entropy or Creation by Chance. Oh well, the best I can do is to state the truth as I see it.

Disorder is the wrong definition for heat, mass, thermodynamics, and entropy. Disorder doesn't correlate with any of these things!

Mass or "heat storage capacity" has a huge amount of order and structure. Mass is actually capable of storing heat, or entropy, or resistance to acceleration. Heat or entropy are the same thing – they are thermodynamics. To classify heat storage capacity or entropy storage capacity as "disorder" is one of the greatest blunders in the whole of human history. It's a deceptive lie that most of us have chosen to believe is true, thanks to the inappropriately named second law of thermodynamics.

Through axiom, fiat, deception, law, and decree, they have taken something that has a correlation coefficient of zero (maximum disorder) and correlated it directly with heat or thermodynamics, in the GREATEST *category error* logic fallacy of all time!

These people literally took something that doesn't correlate with anything, namely "disorder", and they made it correlate with everything by calling it entropy or thermodynamics. It's a logic fallacy. It violates the rules of statistics. Adding insult to injury, in yet another *category error* logic fallacy, they go one step further and turn chance, disorder, entropy, or natural selection into an intelligent causal agent capable of designing and making anything that it sets its mind to – so that they don't have to have some type of Psyche, Intelligence, or God be their designer and creator. It's a logic fallacy. It violates the rules of science. They cheat, lie, and deceive because it gives them what they want most – disorder or entropy or chance as their designer and their creator.

DISORDER CANNOT BE INDEPENDENTLY MANIPULATED, AND ITS EFFECT ON ENTROPY CANNOT BE MEASURED. SCIENCE CANNOT BE DONE WITH DISORDER. IT IS A CATEGORY ERROR LOGIC FALLACY TO DEFINE ENTROPY OR THERMODYNAMICS AS DISORDER. DISORDER DOES NOT CAUSE HEAT! DISORDER DOESN'T ELIMINATE HEAT EITHER! THERE'S NO CORRELATION WHATSOEVER BETWEEN DISORDER AND THERMODYNAMICS! STATISTICALLY, THERE IS NO CORRELATION BETWEEN DISORDER AND ANYTHING. THE ATOMS AND MOLECULES REMAIN PERFECTLY ORGANIZED AND PERFECTLY FUNCTIONAL EVEN AT THERMAL EQUILIBRIUM, OR AT ABSOLUTE ZERO, OR IN A STATE OF MAXIMUM DISORDER. DISORDER DOESN'T TOUCH NOR CHANGE HEAT OR THERMODYNAMICS.

Therefore, disorder and thermodynamics ARE NOT causally related! There's absolutely NO correlation whatsoever between disorder and thermodynamics. There is NO correlation between maximum disorder and anything. There can't be according to rules of statistics. Consequently, there is NO causal relationship between entropy and disorder. There's no correlation between the two. Maximum disorder, by definition, has a correlation coefficient of zero ($r = 0.0$), which means that it doesn't correlate with anything. That also means that entropy and disorder are NOT causally related. There's NO correlation between the two. There can't be! It's a *category error* logic fallacy to classify entropy, heat, mass, or thermodynamics as "disorder". They are not the same thing at all. They are not causally related.

Correlating or equating entropy with disorder is a fraud and a scam – it isn't permitted according to the rules of statistics and common sense. The second law of thermodynamics is based exclusively upon a category error logic fallacy and a statistical fallacy as well.

The test-retest reliability is nil when it comes to disorder and "entropy defined as disorder". Amounts of heat can be correlated with temperature. There is a direct correlation between heat and temperature. There is absolutely NO correlation between heat, temperature, thermodynamics, and "entropy defined as disorder". There is NO correlation between disorder and thermodynamics! Thermodynamics has structure or order or potential; and, disorder, by definition, has NONE.

The "disorder" definition for entropy contradicts or falsifies the other definitions for entropy. Think about it logically. The "disorder" definition for entropy is represented by a volume of gas with its atoms and molecules randomly distributed throughout the volume. Disorder is represented by a cannister of gas whose atoms or molecules are homogenized and effectively have a correlation coefficient of zero. That's maximum disorder or maximum entropy according to the "disorder" definition for entropy.

Yet in the next breath, the same physicists will tell you that black holes and neutron stars contain MOST of the entropy in our universe. So, which is it?

What do we see the atoms and molecules doing when it comes to a black hole or a neutron star? Are the atoms randomly distributed throughout the volume of the black hole or the neutron star? Or are the atoms highly compacted and organized? Are the atoms inert in some kind of gaseous homogenous heat death, or are the atoms doing massive amounts of work producing huge gravity wells and sucking in everything around them?

You see, entropy defined as a black hole or a neutron star FALSIFIES or is MUTUALLY EXCLUSIVE to entropy defined as the maximum disorder of a gas in a cannister. They are NOT the same thing at all. They are two completely different things.

Likewise, entropy defined as maximum disorder is something completely different than entropy defined as heat or thermodynamics. Again, what do we see when we encounter a pre-made hot reservoir and a pre-made cold reservoir? Something or Someone has organized them in advance to be that way. One of those reservoirs is storing a lot of heat within its mass. ALL of that heat and mass required planning and work on the part of Someone or Something. It didn't just spontaneously generate from nothing out of thin air as the Naturalists and Atheists claim. A lot of order and organization and work went into the making of that hot reservoir, so that there could be some entropy or an exchange of heat from the hot reservoir to the cold reservoir.

Entropy defined as some type of heat or entropy defined as some type of thermodynamics – entropy defined as something measurable – has NO correlation whatsoever with disorder. Disorder, by definition, has a correlation coefficient of zero! It's not measurable, because it doesn't correlate with anything! In contrast, there's an actual linear or curvilinear relationship between heat, temperature, and "entropy defined as thermodynamics" depending upon which heat-based equation for entropy that you choose to use. A correlation can be established between entropy, heat, temperature, mass, and thermodynamics! NO correlation can be made between disorder and entropy!

They define entropy as disorder – they define entropy as something that has a correlation coefficient of zero – so that they can then turn around and correlate entropy with anything that they want it to be. If they want to correlate entropy with a black hole, they just do so. If they want to correlate entropy with a sun, they just do so. I've seen them do it. If they want to correlate entropy with heat or thermodynamics, they just do so. If they want to correlate entropy with thermal equilibrium, absolute zero, or heat death, they just do so. If they want to correlate entropy with natural selection or creation by chance, they just do so. If they want to correlate entropy with design and creation, they just do so. If they want to correlate entropy or disorder with all of the energy in the universe, they just do so. The Standard International definition and equation for entropy correlates or equates entropy with ALL of the energy in the universe. Because the "disorder" definition correlates with nothing, they end up correlating disorder or entropy with anything and everything. They pick anything they want and correlate it with entropy.

They correlate entropy or disorder with anything that they want to correlate it with, which is why entropy ends up having a couple dozen different contradictory mutually exclusive definitions.

I listed and discussed sixteen different definitions for entropy in my book, "Quantum Mechanics from a Spiritual Perspective". My initial discoveries into entropy were written in that book. This book is a continuation of what I discovered there in that book.

https://www.amazon.com/gp/product/B01J023TGU/

https://www.amazon.com/gp/product/1521132380/

Since disorder doesn't correlate with anything, they can make disorder or entropy correlate with everything and get away with it. They get away with it because nobody catches onto the deception or the lie. They claim to be scientists, so nobody questions them. Instead, everyone wants to be in sync with them. Nobody wants to stand out and be alone. So, everyone jumps on the bandwagon, defines entropy as disorder, so that entropy can then be anything that we want it to be.

Defining entropy as disorder is the statistically impossible definition for entropy. There's NO correlation between disorder and thermodynamics; and therefore, there is NO correlation between disorder and entropy. Whenever entropy is defined as some type of thermodynamics, it actually falsifies the entropy that is defined as disorder. The two are mutually exclusive and are only forced together by axiom or law or decree. In other words, the second law of thermodynamics defined as "disorder" has absolutely nothing to do with thermodynamics – the two are mutually exclusive and falsify each other.

Entropy defined as disorder is mutually exclusive to entropy defined as heat or thermodynamics. They are NOT the same thing at all. Any way we look at it, to define entropy as "disorder" is a *category error* logic fallacy. Entropy should be defined as heat, or thermodynamics, or the transfer of heat, because the amount of heat is measurable and definable, which means that heat or thermodynamics can actually be correlated with entropy; whereas, disorder cannot by definition be correlated with anything. Disorder cannot be measured because it has a correlation coefficient of zero. It is perfectly randomized. We would have to organize it to measure it.

It is a logic fallacy – a *category error* logic fallacy – to define entropy as disorder or to define entropy as something that has a permanent correlation coefficient of zero. By axiom, the second law of thermodynamics defines entropy as a correlation coefficient of zero ($r = 0.0$); and then, they go forward and correlate entropy or disorder with everything that they encounter in physics. A true statistician can see the error or the logic fallacy in such a procedure; but, most people can't see it nor understand it because they have no idea what a correlation coefficient of zero actually is. Consequently, these people get away with it, and their deceptions and lies become axioms and laws instead.

Of course, one error leads to another.

After defining entropy as "disorder" or "chance" or "randomness", then they boldly go where they shouldn't go and start calling entropy, chance, or natural selection our

Designer and our Creator. One falsehood leads to another; and before you know it, these people have random fluctuations in entropy or quantum fluctuations designing, creating, and making our physical universe. Or they have random fluctuations in entropy, or chance, or natural selection designing and creating Boltzmann Brains, genes, matching proteins, eyes, species, genomes, and life forms. Most of the obvious falsehoods and lies in science come from natural selection and the second law of thermodynamics. That's how they get their foot in the door.

The second law of thermodynamics is protected. No matter how many different ways we might falsify it, it remains axiomatically true, because they have defined it to be unfalsifiable, untouchable, and automatically true. They cheated and broke the rules of statistics, logic, and science in order to make the second law absolutely true all the time. Consequently, the best we can do is to tell the truth as we see it, and let history choose the eventual victors, because for now it is forbidden to falsify the second law of thermodynamics.

I'm interested in the psychology of these people, because I used to be a materialist, naturalist, nihilist, and atheist as recently as 2013. I'm interested in how these people trick themselves and deceive themselves. I'm interested in studying and learning how I was tricked and deceived.

When I discovered that disorder is represented graphically as a correlation coefficient of zero, I had another toy to play with to determine what was done to me and what I did to myself.

Although this can be hard to put into words due to the fact that we are dealing with statistics, once I realized that "disorder" is represented graphically as a correlation coefficient of zero, then I finally had a graphical representation of "entropy defined as disorder". Then, I could finally see the trick or the ruse, because it became obvious that they axiomatically defined every type of disorder as "entropy" in order to make the second law of thermodynamics true. Everything can made to be disordered or seem to be disordered. Once you can see and understand this fact or reality, then you can finally see how they get entropy both coming and going. Everything in the universe has been defined as entropy or correlated with entropy. Literally everything! I have seen it all! Nothing has been exempted from being defined as entropy. Nothing!

Then you begin to see how Sal Khan derived his claim that the sun has more entropy than the heat-dead moon because the sun has way more potential states of being or way much more random disorder than the moon. In other words, Sal Khan defined entropy as microstates, and since the sun has more microstates, or more potential, or more random disorder than the moon, the sun has more entropy.

Then you also begin to see how Matt at PBS Spacetime was able to say such things as "black holes contain most of the entropy in our universe" and "entropy is heat-death or a temperature near absolute zero where NO work can be done". There he has defined entropy as heat-death, no motion, no work, or the end of our universe.

You can literally have any definition you want for entropy and get away with it. This is possible because they set up the second law of thermodynamics so that entropy can literally be defined as anything you want it to be. The whole thing was set up to produce an infinite and ever-increasing amount of entropy. The only thing you can't have – the only thing they forbid – is a decrease in entropy. But you can have entropy be anything you want it to be, so long as it's always increasing.

There is at least a couple dozen different, contradictory, and mutually exclusive definitions for entropy on the books, because entropy doesn't correlate with anything, which means that they can make it correlate with anything that they want to correlate it with. Sal Khan repeatedly correlates entropy with the sun or the big bang or maximum heat, and Matt O'Dowd repeatedly correlates or equates entropy with heat death, black holes, absolute zero temperatures, and the end of our universe. They are able to do this because entropy defined as disorder literally correlates with nothing, which means that they can get away with it and do get away with it whenever and wherever they correlate entropy with anything. Everything seems to have a bit of disorder, does it not? So logically, everything is going to have a bit of entropy. That's the way it goes. Entropy is one size fits all. You can literally define anything in this universe as "entropy". There is no limit to what you can define as entropy.

They can do this because there is NO standardized definition for entropy. Entropy can literally be anything that you want it to be. The Standards International literally defines entropy as ALL of the energy in the universe. All of the energy or quanta in the universe can also be made to seem completely disorganized if one chooses to look at it that way. Because disorder doesn't correlate with anything, there is literally no limit to what you can choose to correlate or equate entropy or disorder with, so long as entropy is defined as any type of disorder and so long as entropy is always increasing.

Random atoms in the sun, random atoms in a gas, and no motion and no work at heat death can all be defined as disorder, so they can all be turned into entropy or called entropy because there is literally no limit to what you can define as disorder. Unless there is perfect correlation – which never happens in statistical sampling – every system has a little bit of disorder. It's a coup! They can't lose! By defining entropy axiomatically as "disorder", they literally defined entropy as everything in our universe. Entropy can be anything you want it to be. They get away with it because we have no idea that that's what they are actually doing and no idea that it's wrong to do so, scientifically and statistically. Correlating something like entropy with something that has zero correlation such as disorder is the ultimate error in statistics and scientific experimentation; but, they get away with it because they have granted themselves a *special pleading* or a *special exemption*.

The Materialists, Naturalists, Darwinists, Nihilists, Behaviorists, and Atheists are automatically exempt from the rules of science and logic and statistics because it's obvious that they have the truth and that everyone else in the universe is deluded and irrational. I have actually had some of them tell me that it's okay for them to lie and cheat and deceive, because they are teaching us the truth and are the only ones who have the truth. When the Darwinists lie to us and deceive us they are doing it for our own good so that we will finally know the truth. The ends justify the means. It's okay for them to lie to us as long as we finally get the truth in the end; and, they define The Truth axiomatically as Materialism, Naturalism, Darwinism, Behaviorism, Physicalism, Determinism, Nihilism, Entropy, and Atheism.

In summary, defining entropy as "disorder" or a "correlation coefficient of zero" is the wrong definition for entropy, and the wrong definition for anything, because maximum disorder cannot be correlated with anything at all. Anything that produces a correlation coefficient of zero cannot be correlated with anything at all. That's the way that statistics really works! It is really simple to understand.

The reason that they can't understand it and have never thought about it is because they are determined to define entropy as disorder, chance, or randomness, so that they can then have entropy, or disorder, or chance be our Designer and Creator. Anything but God. They are willing to believe in anything, but God. They would rather have a correlation coefficient of zero be their designer and creator than a loving Heavenly Father who made this world for them and wants the best for them.

That's just the way they are.

These people literally define entropy and the second law of thermodynamics as "maximum disorder" or as something that has a correlation coefficient of zero; and then, in the GREATEST logic error of all time, they correlate that "maximum disorder" directly with entropy or thermodynamics, and then they call it the second law of thermodynamics. The ever-increasing disorder of the second law of thermodynamics or "entropy defined as disorder" is the greatest logic fallacy ever devised by the mind of man. Yet, we fall for it every time because we desperately want it to be true. And before long, it becomes Creation by Chance, Creation by Entropy, Creation by Natural Selection, or Creation by Death – and then, it becomes our Designer, Creator, and God.

The science community at-large, comprised mostly of atheists and naturalists, will NEVER embrace the falsification of the second law of thermodynamics because it falsifies their deceptions and their lies. The second law of thermodynamics isn't going anywhere. It's perfectly safe. The second law of thermodynamics will always be true, no matter how many different ways we might find to falsify it. The second law of thermodynamics has been enshrined as law and truth in the science community, and nothing will ever change that. We can have fun and falsify the second law of thermodynamics dozens of different ways; yet, it will still be axiomatically true when we are done. Nothing will ever change the truthfulness of the second law of thermodynamics, in our science community, now that the second law has been declared to be unfalsifiable and axiomatically true.

The second law of thermodynamics is protected. It's untouchable. So, we can only attack it or falsify it philosophically, because it will always be true no matter how many different ways we might find to falsify it. Nonetheless, falsifying the second law of thermodynamics in private can be a lot of fun and generate a lot of interesting thought experiments for a theoretical physicist like me. I like it. I find it interesting and thought provoking. I get better and better at it as the years go by.

**There's NO Correlation between Natural Selection and Our Genes**

In a similar vein, while studying science I have observed that there is NO correlation whatsoever between natural selection and our genes. Natural selection doesn't touch our genes. It can't.

Natural selection is the "death" definition for entropy. Natural selection is Creation by Entropy, Creation by Chance, or Creation by Death. Natural selection doesn't do anything except sit around and wait for you to die. When you get selected against, you die. It's that simple. Dying doesn't do anything to change our genome. In fact, a lot of our genome survives the death process, for a while at least. Selection of any kind means that you die, or it means that you were never born in the first place (sexual selection). Natural selection doesn't touch our genes. There's NO correlation between our genomes and natural selection.

The thing that randomizes our genes and produces the occasional mutation is in fact a process that is called genetic recombination. During the production of sperm and egg, the program was designed to shuffle the genes. Because there are thousands of genes that need to be shuffled during genetic recombination, errors are often introduced into the genome during the process. Genetic recombination is the thing that touches and influences our genes – not natural selection.

Once again, design and creation by natural selection is a non-starter, because there is NO correlation and therefore NO causal relationship between natural selection and our genes. Natural selection, or death, or disorder, or entropy has NO physical mechanism for interacting with and controlling our genes. There's no correlation or causal relationship between natural selection and our genes.

It is just another scam that's been perpetrated by the Materialists, Naturalists, Darwinists, Nihilists, Behaviorists, and Atheists. We fall for it because these people claim to be scientists, when they are not. They have NO interest in observation, experience, or truth. They are dogmatic religious fanatics whose only purpose is to convince us that the non-physical does not exist.

## Disorder Doesn't Really Exist in Our Part of the Multiverse

So, in summary, whether we are talking about science, physics, thermodynamics, heat, statistics, philosophy, law, or logic, WE OBSERVE that they broke every rule in the book when they axiomatically defined entropy and the second law of thermodynamics as "disorder", "randomness", "chaos", "chance", or as something that has a correlation coefficient of zero. These people literally took something that correlates with nothing; and through fiat or decree, they made it correlate with everything. They cheated in order to prove their point; and, we blindly follow them and believe them, because we have chosen to trust them and have chosen to believe that they know what they are doing. We believe them because we desperately want their deceptions and lies to be true.

However, choosing to define entropy as "maximum disorder" or as a "correlation coefficient of zero" **(r = 0)** is the stupid, impossible, false, falsified, and invalid definition for entropy and the second law of thermodynamics according to statistics, the rules of science, the rules of logic, the rules of common sense, and the observable rules associated with thermodynamics.

Thermodynamics has structure and order. Mass has heat storage capacity! It's organized! Disorder is the wrong definition for heat storage capacity, entropy storage capacity, and thermodynamics! Even at thermal equilibrium, mass still has heat storage capacity. Mass still has order and organization and resistance to acceleration, even when heat or thermodynamics or entropy has gone to zero. Just load it up with some heat, and it's good to go again.

Anytime that Nature's Psyche chooses to convert entropyless and heatless and massless photons into mass, heat, and entropy, Nature's Psyche introduces NEW structure and NEW order into our physical universe. Nature's Psyche establishes a correlation between heat, mass, and entropy where there was NONE before. Nature's Psyche or Intelligence can make order, organization, quantum fields, photons, heat, mass, and physical matter where there was complete chaos or randomness or disorder before – anywhere and anytime that Nature's Psyche chooses to do so.

This is actually what we have experienced and observed every time that Nature or Nature's Psyche has converted or transformed massless, heatless, and entropyless photons, energy, or light into something that has mass or heat storage capacity or resistance to acceleration instead.

$E = mc^2$

The order or organization goes both ways!

Whenever Nature converts or transforms photons or energy or quantum fields into mass, heat storage capacity, heat, and resistance to acceleration, Nature's Psyche creates NEW order and organization or a NEW FORM of energy.

Likewise, whenever Nature converts quarks, electrons, mass, physical matter, heat storage capacity, and resistance to acceleration back into massless, heatless, and entropyless photons instead, once again Nature's Psyche creates NEW order and organization or a NEW FORM of energy.

$E = mc^2$ is a perpetual motion machine; and, it works perfectly both coming and going. Thanks to the Conservation of Energy and $E = mc^2$, there is NO such thing as disorder. Disorder ceased to exist in our part of the universe, when the Gods, the Controlling Psyches, and Nature's Psyche transformed Chaos into the Quantum Fields.

Thanks to the Quantum Fields, our part of the universe is now perfectly syntropic or perfectly ordered and organized – whether Nature is choosing to FORM parts of those quantum fields into photons or choosing to FORM parts of those quantum fields into mass, quarks, electrons, and physical atoms. The quantum order and organization within the Quantum Fields remains, and the perfect syntropy remains, no matter what FORM the energy is transformed into, and no matter how disordered or disorganized the physical matter at the physical level might become. The perfect order of the Quantum Fields remains, whether parts of it are being transformed into physical matter, or parts of that physical matter are being transformed back into photons and quantum fields.

The Quantum Fields are perfect syntropy. The construction of the Quantum Fields was the Big Bang. It brought perfection and order where there was NONE before. The construction of the Quantum Fields made the construction of mass possible; and, the construction of mass made the storage of heat or entropy possible. The construction of mass made thermodynamics possible. No mass, then no heat and no entropy. It's that simple. Entropy is heat or thermodynamics, NOT disorder. Mass is heat storage capacity, or entropy storage capacity. Entropy, or mass, or heat storage capacity has a great deal of structure and order. Mass, or resistance to acceleration, or entropy has a definite structure and purpose.

Disorganized physical matter doesn't touch nor change the syntropic quantum perfection of the Quantum Fields. Remember, disorder at the universal scale was completely eliminated, the very moment that the Gods and Nature's Psyche transformed Chaos into the Quantum Fields. That transformation is permanent. The Quantum Fields are permanent – until the Gods and Nature's Psyche decide to transform the Quantum Fields into something else instead. Should they ever make that decision, there is nothing we can do about it. The Gods, the Controlling Psyches, and Nature's Psyche could dissolve this planet back into the quantum fields from whence it came, anytime they choose to do so; and, there's nothing we could do to stop them.

Remember, the alleged Big Bang Singularity was the eternal and everlasting Conservation of Energy. The Conservation of Energy and Psyche existed long before the first particle of physical matter was designed and made. The alleged Big Bang took place in our part of the universe when the Gods and Nature's Psyche transformed Chaos into the Quantum Fields. Once the Quantum Fields were made, then it became possible to make and sustain physical matter, mass, resistance to acceleration, heat storage capacity, heat, thermodynamics, and entropy. All of these things are impossible without the Quantum Fields, and the Quantum Fields are huge perpetual motion machines thanks to the Conservation of Energy.

Remember, there is NO such thing as disorder where the Quantum Fields are concerned. ALL of the disorder ceased to exist when the Gods or Controlling Psyches designed and made the Quantum Fields! There is NO disorder in our part of the universe thanks to the Quantum Fields! ALL of the energy is available for use all the time at the quantum level thanks to Psyche and the Quantum Fields. Furthermore, ALL of the work is done for free at the quantum level thanks to the Conservation of Energy and Psyche, and thanks to the perpetual motions machines that we call the Quantum Fields.

This is what we have actually experienced and observed. Quantum Field Theory is our best and most verified science on the planet right now, and Quantum Field Theory makes it obvious that there is NO disorder whatsoever at the quantum level in our part of the universe, because ALL of the disorder or chaos has been completely eliminated by the Quantum Fields.

Therefore, defining entropy or heat or thermodynamics as "disorder" is the nonsensical and irrational definition for entropy and the second law of thermodynamics thanks to the Syntropic Entropyless Quantum Fields, Quantum Field Theory, and the Conservation of Energy. Defining entropy as disorder is the Grand Illusion, because disorder doesn't really exist thanks to the perfect syntropy of the Quantum Fields. The second law of thermodynamics is falsified by the perfect syntropic order and organization contained within the Quantum Fields.

I eventually realized that Quantum Field Theory represents THE TRUTH in science, so long as Nature or Nature's Psyche is allowed to be the maker of, the controller of, and the enforcer behind the Quantum Fields. Quantum Fields are obviously massless, heatless, entropyless, and non-physical, so some type of Psyche or Intelligence or Quantum Consciousness is needed to make them and operate them. The Quantum Fields were not designed and made by entropic physical matter as the Naturalists and Atheists claim. According to Quantum Field Theory, the Gods, the Controlling Psyches, or Nature's Psyche had to design and make the Quantum Fields BEFORE the first particle of physical matter could be made and sustained. The Quantum Fields had to come first, or there could be no mass, heat, or physical matter. Nature's Psyche is absolutely needed for the construction and operation of the Quantum Fields. It is in fact Nature's Psyche who collapses the wave function, and Nature's Psyche was collapsing wave functions long before the first particle of physical matter was designed and made.

The Materialists, Naturalists, Darwinists, Nihilists, Behaviorists, and Atheists BY AXIOM dogmatically deny the existence of the massless, heatless, entropyless, and non-physical Quantum Fields, Photons, Quantum Waves, and Psyches or Intelligences. Therefore, the very existence of massless and entropyless Quantum Fields and Photons FALSIFIES Physicalism, Materialism, Naturalism, Darwinism (Creation by Entropy), Nihilism, and Atheism (Creation Ex Nihilo and Creation by Chance). The two ideas are mutually exclusive. If one of them can be demonstrated to be true, then the other has been automatically falsified. The proven and verified existence of massless, heatless, entropyless, and non-physical wave-like omnipresent quantum photons FALSIFIES Materialism, Naturalism, and their derivatives such as Darwinism and Atheism. The false is falsified by the truth; and, the truth is repeatedly experienced and observed.

Psyche is the Ultimate Causal Agent, and each psyche can do whatever it wants to do with the energy that's under its control. This is the LAW of Psyche. Each psyche or intelligence or quantum consciousness can transform the energy under its control into quantum fields, photons, or physical matter anytime and anywhere that it chooses to do so – and in any order it chooses, now that the quantum fields have been made and established. Nature's Psyche can convert or transform photons into physical matter, or physical matter into massless and entropyless photons, or photons into heat, or heat into heatless photons, or massless quantum fields into physical matter, or physical matter into entropyless and massless quantum fields anytime and anywhere that it chooses to do so. Energy is infinitely malleable, and it can be transformed into anything that its controlling psyche wants it to be, anytime and anywhere; and, there's nothing we can do to stop it. This is Quantum Field Theory. This is the truth. And, it falsifies Materialism, Naturalism, Darwinism, Nihilism, Behaviorism, and Atheism.

Science is in desperate need of a NEW and BETTER definition for entropy – a definition that has nothing to do with disorder or a correlation coefficient of zero. Science needs to correlate entropy with something that has actually been experienced and observed. True disorder or chaos has NEVER been experienced nor observed in our part of the universe, ever since the Quantum Fields were designed and made. It's time for scientists to stop defining entropy as "disorder", because disorder doesn't really exist in our part of the universe thanks to the Quantum Fields and the Conservation of Energy and Psyche.

The state that they call "heat death" at the physical level represents maximum efficiency at the quantum level, so it's a *category error* logic fallacy to call "heat death" maximum disorder, because everything remains perfectly ordered and perfectly functional at the quantum level in the Quantum Realm thanks to the perpetual motion machines that we call the Quantum Fields, even if the system has achieved thermal equilibrium or "heat death" or absolute zero at the physical level.

If we ever run out of heat, Nature's Psyche can make some more, anytime and anywhere that it chooses to do so. This is what has actually been experienced and observed. Every time a massless, heatless, and entropyless photon chooses to stop, it transforms itself into mass, resistance to acceleration, and heat instead. It happens all the time, every day, and we never once think about what's really happening, because most of us have convinced ourselves that the massless or the entropyless does not exist. Self-deception works, and it works every time.

We are completely oblivious to the infinite expanse of perfectly ordered syntropic Quantum Fields that were designed and made by the Gods, the Controlling Psyches, and Nature's Psyche because we have chosen instead to believe that these things don't exist. We don't want them to exist, because their verified and proven existence falsifies our pre-chosen beliefs. In the process, science becomes nothing but a figment of our imagination having nothing to do with reality. Then before you know it, we have entropy, chance, or natural selection making Boltzmann Brains, physical universes, physical matter, stars, galaxies, eyes, brains, RNA, DNA, genes, proteins, genomes, and life forms out of thin air spontaneously because we have convinced ourselves that random fluctuations in entropy, chance, or disorder can design and create at will.

In the greatest scam of all time, we scientists set it up statistically so that "correlation coefficients of zero" can organize themselves into anything that they desire, anytime and anywhere that they choose to do so. That's what happens when we deny the truth and prefer the falsehoods instead. We get "science" that doesn't make any logical sense. We get "science" that is nothing but magic. We get "science" that is demonstrably false.

What an interesting discovery!

In the greatest logic fallacy of all time, the Materialists, Naturalists, Darwinists, Nihilists, and Atheists defined entropy as something that has a correlation coefficient of zero; and then, they go one step further and magically have that "perfect disorder" designing and creating eyes, brains, genes, proteins, genomes, physical universes, planets, stars, galaxies, and life forms at will through random fluctuations in entropy. These people literally turn chance, disorder, or entropy into our Designer, Creator, and God. They took something that doesn't correlate with anything – disorder or randomness or chance – and they made it the cause of everything that exists. These people literally turned "no correlation" into causation.

It's the dumbest idea ever made, and then they have the gall to call it science when they are done. These people have turned idolatry into science, and they don't even know it, because they don't have the background necessary to see it and understand it.

My statistics book tells us to be on the lookout for such scams and falsehoods, where the statisticians and scientists and researchers are turning correlation into causation. Well, the Naturalists and Darwinists and Atheists are going one step further and turning "no correlation" into causation. These people are turning disorder and chance into causation. The results are obviously false, glaringly false, for anyone who is willing to look and see.

Finally, when it comes to the second law of thermodynamics or natural selection everything is defined in terms of it, restricted to it, limited by it; and, every science discovery that falsifies it is rejected, eliminated, banned, censored, ridiculed, and destroyed so that they can continue to verify it. It's the greatest con job in history, and most of us have fallen for it hook, line, and sinker. We are up to our gills in it, and there's no way to escape it because the falsehoods produced by it have been axiomized and turned into laws so that they can't be falsified.

Entropy has been defined as a "correlation coefficient of zero" or "maximum disorder" for over a century now; and, most of us still have no idea that it's a ruse and a con, and that creation by disorder or creation by entropy is impossible to begin with.

Now, don't worry about the second law of thermodynamics. It's perfectly safe. It's not going anywhere. The science community made the second law of thermodynamics into an axiom and a law so that it can't be falsified. You can falsify it all you want, and it will still be true. The second law of thermodynamics is untouchable. It's protected. There's nothing in the world that you can do to falsify it because it will always be true. It was designed to be that way, and it will stay that way.

Mark My Words

## The Thermodynamics Version of Entropy Is Also Based upon Fiction

Once we realize that the "disorder" version of entropy is represented graphically by a scatter plot where the correlation coefficient is zero (**r = 0.0**), then we start to understand why "disorder" is the wrong definition for entropy. Disorder doesn't have anything to do with thermodynamics. There is NO correlation between the two!

Consequently, I too turned to the "thermodynamics" version of entropy in an attempt to define and understand entropy. Thermodynamics is the transfer of heat from a hot system to a cold system. Surely here we will find a true and useful definition for entropy.

Shankar, R. (2014). *Fundamentals of Physics: Mechanics, Relativity, and Thermodynamics*. London: Yale University.

On page 412, Shankar effectively states that the closed loop of the Carnot Cycle is the heart of the whole entropy concept. "It tells us there is another state variable lurking around besides energy $U$. Remember, a state variable is one that returns to its initial value when we go around a closed loop."

The second law of thermodynamics and the "thermodynamics" version of entropy are based upon the closed loop of the Carnot cycle; and, they can't see anything wrong with this, but I certainly can.

First of all, they define entropy as a state variable – a conserved variable – just like energy. In other words, the second law of thermodynamics is the Conservation of Entropy, just like the first law of thermodynamics is the Conservation of Energy. The problem is that entropy cannot be a conserved quantity because its amount is constantly changing. Once again, they cheat by playing fast and loose with their definitions. They turn entropy into a conserved variable and make it equal to or better than energy.

The thermodynamics version of entropy is based upon the Carnot Engine and the Carnot Cycle – it has to be, so that entropy can be defined as something that is always conserved.

Yet, the whole thing is nothing but fiction, because a Carnot Engine is physically impossible; and, so is this conserved state variable or closed loop. It's physically impossible for thermodynamics or entropy to return to its initial value, yet they have it doing so through this fictional closed loop and the fictional Carnot Cycle – just so that they can turn entropy into a conserved entity.

The Carnot engine and Carnot Cycle are also based upon something that they call an ideal gas, which is also a fictional entity. There's no such thing as an ideal gas. The whole thermodynamics version of entropy is based upon nothing but fiction. It's nothing but a thought experiment. It's physically impossible, in actuality.

I like falsifying the second law of thermodynamics. It's really easy to do once you have done it. Everything in science falsifies the second law of thermodynamics and creation by entropy or creation by natural selection.

Falsifying the second law of thermodynamics won't change anything, though, because the second law will always remain axiomatically true no matter how many different ways we might find to falsify it. The second law is safe. It's protected. It's been enshrined into axiom and law. And, entropy is much better than energy. Energy is simply conserved. It's amount never changes. In contrast, the amount of entropy is constantly increasing like magic; and, it can never decrease.  Entropy is much better than energy, because entropy is always increasing.  The second law is unfalsifiable and untouchable, so we don't have to worry about it changing anytime soon, no matter how many different ways we might falsify it.

It's good, though, to study the thermodynamics version of entropy.  You can learn a lot about thermodynamics if you do.

https://www.khanacademy.org/science/physics/thermodynamics

https://www.khanacademy.org/science/physics/thermodynamics/temp-kinetic-theory-ideal-gas-law/v/thermodynamics-part-2

https://www.khanacademy.org/science/physics/thermodynamics/temp-kinetic-theory-ideal-gas-law/a/what-is-the-ideal-gas-law

https://www.khanacademy.org/science/physics/thermodynamics/specific-heat-and-heat-transfer/v/specific-heat-and-latent-leat-of-fusion-and-vaporization-2

https://www.khanacademy.org/science/physics/thermodynamics/laws-of-thermodynamics/v/macrostates-and-microstates

You can also figure out for yourself how they trick us and deceive us into believing that the second law of thermodynamics is true.  It's a lot of smoke and mirrors, with no basis in reality or fact.  The thermodynamics version of entropy is based upon the Carnot Cycle and the Carnot Engine, which are physically impossible and nothing but science fiction to begin with.  It's obviously a kludge.  It was designed with the specific purpose of convincing us that the fictional second law of thermodynamics is axiomatically true.

https://www.khanacademy.org/science/physics/thermodynamics/laws-of-thermodynamics/v/carnot-cycle-and-carnot-engine

You see, if you pay careful attention, you eventually realize that any negative entropy or reversal of entropy that they encounter in their quasi-static thought experiments, or any negative entropy or reversal of entropy that happens in real life, gets relabeled and renamed as "work" so that entropy can always be positive and constantly increasing.  It's a scam.  Any negative entropy or reversal of entropy that they encounter gets to be called "work" instead so that the amount of entropy can always be increasing.  By axiom or by law, they don't allow any negative entropy (syntropy) or a reversal of entropy to exist.  Whenever they encounter a reversal of entropy or negative entropy, they call it "work".

In ALL of their thought experiments into the "thermodynamics" version of entropy, they ALWAYS start with a hot reservoir and a cold reservoir for free as a given.  They NEVER once realize that it took a massive reversal of entropy or a whole bunch of negative entropy (syntropy) to make that hot reservoir to begin with.  They simply start with the hot reservoir for free as a given, so that they can then have some entropy (transfer of heat) when they put that hot reservoir into contact with their cold reservoir.

On page 414 of *Fundamentals of Physics: Mechanics, Relativity, and Thermodynamics*, Shankar states: "The refrigerator needs work (**W**)".  In other words, to have a reversal of entropy or negative entropy requires work.  Negative entropy or a reversal of entropy always gets relabeled and called "work", because it requires tons of work to produce the hot system and the cold system in the first place.  In other words, it requires a massive reversal of entropy or a lot of negative entropy (work or syntropy) to produce the hot reservoir and the cold reservoir in the first place.

Like I said, any negative entropy or reversal of entropy that they encounter they simply relabel as "work", so that entropy can always be increasing and so that entropy can never decrease. It's a scam; and, we fall for it easily because we have no idea what they are talking about in the first place.

In ALL of their thought experiments into entropy, they ALWAYS start with the hot reservoir and the cold reservoir for free as a given, so that they NEVER have to deal with the negative entropy or the reversal of entropy that produced the hot reservoir in the first place. They cheat, so that they can make it seem like entropy is always increasing and that the amount of entropy can never decrease. Whenever entropy decreases and goes to zero, they simply call it "work" instead. Tricky, huh?

It's all semantics. It's all fiction. It's the second law of thermodynamics! And, it's axiomatically true, no matter how many different ways we falsify it.

The second law of thermodynamics is Conservation of Entropy. Consequently, they go out of their way to prove that entropy (**S**) is a valid state variable, like energy (**U**) is a valid state variable. They do this so that entropy can become a conserved quantity and so that entropy can always be increasing. They also do this by renaming any negative entropy and calling it "work" instead.

https://www.khanacademy.org/science/physics/thermodynamics/laws-of-thermodynamics/v/proof-s-or-entropy-is-a-valid-state-variable

They go way out of their way to prove that entropy is the same thing as energy, but even better. You see, energy is simply conserved. Its amount remains constant and unchanged; whereas, the amount of entropy is always increasing and can never decrease according to the second law of thermodynamics. Entropy is super-energy. Entropy is ultra-conserved. Entropy is God!

It's very important to them that they prove that entropy is always conserved. They do so through semantics and thought experiments, where ONLY the increasing work or the increasing amount of entropy is allowed to be called "entropy", while every reversal of entropy or decrease in entropy is simply called "work". It's semantics. It's wordplay. It's sleight-of-hand.

I found all this very interesting. I'm a theoretical physicist, and I love to think about these kinds of things; and, I love to discover how they trick us and deceive us too.

I was surprised at how fast, loose, and liberally they play with science, semantics, and philosophy when it comes to entropy and the second law of thermodynamics. In other words, I was surprised to discover how much they cheat in order to make the second law axiomatically true.

In all of their thought experiments into entropy, they always start with a hot reservoir and a cold reservoir for free as a given. They never once realize that it required a massive reversal of entropy or a huge amount of negative entropy (syntropy) in order to produce the hot reservoir in the first place.

So, what's the first hot reservoir that they start with as a given?

Well, it's the Big Bang. They have that thing infinitely hot right from the very beginning so that it can produce a lot of entropy or "heat transfer" thereafter. Yet, these are Materialists, Naturalists, Darwinists, Nihilists, and Atheists that we are talking about here. They NEVER ONCE think about the huge massive reversal of entropy or the huge amount of negative entropy (syntropy or intelligence) that was required to produce that Big Bang or hot reservoir in the first place. The initial production of the hot reservoir or the infinite temperature Big Bang Singularity required a massive reversal of entropy or tons of negative entropy to produce the Big Bang or heat in the first place. This massive reversal of entropy FALSIFIES the second law of thermodynamics, which is why they can't allow it to exist and pretend that it doesn't exist.

The hot reservoir that they provide for free as a given, at the beginning of each of their thought experiments into entropy, falsifies the second law of thermodynamics!

I find it interesting how they lie and cheat in order to make the second law of thermodynamics true.

The second law of thermodynamics will always be true, no matter how many different ways we might falsify it. They went out of their way to make entropy all things to all people. They are religiously devoted to entropy and the second law of thermodynamics, and they worship it as their Creator and their God. According to these people, entropy, natural selection, or chance has to be our Creator because there is no other.

Nobody can decide or determine what entropy really is because it has literally been defined as everything in our universe. Entropy is multi-purpose – one size fits all. Entropy is anything you want it to be, which explains why I was never able to figure out what it truly is. Entropy isn't anything because entropy has been defined as everything and correlated with everything. Everything is entropy according to the Materialists, Naturalists, Darwinists, Nihilists, Behaviorists, and Atheists.

Over time, I have noticed that all of the obvious falsehoods, deceptions, and lies in science stem from the second law of thermodynamics and natural selection, which are effectively Creation by Entropy, Creation by Random Disorder, Creation by Death, or Creation by Chance. The second law and natural selection are obvious falsehoods that have been defined as being axiomatically true so that they can't be falsified.

The second law of thermodynamics is the ultimate feat of legerdemain in the whole of human history. You should be in awe, because no one does it better.

The only public or official version of entropy that's even remotely true is the version based upon the Gibbs Free Energy equations. The Gibbs Free Energy equations allow negative entropy; and, they allow this negative entropy or syntropy to do work and to make things.

https://www.khanacademy.org/science/chemistry/thermodynamics-chemistry/gibbs-free-energy/v/gibbs-free-energy-and-spontaneity

https://www.khanacademy.org/science/chemistry/thermodynamics-chemistry/entropy-chemistry-sal/v/chem20-entropy

For me personally, I consider two different definitions of entropy to be true. They are conservative, which means that they are measurable, definable, observable, and verifiable.

It makes a great deal of sense to define entropy in terms of heat or thermodynamics.

Entropy is a transfer of heat from a hot system to a cold system; and, Nature or Nature's Psyche can fully reverse the process anytime and anywhere that it chooses to do so. In other words, Nature can make a hot reservoir out of zero temperature massless, entropyless, and heatless Quantum Fields anywhere and anytime it wants, according to the Big Bang Theory and according to Quantum Field Theory! Likewise, Nature or Nature's Psyche can transform mass, heat, and entropy back into massless, heatless, and entropyless Quantum Fields anytime and anywhere that it chooses, according to Quantum Field Theory.

Quantum Field Theory falsifies the Second Law of Thermodynamics. The Big Bang Theory, if true, also falsifies the Second Law of Thermodynamics. Any reversal of entropy or negative entropy (syntropy) FALSIFIES the second law of thermodynamics.

Entropy is also mass, heat storage capacity, friction, or resistance to acceleration. No mass, then no heat storage capacity. No mass, then no heat. No heat storage capacity or no resistance to acceleration, then NO ENTROPY. It's that simple.

Entropy defined as "heat transfer" or thermodynamics is obviously NOT constantly increasing. Instead, at the physical level, heat (after it has been made by Nature's Psyche) seems to be constantly decreasing. In other words, thermodynamics or the constant decrease in heat (entropy) FALSIFIES the second law of thermodynamics. The second law of thermodynamics is a blatant and obvious lie. Entropy or heat isn't constantly increasing. According to observation and experiment, entropy or the amount of heat is constantly decreasing. The truth is the exact opposite of the claims made by the second law of thermodynamics.

I find this interesting.

The amount of heat (entropy) is constantly decreasing and going towards zero, which explains why they had to redefine entropy as "disorder" instead, so that they could have the disorder constantly increasing, so that the second law of thermodynamics could actually be true. But, defining entropy as "disorder" also falsifies the second law of thermodynamics, because it is impossible to correlate disorder with anything, including heat or thermodynamics.

They cheat in order to force the second law of thermodynamics to be true. Study what they say carefully, and you will see that it is so.

On page 419 of *Fundamentals of Physics: Mechanics, Relativity, and Thermodynamics*, Shankar states: "**Notice that energy is conserved but entropy increases**". Entropy is magic. Entropy is better than energy because entropy is always increasing; whereas, energy is simply conserved or constantly remaining the same. Energy is boring. It just stays the same. It's the entropy that's dynamic and alive and constantly growing and increasing. Entropy is God.

On page 418 of *Fundamentals of Physics: Mechanics, Relativity, and Thermodynamics*, Shankar states: "**We can find entropy change anyway we like. We usually pick the easiest path. Now, I'm ready to state the mega law that enforces all that. The second law: the law of increasing entropy. $dS \geq 0$ for the universe. There you have it. That's the great law: *Any process that reduces the entropy of the universe is forbidden.*"**

Entropy for the universe! There you have it – the mega law – the great law – the law that determines your destiny and the destiny of our universe; and, the whole thing is a fiction and a fraud.

It's Nature or Nature's Psyche who enforces everything, NOT entropy and NOT the second law of thermodynamics. Entropy or natural selection can't enforce anything because entropy or natural selection doesn't exist as a person, place, thing, or living entity. It's the LAW of Psyche that enforces everything at the quantum level in the Quantum Realm; and, mass or the physical was based exclusively upon the design and creation of the Quantum Fields by Nature or Nature's Psyche. No Quantum Fields, then no mass and no physical matter. No Nature's Psyche, then NO enforcement of the physical laws and NO quantum fields. No mass, then no heat and no entropy. It would all be random chaos instead, just like it was before the Gods and Nature's Psyche organized the Quantum Fields in our part of the multiverse.

The Gods, Controlling Psyches, or Nature's Psyche had to design and make the Quantum Fields first, before they could make and sustain mass, heat, and physical matter. This is what Quantum Field Theory tells us. Physical matter didn't make the massless, entropyless, and heatless Quantum Fields. Psyche did. Physical matter didn't exist and couldn't exist until after the massless, heatless, and entropyless Quantum Fields were designed and made.

Entropy doesn't determine anything; yet, here he's saying that entropy or the second law of thermodynamics determines everything. So, which is it? Which one is demonstrably true, and which one has been falsified?

Well, the heat-based or thermodynamics version of the second law (or entropy) is also based upon nothing but fiction. It's based upon a lot of things that are physically impossible such as the Carnot Cycle and Carnot Engines. The second law is based upon the Conservation of Entropy, even though the amount of heat or entropy is constantly decreasing in a thermodynamic system. Nature or Nature's Psyche makes the heat and the mass in the first place, and then that heat is constantly decreasing and constantly being absorbed back into the massless, heatless, and entropyless Quantum Fields from whence it originally came.

These people deny the existence of anything that reverses entropy or reduces the amount of entropy in this universe. The second law of thermodynamics is based upon the *elimination of evidence* logic fallacy. These people eliminate all of the evidence that falsifies the second law of thermodynamics. These people cheat, hide evidence, delete evidence, remove evidence, and rename evidence in order to prove that the second law of thermodynamics is true.

What is it that Shankar said above? Let me paraphrase and emphasize the message, in case you missed it.

**"Notice that energy is simply conserved, while the amount of entropy is constantly increasing. Entropy is more powerful and potent than energy. You can define entropy any way you like. Pick the easiest and most obvious definition. Just remember, it is forbidden to falsify the second law of thermodynamics. It's the great law or the mega law. It enforces all. It is forbidden to reduce the amount of entropy in our universe. The reversal of $E = mc^2$ is forbidden. It is only allowed to go one way – the way that produces ever-increasing entropy. Therefore, the Big Bang is also forbidden, because it reversed and eliminated a lot of entropy. Anything that reverses entropy or reduces entropy is axiomatically forbidden. The conversion or transformation of mass, heat, and entropy into massless, heatless, and entropyless photons or quantum fields is forbidden because it reverses entropy and eliminates entropy. The Gibbs Free Energy equations are forbidden because they allow negative entropy or a reversal of entropy. Nature and Quantum Fields are forbidden because together they are reversing and eliminating tons of entropy all the time. You can't have anything that reverses entropy or reduces entropy. It's forbidden! You can have any color you want so long as it's entropy."**

It is forbidden to falsify the second law of thermodynamics. What I'm doing here is forbidden and labeled as pseudo-science because I'm constantly and successfully falsifying the second law of thermodynamics. You can't do that. It's forbidden. You can't do what I'm doing and get away with it, because they won't let you.

The second law is protected by decree. The second law is sacrosanct. It's untouchable. The second law is declared to be unfalsifiable and axiomatically true. The second law is granted a *special pleading* – an exemption from the rules of logic and science. These people employ selective amnesia in order to enforce the truthfulness of the second law of thermodynamics. These people never once realize that it took a massive reversal of entropy or a lot of negative entropy (syntropy) to make the hot reservoir in the first place. They just start with the hot reservoir for free as a given so that entropy can then always be positive and constantly increasing. They cheat. The second law is propped up by a deception – they always start with the hot reservoir for free as a given, because the construction of that hot reservoir in the first place required a massive reversal of entropy, and that's forbidden.

The second law of thermodynamics is a fictional concept that they created with statistics and thought experiments, which means that we can also use thought experiments and statistics to falsify it. It's really easy to do once we know how.

The second law of thermodynamics is obviously false because everything that we have ever experienced and observed falsifies it; yet, the second law will remain axiomatically true while the Darwinists, Naturalists, and Atheists control our public schools. This is also what we have experienced and observed.

Mark My Words

## Entropy Defined as Death or Natural Selection

By turning chance, natural selection, entropy, or death into our Designer and our Creator, Darwinism or the Theory of Evolution was a huge step back into the Dark Ages turning our modern science into falsehoods, deceptions, and lies instead.

The falsehoods and lies in biology can be traced directly to Creation by Chance, Creation by Entropy, or Design and Creation by Natural Selection. Darwinism, Spontaneous Generation, Abiogenesis, Chemical Evolution, and Macro-Evolution were falsified in 1859 by Louis Pasteur – the very same year that Charles Darwin published "On the Origin of Species by Means of Natural Selection". We KNEW the truth before Charles Darwin introduced the Theory of Evolution to the world; but, since then, most of our scientists have deliberately embraced the falsehoods instead by turning Chance or Entropy or Natural Selection or Death into our Designer, our Creator, and our God.

Darwinism, Macro-Evolution, Chemical Evolution, Creation by Chance, Creation by Entropy, and the Theory of Evolution are our modern-day form of idolatry – the worship of false gods, or the worship of Creation by Stones or Creation by Rocks.

Quantum Mechanics and Quantum Field Theory revealed to us that it is in fact Nature or Nature's Psyche who collapses the wave function, and NOT the human psyche, and NOT the stones and rocks as the Materialists and Naturalists claim. The collapse of the wave function doesn't happen by chance. It happens by Intelligent Choice. Only Psyche, Soul, Life Force, Quantum Consciousness, or Intelligence can do choice. Physical matter by itself can't do intelligence or choice, because mass or physical matter is the result or the product of a choice, NOT the choice itself. According to Quantum Field Theory, particles, including physical particles, are chosen into existence by Nature or Nature's Psyche. Particles, including physical particles, can also be chosen out of existence or dissolved back into the Quantum Fields from whence they came by Nature or Nature's Psyche as well.

$E = mc^2$. It works both ways – both coming and going! Nature or Nature's Psyche can form particles including physical particles out of massless, heatless, and entropyless quantum fields anytime and anywhere that it chooses to do so. It's Nature's Psyche who collapses the wave function. Under the direction of the Gods or the Controlling Psyches, it was Nature's Psyche who made the quantum fields in the first place. The Gods had to design and make the quantum fields BEFORE physical matter could be made and sustained. This is what Quantum Field Theory is trying to teach us.

Likewise, according to Quantum Field Theory and $E = mc^2$, Nature or Nature's Psyche can disassemble or dissolve or transform the physical matter, mass, heat, and entropy back into the massless, heatless, and entropyless quantum fields anywhere and anytime that it chooses to do so. This is what Quantum Field Theory is trying to teach us.

We have the truth on one hand, and we have the Theory of Evolution on the other hand. They falsify each other, which means that they are mutually exclusive.

## Quantum Field Theory Falsifies the Second Law of Thermodynamics

The whole purpose of Science 2.0 is to find the truth within science. This process starts by identifying and eliminating from science anything that is demonstrably false – anything that has never been experienced or observed, or anything that can never be experienced or observed.

That means that we typically start with Creation by Chance, Creation by Natural Selection, Creation by Abiogenesis, Creation by Spontaneous Generation, or Creation by Entropy. They are all the same thing in the end, and they are all nothing but pseudoscience. They are false philosophies or false religions masquerading as science. Natural selection will never be observed designing and creating anything in a laboratory because it can't. Natural Selection is, by definition, Creation by Entropy or Creation by Death; and, we all instinctively know that entropy or death cannot design and create anything at all.

Spontaneous Generation, Abiogenesis, and therefore Macro-Evolution were falsified in 1859 by Louis Pasteur, the same year that Charles Darwin published "On the Origin of Species". The theory of evolution was falsified the very same year that it was revealed to the world and became a thing in the minds of its creators. We've known the truth all the way along, and Darwinism or the theory of evolution was a replacement of that truth with a bunch of different falsehoods, deceptions, and lies.

The whole purpose of Statistics and the Scientific Method is to eliminate chance as our designer and our creator. Creation by Chance, or Creation by Entropy, or Creation by Natural Selection falsifies or defies the whole purpose of Statistics and the Scientific Method. The two are mutually exclusive. If we want to find and know the truth, we must be willing to eliminate the falsehoods from science. Most scientists aren't willing to do that, because most of them have defined Chance or Entropy or Random Luck as our designer and our creator in complete violation of the rules of Statistics, Logic, Rationality, and the Scientific Methods.

The whole purpose of Science, Statistics, and the Scientific Methods is to eliminate Chance, or Entropy, or Random Luck as our designer and our creator; but instead, the Darwinists, Naturalists, Physicalists, Materialists, Behaviorists, Determinists, Nihilists, and Atheists have defined and enshrined Chance or Entropy or Random Luck axiomatically as our Designer, our Creator, and our God. They turned it into a LAW so that it can't be falsified.

I falsified Materialism, Naturalism, Darwinism, Nihilism, and Atheism in 2016. In 2019, I started finding dozens of different ways to falsify the Second Law of Thermodynamics.

It took a while, but after I encountered Quantum Field Theory, I slowly realized that Quantum Field Theory represents the truth in Science, as long as we are permitted to go with Paul Dirac's observation that it is in fact Nature (or Nature's Psyche) who collapses the wave function, and not the human observer or the human psyche.

Paul Dirac and Henry P. Stapp should be given the Nobel Prize for their observation that it is Nature who is collapsing the wave function. I should be given the Nobel Prize along with them for discovering that it is Nature's Psyche or the Intelligences within the Quantum Fields who are making the physical atoms, firing the neurons in our brains, mapping our physical brain, designing and making the synapses in our brain and the meaning of those synapses, storing our memories, and collapsing the wave functions that need to be collapsed. Nature or Nature's Psyche is controlling it all. The human psyche isn't consciously aware of any of this stuff. Nature's Psyche was collapsing wave functions long before the first particle of physical matter or mass was designed and made.

Science, Quantum Field Theory, $E = mc^2$, and the Equations for Entropy FALSIFY the second law of thermodynamics. This is probably my greatest scientific discovery. Being able to falsify the second law of thermodynamics is huge, if one wants to find and know the truth about how things really work.

**S = Q/ΔT**

**Q = mcΔT**

**S = mc**

Entropy (**S**) is comprised of heat (**Q**), mass (**m**), and heat storage capacity (**c**). Mass has heat storage capacity (**mc**); and, entropy (**S**) is heat storage capacity, according to these equations for entropy. The temperatures (**T**) cancel out. Entropy remains as a component of heat and mass, no matter what the temperature might be. Thermodynamics is an exchange of heat from a hot reservoir to a cold reservoir.

These equations reveal that at absolute zero (**T = 0 Kelvin**), heat completely ceases to exist, which means that entropy completely ceases to exist along with it. In the massless, heatless, and entropyless Quantum Fields at absolute zero, mass ceases to exist which means that heat ceases to exist, heat storage capacity ceases to exist, and entropy ceases to exist. The Quantum Fields and Quantum Field Theory falsify the second law of thermodynamics. For me personally, this is the greatest scientific discovery of all time. These equations for entropy define what entropy truly is and what entropy is truly comprised of. These equations also tell us how and when entropy goes to zero and ceases to exist.

It doesn't take a genius to realize that when mass goes to zero (**m → 0**), then heat goes to zero along with it (**Q → 0**). Remove all the mass or resistance to acceleration, then you remove mass's heat storage capacity (**mc → 0**), and you eliminate all the heat (**Q → 0**) and all the entropy along with it (**S → 0**)!

It's there in plain sight simply waiting to be discovered by someone who is willing to look and see. Entropy (**S**) is mass (**m**) or heat storage capacity (**mc**). No heat storage capacity or no mass, then NO entropy! Whenever mass ceases to exist, then the associated entropy completely ceases to exist along with it. It's that simple. It also falsifies the second law of thermodynamics which claims that the amount of entropy is always increasing and that the amount of entropy in the universe can never decrease.

The second law of thermodynamics provides us with the wrong definition for entropy. The second law turns "entropy" into some type of mystical, supernatural, magical entity that is constantly increasing and can never decrease in complete violation of the first law of thermodynamics or the Conservation of Energy and Psyche which states that the amount of energy can never increase and that the amount of energy can never decrease. By definition and by axiom, entropy is always increasing, whereas energy is simply conserved or remaining constant and the same. Entropy is supernatural. Entropy is a God. In fact, entropy is better than God, because entropy can do Creation Ex Nihilo or the creation of something from nothing which God can't even do. God is not a magician. God is a scientist. It's entropy that's the magician. It's entropy or natural selection that's imbued or infused with the ability to create something from nothing.

Once we realize that the massless, heatless, and entropyless Quantum Fields falsify the second law of thermodynamics, then we quickly observe that quantum waves (virtual particles), magnetism, gravity, dark energy, tachyons, photons, and psyches or intelligences also falsify the second law of thermodynamics. Everything massless, heatless, and entropyless FALSIFIES the second law of thermodynamics!

The instantaneous infinite acceleration of a photon from zero to the speed-of-light is physically impossible, which means that the photon has to shed its mass or its physicality or its resistance to acceleration, in order to achieve the speed-of-light. A photon must "jettison" its mass or resistance to acceleration by transforming that energy into infinite acceleration instead. According to the equation for heat and the heat-based equation for entropy, when mass goes to zero, so does heat and entropy! Entropy is constantly going to zero and ceasing to exist all the time, according to Quantum Field Theory! The falsification of the second law of thermodynamics is there all over the place in the Feynman Diagrams, and we can't even see it because we aren't looking for it!

https://en.wikipedia.org/wiki/Feynman_diagram

The only reason they don't discover this truth is because they don't allow themselves to discover this truth – they don't allow themselves to falsify the second law of thermodynamics. But you have to be willing to falsify and eliminate everything that is false, if you truly want to find and know the truth. It's that simple. The truth can't be built upon falsehoods, deceptions, and lies because something important will always be missing. In this case, due to the deception or the lie, the massless, the heatless, and the entropyless will always be missing because the second law of thermodynamics doesn't permit them to exist.

When mass ceases to exist, so does heat and entropy. Mass, heat, and entropy are constantly going to zero, according to Quantum Field Theory, which means that Quantum Field Theory FALSIFIES the second law of thermodynamics which states that the amount of entropy is always increasing and that the amount of entropy in the universe can never decrease. The false is falsified by the truth. In Science, the truth is represented by Quantum Mechanics, Psyche or Quantum Consciousness, and Quantum Field Theory. In contrast, the falsehoods within science are contained mostly within Materialism, Naturalism, Darwinism (Creation by Entropy), and Atheism (Creation Ex Nihilo or Creation by Chance), as well as the Second Law of Thermodynamics. These people believe that only Entropy and Physical Matter exist, which means that the discovery of the massless, heatless, and entropyless FALSIFIES their pre-chosen religious beliefs.

When mass goes to zero, heat goes to zero, which means that the amount of entropy goes to zero along with it. Whenever mass or heat is transformed by Nature or Nature's Psyche into massless, heatless, and entropyless photons, quantum waves, or quantum fields, the second law of thermodynamics is falsified. It's that simple. $E = mc^2$ works both coming and going. But, the Naturalists, Darwinists, and Atheists don't permit mass, heat, and entropy to be converted or transformed back into massless, heatless, and entropyless photons or quantum fields or virtual particles (quantum waves) – in complete violation of $E = mc^2$ and Quantum Field Theory. The truth in Science falsifies Materialism, Naturalism, and their derivatives such as Nihilism, Behaviorism, Determinism, Darwinism, Atheism, and the Second Law of Thermodynamics.

Remember, it's Nature's Psyche who collapses the wave function, not the human psyche. Remember, it is Nature or Nature's Psyche who transforms mass, heat, and entropy back into massless, heatless, and entropyless quantum fields, according to Quantum Field Theory. The Big Bang Theory, if true, also falsifies the second law of thermodynamics! The Big Bang, if it happened, represents a huge massive reversal of entropy or the manufacture and construction and organization of a huge amount of syntropy. The false is falsified by the truth; and, the truth is repeatedly experienced and observed.

Truth is confirmed by truth.

Technically, the Big Bang wasn't necessary. What was absolutely necessary was the organization or construction of the Quantum Fields from the Chaos that existed before the Gods, Controlling Psyches, and Nature's Psyche organized the Quantum Fields in our part of the multiverse. The Gods or Controlling Psyches had to design and make the Quantum Fields BEFORE physical matter, or mass, or resistance to acceleration could be made and sustained.

The construction of the Quantum Fields was the Big Bang in our part of the universe; and, it didn't require tons of heat to do the job, either. In fact, mass or heat or heat storage capacity didn't exist and couldn't exist until AFTER the Quantum Fields were designed and organized. Mass, heat, and entropy (heat storage capacity) are made from massless, heatless, and entropyless Quantum Fields all the time according to Quantum Field Theory. Furthermore, mass, heat, and entropy are constantly being transformed back into massless, heatless, and entropyless Quantum Fields or Photons or Quantum Waves (Virtual Particles) all the time as well, according to Quantum Field Theory.

Remember, no mass then no entropy. Whenever mass goes to zero and ceases to exist, entropy goes to zero and ceases to exist along with it. Quantum Field Theory, photons, dark energy, and $E = mc^2$ FALSIFY the second law of thermodynamics.

Where Science is concerned, the obvious flaw in the second law of thermodynamics is that it deliberately excludes ALL of the massless, heatless, and entropyless quantum fields, quantum waves, psyches or intelligences, forces, fields, gravity, magnetism, dark energy, and photons – which comprise most of our universe. The second law is flawed because it deliberately excludes the evidence that falsifies it. It deliberately excludes ALL of the massless and entropyless FORMS of energy or light that exist in great abundance throughout our universe.

The second law of thermodynamics is flawed and grossly incomplete because it is Newtonian Physics, Classical Physics, Materialism, Naturalism, Darwinism, Nihilism, and Atheism. When it comes to the second law of thermodynamics and science in general, it is its Physicalism that falsifies it, because it completely excludes Quantum Field Theory and Quantum Mechanics and the Entropyless Forms of Energy and Light. The second law of thermodynamics can't possibly be true because it deliberately excludes all of the evidence that falsifies it. The second law can't be true because it's just too good to be true.

Mark My Words

## The Truth Is in Plain Sight

The truth is in plain sight where nobody can see it, because they aren't looking for it or don't want it to be true.

Nowadays, we can watch chemistry experiments online for free on YouTube.

When they combine two chemicals together in their chemistry experiments online, the stuff either explodes or it freezes their hands off.

When the stuff explodes, it's generating heat or thermodynamics in real time from room temperature components. The heat seems to be coming from nowhere. Well, it's coming from the massless, entropyless, and heatless Quantum Fields according to Quantum Field Theory. They take room temperature uranium or plutonium, get enough of it to achieve critical mass, and suddenly the stuff is generating NEW heat or thermodynamics seemingly from nothing. Well, it's actually transforming massless, heatless, and entropyless Quantum Fields into heat, mass, and heat storage capacity instead.

Whenever a photon chooses to stop, it literally transforms its massless, entropyless, and heatless infinite acceleration into mass, heat, heat storage capacity, entropy, and resistance to acceleration instead. It's happening all the time every day!

When the chemical reaction freezes their hands off, the stuff is actually returning towards the massless, entropyless, and heatless absolute-zero temperature of the Quantum Fields from whence it originally came. The heat or thermodynamics or entropy is literally ceasing to exist as it's being absorbed back into the heatless, massless, and entropyless absolute-zero temperature Quantum Fields from whence it originally came.

The making of mass, heat, thermodynamics, and entropy from massless, entropyless, and heatless Quantum Fields is only one half of Quantum Field Theory and $E = mc^2$. The transformation of heat, mass, and entropy back into massless, heatless, and entropyless photons, quantum waves (virtual particles), dark energy, and quantum fields is the other half of Quantum Field Theory or $E = mc^2$.

The second law of thermodynamics denies the existence of the other half. The second law denies the existence of the massless, the heatless, and the entropyless, which comprises most of our physical universe.

Mark My Words

## Using Statistics to Falsify the Second Law of Thermodynamics

In the summer of 2019, I used my college statistics course to falsify the second law of thermodynamics, and their attempts to turn disorder, entropy, or chance into a causal agent or a causal force.

The whole statistics course falsifies the second law of thermodynamics, if allowed to do so. This is the harder and more conceptual way to falsify the second law of thermodynamics, but it produces the same results in the end.

If something (like the second law of thermodynamics) is obviously false, there will in fact be many different ways to successfully falsify it. Materialism, Physicalism, Naturalism, Darwinism, Nihilism, Atheism, Behaviorism, Determinism, and the Second Law of Thermodynamics are obviously false, which means that there are hundreds or even thousands of different ways to falsify them. Quantum Mechanics, Quantum Field Theory, and Entropyless Massless Photons FALSIFY these false religions or false philosophies; but, so does statistics, physics, and the other types of logic and math. You just have to train yourself to look for it, and before long, you start to find it.

Falsifying the pseudoscience has become one of my hobbies or specialties. I seem to enjoy it a great deal, and I obviously like writing about it too. It is forbidden to falsify natural selection, the theory of evolution, and the second law of thermodynamics, which is part of the thing that makes it so much fun. It's undiscovered country. It's forbidden science. It's looking where they don't want you to look and discovering what they don't want you to discover. Every day I discover something new, simply by looking where I'm not supposed to look.

In this essay, I intend to list some of the things in my statistics course that I encountered that falsified the second law of thermodynamics in one way or another. Occasionally, something would jump right off the page and say, "Look at me! Here's another way for you to falsify the second law of thermodynamics or 'creation by chance and entropy'." Anytime these people turn chance or entropy into our designer and our creator, the rational or the logical within you is assaulted, and we find ourselves looking at something that is obviously false. Atheism of any kind is based upon a refusal to look at the evidence that falsifies it. That's why it's so much fun to find the evidence that falsifies it, because sometimes you find yourself doing science that has never been done before.

Mark My Words

## Entropy Defined as Maximum Disorder

We are forbidden by the Naturalists and Atheists to talk about it, think about it, or question it; but, most of us instinctively KNOW that there is something fundamentally wrong with the second law of thermodynamics. If the second law were actually true, then NONE of us would exist, physical matter wouldn't exist, black holes and neutrons stars wouldn't exist, and suns and galaxies wouldn't exist, because the Quantum Fields wouldn't exist. If the amount of disorder in the universe were actually increasing as the Materialists and Naturalists claim, then whole planets, stars, and galaxies would be dissolving into chaos right before our very eyes in real time. Your physical body would soon dissolve back into the Chaos Realm from whence it came, if the amount of disorder in our universe was constantly increasing as the Naturalists and Atheists claim.

The FACT that disorder by definition doesn't correlate with anything is statistical proof and scientific proof that disorder is the wrong definition for anything, especially something as organized as mass, heat, physical matter, and thermodynamics.

Disorder is the WRONG definition for anything so long as the Quantum Fields exist, because the Quantum Fields are perfect order, perfect syntropy, perfectly conserved, and perfectly functional perpetual motion machines even when the physical matter in a system is at heat death, at thermal equilibrium, or maximally randomized and homogenized. The second law of thermodynamics can't possibly be true so long as the Quantum Fields exist.

We can't run experiments on disorder, because the simple act of designing and running an experiment introduces order, or structure, or purpose into the system and also because experiments are by nature an attempt to measure the amount of order within a system or within a couple of different systems. Disorder can't be measured. It's the order that we see around us that we are in fact detecting and measuring. Information of any kind is ORDER. Anytime we gain information from a science experiment, we have in fact detected a certain amount of order or organization within the system. If the system were nothing but pure disorder or chaos, there would be nothing to measure. The fact that we can measure atoms, pressure, temperature, and volume within a container of gas tells us clearly and conclusively that there is a huge amount of order within that gas, even if that gas is maximally homogenized or the atoms are maximally and evenly dispersed.

Disorder is the WRONG definition for entropy and the second law of thermodynamics, because ALL of the order remains within the atoms and the quantum fields that make up spacetime, even if thermal equilibrium or perfect homogenization or heat death takes place at the physical level within the system. The fact that there is statistically NO correlation between disorder and anything else tells us clearly and conclusively that "disorder" is the WRONG definition for entropy and the second law of thermodynamics – or the unscientific definition for entropy and the second law of thermodynamics. We can't run science experiments on chaos or disorder. We would have to organize it first in order to do so.

The ONLY thing we can do with disorder is to run thought experiments, which is how the second law of thermodynamics was created in the first place. The second law is pure philosophy, pure metaphysics, and pure theory with absolutely NO experimental backing whatsoever. Perfect disorder or maximum disorder is impossible to achieve so long as the Quantum Fields exist.

The second law of thermodynamics defines entropy axiomatically as "disorder". Ironically, disorder has absolutely nothing to do with thermodynamics, mass, or heat. It is a *category error* logic fallacy to define entropy or thermodynamics as "disorder". There's NO correlation between the two. They have nothing to do with each other at all.

The FACT, that there is NO correlation between "disorder" and the perfectly organized Quantum Fields, also FALSIFIES the second law of thermodynamics which claims that the amount of entropy or disorder in the universe is always increasing and can never decrease. The Quantum Fields represent a permanent decrease in disorder or a permanent elimination of disorder. The Quantum Fields are Pure Syntropy. Syntropy is the opposite of disorder and entropy. There's nothing we can do to destroy the Quantum Fields because they are always being conserved by Nature's Psyche. It's Nature's Psyche who collapses the wave function. Every psyche has a certain amount of energy that's under its control; and, that controlling psyche can form or transform the energy under its control into anything that it wants that energy to be, anytime and anywhere that it chooses to do so. There's nothing we can do about it.

When the Gods and Nature's Psyche in partnership designed and made the Quantum Fields, the second law of thermodynamics was automatically and perpetually FALSIFIED. The Gods and Nature's Psyche would have to destroy the Quantum Fields in order to make the second law of thermodynamics true. The second law hasn't been possible and hasn't been true, ever since the Gods and the Controlling Psyches within Nature designed and made the Quantum Fields.

The only way to have maximum disorder at the quantum level is to return to the pre-Big-Bang or pre-Quantum-Field ERA when there was NO LAW, no Standard Model, no quantum fields, no physical matter, and NO God in our part of the multiverse. Quantum Fields represent perfect order or perfect syntropy at the quantum level. The Quantum Fields are the ultimate perpetual motion machines. The Quantum Fields represent perfect order and organization. Maximum Disorder or the Chaos Realm is impossible to achieve so long as the Quantum Fields exist. The second law of thermodynamics can't possibly be true, as long as the Quantum Fields exist.

Disorder is the wrong definition for entropy, because disorder no longer exists at the quantum level thanks to the Matrix of Quantum Fields that fill our part of the multiverse. Furthermore, the particles that are made from the energies in the Quantum Fields represent increasing amounts of order – NOT disorder. Nature's Psyche makes increasing ORDER at will from the energies within the perfectly ordered and organized Quantum Fields. Maximum Disorder is impossible to attain so long as the Quantum Fields exist. The second law of thermodynamics can't possibly be true so long as the Quantum Fields exist.

The ONLY way to attain Maximum Disorder is to remove God, God's Laws, the Quantum Fields, God's Standard Model, the Law-Giver, and the Law Enforcers (Nature's Psyches) from our part of the multiverse and return the whole thing back to the Chaos Realm from whence it originally came. Maximum Disorder is impossible to achieve so long as the Quantum Fields exist. The Quantum Fields represent perfect syntropy or perfect order. The Quantum Fields function flawlessly at absolute zero temperature in a completely massless, heatless, and entropyless state. The Quantum Fields are perpetual motion machines that are based exclusively on Exergy and the Conservation of Energy. The Quantum Fields are perfect order and perfect organization at the quantum level. Their very existence falsifies the second law of thermodynamics, which claims that the amount of entropy or disorder in our universe is always increasing and can never decrease. The second law of thermodynamics can't possibly be true so long as the Quantum Fields exist.

The false is falsified by the truth, and the truth is constantly experienced and observed. The effects of the massless, heatless, and entropyless Quantum Fields are constantly being experienced and observed. The Quantum Fields ARE perpetual motion. It can't get any more organized than that! Quantum Field Theory is our best and most proven scientific theory of all time. Quantum Field Theory trumps and falsifies the second law of thermodynamics, because Quantum Field Theory is based upon the Quantum Law of Information Conservation and the Quantum LAW of Energy and Psyche Conservation. Information and Conservation FALSIFY the second law's "disorder definition" for entropy. Disorder does NOT exist and cannot exist at the quantum level so long as the Quantum Fields, and Quantum Information, and Quantum Consciousness exist. The construction of particles by Nature or Nature's Psyche represents increasing amounts of order, NOT increasing disorder as the second law of thermodynamics erroneously contends.

There's NO such thing as disorder as long as the Quantum Fields exist. The Quantum Fields would have to be dissolved or disassembled, the physical matter would have to be dissolved and disassembled, and God or the Law-Giver would have to cease to exist in order for our part of the multiverse to return back to Chaos or Maximum Disorder so that the second law of thermodynamics could actually be true. The second law of thermodynamics can't possibly be true so long as the Quantum Fields exist. You would have to get God and Nature's Psyche to destroy the Quantum Fields in order to make the second law of thermodynamics true.

Good luck with that!

The Biblical God Jesus Christ and Nature's Psyche (or the Controlling Psyches within Nature) have formed a partnership and have FORMED the Quantum Fields; and, there's nothing we mere mortals can do to change that. It is Nature's Psyche who collapses the wave function; and, it is the Biblical God Jesus Christ and God the Father who taught the Controlling Psyches within Nature how to organize themselves into quantum fields and physical matter. There's nothing that we human beings can do to change that or influence that, except to get ourselves into partnership with the Biblical God Jesus Christ.

That's just the way it is. So long as the quantum fields remain in existence, the second law of thermodynamics will NEVER be true. Nature's Psyche can and will eliminate mass, heat storage capacity, heat, and entropy anywhere and anytime that Nature's Psyche chooses to do so. Likewise, Nature's Psyche can and will eliminate disorder and entropy anywhere and anytime that it chooses to do so.

So long as Nature's Psyche and the Quantum Fields exist, **$E = mc^2$** will function flawlessly both directions, both coming and going, whenever and wherever the Gods and Nature's Psyche see fit to manipulate and control this equation. The Gods and Nature's Psyche can make new mass and new heat anywhere, at any time they decide to make new mass and new heat, thanks to the Quantum Fields that they have organized and made. Likewise, according to Quantum Field Theory and $E = mc^2$, the Gods and Nature's Psyche can TRANSFORM mass, heat, heat storage capacity, resistance to acceleration, and entropy BACK INTO massless, heatless, and entropyless photons, quantum waves (virtual particles), and quantum fields anytime and anywhere that Nature's Psyche or the Gods choose to do so; and, there's nothing we can do to stop them.

The Gods and the Controlling Psyches within Nature have formed a partnership; and, there's nothing we can do to change that. The Gods and Nature's Psyche can dissolve your physical body, this earth, our sun, our solar system, or our galaxy back into the quantum fields right here right now if they were to choose to do so; and, there's nothing we could do to stop them. We exist at their pleasure. Humbling, is it not?

The ONLY thing they cannot destroy is your Psyche, your Quantum Consciousness, your Life Force, or your Intelligence. It has always existed, and it will always exist. Everything else, the Gods and Nature's Psyche can dissolve at will, including the Quantum Fields and your Spirit Body and your Physical Body.

Fascinating, is it not.

It's interesting what starts to come to the surface after we have eliminated many of the falsehoods that have been enshrined or axiomized into science. If we successfully eliminate everything that is false and everything that has been falsified, then ONLY the truth will remain. ONLY what has been experienced and observed will remain. Energy and Psyche will remain. When it comes to energy and psyche, there's nothing we can do to make more, and there's nothing we can do to eliminate them either. They have always existed, and they will always exist, which means that they will always be TRUE.

Mark My Words

## Entropy Defined as an Infinitely Increasing Superfluid

Energy is a conserved substance. That means that its amount can NEVER increase, and its amount can NEVER decrease. Energy cannot be made, and it cannot be destroyed. That is what it means to be conserved.

The first law of thermodynamics states that energy is conserved, and this universal law is called the Conservation of Energy. Since Psyche or Intelligence or Quantum Consciousness has been experienced and observed by out-of-body travelers, I extended the Conservation of Energy to include the Conservation of Psyche, because they exist at the same level and are the same type of stuff. Every psyche has a certain amount of energy that's under its control, and that controlling psyche can form or transform the energy under its control into anything that it wants that energy to be, anytime and anywhere that it chooses to do so. Psyche is the "brains" or the "intelligence" or the "chooser" within energy.

By axiom, by law, and by decree, the Materialists, Naturalists, Darwinists, Nihilists, and Atheists turned entropy into a superfluid or a super-substance whose amount is always increasing and whose amount can never decrease. Through the second law of thermodynamics, entropy is granted a *special pleading* or a *special exemption* so that it can be the ONLY substance in the universe capable of increasing in amount while simultaneously never decreasing in amount.

Atheism is Creation Ex Nihilo – the creation of something from nothing by nothing. Atheism is magic. The Materialists and Naturalists turned entropy into Creation Ex Nihilo or magic. Contrary to all the rules of logic, statistics, math, and science, the amount of entropy is allowed to increase constantly while its overall amount is never allowed to decrease. The stuff is super-charged and super-conserved. Through the second law of thermodynamics and by *special pleading*, the Materialists, Naturalists, and Atheists turned entropy into a super-substance capable of generating out of thin air from nothing, so that its amount can constantly be increasing in complete violation of the first law of thermodynamics or the Conservation of Energy and Psyche. This is called a *category error* logic fallacy. Through the second law of thermodynamics, entropy is *too good to be true*, which means that it can't possibly be true.

Through a *special pleading* or a *special exemption* logic fallacy, they turned entropy into the ONLY substance in the universe that is constantly generating out of thin air from nothing so that its amount can be constantly increasing while never decreasing. It's magic. It's Creation Ex Nihilo. It's Atheism.

Notice carefully: We can't do statistics, math, science, and logic with Creation Ex Nihilo, magic, or "entropy defined as an ever-increasing superfluid made from nothing". It defies all logic, rationality, and common sense to try to do so. There's NO basis. Even when entropy is simply defined as all of the energy in the universe, or all of the heat in the universe, or all of the disorder in the universe, as it is typically defined, once again there is NO basis for determining what entropy is not whenever entropy is defined as everything in our universe. You can't to math, statistics, science, and logic when you have no basis for comparison.

Through the second law of thermodynamics, entropy is *infinite* and *too good to be true*, which is a fatal flaw and a huge logic fallacy when it comes time to measure it, science it, math it, explain it, and run statistics on it.

There's NO basis for comparison when entropy is defined as all the energy in the universe, all the heat in the universe, all the disorder in the universe, or the creation of something from nothing. The way that entropy is defined within the second law of thermodynamics breaks parsimony, breaks logic, breaks rationality, breaks science, breaks the math, breaks the statistics, and breaks common sense as well. Not even God can do Creation Ex Nihilo; but, entropy can. In complete violation of the first law of thermodynamics, entropy is turned into a magical superfluid capable of constantly increasing in amount from thin air while simultaneously never decreasing in amount. Through the second law of thermodynamics, entropy is turned into creation ex nihilo and defined as such. They turned entropy into the ultimate perpetual motion machine, actually capable of putting out more than what it takes in. They turned entropy into creation ex nihilo. Entropy is better than energy! Entropy is better than God.

The second law of thermodynamics is obviously false; but, it has been granted a *special pleading* logic fallacy or a *special exemption* so that it can be axiomatically true and remain automatically true all the time. In other words, we are forbidden to falsify the second law of thermodynamics, because it has been defined and set into law by decree as being unfalsifiable and axiomatically true. The second law of thermodynamics has been granted a *special exemption* from the rules of statistics, math, logic, and science. Entropy can be anything you want it to be, and it doesn't have to make sense, because it's automatically true by decree or by law.

Mark My Words


**Creation by Entropy**

https://evolution-is-entropy.com/

There are literally dozens of different definitions for entropy. The Materialists, Naturalists, Darwinists, Nihilists, and Atheists believe that ONLY physical matter, entropy, death, and chance exist. Consequently, the mechanism behind evolution has to be one of these. Natural selection is creation by death. Therefore, evolution in general ends up being creation by chance or creation by entropy. These are falsified definitions for entropy. Science as a whole is in desperate need of a couple of true definitions for entropy.

Evolution of any kind is Creation by Chance or Creation by Entropy. The Darwinists only believe in physical matter, entropy, death, and chance.

A look at Boltzmann Brains quickly reveals that Boltzmann intended for entropy to be our Designer, our Creator, and our God.

https://en.wikipedia.org/wiki/Boltzmann_brain

https://www.youtube.com/watch?v=nhy4Z_32kQo

Boltzmann Brains are Creation by Entropy or Creation by Chance. The theory of evolution is Creation by Entropy or Creation by Chance. Creation by chance or creation by entropy is a perversion of statistics. Statistics has been defined as the art of lying with mathematics; and, the Boltzmann Brain is the prime example.

Boltzmann is the creator of the "disorder or chance" definition for entropy. Through the "disorder" equation and definition for entropy and the second law of thermodynamics, Boltzmann tried to turn everything in this universe into Creation by Entropy, Creation by Disorder, Creation by Chaos, or Creation by Chance. The "disorder definition for entropy" is based upon a huge *category error* logic fallacy, because disorder has absolutely nothing to do with thermodynamics. Yet, Boltzmann erroneously defined entropy as a Law of Disorder.

Again, his whole purpose was to turn entropy or chance into our Designer, our Creator, and our God.

https://en.wikipedia.org/wiki/Ludwig_Boltzmann#Second_thermodynamics_law_as_a_law_of_disorder

Following Boltzmann's faulty example, atheist Stephen Hawking at times described the Big Bang as a "quantum fluctuation" in entropy. Others have stated that the Big Bang was a random fluctuation in entropy – essentially a Boltzmann Brain. Many of these geniuses have defined the Big Bang as a universe-sized Boltzmann Brain. Garbage in; garbage out. Through the theory of evolution, Charles Darwin turned biology into Creation by Entropy or Creation by Chance. Creation by natural selection is Creation by Death.

All of these are Creation by Entropy or Creation by Chance. All of these are an example of what statistics was originally designed to prevent. All of these are a perversion of statistics, common sense, observational experiences, and the Rules of Science. Garbage in; garbage out.

If these geniuses are permitted to label such stupid ideas as "science", then we have every right in the universe to falsify their ideas and demonstrate why they are wrong. I used to be a materialist, naturalist, nihilist, and atheist. I erroneously believed that these geniuses knew what they were talking about, were extremely smart, had made their case, and had met their burden of proof. I was wrong! I had no idea how superstitious these people really were until I started to study their theories and their work. Creation Ex Nihilo (Atheism), Creation by Entropy, and Creation by Chance are nothing but supernatural, superstitious magic and wishful thinking. They have nothing to do with reality or truth.

I'm interested in trying to eliminate some of the dumb ideas from science. I want science to be about the truth, and not magic or wishful thinking.

The second law of thermodynamics is Creation by Chance, Creation Ex Nihilo, and Creation by Entropy. The theory of evolution and natural selection are also Creation by Chance, Creation Ex Nihilo, or Creation by Entropy. Natural Selection can also be defined as Creation by Death.

How potent is chance as a creative force?

We can calculate the probability of chance events. Let's delve into it and see what we can learn.

If you have ever looked at a binomial distribution table at the back of a statistics book, you soon realize that with a fair coin the probability of tossing 15 or more heads in a row is effectively ZERO. There is very little CHANCE of ever doing so. At 15 and more trials, the probability of tossing 15 heads in a row effectively drops to ZERO. It can't be done by chance alone. It can only be done by deliberate intelligence. You can defy chance by deliberately and intelligently stacking the deck and loading the dice.

http://www.math.odu.edu/stat130/binomial-tables.pdf

http://www.sjsu.edu/people/saul.cohn/courses/stats/s0/BinomialProbabTable.pdf

In order to make a protein that is 150 amino acids long by entropy or by chance, CHANCE has to toss a twenty-one-sided amino acid dice correctly 150 times in a row. Its probability of doing so is ZERO. Chance can't even reliably toss 15 heads in a row, let alone tossing a twenty-one-sided dice correctly 15 times in a row. It can't be done. The probability of doing so is zero! Yet, to make the simplest of proteins as well as the matching gene to go along with each protein, CHANCE would have to toss that twenty-one-sided dice correctly at least 44 times in a row.

Yeast proteins are on average 466 amino acids long. The smallest human protein is 44 amino acids. The longest human protein is Titin with 34,350 amino acids. With some of the proteins, CHANCE would have to toss that twenty-one-sided dice correctly over 30,000 times in a row. There's NO CHANCE for it to ever do so. It can't be done. It's physically impossible. It's statistically impossible. CHANCE prevents it from happening. Chance can't even make the smallest human protein by chance alone. A protein, 44 amino acids long, is beyond the range of chance. It is so improbable that it is impossible. Chance cannot produce such a thing!

The Materialists, Naturalists, Darwinists, Nihilists, and Atheists ONLY believe in physical matter, entropy, death, and chance. Evolution is literally Creation by Entropy or Creation by Chance. Natural selection is Creation by Death. According to these people, evolution is entropy (entropy defined as chance); and, entropy or natural selection or chance designed and created everything in this universe. Under the control of these people, the second law of thermodynamics was designed to be Creation by Entropy or Creation by Chance. These people only believe in physical matter, entropy, death, and chance. Consequently, these people want entropy, chance, natural selection, or the second law of thermodynamics to be our Designer and our Creator and our God. That is what they want, is it not?

https://evolution-is-entropy.com/

When it comes to the Materialists, Naturalists, Darwinists, and Atheists who believe in Creation Ex Nihilo, Creation by Chance, Creation by Natural Selection, or Creation by Entropy, their chosen beliefs are self-defeating and self-falsifying because chance cannot design and create anything of significance. Chance is very limited in its range or scope. Chance is how we produce disorder; consequently, anything that reverses or eliminates disorder is deliberately done by some type of intelligence or psyche instead, because that's the only way it can be done.

Furthermore, the different versions of the Theory of Evolution are also Creation by Rocks. Whether these people realize it or not, they believe that the rocks or the stones designed and created everything that exists including your brain and your genome. That's physically impossible and statistically impossible, but that's precisely what these people have chosen to believe, is it not? The theory of evolution, creation by entropy, or creation by chance is the last superstition. It is our modern-day form of idolatry – the worship of a false god. It is the worship of stones, chance, disorder, chaos, and entropy.

These people are deathly afraid of consciousness, psyche, or intelligence. They are mortified by the idea that some part of them is going to live forever. They desire annihilation – the cessation of existence. I KNOW because I used to be one of them. When I was an Atheist, I wanted to die and cease to exist. The idea that God or the Universe would someday hold me accountable for my sins scared me to death.

Even though I didn't have a true idea of what the Materialists, Naturalists, Nihilists, and Atheists are and believe at their core, back in 2012, looking back to that time in my life, it is obvious that I was one of them. I used to be a materialist, naturalist, nihilist, and atheist as recently as 2013.

I spent a year around 2015 to 2016 trying to figure out what the Materialists, Naturalists, Darwinists, Nihilists, Behaviorists, Determinists, Physical Reductionists, and Atheists truly believe and who they really are. I wrote a few books on the subject at the time.

They believe in physical matter. They believe in death because they can see the dead physical body. Even though they don't know what it is, they also believe in entropy. They are religiously devoted to entropy, chaos, disorder, or heat death. Furthermore, on the mystical and supernatural side of the equation, these people believe in Creation Ex Nihilo, Creation by Chance, or Magic. These people are superstitious. These people effectively define "entropy" as Creation by Chance or Creation Ex Nihilo. Just look up Boltzmann Brains, Natural Selection, Random Fluctuations in Entropy, and Quantum Fluctuations if you don't believe me. These people definitely believe in Creation by Entropy, Creation by Chance, and Creation Ex Nihilo. Through a process of elimination, we quickly observe that evolution ends up being Creation by Chance, Creation by Entropy, Creation by Death, or Creation Ex Nihilo.

These people really truly believe that entropy, natural selection, or chance designed and created everything that exists in this universe.

A simple binomial distribution table at the back of a statistics textbook falsifies their beliefs. There's NO CHANCE in this universe that these people could possibly be right. Statistics was actually designed or made to eliminate chance as a causal explanation in Science and the Scientific Methods; yet, the Darwinists, Nihilists, Naturalists, Materialists, and Atheists make chance the cause of everything that exists in our universe. That's very unscientific and illogical of them.

The whole purpose of statistics is to remove chance as a causal explanation in science; yet, the Materialists, Naturalists, Darwinists, and Atheists use chance as the cause and the explanation of everything that exists in this universe. There's a huge disconnect taking place here, and these people don't even know it, because they never think about it. These people refuse to look at the evidence that falsifies their beliefs. I KNOW because I used to be one of them.

The Materialists, Naturalists, Darwinists, Nihilists, Behaviorists, Determinists, Physicalists, and Atheists have placed all of their hope into Creation by Chance or Creation by Entropy or Creation Ex Nihilo. They never once realize that the probability of tossing heads fifteen or more times in a row, with a fair coin or with pure chance, is effectively ZERO. They don't realize how absurd their claims are when they start claiming that CHANCE tossed that twenty-one-sided amino acid dice correctly thousands of times in a row, over billions of years of time, trillions of trillions of different times, in order to make all of the 10 million different types of proteins that exist, along with their matching genes – all by chance alone. Such a thing is statistically impossible. Such a thing is physically impossible. God would have to create an ocean full of amino acids and keep that ocean of amino acids in existence for trillions of years; and yet, given that full ocean of the correct types of amino acids and an infinite amount of time, the simplest of proteins will still NEVER form by chance alone. Chance will prevent them from doing so. That's the way that chance really works.

Chance doesn't cause things to happen. Chance prevents things from happening instead. Chance prevents CHANCE from making sequences or proteins longer than about 15 amino acids in length, which means that the proteins and their matching genes are way too long to have been produced by chance alone.

Chance will prevent CHANCE from making a protein or its matching gene. Chance will prevent chance from making brand new life forms from existing life forms. Genetics, natural selection (death), and chance (entropy defined as chance) prevent macro-evolution from happening in the lab and in the wild.

Macro-evolution is two cats giving birth to a dog, or two chimp-like ancestors giving birth to a chimpanzee and a human. Genetics, sexual selection, and chance (entropy defined as chance) prevents macro-evolution from happening. The only way that macro-evolution can happen for real is through genetic engineering, performed by an intelligent being who knows what he is doing. There's NO CHANCE that it can happen by chance, by entropy, by disorder, by random luck, by natural selection, by sexual selection, by breeding, or by wishful thinking. Something like macro-evolution, the construction of a genome, the production of a car or a computer, the production of the quantum fields, the making of physical matter from energy, the making of a planet or a star, or the design and construction of a life form has to be deliberately done by some type of Psyche, Consciousness, Intelligence, or Life Force because the is NO CHANCE that chance or random fluctuations in entropy are ever going to do the job.

The Darwinists, Naturalists, and Atheists have chosen to believe in Creation by Entropy or Creation by Chance; but, the probability of tossing 15 or more heads in a row is effectively ZERO according to a binomial distribution table. Chance can't reliably do 15 amino acids in a row in a correct sequence, let alone a 150 or a thousand. Chance works to prevent it from happening. It is physically and statistically impossible for CHANCE to toss a twenty-one-sided dice correctly 15 times in a row, let alone the 150 to 30,000 times in a row that it would need to toss the dice correctly in order to make the different proteins that exist.

It can't be done, which means that it wasn't done.

Creation by Chance or Creation by Entropy is a statistical impossibility; and, the Theory of Evolution in its different forms is precisely Creation by Entropy or Creation by Chance. Entropy or chance is the ONLY "creative mechanism" that these people actually believe in. They certainly don't believe in Psyche, or God, or Intelligence.

We can calculate the probability of chance events.

Chance has a bit of a chance for making something that is fifteen sequences long; but, at the tail ends of a distribution and at the final end of a series of trials that's fifteen or more in length, the chance of CHANCE actually doing something for us drops exponentially to ZERO. Chance has limits, and when chance reaches its limits it reaches zero probability, zero chance, and the physically impossible. There's NO CHANCE in this universe that disorder, random luck, entropy, natural selection, or chance can make a functional protein, let alone the matching gene to go along with it. It can't be done. There's no chance in this universe that CHANCE can make a functional atom out of a bunch of different random quantum waves. It can't be done.

Some of the amino acids have a short enough sequence that it's realistically possible and probable to produce them by chance alone. However, the sequence length of the proteins and their matching genes is too long for them to ever be produced by chance.

There's NO CHANCE in this universe that entropy, disorder, random luck, or chance can design and create the simplest protein or gene that exists, let alone the genomes and life forms that exist. It is impossible for chance to produce anything longer than 15 units in length with any kind of statistical or measurable reliability. It's physically and statistically impossible for chance to make any type of protein that exists, let alone the matching gene to go along with it. You can produce some of the amino acids and polypeptides by chance, but it is impossible to produce proteins and genes by chance. That binomial distribution table in the back of our statistics books tells us that it is so. It tells us why it is so. Chance just isn't that reliable. Chance has limits!

There's only so much in this universe that can be done by chance; and, that list is very short. The longer the sequence, the more improbable and impossible it becomes for chance to do it alone all by itself. The drop-off towards zero chance at the tail ends of a distribution is exponential. There's effectively zero probability or zero chance that CHANCE, or luck, or random disorder, or natural selection, or "entropy" can produce a protein longer that fifteen amino acids simply by chance alone. It just can't be done, which means that it wasn't done that way.

According to a binomial distribution table, chance effectively prevents us from tossing fifteen heads in a row with a fair coin. In order to toss 15 heads in a row, that coin has to be weighted or manipulated so that it most likely comes up heads. Once the coin has been manipulated, then we are NO longer dealing with chance. Likewise, once the Gods have put the correct amino acid in the right spot within a protein, then we are NO longer dealing with chance.

Chance cannot put the correct amino acid in the correct spot more than about 15 times in a row. Technically, it is much lower than that with a twenty-one-sided dice, but we are trying to give chance the benefit of the doubt here. Fifteen times in a row is about the best that chance can do in a binomial distribution. Beyond that, we are no longer dealing with chance. When we start seeing correct functional sequences that are 150 amino acids long or 30,000 amino acids long, then we KNOW for an absolute fact that some type of intelligence or consciousness or life force was involved, because we KNOW for a fact that we are no longer dealing with chance. We use the laws of probability to eliminate chance as a causal force. Creating proteins by chance is impossible.

https://www.youtube.com/watch?v=W1_KEVaCyaA

Intelligence or psyche or quantum consciousness can put the correct amino acid in the right spot every time, if it chooses to do so. Chance cannot. Chance can only do about 15 heads in a row before it starts to run out of luck. At its ultimate best, chance can only do about fifteen amino acids in a row before it runs out of luck. In contrast, Psyche, or Quantum Consciousness, or Intelligence can do the physically impossible at will. Only Psyche can make choices at the quantum level and get things done at the quantum level, where there is NO physical matter and when there was NO physical matter. There's NO CHANCE that physical matter can make or do anything at the quantum level or the psyche level, because physical matter is limited and therefore prevented from doing so. There's only so much that chance and physical matter can do in this universe, and beyond that, they become totally ineffective.

For me personally, this is one of the most convincing Scientific Proofs of God's Existence. Your genome is God's signature, and that genome or signature is written on every cell in your body.

Remember the Gods, the Controlling Psyches, and Nature's Psyche can do the physically impossible at will. Whenever intelligence or psyche or consciousness is involved, we are no longer looking at chance but at deliberate choices and actions instead. This is science and statistics at its very best, pointing us directly to the truth of the matter. Chance, entropy, natural selection, death, and evolution cannot design and create anything at all that requires the correct placement of more than about fifteen items in a sequence. Proteins and genes are beyond the reach of chance, entropy, disorder, death, natural selection, and evolution.

The Gods had to design and pre-make genomes and molecular machines in order to reliably produce genes and proteins at will. That's something that chance can never do. Ask yourself who is driving or operating those molecular machines, and it quickly becomes clear that it's NOT entropy, chance, natural selection, evolution, or physical matter. It's something much more powerful and reliable instead.

Mark My Words

## The Null Hypothesis

I found the following figure helpful for understanding the null hypothesis, statistics, and their relationship to the scientific method.

Adapted from Pagano, R.R. (2010). *Understanding Statistics in the Behavioral Sciences* (9th ed.). Belmont, CA: Wadsworth. (p. 269.)

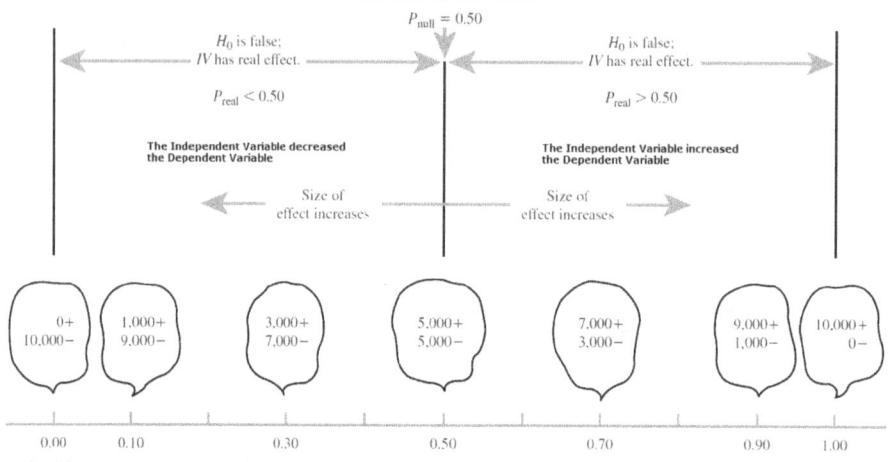

Adapted from: Pagano, R.R. (2010). *Understanding Statistics in the Behavioral Sciences* (9th ed.). Belmont, CA: Wadsworth.

The null hypothesis in a science experiment and statistics states that the independent variable had NO EFFECT on the dependent variable. If the independent variable had an effect on the dependent variable, then the next step is to try to determine the size and the direction of the effect.

It's helpful to have a handle on this, when we start discussing whether the independent variable has the "power" to produce a real effect on the dependent variable, or not. This is science, the scientific method, and statistics in action. We have to understand this, if we want to know how scientists use statistics to do science and science experiments.

In this science experiment, their sample size was 10,000, and they used the sign test. They tested whether the effect of the independent variable was positive or negative.

In the middle there, we have the **Null Hypothesis**, where there are 5,000 positive effects and 5,000 negative effects. The results are fifty/fifty, which is synonymous with pure chance. $P_{null}$, the probability of a null, or the null hypothesis is always 0.50. The null hypothesis is fifty/fifty, or pure chance.

Whenever the results of a science experiment or statistical analysis are fifty/fifty, then they conclude that the results were caused by chance alone. Fifty/fifty is pure chance when gambling as well. When the results of the experiment are fifty/fifty, pure chance, or the null hypothesis, then they conclude that the independent variable didn't have an effect on the dependent variable. It doesn't have to be exactly 50/50 either. The size of the effect is miniscule if the results of the science experiment are hovering around the null hypothesis.

We are looking for a real effect, a sizable effect, or a significant effect, if we are to conclude that the independent variable is causing the dependent variable.

Looking at the percentages at the bottom of the table above, we observe that when the percentage comes within the 0.00 to 0.05 range, then the independent variable had a significant effect on the dependent variable and decreased the dependent variable a lot. If the percentage comes within the 0.95 to 1.00 range, then we conclude that the independent variable had a significant effect on the dependent variable and increased the dependent variable a lot.

Through science experiments, we are trying to establish causality, and thereby remove chance as the "causal explanation". We are trying to reject chance or the null hypothesis. In science and statistics, chance is NOT a valid cause! That's why Materialism, Naturalism, Darwinism, Nihilism, Atheism, and the Second Law of Thermodynamics are invalid sciences or pseudo-sciences, because they use chance as their causal mechanism and are in fact Creation by Chance or Creation Ex Nihilo.

We cannot use chance, randomness, chaos, or disorder in science as our causal explanation and still remain within the realm of science. That's why Materialism, Naturalism, Darwinism, and Atheism are pseudo-sciences or religions masquerading as science, because they are Creation by Chance and use chance as their causal mechanism and make the claim that chance created everything that exists in this universe. Creation by Chance isn't real. Creation by Chance is the null hypothesis, which was produced by chance alone. Chance has no discernable effect on anything. These people literally turned chance into their independent variable and then they made chance the cause of everything that exists in this universe.

Creation by Chance or the Null Hypothesis is precisely what we are trying to falsify and eliminate through science and our science experiments. Creation by Chance or "chance causality" is the Null Hypothesis. Materialism, Naturalism, and their derivatives such as Darwinism and Atheism ARE Creation by Chance or the Null Hypothesis. The Null Hypothesis tells us that we should eliminate Materialism, Naturalism, Darwinism, Nihilism, Atheism, the Second Law of Thermodynamics, and every other form of Creation by Chance from science. So, why aren't we doing so?

It's because the Materialists, Naturalists, Darwinists, and Atheists have granted themselves a *special pleading* or a *special exemption* so that they don't have to follow the rules of science and don't have to prove their theories true through scientific experimentation and statistical analysis. Instead, these people unilaterally turned their theories and ideas into axioms and laws, and forbid us from falsifying their ideas and theories, so that their falsehoods will always be true.

They cheat!

That's how they do science.

You will never see statistical analysis, the null hypothesis, science, the scientific method, and science experiments presented in quite this manner within our public schools because it falsifies Materialism, Naturalism, Darwinism, and even Nihilism and Atheism. Our college professors have to at least make it seem as if the theory of evolution is true, or they will be fired from their jobs. They are not free to tell us the truth.

Our public schools don't reward us for falsifying the second law of thermodynamics, the theory of evolution, or creation by chance. They punish us instead whenever we do.

The whole purpose of our public school system is to indoctrinate us into Materialism, Naturalism, Darwinism, Nihilism, Behaviorism, Determinism, Atheism, and all the other forms of Creation by Chance. If you are not an atheist by the time you graduate from college, then your college professors have failed in their mission and their purpose. They have no other reason for being there. That's how far our society has deteriorated in recent decades.

You can't falsify or reject the Null Hypothesis, whenever the Null Hypothesis is Materialism, Naturalism, Darwinism, Nihilism, Atheism, the Theory of Evolution, the Second Law of Thermodynamics, or any other form of Creation by Chance. By axiom, by law, by edict, and by decree, you have to retain the Null Hypothesis whenever the Null Hypothesis is Materialism, Naturalism, or one of their derivatives such as Darwinism and Atheism. That's the way these people operate. They cheat! They call it peer review. It's censorship and intimidation, which is why I work where they can't get at me.

Thanks to the Atheists, we don't have freedom of speech and freedom to teach in our public schools, but so far, that doesn't seem to apply where my books are concerned. Here, I'm permitted to say anything that comes to mind. Here, I'm permitted to teach the truth if I want to do so.

Mark My Words

### Disorder or Chance Has No Power

In statistics, "power" is the probability that the results of an experiment will allow rejection of the null hypothesis if the independent variable has a real effect on the dependent variable or the measured variable. The "power" of an experiment is a measure of the sensitivity of the experiment to detect a real effect from the independent variable.

This concept didn't make any sense to me at first. In statistics, "power" is the likelihood of you rejecting the null if the null is really false – or, in other words, saying there is an effect when there is an effect. In the physical world, power causes things to happen. "Power" is an attempt to measure how much the independent variable is CAUSING the dependent variable to happen. The power goes up the further away you get from the null hypothesis, or 50 percent, or pure chance alone. Power is the probability of rejecting the null hypothesis when the null hypothesis is false.

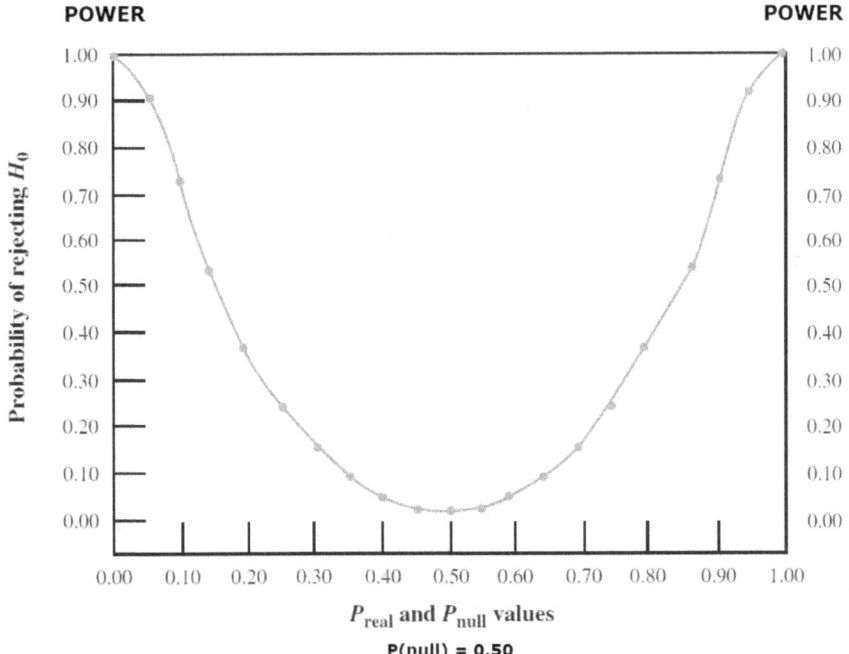

P(null) = 0.50

**POWER is the probability that the null hypothesis will be rejected when the null hypothesis is false. Notice that POWER goes to certainty (1.00) when P(real) is equal to 0.00 or 1.00. POWER goes up the further away it gets from the null hypothesis.**

When P(real) is equal to P(null), **POWER is non-existent or has gone to zero. P(null) is always equal to 0.50, fifty percent, fifty/fifty, or pure chance. The null hypothesis assumes that chance alone is at work. The further away from the null, the more significant the results.**

Adapted from Pagano, R.R. (2010). *Understanding Statistics in the Behavioral Sciences* (9th ed.). Belmont, CA: Wadsworth.

Adapted from Pagano, R.R. (2010). *Understanding Statistics in the Behavioral Sciences* (9th ed.). Belmont, CA: Wadsworth. (p. 275.)

The whole goal of a science experiment is to reject the Null Hypothesis and to get as far away from chance as you can possibly get. The purpose of statistics and science is to eliminate chance as the underlying cause, as best we can. In true science and in statistics, CHANCE is an invalid causal force or an invalid cause. Chance can't be used as our causal mechanism in statistics and True Science.

What we needed in science and statistics is a RULE for determining when the obtained probability is small enough to reject chance as an underlying cause. The goal of statistics and science is to reject chance as the underlying cause. In contrast, the goal of the theory of evolution and the second law of thermodynamics is to turn chance into the cause! These two goals are mutually exclusive, which means that the theory of evolution, the second law of thermodynamics, entropy defined as disorder or chance, chance causality, and creation by chance are NOT science. Creation by chance is wishful thinking, philosophy, magic, religion, and dogma. Creation by chance has no basis in science or reality.

The Null Hypothesis states that the results of the science experiment were caused by chance alone. Science experiments exist to reject the null hypothesis, not to retain it or enshrine it. Materialism, Naturalism, Darwinism, Nihilism, Atheism, Entropy as Disorder or Chance, the Second Law of Thermodynamics, and Creation by Chance ARE the Null Hypothesis; and, these things are precisely what our science experiments are trying to falsify, reject, and eliminate from science.

The null hypothesis in a science experiment states that the results of the experiment were caused by chance or created by chance, which means that the independent variable had NO EFFECT on the dependent variable. That means that there is NO causal relationship between the variables, and that means that your alternative hypothesis is most likely false.

The Null Hypothesis ($P_{null} = 0.50$) assumes that chance alone is at work. Chance can't produce a sizable effect. The Null Hypothesis is "chance causality", "caused by chance", or "created by chance". The Null Hypothesis is Creation by Chance or Chance Causality. The Null Hypothesis is "caused" by chance. In a science experiment, we are trying to reject the Null Hypothesis – we are trying to reject chance, chance causality, or creation by chance. We are trying to establish True Causality. We establish causality in a science experiment by rejecting chance or by rejecting the null hypothesis. Rejecting the null hypothesis is the same thing as rejecting CHANCE as our causal mechanism. True Science rejects chance as a causal mechanism and seeks for true causality instead. Chance can't cause things to happen. If it's causing things to happen, then it is no longer chance!

The null hypothesis predicts that the independent variable has NO EFFECT on the dependent variable. If you can statistically reject the null hypothesis, then you can realistically assume that the independent variable had some type of effect on the dependent variable. Then the next goal is to measure the size of the effect.

The further your effect is from 0.50, the null hypothesis or pure chance, the greater the size of the effect from your science experiment. The probability of the null hypothesis or pure chance is always 0.50. $P_{null}$ is always equal to 0.50. This is fifty/fifty, or pure chance. If your science experiment is producing something besides the null hypothesis or pure chance, then you are detecting a Real Effect. $P_{real}$ is < 0.50, or $P_{real}$ is > 0.50. We can reject the null hypothesis or chance causality or creation by chance, and we can conclude that the independent variable had a real effect on the dependent variable, if we are detecting something besides pure chance or the null hypothesis. The further $P_{real}$ is from

0.50 or pure chance, the greater the size of the effect. This is the scientific method and statistics in action.

In statistics and a true science experiment, a "real effect" is an effect of the independent variable that produces a change in the dependent variable. The goal is to establish causality. The goal is to demonstrate that the independent variable causes the dependent variable to change. The goal of a science experiment is to reject the null hypothesis or to reject chance as our causal mechanism. This is how we establish causality in science – by manipulating the independent variable and measuring its effect on the dependent variable. The purpose of a science experiment is to determine if the independent variable has the power or the ability to produce a real effect on the dependent variable. The higher the "power", the more sensitive the experiment to detect a real effect from the independent variable. In contrast, when the results of the science experiment are null and the independent variable has no effect, there is no "power" because the results of the science experiment were cause by chance alone and $P_{real}$ is close to $P_{null}$, which is 0.50.

In a science experiment, we retain the null hypothesis whenever we fail to detect any real effect from the independent variable on the dependent variable. When we retain the null hypothesis, we conclude that the independent variable (chance) had no effect on the dependent variable (heat, genomes, or thermodynamics).

When "disorder" or "entropy defined as disorder" is the independent variable in a science experiment, that pure disorder or pure randomization has NO CHANCE whatsoever of producing a real effect, or a measurable effect, or a change in the dependent variable. Once it starts to produce an effect, then it is no longer disorder! Whenever we manipulate disorder, it ceases to be disorder. Whenever we manipulate chance, it ceases to be chance. Manipulating disorder and chance eliminates their status as disorder and chance, and it turns them into some type of order or organization instead.

You see, in a true science experiment, they manipulate the independent variable and then observe to see if that manipulation has any effect on the dependent variable. However, whenever we start to manipulate "disorder" then it is no longer disorder but becomes some type of order instead. The effect of manipulating disorder is to produce some type of order! When a system has magically achieved maximum disorder, it has no place to go but up into some type of order. Contrary to the claims made by the second law of thermodynamics, everything in our universe tends towards or strives towards a certain amount of order and organization. Life, consciousness, psyche, or intelligence finds a way to prevail. Psyche, or Quantum Consciousness, or Intelligence is a type of order or organization to begin with; and, it only goes up from there.

In a true science experiment, there's no way to manipulate disorder to see what types of effects it produces on the dependent variable, because once you start to manipulate the disorder, then it is no longer disorder but has become some type of order instead. This truth, along with all of the order and organization that we see all around us, falsifies the second law of thermodynamics which claims that the amount of disorder in our universe is constantly increasing. We don't observe constantly increasing disorder in our universe. Instead, we observe conserved order and organization. Once it has been organized, it seems to stay organized. Once the protons and neutrons were organized or created by Nature or Nature's Psyche, they have stayed organized ever since. Once the quantum fields were designed and organized by the Gods and Nature's Psyche, they have stayed organized ever since. Psyche, or Intelligence, or Quantum Consciousness is some type of order or organization. At the quantum level, everything is organized, and everything is conserved. The verified and proven existence of the Quantum Fields falsifies the second law of thermodynamics which claims that they shouldn't exist. Wherever

Intelligence, Psyche, or Consciousness is involved, chance and disorder cease to exist. This is what we have actually experienced and observed.

If the independent variable does not produce a change in the dependent variable, it has no effect; and, we say that the independent variable does not have a Real Effect. Then we retain the null hypothesis and conclude that the independent variable didn't have a significant or measurable effect on the dependent variable. By retaining the null hypothesis, we conclude that our alternative hypothesis (the second law of thermodynamics or the theory of evolution) is FALSE; and, we are supposed to toss it out and start over with something new, because the results of our science experiment can be attributed to chance alone, because there was no Real Effect from the independent variable (chance) on the dependent variable (genomes, life forms, eyes, planets, stars, heat, thermodynamics, galaxies) whatsoever. This is the scientific method and a science experiment in action. We are supposed to toss out our alternative hypothesis (the second law of thermodynamics or the theory of evolution) if it was caused by chance alone.

By definition, in principle or practice, Materialism, Naturalism, Darwinism, Nihilism, Atheism, and the Second Law of Thermodynamics ARE Creation by Chance or Creation Ex Nihilo, which automatically falsifies them because they are the Null Hypothesis and have NO real effect. Disorder doesn't cause heat, and heat doesn't cause disorder. There's no correlation whatsoever between disorder and thermodynamics. Likewise, natural selection doesn't touch our genes! Natural selection doesn't do anything except wait for us to die.

When disorder or chance is the independent variable, we quickly observe that the independent variable has NO EFFECT on the dependent variable. When heat, thermodynamics, or entropy is the dependent variable, we observe that disorder or chance has no effect on them. We observe that the "disorder definition" for entropy is false, because disorder or chance can't have a measurable effect on anything. Once chance or disorder start producing an effect, then they are no longer disorder or chance, but have become some type of order or organization instead.

We retain the null hypothesis and reject the alternative hypothesis in a science experiment whenever disorder or chance is the independent variable and something like heat, thermodynamics, genomes, stars, or entropy is the dependent variable, because the independent variable has NO EFFECT on the dependent variable.

Varying the amount of disorder or chance has NO REAL EFFECT on the amount of heat or entropy in a system, because we can't vary disorder and have it remain in existence as disorder. Disorder has NO POWER to increase heat or decrease heat. Disorder has absolutely nothing to do with heat, thermodynamics, or entropy. They are two completely different things!

If disorder is the independent variable, then varying the amount of disorder has NO EFFECT on thermodynamics or the amount of heat in a system. Varying the amount of disorder (scrambling the atoms and molecules) has NO REAL EFFECT on heat storage capacity, mass, or entropy (mass's heat storage capacity). If there is heat or thermodynamics, the heat remains no matter how scrambled or disorganized the atoms and molecules might become. There's NO correlation whatsoever between disorder and heat. The heatless, massless, and entropyless Quantum Fields function perfectly at maximum efficiency at absolute zero in the state that the physicists call "heat death". The Quantum Fields are perfectly organized perpetual motion machines. They are pure syntropy, or pure exergy, or pure order and organization; and, they exist and function perfectly in the complete absence of heat. Furthermore, whenever we start to manipulate disorder or chance, then we introduce some level of order or organization into them instead, and then they cease to be disorder and chance.

We can't use disorder or chance as the causal mechanism in our science experiments, as the Materialists, Naturalists, Darwinists, and Atheists do. Creation by Chance is an oxymoron – a contradiction in terms; and, Materialism, Naturalism, Darwinism, Nihilism, Atheism, and the Second Law of Thermodynamics are precisely that – Creation by Chance.

**Remember, disorder or chance has NO POWER to produce a REAL EFFECT! Once they start producing an effect, then they are no longer disorder or chance.**

It's Nature's Psyche who makes the quantum waves in the first place. It's Nature's Psyche who collapses the wave functions and fires the neurons. It's the only thing that can. The Obvious Law of Physics states that the smaller dwells within and controls the larger. Psyche or Intelligence or Quantum Consciousness is the ONLY thing that we know of that is small enough to dwell within and control the quarks and the electrons. Chance or disorder isn't making the quantum waves and firing the neurons in our brains. It's Nature's Psyche or the Controlling Psyches within Nature who is doing that for us. Have you ever had a brain zap? That's what happens whenever "chance" fires your neurons. You KNOW that you didn't do it because it happened by chance or it happened randomly. Whenever the human psyche asks Nature's Psyche to fire the neurons and it happens as requested, then everything functions normally, and we KNOW it. We KNOW that we were instrumental in making it happen! We KNOW that we caused it to happen.

Notice that in a true science experiment, it is always some type of intelligent being who is manipulating the independent variable in order to see if that independent variable has any effect on the dependent variable. You can't do a true science experiment without some type of Intelligence, Consciousness, Psyche, or Chooser at the helm. Nothing of significance ever happens without some type of Psyche or Intelligence choosing to make it happen. The purpose of science, experiments, and the scientific methods is to detect significant effects or causality. Only psyche or intelligence can do causality or choice at the quantum level. Science experiments, manipulation, order, quantum fields, and measurable results are chosen into existence by some type of psyche, intelligence, or quantum consciousness, whether we are talking about the quantum level or the physical level.

Whenever we are dealing with disorder as the independent variable, there's no way to manipulate that disorder so as to test the effects of that manipulation on a dependent variable. Once we start manipulating disorder, then it is no longer disorder but has become the dependent variable or the result instead. Once the Gods start manipulating chaos, or entropy, or random disorder, then it is no longer disorder but becomes some type of order instead.

The null hypothesis in a science experiment states that the results of the experiment were caused by chance or created by chance, which means that the independent variable had NO EFFECT on the dependent variable. That means that there is NO causal relationship between the variables, and that means that your alternative hypothesis is most likely false. You can try to increase the power of the experiment by increasing the size of the sample; but, if the independent variable (chance or disorder) has NO EFFECT on the dependent variable (genomes, stars, proteins, or life forms), increasing the size of the sample to include the entire universe will still have NO EFFECT.

Whenever we start detecting disorder or chance in a science experiment, we are in truth proving that the independent variable has NO EFFECT on the dependent variable. Disorder or chance IS the null hypothesis, or the lack of effect, that we are actually trying to identify, falsify, and eliminate from Science. Chance or disorder IS the null hypothesis, and everything else represents some type of order or some size of effect. Whenever disorder or chance is the independent variable, it is automatically and axiomatically impossible for that independent variable to have a real measurable detectable on any dependent variable, because chance or disorder is precisely the thing that we are trying to eliminate from science by running the science experiment in the first place.

We are using science and science experiments to establish causality by eliminating chance or disorder as the cause. The purpose of science and science experiments is to establish causality, and chance or disorder cannot be the cause of anything according to the Rules of Science, Statistics, and the Philosophy of Science. The purpose of science and statistics is to eliminate chance, or disorder, or random luck as the cause of what we are observing.

So, what do we observe?

We observe the Materialists, Naturalists, Darwinists, Nihilists, and Atheists using disorder, entropy, chaos, or chance as the cause of everything that exists in this universe. This completely violates the purpose for which science, and statistics, and the scientific methods were made. These people do this kind of stuff because they don't understand the Philosophy of Science, and they have no idea how science, experiments, and statistics are supposed to be used in science to eliminate chance as a causal agent or chance as a causal source. If you understand statistics, science, research, experiments, or logic in the first place, then you automatically KNOW why creation by chance, or creation by entropy, or the second law of thermodynamics is Bad Science.

Disorder has NO real effect on anything, but anything that actually exists can have a real effect on disorder by manipulating it, changing it, or eliminating it. Anything that actually exists has the power to have a real effect on the amount of disorder within a system by decreasing the amount of disorder in a system. Power is the ability of an independent variable to have a real effect on the dependent variable or the amount of disorder within a given system. When manipulating an independent variable, we observe a real effect whenever the amount of order is increased within the system.

It is impossible to run a science experiment with disorder or chance as the independent variable, because when you start manipulating that independent variable, then you are no longer looking at chance or disorder but some kind of manipulation or organization instead. It is impossible to run a science experiment where disorder or chance is the dependent variable because once you start to measure an effect, then you are no longer looking at chance or disorder but some type of causality instead. Whenever you detect disorder or chance within your science experiment, then you automatically KNOW that your independent variable had NO EFFECT on your dependent variable. That's why chance and disorder can never be your dependent variable or your independent variable. There's no way that you can manipulate and measure chance and disorder, because they cease to be chance and disorder whenever you measure them or manipulate them.

The Null Hypothesis ($P_{null} = 0.50$) assumes that chance alone is at work. Chance can't produce a sizable effect. The Null Hypothesis is "chance causality", "caused by chance", or "created by chance". The Null Hypothesis is Creation by Chance.

Materialism, Naturalism, and their derivatives ARE the Null Hypothesis because they ARE Creation by Chance. Materialism, Naturalism, Darwinism, Nihilism, Behaviorism, Determinism, Physical Reductionism, Atheism, and the Second Law of Thermodynamics ARE Creation by Chance or the Null Hypothesis. They ARE the thing that we are trying to falsify and eliminate from science through our science experiments!

Chance has no measurable effect. Chance has no sizable effect, because chance is the null hypothesis. Whenever our results are fifty/fifty, we assume that chance alone is at work. We conclude that our independent variable had NO EFFECT on our dependent variable. This is how science, science experiments, statistics, and the null hypothesis work. They try to eliminate chance as the cause of the scientific data or the science experiment.

Disorder, or chance, or the null hypothesis means that your independent variable had no effect on your dependent variable. The null hypothesis in a science experiment states that the results of the experiment were caused by chance or created by chance, which means that the independent variable had NO EFFECT on the dependent variable. That means that there is NO causal relationship between the variables, and that means that your alternative hypothesis is most likely false.

The ONLY thing that chance or disorder can reliably produce is the null hypothesis where the probability is 50 percent or 0.50. The ONLY thing that chance or disorder can reliably produce is a correlation coefficient ($r$) of 0.00. If anything else starts to come out of the equations, then you are looking at some type of order or organization; and then, you are trying to measure or determine the size of the effect. Creation by Disorder and Creation by Chance are precisely what we are trying to falsify and eliminate from science through our science experiments. We are trying to eliminate their faulty attempts to use correlation or chance as causality. Chance, disorder, or "entropy defined as chance or disorder" cannot be a valid cause. There's NO way to run a valid science experiment with chance or disorder as your independent variable, because chance and disorder are precisely what you are trying to eliminate as your independent variable. Furthermore, whenever we start to manipulate chance or disorder, they cease to be chance or disorder, and become some type of order or organization instead.

Creation by Disorder, Creation by Entropy, and Creation by Chance are unscientific; and, the second law of thermodynamics and the theory of evolution are precisely Creation by Disorder, Creation by Death, Creation by Entropy, or Creation by Chance.

Creation by Entropy, Creation by Natural Selection, Creation by Disorder, Creation by Death, Creation from Nothing, and Creation by Chance are bad philosophy, bad statistics, bad logic, and bad science. They are obviously false because they are unscientific and violate the rules of science, the scientific method, the null hypothesis, and the rules of statistics and logic.

Chaos or disorder completely ceased to exist in our part of the multiverse the very moment that the Gods, the Controlling Psyches, and Nature's Psyche made the quantum fields. The quantum fields are perfect syntropic order. The quantum fields are the ultimate perpetual motion machines. After the quantum fields were made from energy, they have been conserved. The construction of the quantum fields had a real effect on the amount of disorder in our part of the multiverse. Making the quantum fields completely eliminated the disorder at the quantum level. The second law of thermodynamics, creation by entropy, creation by randomization, or creation by disorder will NEVER be true as long as the quantum fields exist. Remember, the second law of thermodynamics will never be true so long as the quantum fields exist. You would have to completely destroy the quantum fields in order to make the second law of thermodynamics true. If you can do that, then you are a God. If you are a God, then you can make planets, stars, genomes, and galaxies at will.

In all of their science experiments and thought experiments, the Materialists, Naturalists, Darwinists, Nihilists, and Atheists jump straight to the conclusion that disorder or chance produces a real effect. They are wrong. A real effect is judged by the "decrease in disorder" that takes place after we have manipulated the system. Manipulating a system decreases the amount of disorder in that system. The Gods, the Controlling Psyches, and Nature's Psyche manipulated the Chaos Realm and completely eliminated the disorder at the quantum level, in our part of the multiverse, when they designed and made the Quantum

Fields. Manipulating the independent variables introduces some level of order into a system. The intent is to produce a measurable or detectable effect.

If you detect disorder and chance with your science experiment, then you have proven that your independent variable had no effect on your dependent variable. You have proven that chance, disorder, or "entropy defined as chance and disorder" cannot make things happen or function as a Causal Intelligence.

You have got to understand what these people believe before you can see what's wrong with it. When it comes to the Naturalists, Darwinists, and Atheists, what these people have chosen to believe in is the exact opposite of the observable truth. They really truly believe that disorder or chance can have a real effect on everything that exists in this universe. They are wrong, but they don't know it. The truth is the exact opposite of that. Everything that truly exists in this universe can have a real effect on decreasing the amount of disorder in our universe. Nature or Nature's Psyche can decrease and eliminate disorder at will. Nature's Psyche can do the physically impossible at will. This is what has actually been experienced and observed. Nature's Psyche is the thing that collapses the wave function, and Nature's Psyche is the thing that makes the quantum waves in the first place.

The second law of thermodynamics will never be true as long as the quantum fields exist. The second law will never be true as long as quantum mechanics exists. We will never have true disorder or true chaos in our part of the multiverse as long as the quantum fields exist. That's just the way it is, because that's the way it was made to be.

### Messing with Chance

Chance doesn't produce a measurable nor detectable change in the dependent variable. CHANCE is represented by the control group in a science experiment. The control group was created by a random sample, which means that it was produced by chance alone. The "control group" or "no causality" or "no change in the dependent variable" is what we get from chance alone. By definition, in principle, CHANCE cannot produce a change in ANY dependent variable. If it is producing some type of change in the dependent variable, then it is NO longer chance but has become some type of manipulation or causality instead. Creation by Chance, Chance Causality, the Control Group, or the Null Hypothesis is precisely what we are trying to falsify and eliminate from science by running a science experiment to begin with.

Materialism, Naturalism, Darwinism, Nihilism, Atheism, the Theory of Evolution, Creation by Disorder, Creation by Chaos, and the Second Law of Thermodynamics ARE the Null Hypothesis, the Control Group, or Creation by Chance Alone. They are precisely what we are trying to falsify and eliminate from science by doing science experiments in the first place. These atheistic and naturalistic philosophies are automatically FALSE because they are the Null Hypothesis and Creation by Chance.

A lot of this is semantics, but it does have real world applications. For example, in statistics we say that the Null Hypothesis or Creation by Chance is never proven to be true. It's been said that you can't prove anything in science. That claim is NOT true. You can use science and the scientific methods to prove that your theories and ideas are false. In other words, we can USE science, statistics, the scientific method, and science experiments to PROVE that Materialism, Naturalism, Darwinism, Nihilism, Atheism, the Theory of Evolution, the Second Law of Thermodynamics, the Null Hypothesis, and Creation by Chance ARE FALSE. Whenever we falsify these theories, philosophies, and ideas we have in fact PROVEN that they are false. That's the way Science works, whether we know it or not.

Science is not as gimped nor crippled as the Darwinists, Naturalists, and Atheists claim that it is. Throughout all of my different books, I have used Science, Observations, Experiences, Science Experiments, and the Scientific Method to FALSIFY Materialism, Naturalism, Darwinism, Nihilism, Atheism, the Theory of Evolution, the Second Law of Thermodynamics, Creation Ex Nihilo, and Creation by Chance. That's what Science was designed to do. I have used Science TO PROVE that these things are false. That's what scientists are supposed to do; and, I'm a scientist.

Creation by Chance IS the Null Hypothesis or the Control Group, which means that it is automatically false. Chance is an invalid causal mechanism in statistics and science. Since Materialism, Naturalism, and their derivatives such as Darwinism and Atheism ARE Creation by Chance, they are automatically FALSE. By using CHANCE as their independent variable, as the theory of evolution and the second law of thermodynamics do, they automatically PROVE that "evolution" and "entropy defined as disorder" have NO causal effect on ANY dependent variable that we might choose to examine. These things are impotent because they are Creation by Chance. The theory of evolution and the second law of thermodynamics are self-defeating and self-falsifying because they are in fact the Null Hypothesis or Creation by Chance. They are automatically false according to Science and Statistics because they are Creation by Chance or caused by chance alone. I'm no longer afraid to falsify the things that are obviously false. The theory of evolution and the second law of thermodynamics and creation by chance are obviously false. They should be falsified and eliminated from science.

Since Materialism, Naturalism, Darwinism, Nihilism, Atheism, the Theory of Evolution, and the Second Law of Thermodynamics ARE the Null Hypothesis or Creation by Chance, the Materialists, Naturalists, and Darwinists ALWAYS FAIL to meet their burden of proof. There's no way for them to demonstrate or prove that Creation by Chance happens in the lab and in the wild, therefore their theories, ideas, and interpretations of scientific evidence will always be false. There will NEVER be any evidence supporting Creation by Chance, because there can't be. Once an intelligent being starts manipulating chance or disorder, then it is no longer chance or disorder, but has become some type of order or organization instead. Creation by Chance is automatically false according to Statistics, Science, and the Philosophy of Science. Consequently, when it comes to the Darwinists and Naturalists and Atheists, their theories and ideas have to be taken on blind faith alone as being true because Design and Creation by Chance can NEVER be experienced NOR observed.

I used to be a materialist, naturalist, nihilist, and atheist; but, I no longer have faith enough to be an atheist. I KNOW too much now to be successfully swayed by their deceptions and their lies. Yes, there was a time when I was ignorant enough to fall for the ruse, but that time has passed.

There are other ways to mess with chance and produce faulty "science" and faulty interpretations as a result. Let's mention another one of them.

### There Are Dozens of Different Falsified Definitions for Entropy

There are literally dozens of different, contradictory, mutually exclusive, self-defeating definitions for entropy on the books. The Creation by Chance or Creation by Disorder definition for entropy is one of the worst, in that it's the most easily falsified; but, there are many that are equally as bad.

On Khan Academy, I watched hours of video where they did an excellent job describing the theory behind the second law of thermodynamics. This is the macrostate version of entropy that shows up in some of our physics textbooks.

https://www.khanacademy.org/science/chemistry/thermodynamics-chemistry

https://www.khanacademy.org/science/physics/thermodynamics

https://www.khanacademy.org/science/physics/thermodynamics/laws-of-thermodynamics/v/macrostates-and-microstates

https://www.khanacademy.org/science/physics/thermodynamics/laws-of-thermodynamics/v/more-on-entropy

The whole thing is nothing but a thought experiment; and, careful examination of these videos reveal that the second law of thermodynamics is based exclusively on the fictional and physically impossible Carnot Engine perpetual motion machine consisting of an infinitely sized, cold, absolute zero temperature reservoir and a frictionless piston filled with a fictional pre-heated ideal gas – where some invisible intangible mystical entity is slowly removing and then adding back grains of sand so as to maintain thermal equilibrium within the piston all along the way – just so that entropy can then be declared to be a macrostate variable or a real boy. They also start with a gravity field or an earth as an entropyless given so that the sand will have a real effect on the frictionless piston.

Maxwell's demon is nothing compared to all the science fiction that they are using as the basis for the second law of thermodynamics and the macrostate version of entropy, where entropy is defined as a macrostate variable! The macrostate version entropy and the second law is nothing but science fiction.

The theory behind the macrostate version of entropy or the second law of thermodynamics is based exclusively on a fictional ideal gas, the physically impossible Carnot Engine perpetual motion machine, a premade or God-made frictionless piston cannister of heated gas, a premade or God-made infinite cold reservoir, a pre-made God-given gravity field and earth, a pre-made or God-given amount of sand, and some type of invisible, frictionless, and entropyless God or demon who is slowly adding and removing one grain of sand from the top of that frictionless and entropyless piston so as to maintain constant thermal equilibrium, in order to make it seem like entropy is a macrostate variable or some type of real thing.

This is nothing but science fiction, and it's physically impossible. There's no such thing as an ideal gas. There's no such thing as a Carnot Engine perpetual motion machine. And, according to the Materialists, Naturalists, and Atheists, there is no mystical, intangible, invisible, supernatural entropyless demon or God to add and remove grains of sand from that piston one grain at a time so as to maintain perfect thermal equilibrium all along the way. It's nothing but science fiction. It's nothing but a thought experiment. It has no basis in reality. The whole second law of thermodynamics is physically impossible. It can't exist and it doesn't exist at the physical level. It can only exist at the quantum level among the Quantum Fields, where there is an actual God or Psyche who can manipulate the energies at will and remove sand from a piston one grain at a time.

The "power" of an experiment is a measure of the sensitivity of the experiment to detect a real effect of the independent variable. Disorder as the independent variable has NO real effect on anything, because it is impossible to manipulate the amount of disorder within a system without producing some type of order instead. We manipulate disorder or chaos in order to produce some type of order. Manipulation or experimentation is the production of order or the testing and rearrangement of an order that already exists.

Measurement of a dependent variable is in actuality a measurement of the amount of order within a system. Defining entropy or thermodynamics as disorder is counterproductive and counterintuitive, because physical matter and thermodynamics or heat represent a great deal of order and organization within a physical system.

All of the theory behind the macrostate version of entropy falsifies the "disorder definition for entropy" and the second law of thermodynamics by showing us clearly and conclusively that Creation by Disorder or Creation by Entropy is physically impossible. Perpetual motion machines, such as the second law of thermodynamics, are impossible at the physical level. Creation of any kind requires some type of psyche, intelligence, or God.

The Carnot Engine thought experiment that they used to produce the macrostate version of entropy and the second law required some type of psyche, or intelligence, or consciousness, or God in order to come up with it in the first place. We human beings are Gods. We create things within our minds, and we also organize physical matter in order to make things as well. It's NOT our physical matter that makes us Gods. It's our intelligence, our consciousness, or our human psyche that makes us Gods and gives us the ability to design and create, both at the quantum level and the physical level.

In the macrostate version of entropy or the second law of thermodynamics, they turned logic and rationality upside-down to produce something that is physically impossible out of thin air from nothing. They start with the hot piston of gas, an ideal gas, and an infinite cold reservoir as a given. They never once realize that it took a huge reversal of entropy or a huge reversal of disorder to make that hot reservoir or that hot frictionless piston in the first place. They simply start with it as an entropyless given. They start with a God-given earth and gravity field, so that they have an "entropyless" force to drive their Carnot Engine and thought experiment.

Then they go one step further and have some entropyless, frictionless, invisible, and intangible God or demon adding grains of sand to the piston so as to maintain thermal equilibrium and the illusion that entropy is a macrostate variable. They are trying to turn entropy into a tangible substance or entity, and they are using the intangible, the quantum, the supernatural, and the physically impossible in order to do so. They never once realize that once they start manipulating the system by adding grains of sand to the piston, they are no longer dealing with "disorder" but are in fact dealing with some type of syntropy, order, deliberation, organization, manipulation, or intelligence instead.

It's amazing what these people are willing to believe in – just so that they can eliminate God, Psyche, Consciousness, or Intelligence from the equation and their belief system. They never once realize that it took some type of Psyche, Quantum Consciousness, Intelligence, or God to produce the science experiment, the thought experiment, the order, the piston, the hot and cold reservoirs, the grains of sand, the physical matter, the gravity field, the thoughts, the quantum waves, and the quantum fields in the first place. They simply start with all of these fictional or physically impossible things as a given and jump straight to the conclusion that it required no entropy, no work, and no reversal of entropy to produce them in the first place. That's how they were able to produce the second law of thermodynamics. They used science fiction in order to do so. They jumped straight to the conclusion that disorder or chance has the power or the ability to produce a real effect.

The second law of thermodynamics and the macrostate version of entropy is nothing but science fiction, and they have no idea that it is so, because these people have chosen to believe that entropy is God; and, no amount of evidence will convince them otherwise.

Mark My Words

## The Null Hypothesis Falsifies Naturalism, Darwinism, and Physicalism

The Null Hypothesis falsifies Creation by Chance, Chance Causality, or anything that is caused by chance alone or produced by chance alone.

Pagano, R.R. (2010). *Understanding Statistics in the Behavioral Sciences* (9th ed.). Belmont, CA: Wadsworth.

**If the null hypothesis ($H_0$) is false, then chance does not account for the results. Strictly speaking, this means that something systematic differs between the two groups. Ideally, the only systematic difference is due to the independent variable. Thus, we say that if the null hypothesis ($H_0$) is false, the alternative hypothesis ($H_1$) must be true. Practically speaking, however, the reader should be aware that it is hard to do the perfect experiment. Consequently, in addition to the alternative hypothesis, there are often additional possible explanations of the systematic difference. Therefore, when we say, "we accept $H_1$," you should be aware that there may be additional explanations of the systematic difference.** (*Understanding Statistics*, p. 265.)

Whenever the Null Hypothesis is true, then chance accounts for the results of the science experiment, and the Alternative Hypothesis has been proven false or falsified. Every Alternative Hypothesis that is caused by chance alone is automatically false and automatically falsified. Why? It's because every Alternative Hypothesis that is Caused by Chance, Created by Chance, or is Caused by Chance Alone IS IN FACT the Null Hypothesis; and, when the Null Hypothesis is true, then we KNOW that there is NO correlation and NO causal relationship between the independent variable and the dependent variable. The Null Hypothesis is Creation by Chance or produced by chance alone. Whenever the Alternative Hypothesis is the same thing as the Null Hypothesis, then the Alternative Hypothesis is automatically false, has been falsified, and has been proven false.

By definition and in principle or practice, Materialism, Naturalism, Darwinism, Naturalism, Atheism, Chemical Evolution, Creation by Disorder or Entropy, Creation by Death (natural selection), Creation Ex Nihilo (abiogenesis), Macro-Evolution (spontaneous generation or abiogenesis), the Theory of Evolution, and the Second Law of Thermodynamics ARE the Null Hypothesis or Creation by Chance because they were allegedly produced by chance alone. Since Naturalism and Darwinism were produced by chance alone, that makes them automatically false because they are the Null Hypothesis, Chance Causality, or Creation by Chance.

The whole of science was designed to identify and eliminate Creation by Chance or Chance Causality from science. The whole of science was designed to falsify and reject the Null Hypothesis; and, since Materialism, Naturalism, Darwinism, Nihilism, Atheism, the Theory of Evolution, and the Second Law of Thermodynamics ARE the Null Hypothesis, they have automatically been falsified by statistics and the scientific method. The whole of science was designed to falsify the Theory of Evolution and the Second Law of Thermodynamics. Likewise, any Alternative Hypothesis derived from Naturalism and Darwinism is also automatically false, because the Alternative Hypothesis is always false whenever the Null Hypothesis or Chance Causality is the only explanation for the results of the science experiment. The fact that Darwinism, Naturalism, and their derivatives are caused by chance alone makes them the Null Hypothesis and automatically falsifies them. Technically, Darwinism and Naturalism can't be used as the Alternative Hypothesis because they have already been falsified due to the fact that they are the Null Hypothesis or Creation by Chance Alone.

Whenever the Alternative Hypothesis is the Null Hypothesis or Creation by Chance, then the Alternative Hypothesis is automatically false. We can't accept the Alternative Hypothesis when the Null Hypothesis is true. Whenever the results of the science experiment are produced by chance alone, then the Alternative Hypothesis is always false.

Whether we use them as the Null Hypothesis or the Alternative Hypothesis, Materialism, Naturalism, Darwinism, Nihilism, Atheism, the Theory of Evolution, and the Second Law of Thermodynamics are always false because they were produced by chance alone. That's the way that science is supposed to work. Science was designed to eliminate these things from science. Science was designed to eliminate Creation by Chance, the Theory of Evolution, and the Second Law of Thermodynamics from science. We are supposed to toss out Creation by Chance, the Theory of Evolution, and the Second Law of Thermodynamics because they are the Null Hypothesis, which makes any Alternative Hypothesis derived from them automatically false as well.

Furthermore, even if we are able to demonstrate that the Null Hypothesis is false and establish some type of causality instead, that doesn't mean that mean that we got the correct interpretation or the correct definition for our Alternative Hypothesis. Our Alternative Hypothesis could be completely false when the Null Hypothesis is false simply by getting the wrong interpretation for the Alternative Hypothesis. A faulty interpretation of the scientific data produces a false result, even when the Alternative Hypothesis seems to be true. Science is supposed to be conservative. Just because we were able to reject our Null Hypothesis and establish some level of causality between the independent variable and the dependent variable, that doesn't make our personal interpretation of the data or our pre-chosen Alternative Hypothesis automatically true. If our Alternative Hypothesis is Creation by Chance or Chance Causality, then it is automatically false, whether we are able to reject the Null Hypothesis or not, because Creation by Chance is always a faulty and false interpretation of scientific data and scientific evidence. That's the way science and science experiments are supposed to work. They were designed to falsify and eliminate Materialism, Naturalism, Darwinism, Nihilism, Atheism, Creation by Chance, Creation Ex Nihilo, the Theory of Evolution, Creation by Disorder, and the Second Law of Thermodynamics from science. These falsified philosophies are supposed to be tossed out of science because they were caused by chance alone and ARE the Null Hypothesis.

### When the Null Hypothesis Is True, the Alternative Hypothesis Is False

The null hypothesis in a science experiment states that the results of the experiment were caused by chance or created by chance, which means that the independent variable had NO EFFECT on the dependent variable. That means that there is NO causal relationship between the variables, and that means that your alternative hypothesis is most likely false.

In any science experiment, the null hypothesis is that "the results of the experiment were caused exclusively by chance". The null hypothesis is chance causality, where chance is the cause of the experiment's results.

The whole purpose of a science experiment is to reject the null hypothesis, because statisticians KNOW that chance cannot be a valid cause for anything. If the results of the science experiment can be attributed to chance alone, then you retain the null hypothesis and conclude that the results of the science experiment were caused by chance and conclude that the independent variable had no effect on the dependent variable. That's how science experiments and statistics are supposed to work.

If you can reject the null hypothesis, then you have rejected chance as the cause of the results of the experiment. Then you KNOW that the independent variable had a real effect on the dependent variable. You have established causality. Chance cannot function as a causal agent or a causal source in a science experiment. If it can be demonstrated that chance caused the results of your science experiment, then you retain the null hypothesis and conclude that the independent variable had NO EFFECT on the dependent variable.

That's how science experiments work. Their goal is to eliminate chance as the cause of the results and to establish the independent variable as the cause of the dependent variable instead.

Without running any science experiments whatsoever, the Materialists, Naturalists, Darwinists, Nihilists, Behaviorists, Determinists, Physical Reductionists, and Atheists jump straight to the conclusion that chance, random disorder, chaos, entropy, or natural selection is the cause of everything that exists. That's NOT science! That's religion and dogma and blind faith. It takes a huge, monumental, extravagant amount of blind faith to be a Naturalist, Darwinist, or Atheist. I no longer have faith enough to be an atheist. I understand science, statistics, methodology, logic, and science experiments way too much now to remain a materialist, naturalist, nihilist, and atheist. I'm no longer willing to make chance the cause of everything that exists in this universe. The null hypothesis, statistics, the scientific methods, the rules of science, and science experiments won't allow me to do so.

The null hypothesis or chance causality falsifies Materialism, Naturalism (Creation by Chance), Darwinism (Creation by Entropy), Nihilism, and Atheism (Creation Ex Nihilo). The null hypothesis or chance causality also falsifies "entropy defined as disorder" or "entropy defined as chance".

Creation by Chance IS the null hypothesis, and the null hypothesis is precisely what we are trying to reject and eliminate in Science and through our science experiments. The whole purpose of a science experiment is to reject the null hypothesis or Creation by Chance. In other words, the whole purpose of a science experiment is to reject Materialism, Naturalism, Darwinism, Nihilism, Behaviorism, Determinism, Physical Reductionism, and Atheism which are ALL Creation by Chance. If you are not rejecting or falsifying Naturalism, Darwinism, Atheism, or Creation by Chance with your science experiments, then you aren't doing your science experiments correctly and you don't understand statistics and the null hypothesis.

The Materialists, Darwinists, and Naturalists NEVER think about science experiments and the null hypothesis in this manner because these people rely exclusively on chance as the causal explanation or the causal mechanism for all of the "science" that they do. Consequently, these people grant themselves a *special exemption* or a *special pleading* so that they don't have to follow the Rules of Science and the Rules of Statistics, and don't have to run science experiments, in order to prove that their theories or beliefs are true. These people declare their beliefs or theories to be axiomatically true or automatically true, and then they shift the burden to others to prove that they are false.

Well, it's easy to falsify their beliefs once you catch onto the scam. All you have to do is go out and falsify their beliefs with observations, and experiences, and science experiments that prove that these people are wrong. The proven and verified existence of light, photons, quantum tunneling, action at a distance, non-locality, gravity, radio waves, microwaves, magnetic waves, space, time, quantum waves, x-rays, dark energy, psyche, and quantum fields FALSIFIES Materialism, Naturalism, and their derivatives such as Darwinism and Atheism.

It's that simple.

The false is falsified by the truth; and, the truth is repeatedly and constantly observed. There is NO observational or experiential support for Materialism, Naturalism, Darwinism, Nihilism, and Atheism which claim that the non-physical does not exist. The verified and proven existence of anything non-physical FALSIFIES their beliefs. There is NO experiential or observational support for creation by death (natural selection), creation by chance (Naturalism and Darwinism), creation by entropy (Materialism or Darwinism), or creation ex nihilo (Atheism). Materialism, Naturalism, Darwinism, Nihilism, Atheism, the Second Law of Thermodynamics, and the Theory of Evolution ARE creation by chance. Are they not? There is no other way to define them.

Remember, the whole purpose of a science experiment is to reject the null hypothesis or to reject creation by chance. The purpose of a science experiment is to establish causality, and chance cannot be that cause.

Once you have successfully eliminated everything that is false, then only the truth will remain. The truths that remain will in fact end up being the things that are actually being experienced and observed. We simply don't observe creation by chance, creation by entropy, creation by disorder, creation by chaos, creation by death, or creation ex nihilo in nature and in our laboratories. The reason we don't observe these things is because they are impossible. Chance cannot design and create.

Mark My Words

## Identifying Reality and Truth

The purpose of statistics, math, and science is to help us to understand reality. If it isn't real – if it hasn't been experienced and observed – then it isn't reality and it can't possibly be true.

Pagano, R.R. (2010). *Understanding Statistics in the Behavioral Sciences* (9th ed.). Belmont, CA: Wadsworth.

**Statistics uses probability, logic, and mathematics as ways of determining whether or not observations made in the real world or laboratory are due to random happenstance or perhaps due to an orderly effect one variable has on another. Separating happenstance, or chance, from cause and effect is the task of science, and statistics is a tool to accomplish that end. Occasionally, data will be so clear that the use of statistical analysis isn't necessary. Occasionally, data will be so garbled that no statistics can meaningfully be applied to it to answer any reasonable question. But I will demonstrate that most often statistics is useful in determining whether it is legitimate to conclude that an orderly effect has occurred. If so, statistical analysis can also provide an estimate of the size of the effect.** (*Understanding Statistics*, p. xxvii.)

Happenstance or chance is an invalid causal mechanism in statistics and science. The purpose of statistics is to eliminate chance as the cause of what we are observing. In science and through science experiments, we are trying to determine the true cause of each effect and each event.

The null hypothesis in a science experiment states that the results of the experiment were caused by chance or created by chance, which means that the independent variable had NO EFFECT on the dependent variable. That means that there is NO causal relationship between the variables, and that means that your alternative hypothesis is most likely false.

The null hypothesis in science and statistics is in fact "chance causality" or "creation by chance". The null hypothesis states that the results we are observing from our science experiment were caused by chance alone.

The null hypothesis is that chance produced the results that we have observed. The null hypothesis states that the independent variable in a science experiment has no effect on the dependent variable. In other words, the data from the experiment are telling us that the results of the experiment were caused by chance alone. If chance is the "cause" of the results of a science experiment, then we KNOW that the independent variable had no effect on the dependent variable and that there is no correlation between the two variables, which means that there is no cause and effect relationship between the two variables. This is how statistics and the scientific method are used to eliminate chance from science and science experiments as a causal explanation of what we are observing.

Chance and "entropy defined as chance" cannot be the true cause of anything, according to statistics and science. Yet, there are self-proclaimed "scientists" who make chance or entropy or natural selection the cause of everything that exists.

There's a reason why the good scientists use statistics in science. They are trying to determine what is real and what is fake. There's a reason why good scientists run science experiments. They are trying to determine the true cause of an effect or the true cause of an event. It's that simple.

Choosing a false interpretation of the scientific data will NEVER give us a true science in the end. We need to have true interpretations of scientific evidence, if we want to find and know the truth. If the results of our science experiment were caused by chance alone, then WE KNOW that our independent variable had no effect on our dependent variable. Scientific data, science experiments, and the scientific method is one of the ways that we can find and know the truth, BUT ONLY if we are willing to eliminate chance as a causal explanation for what we are observing and experiencing.

An atheist friend told me not to tell him the statistics. I thought that was fascinating. He preferred to remain in the dark. He didn't want to know the truth. Atheism of any kind is based upon the rejection or the destruction of the evidence that falsifies it. That's how these people do science. They destroy or hide the evidence that falsifies their beliefs. Ironically, Materialism, Naturalism, Darwinism, and Atheism – Creation by Chance or Creation by Entropy – are based upon wishful thinking, with NO evidence whatsoever to support them.

On page 4 of Pagano's book, he states: "**Historically, humankind has employed four methods to acquire knowledge. They are authority, rationalism, intuition, and the scientific method.**"

He and everyone else has overlooked the most important one – and the most powerful one. We find and know the truth through observation and experience. This "new way" of acquiring knowledge was discovered through my research into Science 2.0. One day I realized that the quickest, most effective, most reliable, and most believable way to find and know the truth is to experience it or observe it for yourself, or to choose to trust someone who has.

Okay, he has "experience" buried in rationalism. Rationalism uses reason alone, and then he gives an example where reason alone or logic alone can and does fail.

On page 5, Pagano states: "**In this situation, there are two reasonable explanations of the phenomenon. Hence, reason alone is inadequate in distinguishing between them. We must resort to experience.**"

This is an extremely important truth in science, scientific experimentation, and the philosophy of science. When reason fails us, we MUST resort to observation and experience in order to break the tie. Remember, whenever reason, rationality, logic, or science fails us, we MUST resort to observations and experiences to point us to the truth.

Lived experiences are the best and most reliable way to find and know the truth. I discovered "lived experiences" or phenomenology while studying personality theory, the philosophy of science, and while developing the Ultimate Model of Reality or a Psyche Ontology.

The quickest, most efficient, and most believable way to find and know the truth is to live it, observe it, and experience it for yourself. The false is falsified by the truth, and the truth is repeatedly and constantly experienced and observed.

I've learned not to trust the Materialists, Naturalists, Darwinists, Nihilists, Behaviorists, Determinists, Physical Reductionists, and Atheists. These people have lied to me one too many times and have called their lies "science". These people refuse to look at evidence, and then they proclaim themselves as the "Brights" or as the most intelligent and rational and logical of people on the planet. They are deceiving themselves, and they don't even know it. Self-deception works and it works every time, especially when it comes to PhD scientists who believe that they know everything already. The more I KNOW, the more I realize what I have yet to learn.

"When using the *method of authority*, something is considered true because of tradition or because some person of distinction says it is true. Thus, we may believe in the theory of evolution because our distinguished professors tell us it is true. We frequently accept a large amount of information on the basis of authority, if for no other reason than we do not have the time or the expertise to check it out firsthand. For example, I believe, on the basis of physics authorities, that electrons exist, but I have never seen one." (*Understanding Statistics*, p. 4.)

*Authoritarianism* or *Credentialism* is a logic fallacy. Belief based upon authority alone is bound to fail, especially when those "authorities" are Materialists, Naturalists, Darwinists, Nihilists, and Atheist. I no longer trust the Materialists, Naturalists, Darwinists, and Atheists because they use their PhD degrees to "prove" that their falsified theories, and ideas, and beliefs are true. They rely exclusively on axioms, decrees, laws, and declarations of authority to establish the truthfulness of their falsified beliefs.

I believe on the basis of science experiments and scientific observations that Action at a Distance, Quantum Waves, Virtual Particles, Quantum Tunneling, Non-Locality, Non-Physicality, the Quantum Zeno Effect, $E = mc^2$, Quantum Consciousness, Magnetism, Gravity, Radio Waves, Microwaves, Dark Energy, Massless and Entropyless Photons, and Quantum Fields are real and truly exist. All of these non-physical, or non-local, or quantum things have been experienced and observed. Therefore, WE KNOW that they are real and truly exist. Consequently, I no longer believe the Materialists, Naturalists, Darwinists, and Atheists whenever these people try to tell us that these things don't exist or try to tell us that the non-physical does not exist. WE KNOW that these people are wrong.

These people are always wrong, whenever they start claiming on the basis of their authority alone that the non-physical, the spiritual, the quantum, the transdimensional, or the non-local does not exist. The Materialists, Naturalists, Darwinists, Nihilists, and Atheists don't want to hear about or study anything that is non-physical in nature and origin, because the non-physical or the non-local falsifies their pre-chosen beliefs. Atheism of any kind is based exclusively on a refusal to look at the evidence that falsifies their beliefs. Atheism is a refusal to look at evidence. Atheism is Creation Ex Nihilo – the creation of something from nothing by nothing. It's magic, not science!

On page 5, Pagano states: "**Knowledge is also acquired through intuition. By intuition, we mean that sudden insight, the clarifying idea that springs into consciousness all at once as a whole. It is not arrived at by reason. On the contrary, the idea often seems to occur after conscious reasoning has failed.**"

The Christians in general call this the Gift of the Holy Ghost or revelation from God. Others have called it "conscience" or "the Light of Christ". I started having many flashes of insight after I falsified My Materialism, My Naturalism, My Nihilism, and My Atheism and started looking for the truth instead. Anyone who is sincerely looking for the truth will start to have flashes of insight out-of-the-blue. Start asking God questions, and He will find a way to answer them. There seems to be an eternal rule that states that God is not permitted to answer our questions until after we have asked them. Asking God questions gives Him permission to answer them.

Yes, when reason or rationality fails, as it so often does, then it's time to resort to some other method if one really wants to find and know the truth. Reason, rationality, or logic has failed me many times; so, I started asking questions and started receiving flashes of insight as a result. All of the different books that I have written are based upon the flashes of insight that I have received after I started actively looking for the truth. I eventually discovered that the false is falsified by the truth – by the things that have been experienced and observed – and that the truth is constantly and repeatedly experienced and observed. This is a powerful litmus test that always seems to work.

In the summer of 2019, my statistics class and my statistics book gave me many different flashes of insight, because I was actively looking for and wanting to find the truth.

Starting on page 6 of his book, Pagano effectively equates statistics and the scientific method. He makes statistical analysis an essential and integral part of the scientific method. He is right to do so. Science experiments or the scientific method is indeed a good way to find and know the truth, BUT ONLY if one gets the true interpretation of the scientific evidence that results from the science experiment. If one gets a false interpretation of the evidence, then the results of the science experiment end up being worthless. It actually leads us astray!

The Darwinists are constantly talking about our caveman days. I've never had any caveman days. What about you? How long have you spent living in a cave? Adam and Eve didn't have any caveman days, either. The Darwinists are constantly saying that it took evolution millions of years to make our eyes and our brains. That means that we were blind, stupid, and dumb for millions of years before evolution made our eyes and brains. So, how did we survive? We humans are animals. So, how did we survive for millions of years without eyes and brains while evolution was making our eyes and our brains? Since we were blind and stupid vegetables for millions of years, how did evolution get the know-how and the ability to make our eyes and our brains? And, why hasn't evolution made eyes and brains for the vegetables that currently exist? It's because evolution or chance alone can't make eyes, brains, planes, computers, cars, spaceships, genomes, proteins, and skyscrapers. It's that simple.

**"Using data as the basis for truth assertions is an important part of the scientific process. Sometimes, however, data are reported in a manner that is distorted, leading to false conclusions rather than truth."** (*Understanding Statistics*, p. 16.)

That's what happens with the theory of evolution. The theory of evolution is Creation by Chance, or Creation by Disorder, or Creation by Entropy. Consequently, these people literally present correlation as causation every time they do a science experiment. Because chance doesn't correlate with anything, these people make chance correlate with everything, so that chance can then be the cause of everything.

**"Urban legends are anecdotes so engaging, entertaining, anxiety-affirming, or prejudice-confirming that people repeat them as true, even when they are not."** (*Understanding Statistics*, p. 16.)

The theory of evolution is an urban legend! Creation by Chance is an urban legend!

Based exclusively on observational studies in the complete absence of experimental proof, the Darwinists and Naturalists and Physicists jumped straight to the conclusion that CHANCE caused everything that exists. That's the very definition of Bad Science!

**Commentary pages are the soft underbelly of American journalism. Their writers, however self-interested, are held to a different, which is to say lower, standard of proof because of their presumed expertise. In fact, they are responsible for regularly injecting false information of this sort into our public discourse.** (*Understanding Statistics*, p. 17.)

The theory of evolution or Creation by Chance violates the rules of statistics and common sense. Richard Dawkins and the New Atheists are held to a lower standard of proof because of their presumed or self-declared expertise. They call themselves the Brights so that they don't have to provide proof or evidence that their theories and ideas are correct.

A science experiment takes place when we manipulate an independent variable and observe its effect on a dependent variable. How can you manipulate natural selection, chance, or evolution into designing and creating and making genes, proteins, genomes, and life forms from scratch? They've tried to do so in laboratories, and it doesn't work. Chance, evolution, and natural selection cannot design and create anything. It doesn't work in the laboratory that we call Nature, either! Natural selection doesn't do anything except sit around and wait for us to die. Natural selection is not only Creation by Chance and Creation by Entropy, but natural selection is also Creation by Death. These people literally have chance, death, and entropy creating genes, proteins, genomes, eyes, brains, and life forms out of thin air from nothing all the time. The different types of evolution are Creation by Chance or Creation Ex Nihilo.

In contrast, the scientific method makes an attempt to provide evidence based upon observation, lived experiences, and collected data. Even then, interpretation or inference is the fatal FLAW in the scientific method! Creation by Chance or Chance Causality is the flawed and false interpretation of scientific data, whether we are talking about physics, chemistry, biology, origins, or the second law of thermodynamics.

We have the truth, and then we have everything else that is false. Creation by Chance or Chance Causality in all of its different forms will always be false. There's no way to run the science experiment and prove that chance caused the results that we are observing. By definition, in principle or in practice, CHANCE cannot do causality; and, there's no way to prove that it can.

Experiments into chemical evolution and macro-evolution have falsified the theory of evolution. Evolution is Creation by Entropy or Creation by Chance. Using CHANCE as their causal mechanism automatically falsifies Materialism, Naturalism, Darwinism, Nihilism, and Atheism according to the rules of science and the rules of statistics. Macro-evolution or Creation by Chance or Creation by Entropy is physically impossible, because it is prevented from happening by genetics.

Pagano says that science experiments will expose incorrect hunches or incorrect interpretations. This is ONLY true if one can actually run an experiment. If it is impossible to run the experiment, as it is with Materialism, Naturalism, Darwinism, and Atheism which claim that the non-physical does not exist, then there is NO way to expose their incorrect hunches or their incorrect interpretations.

There's NO way to run a science experiment demonstrating or proving **the non-existence** of God, dark matter or spirit matter, dark energy, action at a distance, quantum mechanisms, gravity, photons, magnetism, light, radio waves, quantum waves, quantum tunneling, non-locality, quantum entanglement, transdimensionality, x-rays, microwaves, quantum consciousness or psyche, virtual particles, space, time, energy, quantum fields, or the other non-physical and immaterial things that have been experienced and observed. In fact, ALL of our science experiments have FALSIFIED the claim that the non-physical or the immaterial does not exist. ALL of our science experiments have FALSIFIED Creation by Chance, Creation by Entropy (the second law of thermodynamics), Creation by Natural Selection, Classical Realism, Materialism, Naturalism, Darwinism, Nihilism, Determinism, Physical Reductionism, Behaviorism, and Atheism (creation ex nihilo) which claim that the non-physical does not exist.

An important aspect of this (scientific) methodology is that the experimenter can hold incorrect hunches, and the data will expose them. The hunches can then be revised in light of the data and retested. This methodology, although sometimes painstakingly slow, has a self-correcting feature that, over the long run, has a high probability of yielding truth. Since in this textbook we emphasize statistical analysis rather than experimental design, we cannot spend a great deal of time discussing the design of experiments. Nevertheless, some experimental design will be covered because it is so intertwined with statistical analysis. (*Understanding Statistics*, p. 6.)

Through statistics, the data is supposed to expose incorrect hunches or false interpretations of the data, unless it is impossible to run the science experiment in the first place. If you can't run the science experiment, then you are left with reason alone or intuition alone as your source of knowledge and truth. When it comes to the Physicalists, Darwinists, Naturalists, and Atheists, their "intuition" is based upon wishful thinking and jumping to conclusions, because there is no way for them to run the science experiment and prove that the non-physical does not exist. In fact, the proven and verified existence of anything non-physical FALSIFIES their pre-chosen beliefs. The false is falsified by the truth; and, the truth is constantly and repeatedly experienced and observed!

Although the scientific method uses both reasoning and intuition for establishing truth, its reliance on objective assessment is what differentiates this method from the others. At the heart of science lies the scientific experiment. The method of science is rather straightforward. By some means, usually by reasoning deductively from existing theory or inductively from existing facts or through intuition, the scientist arrives at a hypothesis about some feature of reality. He or she then designs an experiment to objectively test the hypothesis. The data from the experiment are then analyzed statistically, and the hypothesis is either supported or rejected.

The feature of overriding importance in this methodology is that no matter what the scientist believes is true regarding the hypothesis under study, the experiment provides the basis for an objective evaluation of the hypothesis. The data from the experiment force a conclusion consonant with reality. Thus, scientific methodology has a built-in safeguard for ensuring that truth assertions of any sort about reality must conform to what is demonstrated to be objectively true about the phenomena before the assertions are given the status of scientific truth. (*Understanding Statistics*, p. 6.)

With Materialism, Naturalism, Darwinism, Nihilism, Behaviorism, Determinism, Physical Reductionism, Classical Realism, the Theory of Evolution, Creation Ex Nihilo, Creation by Entropy, Creation by Chance, Creation by Rocks, Creation by Death, and Atheism, there is NO way to run the science experiment that matters most. There is no way to run the science experiment that demonstrates clearly and conclusively that the non-physical or the immaterial does not exist. In fact, the exact opposite holds true. The whole of science demonstrates clearly and conclusively that the non-physical or the immaterial does indeed exist. In fact, over 95% of our universe is non-physical or immaterial in nature and origin according to the science associated with dark matter and dark energy and according to the science associated with Quantum Field Theory. The false is falsified by the truth, and the truth is repeatedly experienced and observed. Materialism, Naturalism, Darwinism, Atheism, Creation by Chance, and Creation by Entropy (the second law of thermodynamics) are constantly and repeatedly falsified by all the different things that we have experienced and observed.

The final step of the scientific method that Pagano left out is the ongoing search for the best explanation or the true interpretation of the scientific data that we have gotten from our science experiments. If we never find the true interpretation for that data or the true meaning of that science, then we will NEVER know the truth. It's that simple.

The purpose of statistics and the purpose of science is to find and know the truth. In contrast, the Materialists, Naturalists, Darwinists, Nihilists, and Atheists in the name of science actively try to hide the truth and the evidence from us so that we will never find or know the truth. These people actively and deliberately hide, suppress, ban, censor, eliminate, destroy, and burn any evidence that falsifies their pre-chosen beliefs. That's how these people do science. *Destroying evidence* or *rejecting evidence* is a logic fallacy; and, this is one of the logic fallacies upon which Materialism and Naturalism and Darwinism are based. These people hide the evidence and hide from the evidence that falsifies their beliefs. I KNOW because I used to be one of them. I used to be a materialist, naturalist, nihilist, and atheist. At that point in my life, I was hiding from evidence – the evidence that falsified my beliefs.

Well, I'm no longer hiding from evidence. In fact, Science 2.0 allows ALL of the evidence into evidence, and then it tries to pursue a preponderance of that evidence. I do this because I KNOW that the fastest, most efficient, most reliable, and most believable way to find and know the truth is to live it, experience it, observe it, and own it for yourself – or to choose to trust someone who has.

Mark My Words

## Inferential Statistics Falsifies the Second Law of Thermodynamics

It's the same thing over and over again, just from different perspectives. Truth verifies truth. The false is falsified by the truth, and the truth is constantly and repeatedly experienced and observed.

Everything falsifies the second law of thermodynamics, which claims that the total amount of entropy or disorder in our universe is constantly increasing and that it can never decrease or go to zero.

The people who created the second law of thermodynamics were trying to make disorder, chaos, or random chance our Designer and our Creator. The second law of thermodynamics was defined axiomatically as Creation by Chance, or Creation by Disorder, or Creation by Entropy. The second law of thermodynamics is Creation Ex Nihilo. The second law is Creation by Chance.

The people who made the second law of thermodynamics were trying to falsify the first law of thermodynamics, which is Creation by Psyche or Creation by Intelligence, or more commonly known as the Conservation of Energy and Psyche. The people who made the second law of thermodynamics don't want Psyche, Intelligence, or God to be our designer and our creator. Consequently, they invented Creation by Chance instead. They made CHANCE our designer, creator, and God.

Once we know that the second law of thermodynamics is Creation by Chance, then we automatically KNOW that the whole of inferential statistics falsifies the second law of thermodynamics, because the Null Hypothesis in any science experiment was caused by chance alone. The second law of thermodynamics IS the Null Hypothesis or Creation by Chance, because it is caused by chance alone.

Likewise, the theory of evolution IS the Null Hypothesis, because it resulted from chance alone. In inferential statistics, anything that results from chance alone ends up being Creation by Chance or the Null Hypothesis. Materialism (Creation by Chance Alignments of Matter), Naturalism (Nature Created by Chance Alone), Darwinism (Creation by Chance), Creation by Natural Selection (Creation by Death or by Bad Luck), Chemical Evolution (Creation by the Random Chance Alignment of Atoms and Molecules), Nihilism (Creation without Intelligence or Psyche), Atheism (Creation Ex Nihilo), the Theory of Evolution (Creation by Blind Luck), the Second Law of Thermodynamics (Creation by Random Disorder or Chaos) ARE ALL the Null Hypothesis because by definition and by axiom they resulted from chance alone. They are ALL the Null Hypothesis or Creation by Chance Alone.

Every science experiment is designed to reject the Null Hypothesis, which states that the results of the experiment were caused by chance alone. The purpose of a science experiment is to establish causality, by eliminating chance as the cause of the results that we are seeing. The second law of thermodynamics is Creation by Chance, which means that the second law is in fact the Null Hypothesis and is caused by chance alone. That makes any alternative hypothesis derived from the second law of thermodynamics automatically false, because the second law IS the Null Hypothesis or Creation by Chance and IS therefore caused by chance alone. In science through science experiments we are trying to reject, eliminate, and falsify everything that is caused by chance alone or that resulted from chance alone, which means that we are trying to reject, eliminate, and falsify the Null Hypothesis or Creation by Chance or Chance Causality.

The second law of thermodynamics, the theory of evolution, materialism, naturalism, atheism or creation ex nihilo, chance causality, and creation by chance are precisely what we are trying to falsify and eliminate from science by doing science experiments in the first place. The whole of inferential statistics was designed to identify and reject the Null Hypothesis, or Creation by Chance, or the Second Law of Thermodynamics, or Chemical Evolution, or the Theory of Evolution. The whole scientific method, inferential statistics, research methodology, science experiments, and science program was designed to identify and eliminate the Null Hypothesis, Creation by Chance, the Second Law of Thermodynamics, Chemical Evolution, and the Theory of Evolution from science.

This is one of the highlights of the Philosophy of Science, which is the study of how Science should be done. We should identify and eliminate from science anything and everything that is the Null Hypothesis, or Creation by Chance, or Caused by Chance Alone. If it resulted from chance alone, then it is automatically false and should be eliminated from science, because there is NO WAY to establish causality with something like the theory of evolution that was caused by chance alone. In other words, when it comes to Science, CHANCE is an invalid cause. Yet, we observe the Materialists, Naturalists, Darwinists, Nihilists, Behaviorists, Determinists, Physical Reductionists, Classical Realists, and Atheists making CHANCE the cause of everything that exists in this universe in complete violation of statistics and science. They are bad scientists doing Bad Science because they are deliberately using CHANCE as the cause of everything that exists and as the cause of everything that they do in science. It's just Bad Science. It's also Bad Philosophy.

Not only do inferential statistics, science experiments, science, and the null hypothesis FALSIFY the second law of thermodynamics, they also FALSIFY the theory of evolution. They falsify every theory or idea that is caused by chance alone or that resulted from chance alone. That's what the Null Hypothesis is all about. The Null Hypothesis is trying to identify and eliminate ALL of the chance from a science experiment and establish true causality instead. That's why they run science experiments in the first place. They are trying to eliminate false ideas and false theories from science, such as the second law of thermodynamics and the theory of evolution that were caused by chance alone or that resulted from chance alone, so that they can establish true causality instead. The whole purpose of a science experiment is to eliminate chance and to find the true cause instead. The whole purpose of science experiment is to identify and eliminate Chance Causality or Creation by Chance from science as a whole. The whole purpose of science and science experiments is to identify and eliminate the second law of thermodynamics, the theory of evolution, chemical evolution, creation by chance, and all other types of "chance causality" from science, so that we can find the true cause of things instead.

This is the pinnacle of the Philosophy of Science, because the whole purpose of Science is to find the truth and to identify the true cause of the things that we experience and observe. We do that in part by experiencing, observing, and living the truth; and, we do that in part by identifying and eliminating from science everything that was caused by chance alone, everything that is false, and everything that has been given a false or falsified cause such as "chance causality" or the null hypothesis.

There are many other things in statistics, science experiments, and the associated scientific method that FALSIFY the second law of thermodynamics.

**Zero Correlation Falsifies Entropy Defined as Disorder**

This next one comes from descriptive statistics.

Boltzmann defined entropy as "disorder", or "chaos", or a random distribution of gas particles in space. Thanks to Boltzmann, the second law of thermodynamics states that the amount of entropy or disorder in our universe is constantly increasing and that it can never decrease. This is NOT what we experience and observe!

It was obvious to me that disorder or chaos or randomness is best represented graphically as a correlation coefficient of zero (**r = 0.0**).

If you know anything about statistics and correlation, then you KNOW that true disorder has absolutely NO relationship with anything else. True disorder doesn't correlate with anything!

Entropy defined as disorder doesn't correlate with anything. It doesn't match with anything. It certainly doesn't correlate with heat or thermodynamics! Entropy defined as disorder doesn't have anything to do with heat or thermodynamics. Heat and thermodynamics represent a great deal of order and organization. A huge amount of order, organization, intelligence, and syntropy went into the design and construction of mass's heat storage capacity. There is NO correlation between disorder and thermodynamics.

Defining entropy or thermodynamics as "disorder" or "chaos" is an oxymoron – a contradiction in terms – as well as a *category error* logic fallacy. They are not the same thing at all. Disorder and thermodynamics are NOT in the same category at all. Disorder doesn't cause heat, and disorder doesn't eliminate heat. You can have what seems to be a lot of disorder or chaos inside of a sun and produce a ton of heat and lots of thermodynamics; or, you can have perfect syntropy and perfect ORDER in the heatless, massless, entropyless, absolute-zero temperature quantum fields that are perfectly organized yet have NO heat. There is NO correlation between disorder and heat.

Thermodynamics is defined as a transfer of heat from a hot reservoir to a cold reservoir. Those pre-given reservoirs represent a huge amount of pre-given order and organization.

There is NO correlation between disorder and thermodynamics. Therefore, it is faulty and false and wrong to define entropy or thermodynamics as "disorder". They have absolutely nothing to do with each other. Disorder is the false definition or the wrong definition for entropy and thermodynamics, because disorder doesn't correlate with anything – including heat, entropy (mass's storage capacity), or thermodynamics.

Entropy defined as disorder is a con, or a scam, or a hoax; and, it should be eliminated from science because it has nothing to do with reality. We have not had true disorder or true chaos in our part of the multiverse ever since the Gods designed and created the Quantum Fields. The "disorder definition for entropy and thermodynamics" will NEVER be true as long as the Quantum Fields exist. Likewise, the second law of thermodynamics will NEVER be true so long as the Quantum Fields exist. You would have to destroy the Quantum Fields in order to make the second law of thermodynamics true. The Quantum Fields are perpetual motion machines. The Quantum Fields represent perfect, perpetual, and eternal ORDER and organization. Quantum Field Theory and the Quantum Fields falsify the second law of thermodynamics and the disorder definition for entropy. The two are mutually exclusive, which means that if one of them can be demonstrated to be true, then the other one has automatically been falsified.

The fact that disorder doesn't correlate with anything and has a correlation coefficient of zero tells us which one is false. There is NO correlation between disorder and thermodynamics, which actually FALSIFIES the inappropriately named second law of thermodynamics. They made a law out of something that doesn't correlate with anything.

Go figure!

It's one of the greatest scams in the whole of science, and we fell for it – hook, line, and sinker. It pulled us right out of science into the mystical and the magical and the impossible, and we didn't even know it. Yet immediately, before we even knew or understood what was happening, Boltzmann, Hawking, and Darwin had entropy or disorder or "random fluctuations in whatever" creating Boltzmann brains, planets, stars, galaxies, genomes, and life forms out of thin air from nothing like magic. These people pulled us out of science and dumped us into superstition, creation by chance, and creation ex nihilo instead. We fell for it because we desperately wanted it to be true; but, wanting it to be true doesn't make it true. Science should be based upon observation and experience, and NOT the wishful thinking of the Materialists, Naturalists, Darwinists, and Atheists.

The false is falsified by the truth, and the truth is repeatedly experienced and observed by someone somewhere, whether we know it or not. Creation by disorder, or creation from nothing, or creation by chaos, or creation by chance has NEVER been experienced NOR observed. It's impossible! Random chaos or random chance cannot design nor create anything. It can only destroy or produce a correlation coefficient of zero. Once a system magically achieves perfect disorder or a correlation coefficient of zero, then it has no place to go but up into some kind of order or organization instead; but even then, ONLY Psyche or Intelligence or Quantum Consciousness can design and create, whether we are talking about the quantum level or the physical level.

Thanks to the quantum fields and quantum mechanics, particles, atoms, and molecules are telepathically connected at the quantum level and interacting with each other at the quantum level, no matter how randomized or chaotic they might seem to be at the physical level. We will NEVER have true disorder and chaos in our part of the multiverse so long as the quantum fields exist.

Should we ever run out of heat, Nature's Psyche can make some more, anytime and anywhere that it chooses to do so. Nature's Psyche is constantly transforming massless, heatless, chargeless, and entropyless photons or quantum waves into mass, heat, and mass's heat storage capacity (entropy) all the time. It happens every time a photon chooses to stop and land on your skin. A miniature Big Bang takes place every time a photon chooses to stop and lands on your skin. That massless, heatless, chargeless, and entropyless photon or quantum wave can transform itself into anything that it wants to be – including mass, heat, or mass's heat storage capacity (entropy).

Likewise, Nature's Psyche is constantly transforming mass, heat, and entropy into massless, heatless, chargeless, and entropyless infinite acceleration, photons, and quantum waves all the time. It's happening right now within our sun. This is the perpetual motion cycle, $E = mc^2$; and, the whole thing is conserved thanks to the Conservation of Energy and Psyche and the Conservation of Quantum Information. The second law of thermodynamics will NEVER be true as long as the perpetual motion cycle exists. The second law of thermodynamics will NEVER be true as long as $E = mc^2$ is true.

**True Disorder Doesn't Correlate with Anything**

https://ultimate-model-of-reality.com/true-disorder-doesnt-correlate-with-anything/

As I write this on 26AUG2019, I have been falsifying the second law of thermodynamics for a couple of years now. I've gotten better at it as the months have gone by.

While taking a Statistics class in college during the summer of 2019, I immediately recognized that "disorder" is best modeled graphically as a correlation coefficient of zero. I immediately realized that True Disorder doesn't correlate with anything. This was a monumentally significant scientific discovery.

I soon realized that "entropy defined as disorder" does NOT correlate with heat. Disorder doesn't make heat, and disorder doesn't eliminate heat. There is NO correlation whatsoever between disorder and heat, which means that there is NO correlation or relationship between disorder and thermodynamics. Therefore, the "disorder definition for entropy" is the wrong definition for entropy. Disorder has nothing to do with thermodynamics or entropy, which is the transfer of heat from a hot system to a cold system.

The heatless, massless, entropyless Quantum Fields are perfectly ordered and organized perpetual motion machines, and they function flawlessly and endlessly at absolute zero temperature for all eternity. The Quantum Fields don't need heat as a lubricant to make them work. Energy, Quantum Fields, and Psyche are perpetual motion machines. They have been working perfectly for all eternity in the complete absence of heat at absolute zero temperature. There's NO correlation between order and heat.

Many people have stated that our sun represents maximum disorder or the complete randomization of atoms and molecules. It is obvious that our sun has a lot of mass. Mass IS heat storage capacity, or resistance to acceleration. Again, we observe that there is NO correlation between disorder and heat. Something that contains a lot of randomized disorder can also contain a lot of heat storage capacity or mass.

"Disorder" is the worthless and falsified definition for thermodynamics, heat, and entropy. It's time that we toss it out of science and find something better to take its place.

This wonderful discovery came as a result of my observation that "disorder" is best represented statistically and graphically as a correlation coefficient of zero.

I wrote the following notes in the margin of my Statistics book while I was making this amazing scientific discovery and falsifying the second law of thermodynamics once again.

Disorder does NOT cause heat. Therefore, disorder and thermodynamics are NOT causally related. That means that disorder and entropy are NOT causally related. There's NO correlation between the two.

The test-retest reliability is nil when it comes to disorder and entropy defined as disorder. If a correlation does not exist between the two variables, a causal relationship can be ruled out. That means that there is NO correlation between disorder and thermodynamics, so a causal relationship can be ruled out. In fact, there can be NO causal relationship between disorder and anything. If there is some type of correlation between two variables, then that means that some type of order or organization exists between the two variables. True disorder doesn't correlate with anything, which means that disorder or random chance can't be the cause of anything. That's just the way it is.

There's NO correlation between disorder (or entropy defined as disorder) and anything else, which means that a causal relationship between disorder and anything else can be ruled out and tossed out. Varying the amount of disorder in a system has absolutely NO effect on the amount of heat or thermodynamics within a system. In fact, whenever a Psyche or Intelligence varies the amount of disorder in a system, it introduces some type of order or organization into the system instead. Manipulating disorder is how we produce order. By making "disorder" our independent variable, whenever we manipulate that "disorder" or independent variable, we always introduce some type of order or organization into the system as a result. Order and disorder are mutually exclusive, and it's impossible to get perfect disorder out of any system. We will NEVER have true disorder or true chaos in our part of the multiverse as long as the Quantum Fields exist. The Quantum Fields are perfect, conserved, syntropic, exergic, massless, entropyless, non-physical perpetual motion machines. The Quantum Fields will never get old and never wear out because they are made from energy, and energy is always conserved. The second law of thermodynamics has NEVER been true ever since the Gods and Nature's Psyche (the Controlling Psyches within Nature or Energy) designed and made the Quantum Fields.

Psyche or Intelligence or Knowledge or Science produces order and organization. It's really hard to get perfect disorder or a perfect correlation coefficient of zero out of a system, because everything in our universe tends towards or strives towards order and organization thanks to Psyche and the Conservation of Energy. The reality of what we experience and observe FALSIFIES the second law of thermodynamics which claims that all of this order and organization shouldn't exist. The fact that you exist and are constantly making things is scientific proof that the second law of thermodynamics is false.

The second law of thermodynamics makes the erroneous claim that the total amount of disorder (or entropy) in our universe is constantly increasing and can never decrease. This is NOT what we experience and observe!

Why?

It's because the second law of thermodynamics and entropy defined as disorder don't correlate with anything! They don't cause anything to happen! Where there is NO correlation, causation has been ruled out. By definition and by observation, disorder or random chance doesn't cause things to happen! It can't!

When correlation is equal to zero, there's NO way to predict anything. We can't use maximum disorder to predict anything. With zero correlation, there's no chance that what we are observing is either a cause or an effect. Yet, the Materialists, Naturalists, Darwinists, Nihilists, and Atheists axiomatically state that "disorder" causes entropy and that entropy is a natural effect of disorder. They are wrong. Disorder doesn't cause heat, and heat doesn't cause disorder. There's NO correlation between the two! The Materialists, Naturalists, and Darwinists erroneously state that chance, random chaos, disorder, or entropy designed and made everything that exists in this universe. They are wrong. Disorder, chance, or entropy can't design and create anything.

These people literally correlate disorder with thermodynamics thereby giving us the false impression that thermodynamics causes disorder, or even the more erroneous impression that disorder causes thermodynamics.

Correlation does not imply causation, except when you are dealing with entropy and natural selection. They are exempt. The Materialists, Naturalists, Darwinists, Nihilists, and Atheists grant themselves a *special pleading* or a *special exemption* so that they don't have to follow the rules of statistics and science in order to demonstrate and prove that their ideas are true. They just *jump straight to the conclusion* that Creation by Chance is true, and they make it a LAW by calling it the second law of thermodynamics or the law of evolution. *Special pleadings*, *special exemptions*, *category errors*, *begging the question*, and *jumping to conclusions* are the logic fallacies upon with the second law of thermodynamics and the theory of evolution are based.

Categorizing "disorder" as heat or thermodynamics is a *category error* logic fallacy, because they are NOT the same thing at all.

True disorder produces a correlation coefficient of zero, because by definition true disorder doesn't correlate with anything. The relationship between disorder and thermodynamics is non-existent, because there is NO relationship or NO correlation between thermodynamics and disorder. Thermodynamics of any kind represents a great deal of order and organization to begin with, whether the associated heat is moving towards zero or not. Even at absolute zero, mass continues to exist, and mass's heat storage capacity (entropy) continues to exist, and mass's resistance to acceleration continues to exist thanks to Quantum Mechanics, the Conservation of Energy, and the massless entropyless Quantum Fields.

Physical systems are constantly achieving thermal equilibrium, yet they continue to function just fine anyway. At "heat death" or absolute zero, the physical matter continues to exist, and the quantum fields continue to exist, and everything continues to function perfectly. We can't use "disorder" or the second law of thermodynamics to predict proton decay, an ever-encroaching gray goo, nor everything coming to a screeching halt at the quantum level, because Quantum Field Theory and the Perpetual Motion Cycle $E = mc^2$ predict the exact opposite. All that we experience and observe is constantly conserved order and organization thanks to the Quantum Fields, the Conservation of Energy and Psyche, the Conservation of Quantum Information within Psyche, and the Perpetual Motion Cycle $E = mc^2$. The thing that we experience and observe is the thing that's actually real and true.

The different equations for entropy defined entropy axiomatically as "all of the energy in the universe" or "all of the heat in the universe" or "all of the disorder in the universe". There's NO scaling and NO interval or ratio possible when they define entropy as all of the energy in the universe or as all of the disorder in the universe. They cheat. They define entropy as infinite to begin with, and then they find ways to make some more by dividing by some temperature close to zero. It's a scam. It's a con. It's a ruse. It's physically impossible! The various different equations for entropy FALSIFY the second law of thermodynamics. Everything falsifies the second law of thermodynamics!

We cannot systematically vary disorder so as to measure its effect on entropy or thermodynamics! Once we start to manipulate disorder, then it ceases to be disorder and becomes some type of order or organization instead. Intelligent manipulation, or psychic manipulation at the quantum level, naturally produces order and organization. That's why we manipulate things in the first place, both at the physical level and at the quantum level. We are trying to bring order out of chaos. When the Gods and Controlling Psyches created the Quantum Fields, they brought permanent order and organization to our part of the Chaos Realm. They brought order out of chaos. Only psyche or intelligence can do that. The second law of thermodynamics will NEVER be true as long as the quantum fields exist. You would have to destroy the quantum fields in order to make the second law of thermodynamics even remotely true. And, even at maximum disorder, the energy and the psyches continue to exist and continue to function perfectly anyway.

You would be surprised how frequently we encounter individuals concluding causation when the data are only correlational. That's precisely what happened with the theory of evolution. They assume causation when in fact the evidence is only correlational at best. Varying genomes randomly in a laboratory doesn't produce new, different, interesting, and functional life forms. Varying genomes randomly in a laboratory or in the field causes cancer and death and extinction. There's NO correlation between random mutations and the causation of life. There is NO correlation between natural selection or death, and the origin of life.

Likewise, with entropy, they assume causation when there is no correlation whatsoever. The Materialists, Naturalists, Darwinists, Nihilists, and Atheists assume that random disorder or entropy causes planets, stars, galaxies, big bangs, genomes, genes, proteins, eyes, brains, and life forms to appear out of thin air at will. They are wrong. There is no correlation between random disorder and the existence of planets, starts, galaxies, and genomes. Random disorder doesn't produce heat, and it doesn't produce genomes either, because there is NO correlation between disorder and anything else that exists in this universe. Stay on the lookout for these kinds of frauds and deceptions. You will see these falsehoods and errors built into "science" at a fundamental level, because the Materialist and Naturalists and Atheists who control our public schools have no difficulty concluding that correlation as well as no-correlation can produce anything that it sets its mind to. These people have chosen to believe that chance produced everything that exists in this universe. They are wrong, but there's no way to convince them that they are wrong.

The second law of thermodynamics and the theory of evolution are the greatest scams in human history, because they are "chance causality" or are caused by chance alone. The second law of thermodynamics and the theory of evolution are Creation by Chance or the Null Hypothesis – they are produced by chance alone. They ARE what we are trying to falsify and eliminate from Science by doing science experiments in the first place. In Science and Statistics, we are trying to find the True Cause of each event, and chance can NEVER be it.

Perfect disorder or randomness produces a correlation coefficient of zero. There's NO correlation between randomly distributed atoms in a gas. Therefore, randomly distributed atoms in a gas cannot be the cause of anything, which means that random clouds of gas cannot be the cause of planets, stars, and galaxies. We have to find some other mechanism for the causal origin of planets, stars, galaxies, genomes, proteins, and life forms BESIDES randomly distributed atoms in a gas cloud. Creation by Disorder, Creation by Entropy, or Creation by Chance is the dumbest idea that we have ever erroneously labeled as "science". Creation by Chance or Creation Ex Nihilo is the stupidest idea every created by the mind of man. That's where the philosophies of men fall down and die. Creation by Chance is philosophy or wishful thinking, NOT science. The second law of thermodynamics, entropy defined as disorder, and the theory of evolution ARE Creation by Chance. They ARE the Null Hypothesis, which means that they are automatically false and have automatically been falsified.

These truths and Science 2.0 will never be adopted nor used by our science community during my lifetime, because they falsify Materialism, Naturalism, Darwinism, Nihilism, Atheism, the Theory of Evolution, and the Second Law of Thermodynamics which are the Null Hypothesis or Creation by Chance.

However, in the centuries to come, when our scientists finally figure out what's really going on in our universe at the non-physical level or the quantum level, then they will vote to replace the falsified second law of thermodynamics with Quantum Field Theory and the Perpetual Motion Cycle $E = mc^2$ instead. This will never happen, though, while these people continue to correlate disorder with causation as they are currently doing. So long as they continue to correlate chance with causation, they will never find and know the truth. That's just the way it is.

Mark My Words

**Creation by Chance Is the Null Hypothesis**

https://origin-science.org/creation-by-chance-is-the-null-hypothesis/

The realization that Creation by Chance is the Null Hypothesis is one of my most favorite and interesting discoveries in science and statistics. It changed the way that I look at reality and science. I'm a mathematician, statistician, and quantum theorist, so I find these probability computations extremely convincing. I KNOW that they are true because I believe in the math and understand the math.

While writing the book on Syntropy, I one day realized that Evolution is Entropy, particularly the Creation by Disorder definition for entropy or the Creation by Chance definition for entropy. There are dozens of different mutually exclusive definitions for entropy, which is why nobody seems to know what entropy is. But, genetic entropy or the deterioration of the human genome is a real phenomenon. Our genome was designed with backup genes in place, but as our human genome continues to deteriorate, the human race will become sterile and eventually go extinct. The experts who have been studying genetic entropy and gauging the rate of deterioration have concluded that the human genome has a lifespan or a shelf life of about 9,000 years total, before the collective human genome is so damaged that it can no longer produce viable offspring.

That kind of suggests that something unseen and supernatural has been propping up and sustaining the human genome, if the current rate of measurable deterioration is any indication.

Genetic entropy is one of the only parts or versions of the second law of thermodynamics that is demonstrably true, and genetic entropy falsifies the theory of evolution, chemical evolution, abiogenesis, spontaneous generation, macro-evolution, and creation by chance.

The other part or version of the second law of thermodynamics that is demonstrably true is its claim that perpetual motion machines are physically impossible at the physical level, because ALL of the perpetual motion machines exist at the quantum level outside our reach or purview as fallen physical mortal beings. The Perpetual Motion Cycle $E = mc^2$ is the only "perpetual motion machine" that we have some influence over. We have ways of assisting nature in converting mass, heat, and entropy into usable energy or exergy or light, and we have ways of transforming energy into mass or physical matter within our particle accelerators. But, the massless and entropyless perpetual motion machines that we call the Quantum Fields seem to be completely beyond our reach as physical beings. Only the Gods and Nature's Psyche seem to have control over the quantum waves, photons, and quantum fields.

It has been observed that Psyche, Intelligence, or Quantum Consciousness can design, create, and organize AT WILL both at the quantum level and the physical level; whereas, disorder or chance cannot. By definition, disorder or chance doesn't have a will, which means that it doesn't have a way to design and create anything at all. That's just the way it is. This is what has actually been experienced and observed. The false is falsified by the truth, and the truth is constantly and repeatedly experienced and observed. Science is observation and experience after all – or it should be.

Creation by Chance has never been experienced nor observed. According to any binomial distribution table, it's impossible to get more than fifteen heads in a row before CHANCE runs out of luck. Your typical average protein is over 400 amino acids long. The sequence has to be perfect, or the protein doesn't work. It's completely impossible to produce proteins by mere chance alone. It can't be done because CHANCE prevents it from happening. It's also impossible to produce the matching genes to go along with those proteins by chance alone. It can't be done, which means that it wasn't done that way.

According to a binomial distribution table, some of the amino acids are short enough or small enough to come about by chance alone, but anything longer than fifteen molecules or fifteen amino acids is impossible to produce by chance alone.

https://origin-science.org/Binomial-Distribution-Table

Just look up the probability of tossing 15 heads in a row with a fair coin (P = .50) on a binomial distribution table, and you can see with your own eyes that the chance of doing so is 0.000. That's already entering into the range of impossibility, and it only gets worse from there. Chance can't reliably produce a pre-chosen sequence fifteen units long, let alone the typical protein that is over 400 amino acids long. Chance can't produce a functional protein, and chance can't produce the matching gene to go along with that protein. ONLY Intelligence, Psyche, or Quantum Consciousness can design and create both at the quantum level and at the physical level.

Your genome is God's Signature, and that signature is written on every cell in your body. The fact that your genome actually exists and the fact that proteins exist ARE Scientific Proof of God's Existence. I had to get used to it, because that's the way it is. Anything that is obviously made obviously has a Maker who made it. Genomes, proteins, and genes were obviously made. It's impossible to produce them by chance alone. They are too long and too complex. Chance only functions reliably when the sequence is extremely short.

According to a binomial distribution table, it is impossible to get fifteen heads in a row or fifteen tails in a row from a fair coin by chance alone. The probability of doing so has effectively dropped to zero. And, that's with a binomial distribution – heads or tails – two possibilities. When it comes to proteins and amino acids, there are 21 different amino acids that are used to make proteins, which means that the sequence that can be produced by chance alone drops significantly lower than a sequence fifteen units long.

Of course, we are dealing with probabilities here, but a correct sequence of amino acids four or five units long is pretty much the best that chance alone can do; and, I'm being extremely generous here, given the fact that there are 21 different amino acids that can be used to form our sequence.

The probability of getting two correct amino acids in a row is $(1/21) * (1/21) =$ "0.0022675737". That's two-tenths of a percent, which some people would consider to be within the realm of possibility. It's highly improbable, but it's possible for chance to get two amino acids in the correct sequence.

The probability of getting three correct amino acids in a row is $(1/21) * (1/21) * (1/21) =$ "0.0001079796998". That's 1/10000, which is still within the realm of detectable possibility according to the binomial distribution table that I listed above, which takes things out to .0001 before the table drops to zero. It's extremely improbable for chance to get three correct amino acids in a row, but the hopeful and those operating on blind faith will claim that it's theoretically possible. "There's still a chance", they will say.

The probability of getting four correct amino acids in a row is $(1/21) * (1/21) * (1/21) * (1/21) =$ "0.000005141890467". That's five chances in a million; and, here we enter into the realm of impossibility. Technically, that's never going to happen, and it gets worse from there. At some point we have to call it a loss, or call it an impossibility; and for me, this is where it starts to happen.

The probability of CHANCE getting five correct amino acids in a row is "2.4485e-7". That's never going to happen. There's just no chance that CHANCE can do the job that needs to be done. So, the best that CHANCE can do by chance alone is a protein that is four or five correct amino acids in a row; and like I said, we are being extremely generous here, giving chance the full benefit of the doubt.

Remember, this is with a God-given tub full of the correct 21 amino acids, and with NOTHING else in the tub besides those correct amino acids. If you add anything else into the tub, then the probability drops to zero even quicker than it would with a tub full of the correct 21 amino acids.

God would have to make all the amino acids in the first place and put them into close proximity to each other so that chance can then combine them into something unusable that is four or five amino acids long. The shortest protein in the human genome is 44 amino acids long, and that's $21^{40}$ more than what chance can do by itself. Think about it. It's true. Chance just isn't all that reliable. The house always wins, which means that the Materialists, Naturalists, Darwinists, and Atheists always lose because they are relying exclusively upon chance alone to get the job done. It's NEVER going to happen!

Creation by Chance is the Null Hypothesis; and, the Null Hypothesis is what we are trying to falsify, eliminate, and reject by doing a science experiment and statistics in the first place. The whole purpose of science, statistics, probability, the scientific method, and science experiments is to eliminate or falsify the Null Hypothesis or Creation by Chance. The whole purpose of science is to falsify and eliminate Materialism, Naturalism, Darwinism, Nihilism, Atheism, the Theory of Evolution, the Second Law of Thermodynamics, and everything else that is Creation by Chance or the Null Hypothesis.

CHANCE did not design and create your genome. God did. CHANCE did not design and create your proteins. God did. The Biblical God Jesus Christ or Jehovah confessed to doing the job. Every protein in your physical body is God's Signature and God's Handiwork. The fact that a protein even exists is Scientific Proof of God's Existence. Proteins and their matching genes cannot be made by chance alone, which means that God or some type of Intelligence made them.

WE KNOW that this is true because Creation by Chance is the Null Hypothesis, and both the theory of evolution and the second law of thermodynamics ARE Creation by Chance or caused by chance alone. The Null Hypothesis or Creation by Chance IS automatically false according to the rules of science and statistics.

If it can be demonstrated that the Null Hypothesis is true, then WE KNOW that there is NO causal relationship between disorder or chance and anything thing else that might exist. If it can be demonstrated that the Null Hypothesis is true, then WE KNOW that there is no causal relationship and no correlation between the independent variable and the dependent variable. Whenever chance is the independent variable as it is in the theory of evolution and the second law of thermodynamics, then we automatically KNOW that chance, the theory of evolution, and the second law of thermodynamics ARE NOT the cause of anything that exists. Chance can NEVER be the cause of the dependent variable! Chance can NEVER be the cause of anything significant. We automatically KNOW that creation by chance, the theory of evolution, creation by random disorder, and the second law of thermodynamics ARE FALSE because they ARE the Null Hypothesis and were produced by chance alone.

Materialism, Naturalism, Darwinism, Nihilism, Atheism, Behaviorism, Physical Reductionism, Determinism, the Theory of Evolution, and the Second Law of Thermodynamics ARE the Null Hypothesis or Creation by Chance, which means that they are automatically FALSE. They cannot be the cause of anything that exists!

Checkmate!

Brutal, is it not?

I used to be a materialist, naturalist, nihilist, and atheist until Science and Statistics convinced me that I was wrong. I had a hard time wrapping my mind around it at first because I had been brainwashed into believing that Materialism, Naturalism, Nihilism, Atheism, and the Second Law of Thermodynamics ARE true. But, everything that exists FALSIFIES the second law of thermodynamic and the theory of evolution because everything that exists FALSIFIES Creation by Chance or the Null Hypothesis.

Mark My Words

**Seeing the Truths Everyone Refuses to Accept**

Thanks to Statistics and Quantum Field Theory, I could finally see and understand what nobody else seems capable of seeing and understanding. I could see what had been done to us. I could see how we had been tricked and deceived into believing that the total amount of disorder in our universe is constantly increasing and constantly conserved, when in fact it is not. We believed these lies because we desperately wanted them to be true. We fell for the stupid ideas in philosophy and science – such as creation by chance, creation ex nihilo, the second law of thermodynamics, and the theory of evolution – because we desperately wanted them to be true, when in fact they are not. Self-deception works, and it works every time. We never question it because we want it to be true, even though it's obviously and demonstrably false. Creation by Chance in any form or disguise is obviously wrong and obviously false and demonstrably false, but we believe in it anyway because we desperately want it to be true. Our *confirmation bias* prevents us from seeing and understanding what's wrong with it.

One of the hallmarks of the Philosophy of Science is the psychological and statistical observation that each and every piece of scientific data is subject to a theoretical infinite number of different, contradictory, and mutually exclusive personal interpretations. The fundamental flaw and weakness of the scientific method consists of all the thousands of different personal interpretations that are being applied to each and every piece of scientific data that we develop. They can't all be true because they falsify each other; and, it's possible that NONE of them are true. It's possible that we still haven't found the true interpretation of Quantum Mechanics, for example. It's obvious that we have developed thousands of interpretations of Quantum Mechanics that are demonstrably and clearly false.

Within this book and my other books, I have developed new and different personal interpretations of scientific evidence that nobody has ever thought of before, because they falsify Materialism, Naturalism, Darwinism, Nihilism, Atheism, the Second Law of Thermodynamics, Creation by Chance, Creation Ex Nihilo, Creation by Disorder, and the Theory of Evolution. The only reason that my personal interpretations of the scientific evidence end up being superior to the others is because they are based exclusively upon what has actually been experienced and observed. Psyche, Quantum Mechanics, and Quantum Field Theory have been experienced and observed. Creation by Chance and Creation Ex Nihilo have NOT been experienced and observed. That's the only difference between the two. They are both personal opinions or personal interpretations of the scientific evidence, but only one of them has actually been experienced and observed. Science 2.0 give preference to what has actually been witnessed, lived, experienced, and observed. That's what makes it superior to the other personal interpretations and philosophies of science that have been presented to the world. I always go with what has been experienced and observed!

Within this and my other books, I have used Statistics, the Scientific Methods, the Philosophy of Science, Logic, Common Sense, as well as the Observations and Experiences of the Human Race to demonstrate clearly and conclusively that up to half of our personal interpretations in modern-day "science" are wrong and false. Anything that is Creation by Chance is automatically wrong and false, especially when chance is being used as a causal explanation or a causal mechanism or a causal force or a causal entity. Creation by Chance is a faulty and falsified personal interpretation of any scientific data that we might develop. This is what Statistics, the Scientific Method, the Null Hypothesis, and the Rules of Scientific Experimentation are trying to teach us, but nobody is listening because nobody wants it to be true.

Creation by Chance and Creation Ex Nihilo are obviously impossible and clearly false. Yet, Boltzmann Brains (Creation by Entropy and Disorder), the falsified aspects of the Big Bang Theory (Creation Ex Nihilo or Creation by Magic or Creation by Chance), the Second Law of Thermodynamics (Creation by Disorder and Entropy and Chance), Chemical Evolution (Creation by Chance and Creation by Disorder), as well as the Theory of Evolution (Creation by Chance and Creation by Death) have become the foundation and the cornerstone of modern-day science. I dare say that half of modern-day science is demonstrably false, totally worthless, and completely wrong because it is in fact Creation by Chance or Creation Ex Nihilo.

Boltzmann, the creator of the disorder definition for entropy, actually tried to turn entropy or disorder into our Designer, our Creator, and our God. His Boltzmann Brain idea is one of the dumbest ideas in science. It's so dumb that the majority of scientists have actually chosen to believe that it is true. Stephen Hawking took it, and ran with it, and soon had "random fluctuations in entropy" or "quantum fluctuations in chance" being our designer and the causal agent behind the Big Bang. Of course, Charles Darwin fed us Creation by Chance or Creation by Death (natural selection) through his theory of evolution. Einstein rejected Action at a Distance (Creation by Psyche or Quantum Consciousness or Intelligence) and tried to force Classical Realism (Creation by Matter and Creation by Chance) onto us instead. Half of the geniuses in science, biology, religion, philosophy, and physics fell for Creation by Chance or Creation Ex Nihilo and tried to make these falsified ideas the cornerstone of modern science. It's still with us to this very day, and we can't get rid of it, because these geniuses desperately want it to be true.

Of course, wanting it to be true doesn't make it true. The only thing that makes it true is if it has actually been experienced and observed. The fact that Creation by Chance and Creation Ex Nihilo (spontaneous generation or abiogenesis) have NEVER been experienced NOR observed is the thing that falsifies them; but, that doesn't matter because the Materialists, Naturalists, Darwinists, Nihilists, and Atheists desperately want them to be true. These people are so desperate to make these falsehoods true that they have actually turned Creation by Chance and Creation Ex Nihilo into a LAW that they now call the second law of thermodynamics. Garbage in. Garbage out. And, that's what modern-day science is all about.

### Death-Bed Repentance

Even at this late date, it's still possible for a scientist like me to find NEW truths or HIDDEN truths by identifying the frauds and exposing the lies. It won't change the way that we do science. Physical matter will continue to be physical matter with all of its limitations and vices. However, it can change for the better the way that we interpret science.

Truth has its own intrinsic reward. My explanatory power has gone through the roof now that I have chosen to include the non-physical Quantum Fields, the massless non-physical Photons, Action at a Distance, Quantum Mechanics, Quantum Field Theory, and Quantum Consciousness or Psyche within science. By bringing Psyche and Quantum Mechanics in to play, we can literally explain everything that comes our way. I haven't found anything that I haven't been able to explain, now that I have embraced the non-physical Quantum Fields and the immaterial non-physical Photons or Psyches as a part of science. Now everything has been explained to my satisfaction. I now KNOW how everything was done and how everything works. Good enough.

As a scientist and quantum theorist, I NEEDED a non-physical and pre-physical mechanism capable of designing, making, and controlling the massless, non-physical, immaterial Quantum Fields BEFORE the first particle of physical matter was designed and made. I now have that mechanism – a perpetual motion machine that we call Psyche, Soul, Intelligence, the Spark, or Quantum Consciousness. I needed a machine capable of working at the quantum level and getting things done at the quantum level, within the massless non-physical quantum waves and quantum fields – and within electrons, quarks, and physical atoms as well. I needed something that is intelligent, conscious, perceptive, sentient, aware, omniscient, telepathic, purposeful, intent, and alive. I needed a real quantum computer, and I needed an infinite singularity. I needed something that has no physical limitations. I needed something that had actually been experienced and observed. The thing we call "Psyche" has actually been experienced and observed. Good enough.

Meanwhile, the Materialists, Naturalists, Darwinists, Nihilists, and Atheists will continue to worship physical matter as if the stuff were God. They have done so for thousands of years, and they will continue to do so for thousands of years to come. Nothing is going to change by showing them the truth. They will remain Naturalists, Darwinists, and Atheists because that's what they want to be. No amount of evidence will convince them that they are wrong. They will simply refuse to look at the evidence. That's what they always do because that's what they have always done. I've been there and done that, so I know how it goes. I used to be a materialist, naturalist, nihilist, and atheist until Science, Observations, or Knowledge convinced me that I was wrong. Now I choose to go with what has been experienced and observed, instead, as any good scientist should do.

Mark My Words

## Everything Falsifies the Second Law of Thermodynamics

So, what's wrong with the second law of thermodynamics?

The second law violates and falsifies $E = mc^2$.

Thanks to the Conservation of Energy, energy is the ultimate perpetual motion machine.

Energy is the mother or the source for ALL of the different perpetual motion machines, including the ones that were made, such as protons and quantum fields. The protons and the quantum fields are being conserved or preserved by Nature or Nature's Psyche. They are perpetual motion machines! The never get old, and they never wear out.

$E = mc^2$ is the perpetual motion cycle. It works perfectly both coming and going, and all the while the energy is conserved. Energy cannot be made, and it cannot be destroyed. Energy has always existed, and it will always exist. Energy is eternal and everlasting. Energy is the ultimate perpetual motion machine. Energy is Syntropy. Energy cannot die or cease to exist. Energy is immortal. Energy will never wear out. Energy is conserved.

The Syntropy has to exist or we wouldn't exist.

According to Quantum Field Theory, particles are born, and particles can die. Particles are made, and particles can be unmade or disassembled. So, ask yourself who is making and then disassembling these particles, or quanta, or quantum waves? Someone has to be making them and disassembling them, or they wouldn't exist, and they wouldn't cease to exist either. We wouldn't exist either.

Particles, or quanta, or quantum waves are MADE by Nature's Psyche. Nature's Psyche can also disassemble and dissolve particles back into the quantum fields from whence they originally came.

Nature's Psyche is the one who collapses the wave function. Nature's Psyche makes the quantum waves or the particles in the first place. Under the tutelage and the instruction of the Gods, Nature's Psyche made the massless, heatless, and entropyless quantum fields.

Nature's Psyche also MAPS quantum functionality onto your physical brain. Nature's Psyche decides the meaning and the purpose of each synapse and neuron in your physical brain. Nature's Psyche makes and MAPS your physical brain. Nature's Psyche made your physical brain, so Nature's Psyche KNOWS what each neuron does and where it is located. Nature's Psyche fires the specific neurons in your brain, whenever you want to move. Nature's Psyche stores your memories at the quantum level. The Human Psyche isn't consciously aware of any of these things. It's Nature or Nature's Psyche who is doing all of these different quantum functions that are needed to keep us alive and make us move.

The memories that show up in your after-death life review are obviously not stored within your physical brain. The memories that survive the death of your physical brain have to be stored someplace else besides your physical brain. Nature's Psyche is handling all of this for us, because the Human Psyche isn't consciously aware of any of these things.

$E = mc^2$ and Quantum Field Theory are the same thing. Quantum fields are perpetual motion machines that were made by Nature or Nature's Psyche. Energy is the ultimate perpetual motion machine. $E = mc^2$ is the perpetual motion cycle. It is eternal and everlasting. It is conserved. It functions perfectly BOTH directions.

Every psyche, intelligence, life force, or quantum consciousness has a certain amount of energy that's under its control. That controlling psyche can form or transform the energy under its control into anything it wants that energy to be, anytime and anywhere that it chooses to do so. This is the LAW of Psyche.

The LAW of Psyche falsifies the second law of thermodynamics or chance causality.

According to Quantum Field Theory, Nature's Psyche can transform the available energy or the exergy under its control into mass, resistance to acceleration, heat, and heat storage capacity (entropy) anytime and anywhere that it chooses to do so. Likewise, according to Quantum Field Theory, Nature or Nature's Psyche can transform physical matter, electrons, quarks, mass, heat, and entropy into massless, heatless, chargeless, and entropyless photons (and quantum waves and quantum fields) anytime and anywhere that it decides to do so; and, there's nothing we can do to stop it.

In the behavioral sciences where decisions and choices are being made by the Human Psyche, correlations between the independent variable and independent variable are typically low because choices are being made and personal preferences are being expressed. The Behaviorists erroneously conclude that the low correlations in the behavioral sciences are being caused by chance – something that they call determinism. They are wrong. The different kinds of Atheists or Determinists are always wrong, because they don't know how Nature or Nature's Psyche truly works.

The correlations are low in the behavioral sciences, especially where human beings are concerned, because of the different choices or preferences that each Human Psyche CHOOSES to express or enact. Unlike the animals, the Human Psyche can break its conditioning at will. The Human Psyche is the ultimate CHOOSER. The Human Psyche decides what it wants to do with the physical matter that's under its control; and, Nature's Psyche collapses the necessary wave functions or fires the appropriate neurons in order to get the job done.

Once we understand the LAW of Psyche and how things truly work at the quantum level among the quantum fields, then we KNOW why Materialism (Behaviorism and Determinism), Naturalism (Supernatural Quantum Fields DO NOT EXIST), Darwinism (Creation by Chance), Nihilism (Spirit and Psyche DO NOT EXIST), Atheism (Creation Ex Nihilo), and the Second Law of Thermodynamics (Creation by Entropy or Creation by Disorder) ARE false.

The second law of thermodynamics is false because it is the Null Hypothesis, Creation by Chance, Chance Causality, or Caused by Chance Alone. The second law of thermodynamics is false because it is a part of Materialism, Naturalism, Darwinism, Nihilism, Atheism, and their derivatives. The second law is false because it violates the Conservation of Energy or the First Law of Thermodynamics by magically having entropy or disorder constantly increasing out of thin air from nothing. The second law of thermodynamics is Creation Ex Nihilo, and there is no such thing as the creation of something from nothing. Not even God can do Creation Ex Nihilo, so why assume that the second law of thermodynamics can?

These truths falsify the second law of thermodynamics. Nature and Nature's Psyche falsifies the second law of thermodynamics. Quantum Field Theory falsifies the second law of thermodynamics. The First Law of Thermodynamics or the Conservation of Energy falsifies the second law of thermodynamics. $E = mc^2$ falsifies the second law of thermodynamics. Everything falsifies the second law of thermodynamics.

The second law of thermodynamics erroneously defines "entropy" and "thermodynamics" as disorder or chaos. Disorder has absolutely no correlation whatsoever with heat or thermodynamics. Through the second law of thermodynamics, the Materialists, Naturalists, Darwinists, Nihilists, and Atheists erroneously worship disorder, chaos, and death. Through the second law, these people reify and deify disorder, chaos, and death. The second law has become Creation by Entropy, Creation by Disorder, Creation by Chaos, Creation by Death, and Creation by Chance.

The second law of thermodynamics was designed to falsify and eliminate the Conservation of Energy or the First Law of Thermodynamics. The second law and $E = mc^2$ are mutually exclusive. They falsify each other. If one of them can be demonstrated to be true, then the other one has automatically been falsified. So, which one is it? Which one is true? Which one has been experienced and observed?

Science 2.0 allows all of the evidence into evidence, and then it pursues a preponderance of that evidence. The whole purpose of Science 2.0 is to identify everything that is false in science and then to eliminate it from science by replacing it with the things that have actually been experienced and observed. The false is falsified by the truth, and the truth is constantly and repeatedly experienced and observed. That's the way science works or should work.

### The Force in Between the Two Objects

In his videos, David SantoPietro keeps saying how confused and flummoxed physicists are when it comes to Action at a Distance, or the ability of charges and gravity to attract each other or repel each other across vast distances of empty space.

#### Electric Field Definition

So, here's a question. **If you got two positive charges, we know they're gonna repel. So, if I put these next to each other, this blue charge would repel the green charge, and vice versa, but how is that working exactly? I mean there's nothing in between these charges. How is the blue charge pushing on the green charge when it's not even touching it?** So, this is kind of a weird thing, right? If I want to push on something in my room, I need to walk up to that thing and actually physically touch it, and then give it a shove. Yet this blue charge seemingly exerts a force on the green charge with just empty space in between, right? There's no strings here. What is the mechanism? So, physicists were kind of embarrassed and concerned about this. So, it was like, all right, we can calculate exactly how much force there would be on each charge, but we don't really know how that is actually working. It doesn't seem to be making any sense that a charge can push on another charge across potentially vast stretches of the universe here. This could be a huge distance and yet somehow this charge over here knows, ah, there's a charge over here pushing on me. How is that possible?

**How is the earth pulling on the moon when there's nothing in between? How is this force at a distance actually being transmitted? How do two objects exert forces on each other across what are potentially vast distances of space?**

https://www.khanacademy.org/science/physics/electric-charge-electric-force-and-voltage/electric-field/v/electric-field-definition

This is THE QUESTION in physics – the whole of physics. How is Action at a Distance done? How are forces transmitted through empty space? How do photons travel through empty space where there is no medium or means of travel? How are immaterial, non-physical, invisible, intangible forces and fields transmitted through empty space?

The reason physicists can't answer this question is because they are Materialists, Naturalists, Darwinists, Nihilists, and Atheists. They have chosen to believe that the non-physical does not exist. Consequently, they never once realize that there is no such thing as empty space. The whole of space, including the 99.999% of "empty space" within a physical atom is filled with a Matrix of invisible, non-physical, intangible, immaterial, non-local, supernatural, syntropic, exergic Quantum Fields. There is no such thing as empty space. Everything is interconnected through the Quantum Fields. The whole observable universe is one large organism consisting of a Matrix of Quantum Fields. It's all enmeshed and connected together at the quantum level. The physical is a drop in the ocean compared to the Matrix of Quantum Fields that have been experienced and observed. At the quantum level, there is NO resistance to acceleration, which means that at the quantum level, everything in our universe is instantaneously connected with everything else thanks to the Quantum Fields.

Yes, omnipresent quantum waves or photons can choose their direction and limit their velocity to something that we call the speed-of-light; but, they don't have to if they don't want to. Quantum waves or photons can be omnipresent when they want to be, and they can quantum tunnel whenever and wherever they choose to do so. Quantum waves or photons can also CHOOSE to stop and transform their energy into mass and heat instead. The Psyche or Intelligence within the quantum wave or photon can do anything that it wants to do with the energy that's under its control. Everything starts to make logical sense, once we choose to bring Psyche and Quantum Mechanics into play.

In contrast, Materialism, Naturalism, Darwinism, Nihilism, Behaviorism, Determinism, Physical Reductionism, Atheism, Classical Realism, and the Second Law of Thermodynamics have NO explanatory power when it comes to the non-physical, the immaterial, the spiritual, or the quantum. These falsified religions or falsified philosophies claim that the non-physical or the quantum does not exist. Rather than trying to explain the quantum or the non-physical, Materialism and Naturalism dismiss, deny, reject, and ignore the quantum or the non-physical.

Remember, once you allow Psyche and Quantum Mechanics in to play, you can literally explain everything that comes your way. Instantly, everything is explained.

As David says, **the one charge KNOWS that the other charge is pushing on it**, and vice versa. In other words, the charges or quanta or particles are intelligent and alive. They each have some type of Psyche or Intelligence within them controlling the energies associated with them. They are conscious and aware of each other across vast distances of space because they are in contact with each other telepathically and telekinetically across vast distances of space thanks to the quantum fields.

Later in this video, David answers his question by invoking an electric field, which is ONE of the many different types of quantum fields.

Faraday and David invoke Quantum Field Theory and an electric field to explain how two particles or charges are interacting with each other at a distance through empty space, **"This positive charge ($Q_1$) creates an electric field everywhere around it at all times whether there's other charges nearby or not. The electric field gets weaker and weaker the farther out you go."**

Now, when **"positive charge number two" ($Q_2$) wanders into this region, it just has to look at its immediate surroundings. At this point right here, it sees this electric field. It senses it and it knows that it is an electric field pointing to the** right. It knows that this electric field is going to cause a force on me to the right or repel me to the right. The electric field causes an electric force. The electric field (or quantum field) is the mediator or the medium through which the two different charges or particles are exerting forces or influences on each other at a distance through what seems to be empty space.

Notice that David, a physicist, uses words that denote or imply Intelligence, Consciousness, Sentience, Life Force, Intention, Purpose, Obedience, Choice, Causality, Perception, and Psyche when talking about this charge, this quantum, or this particle. David has to describe this charge or particle in this manner, because that's the only way to explain what we are experiencing and observing.

This quantum wave, or particle, or charge is sentient. It's alive. It senses its environment. It thinks. It processes information. It makes choices! It has some type of psyche or intelligence within it. These physicists personify and reify Psyche or Intelligence whenever they start talking about what a particle or a charge senses and knows. They turn this particle or charge into a living person or an entity that makes choices. It's the only way to explain what they are observing.

This charge, particle, or quantum KNOWS how it's supposed to behave whenever it is in the presence of other particles that are nearby. That implies that there is some type of Law-Giver or Law-Maker or Standard-Maker, who decided how particles, charges, and quanta are supposed to behave when they are near each other and when they make contact with each other. There has to be a Law-Giver and a standard or a law established by someone; and then, there has to be a Law-Follower who is someone intelligent or psychic at the quantum level who can understand the laws and standards that are in place, and then choose to conform with those laws and standards. Again, whether David realizes it or not, David is talking about some type of Intelligence or Psyche within the charge, the particle, or the quantum who understands the laws and chooses to follow the laws.

The Quantum Fields become the medium and the mediator through which different charges, particles, and quanta interact with each other across vast distances of space. The closer that two particles or quanta are to each other, the stronger the force between them. In the case of electric charges and gravity, the forces decrease the further apart the particles or objects are. However, when it comes to gluons, the attractive force gets stronger the farther apart the quarks become. Each quantum field was programmed and designed by the Gods and Nature's Psyche to produce specific functionality. There was a LAW established by the Lawgivers that provides standards, order, organization, purpose, meaning, and mission to each one of the Quantum Fields and to each of the particles in those fields. Each Quantum Field functions differently, with a different purpose, mission, or law. These differences in functionality required some type of Psyche or Intelligence at the quantum level in order to make them in the first place. They also require intelligent Followers or Psyches at the quantum level to understand and obey the quantum laws. Laws are worthless without intelligent Obeyors. There has to be a Lawmaker, a Law-Enforcer, and a bunch of intelligent and psychic Law-Followers at the quantum level; otherwise, the quantum laws as well as the physical laws would be totally worthless, irrelevant, and ineffective.

These physicists always describe the photons, quanta, quantum waves, charges, particles, electrons, quarks, and physical atoms as if these things were sentient, perceptive, intelligent, conscious, alive, and aware – which they are. These things are connected with each other and communicating with each other telepathically through the Quantum Fields. The Quantum Fields are the universal internet or the cosmic broadcast; and, there is no speed-limit when it comes to the quantum, the spiritual, or the non-physical. The speed-limits come into play when the photons or quantum waves CHOOSE TO STOP and transform themselves into mass, heat, resistance to acceleration, or heat storage capacity (entropy) instead.

Likewise, Nature's Psyche or the psyches within the mass and heat CAN CHOOSE to transform the energy under their control into massless, heatless, chargeless, and entropyless photons, quantum waves, quanta, or quantum fields anytime and anywhere they CHOOSE to do so, according to Quantum Field Theory. This is happening right now within our sun. Where do you think all the light comes from? Nature's Psyche or the controlling psyches within the physical matter in the sun are actively transforming that mass, heat, and entropy into massless, heatless, chargeless, and entropyless photons on a constant basis. The mass, heat, and entropy get transformed into infinite acceleration instead. The photons go from zero to the speed-of-light instantly. That's infinite acceleration. Furthermore, photons can choose to travel at the speed-of-light; or, they can choose to travel at an infinite velocity instead, and quantum tunnel to their destination. It's a choice!

Photons and quantum waves don't have to stop if they don't want to. Photons pass through water and glass all the time. Photons are omnipresent and omniscient while they are quantum waves. Photons KNOW where they are and what they are passing through, and they adjust their behavior accordingly. Photons or quantum waves can pass through our earth, our sun, and a black hole as if these things weren't even there, if a photon or quantum wave wants to do so. A photon doesn't have to stop if it doesn't want to. A photon or quantum wave stops because it CHOOSES to stop.

Photons are conscious, intelligent, sentient, psychic, telepathic, aware, and alive. There's Intelligence or Psyche within every photon, particle, quantum, and quantum wave. A photon has NO physical limitations, until it chooses to stop and transform its energy into mass and heat instead. When a photon chooses to stop, it transforms its infinite acceleration into "resistance to acceleration" (mass), heat, or "mass's heat storage capacity" (entropy) instead. A photon can transform itself into anything that it wants to be. A photon or quantum wave can choose to stop, or a photon can choose to keep going through our earth as if our earth weren't even there. This is what we have actually experienced and observed.

Quantum Field Theory and this video bear witness that these concepts and ideas are real and truly exist.

**Doesn't $Q_2$ also create its own electric field? Wouldn't $Q_2$ also create an electric field everywhere around it? We could call that $E_2$ since it's created by charge too. Doesn't it create its own electric field just like all charges do? Yeah, it does. In fact, it will create an electric field over here next to $Q_1$ and that's how $Q_1$ knows it's supposed to feel the force it feels in the direction that it feels it. So that's how these charges talk to each other. You could think about it that way. The way charges talk to each other is with the electric field. One charge creates an electric field over by the other charge. That charge feels the force. The other charge creates a field by the first charge. That first charge feels the force. So conceptually that's how this electric field works.**

David even quotes the Materialists, Naturalists, Darwinists, Nihilists, and Atheists with their standardized objection or complaint against the non-physical.

**So, at this point you might not be impressed. You might be like, "Are we just making up a story to make ourselves feel better here? Is this just some elaborate fairy tale that makes us so that we don't feel so awkward talking about**

**things exerting forces on each other at a distance? Is there any benefit for doing this?"**

Then David gives the standardized answer, "It makes the math easier." Then he drops the quantum theory or physics theory and switches over to the math instead. These people are embarrassed and constantly making apologies and excuses whenever they discover Quantum Mechanics and Quantum Field Theory, because they don't want the non-physical or the intangible to be real and true, because the verified and proven existence of the Quantum Fields falsifies Materialism, Naturalism, Darwinism, Nihilism, and Atheism which claim that the massless, heatless, chargeless, and entropyless quantum fields DO NOT EXIST. The truth falsifies the false, so these people have to make apologies for the truth, whenever they falsify Materialism and Naturalism and their derivatives such as Darwinism and Atheism. David's apology is that "it makes the math easier".

The Real Answer is that the quantum fields and the quantum waves are Wi-Fi at the quantum level. Just like we have invisible, intangible, non-physical radio waves at the physical level that we can detect and manipulate and transmit and receive, psyches or intelligences have quantum waves at the quantum level that they can detect, manipulate, transmit, receive, process, make, and store. That's how particles communicate with each other at the quantum level – psyche to psyche or telepathically – through the quantum Wi-Fi system that has been organized and made at the quantum level. Psyches are quantum transceivers and quantum computers. It doesn't just make the math easier. It also gives us the truth. It explains everything that has ever been experienced and observed.

We have the truth right before our very eyes, but they refuse to go there, because it falsifies Materialism, Naturalism, Darwinism, Nihilism, and Atheism. They prefer the falsehoods instead.

At the physical level, the radio waves choose to limit themselves to the speed-of-light, which is the standard that has been established for the physical level processes. At the quantum level, the quantum waves have NO speed-limit. The psyches or intelligences at the quantum level can communicate with each other instantly and omnipresently if they choose to do so. They have no speed-limit because they are omnipresent through the quantum fields. The whole thing is instantaneous and omnipresent at the quantum level according to Quantum Mechanics and Quantum Field Theory. The quantum level is just a higher and more powerful order of organization. The quantum level is more efficient because it has NO physical limitations. The quantum level perpetual motion machines make and sustain the less efficient physical level machines and processes. It's that simple.

Each psyche is theoretically capable of storing ALL of the information and knowledge in our universe because it has NO physical limitations. If there is such a thing as an infinite singularity or a true point-particle, Psyche is it.

Once again, by letting Psyche and Quantum Mechanics in to play, we can literally explain everything that comes our way. I have no questions now because I KNOW how everything works at the quantum level. That's the true benefit for allowing Psyche and Quantum Mechanics and Quantum Field Theory into science and physics. Instantly, we can explain everything that we have ever experienced or observed. Our explanatory power goes through the roof to infinity and beyond. That's the true benefit of allowing the massless, heatless, chargeless, and entropyless photons and quantum fields into science and physics. Instantly, we can explain everything that has ever been experienced or observed.

Mark My Words

**Joint Heirs with Christ**

I'm a scientist.

I'm not a prophet, seer, or revelator.

I'm a scientist. I'm a quantum theorist.

I simply want to know how everything works.

However, if we find the truth in science, it will match perfectly with the truths in religion, philosophy, and metaphysics. It will end up being the same truth. Truth bears record of the truth. We use the truth to falsify everything that is false, and the truth is constantly and repeatedly experienced and observed by someone somewhere.

Quantum Field Theory and the Law of Psyche represent the truth in Science. They explain everything that has ever been experienced and observed. There's a massless, intangible, invisible, entropyless order or organization at the quantum level thanks to the Quantum Fields. The Gods work through the Quantum Fields to get things done – both at the physical level and the quantum level. The Quantum Fields sustain our existence as physical beings. Physical matter wouldn't exist without the Quantum Fields.

The Apostle Paul and John the Revelator hint at all of this in the Bible.

**Acts 17: 21-29:**

**21 (For all the Athenians and strangers which were there spent their time in nothing else, but either to tell, or to hear some new thing.)**

**22 Then Paul stood in the midst of Mars' hill, and said, Ye men of Athens, I perceive that in all things ye are too superstitious.**

**23 For as I passed by, and beheld your devotions, I found an altar with this inscription, TO THE UNKNOWN GOD. Whom therefore ye ignorantly worship, him declare I unto you.**

**24 God that made the world and all things therein, seeing that he is Lord of heaven and earth, dwelleth not in temples made with hands;**

**25 Neither is worshipped with men's hands, as though he needed anything, seeing he giveth to all life, and breath, and all things;**

**26 And hath made of one blood all nations of men for to dwell on all the face of the earth, and hath determined the times before appointed, and the bounds of their habitation;**

**27 That they should seek the Lord, if haply they might feel after him, and find him, though he be not far from every one of us:**

**28 For in him we live, and move, and have our being; as certain also of your own poets have said, "For we are also his offspring. "**

**29 Forasmuch then as we are the offspring of God, we ought not to think that the Godhead is like unto gold, or silver, or stone, graven by art and man's device.**

Materialism, Naturalism, Darwinism, Nihilism, Atheism (creation ex nihilo), Creation by Chance (the second law of thermodynamics), and the Theory of Evolution (creation by blind luck) are superstition. In all things, the Naturalists and Darwinists are way too superstitious. These people worship physical matter, entropy, and death. They worship the things that were made by man's hands. These people believe, based on blind faith alone, that everything in this universe was created by chance from nothing by nothing. These people believe in Creation by Rocks, Creation by Chance, Creation by Disorder, Creation by Chaos, Creation by Entropy, and Creation Ex Nihilo. That's the very definition of superstition. I perceive that in all things, the Materialists, Naturalists, Darwinists, and Atheists are too superstitious.

We didn't descend from the apes. That's physically impossible thanks to genetics. We are God's offspring. We descended from the Gods. Our spirit bodies and physical bodies descended from the Gods. We are Gods. Our surname is God. Adam was a son of God, both physically and spiritually. Adam's spirit body and physical body descended from the Gods, and so did we. Our spirit or ghost is comprised of our spirit body and our intelligence or psyche. Our spirit body was made or born, but our psyche or intelligence has always existed and will always exist. These truths have important ramifications for us.

**Revelation 3: 20-22:**

**20 Behold, I stand at the door, and knock: if any man hear my voice, and open the door, I will come in to him, and will sup with him, and he with me.**

**21 To him that overcometh will I grant to sit with me in my throne, even as I also overcame, and am set down with my Father in his throne.**

**22 He that hath an ear, let him hear what the Spirit saith unto the churches.**

Jesus Christ is God, and He is now sitting on his Father's throne. And, what does He say about us? He says that if we overcome our sins and learn to control ourselves as He did, we too shall sit down with Him on our Father's throne. We too shall be Gods. Jesus Christ and God the Father have resurrected physical bodies. They also have Psyches and Spirit Bodies. It's all part of the same package and the same plan. Someday, we too will have resurrected physical bodies, as well as glorified and perfected spirit bodies and intelligences. One day, too, we shall be jacked into the Matrix of Quantum Fields. Our Psyche or Intelligence will be omniscient and omnipresent just like Nature's Psyche and the Quantum Fields.

**Romans 8: 14-18:**

**14 For as many as are led by the Spirit of God, they are the sons of God.**

**15 For ye have not received the spirit of bondage again to fear; but ye have received the Spirit of adoption, whereby we cry, Abba, Father.**

**16 The Spirit itself beareth witness with our spirit, that we are the children of God:**

**17 And if children, then heirs; heirs of God, and joint-heirs with Christ; if so be that we suffer with him, that we may be also glorified together.**

**18 For I reckon that the sufferings of this present time are not worthy to be compared with the glory which shall be revealed in us.**

We are Gods. We are the children of the Gods. Our surname is God. If you can read this, then you know that you are descended from the Gods. We are some of the intelligences or psyches who helped to make the quantum fields and helped to make this earth and the genomes found on this earth. Both our spirit body and our physical body are descended from the Gods.

Paul and the Spirit say that we are heirs, heirs of God, and joint-heirs with Christ who is a God. Can you feel the Spirit testifying to you that this is true? We, too, can become Gods. If we are heirs, then what is it that we are going to inherit? There's no sense being an heir if there is nothing to inherit.

So, what does God the Father have that we don't currently have?

God has the quantum fields. He owns and controls the quantum fields. God made the heavens and the earth. God made the quantum fields. That means that He has some type of contract or arrangement with each of the Psyches or Intelligences within the quantum fields. God gets things done through the quantum fields.

The thing that we call the "Big Bang" took place when the Gods stepped into the Chaos Realm and organized the quantum fields in our part of the multiverse. It was ONLY AFTER the Gods had made the quantum fields that it became possible to make and sustain physical matter, mass, and mass's heat storage capacity (entropy). Physical matter or mass is impossible to make without the quantum fields, according to Quantum Field Theory. The thing that we call the "Big Bang" took place when the Gods made the quantum fields in our part of the multiverse. Given Quantum Field Theory, that is a logical and realistic conclusion. Someone Psyche or Someone Intelligent had to design and make the quantum fields, and the Big Bang seems to be the logical point in time when the quantum fields were made and implemented, because only after the Big Bang was it possible to make and sustain mass and heat storage capacity in our part of the multiverse, according to the Big Bang Theory. It all fits, and it all makes sense.

The Psyches or Intelligences within the energy, and the quantum fields, and the quantum waves CHOOSE to obey God's laws or standards; and that obviously provides them with order, organization, structure, meaning, and purpose in return. Obedience to law provides order, structure, purpose, commonality, unity, meaning, and organization. Every physical atom in this universe is built upon or based upon the same blueprint. It's a universal blueprint that ALL of the psyches or intelligences in the quantum fields KNOW how to read and implement perfectly. The physical atoms are compatible with each other thanks to this universal blueprint or universal law that was designed and made by the Gods. Quantum Laws are worthless without some type of intelligent Law-Obeyor at the quantum level capable of understanding and choosing to obey those Quantum Laws. Without the quantum Law-Obeyor that we call Psyche, all we would have is random chaos in our part of the multiverse just as the second law of thermodynamics predicts, and the second law of thermodynamics would actually be true.

The Gods and Nature's Psyche had to design and make the quantum fields BEFORE they could make and sustain physical matter. If there were no quantum fields, then there would be no physical matter, according to Quantum Field Theory. The thing that we call the Big Bang was in fact the construction of the Quantum Fields in our part of the multiverse. There would be no physical matter, mass, resistance to acceleration, nor mass's heat storage capacity (entropy) without the quantum fields. The Gods and Nature's Psyche can't make and sustain physical matter without the quantum fields. God is light, and God is in the light. God is in the quantum fields. God controls everything through the quantum fields. God is omnipresent and omniscient through the quantum fields. That's the way it works!

The chaos or disorder completely disappeared at the quantum level, once the Intelligences or Psyches within the energy CHOSE to organize themselves into the quantum fields and CHOSE to obey the Quantum Laws or God's Laws. That's when the second law of thermodynamics ceased to be true. Chaos was replaced with order and organization instead, at the quantum level, the very moment that the Quantum Fields were organized and made by Nature or Nature's Psyche in our part of the multiverse. The Gods taught the Intelligences or Psyches how to organize themselves at the quantum level into the Quantum Fields. Now all we have is conserved order, organization, and syntropy at the quantum level thanks to the massless, heatless, chargeless, and entropyless perpetual motion machines that we call the Quantum Fields. The quantum fields were made by God and the Controlling Psyches within Nature; and, the quantum fields have been conserved ever since they were made. The psyches or intelligences who made the quantum fields obviously CHOOSE to conserve them, otherwise, they would cease to exist, and everything would dissolve back into the Chaos from whence it originally came.

The quantum fields are based upon the perpetual motion cycle, which is $E = mc^2$. Quantum Field Theory and $E = mc^2$ are the same thing in the end. The quantum fields are syntropic or conserved. The quantum fields are perpetual motion machines. Alas, the second law of thermodynamics will NEVER be true so long as the quantum fields exist. The second law of thermodynamics will NEVER be true as long as God and Nature's Psyche exist. According to the Big Bang Theory, the second law of thermodynamics ceased to be true the very moment that the quantum fields were made. That's just the way it is. We haven't had true disorder or true chaos in our part of the multiverse ever since the quantum fields were designed and made.

The false is falsified by the truth, and the truth is constantly and repeatedly experienced and observed by someone somewhere.

Just because we can't currently perceive the massless wave-like photons, or quantum waves, or quantum fields with our physical eyes doesn't mean that they don't exist. We can't perceive nor sense radio waves, microwaves, x-rays, gravity waves, chemical bonds, gluons, dark matter or spirit matter, dark energy, or magnetic waves with our physical eyes either; but, that doesn't mean that they don't exist. WE KNOW that they exist by the influence that they have on physical matter. WE KNOW that psyche, intelligence, quantum waves, and quantum fields exist by the influence that they have on physical matter. Psyche has to exist, or there would be no way to control things at the quantum level or the non-physical level. If Psyche, Intelligence, or Consciousness didn't exist, then there would be nothing to know or experience or observe. All would be random chaos instead and we wouldn't exist.

Mark My Words

## Falsifying My Materialism, My Naturalism, and My Atheism

Back in 2012 and 2013, I was a materialist, naturalist, nihilist, and atheist. I truly believed that there was NO scientific proof of God's existence and that there would NEVER be any scientific proof of God's existence.

I considered myself to be a scientist – a completely open-minded scientist. I wanted evidence – NOT philosophical speculation and wishful thinking. I wanted scientific proof of God's existence – safe in the belief that there would never be any such proof. I wasn't interested in being tricked and deceived by a bunch of nutters claiming to have received visions and revelations. I was holding out for scientific proof of God's Existence.

Starting in 2015, science and a couple dozen different scientists began to convince me that I was wrong. In 2016, armed with new knowledge and a new understanding of how things really work, I used Science and the Philosophy of Science to FALSIFY my materialism, my naturalism, my nihilism, and my atheism. The same year, I also used Science and the Philosophy of Science to FALSIFY Darwinism, or Creation by Entropy, or Creation by Death, or Creation by Rocks – more commonly known as Creation by Natural Selection.

I had to break my brainwashing or my conditioning in order to finally discover these new and different scientific truths – truths which clearly and conclusively FALSIFIED My Materialism, My Naturalism, My Nihilism, and My Atheism.

When I finally started looking for it, I started finding Scientific Proof of God's Existence. I started finding Science where God's intervention as the causal source was in fact the BEST, MOST PARSIMONIOUS, and MOST LOGICAL explanation. I wrote a dozen books on the subject at the time.

In 2018, when PBS Space Time introduced me to Quantum Field Theory, it was then that I realized that the Gods or Controlling Psyches had to design and make the couple dozen different Quantum Fields on a universal scale BEFORE physical matter could be made and sustained. No quantum fields, then no physical matter. For me personally, Quantum Field Theory became the most convincing Scientific Proof of God's Existence that I had encountered so far. It became clear and obvious to me that it is Nature or Nature's Psyche who collapses the wave function.

According to the Law of Psyche, every psyche or every intelligence has a certain amount of energy that's under its control. That controlling psyche can form or transform the energy under its control into anything that it wants that energy to be, anytime and anywhere that it chooses to do so. That controlling psyche collapses the wave function where the energies under its control are concerned. It's that simple. Psyches or intelligences were collapsing wave functions long before the first particle of physical matter was designed and made. First things first! Psyches or intelligences are at the same level as energy or exergy or syntropy – they are all subject to the Conservation of Energy, which means that they cannot be made, and they cannot be destroyed. They have always existed, and they will always exist. Energy, exergy, syntropy, psyche, intelligence, and quantum consciousness are at the same level of existence – they are conserved. They cannot be made, and they cannot be destroyed. The just are.

The original construct was nothing but chaos – the Chaos Realm. The Gods entered into all of that chaos in our part of the multiverse with their light, glory, order, and love; and, the Gods provided the intelligences or psyches within that chaos with what we physicists call the Standard Model. Each item on the Standard Model has a quantum field that produces it or makes its existence possible.

The Gods taught the psyches or intelligences in Chaos how to coordinate and organize themselves into the couple dozen different Quantum Fields that are represented by our Standard Model as well as by the different forces and fields that are used to hold together the different particles within a physical atom so that they can function as one. The Gods are the ones who standardized Chaos in our part of the multiverse by teaching the psyches or intelligences how to organize themselves into the different types of quantum fields that have been discovered.

The Gods had to teach the intelligences or psyches in the Chaos Realm how to organize themselves into the different Quantum Fields BEFORE physical matter could be made and sustained. No quantum fields, then no physical matter. It's that simple.

Someone or Something non-physical had to teach the psyches or intelligences in Chaos how to organize their energies into standardized quantum fields, or we still wouldn't have physical matter to this very day. Once I made that realization, I had yet another convincing scientific proof of God's necessity. I found it convincing, because I understood Quantum Field Theory, what it is, and how it works. Quantum Field Theory is the standard model by which the psyches or the intelligences in our part of the multiverse were able to organize the energies under their control into a Matrix of Quantum Fields capable of making and sustaining physical matter.

Eventually, I realized that the thing that we call the Big Bang was in fact the very moment that the Gods organized the psyches and intelligences into a Matrix of Quantum Fields at the quantum level. That's the first time that organized mass, heat, resistance to acceleration, and heat storage capacity (entropy) became possible in our part of the multiverse. Before that point in time – before the Big Bang or the Organization of the Matrix of Quantum Fields – physical matter was impossible in our part of the multiverse, because there were NO standards by which it could be made and sustained.

For me personally, the Big Bang or the Organization of the Quantum Fields ended up being one of the most convincing scientific proofs of God's existence. It wouldn't have happened without the Gods or the Controlling Psyches. Quantum Fields are perpetual motion machines, and they function according to $E = mc^2$, both coming and going. The Gods and the controlling psyches can transform the energies in the massless, heatless, and entropyless quantum fields into mass, heat, and entropy (heat storage capacity) anytime and anywhere that they choose to do so. Remember, each psyche or intelligence chooses what to do with the energy that's under its control. It can do anything it wants to do with the energy that's under its control. Anytime we run out of heat, Nature's Psyche can make some more at will; and, there's nothing we can do to stop it.

Likewise, according to Quantum Field Theory and $E = mc^2$, the Gods and the controlling psyches can transform the mass, heat, and entropy (heat storage capacity) within particles and physical matter back into massless, heatless, and entropyless quantum fields, photons, or quantum waves (virtual particles) anytime and anywhere that they choose to do so. There's nothing we can do to make it happen, and there's nothing we can do to stop it.

The intelligences or psyches in Nature listen to the Gods, not us. It was the Gods who taught them how to organize themselves into quantum fields in the first place. The psyches or intelligences in Nature listen to and obey the Gods, not us. It's Nature or Nature's Psyche who collapses the wave function, and not us. Every psyche or intelligence can do whatever it wants to do with the energy that's under its control. They already chose to obey the Gods and to build a Standardized Matrix of Quantum Fields with the energy that is under their control. The psyches or intelligences in Nature have no incentive from us to do anything else with the energy that's under their control. We are not the ones who taught them how to organize themselves into Quantum Fields. The Gods did.

Remember, it's the psyches or intelligences in Nature who collapse the wave function and choose to do what they want to do with the energy that's under their control. Energy is infinitely malleable. The controlling psyches can transform the energy under their control into anything they want that energy to be, anytime and anywhere they choose to do so. Nature's Psyche under the Word of Command from the Gods could transform our whole galaxy into massless, heatless, and entropyless quantum fields, photons, and quantum waves (virtual particles) instantly right now, if Nature's Psyche or the Controlling Psyches were to choose to do so. There's nothing we can do to make them do it, and there's nothing we can do to stop them from doing it. Only the Gods can give the Word of Command, because it was the Gods who taught the intelligences and the psyches in our part of the multiverse how to make quantum fields, mass, heat storage capacity (entropy), purposeful heat (work), resistance to acceleration, and physical matter in the first place.

According to the Big Bang Theory, Quantum Field Theory, Quantum Mechanics, and $E = mc^2$, Nature or Nature's Psyche can make mass, heat, heat storage capacity, and resistance to acceleration anywhere and anytime that it chooses to do so. It's Nature or Nature's Psyche who collapses the wave function. It is Nature or Nature's Psyche who makes our physical brain and MAPS purpose or specific functions onto our physical brain. It is Nature or Nature's Psyche who collapses the wave functions and fires the neurons in our physical brain. Nature or Nature's Psyche KNOWS which neuron to fire in order to raise your finger off the table because Nature's Psyche is the one who assigned that specific neuron to that specific purpose in the first place. It's Nature's Psyche who MAPS the physical functions onto our physical brain, and MAPS the specific motor neurons to specific purposes, and then fires the specific neurons that make our physical bodies move. The human psyche isn't consciously aware of any of that. It's all done at the quantum level by Nature's Psyche. The Quantum Zeno Effect has taught us that the intelligences or psyches within each physical atom can read our minds and understand our intentions and our desires and can then respond accordingly.

The collapse of the wave functions and the firing of the neurons are all happening at the quantum level under the control of Nature's Psyche, or the intelligences within the physical atoms who are making and controlling the physical atoms. Psyche, intelligence, or quantum consciousness is the only thing that we know of that is capable of existing and working at the quantum level so that the proper wave functions will be collapsed at the quantum level and the appropriate neurons fired at the quantum level. It is the ONLY thing that can make and MAP our physical brains at the quantum level. It is psyche, intelligence, or quantum consciousness who is making and controlling the physical atoms and the quantum particles, NOT the other way around as the Materialists and Naturalists claim.

Science, Quantum Mechanics, and Quantum Field Theory taught me that we Materialists, Naturalists, Darwinists, Nihilists, Behaviorists, Determinists, Physicalists, and Atheists don't know what we are talking about and don't understand the non-physical sciences and how they work. Quantum Mechanics is the science of the interface between the physical and the non-physical. Quantum Field Theory teaches us about many of the different things into which the controlling psyches can transform the energies under their control. Quantum Mechanics is Energy Mechanics or Non-Physical Mechanics. It tries to point us to all the different non-physical things that can be done with energy.

Quantum Field Theory is trying to teach us that Nature's Psyche can take massless, heatless, and entropyless quantum fields, photons, quantum waves (virtual particles) and

TRANSFORM them into mass, heat, heat storage capacity (entropy), and resistance to acceleration anytime and anywhere that the controlling psyches choose to do so. Furthermore, Quantum Field Theory and $E = mc^2$ are trying to tell us that the controlling psyches or Nature's Psyche can TRANSFORM mass, heat, and entropy back into massless, heatless, and entropyless quantum fields, photons, and quantum waves (virtual particles) anytime and anywhere that the controlling psyches choose to do so. There's nothing we can do to make it happen, and there's nothing we can do to prevent it from happening, because it is Nature's Psyche or the controlling psyches who collapse the wave functions and NOT the human observer or the human psyche. Quantum Field Theory makes it obvious that we humans have NO control over the quantum or the non-physical. It is the Gods, Nature's Psyche, and the Controlling Psyches who are running and controlling everything at the quantum level – NOT us.

Mass is resistance to acceleration. Mass is heat storage capacity. No mass, then no heat storage capacity. No heat storage capacity and no heat, then no entropy. The thermodynamics version of entropy is based upon a transfer of heat from a hot reservoir to a cold reservoir. No quantum fields, then NO hot reservoirs and NO cold reservoirs. Organized mass, heat storage capacity, resistance to acceleration, and purposeful heat itself as we know it became possible ONLY AFTER the Gods taught Nature's Psyche how to organize the energies under its control into quantum fields.

No quantum fields, then no mass, no resistance to acceleration, no heat storage capacity, no physical matter, and no entropy. No quantum fields, then no organized thermodynamics and no organized purpose for heat or heat storage capacity. Entropy doesn't exist and can't exist in the massless, heatless, and entropyless Quantum Fields. Entropy is mass's heat storage capacity. No mass, then NO entropy. NO stored heat, then NO entropy.

https://en.wikipedia.org/wiki/Specific_heat_capacity

https://en.wikipedia.org/wiki/Table_of_specific_heat_capacities

**S = Q/ΔT**

**Q = mcΔT**

**S = mc**

In these equations, **mc** is mass's heat storage capacity. Mass (**m**) times specific heat capacity (**c**) is the same thing as entropy (**S**). Entropy (**S**) is mass's heat storage capacity (**mc**) – nothing more and nothing less. It's been there in plain sight all the way along just waiting for someone to discover it and make it known.

Entropy (**S**) is comprised of heat (**Q**), mass (**m**), and heat storage capacity (**c**). Mass has heat storage capacity (**mc**); and, entropy (**S**) is heat storage capacity (**S = mc**), according to these equations for entropy. The temperatures (**T**) cancel out. Entropy remains as a component of heat and mass, no matter what the temperature might be. Thermodynamics is an exchange of heat from a hot reservoir to a cold reservoir. Entropy is mass's heat storage capacity. No mass, then no entropy. Entropy completely ceases to exist whenever mass ceases to exist.

Entropy is the heat storage capacity of mass (**S = mc**). This ends up being the best and most believable definition for entropy. It's the conservative and true definition for entropy. It's the demonstrable and measurable definition for entropy. It's the definition for entropy that I KNOW is true.

These equations reveal that at absolute zero ($T = 0$ **Kelvin**), heat completely ceases to exist, which means that entropy completely ceases to exist along with it. In the massless, heatless, and entropyless Quantum Fields at absolute zero, mass ceases to exist which means that heat ceases to exist, heat storage capacity ceases to exist, and entropy ceases to exist. The Quantum Fields and Quantum Field Theory falsify the second law of thermodynamics. For me personally, this is the greatest scientific discovery of all time. These equations for entropy define what entropy truly is and what entropy is truly comprised of. These equations also tell us how and when entropy goes to zero and ceases to exist.

It doesn't take a genius to realize that when mass goes to zero ($m \rightarrow 0$), then heat goes to zero along with it ($Q \rightarrow 0$). Remove all the mass or resistance to acceleration, then you remove mass's heat storage capacity ($mc \rightarrow 0$); and consequently, you eliminate all the heat ($Q \rightarrow 0$) and all the entropy along with it ($S \rightarrow 0$)! Entropy completely ceases to exist when heat and mass cease to exist. That's why the photons, quantum fields, and quantum waves (virtual particles) are massless, heatless, and entropyless.

Here in the summer of 2019, I'm using Science, the Philosophy of Science, logic and common sense, $E = mc^2$, dark energy, statistics, and the equations for heat and entropy to FALSIFY the second law of thermodynamics and to replace it with something useful and true instead.

The idea that entropy is some type of invisible supernatural substance that is always increasing in amount and can never decrease is nothing but science fiction. Not even energy – the most conserved substance in the universe – can increase in quantity or amount. Yet, entropy – according to the inappropriately named second law of thermodynamics – is always magically increasing in amount and can never decrease on a universal scale. The second law's version of entropy is magical, supernatural, and basically creation from nothing. The stuff just spontaneously generates out of thin air and ends up being the ultimate form of Creation Ex Nihilo. Energy is simply conserved – it's amount cannot increase, and its amount cannot decrease. Yet entropy is constantly being generated out of thin air from nothing so that its amount is always increasing and can never decrease. The second law's version of entropy is the very definition of a superstitious, mystical, magical, mythical, fictional, and supernatural substance that is constantly increasing and constantly being made from nothing in complete violation of common sense and the laws associated with the Conservation of Energy. Imagine my disappointment when I first realized that the second law's version of entropy is nothing but pseudoscience and science fiction. It's magic, not science.

Entropy is mass's heat storage capacity. Entropy is mass's resistance to acceleration. No mass, then no entropy. Whenever Nature's Psyche transforms mass, heat, and entropy back into massless, heatless, and entropyless photons, quantum fields, and quantum waves (virtual particles), the associated entropy completely ceases to exist because the associated heat storage capacity completely ceases to exist. When mass ceases to exist, mass's resistance to acceleration ceases to exist and so does the entropy along with it. It's that simple.

According to the equations for entropy and heat, heat needs someplace where it can be stored; otherwise, it serves no useful purpose. Without purpose, there is nothing but chaos. The Gods gave the psyches and intelligences in the Chaos Realm purpose; and, they organized themselves into a Matrix of Quantum Fields, so that they could then accomplish other purposes. The thing we scientists call the Big Bang was effectively the organization of that Matrix of Quantum Fields. Before that point in time, mass, heat storage capacity, resistance to acceleration, purposeful heat, work, and entropy were impossible to make and sustain. They didn't exist because they couldn't exist. The Gods and the controlling psyches in Nature had to make the Quantum Fields BEFORE mass, useful heat, and physical matter could be made and sustained.  This obvious truth is one of the most powerful and useful scientific discoveries I have ever made.  It explains everything that has ever been experienced or observed.

The truth is hiding in plain sight where nobody can see it because they aren't looking for it and don't want it.  The Materialists, Naturalists, Darwinists, Nihilists, and Atheists don't want the truth, because the truth FALSIFIES their beliefs.  I KNOW because I used to be one of them.  I would still be one of them if Science and a couple dozen different scientists hadn't taught me that I was wrong.

Mark My Words

## Tilting Windmills

Once I was finally willing and able to falsify Naturalism, Darwinism, Physicalism, Nihilism, and Atheism, I soon realized that there are thousands of things in science and physics that need fixing or replacement. However, I'm only one person. I don't have time enough to find and fix everything that's wrong with science and physics. The best I can do is to show other people how to do so.

Nevertheless, I simply realized that I'm tilting windmills. I'm really going nowhere with all of this. Knowing how to find and correct the falsehoods in physics and science isn't going to change anything.

Yes, it's fun to KNOW that you are right, and it's fun to win all the debates taking place online, but eventually one realizes that it is pointless to do so. It doesn't change anything. I stopped debating with the Darwinists, Naturalists, and Atheists online back in 2016 when they started banning me from their websites. These people don't want to find and know the truth. It's a waste of time to even try to give them the truth. They don't want it. They prefer the lies instead. These people delete and destroy the evidence that falsifies their beliefs. Explaining to them why they are wrong doesn't do anything to help them change their beliefs because they want to be wrong. It's discouraging because none of them will ever read a book like this during their entire lives, because they refuse to read and think about anything that falsifies their pre-chosen beliefs. I KNOW, because I used to be a materialist, naturalist, nihilist, and atheist. At the time, I refused to look at, read, or think about anything that falsified my beliefs.

Finding, knowing, and presenting the truth about physics, biology, and science is a pointless and fruitless endeavor in the end. Most people on this planet don't want to find and know the truth. They prefer the status quo instead. That's just the way it is. Most of us wear our ignorance as a badge of honor. I KNOW how it goes, because I used to be a materialist, naturalist, nihilist, and atheist.

After fifty years of thought and research, I was finally willing and able to falsify my materialism, my naturalism, my nihilism, and my atheism. I finally KNOW how everything works. I finally have a true interpretation of Quantum Mechanics. I have found ways to replace Materialism, Naturalism, Darwinism (Creation by Chance and Entropy), Nihilism (Death of the Psyche or the Soul), and Atheism (Creation Ex Nihilo) with the truth. I have a TRUE replacement for each one of these now.

But, having a true interpretation of Quantum Mechanics or knowing the truth about science and physics is NOT going to change the way that we do science. It's going to change the way that we interpret science and physics, but it's not going to change the way that we do science and physics. We fallen mortal beings have taken science and physics about as far as we can take them, I do believe. The next steps that we need to take in science are beyond our reach – mind-over-energy is currently beyond our reach. We have to rely upon the Gods to do that kind of stuff for us, because we are currently blocked from doing mind-over-energy for ourselves. There's no way that we can make quantum waves, quantum fields, and physical matter with our minds. We are blocked from doing so. We can't do that kind of science.

Yes, we now KNOW for a fact that mind-over-matter is real and truly exists. Whether you choose to call it the Placebo Effect, the Quantum Zeno Effect, Faith, Hope, Meditation, or Prayer, we do KNOW for a fact that the Placebo Effect is real, and we actually make allowances for it in our science experiments. However, knowing that the non-physical exists, knowing that the quantum fields and quantum waves exist, knowing that action at a distance exists, and knowing that mind-over-matter exists isn't going to change the way that we do science. It's only going to change the way that we interpret science and physics.

I wrote a few books falsifying the theory of evolution and the second law of thermodynamics, but it's not going to change anything. People will continue to believe in Creation by Entropy, Creation by Chance, and Creation by Natural Selection with blind devotion and a religious zeal. Falsifying their beliefs simply is not enough to change their beliefs. They believe in these falsehoods and lies because they want to believe. No amount of evidence is going to change what they want to believe.

In many respects, I'm simply tilting windmills. I've hit a dead-end.

About the only thing left to science is to figure out how to make ourselves immortal, and when we figure that out, then we are going to either have to find ways to kill each other off or we are going to have to find ways to keep ourselves from starving to death, because physical fallen mortal beings like us consume way too many resources and too much food to be immortal. You just know which one we will choose. We will kill each other in order to prevent ourselves from starving to death. In our current physical format, since we need to eat, immortality would actually be a curse. It's good for the human race as a whole that each one of us has an expiration date. If all of us were immortal, we would soon starve to death.

We would need to be upgraded and turned into quantum fields or perpetual motion machines so that we no longer need to eat, if we truly want to live forever and have the process be a success.

According to Christianity and the Doctrine of Resurrection, the Gods have already figured out how to do that. Thanks to the Atonement of Christ, we each are going to be raised from the dead someday, made immortal, and won't have to eat and sleep in order to survive. The Apostle Paul called it a spiritual body. It's some type of Quantum Body – a body that is made out of quantum fields – a physical body that is a perpetual motion machine – a body that doesn't need to eat and sleep if it doesn't want to. That's the type of physical body that we need, if we want to live forever. We need a physical body that is in fact a perpetual motion machine. We need a physical body that is like the quantum fields. We need a physical body that is indeed a spiritual body or a quantum body. The Apostle Paul knew precisely what he was talking about and what is needed, if we want to be immortal and live forever.

Currently, we don't know how to make a physical body that is a perpetual motion machine. The Gods do. The Gods are the ultimate scientists. The Gods know how to make physical bodies that are perpetual motion machines. We don't.

Only the Gods know how to transfer a psyche, spirit, soul, or consciousness from one physical body to the next. That's something that we human beings haven't figured out how to do. The closest we have ever gotten to that is science fiction and fantasy; and, it NEVER turns out well when we do. Look at something like "Altered Carbon", "Alita Battle Angel", or the "Extinct" television series to get a glimpse of what it's like when human beings figure out how to live forever in their sins. They become some really nasty horrible people. They degrade into animals and monsters. They stop progressing and learning. They become Materialists, Naturalists, Darwinists, Nihilists, and Atheists. Eventually they realize that they have lived too long. We really don't want to live forever in physical bodies like the ones that we currently have. It isn't pretty. It's a living hell!

Look at all the mass shootings that are taking place nowadays. Everyone is now wondering what we can do to solve the problem. Well, we first need to recognize what we did to cause the problem. We deliberately pulled ethics, morality, righteousness, goodness, kindness, compassion, prayer, God, and Christ from our public school systems; and now, fifty years later we are reaping the rewards. We are getting precisely what we created; and, knowing how we created the problem isn't going to change anything, because we are never going to put morality, ethics, goodness, prayer, God, and Christ back into our public school system. We are never going to fix the problem because we don't want to. We are now getting precisely what we designed and created. We are now getting precisely what we wanted in the first place. The shooters feel NO responsibility to anyone or anything. They are atheists just like we designed them to be.

https://en.wikipedia.org/wiki/List_of_mass_shootings_in_the_United_States_in_2019

They've become exactly like our "heroes" in the television series "The Punisher" and "The Blacklist". They have become mass murderers just like we trained them to be, and taught them to be, and conditioned them to be through the television that they watch and the computer games that they play. We have successfully turned them into atheists and monsters just like we intended to do. We are now reaping the rewards for "setting them free" and taking ethics, morality, goodness, prayer, God, and Christ out of their lives. We will never be able to fix the problem, because this is the way we wanted things to be in the first place. We wanted them to be immoral atheists, and now they are. We are never going to put ethics, morality, goodness, self-control, civic responsibility, prayer, God, and Christ back into our public school system because we really don't believe in these things in the first place.

I'm primarily a scientist, but here I venture into social commentary. My bachelor's degree is in psychology, so I'm authorized to speak on the subject if I want to. When it comes to the conscienceless and atheistic shooters, mass murderers, and serial killers, we conditioned them or brainwashed them to be that way according to Behaviorism and the science that we call Psychology. Within our society and within our public schools, we conditioned them and brainwashed them to be atheistic psychopaths. We have been doing this for fifty years. We are now reaping our reward.

It's kind of obvious or self-evident that this is true, and they have done observational studies demonstrating that it is true; but, its truth is denied, nonetheless. The Atheists, Materialists, and Naturalists continue to claim that their television, their music, their lack of prayer, their computer games, their rejection of God, and their pornography had no effect on their violence, anger, hatred, self-control, addictions, murder-rate, level-of-assault, level-of-rape, theft-rate, unbelief, morality, or level-of-evil within their soul. Many of these shooters and mass murderers were bullied and tortured by the atheists (and "Christians") at school. Others have spent a decade of their lives killing and raping human beings on their computer screens. They have become addicted to the violence and the pornography. Yet simultaneously, these people have convinced themselves that evil doesn't exist, at least where they themselves are concerned. They have convinced themselves that everything they do is right.

I, too, am not immune. I have actually experienced both sides of the fence. I used to be a materialist, naturalist, nihilist, and atheist until one day I decided that I didn't like where that road was taking me; and then, I decided to turn around and go the other way.

I now know how the Gods feel, when the gift you try to give someone is rejected by that person because he or she doesn't want it. We human beings as a whole don't want what the Gods have to offer us. We don't want the calm and peace because we also don't want the responsibility that comes along with it. It's discouraging. In America, we built our statue to Liberty, and NOT to Responsibility. We want to be free to murder and kill and rape at will. We don't want to be responsible for the results. Nothing is ever going to change, until we each individually want to change. It's discouraging to know that most of the people on this planet don't want the truth and don't want to change. It's discouraging to know that you aren't going to be able to help people because they don't want your help. That's precisely the way that the Gods feel when it comes to us.

Likewise, knowing the truth about everything is not going to change the way that we do science or live our lives. Until we want to change for the better, we are never going to change for the better. I KNOW. I have already been there and done that. I never changed for the better until after I wanted to change for the better. This is an important observation when it comes to human psychology. We don't adopt the truth and we don't want the truth, until after we have actually decided for ourselves that we need the truth and that we want to find and know the truth. There's no way to help people until they actually want to be helped. Half the time we never find the truth, until after we start wanting it and start searching for it. Yes, some truths impose themselves on us. If we jump off a cliff, we are going to fall, and there's nothing we can do to change that except not to jump in the first place. However, when it comes to the spiritual truths and the non-physical truths, we never discover those until after we go looking for them. Most people don't want the truth. That's just the way it is.

As in life, so in science. We won't have the truth until we actually want the truth. We won't change until we want to change. We won't improve the interpretations that we give to science and scientific data, until we actually want to improve.

On the practical side of the equation, where science is concerned, until we can actually make quantum waves and collapse wave functions with our minds or with the word of command, we will simply continue to do science the way that we currently do it. Until we master mind-over-energy and can actually create quantum fields and physical matter with our minds or through the word of command, we will continue to be fallen mortal beings and will continue to be limited to the physical realm. Until we can actually make physical bodies for ourselves that are perpetual motion machines like the quantum fields, knowing the truth and knowing about quantum field theory isn't going to change the way that we do science. It will only change the way that we interpret science and physics. We are going to have to change or be changed at a fundamental level, if we ever want to have full control over the quantum mechanisms that exist at the quantum level in the Non-Physical Realm or Transdimensional Realm.

Only the Gods can do the things that truly matter. If we truly want to live forever in some type of physical body, we are totally dependent upon the God for the accomplishment of that goal. Alas, when I was an atheist, I didn't want to live forever. I wanted annihilation instead. I wanted to die and cease to exist. That seems to be what most of the atheists want and what most of the atheists truly believe in. They really don't want to live forever in their sins. They are looking for a way out like I used to be. For these people, life has no purpose or meaning. All we are is dust in the wind. Life is just going to cease to exist someday, so why not make it today? Today is a good day to die, as is any other day.

I am a testament or a witness that human beings can change for the better, but I also KNOW that it isn't easy. For me, it has taken years to get some of the falsehoods and rottenness out of my life. The Atonement of Christ can help anyone to change for the better, but it still requires a lot of effort, work, and desire on the part of the person who wants to change. If we don't want to change for the better, then nothing happens. The Atonement of Christ can help us to overcome everything, except for a lack of desire. If we have no desire to change, then there's nothing that God can do to help us change.

As in life, so in science. Until we want scientific truths, we will never have scientific truths, because we will refuse to look for them because we really don't want them. I find

Quantum Field Theory fascinating. It's a great representation of the truth. But it's not going to get us anywhere in the end, because none of us are capable of making quantum fields and quantum waves and quantum particles with our minds or through the word of command, and because we really don't want the truth when it comes to science anyway. We prefer the lies and deceptions instead.

Furthermore, Nature's Psyche doesn't listen to us and obey us. We need the Gods to give those kinds of things to us, if we want mind-over-energy or priesthood power in our lives. That's just the way it is. Thankfully, the Gods don't give power, immortality, or supernatural abilities to fallen mortal people like us who will use that power and those abilities to ruin and destroy the lives of the people around us. The Gods gave us the ability to destroy ourselves, but the Gods are not going to give us the ability to destroy the universe.

Mark My Words

# PART XI — THE MIND OF GOD

The only way to know the Mind of God is for God to tell you or show you what's on His mind. It is my observation that the Biblical God Jesus Christ, Christ Jehovah, has in fact met His burden of proof through a preponderance of the evidence. Satan or Lucifer has done the same.

Jesus Christ in the New Testament is the same individual or person as Jehovah in the Old Testament. In other words, the Bible testifies of or bears witness to the same God. The preponderance of the evidence tells us clearly and conclusively that this particular God does in fact exist. He has been described as the True and Living God – rather than the man-made fictional gods such as evolution, which don't have teeth.

Remember, Science 2.0 allows ALL of the evidence into evidence, including the revelations of God and the revelations from God. Nothing is excluded from Science 2.0. That's why Science 2.0 is a vastly improved version of Science compared with the other "science" which is defined as Materialism, Naturalism, Darwinism, Nihilism, and Atheism because Science 2.0 allows all of the evidence into evidence and these other things don't.

Science 2.0 even allows into evidence ALL of the evidence which the fundamentalist Christians deliberately, and at times knowingly, ridicule and reject.

Consequently, I submit into evidence everything that the Biblical God Jesus Christ has had a hand in writing or producing, because it ALL bears witness of Him – the One who produced it or revealed it.

Again, when it comes to Science 2.0, my only obligation to you is to present the observational evidence and the experiential evidence. Then, I leave it up to you my reader to decide for yourself what to make of it all.

You can point a horse to water, but you can't make him drink.

You can point a man to scripture, but you can't make him think.

## Books that the Biblical God Had a Hand in Producing

I was lucky to find a handful of books that the Biblical God Jesus Christ had a hand in writing or producing.

I have been exposed to a revealed religion from which I have derived a great deal of insight and inspiration. I'm no longer afraid of it, like I used to be when I was a Materialist, Naturalists, Nihilist, Skeptic, and Atheist.

There's lots of observational evidence, experiential evidence, empirical evidence, and eye-witness evidence proving that the Biblical God Jesus Christ does in fact exist.

Hundreds of people have observed, during their Near-Death Experiences (NDEs), that Jesus Christ is the Being of Light and the Being of Love whom the righteous people and the redeemed typically encounter during their NDEs. Jesus Christ has even come and rescued some of these people from hell.

I submit all of these eye-witness accounts into evidence and leave it up to you to decide for yourself what to make of them.

NDE Encounters with Jesus:

https://www.youtube.com/results?search_query=NDE+Jesus

Howard Storm:

https://www.youtube.com/watch?v=Vm647n1360A

https://www.youtube.com/watch?v=UPj4wci_bcI

Dr. Mary Neal:

https://www.youtube.com/watch?v=DX473dF7ChY

Ian McCormack:

https://www.youtube.com/watch?v=HbTAmN4m2lQ

Joe Hadwin:

https://www.youtube.com/watch?v=IOhOynR9Jxq

Another:

https://www.youtube.com/watch?v=N4ut09jDdV0

Remember, unlike Materialism and Naturalism, Science 2.0 allows all of the evidence into evidence. Furthermore, unlike Materialism and Naturalism, Science 2.0 leaves it up to you to interpret the evidence for yourself rather than imposing on you and enforcing upon you an "official interpretation" of the evidence that has been blessed and sanctified and purified by the Materialists and Naturalists who control the review process.

I'm NOT sitting here officiating and reviewing what you are permitted to believe in and what you must reject as being pseudoscience or as being axiomatically and dogmatically false. Under Science 2.0, you are the one who determines what the evidence means to you – there's no official review board or peer review panel to do that for you. There's no censorship process or brainwashing process with Science 2.0.

Since I'm no longer a Materialist, Naturalist, Nihilist, and Atheist, I'm no longer hiding from the evidence. I now allow all of the evidence into evidence, and then I decide for myself how to interpret it and what to make of it. You should do the same. Allow all of the evidence into evidence, and then decide for yourself what to make of it. That's what I finally chose to do.

Instead of hiding from everything that falsifies my pre-chosen beliefs as I used to do when I was an Atheist, I'm now trying to learn something new every day; and, that process sometimes means learning something new about God and how He works.

The following books provide evidence for God's Existence; and since Science 2.0 allows all of the evidence into evidence, I must submit them into evidence as well.

The Holy Bible

https://www.lds.org/scriptures/bible?lang=eng

The Book of Mormon: Another Testament of Jesus Christ

https://www.lds.org/scriptures/bofm?lang=eng

The Doctrine and Covenants

https://www.lds.org/scriptures/dc-testament?lang=eng

The Pearl of Great Price

https://www.lds.org/scriptures/pgp?lang=eng

The Biblical God Jesus Christ had a hand in writing and producing these books. You can decide for yourself what to do with them and what to make of them.

I wish you well!

Mark My Words

Original Version: 31OCT2017
Current Version: 09SEP2019

www.ingramcontent.com/pod-product-compliance
Lightning Source LLC
Chambersburg PA
CBHW072022230526
45466CB00019B/1